NONLINEAR PARTIAL DIFFERENTIAL EQUATIONS AND THEIR APPLICATIONS

Collège de France Seminar Volume XIV

STUDIES IN MATHEMATICS AND ITS APPLICATIONS

VOLUME 31

ELSEVIER

AMSTERDAM – BOSTON – LONDON – NEW YORK – OXFORD – PARIS
SAN DIEGO – SAN FRANCISCO – SINGAPORE – SYDNEY – TOKYO

NONLINEAR PARTIAL DIFFERENTIAL EQUATIONS AND THEIR APPLICATIONS

Collège de France Seminar Volume XIV

Editors

DOINA CIORANESCU

Centre National de la Recherche Scientifique
Laboratoire J.L. Lions
Université Pierre et Marie Curie
Paris, France

and

JACQUES-LOUIS LIONS †

Collège de France
Paris, France

2002

ELSEVIER

AMSTERDAM – BOSTON – LONDON – NEW YORK – OXFORD – PARIS
SAN DIEGO – SAN FRANCISCO – SINGAPORE – SYDNEY – TOKYO

ELSEVIER SCIENCE B.V.
Sara Burgerhartstraat 25
P.O. Box 211, 1000 AE Amsterdam, The Netherlands

First edition 2002

Library of Congress Cataloging in Publication Data
A catalog record from the Library of Congress has been applied for.

British Library Cataloguing in Publication Data
A catalogue record from the British Library has been applied for.

ISBN: 0-444-51103-2
ISSN: 0168-2024

♾ The paper used in this publication meets the requirements of ANSI/NISO Z39.48-1992 (Permanence of Paper).
Printed in The Netherlands.

To the memory of Jacques-Louis Lions

Contents

Preface ix

1 An introduction to critical points for integral functionals 1
D. Arcoya and L. Boccardo

2 A semigroup formulation for electromagnetic waves in dispersive dielectric media 13
H.T. Banks and M.W. Buksas

3 Limite non visqueuse pour les fluides incompressibles axisymétriques 29
J. Ben Ameur and R. Danchin

4 Global properties of some nonlinear parabolic equations 57
M. Ben-Artzi

5 A model for two coupled turbulent flows. Part I: analysis of the system 69
C. Bernardi, T. Chacón Rebollo, R. Lewandowski and F. Murat

6 Détermination de conditions aux limites en mer ouverte par une méthode de contrôle optimal 103
F. Bosseur and P. Orenga

7 Effective diffusion in vanishing viscosity 133
F. Campillo and A. Piatnitski

8 Vibration of a thin plate with a "rough" surface 147
G. Chechkin and D. Cioranescu

9 Anisotropy and dispersion in rotating fluids 171
J.-Y. Chemin, B. Desjardins, I. Gallagher and E. Grenier

10 Integral equations and saddle point formulation for scattering problems 193
F. Collino and B. Despres

11 Existence and uniqueness of a strong solution for nonhomogeneous micropolar fluids 213
C. Conca, R. Gormaz, E. Ortega and M. Rojas

12 Homogenization of Dirichlet minimum problems with conductor type periodically distributed constraints 243
R. De Arcangelis

13 Transport of trapped particles in a surface potential 273
P. Degond

14 Diffusive energy balance models in climatology 297
J.I. Díaz

15 Uniqueness and stability in the Cauchy problem for Maxwell and elasticity systems 329
M. Eller, V. Isakov, G. Nakamura and D. Tataru

16 On the unstable spectrum of the Euler equation 351
S. Friedlander

17 Décomposition en profils des solutions de l'équation des ondes semi linéaire critique à l'extérieur d'un obstacle strictement convexe 367
I. Gallagher and P. Gérard

18 Upwind discretizations of a steady grade-two fluid model in two dimensions 393
V. Girault and L.R. Scott

19 Stability of thin layer approximation of electromagnetic waves scattering by linear and non linear coatings 415
H. Haddar and P. Joly

20 Remarques sur la limite $\alpha \to 0$ pour les fluides de grade 2 457
D. Iftimie

21 Remarks on the Kompaneets equation, a simplified model of the Fokker-Planck equation 469
O. Kavian

22 Singular perturbations without limit in the energy space. Convergence and computation of the associated layers 489
D. Leguillon, E. Sanchez-Palencia and C. de Souza

23 Optimal design of gradient fields with applications to electrostatics 509
R. Lipton and A.P. Velo

24 A blackbox reduced-basis output bound method for noncoercive linear problems 533
Y. Maday, A.T. Patera and D.V. Rovas

25 Simulation of flow in a glass tank 571
V. Nefedov and R.M.M. Mattheij

26 Control localized on thin structures for semilinear parabolic equations 591
P.A. Nguyen and J.-P. Raymond

27 Stabilité des ondes de choc de viscosité qui peuvent être caractéristiques 647
D. Serre

Preface

This volume is the 14th and last one of the series "Nonlinear Partial Differential Equations and their Applications. Collège de France Seminar", which published the texts of the lectures given at the seminars organized by Jacques-Louis Lions, from his election at the Collège de France in 1973 until his retirement in 1998. It was one of the foremost seminars in nonlinear PDE's and their applications during that period.

It is unfortunate that because of his untimely death, on May 17, 2001, Jacques-Louis Lions will not see its publication. This volume is dedicated to his memory.

Studies in Mathematics and its Applications, Vol. 31
D. Cioranescu and J.L. Lions (Editors)

Chapter 1

AN INTRODUCTION TO CRITICAL POINTS FOR INTEGRAL FUNCTIONALS

D. ARCOYA AND L. BOCCARDO

1. Introduction

The study of minima of functionals defined in spaces of functions may be considered one of the keystones of the mathematical analysis. Remind the efforts by the great mathematicians of the last and present century to develop sufficient conditions on the functional for the existence of minimum. This theory is deeply related with the existence of solutions of boundary value problems. Indeed, this connection is estabilished by the well-known Euler-Lagrange equations associated to the functional.

However, there exist boundary value problems for which the associated functional is indefinite, i.e. it is unbounded from below and from above. This means that it has not global extrema and so we have to search the solutions of the problem among the *critical points*, i.e. the points for which the derivative of the functional vanishes.

From the abstract point of view there is a difference between the study of minima and of critical points. Indeed, for the existence of minima we need only assumptions on the functional. On the contrary, we point out that the results of existence of critical points involve additional hypotheses on the regularity of the functional to assure the existence of a *derivative* in some sense. This may explain why the theory of mimima handles classes of functionals with more general hypotheses of smoothness than the critical point theory.

In some papers [4], [5], [6], we overcame this difference by developing a critical point theory for nondifferentiable functionals. We observe explicitely that our model functionals does not involve similar functions to the modulus. In fact, the nondifferentiability of the considered functionals is due to the introduction of some smooth Carathéodory function $A(x, u)$ (as smooth as you want). Specifically, we consider here functionals J defined in $W_0^{1,2}(\Omega)$

($\Omega \subset \mathbb{R}^N$ open, $N > 1$) by

$$J(v) = \int_\Omega A(x,v)|\nabla v|^2\,dx - \int_\Omega F(x,v^+)\,dx,\ \ v \in W_0^{1,2}(\Omega), \tag{1}$$

with $0 < \alpha \leq A(x,z) \leq \beta < \infty$, $|A'_z(x,z)| \leq \gamma$ and $f(x,z) \equiv F'(x,z)$ (derivative respect to z) a subcritical Carathéodory function. Observe that J is only differentiable along directions of $W_0^{1,2}(\Omega) \cap L^\infty(\Omega)$, even for smooth functions A (see [11]).

This note is devoted to present the critical point theory developed in [5] (see also [?, ?, ?, ?, ?, ?, ?]) for functionals which are not differentiable in all directions.

2. A mountain pass theorem for nondifferentiable functionals

Our abstract setting for the functionals J that we study is given by the following asumption:

(H) $(X, \|\cdot\|_X)$ is a Banach space and $Y \subset X$ is a subspace which is a normed space endowed with a norm $\|\cdot\|_Y$. Moreover, $J : X \longrightarrow \mathbb{R}$ is a functional on X such that it is continuous in $(Y, \|\cdot\|_X + \|\cdot\|_Y)$ and satisfies the following hypotheses:

a) J has a directional derivative $\langle J'(u), v\rangle$ at each $u \in X$ through any direction $v \in Y$.

b) For fixed $u \in X$, the function $\langle J'(u), v\rangle$ is linear in $v \in Y$, and for fixed $v \in Y$, the function $\langle J'(u), v\rangle$ is continuous in $u \in X$.

Due to the lack of regularity of the functional, some words are needed in order to establish our definition of critical points.

Definition 2.1 – *A function $u \in X$ is called a critical point of J if*

$$\langle J'(u), v\rangle = 0, \quad \forall v \in Y.$$

In this framework a suitable version of the Ambrosetti-Rabinowitz Theorem has been proved in [5]. Specifically,

Theorem 2.2 – *Assume (H) and that for $e \in Y$,*

$$c = \inf_{\gamma \in \Gamma} \max_{t \in [0,1]} J(\gamma(t)) > c_1 = \max\{J(0), J(e)\}$$

with Γ the set of the continuous paths $\gamma : [0,1] \longrightarrow (Y, \|\cdot\|_X + \|\cdot\|_Y)$, such that $\gamma(0) = 0$ and $\gamma(1) = e$. Suppose, in addition, that J satisfies the condition

(*C*) *Any sequence* $\{u_n\}$ *in the Banach space* Y *satisfying for some* $\{K_n\} \subset \mathbb{R}^+$ *and* $\{\varepsilon_n\} \longrightarrow 0$ *the conditions*

$$\{J(u_n)\} \text{ is bounded,} \tag{2}$$

$$\|u_n\|_Y \leq 2K_n \quad \forall n \in \mathbb{N}, \tag{3}$$

$$|\langle J'(u_n), v\rangle| \leq \varepsilon_n \left[\frac{\|v\|_Y}{K_n} + \|v\|_X\right] \quad \forall v \in Y, \tag{4}$$

possesses a convergent subsequence in X.

Then there exists a nonzero critical point $u \in Y$ *of* J *such that* $J(u) = c$.

Remarks 2.3.

1. The proof of this theorem is done by dividing it into two steps. In the first one, only the geometric hypotheses are used to deduce the existence of a sequence $\{u_n\}$ in Y satisfying for some $\{K_n\} \subset \mathbb{R}^+$ and $\{\varepsilon_n\} \to 0$ the conditions (2)–(4). The proof is then concluded by using condition (C).

2. In this way, condition (C) can be considered as a compactness condition on the functional J, which substitutes in the nondifferentiable case the role done by the well-known Palais-Smale condition in the regular case $Y = X$.

3. This compactness condition is connected with the coercivity of J extending the previous results for C^1 functionals in [11]. To be specifical in [7] we prove

Theorem 2.4 – *In addition to* (H), *assume that* Y *is dense in* X *and that* J *is continuous in* X *and bounded from below. If* J *satisfies condition* (C) *then* J *is coercive, i.e.,*

$$\lim_{\|u\|_X \longrightarrow \infty} J(u) = \infty.$$

3. A simple model

The application of the abstract result quoted in the previous section to the study of the functional J defined in (1) is very technical. In particular, the verification of the condition (C). For this reason, we present here a simple but not natural functional which is not differentiable in $W_0^{1,2}(\Omega)$, but for which the verification of condition (C) has not technical difficulties like for functionals studied in [5], [6]. Specifically, we consider the functional J defined in $W_0^{1,2}(\Omega)$ by setting

$$\left.\begin{array}{c} J(v) = \frac{1}{2}\int_\Omega |\nabla v|^2\,dx + \frac{1}{q}\int_\Omega a(x,v)|\nabla v|^q\,dx - \frac{1}{m}\int_\Omega (v^+)^m\,dx, \\ v \in W_0^{1,2}(\Omega), \end{array}\right\} \tag{5}$$

where $1 \le q < 2 < m < 2^*$ ($2^* = 2N/(N-2)$, if $2 < N$; $2^* = \infty$ if $N \le 2$) and $a : \Omega \times \mathbb{R} \longrightarrow \mathbb{R}$ is a function which is measurable respect to $x \in \Omega$, C^1 with respect to $z \ne 0$ and such that

(a_1) There exist $\beta > \alpha > 0$ such that

$$\alpha \le a(x,z) \le \beta,$$

for almost every $x \in \Omega$ and $z \in \mathbb{R}$.

(a_2) There exists $\gamma > 0$ such that

$$|a'_z(x,z)| \le \gamma, \text{ for almost every } x \in \Omega,\ \forall z > 0,$$

and $a'(x,z) = 0,\ z < 0$.

(a_3) Either

$$a(x,z) \text{ is increasing and concave with respect to } z \ge 0, \tag{6}$$

or

$$a(x,z) \text{ is decreasing and convex with respect to } z \ge 0. \tag{7}$$

Let $X = W_0^{1,2}(\Omega)$, endowed with the usual norm $\|\cdot\|$; $Y = W_0^{1,2}(\Omega) \cap L^{2/(2-q)}(\Omega)$, endowed with the norm $\|\cdot\|_Y = \|\cdot\|_{2/(2-q)}$. By (a_1) and (a_2) the functional J is continuous on X and satisfies (H). We point out that $X = Y$ only for $q \le 1 + \frac{2}{N}$.

Theorem 3.1 [5] – *Assume* $(a_1 - a_3)$ *and* $1 \le q < 2 < m < 2^*$. *Then the functional* J *defined in* (5) *satisfies* (C).

Proof. Let $\{u_n\}$ be a sequence in Y satisfying (2), (3) and (4) for some $\{K_n\} \subset \mathbb{R}^+$ and let $\{\varepsilon_n\} \longrightarrow 0$. We prove that $\{u_n\}$ is bounded in X. Indeed, taking $v = u_n$ as test function in (4), multiplying (2) by m and adding, we obtain

$$\left(\frac{m}{2}-1\right)\int_\Omega |\nabla u_n|^2\,dx - \frac{1}{q}\int_\Omega a'_z(x,u_n)u_n|\nabla u_n|^q\,dx$$
$$+\left(\frac{m}{q}-1\right)\int_\Omega a(x,u_n)|\nabla u_n|^q\,dx \quad\le\quad C_1+\varepsilon_n(2+\|u_n\|).$$

Hence, if $a(x,z)$ satisfies (6), taking into account (a_1), it follows for some $C_2>0$ that

$$\left(\frac{m}{q}-1\right)a(x,z)-\frac{1}{q}a'_z(x,z) \;=\; \frac{1}{q}\left[a(x,z)-a'_z(x,z)z\right]$$
$$+\left(\frac{m}{q}-1-\frac{1}{q}\right)a(x,z)\ge -C_2,$$

and thus we deduce

$$\left(\frac{m}{2}-1\right)\int_\Omega |\nabla u_n|^2\,dx \le C_2\int_\Omega |\nabla u_n|^q\,dx + C_1+\varepsilon_n(2+\|u_n\|),$$

which implies, since $q<2$, that the sequence $\{u_n\}$ is bounded in $W_0^{1,2}(\Omega)$.

On the other hand, if instead of (6), it holds (7), then that the sequence $\{u_n\}$ is bounded follows easily from the fact that $a'_z(x,z)z\le 0$. Therefore the sequence $\{u_n\}$ is bounded in $X=W_0^{1,2}(\Omega)$. Then there exist $u\in X$ and a subsequence (still denoted u_n) such that u_n converges weakly to u.

Now let

$$T_k(z)=\begin{cases} z, & \text{if } |z|\le k\\[2ex] k\dfrac{z}{|z|}, & \text{if } |z|>k.\end{cases}$$

and

$$G_k(z)=z-T_k(z),\ \forall z\in\mathbb{R},\ \ \forall k>0.$$

To conclude the proof, it suffices to prove

Step 1. $\{T_k(u_n)\}\longrightarrow T_k(u)$ in $W_0^{1,2}(\Omega)$, as $n\to\infty$, for every $k\ge R_1$.

Step 2. For every $\delta>0$, there exist $k_0\ge R_1$ and $n_0\in\mathbb{N}$ such that $\|G_k(u_n)\|<\delta$ for every $k\ge k_0$ and $n\ge n_0$.

Indeed, Steps 1 and 2 imply that, given $\delta>0$, there exist $n_1\in\mathbb{N}$ and $k_1\ge R_1$ such that

$$\begin{aligned}\|u_n - u\| &\leq \|u_n - T_{k_1}(u)\| + \|T_{k_1}(u) - u\| \\ &\leq \|T_{k_1}(u_n) - T_{k_1}(u)\| + \|G_{k_1}(u_n)\| + \|T_{k_1}(u) - u\| \\ &\leq 3\delta, \ \forall n \geq n_1,\end{aligned}$$

i.e. $\{u_n\}$ is strongly convergent in $W_0^{1,2}(\Omega)$ to $u \in W_0^{1,2}(\Omega)$.

Step 1. Putting $w_{n,k} = T_k(u_n) - T_k(u)$ as test function in (4), we deduce

$$\begin{aligned}\int_\Omega \nabla u_n \cdot \nabla w_{n,k}\, dx &+ \int_\Omega a(x, u_n)|\nabla u_n|^{q-2}\nabla u_n \cdot \nabla w_{n,k}\, dx \\ &+ \frac{1}{q}\int_\Omega a'_z(x, u_n) w_{n,k}|\nabla u_n|^q\, dx \leq \varepsilon'_n\end{aligned}$$

with $\{\varepsilon'_n\} \longrightarrow \infty$. Remark that

$$\begin{aligned}\int_\Omega a(x, u_n)|\nabla u_n|^{q-2}\nabla u_n \cdot \nabla w_{n,k} dx \geq \int_\Omega a(x, u_n)|\nabla T_k(u)|^{q-2}\nabla T_k(u) \cdot \nabla w_{n,k} dx \\ + \int_{|u_n|>k} a(x, u_n)|\nabla u_n|^{q-2}\nabla u_n \cdot \nabla T_k(u) dx\end{aligned}$$

and the right hand side converges to zero.

Moreover,

$$\int_\Omega a'_z(x, u_n) w_{n,k}|\nabla u_n|^q dx \leq C_1\left(\int_\Omega |\nabla u_n|^2 \nabla u_n dx\right)^{q/2} \|T_k(u_n) - T_k(u)\|_{q/(q-2)}$$

Thus, it follows that the sequence $T_k(u_n)$ is convergent in $W_0^{1,2}(\Omega)$ to $T_k(u)$ for every $k > 0$.

Step 2. The assertion is easily proved by taking $G_k(u_n)$ as test function in (4) and using (a_3). ■

Thanks to the previous lemma, we can prove existence of a nontrivial critical point for the functional J. That is, the existence of a weak solution of the quasilinear Dirichlet problem

$$\left.\begin{aligned}-\Delta u - \text{div}\ \left(a(x, u)|\nabla u|^{q-2}\nabla u\right) + \tfrac{1}{q}a'_z(x, u)|\nabla u|^q = |u|^{m-2}u \\ u \in W_0^{1,2}(\Omega),\ u \geq 0 \text{ in } \Omega\end{aligned}\right\}$$

Theorem 3.2 – *Assume* $(a_1 - a_3)$ *and* $1 \leq q < 2 < m < 2^*$. *Then the functional* J *given by* (5) *has at least a positive critical point.*

Proof. We point out that every nonzero critical point of J is positive. In fact, it is deduced taking $T_k(u^-)$ as test function (note that u may not belong to Y, but $T_k(u^-) \in Y$). In order to show the existence of a nonzero critical point, and following the ideas of Lemma 3.3 in [2], it is easy to check that $u = 0$ is a strict local minimum of J, that is, there exist ρ, $\bar{R}$ such that

$$J(u) \geq \rho > 0 \qquad \text{for } \|u\| = \bar{R} > 0. \tag{8}$$

Moreover, $\lim_{|t|\to\infty} J(t\varphi_1) = -\infty$, being $\varphi_1 > 0$ an eigenfunction associated to the first eigenvalue λ_1 of the homogeneous Dirichlet problem for the laplacian operator with L^2-norm equal to one. Thus, there exists $t_0 > \bar{R}$ such that $J(t_0\varphi_1) < 0$. Thus, letting $e = t_0\varphi_1$ and considering the set Γ of the (continuous) paths $\gamma : [0,1] \longrightarrow \left(W_0^{1,2}(\Omega) \cap L^{2/(2-q)}(\Omega), \|\cdot\| + \|\cdot\|_{2/(2-q)}\right)$ which join 0 and e, i.e. such that $\gamma(0) = 0$ and $\gamma(1) = t_0\varphi_1$, we observe that every $\gamma \in \Gamma$ is continuous from $[0,1]$ to $W_0^{1,2}(\Omega)$, so that, by (8), for every $\gamma \in \Gamma$ there exists $\bar{t} \in [0,1]$ such that

$$\|\gamma(\bar{t})\| = \bar{R}.$$

Hence

$$c \equiv \inf_{\gamma\in\Gamma} \max_{t\in[0,1]} J(\gamma(t)) \geq \rho > \max\{J(0), J(t_0\varphi_1)\} = 0.$$

Then, taking into account Lemma 3.1 and applying Theorem 2.2, we deduce the existence of a critical point $u \in W_0^{1,2}(\Omega) \cap L^{2/(2-q)}(\Omega)$, of J with $J(u) = c > 0$ and thus $u \neq 0$. ■

4. Main examples

The abstract theorem (with $X = W_0^{1,2}(\Omega)$ and $Y = W_0^{1,2}(\Omega) \cap L^\infty(\Omega)$) of the Section 2 is applied now to obtain nonnegative critical points of the functional $J : W_0^{1,2}(\Omega) \longrightarrow \mathbb{R} \cup \{+\infty\}$ defined by

$$J(v) = \int_\Omega A(x,v)|\nabla v|^2\,dx - \int_\Omega F(x,v^+)\,dx, \; v \in W_0^{1,2}(\Omega), \tag{9}$$

i.e. nonnegative solutions of the boundary value problem:

$$\left.\begin{aligned} &u \in W_0^{1,2}(\Omega) \cap L^\infty(\Omega), \\ &-\mathrm{div}(\,A(x,u)\,\nabla u) + \tfrac{1}{2}A'_z(x,u)|\nabla u|^2 = F'_u(x,u) \equiv f(x,u) \end{aligned}\right\} \tag{P}$$

where $f : \Omega \times \mathrm{I\!R} \longrightarrow \mathrm{I\!R}$ is a Carathéodory function with subcritical growth. It is clear that for a solution u of (P) we are meaning

$$\left.\begin{array}{c} u \in W_0^{1,2}(\Omega) \cap L^\infty(\Omega) \\[2ex] \int_\Omega A(x,u)\nabla u \nabla v dx + \frac{1}{2}\int_\Omega A'_z(x,u)|\nabla u|^2 v dx = \int_\Omega f(x,u) v dx \end{array}\right\}$$

for every $v \in W_0^{1,2}(\Omega) \cap L^\infty(\Omega)$.

The hypotheses that we assume on the Carathéodory coefficient $A : \Omega \times \mathrm{I\!R} \to \mathrm{I\!R}$ are the following:

(A_1) There exists $\alpha > 0$ such that

$$\alpha \leq A(x,z),$$

for almost every $x \in \Omega$ and $z \geq 0$.

(A_2) There exists $R_1 > 0$ such that $A'_z(x,z) \geq 0$ for almost every $x \in \Omega$, for every $z \geq R_1$.

(A_3) There exist $m > 2$ and $\alpha_1 > 0$ such that

$$\left(\frac{m-2}{2}\right) A(x,z) - \frac{1}{2} z A'_z(x,z) \geq \alpha_1,$$

for almost every $x \in \Omega$, $z \geq 0$.

Notice that all assumptions on $A(x,z)$ are for $z \geq 0$. In fact, since we are looking for nonnegative solutions of (P) we can assume without loss of generality that $A(x,z)$ is even on z.

On the other hand, we will assume the following conditions on $f(x,z)$:

(f_1) There exist $C_1, C_2 > 0$ such that

$$|f(x,z)| \leq C_1 |z|^\sigma + C_2, \quad \text{a.e. } x \in \Omega, \quad \forall z \in \mathrm{I\!R}^+,$$

with $\sigma + 1 < 2^*$, ($2^* = 2N/(N-2)$ if $2 < N$, and $2^* = \infty$ if $2 \geq N$).

(f_2) There exists $R_2 > 0$ such that

$$0 < mF(x,z) \leq z f(x,z),$$

for almost $x \in \Omega$ and every $z \geq R_2$ (m is the same as in (A_3)).

(f_3) $f(x,|z|) = o(|z|)$ at $z = 0$, uniformly in $x \in \Omega$.

Theorem 4.1 – *Assume* (A_{1-3}), (f_{1-3}) *and that*

$$\lim_{z \to +\infty} \frac{A(x,z)}{z^{\sigma}} = 0, \quad \textit{unif. in } x \in \Omega. \tag{10}$$

Then the problem (P) *has, at least, one nonnegative and nontrivial solution.*

Remarks 4.2.

1. The above theorem is essentially contained in [5]. However, in that paper it is assumed in addition that $A(x,z)$ is bounded from above and its derivative $A'_z(x,z)$ with respect to z is also bounded. In [7], we have seen that these additional hypotheses are not necessary for the existence.

2. The general case of functionals

$$\int_{\Omega} \mathcal{J}(x,v,\nabla v)\,dx - \int_{\Omega} F(x,v^+)\,dx, \; v \in W_0^{1,p}(\Omega), \; (p > 1)$$

could be also handled as in [5]. For simplicity reasons, we just present here the case $p = 2$, $\mathcal{J}(x,v,\nabla v) = A(x,v)|\nabla v|^2$.

3. Some remarks about the meaning of (A_3) and (f_2) may be found in [5, Lemma 3.2 and Remarks 3.1]. ■

Proof of Theorem 4.1. For every $n \in \mathbb{N}$, let h_n be a nondecreasing C^1 function in $[0,\infty)$ satisfying

$$h_n(s) = s, \; \forall s \in [0, n-1],$$

$$h_n(s) \le s, \; \forall s \in (n-1, n),$$

$$h_n(s) = n, \; \forall s \ge n.$$

Consider the coefficients $A_n(x,z) \equiv h_n(A(x,z))$, $x \in \Omega$, $z \in \mathbb{R}$. Clearly, A_n satisfies (A_{1-3}) and, in addition, it is bounded from above with bounded derivative $A'_n(x,z)$ (with respect to z). In this way, if we define the functionals $J_n : W_0^{1,2}(\Omega) \longrightarrow \mathbb{R}$ by setting

$$J_n(v) = \int_{\Omega} A_n(x,v)|\nabla v|^2\,dx - \frac{1}{s+1}\int_{\Omega} F\left(x,v^+\right)\,dx, \; v \in W_0^{1,2}(\Omega),$$

then using (f_{1-2}) and (A_3), it can be seen in a similar way to the one in Section 2 that J_n satisfies (C). Indeed, we have

Lemma 4.3 – *(Compactness condition) Assume* (A_{1-3}) *and* (f_{1-2})*. Then the functional* J_n *satisfies* (C).

Using in addition (f_3) and following the ideas of [2], it is easily seen that J_n satisfies the geometrical hypotheses of Theorem 2.2. Consequently, by it, there exists a nontrivial and nonnegative solution u_n of the problem

$$\left.\begin{array}{r} u_n \in W_0^{1,2}(\Omega) \cap L^\infty(\Omega), \\ -\mathrm{div}(\,A_n(x,u_n)\,\nabla u_n) + \frac{1}{2}A_n'(x,u_n)|\nabla u_n|^2 = f(x,u_n^+). \end{array}\right\} \tag{11}$$

with critical level

$$J_n(u_n) = c_n \equiv \inf_{\gamma\in\Gamma}\max_{t\in[0,1]} J_n(\gamma(t)),$$

where $\Gamma = \{\gamma : [0,1] \longrightarrow W_0^{1,2}(\Omega) \cap L^\infty(\Omega) \,/\, \gamma(0) = 0, \gamma(1) = e_n\}$, $e_n \in W_0^{1,2}(\Omega) \cap L^\infty(\Omega)$ is such that $J_n(e_n) < 0$. Taking into account that $A_n(x,z) \le A(x,z)$ and (10), we observe that

$$J_n(t\varphi_1) \le J(t\varphi_1) = \frac{1}{2}t^{s+1}\left[\in_\Omega \frac{A(x,t\varphi_1)}{t^{s-1}}|\nabla\varphi_1|^2\,dx - \frac{1}{s+1}\int_\Omega \varphi_1^{s+1}\,dx\right] < 0,$$

for all $t \in [t_0,\infty)$ if $t_0 > 0$ is large enough. This allows us to choose $e_n = t_0\varphi_1$ (independent of $n \in \mathbb{N}$). On the other hand, by the Mountain Pass geometry of J_1 there exist $\delta, r > 0$ such that

$$J_n(v) \ge J_1(v) \ge \delta,\ \forall \|v\| \le r,$$

(i.e., roughly speaking, $v = 0$ is a strict local minimum of J_n, uniformly in $n \in \mathbb{N}$). This implies that

$$J_n(u_n) = c_n \ge \delta. \tag{12}$$

We claim that $\{u_n\}$ is bounded in $W_0^{1,2}(\Omega)$. Indeed, using again that $A_n(x,z) \le A(x,z)$, we deduce

$$\begin{aligned} J_n(u_n) &= \inf_{\gamma\in\Gamma}\max_{t\in[0,1]} J_n(\gamma(t)) \\ &\le \inf_{\gamma\in\Gamma}\max_{t\in[0,1]} J(\gamma(t)) \\ &\le \max_{t\in[0,1]} J(tt_0\varphi_1) \equiv C_1. \end{aligned}$$

Subtracting $\dfrac{1}{m}\langle J_n'(u_n), u_n\rangle = 0$ we derive

$$\left(\frac{1}{2}-\frac{1}{m}\right)\int_\Omega A_n(x,u_n)|\nabla u_n|^2\,dx-\frac{1}{2m}\int_\Omega A'_n(x,u_n)u_n|\nabla u_n|^2\,dx\le C_2$$

which, by (A_3), implies that $\|u_n\|$ is bounded, thus proving the claim.

Then, passing to a subsequence, if necessary, we can assume that $\{u_n\}$ is weakly convergent to some $u\in W_0^{1,2}(\Omega)$. Now, we prove that the sequence $\{u_n\}$ is bounded in $L^\infty(\Omega)$. Indeed, we can use $v=G_k(u_n)$, $k>R_1$, as test function in (11) to deduce that $\{u_n\}$ satisfies

$$\|G_k(u_n)\|_{2^*}^2\le C_3\|u\|_r^\sigma\,[\text{meas }\Omega(k)]^{[(2^*-1)r-2^*\sigma]/2^*r},$$

for any $r\ge 2^*$ such that $u\in L^r(\Omega)$.

Hence there is a constant $C_4>0$ such that $\|u_n\|_\infty\le C_4$ (see [14]). Since the sequence $\{u_n\}$ is bounded in $W_0^{1,2}(\Omega)$ and $L^\infty(\Omega)$ and (f_1), we yield that $\{u_n\}$ is compact in $W_0^{1,2}(\Omega)$ by using the results of [9]. Therefore, $\{u_n\}\longrightarrow u$ and $u\in W_0^{1,2}(\Omega)\cap L^\infty(\Omega)$ is a nonnegative critical point of J. In addition, $\{J(u_n)\}\longrightarrow J(u)$ and we get from (12) that $J(u)\ge\delta$ so that $u\ne 0$. Thus u is the weak nontrivial (and nonnegative) solution we were looking for. ■

Remark 4.4. We conclude by noting that in [7] the reader can find more existence results for nonlinearities $f(x,z)$ which are different combinations of concave and convex functions in the quasilinear spirit of [1], [8].

Acknowledgment. This paper was partially presented by the second author at the *Collège de France Seminar* (24.3.1995). Both authors would like to thank to the organizers of the Seminar for having given the opportunity of presenting their work.

References

[1] Ambrosetti, A., Brezis, H. and Cerami, G., Combined effects of concave and convex nonlinearities in some elliptic problems. J. Funct. Anal. 122 (1994), 519–543.

[2] Ambrosetti, A. and Rabinowitz, P. H., Dual variational methods in critical point theory and applications. J. Funct. Anal. 14 (1973), 349–381.

[3] Arcoya, D. and Boccardo, L., Nontrivial solutions to some nonlinear equations via minimization. Variational Methods in Nonlinear Analysis, edited by A. Ambrosetti and K.C. Chang, 49–53, Gordon and Breach Publishers, 1995.

[4] Arcoya, D. and Boccardo, L., A min-max theorem for multiple integrals of the Calculus of Variations and applications. Rend. Mat. Acc. Lincei, s. 9, v. 6, 29–35 (1995).

[5] Arcoya, D. and Boccardo, L., Critical points for multiple integrals of Calculus of Variations. Arch. Rat. Mech. Anal. 134, 3(1996), 249–274.

[6] Arcoya, D. and Boccardo, L., Some remarks on critical point theory for nondifferentiable functionals, to appear in NoDEA.

[7] Arcoya, D. and Orsina, L., Landesman-Lazer conditions and quasilinear elliptic equations, Nonlinear Anal. TMA. 28 (1997), 1623–1632.

[8] Boccardo, L., Escobedo, M. and Peral, I., A Dirichlet problem involving critical exponent. Nonlinear Anal. TMA., 24 (1995), 1639–1648.

[9] Boccardo, L., Murat, F. and Puel, J.P., Existence de solutions faibles pour pour des équations quasilinéaires à croissance quadratique. Res. Notes in Mathematics 84, Pitman, 1983, 19-73.

[10] Canino, A. and Degiovanni, M., Nonsmooth critical point theory and quasilinear elliptic equations, in Top. methods in differential equations and inclusions, Kluwer Academic Publisher, 1995.

[11] Dacorogna, B., Direct Methods in the Calculus of Variations. Springer-Verlag, 1989.

[12] Degiovanni, M. and Marzocchi, M., A critical point theory for nonsmooth functionals. Ann. Mat. Pura Appl. 167 (1994), 73-100.

[13] Pellacci, B. Critical points for non diferentiable functionals, Boll. U.M.I 11-B (1997), 733-749.

[14] Stampacchia, G., Equations elliptiques du second ordre à coefficients discontinus. Les Presses de L'Université du Montréal, 1966.

David Arcoya
Departamento de Análisis Matemático
Universidad de Granada
18071-Granada
Spain
E-mail: darcoya@ugr.es

Lucio Boccardo
Dipartimento di Matematica
Università di Roma 1
Piazza A. Moro 2
00185 Roma
Italy
E-mail: boccardo@mat.uniroma1.it

Studies in Mathematics and its Applications, Vol. 31
D. Cioranescu and J.L. Lions (Editors)

Chapter 2

A SEMIGROUP FORMULATION FOR ELECTRMAGNETIC WAVES IN DISPERSIVE DIELECTRIC MEDIA

H.T. BANKS AND M.W. BUKSAS

1. Introduction

In a forthcoming monograph [2] we have developed a theoretical and computational framework for electromagnetic interrogation of dispersive dielectric media. In that work we show that one can take a time domain variational or weak formulation of Maxwell's equations in dispersive materials and, in the context of inverse problems, use partially reflected polarized microwave pulses to determine both dielectric material properties and geometry of bodies (specifically for plane waves inpinging on slab geometries in paradyms which approximate far field interrogation). This is done in configurations involving either supraconductive reflecting back boundaries or acoustically generated virtual reflectors.

The propagation and reflection of electromagnetic waves in dispersive dielectric media is, of itself, an interesting topic of investigation. As we point out in the next section (and demonstrate computationally in [2]), the underlying dynamical systems are not typical of either standard parabolic or standard hyperbolic (even with the usual dissipation) systems and are hence of mathematical interest. In this short note, we consider the Maxwell system for rather general dispersive dielectric media and show that such systems, under appropriate conditions on the polarization law, generate C_0 semigroup solutions. These results are presented in the context of the 1-dimensional interrogating systems developed in detail in [2] and we invite interested readers to consult that reference for more detailed discussions and development of the underlying model employed here.

2. Modeling of dispersiveness in dielectric media

We begin with time domain Maxwell's equations in second order form (e.g., see [2]) for the electric field $E = E(t, z)$ of 1-dimensional polarized waves

$$\ddot{E} + \frac{1}{\epsilon_0}\ddot{P} + \frac{1}{\epsilon_0}\dot{J}_c - c^2 E'' = -\frac{1}{\epsilon_0}\dot{J}_s \tag{1}$$

where $c = \frac{1}{\sqrt{\epsilon_0 \mu_0}}$ is the speed of light in vacuum, J_c is the conduction current density, J_s is a source current density and P is the electric polarization of the dielectric medium. We assume very general constitutive material laws for the polarization and conductivity given by

$$P(t, z) = (g_p * E)(t, z) = \int_0^t g_p(t - s, z)E(s, z)dz \tag{2}$$

$$J_c(t, z) = (g_c * E)(t, z) = \int_0^t g_c(t - s, z)E(s, z)dz \tag{3}$$

where we have tacitly assumed that $E(t, z) = 0$ for $t < 0$ and that both $g_p(\xi, z)$ and $g_c(\xi, z)$ vanish for $\xi < 0$. With these assumptions, the integrals in equations (2), (3) are equivalent to integration over all of $(-\infty, \infty)$ and thus are indeed convolutions. The displacement susceptibility kernel g_p (also referred to as the dielectric response function(DRF)) and the conductivity susceptibility kernel g_c introduce nonlocality in time in the polarization and conductivity relationships [1], [15] which is equivalent to frequency dependence of the dielctric permittivity ϵ and conductivity σ when using a frequency domain approach. We assume that either P or J_c or both may contain instantaneous (local in time) components by introduction of δ distributions in the kernels g_p and/or g_c, respectively.

A medium is *dispersive* if the phase velocity of plane waves propagating through it depends on the frequency of the waves [16, Chap.7], [10, Chap.8]. To determine the dispersive nature of a medium described by equations (1)-(3) we seek plane-wave solutions of the homogeneous analogue of (1) of the form $E(t, z) = E_0 e^{-i(\omega t \pm \kappa z)}$ which travel in the z direction and have wavelength $\lambda = 2\pi/\kappa$. The phase velocity v_p of these waves is the speed at which planes of constant phase move through the medium. In this case the argument $\omega t - \kappa z$ is constant and

$$v_p = \frac{dz}{dt} = \omega/\kappa. \tag{4}$$

Seeking plane wave solutions of the form $E_0 e^{-i(\omega t \pm \kappa z)}$ in (1) is equivalent to seeking solutions of the form $E_0 e^{\pm i\kappa z}$ in the frequency domain version of (1). Thus we use the Fourier transform in (1) and obtain

$$-\omega^2 \hat{E} - \frac{\omega^2}{\epsilon_0}\hat{P} - \frac{i\omega}{\epsilon_0}\hat{J}_c - c^2 \hat{E}'' = 0 \tag{5}$$

where we have ignored the source term J_s and where the overhat will represent the Fourier transform throughout. Since we see from (2) and (3) that $\hat{P} = \hat{g}_p \hat{E}$ and $\hat{J}_c = \hat{g}_c \hat{E}$, this can be written

$$c^2 \hat{E}'' + \omega^2 (1 + \frac{i\hat{g}_c}{\omega \epsilon_0} + \frac{\hat{g}_p}{\epsilon_0}) \hat{E} = 0. \tag{6}$$

We note that (6) is the generalized Helmholtz equation [16, p. 271]

$$\hat{E}'' + \kappa^2 \hat{E} = 0 \tag{7}$$

with

$$\kappa^2 = \frac{\omega^2}{c^2} (1 + \frac{i\hat{g}_c(\omega)}{\omega \epsilon_0} + \frac{\hat{g}_p(\omega)}{\epsilon_0}) \tag{8}$$

which has solutions $\hat{E}(\omega, z) = E_0 e^{\pm i\kappa(\omega) z}$. It follows that the corresponding time domain solutions are our desired solutions of the form $E(t, z) = E_0 e^{-i(\omega t \pm i\kappa z)}$ where the wavenumber $\kappa = \kappa(\omega)$ will in general depend on the frequency ω. The equation (8) relating the frequency ω and the wavenumber κ of propagating waves is known as the dispersion equation for the medium. In the case of vacuum or free space where $\hat{g}_p = \hat{g}_c = 0$ so that $\kappa = \omega / c$, we obtain the corresponding phase velocity $v_p = c =$ the speed of light as expected. More generally the phase velocity in a dielectric medium with conductivity and polarization is given by

$$v_p = c / \sqrt{1 + i\hat{g}_c / \epsilon_0 \omega + \hat{g}_p / \epsilon_0}. \tag{9}$$

In light of (9) and the definition of a dispersive medium, we see that if either $\hat{g}_c / \omega$ or $\hat{g}_p$ depend on ω, we will have dispersiveness. Several special cases are worthy of note.

For instantaneous conductivity, that is, $g_c(t, z) = \sigma \delta(t)$ so that (3) reduces to Ohm's Law $J_c = \sigma E$, we see that the term $i\hat{g}_c / \epsilon_0 \omega$ becomes $i\sigma / \epsilon_0 \omega$. Thus a medium with simple Ohm's Law conductivity will be dispersive (it is also dissipative in the usual sense since the conductivity term in (1) becomes $\frac{\sigma}{\epsilon_0} \dot{E}$). For instantaneous polarization (often assumed in standard treatments of the Maxwell theory) we find $g_p(t, z) = \epsilon_0 \chi \delta(t)$, where χ is the dielectric susceptibility constant and hence $\hat{g}_p / \epsilon_0 = \chi$ and the medium is *not* dispersive. One must turn to more complicated (and more realistic) models, such as those of Debye or Lorentz, to have a polarization based contribution to dispersiveness in a medium. For the usual Debye polarization model [11, p.386] one has

$$g_p(t) = e^{-t/\tau} \epsilon_0 (\epsilon_s - \epsilon_\infty) / \tau, \quad t > 0, \tag{10}$$

where τ is a relaxation parameter and $\epsilon_s, \epsilon_\infty$ are familiar dielectric constants. In this case one finds

$$\hat{g}_P(\omega) = \epsilon_0(\epsilon_s - \epsilon_\infty)\left[\frac{1 - i\omega\tau}{1 + \tau^2\omega^2}\right]$$

For the Lorentz model [16, p.496] we have

$$g_P(t) = \frac{\epsilon_0\omega_p^2}{\nu_o} e^{-t/2\tau} \sin \nu_0 t, \quad t > 0, \tag{11}$$

where $\nu_0 = \sqrt{\omega_0^2 - 1/4\tau^2}$. In the frequency domain this yields

$$\begin{aligned} \hat{g}_P(\omega) &= \epsilon_0\omega_p^2\left[\frac{4\tau^2}{(1 + i2\tau\omega)^2 + 4\tau^2\nu_0^2}\right] \\ &= \epsilon_0\omega_p^2\nu_0\left[\frac{(\omega_0^2 - \omega^2) - i\omega/\tau}{(\omega_0^2 - \omega^2)^2 + \omega^2/\tau^2}\right] \end{aligned}$$

and again we have a polarization based dispersive medium. Higher order (the Debye and Lorentz models correspond to first and second order, respectively, differential equation models for the polarization P - see [2] and the references therein) models, as well as combinations of such models also lead to dispersion in a medium.

Thus, in summary we see that instantaneous conductivity but not instantaneous polarization yields dispersiveness in a medium. For a polarization contribution to dispersiveness one must include first or higher order polarization models (instantaneous polarization can be correctly viewed as zero order polarization dynamics). For our semigroup presentation in the next section we shall therefore consider the model (1) with instantaneous conductivity and a general (higher order) polarization model given by (2) with $g_P = g$ where the DRF g is assumed smooth in time (i.e., without loss of generality we can assume no instantaneous component for g). Such distributed parameter systems are of interest since they are neither simple hyperbolic nor parabolic in nature.

For simple Ohm's Law conductivity and instantaneous polarization (or no polarization), the system (1) becomes a well understood dissipative or damped hyperbolic system for which a semigroup formulation can readily be found in the research literature on distributed parameter systems. However, for (1) with polarization based dispersiveness, we obtain a system with behavior of solutions that are neither standard hyperbolic (finite speed of wave propagation along characteristics) nor standard parabolic (infinite speed of propagation of disturbances). Indeed for (1) with either Debye or Lorentz polarization, rather fascinating solutions can be observed. These involve the formulation of so-called Brillouin and Sommerfeld precursors where a pulsed excitation (containing waves with a range of frequencies) evolves into waves propagated with different velocities which coalesce into wave "packets" (see Chapter 4 of [2] and [1] and the references therein for discussions of these phenomena).

It is of both mathematical and practical interest to know whether these interesting systems can be described in a semigroup context. The potential advantages afforded by a semigroup formulation are widespread since there is a tremendous literature for control, estimation and identification, and stabilization of systems in a semigroup setting. Results for both stochastic and deterministic control methodologies (in both time domain and frequency domain) including open loop and feedback formulations are abundant [12], [3], [4], [11], [19].

In the next section we present a semigroup formulation of the system (1) with simple Ohm's Law conductivity along with general polarization based dispersiveness generated by polarization laws of the form (2). To be more precise, we take (1) for $t > 0$ and $z \in (0,1)$ with $J_c(t,z) = \sigma(z)E(t,z)$ where $\sigma(z)$ vanishes outside $\Omega \subset (0,1]$. The closed region Ω is some dielectric material region (e.g., a slab or several slab-like regions) containing instantaneous conductivity as well as non trivial polarization of the form (2) with $g_p(t,z) = g(t,z)$ vanishing outside $z \in \Omega$. Using this form of conductivity and polarization in (1), we obtain the system

$$\ddot{E}(t,z) + \frac{1}{\epsilon_0}(\sigma(z) + g(0,z))\dot{E}(t,z) + \frac{1}{\epsilon_0}\dot{g}(0,z)E(t,z)$$

$$+ \int_0^t \frac{1}{\epsilon_0}\ddot{g}(t-s,z)E(s,z)ds - c^2E''(t,z) = -\frac{1}{\epsilon_0}\dot{J}_s(t,z). \tag{12}$$

With this system we take boundary conditions (see [2] for details) that represent a total absorbing boundary at $z = 0$ and a supraconductive boundary at $z = 1$. This can be expressed by

$$\dot{E}(t,0) - cE'(t,0) = 0 \tag{13}$$

$$E(t,1) = 0. \tag{14}$$

With the definitions

$$\alpha(t,z) = \frac{1}{\epsilon_0}\ddot{g}(t,z), \quad \beta(z) = \frac{1}{\epsilon_0}\dot{g}(0,z)$$

$$\gamma(z) = \frac{1}{\epsilon_0}(\sigma(z) + g(0,z)), \quad \mathcal{J}(t,z) = -\frac{1}{\epsilon_0}\dot{J}_s(t,z),$$

we can write equation (12) as

$$\ddot{E} + \gamma\dot{E} + \beta E + \alpha * E - c^2E'' = \mathcal{J} \tag{15}$$

where $\alpha * E$ is the usual convolution

$$\alpha * E(t,z) = \int_0^t \alpha(t-s,z)E(s,z)ds. \tag{16}$$

One can use the boundary conditions (13) - (14) to write (15) in weak or variational form so as to seek solutions $t \to E(t)$ in $V \equiv H_R^1(0,1) = \{\phi \in H^1(0,1) : \phi(1) = 0\}$ in a Gelfand triple setting $V \hookrightarrow H \hookrightarrow V^*$ with pivot space $H = L^2(0,1)$. Under modest regularity assumptions on α, β, γ and $\mathcal{J}$, one can establish existence, uniqueness and continuous dependence (on initial conditions and input) of solutions. Details are given in Chapter 3 of [2].

3. A semigroup formulation

We turn in this section to a semigroup formulation for the dispersive system (12) - (14) or equivalently, (13) - (15), with instantaneous conductivity and general (non instantaneous) polarization.

For our development we assume that $\gamma, \beta \in L^\infty(0,1)$ while $\alpha \in L^\infty((0,T) \times(0,1))$ and α, β, γ vanish outside Ω. We moreover assume that α can be written as $\alpha(t,z) = \alpha_1(t)\alpha_2(z)$ where $0 < \alpha_L \leq \alpha_2(z) \leq \alpha_U$ on $\Omega \subset (0,1]$ for positive constants α_L, α_U, with α_2 vanishing outside Ω. We assume that $t \to \alpha_1(t)$ is positive, monotone nonincreasing, and in $H^1(0,T)$ so that $\dot{\alpha}_1(t) \leq 0$. This monotonicity assumption is typical of the usual assumptions in displacement susceptibility kernels (e.g., see [9, p.102] or [15]). We shall return to discuss this monotonicity requirement further after our semigroup arguments of this section.

We consider the term (16) given by

$$\int_0^t \alpha(t-s)E(s)ds = \int_{-\infty}^t \alpha(t-s)E(s)ds$$

from (15) and note that it can be equivalently written

$$\begin{aligned}\int_{-\infty}^t \alpha(t-s)E(s)ds &= \int_{-\infty}^0 \alpha(-\xi)E(t+\xi)d\xi \\ \approx \int_{-r}^0 \alpha(-\xi)E(t+\xi)d\xi &= \int_{-r}^0 G(\xi)E(t+\xi)d\xi\end{aligned}$$

where $G(\xi) \equiv \alpha(-\xi)$. We denote $G_1(\xi) = \alpha_1(-\xi)$ so that $G(\xi) = G_1(\xi)\alpha_2$.

The approximation is valid for r sufficiently large ($r = \infty$ is permitted) so that $\alpha(t) \approx 0$ for $t > r$. We observe at this point that $\dot{G}_1(\xi) \geq 0$ with $G_1(\xi) > 0$ on $(-r, 0]$.

As introduced in the previous section, we take $V = H^1_R(0,1), H = L^2(0,1)$ with $V \hookrightarrow H \hookrightarrow V^*$. We shall have use of $\tilde{H} = L^2_{\alpha_2}(\Omega)$, the space $L^2(\Omega)$ with weighting function α_2, which is readily seen to be equivalent to $L^2(\Omega)$ due to the upper and lower bounds on $\alpha_2 \in L^\infty(\Omega)$. We shall denote the restriction of functions ϕ in $L^2(0,1)$ to Ω again by ϕ and write $\phi \in L^2(\Omega)$ or $\phi \in L^2_{\alpha_2}(\Omega)$ whenever no confusion will result.

Using the above definitions and approximating, we may write (15) as

$$\ddot{E}(t) + \gamma\dot{E}(t) + \beta E(t) + \int_{-r}^{0} G(\xi)E(t+\xi)d\xi - c^2E''(t) = \mathcal{J}(\sqcup). \tag{17}$$

Using an approach given in [5], [6], [14] and [7] for viscoelastic systems, we define an auxiliary variable $w(t)$ in $W \equiv L^2_{G_1}(-r,0;\tilde{H})$ by $w(t)(\theta) = E(t) - E(t+\theta), -r \leq \theta \leq 0$. Since $G(\theta,z) > 0$ for $\theta \in (-r,0], z \in \Omega$ we may take as an inner product for W the weighted L^2 inner product

$$\langle \eta, w\rangle_W \equiv \int_{-r}^{0} G_1(\theta)\langle \eta(\theta), w(\theta)\rangle_{\tilde{H}} d\theta = \int_{-r}^{0} G_1(\theta)\int_\Omega \alpha_2(z)\eta(\theta,z)w(\theta,z)dz$$

under which W is a Hilbert Space. We note that by our notational convention explained above, we have $w(t) \in W$ for any $E(t,z)$ with $E(\cdot,\cdot) \in L^2_{G_1}(-r,0;H)$. Using a standard shift notation, we may write $w(t) = E(t) - E(t+\theta) = E(t) - E^t(\theta)$ where $E^t(\theta) \equiv E(t+\theta)$ for $-r \leq \theta \leq 0$. Adding and subtracting appropriate terms in (17), we find

$$\ddot{E}(t) + \gamma\dot{E}(t) + \beta E(t) + \int_{-r}^{0} G(\xi)E(t)d\xi - \int_{-r}^{0} G(\xi)\left[E(t) - E^t(\xi)\right]d\xi$$
$$-c^2E''(t) = \mathcal{J}(\sqcup)$$

or, equivalently

$$\ddot{E}(t) + \gamma\dot{E} + (\beta + G_{11})E(t) - \int_{-r}^{0} G(\xi)w(t)(\xi)d\xi - c^2E''(t) = \mathcal{J}(\sqcup) \tag{18}$$

where $G_{11}(z) \equiv \int_{-r}^0 G(\xi)d\xi = \alpha_2(z)\int_{-r}^0 G_1(\xi)d\xi$ and $w(t)(\xi) = E(t) - E^t(\xi)$. We observe that G_{11}, like β, is in $L^2(\Omega)$ and $L^2(0,1)$.

For our semigroup formulation, we consider (18) in the state space $Z = V \times H \times W = H^1_R(0,1) \times L^2(0,1) \times L^2_{G_1}(-r,0,\tilde{H})$ with states $(\phi,\psi,\eta) = (E(t),\dot{E}(t),w(t)) = (E(t),\dot{E}(t),E(t) - E^t(\cdot))$. To define an infinitesimal generator, we begin by defining a fundamental set of component operators. Let $\tilde{A} \in \mathcal{L}(\mathcal{V},\mathcal{V}^*)$ be defined by

$$\tilde{A}\phi \equiv c^2\phi'' - (\beta + G_{11})\phi + c^2\phi'(0)\delta_0 \tag{19}$$

where δ_0 is the Dirac operator $\delta_0\psi = \psi(0)$. Then we find

$$\langle -\tilde{A}\phi, \psi\rangle_{V^*,V} = \langle c^2\phi', \psi'\rangle_H + \langle(\beta + G_{11})\phi, \psi\rangle_H \tag{20}$$

so that it is readily seen that $\tilde{\sigma}_1 : V \times V \mapsto \mathbb{C}$ defined by

$$\tilde{\sigma}_1(\phi, \psi) = \langle -\tilde{A}\phi, \psi\rangle_{V^*,V} \tag{21}$$

is symmetric, V continuous and V coercive (i.e., $\tilde{\sigma}_1(\phi, \phi) \geq c_1|\phi|_V^2 - \lambda_0|\phi|_H^2$ for constants λ_0 and $c_1 > 0$).

We also define operators $B \in \mathcal{L}(\mathcal{V}, \mathcal{V}^*)$ and $\tilde{K} \in \mathcal{L}(\mathcal{W}, \mathcal{H})$ by

$$B\phi = -\gamma\phi - c\phi(0)\delta_0 \tag{22}$$

so that

$$\langle -B\phi, \psi\rangle_{V^*,V} = \langle\gamma\phi, \psi\rangle_H + c\phi(0)\psi(0) \tag{23}$$

and, for $\eta \in W = L^2_{G_1}(-r, 0; \tilde{H})$,

$$(\tilde{K}\eta)(z) = \begin{cases} 0 & z \in [0,1]\backslash\Omega \\ \int_{-r}^0 G(\xi)\eta(\xi)d\xi & z \in \Omega. \end{cases} \tag{24}$$

Since $G(\xi, z) = 0$ for $z \in [0, 1]\backslash\Omega$, we abuse notation and write this as

$$\tilde{K}\eta = \int_{-r}^0 G(\xi)\eta(\xi)d\xi$$

even though, strictly speaking, $\eta(\xi, z)$ is only defined for $z \in \Omega$.

With these definitions and notations, equation (18) can then be written as

$$\begin{aligned}\langle\ddot{E}, \phi\rangle_{V^*,V} + \langle -\tilde{A}E, \phi\rangle_{V^*,V} + \langle -B\dot{E}, \phi\rangle_{V^*,V} \\ + \quad \langle -\tilde{K}(E - E^t), \phi\rangle_{V^*,V} = \langle\mathcal{J}, \phi\rangle_{\mathcal{V}^*,\mathcal{V}}\end{aligned}$$

or

$$\ddot{E}(t) = \tilde{A}E(t) + B\dot{E}(t) + \tilde{K}(E(t) - E^t) + \mathcal{J}(\sqcup) \quad \text{in } \mathcal{V}^*. \tag{25}$$

We rewrite equation (25) as a first order system in the state $\zeta(t) = (E(t), \dot{E}(t), w(t))$ where $w(t) = E(t) - E^t$. To aid in this we introduce another operator

$$D : \text{dom } D \subset W \mapsto W$$

defined on dom $D = \{\eta \in H^1(-r, 0; \tilde{H}) | \eta(0) = 0\}$ by

$$D\eta(\theta) = \frac{\partial}{\partial\theta}\eta(\theta).$$

We then observe that $w(t) = E(t) - E^t$ satisfies

$$\begin{aligned} \dot{w}(t)(\theta) &= \dot{E}(t) - \dot{E}(t+\theta) = \dot{E}(t) - DE^t(\theta) \\ &= \dot{E}(t) + D(E(t) - E^t(\theta)) = \dot{E}(t) + Dw(t)(\theta). \end{aligned}$$

Thus we may formally rewrite (25) as a first order system and adjoin to it the equation

$$\dot{w}(t) = Dw(t) + \dot{E}(t). \tag{26}$$

We then obtain the first order system for $\zeta(t)$ given by

$$\dot{\zeta}(t) = \mathcal{A}\zeta(\sqcup) + \mathcal{F}(\sqcup) \tag{27}$$

where $\mathcal{A}$ given by

$$\mathcal{A} = \begin{pmatrix} 0 & I & 0 \\ \tilde{A} & B & \tilde{K} \\ 0 & I & D \end{pmatrix} \tag{28}$$

is defined on

$$\text{dom } \mathcal{A} = \{(\phi, \psi, \eta) \in \mathcal{Z} | \psi \in \mathcal{V}, \eta \in \text{dom } \mathcal{D}, \tilde{\mathcal{A}}\phi + \mathcal{B}\psi \in \mathcal{H}\}. \tag{29}$$

That is, $\mathcal{A}\oplus = (\phi, \tilde{\mathcal{A}}\phi + \mathcal{B}\psi + \tilde{\mathcal{K}}\eta, \psi + \mathcal{D}\eta)$ for $\Phi = (\phi, \psi, \eta)$ in dom $\mathcal{A}$. The forcing function $\mathcal{F}$ in (27) is given by $\mathcal{F} = \text{col}(\prime, \mathcal{J}, \prime)$. To argue that $\mathcal{A}$ is the infinitesimal generator of a C_0-semigroup, we actually consider the system (27) on an equivlaent space $Z_1 = V_1 \times H \times W$ where V_1 is the space V with equivalent inner product $\langle \phi_1, \phi_2 \rangle_{V_1} = \tilde{\sigma}_1(\phi_1, \phi_2)$ where $\tilde{\sigma}_1$ is the sesquilinear form given in (21). Recall that $\tilde{\sigma}_1$ is symmetric, V continuous and V coercive so that it is topologically equivalent to the V inner product.

We are now ready to prove the following generation theorem.

Theorem – *Suppose that* $\gamma, \beta \in L^\infty(0,1), \alpha \in L^\infty((0,1) \times (0,1))$ *with* α, β, γ *vanishing outside* Ω. *We further assume that* α *can be written* $\alpha(t,z) = \alpha_1(t)\alpha_2(z)$ *where* $\alpha_1 \in H^1(0,T)$ *with* $\alpha_1(t) > 0$, $\dot{\alpha_1}(t) \leq 0$, *and* $0 < \alpha_L \leq \alpha_2(z) \leq \alpha_U$ *for positive constants* α_L, α_U. *Then the operator* $\mathcal{A}$ *defined by (28), (29) is the infinitesimal generator of a* C_0*-semigroup on* Z_1 *and hence on the equivalent space* Z.

Proof. To prove this theorem, we use the Lumer-Phillips theorem ([15, p. 14]). Since Z_1 is a Hilbert space, it suffices to argue that for some λ_0, $\mathcal{A} - \lambda,\mathcal{I}$ is dissipative in Z_1 and $\mathcal{R}(\lambda\mathcal{I} - \mathcal{A}) = \mathcal{Z}_\infty$ for some $\lambda > 0$, where $\mathcal{R}(\lambda\mathcal{I} - \mathcal{A})$ is the range of $\lambda I - \mathcal{A}$. We first argue dissipativeness.

Let $\Phi = (\phi, \psi, \eta) \in \text{dom } \mathcal{A}$. Then

$$\begin{aligned}
\langle A\Phi, \Phi\rangle_{Z_1} &= \langle \psi, \phi\rangle_{V_1} + \langle \tilde{A}\phi + B\psi + \tilde{K}\eta, \psi\rangle_H + \langle \psi + D\eta, \eta\rangle_W \\
&= \langle \psi, \phi\rangle_{V_1} + \langle \tilde{A}\phi + B\psi, \psi\rangle_{V^*,V} + \langle \tilde{K}\eta, \psi\rangle_H + \langle \psi + D\eta, \eta\rangle_W \\
&= \tilde{\sigma}_1(\psi, \phi) - \tilde{\sigma}_1(\phi, \psi) + \langle B\psi, \psi\rangle_H + \langle \tilde{K}\eta, \psi\rangle_H + \langle \psi + D\eta, \eta\rangle_W \\
&= -\langle \gamma\psi, \psi\rangle_H - c|\psi(0)|^2 + \langle \tilde{K}\eta, \psi\rangle_H + \langle \psi + D\eta, \eta\rangle_W \\
&\leq |\gamma|_\infty |\psi|_H^2 + |\langle \tilde{K}\eta, \psi\rangle_H| + |\langle \psi + D\eta, \eta\rangle_W|. \qquad (30)
\end{aligned}$$

We consider estimates for the last two terms in (30) separately. From (24) we have

$$\begin{aligned}
|\langle \tilde{K}\eta, \psi\rangle_H| &= \left|\int_{-r}^0 G_1(\theta)\langle \eta(\theta), \psi\rangle_{\tilde{H}} d\theta\right| \\
&\leq \int_{-r}^0 G_1(\theta)|\eta(\theta)|_{\tilde{H}}|\psi|_{\tilde{H}} d\theta \\
&\leq \frac{1}{2}\int_{-r}^0 G_1(\theta)\{|\eta(\theta)|_{\tilde{H}}^2 + |\psi|_{\tilde{H}}^2\} d\theta \\
&\leq k_1|\eta|_W^2 + k_2|\psi|_H^2.
\end{aligned}$$

Moreover,

$$\begin{aligned}
|\langle \psi, \eta\rangle_W| &\leq \int_{-r}^0 G_1(\theta)|\langle \psi, \eta(\theta)\rangle_{\tilde{H}}| d\theta \\
&\leq \int_{-r}^0 G_1(\theta)\{\frac{1}{2}|\psi|_{\tilde{H}}^2 + \frac{1}{2}|\eta(\theta)|_{\tilde{H}}^2\} d\theta \\
&\leq k_3|\psi|_H^2 + k_4|\eta|_W^2.
\end{aligned}$$

Finally, since $G_1(\theta) \geq 0, \dot{G}_1 \geq 0$, and $\eta \in \text{dom } D$ requires $\eta(0) = 0$, we may argue

$$\begin{aligned}
\langle D\eta, \eta\rangle_W &= \int_{-r}^0 G_1(\theta)\langle D\eta(\theta), \eta(\theta)\rangle_{\tilde{H}} d\theta \\
&= \int_{-r}^0 G_1(\theta)\frac{d}{d\theta}\frac{1}{2}|\eta(\theta)|_{\tilde{H}}^2 d\theta \\
&= \int_{-r}^0 \frac{d}{d\theta}(G_1(\theta)\frac{1}{2}|\eta(\theta)|_{\tilde{H}}^2) d\theta - \int_{-r}^0 \frac{1}{2}(\dot{G}_1(\theta)|\eta(\theta)|_{\tilde{H}}^2) d\theta \\
&= G_1(0)\frac{1}{2}|\eta(0)|_{\tilde{H}}^2 - G_1(-r)\frac{1}{2}|\eta(-r)|_{\tilde{H}}^2 - \frac{1}{2}\int_{-r}^0 \dot{G}_1(\theta)|\eta(\theta)|_{\tilde{H}}^2 d\theta \\
&\leq 0.
\end{aligned}$$

Combining these estimates with (30), we obtain for $\Phi \in \text{dom } \mathcal{A}$

$$\begin{aligned} \langle \mathcal{A}\oplus, \oplus \rangle_{\mathcal{Z}_\infty} &\leq |\gamma|_\infty |\psi|_H^2 + (k_1 + k_4)|\eta|_W^2 + (k_2 + k_3)|\psi|_H^2 \\ &\leq \lambda_0 |\Phi|_{Z_1}^2, \end{aligned}$$

which yields the desired dissipativeness in Z_1.

To establish the range statement, we must argue there exists some $\lambda > 0$ such that for any given $\Psi = (\mu, \nu, \xi)$ in Z, there exists Φ in dom $\mathcal{A}$ satisfying

$$(\lambda I - \mathcal{A})\oplus = \ominus. \tag{31}$$

In view of the definition of $\mathcal{A}$, the equation (31) is equivalent to the system

$$\begin{aligned} \lambda\phi - \psi &= \mu \\ -\tilde{A}\phi + (\lambda - B)\psi - \tilde{K}\eta &= \nu \\ -\psi + (\lambda - D)\eta &= \xi \end{aligned} \tag{32}$$

for $(\phi, \psi, \eta) \in \text{dom } \mathcal{A}$, $(\mu, \nu, \xi) \in \mathcal{Z} = \mathcal{V} \times \mathcal{H} \times \mathcal{W}$. The first equation is the same as $\psi = \lambda\phi - \mu$ while the third can be written as $\eta = (\lambda - D)^{-1}(\xi + \psi) = (\lambda - D)^{-1}(\xi + \lambda\phi - \mu)$. These two equations can be substituted in the second to obtain an equation for ϕ. If this equation can be solved for $\phi \in V$, then the first and third can then be solved for ψ and η, respectively. The equation for ϕ that must be solved is given by

$$-\tilde{A}\phi + (\lambda - B)(\lambda\phi - \mu) - \tilde{K}(\lambda - D)^{-1}(\xi + \lambda\phi - \mu) = \nu$$

or

$$\left[\lambda^2 - \lambda B - \tilde{A} - \tilde{K}(\lambda - D)^{-1}\lambda\right]\phi = (\lambda - B)\mu + \nu + \tilde{K}(\lambda - D)^{-1}(\xi - \mu). \tag{33}$$

If we can invert (33) for $\phi \in V$, then $\psi = \lambda\phi - \mu$ is in V, $\eta = (\lambda - D)^{-1}[\xi + \lambda\phi - \mu]$ is in dom $D \subset W$ and

$$\begin{aligned} \tilde{A}\phi + B\psi &= \lambda^2\phi - \lambda\mu - \nu - \tilde{K}(\lambda - D)^{-1}(\xi + \lambda\phi - \mu) \\ &= \lambda\psi - \nu - \tilde{K}\eta \end{aligned}$$

is in H so that (ϕ, ψ, η) is in dom $\mathcal{A}$ and solves (32).

Thus the range statement reduces to solving (33) for $\phi \in V$. This in turn reduces to invertability of the operator $\lambda^2 - \lambda B - \tilde{A} - \tilde{K}(\lambda - D)^{-1}\lambda$.

We first observe that $(\lambda - D)^{-1} = (1 - e^{\lambda\cdot})/\lambda$ since $(\lambda - D)(1 - e^{\lambda\theta}) = \lambda$ while $\eta(\theta) = \frac{1-e^{\lambda\theta}}{\lambda}[\xi + \lambda\phi - \mu]$ satisfies $\eta(0) = 0$ and hence is in dom D.

Thus, for $\phi \in H$, $\tilde{K}(\lambda - D)^{-1}\lambda$ satisfies

$$\begin{aligned}\langle \tilde{K}(1-e^{\lambda\theta})\phi, \phi\rangle_H &= \int_{-r}^{0} G_1(\theta)(1-e^{\lambda\theta})\langle \phi, \phi\rangle_{\tilde{H}} d\theta \\ &\leq k_5|\phi|_H^2\end{aligned}$$

and

$$\begin{aligned}\langle (\lambda^2 - \lambda B)\phi, \phi\rangle_{V^*,V} &= \langle (\lambda^2 + \lambda\gamma)\phi + \lambda c\phi(0)\delta_0, \phi\rangle_{V^*,V} \\ &= \langle (\lambda^2 + \lambda\gamma)\phi, \phi\rangle_H + \lambda c|\phi(0)|^2 \\ &\geq k_6|\phi|_H^2 \quad \text{for } \lambda \text{ sufficiently large.}\end{aligned}$$

Hence for λ sufficiently large we have

$$\begin{aligned}&\langle (\lambda^2 - \lambda B - \tilde{A} - \tilde{K}(\lambda - D)^{-1}\lambda)\phi, \phi\rangle_{V^*,V} \\ &\quad = \langle (\lambda^2 - \lambda B)\phi, \phi\rangle_{V^*,V} + \tilde{\sigma}_1(\phi,\phi) - \langle \tilde{K}(\lambda - D)^{-1}\lambda\phi, \phi\rangle_H \\ &\quad \geq k_6|\phi|_H^2 + c_1|\phi|_V^2 - \lambda_0|\phi|_H^2 - k_5|\phi|_H^2 \\ &\quad = c_1|\phi|_V^2 + (k_6 - \lambda_0 - k_5)|\phi|_H^2.\end{aligned}$$

Thus if we define the sesquilinear form

$$\sigma_\lambda(\phi,\psi) \equiv \langle (\lambda^2 - \lambda B - \tilde{A} - \tilde{K}(\lambda - D)^{-1}\lambda)\phi, \psi\rangle_{V^*,V},$$

we see that for λ sufficiently large, σ_λ is V coercive and hence, by the Lax-Milgram lemma [20], it is invertible. It follows immediately that (33) is invertible for $\phi \in V$.

This completes the arguments to prove the Theorem. ■

Let $S(t)$ denote the semigroup generated by $\mathcal{A}$ so that solutions to (27) are given by

$$\zeta(t) = S(t)\zeta_0 + \int_0^t S(t-s)\mathcal{F}(\int)\lceil\int. \tag{34}$$

Solutions are clearly continuously dependent on initial data ζ_0 and the nonhomogeneous perturbation $\mathcal{F}$. The first component of $\zeta(t)$ is a generalized solution $E(t)$ of (17). One can now argue that the solution agrees with the unique weak solution obtained in Chapter 3 of [2], by using the arguments in Chapter 4.4 of [8]. Briefly, one argues equivalence for sufficiently regular initial data and nonhomogeneous perturbation. Then density along with continuous dependence is used to extend the equivalence to more general data (see [8] for details).

4. Concluding remarks

In the previous section we presented a semigroup generation theorem under general conditions on the coefficients α, β, γ of (15). The only possibly restrictive condition involved $\alpha(t,z) = \frac{1}{\epsilon_0}\ddot{g}(t,z) = \alpha_1(t)\alpha_2(z)$ where it is required that $\alpha_1(t) > 0, \dot{\alpha_1}(t) \leq 0$. We consider more closely the condition for some common polarization laws.

For Debye polarization in a region Ω, we have $\alpha_1(t) = \frac{1}{\epsilon_0}\ddot{g}_p(t)$ where g_p is given in (10). That is,

$$g_p(t) = \epsilon_0 \frac{(\epsilon_s - \epsilon_\infty)}{\tau} e^{-t/\tau}$$

so that

$$\alpha_1(t) = \frac{1}{\epsilon_0}\ddot{g}_p(t) = \frac{(\epsilon_s - \epsilon_\infty)}{\tau^3} e^{-t/\tau} > 0$$

and

$$\dot{\alpha_1}(t) = -\frac{(\epsilon_s - \epsilon_\infty)}{\tau^4} e^{-t/\tau} < 0.$$

Thus Debye polarization satisfies the conditions of the generation theorem and the associated system generates a C_0 semigroup.

For Lorentz polarization, we have (recall (11))

$$g_p(t) = \frac{\epsilon_0 \omega_p^2}{\nu_0} e^{-t/2\tau} \sin \nu_0 t$$

and hence

$$\alpha_1(t) = \frac{1}{\epsilon_0}\ddot{g}_p(t) = \frac{\omega_p^2}{\nu_0}\left[\left(\frac{1}{2\tau^2} - \omega_0^2\right) \sin \nu_0 t - \frac{\nu_0}{\tau} \cos \nu_0 t\right].$$

We therefore see that it is not possible to conclude that $\alpha_1(t) > 0$ or $\dot{\alpha}_1(t) \leq 0$ so that our generation theorem does *not* guarantee a semigroup representation for systems with a Lorentz polarization law. In spite of this, we do believe that the Lorentz law *does* yield a system with a semigroup representation. We conjecture that the proof of the theorem we present can be modified to weaken the hypothesis on α so as to include Lorentz and other oscillatory (even order) polarization models. We are currently pursuing these ideas.

In closing we point out that the general class of dielectric response functions consisting of a linear combination of decreasing exponentials (essentially multiple Debye mechanisms) suggested for glasseous materials by Hopkinson [15](see the discussion in [9, p.101-103] are included under the theory presented in this note.

Acknowledgment. This research was supported in part by the Air Force Office of Scientific Research under grants AFOSR F49620-98-1-0180 and AFOSR F49620-95-1-0447 and the Department of Energy, under contract W-7405-ENG-36. The authors are grateful to Dr. Richard Albanese, U.S. Air Force Research Laboratory, Brooks AFB, San Antonio, TX, for his continued encouragement and numerous specific technical disscussions throughout the course of the research reported on here and in [2].

References

[1] R. Albanese, J. Penn and R. Medina, Short-rise-time microwave pulse propagation through dispersive biological media, J. of Optical Society of America A, 6:1441–1446, 1989.

[2] H.T. Banks, M.W. Buksas, and T. Lin, Electromagnetic Interrogation of Dielectric Materials, SIAM Frontiers in Applied Mathematics, SIAM, Philadelphia, 2000, to appear.

[3] A. Bensoussan, G. DaPrato, M.C. Delfour, and S.K. Mitter, Representation and Control of Infinite Dimensional Systems, Vol. I, Birkhäuser, Boston, 1992.

[4] A. Bensoussan, G. DaPrato, M.C. Delfour, and S.K. Mitter, Representation and Control of Infinite Dimensional Systems, Vol. II, Birkhäuser, Boston, 1993.

[5] H.T. Banks, R.H. Fabiano and Y. Wang, Estimation of Boltzmann damping coefficients in beam models, In COMCON Conf.on Stabilization of Flexible Structures, 13–35, New York, 1988, Optimization Software, Inc.

[6] H.T. Banks, R.H. Fabiano and Y. Wang, Inverse problem techniques for beams with tip body and time hysteresis damping, Mat. Aplic. Comp., 8:101–118, 1989.

[7] H.T. Banks, N.G. Medhin and Y. Zhang, A mathematical framework for curved active constrained layer structures: Wellposedness and approximation, Num. Func. Analysis Optim., 17:1–22, 1996.

[8] H.T. Banks, R.C. Smith and Y. Wang, Smart Material Structures: Modeling, Estimation and Control, Masson/J. Wiley, Paris/Chichester, 1996.

[9] F. Bloom, Ill-Posed Problems for Integrodifferential Equations in Mechanics and Electromagnetic Theory, Vol. 3 of SIAM Studies in Applied Math, SIAM, Philadelphia, 1981.

[10] D.K. Cheng, Field and Wave Electromagnetics, Addison Wesley, Reading, MA, 1989.

[11] R.F. Curtain and A.J. Pritchard, Infinite-Dimensional Linear Systems Theory, LN in Control and Info. Sci., 8, Springer Verlag, Berlin, 1978.

[12] R.F. Curtain and H.J. Zwart, An Introduction to Infinite - Dimensional Linear Systems Theory, Springer Verlag, New York, 1995.

[13] R.S. Elliott, Electromagnetics: History, Theory and Applications, IEEE Press, New York, 1993.
[14] R.H. Fabiano and K. Ito, Semigroup theory and numerical approximation for equations in linear viscoelasticity, SIAM J. Math. Analysis, 21: 374–393, 1990.
[15] J. Hopkinson, The residual charge of the leyden jar, Phil. Trans. Roy. Soc. London, 167:599–626, 1877.
[16] J.D. Jackson, Classical Electrodynamics, J. Wiley & Sons, New York, 2nd edition, 1975.
[17] A. Pazy, Semigroups of Linear Operators and Applications to Partial Differential Equations, Springer-Verlag, New York, 1983.
[18] J.R. Reitz, F.J. Melford and R.W. Christy, Foundations of Electromagnetic Theory, Addison Wesley, Reading, MA, 1993.
[19] B. van Keulen, H_∞ Control for Distributed Parameter Systems : A State Space Approach, Birkhäuser, Boston, 1993.
[20] J. Wloka, Partial Differential Equations, Cambridge University Press, Cambridge, 1987.

H.T. Banks
Center for Research in Scientific Computation
NC State University
Raleigh, NC. 27695–8205
USA
E-mail: htbanks@unity.ncsu.edu

M.W. Buksas
Los Alamos National Laboratory
T-CNLS MS B258
Los Alamos, NM. 87545
USA
E-mail: mwbuksas@lanl.gov

Studies in Mathematics and its Applications, Vol. 31
D. Cioranescu and J.L. Lions (Editors)

Chapter 3

LIMITE NON VISQUEUSE POUR LES FLUIDES INCOMPRESSIBLES AXISYMETRIQUES

J. BEN AMEUR and R. DANCHIN

Résumé. On s'intéresse à la limite non visqueuse des équations de Navier-Stokes incompressibles tridimensionnelles axisymétriques. On suppose que les données initiales ont des propriétés de régularité stratifiée de type poche de tourbillon. En utilisant la conservation du tourbillon divisé par la distance à l'axe de symétrie (dans le cas non visqueux), on trouve des résultats de convergence pour tout temps similaires à ceux de la dimension deux. En particulier, on a convergence forte au sens de la régularité stratifiée, et le gradient de la vitesse est borné indépendamment de la viscosité. Lorsque les données initiales n'ont pas de régularité stratifiée, on donne une majoration de la vitesse de convergence L^2 en fonction de la viscosité, très proche de celle de [4] pour la dimension deux.

Abstract. We are concerned with the inviscid limit for three-dimensional axisymmetric incompressible flows. The initial data are vortex patches or, more generally, have striated regularity. Using the conservation of the vorticity divided by the distance to the axis of symmetry (in the inviscid case), we gather global convergence results similar to those of dimension two, namely, strong convergence for striated regularity and uniform estimates for the gradient of the velocity. When initial data do not have striated vorticity, we give an upper bound depending on the viscosity for the speed of convergence in L^2 norms. This result is similar to the one stated in [4] for two-dimensional fluids.

Introduction

Considérons le système de Navier-Stokes incompressible en dimension $d = 3$:

$$\begin{cases} \partial_t v_\nu + v_\nu \cdot \nabla v_\nu - \nu \Delta v_\nu = -\nabla p_\nu, \\ \operatorname{div} v_\nu = 0, \\ v_\nu(0) = v^0, \end{cases} \qquad (NS_\nu)$$

où ν, la viscosité, est une constante strictement positive, la vitesse $v_\nu(t,x)$ est un champ de vecteurs sur $\mathbb{R}^3$ dépendant du temps $t \geq 0$ et la pression $p_\nu(t,x)$ est un scalaire. La variable d'espace x décrit $\mathbb{R}^3$ entier et on s'intéresse à l'évolution du fluide pour tout temps t positif.

Formellement, pour ν tendant vers zéro, on obtient les équations d'Euler

$$\begin{cases} \partial_t v_0 + v_0 \cdot \nabla v_0 = -\nabla p_0, \\ \operatorname{div} v_0 = 0, \\ v_0(0) = v^0, \end{cases} \qquad (NS_0)$$

Il est bien connu que, pour un champ de vitesse v^0 "un peu mieux que lipschitzien" et $\nu \geq 0$, le système (NS_ν) est localement bien posé et que v_ν tend vers v_0 fortement lorsque ν tend vers zéro.

Notre résultat de référence, dû à T. Kato, sera le suivant (voir [8]):

Théorème 0.1 - *Soit $s > 5/2$ et v^0 un champ de vecteurs à divergence nulle et à coefficients dans l'espace de Sobolev $H^s(\mathbb{R}^3)$. Alors il existe un temps $T > 0$ tel que pour tout $\nu \geq 0$, le système (NS_ν) admette une unique solution v_ν dans $C([0,T];H^s) \cap C^1([0,T];H^{s-2})$ et tel que de plus, v_ν tende vers v_0 dans $C([0,T];H^s) \cap C^1([0,T];H^{s-2})$ lorsque ν tend vers 0.*

Sans hypothèse supplémentaire sur la donnée initiale, la question de l'existence globale d'une solution régulière reste ouverte mais on dispose du critère d'explosion suivant (voir [1] et [9]):

Théorème 0.2 – *Soit $\nu \geq 0$, $s > 5/2$ et $v_\nu \in C([0,T^\star[;H^s)$ une solution de (NS_ν) n'appartenant pas à $C([0,T^\star];H^s)$. Soit $\omega_\nu = \operatorname{rot} v_\nu$ le vecteur tourbillon associé au champ v_ν. Alors on a*

$$\int_0^{T^\star} \|\omega_\nu(t)\|_{L^\infty}\, dt = +\infty.$$

Ce critère d'explosion permet bien évidemment de retrouver l'existence globale de solutions régulières en dimension deux.

Nous nous intéressons ici à des champs initiaux v^0 axisymétriques, c'est-à-dire de la forme

$$v^0 = v_r^0(r,z)e_r + v_z^0(r,z)e_z$$

où nous avons adopté un système de coordonnées cylindriques (r,θ,z) et noté

$$e_r = (\cos\theta, \sin\theta, 0), \quad e_\theta = (-\sin\theta, \cos\theta, 0), \quad e_z = (0,0,1),$$

les trois vecteurs de base au point $x = (r\cos\theta, r\sin\theta, z)$.

Pour v^0 suffisamment régulière, (NS_ν) conserve cette propriété de symétrie. Le tourbillon ω_ν se réduit alors à $\omega_\nu = \omega_{\nu,\theta}(r,z)e_\theta$. En identifiant le vecteur tourbillon ω_ν au scalaire $\omega_{\nu,\theta}$, on constate que la quantité $\alpha_\nu \stackrel{\text{déf}}{=} \omega_\nu/r$ vérifie

$$(\partial_t + v_\nu \cdot \nabla)\alpha_\nu - \nu(\partial_r^2\alpha_\nu + \partial_z^2\alpha_\nu + \frac{3}{r}\partial_r\alpha_\nu) = 0. \qquad (T_\nu)$$

Lorsque $\nu = 0$, la quantité α_ν est visiblement transportée par le flot ψ_ν de v_ν, et si $\nu > 0$, l'opérateur du second ordre apparaissant dans (T_ν) a le "bon signe". Ceci a permis à M. Ukhovskiì et V. Yudovitch d´établir dans [14] le résultat suivant:

Théorème 0.3 – *Soit $v^0 \in (H^1(\mathbb{R}^3))^3$ un champ de vitesse axisymétrique à divergence nulle. Notons ω^0 le tourbillon initial. On suppose que ω^0, $\omega^0/r \in L^2(\mathbb{R}^3) \cap \mathrm{L}^\infty(\mathbb{R}^3)$. Alors, pour tout $\nu \geq 0$, (NS_ν) admet une unique solution v_ν dans $L^\infty(\mathbb{R}^+; \mathrm{L}^2)$ telle que de plus $\omega_\nu \in L^\infty_{loc}(\mathbb{R}^+; \mathrm{L}^2 \cap \mathrm{L}^\infty)$ et $\omega_\nu/r \in L^\infty(\mathbb{R}^+; \mathrm{L}^2 \cap \mathrm{L}^\infty)$. Cette solution reste axisymétrique pour tout temps.*

En combinant le théorème 0.3 avec les théorèmes 0.1 et 0.2, il est aisé de voir que si $v^0 \in H^s$ $(s > 5/2)$ et $\omega^0/r \in L^2(\mathbb{R}^3) \cap L^\infty(\mathbb{R}^3)$, alors les résultats du théorème 0.1 sont valables pour tout temps. De plus, la solution obtenue reste axisymétrique.

Dans les trois premières parties de cet article, on s'intéresse plus particulièment à des données initiales de type poche de tourbillon.

L'étude de ce genre de structures provient du cas bidimensionnel non visqueux. On parle de poche de tourbillon lorsque ω est la fonction caractéristique d'un domaine borné de $\mathbb{R}^2$. Lorsque $\nu = 0$, le tourbillon est conservé par le flot de la solution. Un résultat de Yudovitch (voir [15]) nous assure alors que la structure de poche de tourbillon est stable pour tout temps: le domaine de départ est simplement transporté par le flot.

Lorsque la frontière du domaine initial a une régularité höldérienne C^r $(r > 1)$, J.-Y. Chemin montre que la solution v_0 de (NS_0) appartient à $L^\infty_{loc}(\mathbb{R}^+; \mathrm{Lip})$ (où Lip désigne l'espace des fonctions bornées et lipschitziennes), et que la régularité C^r de la poche est préservée pour tout temps. Ceci résulte en fait de résultats bien plus généraux de persistance de la régularité stratifiée pour les fluides incompressibles (voir [3] et les références jointes).

Dans [5], le premier auteur s'est intéressé à la généralisation du résultat de J.-Y. Chemin pour les fluides faiblement visqueux. On obtient notamment le résultat suivant:

Théorème 0.4 – *Soit Ω^0 un ouvert borné de $\mathbb{R}^2$ dont la frontière est une courbe simple de classe $C^{1+\epsilon}$ $(\epsilon \in]0,1[)$. Soit v^0 le champ de vitesses à* *divergence nulle et s'annulant à l'infini, de tourbillon* $\omega^0 = 1_{\Omega^0}$. *Alors, pour tout* $\nu \geq 0$, (NS_ν) *admet une unique solution* v_ν *dans* $L^\infty_{loc}(\mathbb{R}^+; \mathrm{Lip}(\mathbb{R}^2))$ *avec donnée initiale* v^0, *et il existe une constante* C *ne dépendant que de* Ω^0 *telle que*

$$\forall \nu \in \mathbb{R}^+, \forall \mathrm{t} \in \mathbb{R}^+, \|\nabla \mathrm{v}_\nu(\mathrm{t})\|_{\mathrm{L}^\infty} \leq \mathrm{C}e^{\mathrm{Ct}}.$$

Notons $\psi_{t,\nu}$ *le flot de* v_ν *à l'instant* t *et* $\Omega_{t,\nu} = \psi_{t,\nu}(\Omega^0)$. *On a les résultats suivants de convergence:*

(i) Pour tout $\epsilon' < \epsilon$, $\partial\Omega_{t,\nu}$ *est une courbe simple de classe* $C^{1+\epsilon'}$. *Plus précisément, si* $\gamma^0 \in C^{1+\epsilon}(S^1; \mathbb{R}^2)$ *est une paramétrisation régulière de* $\partial\Omega^0$, *et si l'on pose* $\gamma_\nu(t) = \psi_{t,\nu}(\gamma^0)$, *alors* $\gamma_\nu(t)$ *est une paramétrisation régulière de* $\partial\Omega_{t,\nu}$. *De plus* $\gamma_\nu \in L^\infty_{loc}(\mathbb{R}^+; C^{1+\epsilon'}(S^1, \mathbb{R}^2))$ *uniformément en* ν *et* γ_ν *tend vers* γ_0 *dans* $L^\infty_{loc}(\mathbb{R}^+; \mathrm{C}^{1+\epsilon'}(\mathrm{S}^1, \mathbb{R}^2))$ *lorsque* ν *tend vers* 0.

(ii) Notons

$$(\Omega_{t,\nu})^c_h = \{x \in \mathbb{R}^2, \, \mathrm{d}(\mathrm{x}, \Omega_{\mathrm{t},\nu}) > \mathrm{h}\}$$

et $(\Omega^c_{t,\nu})_h = \{x \in \Omega_{t,\nu}, \, d(x, \partial\Omega_{t,\nu}) > h\}$. *Alors il existe une constante* C *ne dépendant que de* Ω^0 *et telle que pour* $\nu, t, h > 0$, *on ait*

$$\|\omega_\nu(t)\|_{L^2\left((\Omega_{t,\nu})^c_h\right)} \leq e^{-\frac{h^2}{4\nu t}\exp\left(-4(e^{Ct}-1)\right)} \left\|\omega^0\right\|_{L^2},$$

$$\|\omega_\nu(t) - 1_{\Omega_{t,\nu}}\|_{L^2\left((\Omega^c_{t,\nu})_h\right)}$$

$$\leq 2\left\|\omega^0\right\|_{L^2} \min\left(1, C\frac{(\nu t)^{1/2}}{h} e^{2(e^{Ct}-1)} e^{-\frac{h^2}{32\nu t}\exp\left(-4(e^{Ct}-1)\right)}\right).$$

Remarque 0.1. Pour des raisons techniques, nous avons été amenés dans [5] à utiliser les espaces de Besov $B^r_{a,\infty}$ $(2 \leq a < +\infty)$ pour mesurer la régularité stratifiée lorsque $\nu > 0$.

L'apparente perte de régularité dans $\partial\Omega_{t,\nu}$ n'est en fait due qu'à l'utilisation d'espaces de Hölder dans l'énoncé du théorème 0.4. Elle n'a pas lieu pour des poches de tourbillon à frontière dans $B^{1+\epsilon}_{a,\infty}$. ■

En dimension trois, même pour un fluide non visqueux, il n'y a aucune chance pour que la structure de poche de tourbillon *stricto sensu* soit stable, même à temps petit: le tourbillon n'est pas constant le long des lignes de flot. Pour un fluide axisymétrique non visqueux, on sait cependant que la quantité α_0 est conservée. Ceci a permis à P. Gamblin, X. Saint-Raymond et P. Serfati de prouver des résultats globaux de persistance de structures stratifiées (voir [7], [10] et [11]).

Dans le cas visqueux en dimension quelconque et sans hypothèse particulière de symétrie, on dispose de résultats de convergence de (NS_ν) vers

(NS_0) en un sens qui préserve la régularité stratifiée de type Besov (voir [6]). Ces résultats ne sont bien sûr que *locaux* en temps.

Nous nous proposons de montrer que dans le cas axisymétrique, les résultats de [6] sont globaux. Lorsque la donnée initiale est une *poche de tourbillon axisymétrique* (i. e. $\omega^0 = r1_{\Omega^0}$ avec Ω^0 domaine axisymétrique à frontière $C^{1+\epsilon}$), nous prouvons en sus l'analogue du théorème 0.4, partie *(ii)* (à ceci près que ω_ν est changé en ω_ν/r).

Énonçons le résultat de convergence que nous obtenons pour une telle donnée initiale.

Théorème 0.5 – *Soit $\Omega^0 \subset \mathbb{R}^3$ un ouvert borné à symétrie axiale et à frontière de classe $C^{1+\epsilon}$ ($\epsilon \in]0,1[$). Soit v^0 l'unique champ de vitesses à coefficients H^1, à divergence nulle et de tourbillon $\omega^0 = r1_{\Omega^0}$. Alors, pour tout $\nu \geq 0$, (NS_ν) admet une unique solution v_ν dans $L^\infty_{loc}(\mathbb{R}^+; \mathrm{Lip} \cap \mathrm{H}^1)$ avec donnée initiale v^0, et il existe une constante C ne dépendant que de Ω^0, telle que*

$$\forall \nu \in \mathbb{R}^+, \forall t \in \mathbb{R}^+, \|\nabla v_\nu(t)\|_{L^\infty} \leq C e^{Ct^{\frac{7}{2}} \log(1+t)}.$$

On a de plus les résultats de convergence suivants.

(i) Pour tout $t \geq 0$ et $\epsilon' < \epsilon$, $\partial\Omega_{t,\nu}$ est dans $C^{1+\epsilon'}$. Soit $\{f^0 = 0\}$ une équation non dégénérée de $\partial\Omega^0$ et $f_{t,\nu} = f^0 \circ \psi_{t,\nu}^{-1}$. Alors $\{f_{t,\nu} = 0\}$ est une équation non dégénérée de $\partial\Omega_{t,\nu}$, $f_\nu \in L^\infty_{loc}(\mathbb{R}^+; \mathrm{C}^{1+\epsilon'})$ uniformément en ν et f_ν tend vers f_0 dans $L^\infty_{loc}(\mathbb{R}^+; \mathrm{C}^{1+\epsilon'})$ lorsque ν tend vers 0.

(ii) Il existe une constante C ne dépendant que de Ω^0 et telle que si l'on note $z(t) \stackrel{\text{déf}}{=} \exp\big(-4((1+t)^{Ct^{\frac{7}{2}}} - 1)\big)$, on ait pour tout $\nu, t, h > 0$,

$$\|\frac{\omega_\nu(t)}{r}\|_{L^2\left((\Omega_{t,\nu})^c_h\right)} \leq e^{-\frac{h^2}{20\nu t} z(t)} \|\frac{\omega^0}{r} d\|_{L^2},$$

$$\|\frac{\omega_\nu(t)}{r} - 1_{\Omega_{t,\nu}}\|_{L^2\left((\Omega^c_{t,\nu})_h\right)} \leq 2\|\frac{\omega^0}{r}\|_{L^2} \min\left(1, C\frac{(\nu t)^{1/2}}{h} e^{2((1+t)^{Ct^{\frac{7}{2}}}-1)} e^{-\frac{h^2}{96\nu t} z(t)}\right).$$

Dans la dernière partie de ce travail, on abandonne les hypothèses de régularité stratifiée pour les données initiales et on s'intéresse à la vitesse de convergence des solutions v_ν de (NS_ν) vers celle v_0 de (NS_0) à v^0 fixée vérifiant les hypothèses du théorème 0.3. Le gradient de v_0 n'est alors pas nécessairement borné et peut exploser comme un logarithme au voisinage de certains points. On peut cependant montrer que v_ν converge vers v_0 fortement dans $L^\infty_{loc}(\mathbb{R}^+; L^2)$ avec une vitesse de convergence qui se comporte comme une puissance de ν se dégradant au cours du temps:

Théorème 0.6 - *Soit $v^0 \in H^1$ un champ de vecteur axisymétrique à divergence nulle tel que ω^0, $\omega^0/r \in L^2 \cap L^\infty$. Notons v_ν la solution de (NS_ν) donnée par le théorème* 0.3. *Alors il existe une constante C universelle telle que pour tout temps $T > 0$ et viscosité $\nu \geq 0$ vérifiant*

$$\nu T g^2(T) \leq 2e^{-\frac{4}{3}\exp\left(CT\|v^0\|_{L^2} f(C_0+T\|\alpha^0\|_{L^\infty})\right)},$$

avec

$$f(t) = t^{\frac{5}{2}} \log t, \quad g(t) = \|\omega^0\|_{L^2} \|v^0\|_{L^2}^{-1} + t\,\|\alpha^0\|_{L^\infty}$$

et

$$C_0 = 2 + \frac{\|\omega^0\|_{L^2\cap L^\infty}}{\|v^0\|_{L^2}},$$

on ait

$$\|v_\nu - v_0\|_{L^\infty(0,T;L^2)} \leq \|v^0\|_{L^2}\left(\frac{\nu T}{2} g^2(T)\right)^{\frac{1}{2}\exp\left(-CT\|v^0\|_{L^2} f(C_0+T\|\alpha^0\|_{L^\infty})\right)}. \tag{0.1}$$

Si de plus $\nabla v \in L^1(0,T;L^\infty)$, alors

$$\|v_\nu - v_0\|_{L^\infty(0,T;L^2)} \leq \frac{\nu^{\frac{1}{2}} e^{V(T)}}{2} \min\left(T\,\|\omega^0\|_{L^2}^2, \int_0^T \|v^0\|_{L^2}^2 g^2(\tau) e^{-2V(\tau)} d\tau\right)^{\frac{1}{2}} \tag{0.2}$$

avec $V(t) = \int_0^t \|\nabla v(\tau,\cdot)\|_{L^\infty} d\tau$.

Rappelons que dans [4] J.-Y. Chemin prouve un résultat très similaire pour les fluides bi-dimensionnels. Dans ce dernier cas cependant, la vitesse de convergence est du type $\nu^{\frac{1}{2}\exp(-CT)}$ et non pas $\nu^{\frac{1}{2}\exp(-CT^{\frac{7}{2}}\log(1+T))}$. L'apparition de la puissance $\frac{7}{2}$ ressemble certes à un artéfact technique. On peut cependant la justifier dans une certaine mesure par des considérations de *scaling* (voir la remarque A.1 de l'appendice).

Nous avons adopté le plan suivant:

Dans la première partie, on rappelle la définition des espaces stratifiés et conormaux construits sur des espaces de Besov selon [6]. Ceci permet d'énoncer un théorème général de convergence de (NS_ν) vers (NS_0) dont découle la partie *(i)* du théorème 0.5 (voir théorème 1.1).

La seconde section est consacrée à l'étude de l'équation (T_ν) vérifiée par α_ν. On y prouve un résultat de décroissance exponentielle hors du support de la donnée initiale transporté par le flot. Ceci entraîne l'assertion *(ii)* du théorème 0.5.

Dans la troisième partie, on prouve le théorème 1.1. On indique les quelques modifications à faire à [6] pour avoir un résultat global en temps.

La quatrième partie est dévolue au théorème 0.6.

On donne en appendice un lemme de régularisation utilisé dans la partie 3 et des bornes sur la croissance temporelle des normes L^p du tourbillon pour les fluides axisymétriques.

1. Résultats de convergence pour la régularité stratifiée

1.1. Espaces de Besov et espaces stratifiés

Dans toute cette partie, d désigne un entier strictement positif. Définis-sons un découpage en fréquences dyadiques: la décomposition de Littlewood-Paley.

Proposition 1.1 – *Il existe deux applications radiales à valeurs dans* $[0,1]$, $\chi \in C_0^\infty(\mathcal{B})$ *et* $\varphi \in C_0^\infty(\mathcal{C})$ *(avec* $\mathcal{C} = \left\{x \in \mathbb{R}^d \mid 5/6 \leq |x| \leq 12/5\right\}$ *et* $\mathcal{B} = \left\{x \in \mathbb{R}^d \mid |x| \leq 6/5\right\}$*) telles que :*

$$\forall \xi \in \mathbb{R}^d,\ \chi(\xi) + \sum_{q \in \mathbb{N}} \varphi(2^{-q}\xi) = 1.$$

On définit alors des opérateurs Δ_p et S_p de $\mathcal{L}(\mathcal{S}'(\mathbb{R}^d), C^\infty(\mathbb{R}^d))$ qui correspondent respectivement à des localisations en fréquences voisines de 2^p pour Δ_p et plus petites que 2^p pour S_p.

Plus précisément, soit $h = \mathcal{F}^{-1}\varphi$ et $\tilde{h} = \mathcal{F}^{-1}\chi$. On pose

$$\Delta_p u = 0 \ \text{ si } \ p \leq -2, \qquad \Delta_{-1} u = \chi(D)u = \tilde{h} \star u,$$

$$\Delta_p u = \varphi(2^{-p}D)u = 2^{pd} \int h(2^p y) u(x-y)\, dy \ \text{ si } \ p \geq 0,$$

$$S_p u = \chi(2^{-p}D)u = \sum_{q \leq p-1} \Delta_q u = 2^{pd} \int \tilde{h}(2^p y) u(x-y)\, dy.$$

Définition 1.1 – *Pour* $a \in [1,+\infty]$ *et* $r \in \mathbb{R}$, *on note* $B_a^r(\mathbb{R}^d)$ *l'ensemble des distributions* $u \in \mathcal{S}'(\mathbb{R}^d)$ *telles que*

$$\|u\|_{B_a^r} \stackrel{\text{déf}}{=} \sup_q 2^{qr} \left\|\Delta_q u\right\|_{L^a} < +\infty.$$

Remarque 1.1. On dispose d'une caractérisation des espaces de Besov par différence finie qui permet de faire le lien avec la définition habituelle des espaces de Hölder (voir [13]). En effet, pour $r \in]0,1[$ par exemple, $B^r_a(\mathbb{R}^d)$ est l'ensemble des éléments u de $L^a(\mathbb{R}^d)$ tels que

$$\|u\|_{L^a} + \sup_{0<|h|\leq 1} \frac{\|\tau_h u - u\|_{L^a}}{|h|^r} < +\infty \qquad \text{où} \quad \tau_h u(x) = u(x+h).$$

Remarque 1.2. L'espace $\left(B^r_a(\mathbb{R}^d), \|\cdot\|_{B^r_a}\right)$ est un Banach. C'est une algèbre lorsque $r > d/a$. Dans le cas particulier où $a = +\infty$, on note $C^r \stackrel{\text{déf}}{=} B^r_\infty$. L'espace C^r coïncide avec l'espace de Hölder usuel lorsque r est positif non entier et on a l'injection $B^r_a \hookrightarrow C^{r-\frac{d}{a}}$.

Remarque 1.3. On a le classique résultat d'interpolation suivant:

$$\begin{aligned} &\forall u \in B^s_a(\mathbb{R}^d) \cap B^t_a(\mathbb{R}^d), \quad \forall \theta \in [0,1], \\ &\|u\|_{B_a^{\theta s+(1-\theta)t}} \leq \|u\|_{B^s_a}{}^{\theta} \|u\|_{B^t_a}{}^{1-\theta}. \end{aligned}$$

Nous allons maintenant construire des espaces stratifiés de type Besov. Les notations adoptées ici sont celles de [6]. On les rappelle pour la commodité du lecteur.

Notations. Si X est un champ de vecteurs à coefficients et divergence dans B^r_a, on pose

$$\|X\|_{B^r_a} \stackrel{\text{déf}}{=} \sup_{1\leq i\leq d} \|X^i\|_{B^r_a} \quad \text{et} \quad \widetilde{\|} X\|_{B^r_a} \stackrel{\text{déf}}{=} \|X\|_{B^r_a} + \|\operatorname{div} X\|_{B^r_a}.$$

On définit la dérivée d'une fonction bornée u selon un tel champ X en posant $X(x,D)u = \partial_i(X^i u) - u \operatorname{div} X$. Si $(f_\lambda)_{\lambda\in\Lambda}$ est une famille d'éléments d'un espace de Banach $(B, \|\cdot\|_B)$, on écrit

$$|||f|||_B \stackrel{\text{déf}}{=} \sup_{\lambda\in\Lambda} \|f_\lambda\|_B.$$

Enfin, on notera $A \wedge B$ le produit vectoriel entre deux éléments de $\mathbb{R}^3$.

Définition 1.2 – *Soit $a \in]3,\infty[$, $\epsilon \in]3/a,1[$ et $m \in \mathbb{N}^\star$. On dit qu'une famille $X = (X_i)_{1\leq i\leq m}$ de champs de vecteurs est (ϵ,a)-substantielle si les X_i sont à coefficients et divergence dans $B^\epsilon_a(\mathbb{R}^3)$ et*

$$I(X) \stackrel{\text{déf}}{=} \inf_{x\in\mathbb{R}^3} \sup_{\lambda\neq\mu} |X_\lambda \wedge X_\mu(x)|^{\frac{1}{2}} > 0.$$

On note alors $B_a^\epsilon(X)$ l'espace des fonctions u bornées sur $\mathbb{R}^3$ et telles que $X_i(x,D)u \in B_a^{\epsilon-1}(\mathbb{R}^3)$ pour tout $i \in \{1,\cdots,m\}$, muni de la norme

$$\|u\|_{B^\epsilon_{a,X}} \overset{\text{déf}}{=} \left(\frac{1}{I(X)}\right)\left(\|u\|_{L^\infty}\widetilde{|||}X|||_{B_a^\epsilon} + |||X(x,D)u|||_{B_a^{\epsilon-1}}\right).$$

Comme cas particuliers d'espaces stratifiés, nous allons définir des espaces conormaux par rapport à des surfaces à régularité dans une classe de Besov.

Définition 1.3 – *Soit $a \in [3,+\infty]$ et $r > 3/a$. On dira qu'un fermé Σ de $\mathbb{R}^3$ est une surface de classe B_a^{r+1} s'il existe un voisinage V de Σ et une application $f \in B_a^{r+1}(\mathbb{R}^3)$ telle que ∇f ne s'annule pas sur V et $\Sigma = f^{-1}(0)\cap V$. On dira qu'un champ X à coefficients B_a^r est tangent à Σ si et seulement si $X\cdot\nabla f_{|\Sigma} \equiv 0$. On notera $\mathcal{T}_a^r(\Sigma)$ l'ensemble des champs de classe B_a^r tangents à Σ.*

Définition 1.4 – *Soit $a \in [3,+\infty]$, $\epsilon \in]3/a,1[$ et Σ une surface compacte de classe $B_a^{1+\epsilon}$. On définit alors l'espace conormal $B^\epsilon_{a,\Sigma}$ associé à Σ par*

$$B^\epsilon_{a,\Sigma} = \left\{u \in L^\infty \mid \forall X \in \mathcal{T}_a^\epsilon(\Sigma), \operatorname{div}(Xu) \in B_a^{\epsilon-1}\right\}.$$

Pour terminer, signalons que l'espace $B^\epsilon_{a,\Sigma}$ contient la fonction $r1_\Omega$ où Ω est l'ouvert borné délimité par Σ. Plus généralement, on a le résultat suivant (voir [6]):

Proposition 1.2 – *Soit Ω un ouvert borné à frontière $B_a^{1+\epsilon}$ avec $\epsilon \in]3,+\infty]$ et $\epsilon \in]3/a,1[$. Alors pour toute fonction $\phi \in B_a^\epsilon$, on a $1_\Omega\phi \in B^\epsilon_{a,\Sigma}$.*

1.2. Un résultat de convergence pour la régularité stratifiée ou conormale

Commençons par définir la notion de champ de vecteurs transporté par le flot.

Définition 1.5 – *Soit v un champ de vecteurs lipschitzien et ψ son flot, c'est-à-dire l'unique solution de l'équation*

$$\psi(t,x) = x + \int_0^t v\big(s,\psi(s,x)\big)\,ds.$$

Soit X^0 un champ de vecteurs de B_a^ϵ. On dira que X_t est le champ transporté de X^0 par le flot de v à l'instant t si $X_t(x) = X^0(x,D)\psi_t(\psi_t^{-1}(x))$.

Énonçons notre résultat de convergence de (NS_ν) vers (NS_0) dans toute sa généralité.

Théorème 1.1 – *Soit $a \in]3,+\infty[$ et $\epsilon \in]3/a,1[$. On suppose que le champ de vecteurs v^0 est à coefficients dans $H^1(\mathbb{R}^3)$, à divergence nulle et que son tourbillon ω^0 vérifie $\omega^0/r \in L^2 \cap L^\infty(\mathbb{R}^3)$. On suppose de plus que l'une des deux hypothèses suivantes est vérifiée:*

$(\mathcal{H}_1)$ *ω^0 est dans $B^\epsilon_a(X^0)$ avec $(X^0_\lambda)_{1\leq\lambda\leq m}$ famille (ϵ,a)-substantielle de champs de vecteurs.*

$(\mathcal{H}_2)$ *ω^0 est dans B^ϵ_{a,Σ^0} où Σ^0 est une surface compacte de classe $B^{1+\epsilon}_a$.*

Alors pour tout $\nu \geq 0$, le système (NS_ν) admet une unique solution v_ν dans $L^\infty_{loc}(\mathbb{R}^+;\mathrm{Lip}) \cap \mathrm{C}(\mathbb{R}^+;\mathrm{H}^1)$. En outre, v_ν tend vers v_0 et $\psi_\nu - Id$ tend vers $\psi_0 - Id$ dans $L^\infty_{loc}(\mathbb{R}^+;\mathrm{C}^{1-\eta})$ pour tout $\eta > 0$ et il existe une constante C ne dépendant que des données initiales et telle que

$$\forall \nu \in \mathbb{R}^+,\ \forall \mathrm{t} \in \mathbb{R}^+,\ \|\nabla \mathrm{v}_\nu(\mathrm{t})\|_{\mathrm{L}^\infty} \leq \mathrm{C}e^{\mathrm{C}t^{\frac{7}{2}}\log(1+\mathrm{t})}.$$

Sous l'hypothèse $(\mathcal{H}_1)$, on a de plus le résultat suivant:

Notons $X_{t,\nu}$ la famille transportée par le flot de v_ν. Alors $X_{t,\nu}$ reste (ϵ,a)-substantielle pour tout temps $t \geq 0$ et $\omega_\nu(t)$ reste dans $B^\epsilon_a(X_{t,\nu})$ uniformément en ν. Enfin, Si $\epsilon' < \epsilon$, alors $X^0_\lambda(x,D)\psi_\nu$, $X_{\nu,\lambda}$ et $\operatorname{div} X_{\nu,\lambda}$ tendent respectivement vers $X^0_\lambda(x,D)\psi_0$, $X_{0,\lambda}$ et $\operatorname{div} X_{0,\lambda}$ dans $L^\infty_{loc}(\mathbb{R}^+;\mathrm{B}^{\epsilon'}_{\mathrm{a}})$, et $X_{\nu,\lambda}(x,D)\omega_\nu$ tend vers $X_{0,\lambda}(x,D)\omega_0$ dans $L^\infty_{loc}(\mathbb{R}^+;\mathrm{B}^{\epsilon'-1}_{\mathrm{a}})$.

Sous l'hypothèse $(\mathcal{H}_2)$, on a de plus le résultat suivant:

Pour tout $t \geq 0$ et $\nu \geq 0$, $\Sigma_{t,\nu} \stackrel{\text{déf}}{=} \psi_{t,\nu}(\Sigma^0)$ est une surface compacte de classe $B^{1+\epsilon}_a$, et $\omega_\nu(t)$ est dans $B^\epsilon_{a,\Sigma_{t,\nu}}$. Enfin, $\Sigma_{t,\nu}$ tend vers $\Sigma_{t,0}$ au sens suivant. Soit $\{f^0 = 0\}$ une équation de classe $B^{1+\epsilon}_a$ de Σ^0. Posons $f_{t,\nu} \stackrel{\text{déf}}{=} f^0 \circ \psi^{-1}_{t,\nu}$. Alors $\{f_{t,\nu} = 0\}$ est une équation de $\Sigma_{t,\nu}$ et pour tout $\epsilon' < \epsilon$, f_ν tend vers f_0 dans $L^\infty_{loc}(\mathbb{R}^+;\mathrm{B}^{1+\epsilon'}_{\mathrm{a}})$ lorsque ν tend vers 0.

Remarque 1.4. L'hypothèse $v^0 \in H^1$ est automatiquement vérifiée dès que $\omega^0 \in L^p \cap L^\infty$ pour un $p \leq 6/5$ (c'est le cas par exemple si $\omega^0 \in B^\epsilon_a(X^0)$ est à support compact ou dans L^1).

Remarque 1.5. Lorsque $\nu = 0$, la partie existence et régularité d'une solution pour le système d'Euler s'étend aux espaces de Hölder (voir [7], [10] et [11]). Dans le cas $\nu > 0$, la preuve du théorème 1.1 utilise des estimations uniformes en ν pour les solutions d'une équation de la chaleur avec terme de convection. De telles estimations ne sont pas connues dans le cadre des espaces de Hölder (voir [5] section 4) mais sont vraies dans les B^r_a pour $1 < a < +\infty$.

Remarque 1.6. Si $(\mathcal{H}_2)$ est vérifiée, on peut construire une famille (ϵ,a)-substantielle X^0 constituée de 6 champs de vecteurs tangents à Σ^0 et telle que $\omega^0 \in B^\epsilon_a(X^0)$ (voir [6] partie 5).

Indiquons brièvement pourquoi le *(i)* du théorème 0.5 découle du théorème 1.1.

Supposons que $\omega^0 = r1_{\Omega^0}$ avec Ω^0 ouvert borné de classe $C^{1+\epsilon}$. Comme les fonctions de $C^{1+\epsilon}$ à support compact sont aussi dans tous les $B_a^{1+\epsilon}$, on a $\omega^0 \in B^\epsilon_{a,\partial\Omega^0}$ pour tout $a > 3/\epsilon$ d'après la proposition 1.2. L'hypothèse $(\mathcal{H}_2)$ est donc vérifiée. On conclut en appliquant le théorème 1.1 puis en utilisant à répétition que $B_a^\epsilon \hookrightarrow C^{\epsilon-\frac{3}{a}}$.

2. Un résultat de décroissance exponentielle

Dans cette partie, on étudie le système

$$\begin{cases} (\partial_t + v\cdot\nabla)a - \nu(\partial_r^2 a + \partial_z^2 a + \frac{3}{r}\partial_r a) = 0, \\ a_{|t=0} = a_0, \end{cases} \qquad (T_\nu)$$

que vérifie ω_ν/r avec le champ de vitesses $v = v_\nu$. Nous nous intéressons au cas où a_0 est la fonction caractéristique d'un domaine borné et où v est lipschitzien. Dans le théorème suivant, nous montrons qu'aux échelles spatiales grandes devant $\sqrt{\nu t}$, la fonction $a(t)$ est proche de la fonction caractéristique du domaine transporté par le flot.

Théorème 2.1 – *Soit $\nu > 0$ et $v \in L^\infty_{loc}(\mathbb{R}^+;\mathrm{Lip}(\mathbb{R}^3))$ un champ de vecteurs axisymétrique à divergence nulle. On suppose que $a = a(t,r,z)$ vérifie (T_ν) avec $a_0 = a_0(r,z) \in L^2(\mathbb{R}^3)$. Notons F_0 le support de a_0, ψ le flot de v et $F_t = \psi_t(F_0)$. Posons $(F_t)^c_h = \{x \in \mathbb{R}^3,\ \mathrm{d(x,F_t) > h}\}$, $(F_t^c)_h = \{x \in F_t,\ d(x,\partial F_t) > h\}$ et $V(t) = \int_0^t \|\nabla v(s)\|_{L^\infty}\,ds$. Pour tout $h, t > 0$, on a*

$$\|a(t)\|_{L^2\left((F_t)^c_h\right)} \le e^{-\frac{h^2}{20\nu t}\exp(-4V(t))}\,\|a_0\|_{L^2}. \qquad (2.1)$$

Dans le cas où a_0 est la fonction caractéristique d'un domaine borné F_0, on a de plus

$$\begin{aligned} &\|a(t) - 1_{F_t}\|_{L^2\left((F_t^c)_h\right)} \\ &\quad \le 2\|a_0\|_{L^2}\min\left(1, \left(1 + C\frac{\sqrt{\nu t}}{h}e^{2V(t)}\right)e^{-\frac{h^2}{96\nu t}\exp(-4V(t))}\right), \end{aligned} \qquad (2.2)$$

où C est une constante universelle.

Démonstration. Elle ressemble beaucoup à celle du théorème 1.1 de [5]. La principale différence provient du terme du second ordre dans (T_ν) qui n'est plus un Laplacien “classique”.

Par une méthode d'énergie, on trouve

$$\forall t \in \mathbb{R}^+,\ \|a(t)\|_{L^2}^2 + 2\nu \int_0^t \|\nabla a(s)\|_{L^2}^2\, ds \leq \|a_0\|_{L^2}^2 . \tag{2.3}$$

Soit $\Phi_0 \in C^\infty(\mathbb{R}^3)$ axisymétrique (c'est-à-dire ne dépendant que de r et de z). On pose $\Phi(t,x) = \Phi_0(\psi_t^{-1}(x))$ si bien que Φ est conservée par le flot et reste axisymétrique. Supposons dans un premier temps que $v \in L^\infty_{loc}(\mathbb{R}^+; (\mathcal{S}(\mathbb{R}^3))^3)$ et $a \in L^\infty_{loc}(\mathbb{R}^+; \mathcal{S}(\mathbb{R}^3))$.

D'après (T_ν), Φa vérifie

$$\big(\partial_t + v\cdot\nabla - \nu(\partial_r^2 + \partial_z^2 + \frac{3}{r}\partial_r)\big)(\Phi a) = -\nu a(\partial_r^2 + \partial_z^2 + \frac{3}{r}\partial_r)\Phi - 2\nu \nabla\Phi\cdot\nabla a. \tag{2.4}$$

Supposons de plus que Φ est constante hors d'un compact (qui dépend du temps). Dans ces conditions, $\Phi a(t)$ est dans H^1. Rappelons par ailleurs que pour une fonction $u = u(r,z)$, le Laplacien se réduit à $\Delta u = \partial_r^2 u + \partial_z^2 u + \partial_r u/r$. En prenant le produit scalaire de (2.4) avec $\Phi a(t)$ au sens L^2 et en se souvenant que $\operatorname{div} v = 0$, on trouve

$$\begin{aligned}\frac{1}{2}\frac{d}{dt}\|\Phi a\|_{L^2}^2 &- \nu \int_{\mathbb{R}^3} \Phi a\big(\Delta(\Phi a) + \frac{2}{r}\partial_r(\Phi a)\big)\, dx \\ &= -\nu \int_{\mathbb{R}^3} a^2 \Phi\big(\Delta\Phi + \frac{2}{r}\partial_r(\Phi a)\big)\, dx - 2\nu \int_{\mathbb{R}^3} \Phi a \nabla\Phi\cdot\nabla a\, dx.\end{aligned}$$

En intégrant par parties, on obtient

$$\frac{1}{2}\frac{d}{dt}\big(\|\Phi a\|_{L^2}^2\big) + \nu\|\nabla(\Phi a)\|_{L^2}^2 \leq \nu\Big(\|a\nabla\Phi\|_{L^2}^2 - \int_{\mathbb{R}^3} \frac{a^2}{r}\partial_r\Phi^2\, dx\Big). \tag{2.5}$$

Pour démontrer (2.1), prenons Φ de la forme $\Phi(t,x) = \exp\phi(t,x)$ avec $\phi(t,x) = f_0(\psi^{-1}(t,x))$ et f_0 axisymétrique constante hors d'un compact.

On a visiblement $\|a\nabla\Phi\|_{L^2} \leq \|\nabla\phi\|_{L^\infty}\|a\Phi\|_{L^2}$.

Pour majorer le terme $\displaystyle\int_{\mathbb{R}^3} \frac{a^2}{r}\partial_r\Phi^2\, dx$ (qui n'apparaît pas en dimension deux), on écrit

$$\begin{aligned}\Big|\int_{\mathbb{R}^3} \frac{a^2}{r}\partial_r\Phi^2\, dx\Big| &\leq 2\|\partial_r\phi\|_{L^\infty} \int_{\mathbb{R}^3} \frac{a^2\Phi^2}{r}\, dx, \\ &\leq 4\pi\|\partial_r\phi\|_{L^\infty} \int_{-\infty}^{+\infty}\int_0^{+\infty} a^2\Phi^2\, dr\, dz.\end{aligned} \tag{2.6}$$

Une intégration par parties donne

$$\int_0^{+\infty} a^2\Phi^2\,dr = -2\int_0^{+\infty} a\Phi\partial_r(a\Phi)r\,dr$$

Injectons cette expression dans (2.6) puis appliquons l'inégalité de Cauchy-Schwarz. On trouve

$$\begin{aligned}\left|\int_{\mathbb{R}^3}\frac{a^2}{r}\partial_r\Phi^2\,dx\right| &\leq 4\,\|\partial_r\phi\|_{L^\infty}\,\|a\Phi\|_{L^2}\,\|\partial_r(a\Phi)\|_{L^2}\,,\\ &\leq 4\,\|\partial_r\phi\|_{L^\infty}^2\,\|a\Phi\|_{L^2}^2+\|\partial_r(a\Phi)\|_{L^2}^2\,.\end{aligned}$$

En revenant à (2.5), on obtient

$$\frac{1}{2}\frac{d}{dt}\left(\|\Phi a\|_{L^2}^2\right)\leq 5\nu\,\|\nabla\phi\|_{L^\infty}^2\,\|\Phi a\|_{L^2}^2\,,$$

et le lemme de Gronwall donne donc

$$\|(\Phi a)(t)\|_{L^2}\leq e^{5\nu\|\nabla f_0\|_{L^\infty}^2\int_0^t\|\nabla\psi^{-1}(s)\|_{L^\infty}^2\,ds}\,\|(\Phi a)(0)\|_{L^2}\,.$$

A ce stade, on peut conclure exactement comme dans [5] en supposant d'abord a_0 à support compact et en choisissant $f_0 = \alpha\min\{R, d(x,F_0)\}$ avec $R>0$ et $\alpha>0$ (convenablement régularisée), puis en faisant tendre R vers $+\infty$ et en prenant le "meilleur" α.

Pour démontrer (2.2), posons $w(t,x) = a(t,x) - 1_{F_t}(x)$ et $\Phi(t,x) = \Phi_0(\psi_t^{-1}(x))$ avec $\Phi_0\in C_0^\infty(F_0)$ axisymétrique. Remarquons que 1_{F_t} vérifie $(\partial_t + v\cdot\nabla)1_{F_t} = 0$ et que $\operatorname{Supp}\Phi_t$ est compact dans F_t, ce qui implique $\Phi_t\partial_r w = \Phi_t\partial_r a$. On en déduit donc que

$$\begin{aligned}&\left(\partial_t + v\cdot\nabla - \nu\left(\Delta + \frac{2}{r}\partial_r\right)\right)(\Phi w)\\ &\qquad = -\nu w\left(\Delta+\frac{2}{r}\partial_r\right)\Phi - 2\nu\nabla\Phi\cdot\nabla w.\end{aligned}\tag{2.7}$$

Une méthode d'énergie donne

$$\frac{1}{2}\frac{d}{dt}\left(\|\Phi w\|_{L^2}^2\right)+\nu\,\|\nabla(\Phi w)\|_{L^2}^2\leq\nu\,\|w\nabla\Phi\|_{L^2}^2-\nu\int_{\mathbb{R}^3}\frac{w^2\partial_r\Phi^2}{r}\,dx.\tag{2.8}$$

Soit $h_0>0$ et $\chi_0\in C_0^\infty(F_0)$ axisymétrique valant 1 sur $(F_0^c)_{h_0}$ et à valeurs dans $[0,1]$. On impose de plus que $\|\nabla\chi_0\|_{L^\infty}\leq C_0h_0^{-1}$ où C_0 est une constante universelle. On choisit alors $\Phi_0 = \chi_0^2 e^{f_0}$ avec $f_0 = \alpha d(x, F_0^c)$. Notons $f_t = f_0\circ\psi_t^{-1}$ et $\chi_t = \chi_0\circ\psi_t^{-1}$.

En remarquant que $\left\|\nabla\psi_t^{\pm1}\right\|_{L^\infty} \leq e^{V(t)}$ et que $\|w\|_{L^2} \leq 2\|a_0\|_{L^2}$, on trouve

$$\begin{aligned}\|w\nabla\Phi\|_{L^2}^2 &\leq 8\left\|\chi_t\nabla\chi_t we^{f_t}\right\|_{L^2}^2 + 2\|\Phi w\nabla f_t\|_{L^2}^2, \\ &\leq 2e^{2V(t)}\left(16C_0^2h_0^{-2}e^{2\alpha h_0}\|a_0\|_{L^2}^2 + \alpha^2\|\Phi w\|_{L^2}^2\right).\end{aligned}$$

On a par ailleurs

$$\int_{\mathbb{R}^3}\frac{w^2\partial_r\Phi^2}{r}\,dx = \underbrace{4\int_{\mathbb{R}^3}\frac{w^2\chi_t^3\partial_r\chi_t e^{2f_t}}{r}\,dx}_{I_1} + \underbrace{2\int_{\mathbb{R}^3}\frac{\Phi^2w^2\partial_r f_t}{r}\,dx}_{I_2}.$$

Remarquons que

$$|I_1| \leq 4\left\|\chi_t\partial_r\chi_t e^{2f_t}\right\|_{L^\infty}\int_{\mathbb{R}^3}\frac{\chi_t^2w^2}{r}\,dx.$$

En procédant comme pour majorer $\displaystyle\int_{\mathbb{R}^3}\frac{a^2\Phi^2}{r}\,dx$, on obtient donc

$$\begin{aligned}|I_1| &\leq 8\left\|\chi_t\partial_r\chi_t e^{2f_t}\right\|_{L^\infty}\|\chi_t w\|_{L^2}\|\partial_r(\chi_t w)\|_{L^2}, \\ &\leq 8\left\|\chi_t\partial_r\chi_t e^{2f_t}\right\|_{L^\infty}\|\chi_t w\|_{L^2}(\|\chi_t\partial_r w\|_{L^2} + \|w\partial_r\chi_t\|_{L^2}).\end{aligned} \tag{2.9}$$

Comme $\chi_t\partial_r w = \chi_t\partial_r a$, on en déduit que

$$|I_1| \leq 16C_0h_0^{-1}e^{V(t)}e^{2\alpha h_0}\|a_0\|_{L^2}(\|\partial_r a\|_{L^2} + 2C_0h_0^{-1}e^{V(t)}\|a_0\|_{L^2}). \tag{2.10}$$

Pour le terme I_2, on écrit simplement

$$\begin{aligned}|I_2| &\leq 4\|\partial_r f_t\|_{L^\infty}\|\Phi w\|_{L^2}\|\partial_r(\Phi w)\|_{L^2}, \\ &\leq \|\partial_r(\Phi w)\|_{L^2}^2 + 4\alpha^2e^{2V(t)}\|\Phi w\|_{L^2}^2.\end{aligned} \tag{2.11}$$

En injectant les inégalités (2.10) et (2.11) dans (2.8), on conclut que

$$\begin{aligned}\frac{1}{2\nu}\frac{d}{dt}\|\Phi w\|_{L^2}^2 \leq 6\alpha^2e^{2V(t)}\|\Phi w\|_{L^2}^2 &+ 64C_0^2h_0^{-2}e^{2\alpha h_0}e^{2V(t)}\|a_0\|_{L^2}^2 \\ &+16C_0h_0^{-1}e^{V(t)}e^{2\alpha h_0}\|a_0\|_{L^2}\|\partial_r a\|_{L^2},\end{aligned}$$

d'où

$$\begin{aligned}\frac{1}{2}\frac{d}{dt}\|\Phi w\|_{L^2}^2 &\leq 6\nu\alpha^2e^{2V(t)}\|\Phi w\|_{L^2}^2 \\ &+ 128\nu C_0^2h_0^{-2}e^{2\alpha h_0}e^{2V(t)}\|a_0\|_{L^2}^2 + \nu e^{2\alpha h_0}\|\partial_r a\|_{L^2}^2.\end{aligned} \tag{2.12}$$

Intégrons (2.12) en temps. Grâce à (2.3), il vient

$$\|\Phi w\|_{L^2}^2 \le e^{2\alpha h_0} \|a_0\|_{L^2}^2 \left(1 + 256 C_0^2 h_0^{-2} \nu \int_0^t e^{2V(\tau)}\, d\tau\right) + 12\alpha^2 \nu \int_0^t e^{2V(\tau)} \|\Phi w(\tau)\|_{L^2}^2\, d\tau.$$

Le lemme de Gronwall donne donc

$$\|\Phi w\|_{L^2}^2 \le e^{2\alpha h_0} \|a_0\|_{L^2}^2 \left(e^{12\alpha^2 \nu \int_0^t e^{2V(\tau)}\, d\tau} + \frac{64 C_0^2}{3 h_0^2 \alpha^2}\left(e^{12\alpha^2 \nu \int_0^t e^{2V(\tau)}\, d\tau} - 1\right)\right).$$

On peut alors conclure comme dans [5]. En effet, si l'on prend $h_0 = h e^{-V(t)}/2$, on trouve

$$\|w(t)\|_{L^2((F_t^c)_h)}^2 \le \|a_0\|_{L^2}^2 e^{\alpha h e^{-V(t)}} \left(e^{12\alpha^2 \nu t e^{2V(t)}} + \frac{256 C_0^2 e^{2V(t)}}{3 h^2 \alpha^2}\left(e^{12\alpha^2 \nu t e^{2V(\tau)}} - 1\right)\right).$$

Il ne reste plus qu'à choisir $\alpha = \dfrac{h e^{-3V(t)}}{24 \nu t}$ pour obtenir (2.2). ■

3. Preuve du résultat de convergence pour la régularité stratifiée ou conormale

On traite d'abord le cas où l'hypothèse $(\mathcal{H}_1)$ est vérifiée. On procède en quatre étapes:
1) Preuve d'estimations dans les espaces stratifiés pour les solutions de (NS_ν) lorsque la donnée initiale est régulière.
2) Régularisation de la donnée initiale et preuve d'estimations uniformes pour la solution correspondante.
3) Convergence des solutions régularisées vers la solution de (NS_ν).
4) Convergence de (NS_ν) vers (NS_0).

Première étape: Il s'agit de prouver la proposition suivante.

Proposition 3.1 – *Soit $a \in]3, +\infty[$ et $\epsilon \in]3/a, 1[$. Soit $v^0 \in H^\infty(\mathbb{R}^3)$ un champ de vecteurs axisymétrique à divergence nulle. On suppose en outre que $\alpha^0 \overset{\text{déf}}{=} \omega^0/r \in L^2 \cap L^\infty$. Alors (NS_ν) a une unique solution globale $v \in C(\mathbb{R}^+; \mathrm{H}^\infty)$. De plus si $(X_\lambda^0)_{1\le\lambda\le m}$ est une famille (ϵ, a)-substantielle* *de champs de vecteurs alors la famille X_t transportée par le flot ψ de v reste (ϵ, a)-substantielle pour tout temps, et on a les estimations suivantes (pour une constante C ne dépendant que de ϵ et de a):*

$$\|\omega(t)/r\|_{L^p} \leq \|\alpha^0\|_{L^p} \quad \text{pour tout} \quad p \in [2, +\infty], \tag{3.1}$$

$$\|\omega(t)\|_{L^2} \leq \|\omega^0\|_{L^2} + t\|v^0\|_{L^2}\|\alpha^0\|_{L^\infty}, \tag{3.2}$$

$$\|\omega(t)\|_{L^\infty} \leq \|\omega^0\|_{L^\infty} + C\|v^0\|_{L^2} f\left(2 + \frac{\|\omega^0\|_{L^2\cap L^\infty}}{\|v^0\|_{L^2}} + t\|\alpha^0\|_{L^\infty}\right) \tag{3.3}$$

$$\|\operatorname{div} X_{t,\lambda}\|_{B^\epsilon_a} \leq \|\operatorname{div} X^0_\lambda\|_{B^\epsilon_a} e^{CV(t)}, \tag{3.4}$$

$$\begin{aligned} \|X_{t,\lambda}\|_{B^\epsilon_a} + &\|X^0_\lambda(x,D)\psi(t)\|_{B^\epsilon_a} \\ &\leq C\left(\frac{\|\operatorname{div}(X^0_\lambda \otimes \omega^0)\|_{B^{\epsilon-1}_a}}{\|\omega^0\|_{L^\infty}} + \|X^0_\lambda\|_{B^\epsilon_a}\right)e^{CW(t)}, \end{aligned} \tag{3.5}$$

$$\|\omega(t)\|_{B^\epsilon_{a,X_t}} \leq C\|\omega^0\|_{B^\epsilon_{a,X^0}} e^{CW(t)}, \tag{3.6}$$

avec

$$V(t) = \int_0^t \|\nabla v(\tau)\|_{L^\infty} d\tau, \quad W(t) = \|\omega^0\|_{L^\infty} t + V(t), \quad f(t) = t^{\frac{5}{2}} \log t.$$

Enfin, on a

$$\|\nabla v(t)\|_{L^\infty} \leq CL^0 e^{Ct\|v^0\|_{L^2} f\left(2 + \frac{\|\omega^0\|_{L^2\cap L^\infty}}{\|v^0\|_{L^2}} + t\|\alpha^0\|_{L^\infty}\right)}, \tag{3.7}$$

avec

$$L^0 \stackrel{\text{déf}}{=} \left(\|\omega^0\|_{L^2\cap L^\infty} + \|v^0\|_{L^2}\right) \log\left(e + \frac{\|\omega^0\|_{B^\epsilon_{a,X^0}}}{\|\omega^0\|_{L^2\cap L^\infty} + \|v^0\|_{L^2}}\right).$$

Démonstration. L'existence et l'unicité d'une solution globale v axisymétrique vérifiant entre autres $\omega \in L^\infty_{loc}(\mathbb{R}^+; L^2 \cap L^\infty)$ et $\omega/r \in L^\infty(\mathbb{R}^+; L^2 \cap L^\infty)$ résulte du théorème 0.3. Comme v^0 est de plus dans tous les H^s, les théorèmes 0.1 et 0.2 nous assurent que $v \in C(\mathbb{R}^+; H^\infty)$.

Les inégalités (3.1), (3.2) sont démontrées dans [14]. On renvoie au lemme A.2 pour la preuve de (3.3). Le fait que la famille X_t reste substantielle découle de la proposition 4.1 de [6]. Il en est de même pour les inégalités (3.4), (3.5) et (3.6).

Pour prouver (3.7), on utilise l'estimation stationnaire suivante (voir par exemple [6] proposition 1.3 et les références jointes):

$$\|\nabla v(t)\|_{L^\infty} \leq C \left(\|\omega(t)\|_{L^2 \cap L^\infty} \log \left(e + \frac{\|\omega(t)\|_{B^\epsilon_{a,X_t}}}{\|\omega(t)\|_{L^\infty}} \right) \right). \tag{3.8}$$

Il suffit alors d'injecter les inégalités (3.1), (3.2),(3.3) et (3.6) dans (3.8) puis d'appliquer le lemme de Gronwall pour obtenir l'estimation souhaitée. □

Deuxième étape: On se donne une vitesse initiale vérifiant l'hypothèse $(\mathcal{H}_1)$ du théorème 1.1. On la régularise comme dans [3] à l'aide de la décomposition de Littlewood-Paley radiale donnée dans la proposition 1.1: on pose $v_n^0 = S_n v^0$. Comme v^0 est dans L^2, on a clairement $v_n^0 \in H^\infty$. De plus, grâce au lemme A.1 de l'appendice, on a pour $p \in [2, +\infty]$,

$$\left\| \frac{S_n \omega^0}{r} \right\|_{L^p} \leq C \left\| \alpha^0 \right\|_{L^p},$$

ce qui permet d'appliquer la proposition 3.1.

Pour tout $\nu \geq 0$, on a donc construit une solution régulière $v_{n,\nu}$ de (NS_ν) qui correspond à la donnée initiale v_n^0. Toutes les estimations de la proposition 3.1 s'appliquent, mais les termes de droite dépendent de la donnée initiale donc de n. Il est en fait classique (voir [3] ou [5]) que

$$\|S_n \omega^0\|_{L^2 \cap L^\infty} \leq C \|\omega^0\|_{L^2 \cap L^\infty},$$
$$\|S_n \omega^0\|_{B^\epsilon_{a,X^0}} \leq \|\omega^0\|_{B^\epsilon_{a,X^0}}.$$

On conclut donc au résultat suivant.

Proposition 3.2 – *La solution $v_{n,\nu}$ vérifie toutes les estimations de la proposition* 3.1 *avec une constante C indépendante de n et de ν.*

Troisième étape: La convergence de $v_{n,\nu}$ vers v_ν solution de (NS_ν) vérifiant les propriétés de régularité stratifiée voulues se fait comme dans [5]. On commence par établir que la suite est de Cauchy en petite norme (dans des espaces de Hölder à indice négatif par exemple) puis on interpole avec les estimations uniformes de la proposition 3.2. La limite v_ν vérifie de plus toutes les estimations de la proposition 3.1 avec une constante C *indépendante* de ν.

Quatrième étape: Il s'agit d'étudier la limite non visqueuse de v_ν. D'après l'estimation (0.2) du théorème 0.6, le champ $v_{n,\nu}$ tend vers $v_{n,0}$ dans $L^\infty_{loc}(R^+; L^2)$. On utilise alors les estimations uniformes des étapes 2 et 3, et les résultats de convergence de l'étape 3 pour obtenir les résultats de convergence pour la régularité stratifiée.

Il reste à traiter le cas où $(\mathcal{H}_2)$ est vérifiée. D'après la remarque 1.6, on sait déjà que $(\mathcal{H}_1)$ est vérifiée pour une famille X^0 (ϵ, a)-substantielle bien choisie. On dispose donc d'une unique solution globale vérifiant des propriétés de régularité stratifiée. En appliquant à cette solution les résultats de la partie 5 de [6], on obtient les propriétés de convergence voulues pour la régularité conormale. ■

4. Convergence pour des données initiales sans régularité stratifiée

Cette partie est consacrée à la preuve du théorème 0.6.

Sous les hypothèses $\omega^0, \alpha^0 \in L^2 \cap L^\infty$, l'existence de solutions ayant les propriétés voulues est assurée par le théorème 0.3. En appliquant une méthode d'énergie à $w_\nu \stackrel{\text{déf}}{=} v_\nu - v_0$, on obtient

$$\frac{1}{2}\frac{d}{dt}\|w_\nu\|_{L^2}^2 + \nu\|\nabla w_\nu\|_{L^2}^2 \leq \nu\|\nabla v_0\|_{L^2}\|\nabla w_\nu\|_{L^2} + I(t) \tag{4.1}$$

avec

$$I(t) = \int_{\mathbb{R}^3} |w_\nu(t,x)|^2 |\nabla v_0(t,x)|\, dx.$$

On constate par un calcul direct exploitant l'incompressibilité du champ v_0 que $\|\nabla v_0\|_{L^2} = \|\omega_0\|_{L^2}$. En revenant à (4.1), on trouve donc

$$\frac{1}{2}\frac{d}{dt}\|w_\nu\|_{L^2}^2 \leq \frac{\nu}{4}\|\omega_0\|_{L^2}^2 + I. \tag{4.2}$$

Traitons d'abord le cas très simple où la solution v_0 de (NS_0) est dans $L^1_{loc}(\mathbb{R}^+; Lip)$. On a

$$I \leq \|w_\nu\|_{L^2}^2 \|\nabla v_0\|_{L^\infty}. \tag{4.3}$$

En injectant (4.3) dans (4.2) puis en appliquant le lemme de Gronwall, on en déduit finalement que

$$\|w_\nu(t,\cdot)\|_{L^2}^2 \leq \frac{\nu}{2}\int_0^t \|\omega_0(\tau,\cdot)\|_{L^2}^2 e^{\int_\tau^t \|\nabla v_0(\tau')\|_{L^\infty} d\tau'}\, d\tau.$$

Sans hypothèse particulière de symétrie, une méthode d'énergie appliquée à l'équation du tourbillon

$$\partial_t \omega_0 + v_0 \cdot \nabla \omega_0 = (\omega_0 \cdot \nabla) v_0$$

permet d'obtenir

$$\|\omega_0(t,\cdot)\|_{L^2} \leq \left\|\omega^0\right\|_{L^2} e^{\int_0^t \|\nabla v_0(\tau,\cdot)\|_{L^\infty}\,d\tau}. \tag{4.4}$$

Dans le cas axisymétrique, on dispose aussi de l'inégalité (A.7) de l'annexe. On en déduit (0.2).

Dorénavant, on ne suppose plus que ∇v_0 est borné. L'inégalité (0.1) se montre en adaptant la preuve de [4]. Fixons un réel $a \geq 2$. Rappelons que d'après [12] pages 42 et 250, il existe une constante C universelle telle que

$$\|\nabla v_0\|_{L^a} \leq Ca\,\|\omega_0\|_{L^a}\,. \tag{4.5}$$

En appliquant l'inégalité de Hölder pour majorer I, on obtient donc

$$I \leq Ca\,\|\omega_0\|_{L^a}\,\|w_\nu\|^2_{L^{\frac{2a}{a-1}}} \tag{4.6}$$

En combinant l'inégalité de Sobolev $\|z\|_{L^6} \leq 4\,\|\nabla z\|_{L^2}$ (voir [2] page 162), l'inégalité

$$\|z\|_{L^{\frac{2a}{a-1}}} \leq \|z\|_{L^2}^{1-\frac{3}{2a}}\,\|z\|_{L^6}^{\frac{3}{2a}}, \tag{4.7}$$

le fait que $\|\nabla v_\nu\|_{L^2} = \|\omega_\nu\|_{L^2}$ et l'estimation (A.7), on obtient

$$\|w_\nu\|^2_{L^{\frac{2a}{a-1}}} \leq C(\left\|\omega^0\right\|_{L^2} + t\left\|v^0\right\|_{L^2}\left\|\alpha^0\right\|_{L^\infty})^{\frac{3}{a}}\,\|w_\nu\|_{L^2}^{2-\frac{3}{a}}.$$

Par interpolation entre (A.7) et (A.4), on a

$$\|\omega(t,\cdot)\|_{L^a} \leq \left\|v^0\right\|_{L^2}(C_0 + t\left\|\alpha^0\right\|_{L^\infty})^{\frac{5}{2}-\frac{3}{a}}\left(\log(C_0 + t\left\|\alpha^0\right\|_{L^\infty})\right)^{1-\frac{2}{a}}.$$

En injectant cette dernière inégalité dans (4.6) et en utilisant (4.7), on obtient finalement

$$I(t) \leq Ca\left\|v^0\right\|_{L^2}^{1+\frac{3}{a}}(C_0 + t\left\|\alpha^0\right\|_{L^\infty})^{\frac{5}{2}}\log(C_0 + t\left\|\alpha^0\right\|_{L^\infty})\,\|w_\nu\|_{L^2}^{2-\frac{3}{a}}, \tag{4.8}$$

d'où, en revenant à (4.2) et en utilisant (A.7),

$$\begin{aligned}\frac{d}{dt}\|w_\nu(t,\cdot)\|^2_{L^2} \leq\; & \frac{\nu}{2}(\left\|\omega^0\right\|_{L^2} + t\left\|v^0\right\|_{L^2}\left\|\alpha^0\right\|_{L^\infty})^2 \\ & + Ca\left\|v^0\right\|_{L^2}^{1+\frac{3}{a}} f(C_0 + t\left\|\alpha^0\right\|_{L^\infty})\,\|w_\nu\|_{L^2}^{2-\frac{3}{a}}\end{aligned} \tag{4.9}$$

avec $f(t) = t^{\frac{5}{2}}\log t$.

Fixons un $\delta \in]0, e^{-\frac{4}{3}}[$ et notons

$$x_{\delta,\nu}(t) \stackrel{\text{déf}}{=} \frac{\|w_\nu(t,\cdot)\|_{L^2}^2}{\|v^0\|_{L^2}^2} + \delta.$$

Le fait que $w_\nu \in C(\mathbb{R}^+; L^2)$ et que $w_\nu(0) = 0$ assure l'existence d'un temps $T_{\delta,\nu} > 0$ (que l'on peut supposer maximal) tel que $x_{\delta,\nu} < e^{-\frac{4}{3}}$ sur $[0, T_{\delta,\nu}[$.

Autorisons la "constante" a de l'inégalité (4.9) à dépendre du temps. On a sur l'intervalle $[0, T_{\delta,\nu}[$:

$$\begin{aligned} x'_{\delta,\nu}(t) \leq & \frac{\nu}{2}\left(\frac{\|\omega^0\|_{L^2}}{\|v^0\|_{L^2}} + t\|\alpha^0\|_{L^\infty}\right)^2 \\ & + Ca(t)\|v^0\|_{L^2} f(C_0 + t\|\alpha^0\|_{L^\infty}) x_{\delta,\nu}^{1-\frac{3}{2a}}(t). \end{aligned}$$

Choisissons $a(t) \stackrel{\text{déf}}{=} -\frac{3}{2}\log x_{\delta,\nu}(t)$ (ce qui entraîne bien $a(t) \geq 2$ puisque $x_{\delta,\nu} < e^{-\frac{4}{3}}$). Après intégration, il vient

$$\begin{aligned} x_{\delta,\nu}(t) \leq \delta + & \frac{\nu t}{2}\left(\frac{\|\omega^0\|_{L^2}}{\|v^0\|_{L^2}} + t\|\alpha^0\|_{L^\infty}\right)^2 \\ & + C\|v^0\|_{L^2}\int_0^t f(C_0 + \tau\|\alpha^0\|_{L^\infty})\mu(x_{\delta,\nu}(\tau))\,d\tau \end{aligned}$$

avec $\mu(z) = -z\log z$.

Comme la fonction $z \mapsto -z\log z$ est croissante sur $]0, e^{-1}]$, et que nous avons $x_{\delta,\nu} \in]0, e^{-\frac{4}{3}}[$ sur $[0, T_{\delta,\nu}[$, on peut appliquer le lemme d'Osgood à l'inégalité ci-dessus (voir par exemple [4]).

En remarquant que $z \mapsto \log(-\log z)$ est une primitive de μ^{-1} sur $]0, e^{-1}]$, on en déduit que pour tout $t \in [0, T_{\delta,\nu}[$ tel que

$$\delta + \frac{\nu t}{2}\left(\frac{\|\omega^0\|_{L^2}}{\|v^0\|_{L^2}} + t\|\alpha^0\|_{L^\infty}\right)^2 < e^{-1},$$

on a

$$\begin{aligned} \log\left(-\log\left(\delta + \frac{\nu t}{2}\left(\frac{\|\omega^0\|_{L^2}}{\|v^0\|_{L^2}} + t\|\alpha^0\|_{L^\infty}\right)^2\right)\right) - \log(-\log x_{\delta,\nu}(t)) \\ \leq C\|v^0\|_{L^2}\int_0^t f(C_0 + \tau\|\alpha^0\|_{L^\infty})\,d\tau \end{aligned}$$

d'où

$$x_{\delta,\nu}(t) \leq \left[\delta + \frac{\nu t}{2}\left(\frac{\|\omega^0\|_{L^2}}{\|v^0\|_{L^2}} + t\,\|\alpha^0\|_{L^\infty}\right)^2\right]^{\exp\left(-C\|v^0\|_{L^2} t f(C_0+\tau\|\alpha^0\|_{L^\infty})\right)}.$$

En faisant tendre δ vers 0 et en utilisant un argument de bootstrap, on obtient (0.1). ■

Appendice

On généralise le lemme 3.1 de [10] au cas de régularisations construites à partir de troncatures qui ne sont pas à support compact. On peut ainsi, dans l'étape 3 de la partie précédente, régulariser les données initiales à l'aide d'une décomposition de Littlewood-Paley radiale et réutiliser telles quelles les estimations de [5] sur la régularité stratifiée.

Lemme A.1 – *Soit $\chi \in \mathcal{S}(\mathbb{R}^3)$ radiale. On suppose que ω est le vecteur tourbillon d'un champ de vitesse v axisymétrique. Notons $\chi_n(x) = n^3\chi(nx)$, $\omega_n = \chi_n * \omega$, $\alpha = \omega/r$ et $\alpha_n = \omega_n/r$ où r désigne la distance à l'axe. Alors il existe une constante C ne dépendant que de χ et telle que*

$$\|\alpha_n\|_{L^p} \leq C\,\|\alpha\|_{L^p} \quad \textit{pour} \quad 1 \leq p \leq +\infty.$$

Démonstration. Notons $x = (x_1, x_2, x_3) = (x', x_3)$ et $\bar{x} = (0, 0, x_3)$. Le caractère radial de χ et les propriétés de symétrie du vecteur tourbillon nous assurent que

$$\int \chi_n(\bar{x} - y)\omega(y)\,dy = 0 \quad \text{pour tout} \quad x \in \mathbb{R}^3.$$

On en déduit donc que

$$\alpha_n(x) = \int \underbrace{\phi_n(x)\frac{|y'|}{|x'|}(\chi_n(x-y) - \chi_n(\bar{x}-y))}_{K_n^1(x,y)}\,\alpha(y)\,dy + \int \underbrace{(1-\phi_n(x))\frac{|y'|}{|x'|}\chi_n(x-y)}_{K_n^2(x,y)}\,\alpha(y)\,dy.$$

où l'on a noté $\phi_n(z) \stackrel{\text{déf}}{=} 1_{|z'|\leq n^{-1}}(z)$.

Supposons qu'il existe une constante C indépendante de n et telle que

$$\max\left(\int |K_n^i(x,y)|\,dx, \int |K_n^i(x,y)|\,dy\right) \leq C \quad \text{pour} \quad i=1,2. \qquad (A.1)$$

Alors le lemme de Schur donne $\|\alpha_n\|_{L^p} \leq C\,\|\alpha\|_{L^p}$ pour $1 \leq p \leq +\infty$.

Montrons que (A.1) est effectivement vérifiée. En appliquant la formule de Taylor avec reste intégral, on obtient les inégalités suivantes:

$$\begin{aligned}
\int |K_n^1(x,y)|\,dy &\leq \phi_n(x)\int_0^1\int_{\mathbb{R}^3} |y'|\,\big|\nabla\chi_n\big(\bar{x}+t(x-\bar{x})-y\big)\big|\,dy\,dt,\\
&\leq \phi_n(x)\Big(\int_0^1\int_{\mathbb{R}^3} \big|(\bar{x}+t(x-\bar{x})-y)'\big|\,\big|\nabla\chi_n\big(\bar{x}+t(x-\bar{x})-y\big)\big|\,dy\,dt\\
&\qquad\qquad + \int_0^1\int_{\mathbb{R}^3} t|x'|\,\big|\nabla\chi_n\big(\bar{x}+t(x-\bar{x})-y\big)\big|\,dy\,dt\Big),\\
&\leq \int_{\mathbb{R}^3} |\nabla\chi_n(z)||z'|\,dz + n^{-1}\int_{\mathbb{R}^3} |\nabla\chi_n(z)|\,dz,\\
&\leq \|(1+|.|)\nabla\chi\|_{L^1}.
\end{aligned}$$

Comme $\nabla\chi$ est dans $\mathcal{S}(\mathbb{R}^3)$, il existe une constante C_2 telle que

$$|\nabla\chi(z)| \leq \frac{C_2}{(1+|z'|)^2(1+|z_3|)^2}.$$

Utilisons cette inégalité et la formule de Taylor avec reste intégral. Il vient

$$\begin{aligned}
&\int |K_n^1(x,y)|\,dx\\
&\qquad \leq n^3|y'|\int_0^1\int_{|x'|<n^{-1}}\int_{\mathbb{R}} |\nabla\chi(n(tx'-y'),x_3)|\,dx_3\,dx'\,dt,\\
&\qquad \leq C_2n^3|y'|\left(\int_{\mathbb{R}}\frac{1}{(1+|x_3|)^2}\,dx_3\right) \qquad\qquad (A.2)\\
&\qquad\qquad \times\left(\int_0^1\int_{|x'|<n^{-1}}\frac{1}{(1+n|y'-tx'|)^2}\,dx'\,dt\right).
\end{aligned}$$

Lorsque $|y'| \leq 2n^{-1}$, on écrit simplement que $(1+n|y'-tx'|)^{-2} \leq 1$. De l'inégalité (A.2), on déduit alors aisément que

$$\int |K_n^1(x,y)|\,dx \leq C.$$

Lorsque $|y'| > 2n^{-1}$, on a visiblement $|y' - tx'| \geq |y'|/2$ pour tout $t \in [0,1]$ et $x \in \mathbb{R}^3$ tel que $|x'| \leq n^{-1}$. De (A.2), on déduit donc

$$\int |K_n^1(x,y)|\,dx \leq C\frac{n|y'|}{(1+\frac{n}{2}|y'|)^2} \leq C'.$$

Il ne reste plus maintenant qu'à borner $\int |K_n^2(x,y)|\,dx$ et $\int |K_n^2(x,y)|\,dy$ indépendamment de n. Ceci se fait de la manière suivante:

$$\begin{aligned}\int |K_n^2(x,y)|\,dx &\leq \int (1-\phi_n(x))|\chi_n(x-y)|\,dx \\ &\qquad + \int (1-\phi_n(x))\frac{|x'-y'|}{|x'|}|\chi_n(x-y)|\,dx, \\ &\leq \int |\chi_n(z)|\,dz + n\int |z'||\chi_n(z)|\,dz, \\ &\leq \|(1+|.|)\nabla\chi\|_{L^1}.\end{aligned}$$

Un calcul strictement analogue donne également

$$\int |K_n^2(x,y)|\,dy \leq \|(1+|.|)\nabla\chi\|_{L^1}.$$

A.2. Estimations pour le tourbillon

Dans cette partie, on améliore un peu les estimations de [14] sur la croissance en temps des normes L^p du tourbillon. Nous obtenons le résultat suivant:

Lemme A.2 - *Soit $v \in L^\infty(0,T;H^1)$ un champ de vecteurs axisymétrique solution de (NS_ν) avec une donnée initiale $v^0 \in H^1$ telle que $\omega^0 \in L^2 \cap L^\infty$ et $\alpha^0 \stackrel{\text{déf}}{=} r^{-1}\omega^0 \in L^\infty$. On suppose que $\omega \in L^\infty(0,T;L^2 \cap L^\infty)$ et $\alpha \stackrel{\text{déf}}{=} r^{-1}\omega \in L^\infty(0,T\times\mathbb{R}^3)$. Soit*

$$C_0 \stackrel{\text{déf}}{=} 2 + \|v^0\|_{L^2}^{-1}\|\omega^0\|_{L^2\cap L^\infty}.$$

Il existe une constante C universelle telle que pour $p \in [2,+\infty[$, on ait

$$\|\omega(t)\|_{L^p} \leq \|\omega^0\|_{L^p} + Cpt\|v^0\|_{L^2}\|\alpha^0\|_{L^\infty}(C_0 + t\|\alpha^0\|_{L^\infty})^{\frac{3}{2}-\frac{3}{p}}, \quad (A.3)$$

et, dans le cas $p = +\infty$,

$$\begin{aligned}\|\omega(t)\|_{L^\infty} &\leq \|\omega^0\|_{L^\infty} \\ &\quad + Ct\|v^0\|_{L^2}\|\alpha^0\|_{L^\infty}(C_0+t\|\alpha^0\|_{L^\infty})^{\frac{3}{2}}\log(C_0+t\|\alpha^0\|_{L^\infty}).\end{aligned} \quad (A.4)$$

Démonstration. On reprend la preuve de [14] en utilisant “au mieux” les injections de Sobolev. On peut d'emblée remarquer que, comme $\omega \in L^2 \cap L^\infty$ et $v \in H^1$, le champ v et son tourbillon ω sont dans tous les L^p pour $2 \leq p \leq +\infty$. Il s'agit maintenant d'obtenir des inégalités pour ces normes L^p.

Considérons l'équation du tourbillon en coordonnées cylindriques:

$$\partial_t \omega + v \cdot \nabla \omega - \nu \left(\frac{1}{r} \partial_r (r \partial_r \omega) - \frac{\omega}{r^2} + \partial^2_{zz} \omega \right) = v_r \alpha.$$

Multiplions cette équation par $r\omega|\omega|^{p-2}$ et intégrons en (r, θ, z). Une intégration par partie permet de constater que le terme provenant de la viscosité est positif. Le membre de droite se majore à l'aide de l'inégalité de Hölder. On obtient

$$\frac{1}{p} \frac{d}{dt} \|\omega\|^p_{L^p} \leq \|\alpha\|_{L^\infty} \|v\|_{L^p} \|\omega\|_{L^p}{}^{p-1}$$

d'où, après intégration en temps,

$$\|\omega(t,\cdot)\|_{L^p} \leq \left\|\omega^0\right\|_{L^p} + \int_0^t \|\alpha(\tau,\cdot)\|_{L^\infty} \|v(\tau,\cdot)\|_{L^p} \, d\tau. \qquad (A.5)$$

En faisant tendre p vers l'infini dans (A.5), on obtient aussi

$$\|\omega(t,\cdot)\|_{L^\infty} \leq \left\|\omega^0\right\|_{L^\infty} + \int_0^t \|\alpha(\tau,\cdot)\|_{L^\infty} \|v(\tau,\cdot)\|_{L^\infty} \, d\tau. \qquad (A.6)$$

Lorsque $p = 2$, le fait que $\|v(t,\cdot)\|_{L^2} \leq \left\|v^0\right\|_{L^2}$ et que $\|\alpha(t,\cdot)\|_{L^\infty} \leq \left\|\alpha^0\right\|_{L^\infty}$ (voir [14]) assure que

$$\|\omega(t,\cdot)\|_{L^2} \leq \left\|\omega^0\right\|_{L^2} + t \left\|v^0\right\|_{L^2} \left\|\alpha^0\right\|_{L^\infty}. \qquad (A.7)$$

Traitons d'abord le cas $2 \leq p \leq 6$. Par interpolation, on a

$$\|v\|_{L^p} \leq \|v\|_{L^2}^{\frac{3}{p}-\frac{1}{2}} \|v\|_{L^6}^{\frac{3}{2}-\frac{3}{p}}.$$

En combinant avec l'injection de Sobolev suivante (voir [2] page 162):

$$\|v\|_{L^6} \leq 4 \|\nabla v\|_{L^2}$$

et en remarquant que $\|\omega\|_{L^2} = \|\nabla v\|_{L^2}$, on obtient finalement

$$\|v\|_{L^p} \leq 4\, \|v\|_{L^2}^{\frac{3}{p}-\frac{1}{2}}\, \|\omega\|_{L^2}^{\frac{3}{2}-\frac{3}{p}}. \tag{A.8}$$

Il ne reste plus qu'à reporter cette dernière inégalité combinée avec (A.7) dans (A.5). En utilisant l'inégalité d'énergie $\|v(t,\cdot)\|_{L^2} \leq \|v^0\|_{L^2}$, on trouve finalement

$$\|\omega(t,\cdot)\|_{L^p} \leq \|\omega^0\|_{L^p} + C\, \|v^0\|_{L^2}\, \|\alpha^0\|_{L^\infty}\, t \left(\frac{\|\omega^0\|_{L^2}}{\|v^0\|_{L^2}} + \|\alpha^0\|_{L^\infty}\, t \right)^{\frac{3}{2}-\frac{3}{p}}. \tag{A.9}$$

Traitons maintenant le cas $6 < p < +\infty$. D'après [2], page 162, on a pour $p \geq 3/2$,

$$\|v\|_{L^p} \leq \frac{2p}{3}\, \|\nabla v\|_{L^{\frac{3p}{3+p}}}. \tag{A.10}$$

Dans notre cas, $6 < p < +\infty$, donc $2 < 3p/(3+p) < 3$. Comme, d'après (4.5),

$$\|\nabla v\|_{L^{\frac{3p}{3+p}}} \leq C\, \|\omega\|_{L^{\frac{3p}{3+p}}},$$

on peut appliquer (A.7) pour majorer $\|\nabla v\|_{L^{\frac{3p}{3+p}}}$. On en déduit (A.3).

Traitons maintenant le cas $p = +\infty$. Avec les notations de la partie 1.1, on a

$$\|v\|_{L^\infty} \leq \|\chi(D)v\|_{L^\infty} + \|(1-\chi(D))v\|_{L^\infty} \leq C\, \|v\|_{L^2} + \|(1-\chi(D))v\|_{L^\infty}.$$

Pour majorer le dernier terme, on utilise l'inégalité de Sobolev suivante (voir [2] page 167):

$$\forall \epsilon > 0,\ \|u\|_{L^\infty} \leq C\epsilon^{-1}\, \|\nabla u\|_{L^{3+\epsilon}}.$$

En appliquant cette inégalité puis l'inégalité (4.5) à $(1-\chi(D))v$, on trouve finalement que

$$\|v\|_{L^\infty} \leq C \left(\|v\|_{L^2} + \frac{3+\epsilon}{\epsilon}\, \|\omega\|_{L^{3+\epsilon}} \right).$$

On majore $\|\omega\|_{L^{3+\epsilon}}$ à l'aide de (A.3). Après quelques calculs, on aboutit à

$$\|\omega(t,\cdot)\|_{L^\infty} \leq \|\omega^0\|_{L^\infty} + \frac{C(3+\epsilon)^2 t}{\epsilon}\, \|\alpha^0\|_{L^\infty}\, \|v^0\|_{L^2}\, \left(C_0 + t\, \|\alpha^0\|_{L^\infty}\right)^{\frac{3}{2}+\frac{\epsilon}{3+\epsilon}}.$$

pour Pour finir la preuve de l'inégalité (A.2), il suffit alors de choisir

$$\epsilon = \frac{3}{\log(C_0 + \|\alpha^0\|_{L^\infty}\, t)}. \qquad \blacksquare$$

Remarque A.1. L'inégalité (A.3) est optimale parmi les inégalités

$$\|\omega(t,\cdot)\|_{L^q} \leq C \left\|v^0\right\|_{L^2} \left(C_1 + \left\|\alpha^0\right\|_{L^\infty} t\right)^{\alpha_q}$$

avec une constante C universelle et une constante C_1 ne dépendant que des données initiales mais pas de la viscosité: des considérations de *scaling* montrent que $\alpha_q = 5/2 - 3/q$ est le "bon exposant". Il suffit d'utiliser l'invariance de (NS_ν) par la transformation $v(t,x) \rightarrow \lambda v(\lambda^2 t, \lambda x)$ pour le voir. Dans le cas L^∞, le "bon exposant" serait 5/2. Au logarithme près, c'est ce que nous avons obtenu. Rappelons que dans [14], l'exposant trouvé était 11/4.

Références bibliographiques

[1] J. Beale, T. Kato et A. Majda, Remarks on the breakdown of smooth solutions for the 3-D Euler equations, Comm. in Mathematical Physics, 94 (1984), 61–66 .

[2] H. Brézis, Analyse fonctionnelle. Théorie et applications, Collection Mathématiques Appliquées pour la Maîtrise, Masson, Paris, 1983.

[3] J.-Y. Chemin, Fluides parfaits incompressibles, Astérisque, 230 (1995).

[4] J.-Y. Chemin, A remark on the inviscid limit for two-dimensionnal incompressible fluid, Comm. Part. Diff. Eq., 21 (1996), 1771–1779.

[5] R. Danchin, Poches de tourbillon visqueuses, Journal de Mathématiques Pures et Appliquées, 76 (1997), 609–647.

[6] R. Danchin, Persistance de structures géométriques et limite non visqueuse pour les fluides incompressibles en dimension quelconque, Bull. Société Mathématique de France, 27 (1999), 179–227 .

[7] P. Gamblin et X. Saint Raymond, On three-dimensional vortex patches, Bull. Société Mathématique de France, 123 (1995), 375–424.

[8] T. Kato, Quasi-linear equations of evolution, with applications to partial differential equations, Lecture Notes in Mathematics, 448 (1975) , Springer-Verlag, 25–70.

[9] T. Kato et G. Ponce, Commutator estimates and the Euler and Navier-Stokes equations, Comm. Pure and Appl. Math., 41 (1988), 891–907.

[10] X. Saint Raymond, Remarks on axisymmetric solutions of the incompressible Euler system, Comm. Part. Diff. Eq., 19 (1994), 321–334.

[11] P. Serfati, Régularité stratifiée et équation d'Euler à temps grand, Note aux Comptes-rendus Acad. Sci. Paris, 318 (1994), série 1, 925–928.

[12] E. Stein, Harmonic analysis: real-variable methods, orthogonality and oscillatory integrals, Princeton University Press (1993).

[13] H. Triebel, Theory of function spaces, Birkhauser, 1983.

[14] M. Ukhovskiì and V. Yudovitch: Axially symmetric flows of ideal and viscous fluids filling the whole space, J. Appl. Math.and Mechanics, 32 (1968), 59–69.

[15] V. Yudovitch, Non stationary flows of an ideal incompressible fluid, Jurnal vychislitel'noj mat. i matematiceskoj fiziki, 3 (1963), 1032–1066.

Jamel Ben Ameur
Département de Mathématiques
Université de Sciences de Tunis
1060 Tunis
Tunisie

E-mail: jamel.benameur@ipest.rnu.tn

Raphaël Danchin
Laboratoire d'Analyse Numérique
Université Pierre et Marie Curie
175 rue du Chevaleret
75013 Paris
France

E-mail:danchin@ann.jussieu.fr

Studies in Mathematics and its Applications, Vol. 31
D. Cioranescu and J.L. Lions (Editors)

Chapter 4

GLOBAL PROPERTIES OF SOME NONLINEAR PARABOLIC EQUATIONS

M. BEN-ARTZI

1. Introduction

In this paper we review some recent results concerning the class of nonlinear equations of evolution given by,

$$u_t - \Delta u = \mu|\nabla u|^p, \ \mu \in \mathbb{R}, \ p \geq 1, \tag{1.1}$$

$$u(x,0) = u_0(x), \ x \in \mathbb{R}^n. \tag{1.2}$$

We denote

$$\nabla = \nabla_x = \left(\frac{\partial}{\partial x_1}, \ldots, \frac{\partial}{\partial x_n}\right), \quad \Delta = \sum_{i=1}^{n} \left(\frac{\partial}{\partial x_i}\right)^2.$$

While equations of the type $u_t - \Delta u = u^p$ have been extensively studied (see e.g. [14, 15, 29] and references there), the same is not true for (1.1). We note that most of the results mentioned in this paper apply to the more general case where the right-hand side of (1.1) is replaced by $F(\nabla u)$, with suitable growth conditions on F. Thus, (1.1) can be viewed as a model for a "viscous Hamilton-Jacobi" equation. Indeed, this equation appears naturally in a variety of studies. Some examples include

(a) The one-dimensional case $n = 1$. In this case the equation appears in the study of growth of surfaces and is called as the "generalized KPZ equation" [16, 17, 20, 21].

(b) Still in the one-dimensional case, we take $\mu = -1$ and $p = 2$, thus obtaining the equation $u_t + u_x^2 = u_{xx}$. Differentiating with respect to x and setting $v = u_x$, we get for $v(x,t)$ the equation $v_t + (v^2)_x = v_{xx}$, which is the well-known Burgers equation.

(c) Consider the Navier-Stokes equations in the plane ($n = 2$), which in vorticity form can be written as

$$\xi_t + (\underline{u} \cdot \underline{\nabla})\xi = \nu \Delta \xi,$$

(ξ is the vorticity $\partial_2 u^1 - \partial_1 u^2$ of the velocity field $\underline{u} = (u^1, u^2)$). Suppose we know in advance that $|\underline{u}|$ is bounded. Then ξ satisfies the inequality

$$\xi_t - \nu \Delta \xi \leq C|\nabla \xi|.$$

Thus, the methods used in the study of (1.1) are also applicable in the case of the inequality.

In the following sections we shall discuss the global well-posedness of (1.1) in various spaces and the decay properties of solutions as $t \to +\infty$.

2. Existence of global solutions

Let $C_b^2(\mathbb{R}^n) := C^2(\mathbb{R}^n) \cap W^{2,\infty}(\mathbb{R}^n)$, namely, the space of twice continuously differentiable functions with bounded derivatives. It was proved in [3] that $C_b^2(\mathbb{R}^n)$ is a "persistence space" to classical solutions of (1.1). Namely, we have the theorem.

Theorem 2.1 [3]– *Let $u_0 \in C_b^2(\mathbb{R}^n)$. Then for any $\mu \in \mathbb{R}$, $p \geq 1$, there exists a unique classical solution to (1.1) - (1.2), such that $u(\cdot, t) \in C_b^2(\mathbb{R}^n)$ for all $t \geq 0$ and the mapping*

$$u_0 \in C_b^2(\mathbb{R}^n) \to u \in C(\overline{\mathbb{R}_+}, C_b^2(\mathbb{R}^n))$$

is continuous.

Furthermore, the solution satisfies the following maximum-minimum principles:

$$\sup_{\substack{x \in \mathbb{R}^n \\ t \in (0,T]}} u(x,t) = \sup_{x \in \mathbb{R}^n} u_0(x), \ \inf_{\substack{x \in \mathbb{R}^n \\ t \in (0,T]}} u(x,t) = \inf_{x \in \mathbb{R}^n} u_0(x), \forall T > 0, \tag{2.1}$$

$$\|\nabla u(\cdot, t)\|_{L^\infty(\mathbb{R}^n)} \leq \|\nabla u_0\|_{L^\infty(\mathbb{R}^n)}, \forall t \geq 0. \tag{2.2}$$

In the proof, one shows that the solution exists in a time interval $(0, T]$, where T depends only on $\|\nabla u_0\|_{L^\infty(\mathbb{R}^n)}$. The inequality (2.2) then allows the continuation of the solution to $[T, 2T], \ldots$. We remark that to prove (2.2) the equation (1.1) is differentiated with respect to x_j. Denoting $u_j = \frac{\partial}{\partial x_j} u$, we get

$$\frac{\partial}{\partial t} u_j - \Delta u_j = \sum_{i=1}^{n} \Psi_i(x,t) \frac{\partial u_j}{\partial x_i}, \tag{2.3}$$

where

$$\Psi_i(x,t) = \mu p |\nabla u|^{p-2} \frac{\partial u}{\partial x_i} \in L^\infty(\mathbb{R}^n \times (0,T].$$

However, the solution u_j to the linear parabolic equation (2.3) is not twice continuously differentiable hence some care must be taken in deducing (2.2) from the standard (linear) maximum principle. See the Appendix in [3] for details.

Naturally, our next goal is to investigate the well-posedness of (1.1) in wider spaces of less regular functions, for instance, $L^q(\mathbb{R}^n)$ for suitable exponents q (possibly depending on p). To allow such solutions, the equation (1.1) is first cast in the integral form,

$$u(x,t) = \int_{\mathbb{R}^n} G(x-y,t) u_0(y) dy + \mu \int_0^t \int_{\mathbb{R}^n} G(x-y,t-s) |\nabla u(y,s)|^p dy ds, \tag{2.4}$$

where $G(x,t) = (4\pi t)^{-n/2} \exp(-|x|^2/4t)$ is the heat kernel. Taking ∇_x of Eq. (2.4) and using norms of the type

$$\sup_{t\in(0,T]} t^\alpha \|\nabla u(\cdot,t)\|_{L^r(\mathbb{R}^n)},$$

for suitable r, α, as in [29] one obtains the local (in time) existence of solutions to (2.4) in $L^q(\mathbb{R}^n)$, for certain exponents. Then, by using the regularizing effect of the parabolic equation (2.4) (see also [9] for a direct argument) one shows that the solution $u(\cdot,t) \in C_b^2(\mathbb{R}^n)$ for $t > 0$, hence global existence follows from Theorem 2.1 above. As for uniqueness, we note that the solution was constructed by using "growth norms" of the type

$$\sup_{t\in(0,T]} t^\alpha \|\nabla u(\cdot,t)\|_{L^r(\mathbb{R}^n)}.$$

Thus, a contraction argument yields uniqueness using such norms (the "Kato-Fujita condition" [19]). However, an alternative approach as in [13] gives uniqueness for solutions in classes like $C([0,T]; L^q(\mathbb{R}^n)) \cap C((0,T]; C_b(\mathbb{R}^n))$. The exact exponents are summarized in the following theorem.

Theorem 2.2 [11] – *For $1 \le p < 2$, let*

$$q_c = n\frac{p-1}{2-p},$$

and take any $q \ge \max(1, q_c)$, $q < \infty$ (but $q > 1$ if $q_c = 1$). Then, given any $u_0 \in L^q(\mathbb{R}^n)$ (and any $\mu \in \mathbb{R}$), the equation (2.4) has a unique, global (in time) solution $u \in C([0,\infty), L^q(\mathbb{R}^n))$.

In particular we note that if

$$p > p_c := \frac{n+2}{n+1} \tag{2.5}$$

then $q_c > 1$ and the exponent $q = 1$ is outside the scope of Theorem 2.2. Indeed, as the following claim shows, one cannot expect, for $p > p_c$, to have solutions u of (2.4) for any $u_0 \in L^1(\mathbb{R}^n)$, even under the mildest assumptions on u. In presenting the next claim, there is no attempt at achieving maximal generality.

Claim 2.3 [12] – *Let $p > p_c = (n+2)/(n+1)$ and $\mu = 1$. Introduce, for*

$$0 < \delta < \frac{1}{2}\left(\frac{n+1}{n+2} - \frac{1}{p}\right),$$

the following function:

$$v_\delta(x) = \begin{cases} |x|^{-n+\delta}, & |x| < 1, \\ 0 & |x| \ge 1. \end{cases}$$

Then, given any $T > 0$, there is no solution $u(x,t)$ of (2.4) in $(x,t) \in \mathbb{R}^n \times (0,T]$, where $u_0 = v_\delta$ and such that

$$u \in L^p((0,T); W^{1,p}(\mathbb{R}^n)).$$

Proof. Assume the existence of a solution $u(x,t)$ with the above properties. Since

$$\int_0^T \int_{\mathbb{R}^n} |\nabla u|^p dx dt < \infty,$$

given $\varepsilon > 0$ there exists a sequence $t_j \to 0$ such that

$$\int_{\mathbb{R}^n} |\nabla u(x,t_j)|^p dx < \varepsilon t_j^{-1}, j = 1, 2, \dots, \tag{2.6}$$

which implies, by the Sobolev inequality (observe that $u \geq 0$ in view of (2.4)),

$$\int_{\mathbb{R}^n} u(x,t_j)^{p^*}dx \leq C(\varepsilon t_j^{-1})^{\frac{p^*}{p}}, \frac{1}{p^*} = \frac{1}{p} - \frac{1}{n}. \tag{2.7}$$

Take $0 < \beta < \frac{1}{2}$ (to be specified later) and use Hölder's inequality and (2.7) to get,

$$\int\limits_{|x|<t_j^\beta} u(x,t_j)dx \leq (C\varepsilon t_j^{-1})^{\frac{1}{p}} \cdot (\omega_n t_j^{\beta n})^{1-\frac{1}{p^*}}, \tag{2.8}$$

(ω_n= volume of unit ball).

Now

$$p > \frac{n+2}{n+1} \Longrightarrow n(1-\frac{1}{p^*}) > \frac{2(n+1)}{n+2},$$

so from (2.8),

$$\int\limits_{|x|<t_j^\beta} u(x,t_j)dx \leq C\varepsilon^{1/p} t_j^{-\frac{1}{p}+\frac{2\beta(n+1)}{n+2}}, j = 1,2,\ldots \tag{2.9}$$

Since $p > ((n+2)/(n+1)$ we can choose $\beta < \frac{1}{2}$ such that

$$\eta = -\frac{1}{p} + 2\beta\frac{n+1}{n+2} > 0,$$

hence

$$\int\limits_{|x|<t_j^\beta} u(x,t_j)dx \leq C\varepsilon^{1/p} t_j^\eta \to 0 \text{ as } \quad j \to \infty. \tag{2.10}$$

Let $\tilde{u}(x,t) = G(x,t) * u_0$ be the solution to the heat equation with the same initial data. Clearly $u(x,t) \geq \tilde{u}(x,t)$. We have, for $t > 0$,

$$\begin{aligned}
\int\limits_{|x|>t^\beta} \tilde{u}(x,t)dx &= \int\limits_{|x|>t^\beta}\int\limits_{\mathbb{R}^n} G(x-y,t)u_0(y)dydx \\
&= \int\limits_{|x|>t^\beta}\int\limits_{|y|<\frac{1}{2}t^\beta} + \int\limits_{|x|>t^\beta}\int\limits_{|y|>\frac{1}{2}t^\beta} G(x-y,t)u_0(y)dydx \\
&\leq \int\limits_{|\xi|>\frac{1}{2}t^\beta} G(\xi,t)d\xi \cdot \|u_0\|_{L^1(\mathbb{R}^n)} + \int\limits_{|y|>\frac{1}{2}t^\beta} u_0(y)dy.
\end{aligned} \tag{2.11}$$

Since $\beta < \frac{1}{2}$, we have

$$\int\limits_{|\xi|>\frac{1}{2}t^\beta} G(\xi,t)d\xi = O(t^N) \text{ as } t \to 0, N = 1,2,\ldots$$

and, for $u_0 = v_\delta$

$$\int\limits_{|y|>\frac{1}{2}t^\beta} u_0(y)dy = (1 - 2^{-\delta}t^{\beta\delta})\|u_0\|_{L^1(\mathbb{R}^n)}, t < 1, \tag{2.13}$$

so, since $\|\tilde{u}(\cdot,t)\|_{L^1(\mathbb{R}^n)} = \|u_0\|_{L^1(\mathbb{R}^n)}$, we conclude that

$$\int\limits_{|x|<t^\beta} \tilde{u}(x,t)dx = 2^{-\delta}t^{\beta\delta}\|u_0\|_{L^1(\mathbb{R}^n)} + O(t^N) \text{ as } t \to 0.$$

Setting $t = t_j$ and comparing with (2.10) we get, for $j = 1,2,\ldots$,

$$C\varepsilon^{1/p}t_j^\eta \geq 2^{-\delta}t_j^{\beta\delta}\|u_0\||_{L^1(\mathbb{R}^n)} + O(t_j^N) \tag{2.14}$$

which is a contradiction by the choice of δ, since $\beta < \frac{1}{2}$ can be chosen such that $\beta\delta < \eta$. ■

Remark 2.4. In view of the last claim, one may ask, in the case $p > p_c, \mu = 1$, what is the set of initial data $u_0(x) \in L^1(\mathbb{R}^n)$ for which a solution to (2.4) does exist. Theorem 2.2 implies that this set contains all $u_0 \in L^1(\mathbb{R}^n) \cap L^q(\mathbb{R}^n)$, $q \geq q_c$, and in particular, all $u_0(x) \in L^1(\mathbb{R}^n) \cap L^\infty(\mathbb{R}^n)$. However, Claim 2.3 says this set is not all of $L^1(\mathbb{R}^n)$. The situation is still not clear for $\mu = -1$. On the other hand, if $\mu = 1$ and $p \geq 2$, Claim 2.3 can be strengthened as it is shown in Proposition 2.5 below.

Proposition 2.5 [11] – *Let $u(x,t)$ be a classical solution of (1.1), with $\mu = 1, p \geq 2$, in a strip $\mathbb{R}^n \times (0,T)$. Assume that*

$$\lim_{t\to 0} u(\cdot,t) = u_0 \quad \textit{in } L^1_{loc}(\mathbb{R}^n).$$

Then

$$\exp(u_0) \in L^1_{loc}(\mathbb{R}^n).$$

Finally, while (for $p > p_c, \mu = 1$) existence is not guaranteed for all $u_0 \in L^q(\mathbb{R}^n), 1 \leq q < q_c$, uniqueness can also fail, as the following theorem shows:

Theorem 2.6 [11] – *Assume* $2 > p > p_c$ *and let* $1 \leq q < q_c$ *and* $\mu = 1$. *Then, for* $u_0 = 0$, *there exists a positive solution* u *to (2.4). In fact,* u *is self-similar,*

$$u(x,t) = t^{-k} U(|x|t^{-\frac{1}{2}}), \quad k = \frac{2-p}{2(p-1)},$$

where $U = U(r) \in C^2([0,\infty))$.

Remark 2.7. The case of a coupled system of equations of the type (1.1) was treated in [4]. ■

3. Further extensions. The case $\mu = -1$

We consider here some further results for solutions of Eq. (1.1) (or (2.4)) under the assumptions that $\mu = -1$ and $u_0 \geq 0$. The maximum-minimum principle guarantees that the solution u is nonnegative and is majorized by the corresponding solution of the heat equation. In this case, the subcritical part of Theorem 2.2 has been extended by Benachour and Laurencot [5] to include positive bounded measures, as follows.

Theorem 3.1 [5] – *Let*

$$1 < p < p_c = \frac{n+2}{n+1}, \quad \mu = -1,$$

and $u_0 \in M_b^+(\mathbb{R}^n)$ (= *the space of positive bounded Borel measures). Then there exists a unique weak solution (in the sense of (2.4))* u *such that,*

$$u \in C((0,\infty); L^1(\mathbb{R}^n)) \cap L^p_{loc}((0,\infty); W^{1,p}(\mathbb{R}^n)).$$

Remark 3.2 (a). We refer to [5] for a precise definition of a "weak solution". Also, for the uniqueness a "growth condition" (as $t \to 0$) of the "Kato-Fujita" type is required, as in the discussion preceding Theorem 2.2 above.

(b). The case $p = 1$ (and $\mu = -1$) was treated in [7], by probabilistic methods, producing a spherically symmetric solution for any initial data $u_0(x)$ which is a "profiled" spherically symmetric bounded positive measure.

(c). The more general equation

$$u_t - \Delta u = -a(x) u^q (\nabla u)^p, u_0 \geq 0,$$

was treated in [23]. ■

The supercritical case ($p \geq p_c$) is more difficult. Clearly, the method of proof of Claim 2.3 does not work here and the question whether or not the equation is well-posed in $L^1(\mathbb{R}^n)$ remains an open problem. However, Benachour and Laurencot [5] have managed to prove the non-existence of "source-type" solutions, namely, solutions that converge (in the sense of distributions) to a multiple of the Delta-function. The exact formulation of the theorem is as follows.

Theorem 3.3 [5]– *Let $M > 0, T > 0,\ p \geq p_c$. There is no function $u \in L^\infty((0,T); L^1(\mathbb{R}^n)) \cap L^p((0,T); W^{1,p}(\mathbb{R}^n))$ such that $u_t - \Delta u = -|\nabla u|^p$ in $\mathcal{D}'(\mathbb{R}^n \times (0,T))$ and*

$$\lim_{t\to 0} \int_{\mathbb{R}^n} u(x,t)\Psi dx = M\Psi(o), \forall \Psi \in C_0^\infty(\mathbb{R}^n).$$

4. Decay as $t \to +\infty$

Let us go back to classical (say, as in Theorem 2.1) solutions to (1.1), where we assume now that $u_0 \geq 0$ and $\mu = -1$. Then the solution $u(x,t)$ is nonnegative and an integration of (1.1) shows that if, in addition, $u_0 \in L^1(\mathbb{R}^n)$ then $u(\cdot,t) \in L^1$ for all $t \geq 0$ and the nonnegative function $I(t) = \int_{\mathbb{R}^n} u(x,t)dx$ is nonincreasing. Thus, the limit $I_\infty = \lim_{t\to\infty} I(t) \geq 0$ always exists. It is interesting that the question whether or not $I_\infty = 0$ is determined uniquely by $p_c = (n+2)/(n+1)$, the same critical value as in the previous sections. We have the following theorem.

Theorem 4.1 [10] – *Let $0 \leq u_0 \in C_b^2(\mathbb{R}^n) \cap L^1(\mathbb{R}^n), u_0 \not\equiv 0$. Let $u(x,t)$ be the solution to (1.1), with $\mu = -1$. Then*

$$I_\infty > 0 \Leftrightarrow p > p_c = \frac{n+2}{n+1}.$$

Remark 4.2. As was seen in Theorem 2.2, the well-posedness of (1.1) in $L^1(\mathbb{R}^n)$ was also linked to the same critical index p_c. However, there is yet no direct argument connecting this well-posedness (essentially a short-time feature) with the long-time decay as expressed in Theorem 4.1.

Remark 4.3. In the case $p < p_c$ the equation is well-posed in $L^1(\mathbb{R}^n)$. Then, as in the discussion preceding Theorem 2.2, if $0 \leq u_0 \in L^1(\mathbb{R}^n)$ (and $\mu = -1$), it follows that $u(\cdot,t) \in C_b^2(\mathbb{R}^n) \cap L^1(\mathbb{R}^n)$ for $t > 0$. Hence, Theorem 4.1 is applicable also, in the subcritical case, to all $0 \leq u_0 \in L^1(\mathbb{R}^n)$.

Remark 4.4 In the case $p \leq p_c$, the rate of decay of $I(t)$ to zero becomes slower as p approaches p_c. More precisely, let $1 < p \leq p_c$ and

$$\alpha > \frac{2-p}{2(p-1)} - \frac{n}{2}.$$

Then [10] $I(t) \leq Ct^{-\alpha}$ (for all sufficiently large t) implies $u_0 = 0$. In particular, if $p = p_c$ then $I(t)$ cannot decay like $t^{-\alpha}$ for any $\alpha > 0$. On the other hand, if $p = 1$ and u_0 is compactly supported then, for some $A, \theta > 0$ we have (see [3]),

$$\sup_{0 \leq t < \infty} \exp(At^\theta) I(t) < \infty.$$

Acknowledgement. The author is deeply indebted to L. Amour, J. Goodman, H. Koch, A. Levy, Ph. Souplet and F. Weissler for pleasant collaborations and useful discussions. It is also a pleasure to thank S. Benachour for interesting correspondence and exchange of ideas.

References

[1] N. Alaa, Solutions faibles d'équations paraboliques quasilinéaires avec données initiales mesures, Ann. Math. Blaise Pascal 3 (1996), 1–15.

[2] L. Alfonsi and F. B. Weissler, Blow-up in $\mathbb{R}^n$ for a parabolic equation with a damping nonlinear gradient term, in Progress in Nonlinear Differential Equations, (N.G. Lloyd et al Eds.), Birkhäuser 1992.

[3] L. Amour and M. Ben-Artzi, Global existence and decay for viscous Hamilton-Jacobi equations, Nonlinear Anal. TMA 31 (1998), 621–628.

[4] L. Amour and T. Raoux, Existence et décroissance en norme L^1 des solutions d'un système parabolique semi-linéaire, Note CRAS Paris, Série I , 329 (1999), 367-370.

[5] S. Benachour and Ph. Laurencot, Global solutions to viscous Hamilton-Jacobi equation with irregular data, Comm. PDE 24 (1999), 1999–2021.

[6] S. Benachour and Ph. Laurencot, "Solutions très singulières" d'une équation parabolique non linéaire avec absorption, Note CRAS Paris, Série I Math. 328 (1999), 215–220.

[7] S. Benachour, B. Roynette and P. Vallois, Asymptotic estimates of solutions of $u_t - \frac{1}{2}\Delta u = -|\nabla u|$ in $\mathbb{R}_+ \times \mathbb{R}^d, d \geq 2$, J. Func. Anal. 144 (1997), 301–324.

[8] M. Ben-Artzi, Global existence and decay for a nonlinear parabolic equation, Nonlinear Anal. TMA 19 (1992), 763–768.

[9] M. Ben-Artzi, J. Goodman and A. Levy, Remarks on a nonlinear parabolic equation, Trans. Amer. Math. Soc. 352 (2000), 731–751.

[10] M. Ben-Artzi and H. Koch, Decay of mass for a semilinear parabolic equation, Comm. PDE 24 (1999), 869–881.

[11] M. Ben-Artzi, Ph. Souplet and F.B. Weissler, Sur la non-existence et la non-unicité des solutions du problème de Cauchy pour une équation parabolique semi-linéaire, Note CRAS Paris, Série I, 329 (1999), 371–376.

[12] M. Ben-Artzi, Ph. Souplet and F.B. Weissler, The local theory for viscous Hamilton-Jacobi equations in Lebesgue spaces, J. Math. Pures et Appliquées, 81 (2002) to appear

[13] H. Brezis, Remarks on the preceding paper by M. Ben-Artzi, "Global solutions of two-dimensional Navier-Stokes and Euler Equations", Arch. Rat. Mech. Anal. 128 (1994), 359–360.

[14] H. Brezis and A. Friedman, Nonlinear parabolic equations involving measures as initial conditions, J. Math. Pures Appl. IX, Ser. 62 (1983), 73–97.

[15] H. Brezis, L. Peletier and D. Terman, A very singular solution of the heat equation with absorption, Arch. Rat. Mech. Anal. 95 (1986), 185–219.

[16] B. Gilding, M. Guedda and R. Kersner, The Cauchy problem for $u_t = \Delta u + |\nabla|^q$, Preprint 1998.

[17] M. Guedda, R. Kersner and L. Veron, On self-similar-type solutions to the generalized KPZ equation, Preprint 1998.

[18] A. Haraux and F.B. Weissler, Non-uniqueness for a semilinear initial value problem, Indiana Univ. Math. J. 31 (1982), 167–189.

[19] T. Kato and H. Fujita, On the nonstationary Navier-Stokes system, Rend. Sem. Math. Univ. Padova 32 (1962), 243–260.

[20] J. Krug and H. Spohn, Universality classes for deterministic surface growth, Phys. Rev. A 38 (1988), 4271–4283.

[21] J. Krug and H. Spohn, Kinetic roughening of growing surfaces, in "Solids far from equilibrium", C. Godreche (Ed.), Cambridge Univ. Press, 1991, 479–582.

[22] P.L. Lions, Generalized solutions of Hamilton-Jacobi Equations, Pitman Research Notes in Mathematics, 69, 1982.

[23] R.G. Pinsky, Decay of mass for the equation $u_t = \Delta u - a(x)u^p|\nabla u|^q$, J. Diff. Eqs. 165 (2000), 1–23

[24] S. Snoussi, S. Tayachi and F.B. Weissler, Asymptotically self-similar global solutions of semilinear parabolic equations with nonlinear gradient terms, Proc. Royal Soc. Edinburgh, 129A (1999), 1291–1307.

[25] Ph. Souplet, Résultats d'explosion en temps fini pour une èquation de la chaleur non linéaire, Note C. R. Acad. Sci. Paris, Série I. 321 (1995), 721–726.

[26] Ph. Souplet, Geometry of unbounded domains, Poincaré inequalities and stability in semilinear parabolic equations, Comm. PDE, 24 (1999), 951–973.

[27] Ph. Souplet and F.B. Weissler, Poincaré's inequality and global solutions of a nonlinear parabolic equation, Ann. Inst. H. Poincaré, Analyse non linéaire, 16, 3 (1999), 337–373.

[28] S. Tayachi, Forward self-similar solutions of a semilinear parabolic equation with a nonlinear gradient term, Diff. and Integral Equations, 9 (1996), 1107–1117.

[29] F.B. Weissler, Local existence and nonexistence for semilinear parabolic equations in L^p, Indiana Univ. Math. J. 29 (1980), 79–102.

M. Ben-Artzi
Institute of Mathematics
Hebrew University
Jerusalem 91904
Israel
E-mail: mbartzi@math.huji.ac.il

Studies in Mathematics and its Applications, Vol. 31
D. Cioranescu and J.L. Lions (Editors)

Chapter 5

A MODEL FOR TWO COUPLED TURBULENT FLUIDS PART I: ANALYSIS OF THE SYSTEM

C. Bernardi, T. Chacón Rebollo, R. Lewandowski and F. Murat

1. Introduction.

This paper is devoted to the analysis of the following system which models two stationary turbulent fluids coupled by boundary conditions on the interface:

$$\begin{cases} -\mathrm{div}\,\big(\alpha_i(k_i)\,\nabla \boldsymbol{u}_i\big) + \mathbf{grad}\,p_i = \boldsymbol{f}_i & \text{in } \Omega_i,\ 1 \le i \le 2, \\ \mathrm{div}\,\boldsymbol{u}_i = 0 & \text{in } \Omega_i,\ 1 \le i \le 2, \\ -\mathrm{div}\,\big(\gamma_i(k_i)\,\nabla k_i\big) = \alpha_i(k_i)\,|\nabla \boldsymbol{u}_i|^2 & \text{in } \Omega_i,\ 1 \le i \le 2, \\ \boldsymbol{u}_i = \boldsymbol{0} & \text{on } \Gamma_i,\ 1 \le i \le 2, \\ k_i = 0 & \text{on } \Gamma_i,\ 1 \le i \le 2, \\ \alpha_i(k_i)\,\partial_{n_i}\boldsymbol{u}_i - p_i\,\boldsymbol{n}_i + (\boldsymbol{u}_i - \boldsymbol{u}_j)\,|\boldsymbol{u}_i - \boldsymbol{u}_j| = \boldsymbol{0} & \\ \qquad\qquad \text{on } \Gamma, & 1 \le i \ne j \le 2, \\ k_i = |\boldsymbol{u}_1 - \boldsymbol{u}_2|^2 & \text{on } \Gamma,\ 1 \le i \le 2, \end{cases} \tag{1.1}$$

where each triple $(\boldsymbol{u}_i, k_i, p_i)$ is defined in the domain Ω_i, $1 \le i \le 2$.

In what follows, Ω_1 and Ω_2 stand for disjoint bounded domains in $\mathbb{R}^d$, $d = 2$ or 3, which are either convex or of class $\mathcal{C}^{1,1}$. The generic point in $\mathbb{R}^2$, resp. in $\mathbb{R}^3$, is denoted by $\boldsymbol{x} = (x, z)$, resp. $\boldsymbol{x} = (x, y, z)$. We assume for simplicity that the interface $\Gamma = \partial\Omega_1 \cap \partial\Omega_2$ coincides with the intersection of both $\overline{\Omega}_1$ and $\overline{\Omega}_2$ with the hyperplane $z = 0$, while Ω_1 and Ω_2 are contained in the half-spaces $z > 0$ and $z < 0$ respectively, see the following figure where each Γ_i is equal to $\partial\Omega_i \setminus \Gamma$. Note also that, in the physical context, the heights of the domains are much smaller than their horizontal diameters.

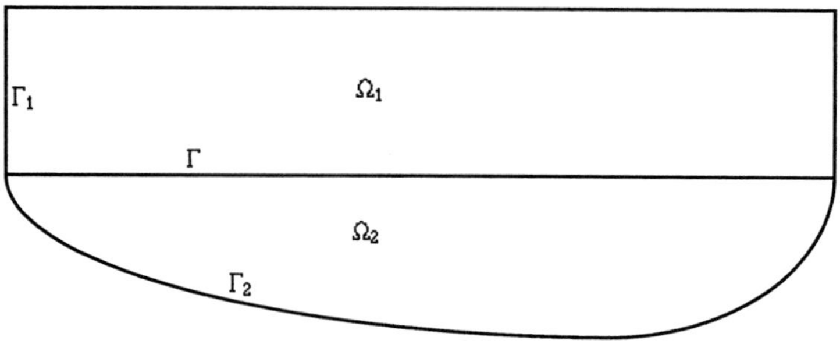

Figure 1

The vector field $\boldsymbol{u}_i$ stands for the velocity of a turbulent fluid in Ω_i, p_i represents its pressure and k_i its turbulent kinetic energy (TKE in what follows). The quantity $\alpha_i(k_i)$ is the eddy viscosity, and we shall assume throughout this paper that the functions α_i and γ_i satisfy

$$\begin{cases} \alpha_i \in C^0(\mathbb{R}) \cap L^\infty(\mathbb{R}) \quad \text{and} \quad \forall k \in \mathbb{R}, \quad \alpha_i(k) \geq \nu, \qquad 1 \leq i \leq 2, \\ \gamma_i \in C^0(\mathbb{R}) \cap L^\infty(\mathbb{R}) \quad \text{and} \quad \forall k \in \mathbb{R}, \quad \gamma_i(k) \geq \nu, \qquad 1 \leq i \leq 2, \end{cases} \tag{1.2}$$

for some positive constant ν. The functions $\boldsymbol{f}_i$ are given, with

$$\boldsymbol{f}_i \in L^2(\Omega_i)^d, \qquad 1 \leq i \leq 2. \tag{1.3}$$

System (1.1) is motivated by the coupling of two turbulent fluids F_i, $i = 1$ and 2, such as in the framework ocean/atmosphere or in the case of two layers of a stratified fluid (see e.g. [13, Chap. 1 & 3] or [17]). These fluids F_i are coupled through the interface condition in the sixth line of (1.1), on their common boundary Γ (which is supposed to be fixed). Indeed, we assume that the so-called "rigid lid hypothesis" holds, an hypothesis which is standard in geophysics and oceanography. Actually, Γ is a mean interface and the values of $\boldsymbol{u}_i$, p_i and k_i on Γ are in fact mean values of the velocity, pressure and TKE. So, the turbulent mixed layer of the two turbulent fluids is modelized by the sixth and seventh lines in (1.1) which summarize the information related to a realistic interface ocean/atmosphere (see e.g. [13, §1.4] for more details about this modelization).

The main goal of this paper is to prove the existence of a solution $(\boldsymbol{u}_i, k_i, p_i)_{1 \le i \le 2}$ of system (1.1) (see Corollary 5.3). We start by giving a sense to the equations. The two first lines of system (1.1) are the Stokes equations in Ω_i, equipped with the eddy viscosity $\alpha_i(k_i)$ which is the quantity of interest. The third line is a scalar equation which allows one to compute the k_i.

Following the ideas of [15], one can write a weak mixed formulation of the first two lines (see also [17] for the case of coupling ocean/atmosphere without turbulence). The third line and the corresponding boundary conditions are more complex: the main difficulty comes from the fact that the right-hand side only belongs to $L^1(\Omega_i)$. In the case of homogeneous boundary conditions and with only one turbulent fluid, such type of equations has already been studied (see e.g. [4], [10], [13, Chap. 4], [14]). In those references, the equation for the TKE is taken in the renormalized sense of Lions and Murat (see [18], [20], [21]), or in the equivalent entropy sense of Benilan *et al.* (see [3]), and *a priori* estimates of Boccardo–Gallouët type [5] are used. However, because of the boundary conditions at the interface Γ, this renormalization does not seem an easy way for the study of the TKE equations in the present problem, and one cannot hope to use directly the results of [5]. For this reason, we make Kirchoff's change of unknown in order to replace the operator div $\big(\gamma_i(k_i)\, \nabla\big)$ by a simpler Laplace operator. We consider the corresponding new equation in the sense of transposition, according to the ideas of Stampacchia [23] and of Lions and Magenes [16, Chap. 2, §6]. We are then able to prove the existence of a solution of the global system (1.1).

Under some rather restrictive assumptions on the variations of the functions α_i, we prove a uniqueness result for the solution of (1.1). Some further regularity properties of this solution are also derived with the same assumptions, when the domains Ω_i are rectangles.

Most results of this paper have been announced in [1], however with a slightly different proof relying on Leray–Schauder fixed point theorem. The discretization of system (1.1) by spectral and finite element methods is presently under consideration, from both points of view of numerical analysis and experiments. It must also be observed that the present analysis can be extended to slightly different models,

- by adding convection terms in the momentum equations for the velocities,

- by replacing the transmission conditions for the velocities on Γ (sixth line in (1.1)) by Manning's law

$$\begin{cases} \alpha_i(k_i)\,\partial_{n_i}\boldsymbol{u}_{iH} + (\boldsymbol{u}_{iH} - \boldsymbol{u}_{jH})\,|\boldsymbol{u}_i - \boldsymbol{u}_j| = 0 & \text{on } \Gamma,\ 1 \le i \ne j \le 2, \\ u_{iV} = 0 & \text{on } \Gamma,\ 1 \le i \le 2, \end{cases}$$

where $\boldsymbol{u}_{iH}$ and u_{iV} respectively stand for the horizontal and vertical components of the velocities,
- by adding the term due to Coriolis acceleration.

An outline of the paper is as follows. In Section 2, we write a system which is equivalent to (1.1) through a change of unknowns. Sections 3 and 4 are devoted to the proof of the existence of a solution for the equations on the velocities and the turbulent energies, respectively. In Section 5, we state and establish our main result, namely that the global system (1.1) admits a solution. In Section 6, we prove the conditional uniqueness result and, in Section 7, we derive the optimal regularity of this solution when the domains Ω_i are rectangles.

2. Transformation of the system.

Let us define the functions G_i, $1 \le i \le 2$, by

$$G_i(k) = \int_0^k \gamma_i(\kappa)\,d\kappa. \tag{2.1}$$

In view of (1.2), the G_i are increasing functions of class $\mathcal{C}^1$, so that they admit an inverse G_i^{-1} from $\mathbb{R}$ into $\mathbb{R}$. Moreover, the functions $\tilde{\alpha}_i$, $i = 1$ and 2, defined by

$$\tilde{\alpha}_i = \alpha_i \circ G_i^{-1}, \tag{2.2}$$

satisfy the same properties as the α_i, namely

$$\tilde{\alpha}_i \in \mathcal{C}^0(\mathbb{R}) \cap L^\infty(\mathbb{R}) \quad \text{and} \quad \forall \ell \in \mathbb{R}, \quad \tilde{\alpha}_i(\ell) \ge \nu, \quad 1 \le i \le 2. \tag{2.3}$$

The idea is to introduce the new unknowns ℓ_i by

$$\ell_i = G_i(k_i), \qquad 1 \le i \le 2. \tag{2.4}$$

From the formula $\nabla \ell_i = \gamma_i(k_i)\,\nabla k_i$, it is readily checked that system (1.1) is equivalent to

$$\begin{cases} -\mathrm{div}\,\big(\tilde{\alpha}_i(\ell_i)\,\nabla \boldsymbol{u}_i\big) + \mathbf{grad}\,p_i = \boldsymbol{f}_i & \text{in } \Omega_i,\ 1 \le i \le 2, \\ \mathrm{div}\,\boldsymbol{u}_i = 0 & \text{in } \Omega_i,\ 1 \le i \le 2, \\ -\Delta \ell_i = \tilde{\alpha}_i(\ell_i)\,|\nabla \boldsymbol{u}_i|^2 & \text{in } \Omega_i,\ 1 \le i \le 2, \\ \boldsymbol{u}_i = \boldsymbol{0} & \text{on } \Gamma_i,\ 1 \le i \le 2, \\ \ell_i = 0 & \text{on } \Gamma_i,\ 1 \le i \le 2, \\ \tilde{\alpha}_i(\ell_i)\,\partial_{n_i}\boldsymbol{u}_i - p_i\,\boldsymbol{n}_i + (\boldsymbol{u}_i - \boldsymbol{u}_j)\,|\boldsymbol{u}_i - \boldsymbol{u}_j| = \boldsymbol{0} & \\ \qquad\qquad\qquad\qquad \text{on } \Gamma, & 1 \le i \ne j \le 2, \\ \ell_i = G_i(|\boldsymbol{u}_1 - \boldsymbol{u}_2|^2) & \text{on } \Gamma,\ 1 \le i \le 2. \end{cases} \tag{2.5}$$

The interest is that now each function ℓ_i is a solution of a nearly standard Laplace equation.

The goal of this paper is to prove the existence of a solution of (2.5), see Theorem 5.2 below. Throughout the paper, we use the spaces $L^p(\Omega_i)$, $1 \le p \le \infty$, and the Sobolev spaces $H^s(\Omega_i)$ and $H^s_0(\Omega_i)$ for any real number s, provided with the standard norm $\|\cdot\|_{H^s(\Omega_i)}$ and semi-norm $|\cdot|_{H^s(\Omega_i)}$, together with their analogues on Γ. We also need the Sobolev spaces $W^{1,p}(\Omega_i)$ and the special space $H^{\frac{1}{2}}_{00}(\Gamma)$, defined e.g. in [16, Chap. 1, Th. 11.7].

3. The equations on the velocities.

Throughout this section, we assume that the functions ℓ_1 and ℓ_2 are given such that

$$\ell_i \in L^1(\Omega_i), \qquad 1 \le i \le 2, \tag{3.1}$$

even if they turn out to be smoother in what follows. For $1 \le i \le 2$, we introduce the spaces

$$X_i = \big\{\boldsymbol{v}_i \in H^1(\Omega_i)^d;\ \boldsymbol{v}_i = \boldsymbol{0} \text{ on } \Gamma_i\big\}. \tag{3.2}$$

We now write correctly (and not only formally) the equation on each velocity

$\boldsymbol{u}_i$ through its variational formulation:

Find $(\boldsymbol{u}_i, p_i)$, $1 \le i \le 2$, *in* $X_i \times L^2(\Omega_i)$ *such that, for* $1 \le i \ne j \le 2$*:*

$$\begin{aligned} \forall \boldsymbol{v}_i \in X_i, \quad & \int_{\Omega_i} \tilde{\alpha}_i(\ell_i)\, \nabla \boldsymbol{u}_i \cdot \nabla \boldsymbol{v}_i \, d\boldsymbol{x} + b_i(\boldsymbol{v}_i, p_i) \\ & \qquad + \int_\Gamma |\boldsymbol{u}_i - \boldsymbol{u}_j|\, (\boldsymbol{u}_i - \boldsymbol{u}_j) \cdot \boldsymbol{v}_i \, d\tau = \int_{\Omega_i} \boldsymbol{f}_i \cdot \boldsymbol{v}_i \, d\boldsymbol{x}, \\ \forall q_i \in L^2(\Omega_i), \quad & b_i(\boldsymbol{u}_i, q_i) = 0, \end{aligned} \tag{3.3}$$

where the form $b_i(\cdot,\cdot)$ is defined by

$$b_i(\boldsymbol{v}_i, q_i) = -\int_{\Omega_i} q_i (\operatorname{div} \boldsymbol{v}_i)\, d\boldsymbol{x}. \tag{3.4}$$

Note that the first bilinear form in (3.3) depends on ℓ_i.

As standard for the Stokes problem, we consider the kernel

$$V_i = \{\boldsymbol{v}_i \in X_i;\ \operatorname{div} \boldsymbol{v}_i = 0 \text{ in } \Omega_i\},$$

and we observe that, for each solution $(\boldsymbol{u}_i, p_i)$ of problem (3.3), the velocity $\boldsymbol{u}_i$ is a solution of:

Find $\boldsymbol{u}_i$ *in* V_i, $1 \le i \le 2$, *such that, for* $1 \le i \ne j \le 2$*:*

$$\begin{aligned} \forall \boldsymbol{v}_i \in V_i, \quad & \int_{\Omega_i} \tilde{\alpha}_i(\ell_i)\, \nabla \boldsymbol{u}_i \cdot \nabla \boldsymbol{v}_i \, d\boldsymbol{x} \\ & \qquad + \int_\Gamma |\boldsymbol{u}_i - \boldsymbol{u}_j|\, (\boldsymbol{u}_i - \boldsymbol{u}_j) \cdot \boldsymbol{v}_i \, d\tau = \int_{\Omega_i} \boldsymbol{f}_i \cdot \boldsymbol{v}_i \, d\boldsymbol{x}. \end{aligned} \tag{3.5}$$

The converse property relies on the following inf-sup condition of Babuška and Brezzi type.

Lemma 3.1. – *For* $i = 1$ *and* 2, *there exists a constant* $\beta_i > 0$ *such that*

$$\forall q_i \in L^2(\Omega_i), \quad \sup_{\boldsymbol{v}_i \in X_i} \frac{b_i(\boldsymbol{v}_i, q_i)}{\|\boldsymbol{v}_i\|_{H^1(\Omega_i)^d}} \ge \beta_i \, \|q_i\|_{L^2(\Omega_i)}. \tag{3.6}$$

Proof. The argument here is due to Boland and Nicolaides [6]. Let us write any function q_i in $L^2(\Omega_i)$ as

$$q_i = \tilde{q}_i + \overline{q}_i, \qquad \text{with} \quad \overline{q}_i = \frac{1}{\operatorname{meas}(\Omega_i)} \int_{\Omega_i} q_i \, d\boldsymbol{x}.$$

Indeed, from the standard inf-sup condition [11, Chap. I, Cor. 2.4], there exists a function $\tilde{\boldsymbol{v}}_i$ in $H_0^1(\Omega_i)^d$ such that

$$\operatorname{div} \tilde{\boldsymbol{v}}_i = -\tilde{q}_i \quad \text{and} \quad \|\tilde{\boldsymbol{v}}_i\|_{H^1(\Omega_i)^d} \leq c_i \, \|\tilde{q}_i\|_{L^2(\Omega_i)},$$

where the constant c_i only depends on the geometry of Ω_i. On the other hand, there exist an open interval or disk $\mathcal{D}$ such that its closure is contained in the interior of Γ and positive real numbers ε_i such that the cylinders $\mathcal{C}_1 = \mathcal{D}\times]0, \varepsilon_1[$ and $\mathcal{C}_2 = \mathcal{D}\times]-\varepsilon_2, 0[$ are contained in Ω_1 and Ω_2, respectively. If ρ denotes a smooth function that is positive on $\mathcal{D}$ and vanishes on $\partial\mathcal{D}$, the function with horizontal components equal to zero and vertical component v_{iz} defined on the cylinder $\mathcal{C}_i$ by (in dimension $d = 3$ for instance)

$$v_{iz}(x, y, z) = \overline{q}_i \, \rho(x, y) \, (\pm\varepsilon_i - z),$$

can be extended by zero into a function $\overline{\boldsymbol{v}}_i$ of X_i which satisfies

$$b_i(\overline{\boldsymbol{v}}_i, \overline{q}_i) = c_i' \, \|\overline{q}_i\|_{L^2(\Omega_i)}^2 \quad \text{and} \quad \|\overline{\boldsymbol{v}}_i\|_{H^1(\Omega_i)^d} \leq c_i'' \, \|\overline{q}_i\|_{L^2(\Omega_i)}.$$

Finally, taking $\boldsymbol{v}_i = \tilde{\boldsymbol{v}}_i + \mu_i \, \overline{\boldsymbol{v}}_i$ yields

$$\begin{aligned} b_i(\boldsymbol{v}_i, q_i) &= \|\tilde{q}_i\|_{L^2(\Omega_i)}^2 + \mu_i c_i' \, \|\overline{q}_i\|_{L^2(\Omega_i)}^2 + \mu_i \, b_i(\overline{\boldsymbol{v}}_i, \tilde{q}_i) \\ &\geq \|\tilde{q}_i\|_{L^2(\Omega_i)}^2 + \mu_i c_i' \, \|\overline{q}_i\|_{L^2(\Omega_i)}^2 - \mu_i c_i'' \, \|\tilde{q}_i\|_{L^2(\Omega_i)} \|\overline{q}_i\|_{L^2(\Omega_i)} \\ &\geq \frac{1}{2} \|\tilde{q}_i\|_{L^2(\Omega_i)}^2 + \mu_i \big(c_i' - \frac{\mu_i c_i''^2}{2}\big) \, \|\overline{q}_i\|_{L^2(\Omega_i)}^2, \end{aligned}$$

and also

$$\|\boldsymbol{v}_i\|_{H^1(\Omega_i)^d} \leq c_i \, \|\tilde{q}_i\|_{L^2(\Omega_i)} + \mu_i c_i'' \, \|\overline{q}_i\|_{L^2(\Omega_i)} \leq \big(c_i^2 + \mu_i^2 c_i''^2\big)^{\frac{1}{2}} \, \|q_i\|_{L^2(\Omega_i)}.$$

So choosing $\mu_i = \frac{c_i'}{c_i''^2}$ leads to the desired result. ∎

The next corollary is now a direct consequence of Lemma 3.1, see [11, Chap. I, Lemma 4.1].

Corollary 3.2. – *For $i = 1$ and 2 and for any data $\boldsymbol{f}_i$ in $L^2(\Omega_i)^d$,*
(i) for any solution $(\boldsymbol{u}_i, p_i)$ of problem (3.3), the velocity $\boldsymbol{u}_i$ is a solution of problem (3.5),
(ii) for any solution $\boldsymbol{u}_i$ of problem (3.5), there exists a unique pressure p_i in $L^2(\Omega_i)$ such that the pair $(\boldsymbol{u}_i, p_i)$ is a solution of problem (3.3).

We now prove the existence of a solution of problem (3.5). We begin with an *a priori* estimate.

Lemma 3.3. – *For every ℓ_i in $L^1(\Omega_i)$ and $\boldsymbol{f}_i$ in $L^2(\Omega_i)^d$, $1 \le i \le 2$, any solution $(\boldsymbol{u}_1, \boldsymbol{u}_2)$ of problem* (3.5) *satisfies*

$$\|\boldsymbol{u}_1\|_{H^1(\Omega_1)^d} + \|\boldsymbol{u}_2\|_{H^1(\Omega_2)^d} \le \frac{c}{\nu}\big(\|\boldsymbol{f}_1\|_{L^2(\Omega_1)^d} + \|\boldsymbol{f}_2\|_{L^2(\Omega_2)^d}\big). \tag{3.7}$$

Proof. Taking $\boldsymbol{v}_i$ equal to $\boldsymbol{u}_i$ in (3.5), using (2.3) and summing up on i give

$$\nu\left(|\boldsymbol{u}_1|^2_{H^1(\Omega_1)^d} + |\boldsymbol{u}_2|^2_{H^1(\Omega_2)^d}\right) + \int_\Gamma |\boldsymbol{u}_i - \boldsymbol{u}_j|^3 \, d\tau \le \int_{\Omega_1} \boldsymbol{f}_1 \cdot \boldsymbol{u}_1 \, d\boldsymbol{x} + \int_{\Omega_2} \boldsymbol{f}_2 \cdot \boldsymbol{u}_2 \, d\boldsymbol{x}.$$

Since the term integrated on Γ is nonnegative, the desired estimate follows from the Cauchy–Schwarz and Poincaré–Friedrichs inequalities. ■

Proposition 3.4. – *For every ℓ_i in $L^1(\Omega_i)$ and $\boldsymbol{f}_i$ in $L^2(\Omega_i)^d$, $1 \le i \le 2$, problem* (3.5) *has a solution $(\boldsymbol{u}_1, \boldsymbol{u}_2)$. Moreover, this solution satisfies* (3.7).

Proof. Since both spaces V_1 and V_2 are separable (indeed, they are closed subspaces of the spaces $H^1(\Omega_i)$ which are separable), there exist increasing sequences of finite-dimensional Hilbert subspaces V_i^m of V_i, $i = 1$ and 2, such that

$$V_i = \bigcup_{m \ge 0} V_i^m, \qquad 1 \le i \le 2.$$

We define a mapping Φ_m from $V_1^m \times V_2^m$ into itself by

$$\begin{aligned}
&\forall (\boldsymbol{u}_1, \boldsymbol{u}_2) \in V_1^m \times V_2^m, \quad \forall (\boldsymbol{v}_1, \boldsymbol{v}_2) \in V_1^m \times V_2^m,\\
&\big(\Phi_m(\boldsymbol{u}_1, \boldsymbol{u}_2), (\boldsymbol{v}_1, \boldsymbol{v}_2)\big)\\
&\quad = \int_{\Omega_1} \tilde{\alpha}_1(\ell_1)\, \nabla \boldsymbol{u}_1 \cdot \nabla \boldsymbol{v}_1 \, d\boldsymbol{x} + \int_{\Omega_2} \tilde{\alpha}_2(\ell_2)\, \nabla \boldsymbol{u}_2 \cdot \nabla \boldsymbol{v}_2 \, d\boldsymbol{x}\\
&\qquad + \int_\Gamma |\boldsymbol{u}_1 - \boldsymbol{u}_2|\, (\boldsymbol{u}_1 - \boldsymbol{u}_2)\, (\boldsymbol{v}_1 - \boldsymbol{v}_2) \, d\tau\\
&\qquad - \int_{\Omega_1} \boldsymbol{f}_1 \cdot \boldsymbol{v}_1 \, d\boldsymbol{x} - \int_{\Omega_2} \boldsymbol{f}_2 \cdot \boldsymbol{v}_2 \, d\boldsymbol{x},
\end{aligned} \tag{3.8}$$

where $(\cdot, \cdot)$ stands for the scalar product on $V_1 \times V_2$. Since the function $\tilde{\alpha}_i$ is bounded and the traces of functions in V_i on Γ belong to $H^{\frac{1}{2}}(\Gamma)^d$, hence at least to $L^4(\Gamma)^d$ from the Sobolev embedding (recall that $d \le 3$), each mapping Φ_m is well-defined and continuous on $V_1^m \times V_2^m$. Moreover, as for estimate (3.7), it is readily checked that

$$\begin{aligned}\big(\Phi_m(\boldsymbol{u}_1, \boldsymbol{u}_2), (\boldsymbol{u}_1, \boldsymbol{u}_2)\big) \\ &\ge \nu \left(|\boldsymbol{u}_1|^2_{H^1(\Omega_1)^d} + |\boldsymbol{u}_2|^2_{H^1(\Omega_2)^d}\right) \\ &\quad - c \left(\|\boldsymbol{f}_1\|^2_{L^2(\Omega_1)^d} + \|\boldsymbol{f}_2\|^2_{L^2(\Omega_2)^d}\right)^{\frac{1}{2}} \left(|\boldsymbol{u}_1|^2_{H^1(\Omega_1)^d} + |\boldsymbol{u}_2|^2_{H^1(\Omega_2)^d}\right)^{\frac{1}{2}}.\end{aligned}$$

So the right-hand side is nonnegative on the sphere of radius μ defined by

$$\mu = \frac{c}{\nu} \left(\|\boldsymbol{f}_1\|^2_{L^2(\Omega_1)^d} + \|\boldsymbol{f}_2\|^2_{L^2(\Omega_2)^d}\right)^{\frac{1}{2}}.$$

Applying Brouwer's fixed point theorem (see e.g. [11, Chap. IV, Cor. 1.1]) yields the existence of a pair $(\boldsymbol{u}_1^m, \boldsymbol{u}_2^m)$ in $V_1^m \times V_2^m$, with norm less than μ, such that

$$\Phi_m(\boldsymbol{u}_1^m, \boldsymbol{u}_2^m) = 0.$$

Since the sequence $(\boldsymbol{u}_1^m, \boldsymbol{u}_2^m)_m$ is bounded by μ in $V_1 \times V_2$, there exists a subsequence which converges weakly to $(\boldsymbol{u}_1, \boldsymbol{u}_2)$ in $V_1 \times V_2$. Using the compactness of the imbedding of $H^{\frac{1}{2}}(\Gamma)$ into $L^3(\Gamma)$, we obtain that this pair $(\boldsymbol{u}_1, \boldsymbol{u}_2)$ is a solution of problem (3.5). ■

We conclude the study of problem (3.5) by a uniqueness result.

Proposition 3.5. – *For every ℓ_i in $L^1(\Omega_i)$ and $\boldsymbol{f}_i$ in $L^2(\Omega_i)^d$, $1 \le i \le 2$, the solution $(\boldsymbol{u}_1, \boldsymbol{u}_2)$ of problem (3.5) is unique.*

Proof. Let $(\boldsymbol{u}_1, \boldsymbol{u}_2)$ and $(\overline{\boldsymbol{u}}_1, \overline{\boldsymbol{u}}_2)$ be two solutions of this problem. By setting $\boldsymbol{w}_i = \boldsymbol{u}_i - \overline{\boldsymbol{u}}_i$, $1 \le i \le 2$, we observe that each $\boldsymbol{w}_i$ belongs to V_i and satisfies for $1 \le i \le 2$

$$\begin{aligned}\forall \boldsymbol{v}_i \in V_i, \quad &\int_{\Omega_i} \tilde{\alpha}_i(\ell_i)\, \nabla \boldsymbol{w}_i \cdot \nabla \boldsymbol{v}_i \, d\boldsymbol{x} \\ &+ \int_\Gamma \big(|\boldsymbol{u}_i - \boldsymbol{u}_j|\,(\boldsymbol{u}_i - \boldsymbol{u}_j) - |\overline{\boldsymbol{u}}_i - \overline{\boldsymbol{u}}_j|\,(\overline{\boldsymbol{u}}_i - \overline{\boldsymbol{u}}_j)\big)\, \boldsymbol{v}_i \, d\tau = 0.\end{aligned}$$

Taking $\boldsymbol{v}_i$ equal to $\boldsymbol{w}_i$ in this equation and summing on i yield, since $\tilde{\alpha}_i$ is $\ge \nu$,

$$\begin{aligned}\nu &\left(|\boldsymbol{w}_1|^2_{H^1(\Omega_1)^d} + |\boldsymbol{w}_2|^2_{H^1(\Omega_2)^d}\right) \\ &+ \int_\Gamma \big(|\boldsymbol{u}_1 - \boldsymbol{u}_2|\,(\boldsymbol{u}_1 - \boldsymbol{u}_2) - |\overline{\boldsymbol{u}}_1 - \overline{\boldsymbol{u}}_2|\,(\overline{\boldsymbol{u}}_1 - \overline{\boldsymbol{u}}_2)\big) \\ &\qquad \big((\boldsymbol{u}_1 - \boldsymbol{u}_2) - (\overline{\boldsymbol{u}}_1 - \overline{\boldsymbol{u}}_2)\big)\, d\tau \le 0.\end{aligned}$$

From the inequality, valid for all real numbers λ and $\overline{\lambda}$,

$$(|\lambda|\,\lambda - |\overline{\lambda}|\,\overline{\lambda})\,(\lambda - \overline{\lambda}) \geq 0, \tag{3.9}$$

we deduce that both $|\boldsymbol{w}_i|_{H^1(\Omega_i)^d}$ are zero, whence the result. ∎

Note as a conclusion of this section that, for any ℓ_i in $L^1(\Omega_i)$ and $\boldsymbol{f}_i$ in $L^2(\Omega_i)^d$, system (3.3) has a unique solution $\big((\boldsymbol{u}_1, p_1), (\boldsymbol{u}_2, p_2)\big)$ in the space $\big(X_1 \times L^2(\Omega_1)\big) \times \big(X_2 \times L^2(\Omega_2)\big)$, which is bounded as a function of the norms of the data $\boldsymbol{f}_i$.

4. The equations on the turbulent kinetic energies (TKE).

Throughout this section, we assume that the functions $\boldsymbol{u}_1$ and $\boldsymbol{u}_2$ are given such that

$$\boldsymbol{u}_i \in H^1(\Omega_i)^d, \qquad 1 \leq i \leq 2. \tag{4.1}$$

The correct formulation of the equations on the TKE is by transposition, following the ideas of Stampacchia [23] and Lions and Magenes [16].

Let us first perform a formal computation. If φ_i, $1 \leq i \leq 2$, are functions in $\mathcal{C}^\infty(\overline{\Omega}_i)$ which vanish on $\partial\Omega_i$, we obtain by multiplying the third line of (2.5) by these functions and integrating twice by parts

$$-\int_{\Omega_i} \ell_i\, \Delta\varphi_i \, d\boldsymbol{x} = -\int_\Gamma G_i(|\boldsymbol{u}_1 - \boldsymbol{u}_2|^2)\, \partial_{n_i}\varphi_i \, d\tau + \int_{\Omega_i} \tilde{\alpha}_i(\ell_i)\, |\nabla \boldsymbol{u}_i|^2\, \varphi_i \, d\boldsymbol{x}.$$

So, from now on, we look for a solution of the following problem:

Find ℓ_i in $L^2(\Omega_i)$, $1 \leq i \leq 2$, such that, for $1 \leq i \leq 2$:

$\forall \varphi_i \in H^2(\Omega_i) \cap H^1_0(\Omega_i)$,

$$\begin{aligned} -\int_{\Omega_i} \ell_i\, \Delta\varphi_i \, d\boldsymbol{x} = -\int_\Gamma G_i(|\boldsymbol{u}_1 - \boldsymbol{u}_2|^2)\, \partial_{n_i}\varphi_i \, d\tau \\ + \int_{\Omega_i} \tilde{\alpha}_i(\ell_i)\, |\nabla \boldsymbol{u}_i|^2\, \varphi_i \, d\boldsymbol{x}. \end{aligned} \tag{4.2}$$

Our existence proof relies on the fact that, since the domain Ω_i is convex or of class $\mathcal{C}^{1,1}$, the Laplace operator $\mathcal{L}_i$ which, with data g_i in $H^{-1}(\Omega)$, associates the solution $\varphi_i = \mathcal{L}_i g_i$ in $H^1_0(\Omega_i)$ of the problem

$$\begin{cases} -\Delta\varphi_i = g_i & \text{in } \Omega_i, \\ \varphi_i = 0 & \text{on } \partial\Omega_i, \end{cases} \tag{4.3}$$

is continuous from $L^2(\Omega_i)$ into $H^2(\Omega_i)$ (see e.g. [12, Thm. 3.2.1.2]). Combining this property with an interpolation argument [16, Chap. 1, Th. 5.1] leads to the following result.

Lemma 4.1. – *Let t be a real number, $1 \leq t \leq 2$. For $1 \leq i \leq 2$, the operator $\mathcal{L}_i$ is continuous from $H^{t-2}(\Omega_i)$ into $H^t(\Omega_i)$ and satisfies*

$$\forall g_i \in H^{t-2}(\Omega_i), \quad \|\mathcal{L}_i g_i\|_{H^t(\Omega_i)} \leq c \, \|g_i\|_{H^{t-2}(\Omega_i)}, \tag{4.4}$$

for a constant c independent of t.

We are now in a position to prove the *a priori* estimate.

Lemma 4.2. – *Let s be a real number, $0 < s < \frac{1}{2}$. For $1 \leq i \leq 2$ and for every pair $(\boldsymbol{u}_1, \boldsymbol{u}_2)$ in $H^1(\Omega_1)^d \times H^1(\Omega_2)^d$, there exists a constant c depending on s and on the maximum of α_i and γ_i such that every solution ℓ_i of equation (4.2) satisfies*

$$\|\ell_i\|_{H^s(\Omega_i)} \leq c \left(\|\boldsymbol{u}_1\|^2_{H^1(\Omega_1)^d} + \|\boldsymbol{u}_2\|^2_{H^1(\Omega_2)^d} \right). \tag{4.5}$$

Proof. Let us introduce the linear form

$$F_i \, \varphi_i = - \int_\Gamma G_i(|\boldsymbol{u}_1 - \boldsymbol{u}_2|^2) \, \partial_{n_i} \varphi_i \, d\tau + \int_{\Omega_i} \tilde{\alpha}_i(\ell_i) \, |\nabla \boldsymbol{u}_i|^2 \, \varphi_i \, d\boldsymbol{x},$$

and check that it is continuous on $H^{2-s}(\Omega_i)$. Firstly, since γ_i is bounded, it follows from the definition (2.1) of G_i that

$$G_i(|\boldsymbol{u}_1 - \boldsymbol{u}_2|^2) \leq c \, |\boldsymbol{u}_1 - \boldsymbol{u}_2|^2.$$

Moreover the trace $\boldsymbol{u}_1 - \boldsymbol{u}_2$ belongs to $H^{\frac{1}{2}}(\Gamma)^d$, hence to $L^4(\Gamma)^d$ thanks to the Sobolev imbedding in dimension $d \leq 3$. On the other hand, since Γ as a part of an hyperplane is smooth and φ_i belongs to $H^{2-s}(\Omega_i)$, $\partial_{n_i} \varphi_i$ belongs to $H^{\frac{1}{2}-s}(\Gamma)$. So we have

$$\left| \int_\Gamma G_i(|\boldsymbol{u}_1 - \boldsymbol{u}_2|^2) \, \partial_{n_i} \varphi_i \, d\tau \right| \leq c \left(\int_\Gamma |\boldsymbol{u}_1 - \boldsymbol{u}_2|^4 \, d\tau \right)^{\frac{1}{2}} \|\partial_{n_i} \varphi_i\|_{L^2(\Gamma)}.$$

Similarly, since $\tilde{\alpha}_i$ is bounded and φ_i belongs to $H^{2-s}(\Omega_i)$, hence to $L^\infty(\Omega_i)$, we obtain

$$\left| \int_{\Omega_i} \tilde{\alpha}_i(\ell_i) \, |\nabla \boldsymbol{u}_i|^2 \, \varphi_i \, d\boldsymbol{x} \right| \leq c \left(\|\boldsymbol{u}_1\|^2_{H^1(\Omega_1)^d} + \|\boldsymbol{u}_2\|^2_{H^1(\Omega_2)^d} \right) \|\varphi_i\|_{L^\infty(\Omega_i)}.$$

Combining all this leads to

$$F_i \, \varphi_i \leq c \left(\|\boldsymbol{u}_1\|^2_{H^1(\Omega_1)^d} + \|\boldsymbol{u}_2\|^2_{H^1(\Omega_2)^d} \right) \|\varphi_i\|_{H^{2-s}(\Omega_i)}. \tag{4.6}$$

When combined with (4.4) for $t = 2 - s$, this yields that the linear form $F_i \circ \mathcal{L}_i$ is continuous on $H^{-s}(\Omega_i)$. Note [16, Chap. 1, Thm 11.1] that, for $0 \leq s < \frac{1}{2}$, the dual space of $H^s(\Omega_i)$ coincides with $H^{-s}(\Omega_i)$ and that these spaces are reflexive. So any function ℓ_i satisfying (4.2), hence

$$\forall g_i \in H^{-s}(\Omega_i), \quad < g_i, \ell_i >= F_i \circ \mathcal{L}_i\, g_i,$$

belongs to $H^s(\Omega_i)$ and satisfies (4.5). ■

From Lemma 4.2, we derive that equation (4.2) with the integral in the left-hand side replaced by the duality pairing is satisfied for any φ_i in $H^{2-s}(\Omega_i) \cap H_0^1(\Omega_i)$. Proving the existence result is now easy, by using the same arguments as for Proposition 3.4, with a further regularization of the data.

Proposition 4.3. – *For $1 \leq i \leq 2$, and for any pair $(\boldsymbol{u}_1, \boldsymbol{u}_2)$ in $X_1 \times X_2$, problem (4.2) has a solution ℓ_i. Moreover, this solution belongs to $H^s(\Omega_i)$, for every $s < \frac{1}{2}$.*

Proof. We firstly consider the problem with more regular data (λ_i, ρ_i) in $L^2(\Omega_i) \times H_{00}^{\frac{1}{2}}(\Gamma)$:

$$\begin{cases} -\Delta \ell_i = \tilde{\alpha}_i(\ell_i)\, \lambda_i & \text{in } \Omega_i, \\ \ell_i = 0 & \text{on } \Gamma_i, \\ \ell_i = \rho_i & \text{on } \Gamma. \end{cases} \tag{4.7}$$

Let us still denote by ρ_i a function in $H^1(\Omega_i)$, the trace of which vanishes on Γ_i and coincides with ρ_i on Γ. The idea is to set $\ell_i^0 = \ell_i - \rho_i$. So, we now consider the problem:

Find ℓ_i^0 in $H_0^1(\Omega_i)$, $1 \leq i \leq 2$, such that:

$$\forall g_i \in H_0^1(\Omega_i), \quad \int_{\Omega_i} \nabla \ell_i^0 \cdot \nabla g_i \, d\boldsymbol{x} = \int_{\Omega_i} \tilde{\alpha}_i(\ell_i^0 + \rho_i)\, \lambda_i\, g_i \, d\boldsymbol{x} - \int_{\Omega_i} \nabla \rho_i \cdot \nabla g_i \, d\boldsymbol{x}. \tag{4.8}$$

Since $H_0^1(\Omega_i)$ is separable, there exists an increasing sequence of finite-dimensional Hilbert subspaces Z_i^m of $H_0^1(\Omega_i)$ such that

$$H_0^1(\Omega_i) = \bigcup_{m \geq 0} Z_i^m.$$

We then define a mapping Ψ_{im} from Z_i^m into itself by

$$\forall \ell_i^0 \in Z_i^m, \forall g_i \in Z_i^m,$$

$$(\Psi_{im}(\ell_i^0), g_i) = \int_{\Omega_i} \nabla \ell_i^0 \cdot \nabla g_i \, d\boldsymbol{x} - \int_{\Omega_i} \tilde{\alpha}_i(\ell_i^0 + \rho_i) \, \lambda_i \, g_i \, d\boldsymbol{x} + \int_{\Omega_i} \nabla \rho_i \cdot \nabla g_i \, d\boldsymbol{x}. \tag{4.9}$$

It is readily checked that

$$(\Psi_{im}(\ell_i^0), \ell_i^0) \geq |\ell_i^0|^2_{H^1(\Omega_i)} - (c \, \|\lambda_i\|_{L^2(\Omega_i)} + |\rho_i|_{H^1(\Omega_i)}) \, |\ell_i^0|_{H^1(\Omega_i)},$$

where c only depends on the maximum of the function $\tilde{\alpha}_i$ and on the Poincaré–Friedrichs constant. So, this quantity is nonnegative when the function ℓ_i^0 belongs to a sphere with appropriate radius

$$\mu_i = c \, \|\lambda_i\|_{L^2(\Omega_i)} + |\rho_i|_{H^1(\Omega_i)}.$$

Using once more Brouwer's fixed point theorem (see e.g. [11, Chap. IV, Cor. 1.1]) yields the existence of a solution ℓ_i^{0m} of the equation $\Psi_{im}(\ell_i^{0m}) = 0$ with norm less than μ_i.

Since the sequence $(\ell_i^{0m})_{m \geq 0}$ is bounded in $H^1(\Omega_i)$, it admits a subsequence which tends to ℓ_i^0 weakly in $H^1(\Omega_i)$. Due to the compactness of $H^1(\Omega_i)$ into $L^2(\Omega_i)$, this subsequence converges strongly in $L^2(\Omega_i)$. Hence by the inverse Lebesgue theorem (see e.g. [7, Th. IV.9]), there exists a further subsequence, still denoted by $(\ell_i^{0m})_{m \geq 0}$, which tends to ℓ_i^0 a.e. on Ω_i. Since the function $\tilde{\alpha}_i$ is continuous and bounded, the subsequence $(\tilde{\alpha}_i(\ell_i^{0m} + \rho_i) \, g_i)_{m \geq 0}$ is dominated by $c \, g_i$ which belongs to $L^2(\Omega_i)$, and tends to $\tilde{\alpha}_i(\ell_i^0 + \rho_i) \, g_i$ a.e. in Ω_i, hence in $L^2(\Omega_i)$. So, the function ℓ_i^0 is a solution of (4.8) and finally the function $\ell_i = \ell_i^0 + \rho_i$ is a solution of (4.7).

Since we have proved the existence of a solution of problem (4.7), we now pass to the case of real data. We approximate each $\boldsymbol{u}_i$ in X_i, $1 \leq i \leq 2$, by a sequence $(\boldsymbol{u}_i^n)_{n \geq 0}$ of $\mathcal{C}^\infty(\overline{\Omega}_i)^d \cap X_i$ which converges towards $\boldsymbol{u}_i$ in $H^1(\Omega_i)^d$. Then, it is readily checked that the functions $\lambda_i^n = |\nabla \boldsymbol{u}_i^n|^2$, resp. $\rho_i^n = G_i(|\boldsymbol{u}_1^n - \boldsymbol{u}_2^n|^2)$, belong to $L^2(\Omega_i)$, resp. $H_{00}^{\frac{1}{2}}(\Gamma)$ (see [16, Chap. 1, Th. 11.7] for the characterization of this space). Let ℓ_i^n be a solution of problem (4.7) with data $\lambda_i = \lambda_i^n$ and $\rho_i = \rho_i^n$. Since it belongs to $H^1(\Omega_i)$, by integration by parts, it is readily checked that it satisfies

$$\forall \varphi_i \in H^2(\Omega_i) \cap H_0^1(\Omega_i),$$

$$-\int_{\Omega_i} \ell_i^n \, \Delta \varphi_i \, d\boldsymbol{x} = -\int_\Gamma G_i(|\boldsymbol{u}_1^n - \boldsymbol{u}_2^n|^2) \, \partial_{n_i} \varphi_i \, d\tau + \int_{\Omega_i} \tilde{\alpha}_i(\ell_i^n) \, |\nabla \boldsymbol{u}_i^n|^2 \, \varphi_i \, d\boldsymbol{x}.$$

Let now s_0 be such that $0 < s_0 < \frac{1}{2}$. From Lemma 4.2, the sequence $(\ell_i^n)_{n\geq 0}$ is bounded in $H^{s_0}(\Omega_i)$, so that it admits a subsequence which converges to ℓ_i weakly in $H^{s_0}(\Omega_i)$ (and in fact in all $H^s(\Omega_i)$, $s < \frac{1}{2}$). From the compactness of $H^{s_0}(\Omega_i)$ into $L^2(\Omega_i)$ and the boundedness of the function $\tilde{\alpha}_i$, by the same arguments as previously, we deduce the existence of subsequences, still denoted by $(\ell_i^n)_{n\geq 0}$ and $(\boldsymbol{u}_i^n)_{n\geq 0}$, such that the sequence $(\tilde{\alpha}_i(\ell_i^n)\,|\nabla \boldsymbol{u}_i^n|^2)_{n\geq 0}$ is dominated by a function of $L^1(\Omega_i)$ and tends to $\tilde{\alpha}_i(\ell_i)\,|\nabla \boldsymbol{u}_i|^2$ a.e. in Ω_i, hence in $L^1(\Omega_i)$. Similarly, there exists a subsequence $(G_i(|\boldsymbol{u}_1^n - \boldsymbol{u}_2^n|^2))_{n\geq 0}$ which tends to the term $G_i(|\boldsymbol{u}_1 - \boldsymbol{u}_2|^2)$ in $L^2(\Gamma)$. So, the limit ℓ_i is a solution of (4.2), which completes the proof. ∎

Remark. Consider a solution ℓ_i of problem (4.2) which further belongs to $H^1(\Omega_i)$. Then, if the pair $(\boldsymbol{u}_1, \boldsymbol{u}_2)$ belongs to $W^{1,\frac{12}{5}}(\Omega_1)^d \times W^{1,\frac{12}{5}}(\Omega_2)^d$, it solves the following variational problem:

$$\begin{aligned}
&\textit{Find } \ell_i \textit{ in } H^1(\Omega_i),\ 1 \leq i \leq 2, \textit{ with}\\
&\qquad \ell_i = 0 \quad \textit{on } \Gamma_i \qquad \textit{and} \qquad \ell_i = G_i(|\boldsymbol{u}_1 - \boldsymbol{u}_2|^2) \quad \textit{on } \Gamma,\\
&\textit{such that, for } 1 \leq i \leq 2:\\
&\forall \varphi_i \in H_0^1(\Omega_i), \quad \int_{\Omega_i} \nabla \ell_i \cdot \nabla \varphi_i \, d\boldsymbol{x} = \int_{\Omega_i} \tilde{\alpha}_i(\ell_i)\,|\nabla \boldsymbol{u}_i|^2\, \varphi_i \, d\boldsymbol{x}.
\end{aligned} \tag{4.10}$$

Note that (4.10) is now a classical formulation, which could be discretized in an easy way.

5. The global system.

Proving the existence of a solution for the full system (2.5) follows from the same arguments as previously, however for technical reasons we begin by working with a truncated problem.

For each positive integer n, we introduce the function T_n defined from $\mathbb{R}$ onto $\mathbb{R}$ by

$$T_n(x) = \begin{cases} -n & \text{if } x \leq -n, \\ x & \text{if } -n \leq x \leq n, \\ n & \text{if } x \geq n, \end{cases}$$

and we consider the problem

$$
\begin{cases}
-\mathrm{div}\,\big(\tilde{\alpha}_i(\ell_i)\,\nabla \boldsymbol{u}_i\big) + \mathbf{grad}\, p_i = \boldsymbol{f}_i & \text{in } \Omega_i,\ 1 \le i \le 2, \\
\mathrm{div}\, \boldsymbol{u}_i = 0 & \text{in } \Omega_i,\ 1 \le i \le 2, \\
-\Delta \ell_i = T_n\big(\tilde{\alpha}_i(\ell_i)\,|\nabla \boldsymbol{u}_i|^2\big) & \text{in } \Omega_i,\ 1 \le i \le 2, \\
\boldsymbol{u}_i = \boldsymbol{0} & \text{on } \Gamma_i,\ 1 \le i \le 2, \\
\ell_i = 0 & \text{on } \Gamma_i,\ 1 \le i \le 2, \\
\tilde{\alpha}_i(\ell_i)\,\partial_{n_i} \boldsymbol{u}_i - p_i\, \boldsymbol{n}_i + (\boldsymbol{u}_i - \boldsymbol{u}_j)\,|\boldsymbol{u}_i - \boldsymbol{u}_j| = \boldsymbol{0} & \\
\qquad\qquad\qquad\qquad \text{on } \Gamma, & 1 \le i \ne j \le 2, \\
\ell_i = T_n\big(G_i(|\boldsymbol{u}_1 - \boldsymbol{u}_2|^2)\big) & \text{on } \Gamma,\ 1 \le i \le 2.
\end{cases}
\tag{5.1}
$$

We first write its reduced variational formulation (where the word "reduced" means that the pressures p_i do not appear in it):

$$
\begin{array}{l}
\textit{Find } \boldsymbol{u}_i \textit{ in } V_i,\ 1 \le i \le 2, \textit{ such that, for } 1 \le i \ne j \le 2\textit{:} \\
\forall \boldsymbol{v}_i \in V_i, \quad \displaystyle\int_{\Omega_i} \tilde{\alpha}_i(\ell_i)\,\nabla \boldsymbol{u}_i \cdot \nabla \boldsymbol{v}_i\, d\boldsymbol{x} \\
\qquad\qquad + \displaystyle\int_\Gamma |\boldsymbol{u}_i - \boldsymbol{u}_j|\,(\boldsymbol{u}_i - \boldsymbol{u}_j) \cdot \boldsymbol{v}_i\, d\tau = \int_{\Omega_i} \boldsymbol{f}_i \cdot \boldsymbol{v}_i\, d\boldsymbol{x}. \\
\textit{Find } \ell_i \textit{ in } H^1(\Omega_i),\ 1 \le i \le 2, \textit{ with} \\
\qquad \ell_i = 0 \quad \textit{on } \Gamma_i \qquad \textit{and} \qquad \ell_i = T_n\big(G_i(|\boldsymbol{u}_1 - \boldsymbol{u}_2|^2)\big) \quad \textit{on } \Gamma, \\
\textit{such that, for } 1 \le i \le 2\textit{:} \\
\forall \varphi_i \in H^1_0(\Omega_i), \quad \displaystyle\int_{\Omega_i} \nabla \ell_i \cdot \nabla \varphi_i\, d\boldsymbol{x} = \int_{\Omega_i} T_n\big(\tilde{\alpha}_i(\ell_i)\,|\nabla \boldsymbol{u}_i|^2\big)\, \varphi_i\, d\boldsymbol{x}.
\end{array}
\tag{5.2}
$$

The next lemma states that problem (5.1) is well-posed.

Lemma 5.1. – *For any positive integer n, for any pair $(\boldsymbol{f}_1, \boldsymbol{f}_2)$ in $L^2(\Omega_1)^d \times L^2(\Omega_2)^d$, problem (5.1) admits the variational formulation (5.2). System (5.2) has a solution (U_1, U_2) with each $U_i = (\boldsymbol{u}_i, \ell_i)$ in $V_i \times H^1(\Omega_i)$. Moreover the functions ℓ_1 and ℓ_2 are nonnegative.*

The proof of this lemma is rather complex, it is performed in five steps.

Proof (I). Liftings of traces.
For $i = 1$ and 2, we introduce the continuous lifting operator L_i from $H_{00}^{\frac{1}{2}}(\Gamma)$ into harmonic functions in X_i (note that the extension by zero of a function in $H_{00}^{\frac{1}{2}}(\Gamma)$ to $\partial\Omega_i$ belongs to $H^{\frac{1}{2}}(\partial\Omega_i)$). Next, with each pair $(\boldsymbol{u}_1, \boldsymbol{u}_2)$ in $X_1 \times X_2$, we associate a function $\rho_i(\boldsymbol{u}_1, \boldsymbol{u}_2)$ defined on Ω_i by

$$\rho_i(\boldsymbol{u}_1, \boldsymbol{u}_2) = T_n\big(G_i(|L_i(\boldsymbol{u}_1 - \boldsymbol{u}_2)|^2)\big).$$

Since the trace of each $\boldsymbol{u}_i$ on Γ belongs to $H_{00}^{\frac{1}{2}}(\Gamma)^d$, the function $\boldsymbol{g}_i = L_i(\boldsymbol{u}_1 - \boldsymbol{u}_2)$ belongs to $H^1(\Omega_i)^d$ and satisfies

$$\|\boldsymbol{g}_i\|_{H^1(\Omega_i)^d} \le c\left(\|\boldsymbol{u}_1\|^2_{H^1(\Omega_1)^d} + \|\boldsymbol{u}_2\|^2_{H^1(\Omega_2)^d}\right)^{\frac{1}{2}}.$$

Let us now prove that the function $T_n\big(G_i(|\boldsymbol{g}_i|^2)\big)$ belongs to $H^1(\Omega_i)$. We have [23, Lemme 1.1]

$$\textbf{grad}\; T_n\big(G_i(|\boldsymbol{g}_i|^2)\big) = 2T_n'\big(G_i(|\boldsymbol{g}_i|^2)\big)\,\gamma_i(|\boldsymbol{g}_i|^2)\,\textbf{grad}\;\boldsymbol{g}_i\,\cdot\,\boldsymbol{g}_i.$$

Note that γ_i is bounded and that $\textbf{grad}\;\boldsymbol{g}_i$ belongs at most to $L^2(\Omega_i)^{d^2}$. Moreover, $T_n'\big(G_i(|\boldsymbol{g}_i|^2)\big)$ is either 0 or 1, and, when it is not zero,

$$\nu\,|\boldsymbol{g}_i|^2 \le \int_0^{|\boldsymbol{g}_i|^2} \gamma_i(\kappa)\,d\kappa = G_i(|\boldsymbol{g}_i|^2) \le n,$$

so that $T_n'\big(G_i(|\boldsymbol{g}_i|^2)\big)\,\boldsymbol{g}_i$ is bounded and $T_n'\big(G_i(|\boldsymbol{g}_i|^2)\big)\,\gamma_i(|\boldsymbol{g}_i|^2)\,\textbf{grad}\;\boldsymbol{g}_i\,\cdot\,\boldsymbol{g}_i$ belongs to $L^2(\Omega_i)^d$. As a consequence, the function $\rho_i(\boldsymbol{u}_1, \boldsymbol{u}_2)$ belongs to $H^1(\Omega_i)$ and satisfies

$$\|\rho_i(\boldsymbol{u}_1, \boldsymbol{u}_2)\|_{H^1(\Omega_i)} \le c\,n^{\frac{1}{2}}\left(\|\boldsymbol{u}_1\|^2_{H^1(\Omega_1)^d} + \|\boldsymbol{u}_2\|^2_{H^1(\Omega_2)^d}\right)^{\frac{1}{2}}. \qquad (5.3)$$

Next we work with the new unknowns $(\boldsymbol{u}_i, \ell_i^0 = \ell_i - \rho_i(\boldsymbol{u}_1, \boldsymbol{u}_2))$ in $V_i \times H_0^1(\Omega_i)$.

Proof (II). Existence of a solution for a finite-dimensional system.
For $i = 1$ and 2, as in the proof of Proposition 3.4, we introduce an increasing sequence of finite-dimensional Hilbert subspaces V_i^m of V_i, $i = 1$ and 2, such that

$$V_i = \bigcup_{m \ge 0} V_i^m, \qquad 1 \le i \le 2,$$

and, as in the proof of Proposition 4.3, we introduce an increasing sequence of finite-dimensional Hilbert subspaces Z_i^m of $H_0^1(\Omega_i)$ such that

$$H_0^1(\Omega_i) = \bigcup_{m \geq 0} Z_i^m.$$

We now define the mapping Φ_m from $V_1^m \times V_2^m$ into itself by

$$\begin{aligned}
&\forall (\boldsymbol{u}_1, \boldsymbol{u}_2) \in V_1^m \times V_2^m, \quad \forall (\boldsymbol{v}_1, \boldsymbol{v}_2) \in V_1^m \times V_2^m,\\
&\big(\Phi_m(\boldsymbol{u}_1, \boldsymbol{u}_2), (\boldsymbol{v}_1, \boldsymbol{v}_2)\big) = \int_{\Omega_1} \tilde{\alpha}_1(\ell_1^0 + \rho_1(\boldsymbol{u}_1^m, \boldsymbol{u}_2^m))\, \nabla \boldsymbol{u}_1 \cdot \nabla \boldsymbol{v}_1 \, d\boldsymbol{x}\\
&\qquad + \int_{\Omega_2} \tilde{\alpha}_2(\ell_i^0 + \rho_2(\boldsymbol{u}_1^m, \boldsymbol{u}_2^m))\, \nabla \boldsymbol{u}_2 \cdot \nabla \boldsymbol{v}_2 \, d\boldsymbol{x}\\
&\qquad + \int_{\Gamma} |\boldsymbol{u}_1 - \boldsymbol{u}_2|\, (\boldsymbol{u}_1 - \boldsymbol{u}_2)\, (\boldsymbol{v}_1 - \boldsymbol{v}_2)\, d\tau\\
&\qquad - \int_{\Omega_1} \boldsymbol{f}_1 \cdot \boldsymbol{v}_1 \, d\boldsymbol{x} - \int_{\Omega_2} \boldsymbol{f}_2 \cdot \boldsymbol{v}_2 \, d\boldsymbol{x},
\end{aligned}$$

and, similarly, for $i = 1$ and 2, a mapping Ψ_{im} from Z_i^m into itself by

$$\begin{aligned}
&\forall \ell_i^0 \in Z_i^m, \forall g_i \in Z_i^m,\\
&\big(\Psi_{im}(\ell_i^0), g_i\big) = \int_{\Omega_i} \nabla \ell_i^0 \cdot \nabla g_i \, d\boldsymbol{x}\\
&\qquad - \int_{\Omega_i} T_n\big(\tilde{\alpha}_i(\ell_i^0 + \rho_i(\boldsymbol{u}_1, \boldsymbol{u}_2))\, |\nabla \boldsymbol{u}_i|^2\big)\, g_i \, d\boldsymbol{x}\\
&\qquad + \int_{\Omega_i} \nabla \rho_i(\boldsymbol{u}_1, \boldsymbol{u}_2) \cdot \nabla g_i \, d\boldsymbol{x}.
\end{aligned}$$

Finally, we consider the mapping Ξ_m from $V_1^m \times Z_1^m \times V_2^m \times Z_2^m$ defined by

$$\begin{aligned}
&\forall (\boldsymbol{u}_1, \ell_1^0, \boldsymbol{u}_2, \ell_2^0) \in V_1^m \times Z_1^m \times V_2^m \times Z_2^m,\\
&\qquad\qquad \forall (\boldsymbol{v}_1, g_1, \boldsymbol{v}_2, g_2) \in V_1^m \times Z_1^m \times V_2^m \times Z_2^m,\\
&\big(\Xi_m(\boldsymbol{u}_1, \ell_1^0, \boldsymbol{u}_2, \ell_2^0), (\boldsymbol{v}_1, g_1, \boldsymbol{v}_2, g_2)\big)\\
&\qquad = \big(\Phi_m(\boldsymbol{u}_1, \boldsymbol{u}_2), (\boldsymbol{v}_1, \boldsymbol{v}_2)\big) + \big(\Psi_{1m}(\ell_1^0), g_1\big) + \big(\Psi_{2m}(\ell_2^0), g_2\big).
\end{aligned}$$

Taking $(\boldsymbol{u}_1, \boldsymbol{u}_2)$ on the sphere with radius μ, for the same μ as in the proof of Proposition 3.4, yields that $\Phi_m\big((\boldsymbol{u}_1, \boldsymbol{u}_2), (\boldsymbol{u}_1, \boldsymbol{u}_2)\big)$ is nonnegative. This combined with the arguments of the proof of Proposition 4.3 implies

$$\begin{aligned}
&\big(\Xi_m(\boldsymbol{u}_1, \ell_1^0, \boldsymbol{u}_2, \ell_2^0), (\boldsymbol{u}_1, \ell_1^0, \boldsymbol{u}_2, \ell_2^0)\big)\\
&\qquad \geq \sum_{i=1}^{2} \big(|\ell_i^0|^2_{H^1(\Omega_i)} - \big(c'\, n + |\rho_i(\boldsymbol{u}_1, \boldsymbol{u}_2)|_{H^1(\Omega_i)}\big)\, |\ell_i^0|_{H^1(\Omega_i)}\big).
\end{aligned}$$

So using the bound (5.3) for $\|\rho_i(\boldsymbol{u}_1, \boldsymbol{u}_2)\|_{H^1(\Omega_i)}$ and taking each ℓ_i^0 on the sphere with radius

$$\mu_i = c' \, n + c n^{\frac{1}{2}} \, \mu,$$

we deduce that the previous quantity is nonnegative. Thus, applying Brouwer's fixed point theorem yields the existence of a solution $(\boldsymbol{u}_1^m, \ell_1^{0m}, \boldsymbol{u}_2^m, \ell_2^{0m})$ of

$$\Xi_m(\boldsymbol{u}_1^m, \ell_1^{0m}, \boldsymbol{u}_2^m, \ell_2^{0m}) = 0, \tag{5.4}$$

which satisfies

$$\|\boldsymbol{u}_1^m\|_{H^1(\Omega_1)^d} + \|\boldsymbol{u}_2^m\|_{H^1(\Omega_2)^d} \leq \mu, \quad \|\ell_i^{0m}\|_{H^1(\Omega_i)} \leq \mu_i, \quad i = 1 \text{ and } 2.$$

Since the sequence $(\boldsymbol{u}_1^m, \ell_1^{0m}, \boldsymbol{u}_2^m, \ell_2^{0m})_m$ is bounded, there exists a subsequence, still denoted by $(\boldsymbol{u}_1^m, \ell_1^{0m}, \boldsymbol{u}_2^m, \ell_2^{0m})_m$ for simplicity, which converges to $(\boldsymbol{u}_1, \ell_1^0, \boldsymbol{u}_2, \ell_2^0)$ weakly in $V_1 \times H_0^1(\Omega_1) \times V_2 \times H_0^1(\Omega_2)$. Next, for a fixed $(\boldsymbol{v}_1, g_1, \boldsymbol{v}_2, g_2)$, we pass to the limit in problem (5.4).

Proof (III). The limit on the equations for the velocities.
We start from the equation $\Phi_m(\boldsymbol{u}_1^m, \boldsymbol{u}_2^m) = 0$. For $1 \leq i \leq 2$, there exists a subsequence still denoted by $(\ell_i^{0m})_m$ which converges to ℓ_i^0 strongly in $L^2(\Omega_i)$, hence a.e. in Ω_i. On the other hand, due to the continuity of L_i, the sequence $\big(\boldsymbol{g}_i^m = L_i(\boldsymbol{u}_1^m - \boldsymbol{u}_2^m)\big)_m$ converges to $L_i(\boldsymbol{u}_1 - \boldsymbol{u}_2)$ weakly in $H^1(\Omega_i)$. Due to the formula

$$\begin{aligned}\textbf{grad } \rho_i^m &= \textbf{grad } T_n\big(G_i(|\boldsymbol{g}_i^m|^2)\big) \\ &= 2T_n'\big(G_i(|\boldsymbol{g}_i^m|^2)\big)\, \gamma_i(|\boldsymbol{g}_i^m|^2)\, \textbf{grad } \boldsymbol{g}_i^m \cdot \boldsymbol{g}_i^m,\end{aligned}$$

there exists another subsequence $(\rho_i(\boldsymbol{u}_1^m, \boldsymbol{u}_2^m))_m$ which converges to the function $\rho_i(\boldsymbol{u}_1, \boldsymbol{u}_2)$ weakly in $H^1(\Omega_i)^d$ and strongly in $L^2(\Omega_i)^d$. So, the corresponding subsequence $(\ell_i^{0m} + \rho_i(\boldsymbol{u}_1^m, \boldsymbol{u}_2^m))_m$ converges to $\ell_1^0 + \rho_i(\boldsymbol{u}_1, \boldsymbol{u}_2)$ a.e. in Ω_i and, since $\tilde{\alpha}_i$ is continuous and bounded, for all fixed $\boldsymbol{v}_i$ in X_i, the sequence $(\tilde{\alpha}_i(\ell_i^{0m} + \rho_i(\boldsymbol{u}_1^m, \boldsymbol{u}_2^m))\, \nabla \boldsymbol{v}_i)_m$ tends to $\tilde{\alpha}_i(\ell_i^0 + \rho_i(\boldsymbol{u}_1, \boldsymbol{u}_2))\, \nabla \boldsymbol{v}_i$ a.e. in Ω_i and is bounded in $L^2(\Omega_i)^{d^2}$, hence converges strongly in $L^2(\Omega_i)^{d^2}$. This yields the convergence of the first two integrals in the definition of $\Phi_m(\cdot, \cdot)$: for $i = 1$ and 2,

$$\lim_{m \to \infty} \int_{\Omega_i} \tilde{\alpha}_i(\ell_i^{0m} + \rho_i(\boldsymbol{u}_1^m, \boldsymbol{u}_2^m))\, \nabla \boldsymbol{u}_i^m \cdot \nabla \boldsymbol{v}_i \, d\boldsymbol{x}$$
$$= \int_{\Omega_i} \tilde{\alpha}_i(\ell_i^0 + \rho_i(\boldsymbol{u}_1, \boldsymbol{u}_2))\, \nabla \boldsymbol{u}_i \cdot \nabla \boldsymbol{v}_i \, d\boldsymbol{x}.$$

The convergence of the third one follows from the compact imbedding of $H^{\frac{1}{2}}(\Gamma)$ into $L^3(\Gamma)$:

$$\lim_{m\to\infty}\int_\Gamma |\boldsymbol{u}_1^m - \boldsymbol{u}_2^m|\,(\boldsymbol{u}_1^m - \boldsymbol{u}_2^m)\,(\boldsymbol{v}_1 - \boldsymbol{v}_2)\,d\tau$$
$$= \int_\Gamma |\boldsymbol{u}_1 - \boldsymbol{u}_2|\,(\boldsymbol{u}_1 - \boldsymbol{u}_2)\,(\boldsymbol{v}_1 - \boldsymbol{v}_2)\,d\tau.$$

Combining all this implies that the desired equation is satisfied by $(\boldsymbol{u}_1, \boldsymbol{u}_2)$.

Proof (IV). A stronger convergence result.
For $i = 1$ and 2, we start from the formula

$$\int_{\Omega_i} \tilde{\alpha}_i(\ell_i^{0m} + \rho_i(\boldsymbol{u}_1^m, \boldsymbol{u}_2^m))\,\nabla(\boldsymbol{u}_i^m - \boldsymbol{u}_i)\cdot\nabla(\boldsymbol{u}_i^m - \boldsymbol{u}_i)\,d\boldsymbol{x}$$
$$= \int_{\Omega_i} \tilde{\alpha}_i(\ell_i^{0m} + \rho_i(\boldsymbol{u}_1^m, \boldsymbol{u}_2^m))\,\nabla\boldsymbol{u}_i^m\cdot\nabla\boldsymbol{u}_i^m\,d\boldsymbol{x}$$
$$- 2\int_{\Omega_i} \tilde{\alpha}_i(\ell_i^{0m} + \rho_i(\boldsymbol{u}_1^m, \boldsymbol{u}_2^m))\,\nabla\boldsymbol{u}_i^m\cdot\nabla\boldsymbol{u}_i\,d\boldsymbol{x}$$
$$+ \int_{\Omega_i} \tilde{\alpha}_i(\ell_i^{0m} + \rho_i(\boldsymbol{u}_1^m, \boldsymbol{u}_2^m))\,\nabla\boldsymbol{u}_i\cdot\nabla\boldsymbol{u}_i\,d\boldsymbol{x}.$$

From equation (5.4), the first term in the right-hand side is equal to

$$-\int_\Gamma |\boldsymbol{u}_i^m - \boldsymbol{u}_j^m|\,(\boldsymbol{u}_i^m - \boldsymbol{u}_j^m)\cdot\boldsymbol{u}_i^m\,d\tau + \int_{\Omega_i} \boldsymbol{f}_i\cdot\boldsymbol{u}_i^m,$$

and using once more the compactness of $H^{\frac{1}{2}}(\Gamma)$ into $L^3(\Gamma)$ implies its convergence. The convergence of the second and third term follows from the weak convergence of $(\nabla\boldsymbol{u}_i^m)_m$ in $L^2(\Omega_i)^{d^2}$ and the strong convergence of $(\tilde{\alpha}_i(\ell_i^{0m} + \rho_i(\boldsymbol{u}_1^m, \boldsymbol{u}_2^m))\,\nabla\boldsymbol{v}_i)_m$ in $L^2(\Omega_i)$. So, we obtain that

$$\lim_{m\to\infty}\int_{\Omega_i} \tilde{\alpha}_i(\ell_i^{0m} + \rho_i(\boldsymbol{u}_1^m, \boldsymbol{u}_2^m))\,\nabla(\boldsymbol{u}_i^m - \boldsymbol{u}_i)\cdot\nabla(\boldsymbol{u}_i^m - \boldsymbol{u}_i)\,d\boldsymbol{x} = 0,$$

which yields the strong convergence of $(\boldsymbol{u}_i^m)_m$ towards $\boldsymbol{u}_i$ in $H^1(\Omega_i)^d$.

Proof (V). The limit on the equations for the TKE.
Next, we consider each equation $\Psi_m(\ell_i^{0m}) = 0$. As previously, there exists a subsequence $(\tilde{\alpha}_i(\ell_i^{0m} + \rho_i(\boldsymbol{u}_1^m, \boldsymbol{u}_2^m)))_m$ which tends to $\tilde{\alpha}_i(\ell_i^0 + \rho_i(\boldsymbol{u}_1, \boldsymbol{u}_2))$ strongly in $L^2(\Omega_i)$. From part IV of the proof, there also exists a subsequence $(|\nabla\boldsymbol{u}_i^m|^2)_m$ which tends to $|\nabla\boldsymbol{u}_i|^2$ strongly in $L^1(\Omega_i)$. Hence, the

sequence $(\tilde{\alpha}_i(\ell_i^{0m} + \rho_i(\boldsymbol{u}_1^m, \boldsymbol{u}_2^m))\,|\nabla \boldsymbol{u}_i^m|^2)_m$ converges a.e. in Ω_i and, since T_n is continuous and bounded, the sequence

$$(T_n(\tilde{\alpha}_i(\ell_i^{0m} + \rho_i(\boldsymbol{u}_1^m, \boldsymbol{u}_2^m))\,|\nabla \boldsymbol{u}_i^m|^2))_m,$$

converges towards $T_n(\tilde{\alpha}_i(\ell_i^0 + \rho_i(\boldsymbol{u}_1, \boldsymbol{u}_2))\,|\nabla \boldsymbol{u}_i|^2)$ strongly in $L^2(\Omega_i)$. This yields

$$\lim_{m\to\infty} \int_{\Omega_i} T_n\big(\tilde{\alpha}_i(\ell_i^{0m} + \rho_i(\boldsymbol{u}_1^m, \boldsymbol{u}_2^m))\,|\nabla \boldsymbol{u}_i^m|^2\big)\, g_i\, d\boldsymbol{x} \\ = \int_{\Omega_i} T_n\big(\tilde{\alpha}_i(\ell_i^0 + \rho_i(\boldsymbol{u}_1, \boldsymbol{u}_2))\,|\nabla \boldsymbol{u}_i|^2\big)\, g_i\, d\boldsymbol{x}.$$

Also, from the weak convergence of a subsequence $(\rho_i(\boldsymbol{u}_1^m, \boldsymbol{u}_2^m))_m$ to $\rho_i(\boldsymbol{u}_1, \boldsymbol{u}_2)$ in $H^1(\Omega_i)^d$, we deduce

$$\lim_{m\to\infty} \int_{\Omega_i} \nabla \rho_i(\boldsymbol{u}_1^m, \boldsymbol{u}_2^m) \cdot \nabla g_i\, d\boldsymbol{x} = \int_{\Omega_i} \nabla \rho_i(\boldsymbol{u}_1, \boldsymbol{u}_2) \cdot \nabla g_i\, d\boldsymbol{x}.$$

So the desired equation is satisfied by ℓ_i. Finally, the nonnegativity of the ℓ_i follows from the standard maximum principle [7, Prop. IX.29].

We are now in a position to state the main result of this section. There also, we write the reduced variational formulation of system (2.5), where the equation on the ℓ_i has now the same "transposed" form as in Section 4:

Find $\boldsymbol{u}_i$ in V_i, $1 \le i \le 2$, such that, for $1 \le i \ne j \le 2$:

$$\forall \boldsymbol{v}_i \in V_i, \quad \int_{\Omega_i} \tilde{\alpha}_i(\ell_i)\, \nabla \boldsymbol{u}_i \cdot \nabla \boldsymbol{v}_i\, d\boldsymbol{x} \\ + \int_\Gamma |\boldsymbol{u}_i - \boldsymbol{u}_j|\,(\boldsymbol{u}_i - \boldsymbol{u}_j) \cdot \boldsymbol{v}_i\, d\tau = \int_{\Omega_i} \boldsymbol{f}_i \cdot \boldsymbol{v}_i\, d\boldsymbol{x}.$$

Find ℓ_i in $L^2(\Omega_i)$, $1 \le i \le 2$, such that, for $1 \le i \le 2$: (5.5)

$$\forall \varphi_i \in H^2(\Omega_i) \cap H_0^1(\Omega_i), \\ -\int_{\Omega_i} \ell_i\, \Delta \varphi_i\, d\boldsymbol{x} = -\int_\Gamma G_i(|\boldsymbol{u}_1 - \boldsymbol{u}_2|^2)\, \partial_{n_i} \varphi_i\, d\tau \\ + \int_{\Omega_i} \tilde{\alpha}_i(\ell_i)\, |\nabla \boldsymbol{u}_i|^2\, \varphi_i\, d\boldsymbol{x}.$$

Theorem 5.2. – *For any $\boldsymbol{f}_i$ in $L^2(\Omega_i)^d$, $i = 1$ or 2, problem (2.5) admits the formulation (5.5). System (5.5) has a solution (U_1, U_2) with each*

$U_i = (\boldsymbol{u}_i, \ell_i)$ in $X_i \times L^2(\Omega_i)$. Moreover, each function ℓ_i, $i = 1$ and 2, is nonnegative and belongs to $H^s(\Omega_i)$ for all $s < \frac{1}{2}$, and this solution satisfies (3.7) and (4.5).

Proof. For each integer n, let us now denote by (U_1^n, U_2^n), with $U_i^n = (\boldsymbol{u}_i^n, \ell_i^n)$, a solution of problem (5.1) (its existence is proven in Lemma 5.1). As previously, see (3.7) and (4.5), it satisfies, for a fixed number $s < \frac{1}{2}$ and for a constant c independent of n,

$$\|\boldsymbol{u}_1^n\|_{H^1(\Omega_1)^d} + \|\boldsymbol{u}_2^n\|_{H^1(\Omega_2)^d} \le \frac{c}{\nu}\big(\|\boldsymbol{f}_1\|_{L^2(\Omega_1)^d} + \|\boldsymbol{f}_2\|_{L^2(\Omega_2)^d}\big),$$
$$\|\ell_i^n\|_{H^s(\Omega_i)} \le c\big(\|\boldsymbol{u}_1^n\|^2_{H^1(\Omega_1)^d} + \|\boldsymbol{u}_2^n\|^2_{H^1(\Omega_2)^d}\big), \quad i = 1 \text{ and } 2.$$

So, there exists a subsequence, still denoted by $(\boldsymbol{u}_1^n, \ell_1^n, \boldsymbol{u}_2^n, \ell_2^n)_n$, which converges towards $(\boldsymbol{u}_1, \ell_1, \boldsymbol{u}_2, \ell_2)$ weakly in $V_1 \times H^s(\Omega_1) \times V_2 \times H^s(\Omega_2)$. We must now prove that $(\boldsymbol{u}_1, \ell_1, \boldsymbol{u}_2, \ell_2)$ satisfies (5.5), which is performed in three steps.
1) We start from the variational formulation, for $i = 1$ and 2:

$$\forall \boldsymbol{v}_i \in V_i, \quad \int_{\Omega_i} \tilde{\alpha}_i(\ell_i^n)\, \nabla \boldsymbol{u}_i^n \cdot \nabla \boldsymbol{v}_i \, d\boldsymbol{x} + \int_\Gamma (\boldsymbol{u}_i^n - \boldsymbol{u}_j^n)\, |\boldsymbol{u}_i^n - \boldsymbol{u}_j^n| \, \boldsymbol{v}_i \, d\tau = < \boldsymbol{f}_i, \boldsymbol{v}_i > . \tag{5.6}$$

Next, from the convergence properties of the sequence $(\ell_i^n)_n$, there exists a subsequence, still denoted by $(\ell_i^n)_n$, which converges to ℓ_i strongly in $L^2(\Omega_i)$ and a.e. in Ω_i. So, thanks to the continuity and boundedness of the function $\tilde{\alpha}_i$, for any fixed $\boldsymbol{v}_i$ in V_i, the sequence $(\tilde{\alpha}_i(\ell_i^n)\, \nabla \boldsymbol{v}_i)_n$ tends to $\tilde{\alpha}_i(\ell_i)\, \nabla \boldsymbol{v}_i$ a.e. in Ω_i and is bounded in $L^2(\Omega_i)^{d^2}$ by $c\, \|\nabla \boldsymbol{v}_i\|_{L^2(\Omega_i)^{d^2}}$, hence it converges strongly in $L^2(\Omega_i)^{d^2}$. Since $(\nabla \boldsymbol{u}_i^n)_n$ converges to $\nabla \boldsymbol{u}_i$ weakly in $L^2(\Omega_i)^{d^2}$, this yields

$$\lim_{n\to\infty} \int_{\Omega_i} \tilde{\alpha}_i(\ell_i^n)\, \nabla \boldsymbol{u}_i^n \cdot \nabla \boldsymbol{v}_i \, d\boldsymbol{x} = \int_{\Omega_i} \tilde{\alpha}_i(\ell_i)\, \nabla \boldsymbol{u}_i \cdot \nabla \boldsymbol{v}_i \, d\boldsymbol{x}.$$

Moreover, due to the compactness of the embedding of $H^{\frac{1}{2}}(\Gamma)$ into $L^3(\Gamma)$, there exists two subsequences, still denoted by $(\boldsymbol{u}_1^n)_n$ and $(\boldsymbol{u}_2^n)_n$, so that $((\boldsymbol{u}_i^n - \boldsymbol{u}_j^n)\, |\boldsymbol{u}_i^n - \boldsymbol{u}_j^n|)_n$ converges to $(\boldsymbol{u}_i - \boldsymbol{u}_j)\, |\boldsymbol{u}_i - \boldsymbol{u}_j|$ strongly in $L^{\frac{3}{2}}(\Gamma)$. Consequently, $(\boldsymbol{u}_1, \boldsymbol{u}_2)$ satisfies the first equation in (5.5) for $i = 1$ and 2.
2) Taking $\boldsymbol{v}_i = \boldsymbol{u}_i^n$ in (5.2) yields

$$\int_{\Omega_1} \tilde{\alpha}_i(\ell_1^n)\, |\nabla \boldsymbol{u}_1^n|^2 \, d\boldsymbol{x} + \int_{\Omega_2} \tilde{\alpha}_i(\ell_2^n)\, |\nabla \boldsymbol{u}_2^n|^2 \, d\boldsymbol{x} + \int_\Gamma |\boldsymbol{u}_1^n - \boldsymbol{u}_2^n|^3 \, d\tau = < \boldsymbol{f}_1, \boldsymbol{u}_1^n > + < \boldsymbol{f}_2, \boldsymbol{u}_2^n >,$$

so that passing to the limit yields

$$\lim_{n\to\infty} \int_{\Omega_1} \tilde{\alpha}_i(\ell_1^n)\,|\nabla \boldsymbol{u}_1^n|^2\,d\boldsymbol{x} + \int_{\Omega_2} \tilde{\alpha}_i(\ell_2^n)\,|\nabla \boldsymbol{u}_2^n|^2\,d\boldsymbol{x}$$
$$= <\boldsymbol{f}_1, \boldsymbol{u}_1> + <\boldsymbol{f}_2, \boldsymbol{u}_2> - \int_\Gamma |\boldsymbol{u}_1 - \boldsymbol{u}_2|^3\,d\tau.$$

We also derive from the first equation in (5.5) that

$$\int_{\Omega_1} \tilde{\alpha}_i(\ell_1)\,|\nabla \boldsymbol{u}_1|^2\,d\boldsymbol{x} + \int_{\Omega_2} \tilde{\alpha}_i(\ell_2)\,|\nabla \boldsymbol{u}_2|^2\,d\boldsymbol{x}$$
$$= <\boldsymbol{f}_1, \boldsymbol{u}_1> + <\boldsymbol{f}_2, \boldsymbol{u}_2> - \int_\Gamma |\boldsymbol{u}_1 - \boldsymbol{u}_2|^3\,d\tau,$$

whence

$$\begin{aligned}\lim_{n\to\infty} \int_{\Omega_1} \tilde{\alpha}_1(\ell_1^n)\,|\nabla \boldsymbol{u}_1^n|^2\,d\boldsymbol{x} &+ \int_{\Omega_2} \tilde{\alpha}_2(\ell_2^n)\,|\nabla \boldsymbol{u}_2^n|^2\,d\boldsymbol{x} \\ &= \int_{\Omega_1} \tilde{\alpha}_1(\ell_1)\,|\nabla \boldsymbol{u}_1|^2\,d\boldsymbol{x} + \int_{\Omega_2} \tilde{\alpha}_2(\ell_2)\,|\nabla \boldsymbol{u}_2|^2\,d\boldsymbol{x}.\end{aligned} \tag{5.7}$$

On the other hand, let us set: $\boldsymbol{h}_i^n = \sqrt{\tilde{\alpha}_i(\ell_i^n)}\,\nabla \boldsymbol{u}_i^n$. Since the sequence $(\boldsymbol{h}_i^n)_n$ is bounded in $L^2(\Omega_i)^{d^2}$, there exists a subsequence, still denoted by $(\boldsymbol{h}_i^n)_n$, which converges to $\boldsymbol{h}_i$ weakly in $L^2(\Omega_i)^{d^2}$. In order to identify $\boldsymbol{h}_i$, we introduce a function φ in $L^2(\Omega_i)^{d^2}$. Since the previous subsequence $(\sqrt{\tilde{\alpha}_i(\ell_i^n)})_n$ converges to $\sqrt{\tilde{\alpha}_i(\ell_i)}$ a.e. in Ω_i, the subsequence $(\sqrt{\tilde{\alpha}_i(\ell_i^n)}\,\varphi)_n$ also converges to $\sqrt{\tilde{\alpha}_i(\ell_i)}\,\varphi$ a.e. in Ω_i and, since it is obviously bounded by a function in $L^2(\Omega_i)$, it tends to $\sqrt{\tilde{\alpha}_i(\ell_i)}\,\varphi$ in $L^2(\Omega_i)^{d^2}$. Since $(\nabla \boldsymbol{u}_i^n)_n$ converges to $\nabla \boldsymbol{u}_i$ weakly in $L^2(\Omega_i)^{d^2}$, this yields

$$\lim_{n\to\infty} \int_{\Omega_i} \sqrt{\tilde{\alpha}_i(\ell_i^n)}\,\nabla \boldsymbol{u}_i^n \cdot \varphi\,d\boldsymbol{x} = \int_{\Omega_i} \sqrt{\tilde{\alpha}_i(\ell_i)}\,\nabla \boldsymbol{u}_i \cdot \varphi\,d\boldsymbol{x},$$

so that $\boldsymbol{h}_i$ is equal to $\sqrt{\tilde{\alpha}_i(\ell_i)}\,\nabla \boldsymbol{u}_i$. Moreover, from the weak convergence of $(\boldsymbol{h}_i^n)_n$, we deduce that

$$\int_{\Omega_i} \tilde{\alpha}_i(\ell_i)\,|\nabla \boldsymbol{u}_i|^2\,d\boldsymbol{x} \le \liminf_{n\to\infty} \int_{\Omega_i} \tilde{\alpha}_i(\ell_i^n)\,|\nabla \boldsymbol{u}_i^n|^2\,d\boldsymbol{x},$$

and combining this inequality with (5.7) implies

$$\lim_{n\to\infty} \int_{\Omega_i} \tilde{\alpha}_i(\ell_i^n)\,|\nabla \boldsymbol{u}_i^n|^2\,d\boldsymbol{x} = \int_{\Omega_i} \tilde{\alpha}_i(\ell_i)\,|\nabla \boldsymbol{u}_i|^2\,d\boldsymbol{x}, \quad i = 1 \text{ and } 2. \tag{5.8}$$

Equivalently, the sequence $(\sqrt{\tilde{\alpha}_i(\ell_i^n)}\,\nabla \boldsymbol{u}_i^n)_n$ tends to $\sqrt{\tilde{\alpha}_i(\ell_i)}\,\nabla \boldsymbol{u}_i$ strongly in $L^2(\Omega_i)^{d^2}$, so that the sequence $(\tilde{\alpha}_i(\ell_i^n)\,|\nabla \boldsymbol{u}_i^n|^2)_n$ tends to $\tilde{\alpha}_i(\ell_i)\,|\nabla \boldsymbol{u}_i|^2$ strongly in $L^1(\Omega_i)$.
3) We observe that the solution ℓ_i^n, $i = 1$ and 2, of the second equation in (5.2) also satisfies the "transposed" formulation, where φ_i is a smooth enough function on Ω_i:

$$-\int_{\Omega_i} \ell_i^n\,\Delta\varphi_i\,d\boldsymbol{x} = -\int_\Gamma T_n\big(G_i(|\boldsymbol{u}_1^n - \boldsymbol{u}_2^n|^2)\big)\,\partial_{n_i}\varphi_i\,d\tau + \int_{\Omega_i} T_n\big(\tilde{\alpha}_i(\ell_i^n)\,|\nabla \boldsymbol{u}_i^n|^2\big)\,\varphi_i\,d\boldsymbol{x}.$$

The convergence of the last term follows from part 2) of the proof together with the definition of T_n. The convergence $\big(G_i(|\boldsymbol{u}_1^n - \boldsymbol{u}_2^n|^2)\big)_n$ can easily be deduced from the sublinearity of G_i. So, for $i = 1$ and 2, each ℓ_i satisfies the second part of (5.5), which ends the proof. ■

To conclude, we go back to the initial system (1.1) and we write its full variational formulation:

Find $(\boldsymbol{u}_i, p_i)$ *in* $X_i \times L^2(\Omega_i)$, $1 \le i \le 2$, *such that,*
for $1 \le i \ne j \le 2$:

$$\begin{aligned}
\forall \boldsymbol{v}_i \in X_i,\quad & \int_{\Omega_i} \alpha_i(k_i)\,\nabla \boldsymbol{u}_i \cdot \nabla \boldsymbol{v}_i\,d\boldsymbol{x} - \int_{\Omega_i} p_i(\operatorname{div} \boldsymbol{v}_i)\,d\boldsymbol{x} \\
&\quad + \int_\Gamma |\boldsymbol{u}_i - \boldsymbol{u}_j|\,(\boldsymbol{u}_i - \boldsymbol{u}_j)\cdot \boldsymbol{v}_i\,d\tau = \int_{\Omega_i} \boldsymbol{f}_i \cdot \boldsymbol{v}_i\,d\boldsymbol{x}. \\
\forall q_i \in L^2(\Omega_i),\quad & -\int_{\Omega_i} q_i(\operatorname{div} \boldsymbol{u}_i)\,d\boldsymbol{x} = 0,
\end{aligned} \tag{5.9}$$

Find k_i *in* $L^2(\Omega_i)$, $1 \le i \le 2$, *such that, for* $1 \le i \le 2$:
$\forall \varphi_i \in H^2(\Omega_i) \cap H_0^1(\Omega_i)$,

$$-\int_{\Omega_i} G_i(k_i)\,\Delta\varphi_i\,d\boldsymbol{x} = -\int_\Gamma G_i(|\boldsymbol{u}_1 - \boldsymbol{u}_2|^2)\,\partial_{n_i}\varphi_i\,d\tau + \int_{\Omega_i} \alpha_i(k_i)\,|\nabla \boldsymbol{u}_i|^2\,\varphi_i\,d\boldsymbol{x}.$$

Here, the argument is due to [24].

Corollary 5.3. – *For any* $\boldsymbol{f}_i$ *in* $L^2(\Omega_i)^d$, $i = 1$ *or* 2, *system* (1.1) *admits the formulation* (5.9). *System* (5.9) *has a solution* (W_1, W_2) *with each* $W_i =$

$(\boldsymbol{u}_i, p_i, k_i)$ in $X_i \times L^2(\Omega_i) \times L^2(\Omega_i)$. Moreover, each function k_i, $i = 1$ and 2, is nonnegative and belongs to $H^s(\Omega_i)$ for all $s < \frac{1}{2}$.

Proof. Since the existence of p_i in $L^2(\Omega_i)$ is a consequence of Lemma 3.1, it suffices to check that the mapping: $\ell \mapsto k = G_i^{-1}(\ell)$ is continuous from $H^s(\Omega_i)$ into itself. This follows by an interpolation argument: indeed, it is continuous from $L^2(\Omega_i)$ into itself and from $H^1(\Omega_i)$ into itself thanks to the inequalities

$$k \leq \nu^{-1}\,\ell, \quad |\nabla k| \leq \nu^{-1}\,|\nabla \ell|.$$

6. A uniqueness result for smooth solutions.

The aim of this section is to prove that any solution of system (1.1) which is slightly more regular than in the existence theorem, is unique when a further condition holds: the data must be small enough in comparison of the relative variation of the functions α_i. So, we assume that, for some data $\boldsymbol{f}_i$ in $L^2(\Omega_i)^d$, system (1.1) has two solutions $(\boldsymbol{u}_i, p_i, k_i)_{i=1,2}$ and $(\overline{\boldsymbol{u}}_i, \overline{p}_i, \overline{k}_i)_{i=1,2}$, we define the corresponding functions $\ell_i = G_i(k_i)$ and $\overline{\ell}_i = G_i(\overline{k}_i)$ as in Section 2 and we set:

$$\boldsymbol{w}_i = \boldsymbol{u}_i - \overline{\boldsymbol{u}}_i, \qquad m_i = \ell_i - \overline{\ell}_i, \quad i = 1, 2. \tag{6.1}$$

We assume that the functions α_i are continuously differentiable with bounded derivatives and, for simplicity, we also introduce the notation

$$\nu^* = \max_{i=1,2} \sup_{k \in \mathbb{R}} \{\alpha_i(k), \gamma_i(k)\}, \qquad \nu' = \max_{i=1,2} \sup_{k \in \mathbb{R}} |\alpha_i'(k)|.$$

Note that the maximum of $\tilde{\alpha}_i'$ is smaller than $\frac{\nu'}{\nu}$.

In the next two lemmas, we treat separately the equations on the velocities and on the turbulent energies.

Lemma 6.1. – *Assume that the pair $(\overline{\boldsymbol{u}}_1, \overline{\boldsymbol{u}}_2)$ belongs to $W^{1,p}(\Omega_1)^d \times W^{1,p}(\Omega_2)^d$ for a real number $p > 2$, and let q be such that: $\frac{1}{p} + \frac{1}{q} = \frac{1}{2}$. The following estimate holds for the functions $\boldsymbol{w}_i$, $i = 1$ and 2:*

$$\begin{aligned}
&\left(|\boldsymbol{w}_1|^2_{H^1(\Omega_1)^d} + |\boldsymbol{w}_2|^2_{H^1(\Omega_2)^d}\right)^{\frac{1}{2}} \\
&\quad \leq \frac{\nu'}{\nu^2}\left(\|\overline{\boldsymbol{u}}_1\|^p_{W^{1,p}(\Omega_1)^d} + \|\overline{\boldsymbol{u}}_2\|^p_{W^{1,p}(\Omega_2)^d}\right)^{\frac{1}{p}} \\
&\qquad\qquad \left(\|m_1\|^q_{L^q(\Omega_1)} + \|m_2\|^q_{L^q(\Omega_2)}\right)^{\frac{1}{q}}.
\end{aligned} \tag{6.2}$$

Proof. Subtracting formulation (3.5) for $\boldsymbol{u}_i$ and $\overline{\boldsymbol{u}}_i$, we obtain

$$\forall \boldsymbol{v}_i \in V_i, \quad \int_{\Omega_i} \tilde{\alpha}_i(\ell_i)\, \nabla \boldsymbol{w}_i \cdot \nabla \boldsymbol{v}_i \, d\boldsymbol{x} + \int_\Gamma (\boldsymbol{u}_i - \boldsymbol{u}_j)\, |\boldsymbol{u}_i - \boldsymbol{u}_j|\, \boldsymbol{v}_i \, d\tau - \int_\Gamma (\overline{\boldsymbol{u}}_i - \overline{\boldsymbol{u}}_j)\, |\overline{\boldsymbol{u}}_i - \overline{\boldsymbol{u}}_j|\, \boldsymbol{v}_i \, d\tau = - \int_{\Omega_i} \big(\tilde{\alpha}_i(\ell_i) - \tilde{\alpha}_i(\overline{\ell}_i)\big)\, \nabla \overline{\boldsymbol{u}}_i \cdot \nabla \boldsymbol{v}_i \, d\boldsymbol{x}.$$

Since $\boldsymbol{w}_i$ belongs to V_i, we take each $\boldsymbol{v}_i$ equal to $\boldsymbol{w}_i$, we sum up on i and we observe from (3.9) that, as in the proof of Proposition 3.5, the quantity that is integrated on Γ is nonnegative. This yields

$$\sum_{i=1}^{2} \int_{\Omega_i} \tilde{\alpha}_i(\ell_i)\, |\nabla \boldsymbol{w}_i|^2 \, d\boldsymbol{x} \le \sum_{i=1}^{2} \Big| \int_{\Omega_i} \big(\tilde{\alpha}_i(\ell_i) - \tilde{\alpha}_i(\overline{\ell}_i)\big)\, \nabla \overline{\boldsymbol{u}}_i \cdot \nabla \boldsymbol{w}_i \, d\boldsymbol{x} \Big|,$$

whence

$$\nu \Big(\sum_{i=1}^{2} |\boldsymbol{w}_i|^2_{H^1(\Omega_i)^d} \Big) \le \sum_{i=1}^{2} \|\tilde{\alpha}_i(\ell_i) - \tilde{\alpha}_i(\overline{\ell}_i)\|_{L^q(\Omega_i)}\, |\overline{\boldsymbol{u}}_i|_{W^{1,p}(\Omega_i)^d}\, |\boldsymbol{w}_i|_{H^1(\Omega_i)^d}.$$

We also have

$$\|\tilde{\alpha}_i(\ell_i) - \tilde{\alpha}_i(\overline{\ell}_i)\|_{L^q(\Omega_i)} \le \frac{\nu'}{\nu}\, \|m_i\|_{L^q(\Omega_i)},$$

so that the desired estimate follows by applying Hölder's inequality. ∎

Lemma 6.2. – *Assume that the pair $(\overline{\boldsymbol{u}}_1, \overline{\boldsymbol{u}}_2)$ belongs to $W^{1,p}(\Omega_1)^d \times W^{1,p}(\Omega_2)^d$ for a real number $p > 2$, and let q be such that: $\frac{1}{p} + \frac{1}{q} = \frac{1}{2}$. Let s be a real number, $0 < s < \frac{1}{2}$. The following estimate holds for the functions m_i, $i = 1$ and 2:*

$$\|m_i\|_{H^s(\Omega_i)} \le c\, \frac{\nu^*}{\nu} \big(\|\boldsymbol{f}_1\|^2_{L^2(\Omega_1)^d} + \|\boldsymbol{f}_2\|^2_{L^2(\Omega_2)^d} \big)^{\frac{1}{2}} \big(\|\boldsymbol{w}_1\|^2_{H^1(\Omega_1)^d} + \|\boldsymbol{w}_2\|^2_{H^1(\Omega_2)^d} \big)^{\frac{1}{2}} + c'\, \frac{\nu'}{\nu}\, \|\overline{\boldsymbol{u}}_i\|^2_{W^{1,p}(\Omega_i)}\, \|m_i\|_{L^{\frac{q}{2}}(\Omega_i)}. \tag{6.3}$$

Proof. Subtracting formulation (4.2) for ℓ_i and $\overline{\ell}_i$ gives

$\forall \varphi_i \in H^2(\Omega_i) \cap H^1_0(\Omega_i)$,

$$- \int_{\Omega_i} m_i\, \Delta \varphi_i \, d\boldsymbol{x} = - \int_\Gamma \big(G_i(|\boldsymbol{u}_1 - \boldsymbol{u}_2|^2) - G_i(|\overline{\boldsymbol{u}}_1 - \overline{\boldsymbol{u}}_2|^2) \big)\, \partial_{n_i} \varphi_i \, d\tau + \int_{\Omega_i} \big(\tilde{\alpha}_i(\ell_i)\, |\nabla \boldsymbol{u}_i|^2 - \tilde{\alpha}_i(\overline{\ell}_i)\, |\nabla \overline{\boldsymbol{u}}_i|^2 \big)\, \varphi_i \, d\boldsymbol{x}.$$

As in the proof of Lemma 4.2, relying on the analogue of (4.6), with any function g_i in $H^{-s}(\Omega_i)$, we associate the function $\varphi_i = \mathcal{L}_i g_i$ defined by (4.3) and we recall that

$$\|\partial_n \varphi_i\|_{L^2(\Gamma)} + \sup_{\boldsymbol{x}\in\Omega_i} |\varphi_i(\boldsymbol{x})| \leq c_i \|g_i\|_{H^{-s}(\Omega_i)},$$

for a constant c_i only depending on Ω_i. We now estimate the right-hand side of the previous equation for such a φ_i. We observe that

$$|G_i(|\boldsymbol{u}_1 - \boldsymbol{u}_2|^2) - G_i(|\overline{\boldsymbol{u}}_1 - \overline{\boldsymbol{u}}_2|^2)| \leq \nu^* \, |\boldsymbol{u}_1 + \boldsymbol{u}_2 + \overline{\boldsymbol{u}}_1 + \overline{\boldsymbol{u}}_2| \, |\boldsymbol{w}_1 - \boldsymbol{w}_2|,$$

so that denoting by c_i' the norm of the embedding of $H_0^1(\Omega_i)$ into $L^4(\Omega_i)$, we obtain

$$\begin{aligned}
&|\int_\Gamma (G_i(|\boldsymbol{u}_1 - \boldsymbol{u}_2|^2) - G_i(|\overline{\boldsymbol{u}}_1 - \overline{\boldsymbol{u}}_2|^2)) \, \partial_{n_i}\varphi_i \, d\tau| \\
&\qquad \leq c_i c_i'^2 \, \nu^* \, \big(\sum_{i=1}^2 |\boldsymbol{u}_i|^2_{H^1(\Omega_i)^d} + |\overline{\boldsymbol{u}}_i|^2_{H^1(\Omega_i)^d}\big)^{\frac{1}{2}} \\
&\qquad\qquad \big(\sum_{i=1}^2 |\boldsymbol{w}_i|^2_{H^1(\Omega_i)^d}\big)^{\frac{1}{2}} \, \|g_i\|_{H^{-s}(\Omega_i)}.
\end{aligned}$$

The norms $|\boldsymbol{u}_i|^2_{H^1(\Omega_i)^d}$ and $|\overline{\boldsymbol{u}}_i|^2_{H^1(\Omega_i)^d}$ are bounded from (3.7). Similarly, we have

$$|\tilde{\alpha}_i(\ell_i) \, |\nabla \boldsymbol{u}_i|^2 - \tilde{\alpha}_i(\overline{\ell}_i) \, |\nabla \overline{\boldsymbol{u}}_i|^2| \leq \nu^* \, |\nabla(\boldsymbol{u}_i + \overline{\boldsymbol{u}}_i)| \, |\nabla \boldsymbol{w}_i| + \frac{\nu'}{\nu} \, |m_i| \, |\nabla \overline{\boldsymbol{u}}_i|^2,$$

whence

$$\begin{aligned}
&|\int_{\Omega_i} (\tilde{\alpha}_i(\ell_i) \, |\nabla \boldsymbol{u}_i|^2 - \tilde{\alpha}_i(\overline{\ell}_i) \, |\nabla \overline{\boldsymbol{u}}_i|^2) \, \varphi_i \, d\boldsymbol{x}| \\
&\qquad \leq c_i \Big(\nu^* \, \big(\sum_{i=1}^2 |\boldsymbol{u}_i|^2_{H^1(\Omega_i)^d} + |\overline{\boldsymbol{u}}_i|^2_{H^1(\Omega_i)^d}\big)^{\frac{1}{2}} \big(\sum_{i=1}^2 |\boldsymbol{w}_i|^2_{H^1(\Omega_i)^d}\big)^{\frac{1}{2}} \\
&\qquad\qquad + \frac{\nu'}{\nu} \, \|\overline{\boldsymbol{u}}_i\|^2_{W^{1,p}(\Omega_i)} \, \|m_i\|_{L^{\frac{q}{2}}(\Omega_i)}\Big) \|g_i\|_{H^{-s}(\Omega_i)}.
\end{aligned}$$

Combining all this yields the desired result.

We are now in a position to prove the uniqueness result.

Theorem 6.3. – *Assume that the data $\boldsymbol{f}_i$, $i = 1$ and 2, belong to $L^2(\Omega_i)^d$. If system (1.1) admits a solution $(\overline{W}_1, \overline{W}_2)$, with $\overline{W}_i = (\overline{\boldsymbol{u}}_i, \overline{p}_i, \overline{k}_i)$, such that the $\overline{\boldsymbol{u}}_i$, $i = 1$ and 2, belong to $W^{1,p}(\Omega_i)^d$ for a real number $p > 2d$ and that the following condition holds for appropriate constants c and c'*

$$c\,\frac{\nu'}{\nu}\,\overline{\Lambda}\left(\overline{\Lambda} + c'\,\frac{\nu^*}{\nu^2}\left(\|\boldsymbol{f}_1\|^2_{L^2(\Omega_1)^d} + \|\boldsymbol{f}_2\|^2_{L^2(\Omega_2)^d}\right)^{\frac{1}{2}}\right) < 1,$$
$$\text{with} \quad \overline{\Lambda} = \left(\|\overline{\boldsymbol{u}}_1\|^p_{W^{1,p}(\Omega_1)^d} + \|\overline{\boldsymbol{u}}_2\|^p_{W^{1,p}(\Omega_2)^d}\right)^{\frac{1}{p}}, \tag{6.4}$$

then it is the unique solution of system (1.1) in the sense that the pair (W_1, W_2) is the unique solution of problem (5.9).

Proof. Since p is $> 2d$, there exists an $s < \frac{1}{2}$ such that $H^s(\Omega_i)$ is imbedded in $L^q(\Omega_i)$. So replacing the $\|m_i\|_{L^q(\Omega_i)}$ and $\|m_i\|_{L^{\frac{q}{2}}(\Omega_i)}$ in (6.2) and (6.3), respectively, by an appropriate constant times $\|m_i\|_{H^s(\Omega_i)}$, summing up the estimate (6.3) on i and inserting (6.2) in the sum, we derive from (6.4) that both m_i cancel. The complete uniqueness result follows. ∎

So, the uniqueness mainly depends the parameter $\frac{\nu'}{\nu}$ which represents the variation of the function $\tilde{\alpha}_i$. When the α_i are constant, we recover in the previous theorem the nearly obvious and unconditional uniqueness result which also follows from Proposition 3.5.

7. Some further regularity properties in dimension 2.

We intend to study the regularity of the solution of system (1.1) in the simple case of dimension $d = 2$, when the Ω_i are rectangles. Indeed this geometry is the key one for most discretizations. We need several lemmas.

The first one deals with the Stokes problem with mixed boundary conditions, and results from [22, Cor. 4.2], extended to less smooth data thanks to the arguments in [9, Chap. 8]. From now on, we set:

$$s_0 = 1.5946. \tag{7.1}$$

Lemma 7.1. – *For $i = 1$ and 2, if the domain Ω_i is a rectangle, the* *mapping:* $\boldsymbol{f}_i \mapsto (\boldsymbol{v}_i, q_i)$*, where* $(\boldsymbol{v}_i, q_i)$ *is the solution of the Stokes problem*

$$\begin{cases} -\Delta \boldsymbol{v}_i + \mathbf{grad}\, q_i = \boldsymbol{f}_i & \text{in } \Omega_i, \\ \operatorname{div} \boldsymbol{v}_i = 0 & \text{in } \Omega_i, \\ \boldsymbol{v}_i = \mathbf{0} & \text{on } \Gamma_i, \\ \partial_{n_i} \boldsymbol{v}_i - q_i\, \boldsymbol{n}_i = \mathbf{0} & \text{on } \Gamma, \end{cases} \tag{7.2}$$

is continuous from $H^{s-2}(\Omega_i)^2$ *into* $H^s(\Omega_i)^2 \times H^{s-1}(\Omega_i)$ *for all* s, $\frac{3}{2} < s \le s_0$.

The second lemma extends this result to the case where the Neumann boundary conditions are not zero.

Lemma 7.2. – *For* $i = 1$ *and* 2*, if the domain* Ω_i *is a rectangle, the Stokes mapping* $\mathcal{S}_i$: $(\boldsymbol{f}_i, \boldsymbol{g}_i) \mapsto (\boldsymbol{v}_i, q_i)$*, where* $(\boldsymbol{v}_i, q_i)$ *is the solution of the Stokes problem*

$$\begin{cases} -\Delta \boldsymbol{v}_i + \mathbf{grad}\, q_i = \boldsymbol{f}_i & \text{in } \Omega_i, \\ \operatorname{div} \boldsymbol{v}_i = 0 & \text{in } \Omega_i, \\ \boldsymbol{v}_i = \mathbf{0} & \text{on } \Gamma_i, \\ \partial_{n_i} \boldsymbol{v}_i - q_i\, \boldsymbol{n}_i = \boldsymbol{g}_i & \text{on } \Gamma, \end{cases} \tag{7.3}$$

is continuous from $H^{s-2}(\Omega_i)^2 \times H^{s-\frac{3}{2}}(\Gamma)^2$ *into* $H^s(\Omega_i)^2 \times H^{s-1}(\Omega_i)$ *for all* s, $\frac{3}{2} < s \le s_0$.

Proof. We want to construct a divergence-free function $\boldsymbol{w}_i$ and a function r_i with the required regularity such that:

$$\boldsymbol{w}_i = \mathbf{0} \quad \text{on } \Gamma_i \qquad \text{and} \qquad \partial_{n_i} \boldsymbol{w}_i - r_i\, \boldsymbol{n}_i = \boldsymbol{g}_i \quad \text{on } \Gamma,$$

and to apply the previous Lemma 7.1 to the pair $(\boldsymbol{v}_i - \boldsymbol{w}_i, q_i - r_i)$. So, denoting by g_{ix} and g_{iz} the components of $\boldsymbol{g}_i$, we look for a function $\boldsymbol{w}_i$ of the form $\mathbf{curl}\, \psi$, so that the desired boundary conditions are written

$$\psi = \partial_{n_i} \psi = 0 \quad \text{on } \Gamma_i$$
$$\text{and} \qquad \partial_z \psi = \pm \int_a^x (g_{iz} - r_i)(\xi)\, d\xi, \quad \partial_z^2 \psi = \pm(-g_{ix}) \quad \text{on } \Gamma,$$

where $(a, 0)$ stands for the left endpoint of Γ and the sign $\pm$ depends on i. The idea is to take:

$$r_i = \frac{1}{\operatorname{meas}\Gamma} \int_\Gamma g_{iz}(x)\, dx$$

(such a constant is well-defined since g_{iz} belongs to $L^1(\Gamma)$). Indeed, thanks to this choice, the right compatibility conditions are satisfied at the corners of Ω_i. So, the existence of a function ψ in $H^{s+1}(\Omega_i)$ follows from [2, Thm. 2.d.2], together with the desired bound on its norm. This, combined with Lemma 7.1, yields the result. ∎

Let us now check how the previous results extend to smaller values of s, namely $1 < s < \frac{3}{2}$.

Corollary 7.3. – *For $i = 1$ and 2, if the domain Ω_i is a rectangle, the Stokes mapping $\mathcal{S}_i$: $(\boldsymbol{f}_i, \boldsymbol{g}_i) \mapsto (\boldsymbol{v}_i, q_i)$, where $(\boldsymbol{v}_i, q_i)$ is the solution of (7.3), is continuous from $(X_i^{2-s})' \times H^{s-\frac{3}{2}}(\Gamma)^2$ into $H^s(\Omega_i)^2 \times H^{s-1}(\Omega_i)$ for all s, $1 < s < \frac{3}{2}$, where $(X_i^{2-s})'$ stands for the dual space of*

$$X_i^{2-s} = \{\boldsymbol{v}_i \in H^{2-s}(\Omega_i)^2;\ \boldsymbol{v}_i = \mathbf{0} \text{ on } \Gamma_i\}. \tag{7.4}$$

Proof. From Lemma 7.2, the mapping is continuous from $H^{s_0-2}(\Omega_i)^2 \times H^{s_0-\frac{3}{2}}(\Gamma)^2$ into $H^{s_0}(\Omega_i)^2 \times H^{s_0-1}(\Omega_i)$. When writing the variational formulation of problem (7.3), we observe that it is also continuous from $X_i' \times \big(H_{00}^{\frac{1}{2}}(\Gamma)\big)'^2$ (the $'$ denotes the dual space) into $H^1(\Omega)^2 \times L^2(\Omega)$. So, if $H^{s_0-2}(\Omega)^2$ is dense in X_i' and if $H^{s_0-\frac{3}{2}}(\Gamma)$ is dense in $\big(H_{00}^{\frac{1}{2}}(\Gamma)\big)'$, applying the main theorem of interpolation [16, Chap. 1, Th. 5.1] yields that the mapping is continuous from $F_s \times G_s$ into $H^s(\Omega_i)^2 \times H^{s-1}(\Omega_i)$, with:

$$F_s = [H^{s_0-2}(\Omega_i)^2, X_i']_\theta, \qquad G_s = [H^{s_0-\frac{3}{2}}(\Gamma)^2, \big(H_{00}^{\frac{1}{2}}(\Gamma)\big)'^2]_\theta,$$
$$\text{with} \quad \theta = \frac{s_0 - s}{s_0 - 1}.$$

So, it remains to check the density results and to characterize the spaces F_s and G_s.

1) First, since $2 - s_0$ is $< \frac{1}{2}$, $\mathcal{D}(\Omega)$ is dense in $H^{2-s_0}(\Omega)$. Since $\mathcal{D}(\Omega)^2$ is contained in X_i, X_i is dense in $H^{2-s_0}(\Omega)^2$, so that $H^{s_0-2}(\Omega)^2$ is dense in X_i'. Moreover the following characterization holds [16, Chap. 1, Th. 6.2]:

$$F_s = \Big([X_i, H^{2-s_0}(\Omega_i)^2]_{1-\theta}\Big)'.$$

And the interpolate space in the previous line coincides with X_i^{2-s}. This follows from one-dimensional interpolation results, combined with the tensorization property (where the rectangle Ω_i is the product of the two intervals Λ_i^x and Λ_i^z)

$$X_i^{2-s} = H_0^{2-s}(\Lambda_i^x; L^2(\Lambda_i^z)) \cap L^2(\Lambda_i^x; H_*^{2-s}(\Lambda_i^z)),$$

where $H_*^{2-s}(\Lambda_i^z)$ stands for the space of functions in $H^{2-s}(\Lambda_i^z)$ vanishing in the upper endpoint.
2) Next, we observe that $H_{00}^{\frac{1}{2}}(\Gamma)$ is dense in $L^2(\Gamma)$ which is dense in $H^{\frac{3}{2}-s_0}(\Gamma)$, so that $H^{s_0-\frac{3}{2}}(\Gamma)$ is dense in $\big(H_{00}^{\frac{1}{2}}(\Gamma)\big)'$ and the space G_s is well-defined. Moreover, applying [16, Chap. 1, Th. 6.2] gives

$$G_s = \Big([H_{00}^{\frac{1}{2}}(\Gamma)^2, H^{\frac{3}{2}-s_0}(\Gamma)^2]_{1-\theta}\Big)'.$$

Next, relying on [16, Chap. 1, Rem. 12.6], we observe that the interpolate space $[H_{00}^{\frac{1}{2}}(\Gamma), H^{\frac{3}{2}-s_0}(\Gamma)]_{1-\theta}$ coincides with $H_0^{\frac{3}{2}-s}(\Gamma)$, which coincides with $H^{\frac{3}{2}-s}(\Gamma)$ since s is > 1.
This ends the proof. ■

We also need some regularity properties of problem (4.2) when the functions $\boldsymbol{u}_1$ and $\boldsymbol{u}_2$ are smoother than in the existence result of Theorem 5.2.

Lemma 7.4. – *Let s be a real number, $1 < s \leq \frac{3}{2}$. For $i = 1$ or 2, and for any pair $(\boldsymbol{u}_1, \boldsymbol{u}_2)$ in $H^s(\Omega_1)^2 \times H^s(\Omega_2)^2$, the solution ℓ_i of problem (4.2) belongs to $H^s(\Omega_i)$ and satisfies, for a constant κ_i only depending on Ω_i,*

$$\|\ell_i\|_{H^s(\Omega_i)} \leq \kappa_i \left(\|\boldsymbol{u}_1\|^2_{H^s(\Omega_1)^2} + \|\boldsymbol{u}_2\|^2_{H^s(\Omega_2)^2}\right). \tag{7.5}$$

The same properties holds for $\frac{3}{2} < s < 2$ if the functions γ_i are continuously differentiable with bounded derivatives

Proof. Thanks to Lemma 4.1, it suffices to check that $\tilde{\alpha}_i(\ell_i)\,|\nabla \boldsymbol{u}_i|^2$ belongs to $H^{s-2}(\Omega_i)$ and that $G_i(|\boldsymbol{u}_1 - \boldsymbol{u}_2|^2)$, extended by zero on Γ_i, belongs to $H^{s-\frac{1}{2}}(\partial\Omega_i)$.
1) On one hand, the functions $\boldsymbol{u}_i$ belong to $H^s(\Omega_i)^2$, hence to $W^{1,\frac{2}{2-s}}(\Omega_i)^2$ by the Sobolev imbedding theorem. So, $|\nabla \boldsymbol{u}_i|^2$ belongs to $L^{\frac{1}{2-s}}(\Omega_i)$ and, since $\tilde{\alpha}_i$ is bounded, the same property holds for $\tilde{\alpha}_i(\ell_i)\,|\nabla \boldsymbol{u}_i|^2$.
For $s \geq \frac{3}{2}$, we obtain that $\tilde{\alpha}_i(\ell_i)\,|\nabla \boldsymbol{u}_i|^2$ belongs to $L^2(\Omega_i)$, hence to $H^{s-2}(\Omega_i)$. For $s \leq \frac{3}{2}$, since the imbedding of $H^{3-2s}(\Omega_i)$ into

$L^{\frac{1}{s-1}}(\Omega_i)$ yields the imbedding of $L^{\frac{1}{2-s}}(\Omega_i)$ into $H^{2s-3}(\Omega_i)$, we derive that $\tilde{\alpha}_i(\ell_i)\,|\nabla \boldsymbol{u}_i|^2$ belongs to $H^{2s-3}(\Omega_i)$, hence to $H^{s-2}(\Omega_i)$. Moreover, in both cases, we have the following estimate

$$\|\tilde{\alpha}_i(\ell_i)\,|\nabla \boldsymbol{u}_i|^2\|_{H^{s-2}(\Omega_i)} \le c\,\|\boldsymbol{u}_i\|^2_{H^s(\Omega_i)^2}. \tag{7.6}$$

2) On the other hand, since $G_i(|\boldsymbol{u}_1 - \boldsymbol{u}_2|^2)$ vanishes at the two endpoints of Γ, it suffices to check that it belongs to $H^{s-\frac{1}{2}}(\Gamma)$. First, we observe that $\boldsymbol{u}_1 - \boldsymbol{u}_2$ belongs to $H^{s-\frac{1}{2}}(\Gamma)^2$ and, since $H^{s-\frac{1}{2}}(\Gamma)$ is an algebra [12, Thm 1.4.4.2], $|\boldsymbol{u}_1 - \boldsymbol{u}_2|^2$ also belongs to $H^{s-\frac{1}{2}}(\Gamma)$. When s is $\le \frac{3}{2}$, relying on [24] and using similar arguments as for Corollary 5.3, we derive that $G_i(|\boldsymbol{u}_1 - \boldsymbol{u}_2|^2)$ belongs to $H^{s-\frac{1}{2}}(\Gamma)$ and satisfies

$$\|G_i(|\boldsymbol{u}_1 - \boldsymbol{u}_2|^2)\|_{H^{s-\frac{1}{2}}(\Gamma)} \le c\,\big(\|\boldsymbol{u}_1\|^2_{H^s(\Omega_i)^2} + \|\boldsymbol{u}_2\|^2_{H^s(\Omega_i)^2}\big). \tag{7.7}$$

The same result is derived when s is $> \frac{3}{2}$, by applying the same arguments to the derivative of $G_i(|\boldsymbol{u}_1 - \boldsymbol{u}_2|^2)$.
This ends the proof. ■

We are in a position to prove the main result of this section.

Theorem 7.5. – *Assume that the domains Ω_1 and Ω_2 are rectangles and that the functions α_i and γ_i, $i = 1$ and 2, are continuously differentiable with bounded derivatives. Let $\boldsymbol{f}_i$, $i = 1$ and 2, belong to $L^2(\Omega_i)^2$, and let (W_1, W_2) be a solution of system (1.1), with $W_i = (\boldsymbol{u}_i, p_i, k_i)$. If the $\boldsymbol{u}_i$, $i = 1$ and 2, belong to $H^{s^*}(\Omega_i)^2$ for some $s^* > 1$, the W_i, $i = 1$ and 2, belong to $H^{s_0}(\Omega_i)^d \times H^{s_0-1}(\Omega_i) \times H^{s_0}(\Omega_i)$.*

Proof. Let (W_1, W_2) be any solution of problem (1.1). The proof relies on a boot-strap argument and is performed in two steps.
1) From the assumption on the $\boldsymbol{u}_i$, each pair $(\boldsymbol{u}_i, p_i)$ belongs to $H^{s^*}(\Omega_i)^2 \times H^{s^*-1}(\Omega_i)$. So, applying Lemma 7.4 yields that each ℓ_i belongs to $H^{s^*}(\Omega_i)$. Finally, from [24], the function $\frac{p_i}{\tilde{\alpha}_i(\ell_i)}$ also belongs to $H^{s^*-1}(\Omega_i)$.
2) Assume now that both triples $(\boldsymbol{u}_i, \frac{p_i}{\tilde{\alpha}_i(\ell_i)}, \ell_i)$ belong to $H^t(\Omega_i)^2 \times H^{t-1}(\Omega_i) \times H^t(\Omega_i)$, for a real number t, $1 < t \le s_0$. Then the pair $(\boldsymbol{u}_i, q_i)$, with $q_i = \frac{p_i}{\tilde{\alpha}_i(\ell_i)}$, is a solution of the Stokes problem

$$\begin{cases} -\Delta \boldsymbol{u}_i + \mathbf{grad}\, q_i = \frac{f_i}{\tilde{\alpha}_i(\ell_i)} + \frac{\tilde{\alpha}_i'(\ell_i)}{\tilde{\alpha}(\ell_i)} \nabla \ell_i \cdot \nabla \boldsymbol{u}_i - \frac{\tilde{\alpha}_i'(\ell_i)}{\tilde{\alpha}(\ell_i)^2} \nabla \ell_i \, q_i & \text{in } \Omega_i, \\ \operatorname{div} \boldsymbol{u}_i = 0 & \text{in } \Omega_i, \\ \boldsymbol{u}_i = \mathbf{0} & \text{on } \Gamma_i, \\ \partial_{n_i} \boldsymbol{u}_i - q_i \, \boldsymbol{n}_i = -\frac{1}{\tilde{\alpha}_i(\ell_i)} (\boldsymbol{u}_i - \boldsymbol{u}_j)|\boldsymbol{u}_i - \boldsymbol{u}_j| & \text{on } \Gamma, \end{cases} \tag{7.8}$$

In this equation, $\nabla \ell_i$ and q_i belongs to $H^{t-1}(\Omega_i)$, together with each component of $\nabla \boldsymbol{u}_i$. Since $H^{t-1}(\Omega_i)$ is included in $L^{\frac{2}{2-t}}(\Omega_i)$, the right-hand side in the first line of (7.8) belongs to $L^{\frac{1}{2-t}}(\Omega_i)^2$, hence to $L^2(\Omega_i)^2$ when t is $\geq \frac{3}{2}$ and to $(X_i^{3-2t})'$ when t is $< \frac{3}{2}$. It is also clear that the right-hand side in the fourth line of (7.8) belongs to $H^{t-\frac{1}{2}}(\Gamma)^2$, hence to $H^{2t-\frac{3}{2}}(\Gamma)^2$. So using Corollary 7.3 (or Lemma 7.2) yields that $(\boldsymbol{u}_i, q_i)$ belongs to $H^{t^*}(\Omega_i)^2 \times H^{t^*-1}(\Omega_i)$, with $t^* = \min\{2t-1, s_0\}$, and applying Lemma 7.4 yields that ℓ_i belongs to $H^{t^*}(\Omega_i)$.
Iterating n times this argument, where n is the smallest integer such that $s^* + n(s^*-1)$ is $\geq s_0$, we obtain that the triple $(\boldsymbol{u}_i, q_i, \ell_i)$, hence $(\boldsymbol{u}_i, p_i, \ell_i)$, belongs to $H^{s_0}(\Omega_i)^2 \times H^{s_0-1}(\Omega_i) \times H^{s_0}(\Omega_i)$. Then, k_i also belongs to $H^{s_0}(\Omega_i)$.

Remark. A more technical proof, relying on Meyers' argument [19] (see also [8]), allows for replacing , in the statement of Theorem 7.5, the assumption "the $\boldsymbol{u}_i$ belong to $H^{s^*}(\Omega_i)^2$" by the modified one "the k_i (or ℓ_i) belong to $H^{s^*}(\Omega_i)$", also for any $s^* > 1$, however this new assumption seems stronger.

Of course, these regularity properties can be extended to any convex polygons Ω_i, since the corresponding values of s_0 can easily be computed from [22]. By a boot-strap argument, they also hold in the case where convection terms are added in the system. But, when the interface conditions are replaced by Manning's law, the regularity of the solution of the basic Stokes problem, *a fortiori* of the present system, seems unknown.

References

[1] C. Bernardi, T. Chacon, R. Lewandowski and F. Murat, Existence d'une solution pour un modèle de deux fluides turbulents couplés, C. R. Acad. Sc. Paris 328 série I (1999), 993–998.

[2] C. Bernardi, M. Dauge and Y. Maday, Polynomials in weighted Sobolev spaces: basics and trace liftings, Internal Report 92039, Laboratoire d'Analyse Numérique, Université Pierre et Marie Curie, Paris (1992).

[3] P. Benilan, L. Boccardo, T. Gallouët, R. Gariepy, M. Pierre and J. L. Vasquez, An L^1 theory of existence and uniqueness of nonlinear elliptic equations, Ann. Scuola Norm. Sup. Pisa Cl. Sci. 22 (1995), 241–273.

[4] D. Blanchard and H. Redwane, Solutions renormalisées d'équations paraboliques à deux nonlinéarités, C. R. Acad. Sc. Paris 319 série I (1994), 831–835.

[5] L. Boccardo and T. Gallouët, Nonlinear elliptic and parabolic equations involving measure data, J. Funct. Anal. 87 (1989), 149–169.

[6] J. Boland and R. Nicolaides, Stability of finite elements under divergence constraints, SIAM J. Numer. Anal. 20 (1983), 722–731.

[7] H. Brezis, Analyse fonctionnelle, Collection "Mathématiques appliquées pour la maîtrise", Masson (1983).

[8] S. Clain and R. Touzani, Solution of a two-dimensional stationary induction heating problem without boundedness of the coefficients, Modél. Math. et Anal. Numér. 31 (1997), 845–870.

[9] M. Dauge, Elliptic Boundary Value Problems on Corner Domains, Lecture Notes in Mathematics 1341, Springer-Verlag (1988).

[10] T. Gallouët and R. Herbin, Existence of a solution to a coupled elliptic system, Applied Maths Letters 2 (1994), 49–55.

[11] V. Girault and P.-A. Raviart, Finite Element Methods for the Navier–Stokes Equations, Theory and Algorithms, Springer–Verlag (1986).

[12] P. Grisvard, Elliptic Problems in Nonsmooth Domains, Pitman (1985).

[13] R. Lewandowski, Analyse mathématique et océanographie, Collection "Recherches en Mathématiques Appliquées", Masson (1997).

[14] R. Lewandowski, The mathematical analysis of the coupling of a turbulent kinetic energy equation to the Navier-Stokes equation with an eddy viscosity, Nonlinear Analysis TMA 28 (1997), 393–417.

[15] J.-L. Lions, Quelques méthodes de résolution des problèmes aux limites non linéaires, Dunod & Gauthier–Villars (1969).

[16] J.-L. Lions and E. Magenes, Problèmes aux limites non homogènes et applications, Vol. 1, Dunod (1968).

[17] J.-L. Lions, R. Temam and S. Wang, Models for the coupled atmosphere and ocean, *Computational Mechanics Advances* 1 (1993), 1–120.

[18] P.-L. Lions and F. Murat, Solutions renormalisées d'équations elliptiques non linéaires, to appear.

[19] N.G. Meyers, An L^p-estimate for the gradient of solutions of second order elliptic divergence equations, Ann. Sc. Norm. Sup. Pisa 17 (1963), 189–206.

[20] F. Murat, Soluciones renormalizadas de EDP elipticas no lineales, Internal Report 93023, Laboratoire d'Analyse Numérique, Université Pierre et Marie Curie, Paris (1993).

[21] F. Murat, Équations elliptiques non linéaires avec second membre L^1 ou mesure, Actes du 26ème Congrès National d'Analyse Numérique, Les Karellis, France (1994), A12–A24.

[22] M. Orlt and A.-M. Sändig, Regularity of viscous Navier–Stokes flows in nonsmooth domains, Proc. Conf. Boundary Value Problems and Integral Equations in Nonsmooth Domains, M. Costabel, M. Dauge et S. Nicaise eds., Lecture Notes in Pure and Applied Mathematics 167, Dekker (1995), 185–201.

[23] G. Stampacchia, Équations elliptiques du second ordre à coefficients discontinus, Presses de l'Université de Montréal (1965).

[24] L. Tartar, Interpolation non linéaire et régularité, J. Functional Analysis 9 (1972), 469–489.

Christine Bernardi, François Murat
Laboratoire Jacques-Louis Lions
C.N.R.S. & Université Pierre et Marie Curie
Boîte postale 187
4 place Jussieu
75252 Paris Cedex 05
France
E-mail: bernardi@ann.jussieu.fr, murat@ann.jussieu.fr

Tomas Chacón Rebollo
Departamento de Ecuaciones Diferenciales y Análisis Numerico
Universidad de Sevilla
Tarfia s/n
41012 Sevilla
Spain
E-mail: chacon@numer.us.es

Roger Lewandowski
Equipe de Mécanique, IRMAR
Campus de Beaulieu
Université de Rennes 1
35042 Rennes Cedex 03
France
E-mail: lewandow@maths.univ-rennes1.fr

Research partially supported by Spanish Government, MAR97-1055-C02-02 Grant.

Studies in Mathematics and its Applications, Vol. 31
D. Cioranescu and J.L. Lions (Editors)

Chapter 6

DETERMINATION DE CONDITIONS AUX LIMITES EN MER OUVERTE PAR UNE METHODE DE CONTROLE OPTIMAL

F. Bosseur et P. Orenga

Résumé. Nous présentons dans ce travail le principe d'une méthode numérique adaptée à la détermination de conditions aux limites en mer ouverte d'un problème de shallow water. Cette méthode repose essentiellement sur l'utilisation de la théorie du contrôle optimal et de l'assimilation de données. On se propose de reconstituer des conditions aux limites à partir d'un ensemble de mesures disponibles à l'intérieur du domaine d'étude. Le contrôle est effectué sur les conditions aux limites de la vitesse aux frontières ouvertes, en considérant des observations ponctuelles. Si au niveau théorique, nous avons montré des résultats d'existence dans le cas linéaire, nous n'avons pu le faire dans le cas non linéaire. Nous donnons des résultats numériques dans le cas linéaire et non linéaire, tout d'abord sur une géométrie simplifiée pour justifier la méthode numérique, puis sur le cas réel d'une baie.

Abstract. We present here a numerical method adapted to the boundary conditions in open seas for a shallow water problem. Essentially, this method depends on the use of optimal control and data acquisition theory. We propose to reconstruct the boundary conditions from a set of available measurements taken in the interior of the region under study. The control is made on the velocity boundary condition on open boundaries, in considering observations at isolated points. If on the theoretical level we showed existence in the linear case, we could not achieve it in the nonlinear case. We present numerical results in the linear and nonlinear cases, first in a simplified geometry in order to justify the numerical method, then in the real case of a bay.

1. Introduction

Nous étudions dans ce travail un modèle de type *shallow water* qui, dans le cas d'une mer peu stratifiée ou peu profonde, s'avère suffisant pour représenter la dynamique des fluides et peut constituer une étape préalable à une étude plus approfondie des phénomènes par un modèle tridimensionnel. Plusieurs études théoriques et numériques ont été faites sur ce modèle (Orenga, 1991; Chatelon–Orenga, 1997; Bisgambiglia, 1989).

Pour notre part, nous nous proposons d'étudier le cas de domaines avec des frontières ouvertes où se pose souvent le problème de la connaissance des conditions aux limites sur la vitesse u. Celles-ci sont généralement évaluées grâce à des données expérimentales issues de campagnes de mesures, ou à des données calculées à l'aide d'un modèle de plus grande emprise.

Dans le cas de la baie de Calvi, des mesures *in situ* ont pu être effectuées (Norro, 1995) grâce à l'implantation voisine de la station de recherche océanographique STARESO. Toutefois, celles-ci sont en nombre insuffisant et ne nous donnent que peu de renseignements sur les conditions aux limites en mer ouverte à adjoindre au système. De même, les résultats issus des programmes de recherche tels que Medalpex qui donnent des renseignements sur la circulation générale en Méditerranée, ne peuvent prendre en compte les phénomènes à l'échelle de la baie.

C'est dans le but de lever ces indéterminations et d'utiliser au mieux les données disponibles, que nous nous proposons d'utiliser des méthodes d'assimi-lation de données, déjà utilisées avec succès dans de nombreux domaines, comme par exemple les prévisions météorologiques.

Le problème est résolu par la méthode de Galerkin. Afin d'obtenir des conditions aux limites homogènes, on effectue le changement de variable $v = u - w$, où u est la solution et w un relèvement de la condition aux limites. Le contrôle est alors effectué sur w, en utilisant des données ponctuelles issues de mesures effectuées *in situ* ou de résultats de modèles à plus grande échelle. Dans la section 2., nous rappelons brièvement les équations du modèle utilisé. Dans la section 3., nous donnons les équations vérifiées par le contrôle optimal dont nous démontrons, dans le cas linéaire, l'existence et l'unicité dans la section 4. Les sections 5. et 6. sont, quant à elles, consacrées à l'analyse numérique des équations. Nous présentons dans la section 7. un certain nombre de résultats obtenus sur une géométrie simplifiée, constituée par un carré de côté unitaire et qui nous a surtout permis de tester l'efficacité du code implémenté, puis sur le cas plus concret de la baie de Calvi (Corse), où l'on constate des phénomènes d'érosion des côtes dûs principalement à la modification des courants après l'implantation de structures en mer (extension du port, ...).

2. Equations du modèle

Dans un modèle océanique complet, sont prises en considération des variables biochimiques (concentrations chimiques, biomasses, ...) et hydrodynamiques (salinité, température, vitesse et masse volumique). Dans notre cas, nous nous sommes limités à l'étude des variables hydrodynamiques du système.

Pour décrire l'évolution de ces variables, nous utilisons des modèles de mécanique des fluides géophysiques qui se distinguent des problèmes classiques d'écoulement, tels que ceux de Navier-Stokes, par les dimensions des domaines, les échelles temporelles, la loi de conservation de la masse, la faible profondeur et surtout les conditions aux limites. A partir des équations générales de conservation et des différentes hypothèses simplificatrices liées aux propriétés des fluides géophysiques (Nihoul, 1977), on établit les équations du modèle tridimensionnel.

Les équations du modèle que nous utilisons sont quant à elles obtenues en intégrant les équations du modèle tridimensionnel non-stratifié sur la verticale.

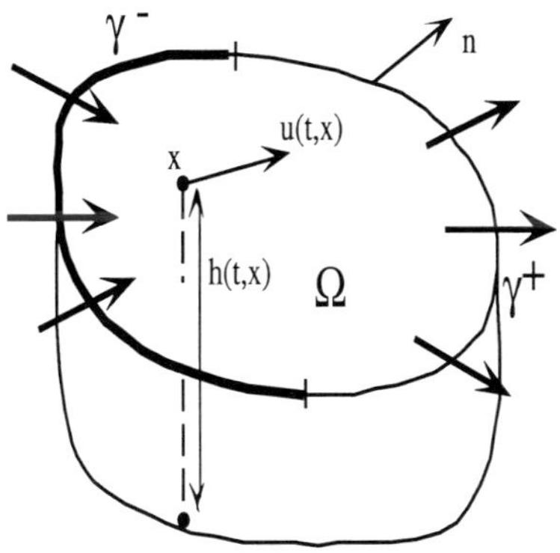

Fig. 2.1: *Domaine d'étude*

- γ_f représente les côtes ($u.n = 0$),
- γ_e les frontières ouvertes ($u.n = G(x, t)$).

De plus, pour les frontières ouvertes, on distingue les parties de la frontière où le fluide est entrant ($G(x,t) < 0$) que l'on notera γ^-, des parties où celui-ci est sortant ($G(x,t) > 0$), alors notées γ^+. Cette différenciation est justifiée par la nécessité de fixer la hauteur d'eau sur la partie de la frontière où le fluide est entrant (Chatelon–Orenga, 1997). On désigne par $u(x,t)$, la fonction de $\Omega\times]0,T[$ à valeur dans $\mathbb{R}^2$ représentant la vitesse à l'intérieur du fluide et $h(x,t)$, la fonction de $\Omega\times]0,T[$ à valeur dans $\mathbb{R}$ désignant la hauteur de la colonne d'eau. On note

- l'opérateur α, qui à $u = (u_1, u_2) \in \mathbb{R}^2$ fait correspondre $(-u_2, u_1) \in \mathbb{R}^2$,
- l'opérateur Rot, qui à une fonction scalaire $q(x,y)$ fait correspondre la fonction vectorielle $\text{Rot}\, q = \left(\frac{\partial q}{\partial y}, -\frac{\partial q}{\partial x}\right)$,
- l'opérateur rot, qui à une fonction vectorielle $u = (u_1, u_2)$ fait correspondre la fonction scalaire $\text{rot}\, u = \left(\frac{\partial u_2}{\partial x} - \frac{\partial u_1}{\partial y}\right)$,

et on introduit

$$
\begin{aligned}
Q &= \Omega\times]0,T[,\\
\Sigma &= \gamma\times]0,T[,\\
\Sigma^- &= \gamma^-\times]0,T[,\\
\Sigma^+ &= \gamma^+\times]0,T[.
\end{aligned}
$$

Les équations du problème de *shallow water* s'écrivent

$$
\begin{cases}
\frac{\partial u}{\partial t} - A\Delta u + \frac{1}{2}\nabla u^2 + \text{rot}\, u\alpha(u) & \\
\quad +Du + \omega\alpha(u) + g\nabla h = f, & \text{dans } Q,\\
\quad u.n = G(x,t), & \text{sur } \Sigma,\\
\quad \text{rot}\, u = 0, & \text{sur } \Sigma,\\
\quad u(t=0) = u_0(x), & \text{dans } \Omega,\\
 & \\
\frac{\partial h}{\partial t} + \text{div}\,(uh) = 0, & \text{dans } Q,\\
\quad h = \mu(x,t), & \text{sur } \Sigma^-,\\
\quad h(t=0) = h_0(x), & \text{dans } \Omega.
\end{cases}
\tag{2.1}
$$

Pour la résolution numérique des équations du modèle, et notamment pour pouvoir utiliser la base spéciale dans la méthode de Galerkin, nous effectuons un changement de variable de manière à nous ramener à des conditions aux limites homogènes.

Ainsi, on pose

$$u = v + w,$$

où

$$w = \nabla p, \tag{2.2}$$

avec p solution du problème

$$
\begin{cases}
-\Delta p = 0 \quad \text{dans } \Omega\\
\\
\dfrac{\partial p}{\partial n} = G \quad \text{sur } \gamma, \ \displaystyle\int_\gamma G\, d\ell = 0
\end{cases}
\tag{2.3}
$$

Les équations du modèle se réécrivent, en remplaçant u par $(v + w)$

$$
(\mathcal{P}) \left\{
\begin{array}{ll}
\frac{\partial v}{\partial t} - A\Delta v + \frac{1}{2}\nabla v^2 + \mathrm{grad}\,(v.w) + \mathrm{rot}\ v\alpha(v) & \\
\quad + \mathrm{rot}\ v\alpha(w) + Dv + \omega\alpha(v) + g\nabla h & \\
\quad\quad = f - \frac{\partial w}{\partial t} - \frac{1}{2}\nabla w^2 - Dw - \omega\alpha(w) & \text{dans } Q, \\
v.n = 0, & \text{sur } \Sigma, \\
\mathrm{rot}\ v = 0, & \text{sur } \Sigma, \\
v(t=0) = u_0(x) - w(t=0), & \text{dans } \Omega, \\
 & \\
\frac{\partial h}{\partial t} + \mathrm{div}\,(vh) = -\,\mathrm{div}\,(wh), & \text{dans } Q, \\
h = \mu(x,t), & \text{sur } \Sigma^-, \\
h(t=0) = h_0(x), & \text{dans } \Omega.
\end{array}
\right. \tag{2.4}
$$

Des résultats théoriques d'existence et de régularité de solutions de ce problème sont donnés dans Chatelon–Orenga (1997). Dans la suite, où les résultats théoriques sont démontrés dans le cas linéaire, nous considérons le problème $(\mathcal{P})$ linéarisé noté $(\mathcal{P})_\ell$

$$
(\mathcal{P})_\ell \left\{
\begin{array}{ll}
\frac{\partial v}{\partial t} - A\Delta v + Dv + \omega\alpha(v) + g\nabla h & \\
\quad\quad = f - \frac{\partial w}{\partial t} - Dw - \omega\alpha(w) & \text{dans } Q, \\
v.n = 0, & \text{sur } \Sigma, \\
\mathrm{rot}\ v = 0, & \text{sur } \Sigma, \\
v(t=0) = u_0(x) - w(t=0), & \text{dans } \Omega, \\
 & \\
\frac{\partial h}{\partial t} + \underline{h}\,\mathrm{div}\ v = 0, & \text{dans } Q, \\
h(t=0) = h_0(x), & \text{dans } \Omega.
\end{array}
\right. \tag{2.5}
$$

où $\underline{h}$ représente la hauteur moyenne sur le domaine, ne dépendant pas du temps.

3. Le modèle adjoint

Le principe de base du contrôle optimal est la minimisation d'une fonction coût J mesurant les écarts entre la solution calculée et un ensemble d'observations disponibles.

Ainsi, si l'on note X, l'espace des contrôles (généralement un espace de Hilbert), X_{ad}, un convexe fermé de X et $w \in X$, la variable de contrôle, le problème s'écrit

trouver $w_o \in X_{ad}$ *réalisant*

$$\inf_{w \in X_{ad}} J(w).$$

En pratique, la minimisation de la fonctionnelle $J(w)$ nécessite la connaissance du gradient de J par rapport aux variables de contrôle. Parmi les différentes méthodes de détermination de ce gradient, l'utilisation des équations adjointes du modèle linéaire tangent semble être la plus intéressante, notamment au niveau numérique (Lions, 1968; Talagrand–Courtier, 1987).

Puisque l'on considère ici des observations ponctuelles, il est essentiel d'exiger comme condition préalable que les fonctions recherchées soient au moins continues sur le domaine d'étude. Dans notre cas, où la dimension du domaine est égale à deux, le théorème de plongement de Sobolev montre que ces fonctions doivent être recherchées naturellement dans $H^2(\Omega)$. Compte tenu du type de relèvement considéré, l'espace des contrôles est donc donné par $L^2(0,T;W)$, où

$$W = \left\{ w \in H^2(\Omega)^2; w = \nabla p,\ \Delta p = 0,\ \frac{\partial p}{\partial n} = 0 \text{ sur } \gamma_f \right\}.$$

Considérons $x_1, \dots x_m$, des points de Ω, où m représente le nombre d'observa-tions disponibles. On définit la fonctionnelle $J(w)$ par

$$J(w) = \left\| C(v+w) - u_d \right\|_H^2 + \epsilon \left\| w \right\|_{L^2(0,T;W)}^2 , \tag{3.1}$$

où C est l'opérateur d'observabilité (défini ci-après) et $H = \left(L^2(\mathbb{R})\right)^m$ l'espace des observations.

Si les données du problème sont suffisamment régulières, il a été démontré dans Chatelon–Orenga (1997) que la vitesse v appartient au moins à $L^2\big(0,T;V\big)$, où

$$V = \left\{ \varphi \in H^2(\Omega)^2; \varphi.n = 0,\ \text{rot } \varphi = 0 \text{ sur } \gamma \right\}.$$

L'opérateur d'observabilité C est donc défini par

$$\begin{array}{rccc} C: & L^2\big(0,T;V \cup W\big) & \longrightarrow & \left(L^2(\mathbb{R})\right)^m \\ & v & \longmapsto & \begin{pmatrix} v(x_1) \\ \vdots \\ v(x_m) \end{pmatrix}. \end{array}$$

Notons enfin la vitesse désirée, $u_d \in H$, sous la forme

$$u_d = \begin{pmatrix} u_d(x_1) \\ \vdots \\ u_d(x_m) \end{pmatrix} = \begin{pmatrix} u_{d_1} \\ \vdots \\ u_{d_m} \end{pmatrix}.$$

Alors, (3.1) se réécrit sous la forme

$$J(w) = \sum_{j=1}^{m} \int_0^T \left|v(x^j, t; w) + w(x^j) - u_{d_j}(t)\right|^2 dt + \epsilon \left\|w\right\|^2_{L^2(0,T;W)}. \quad (3.2)$$

Expression du problème adjoint

Les équations adjointes sont obtenues formellement en multipliant les équa-tions du modèle linéaire tangent[1] par les variables adjointes, notées v^* et h^*, puis en intégrant sur $Q = \Omega \times]0, T[$.

On obtient alors le problème $(\mathcal{P}^*)$ suivant:

$$\left\{\begin{array}{ll} -\frac{\partial v^*}{\partial t} - A\Delta v^* - v.\operatorname{div} v^* - w \operatorname{div} v^* + \operatorname{rot}(v\alpha(v^*)) & \\ + \operatorname{rot}(w\alpha(v^*)) + v^*\alpha(\operatorname{rot} v) - h \operatorname{grad} h^* - \omega\alpha(v^*) + Dv^* & \\ \quad = \sum_{j=1}^m \big(v(x^j, t; w) + w(x_j) - u_{d_j}(t)\big) \otimes \delta(x - x^j) & \text{dans } Q, \\ v^*.n = 0, & \text{sur } \Sigma, \\ \operatorname{rot} v^* = 0, & \text{sur } \Sigma, \\ v^*(t = T) = 0, & \text{dans } \Omega. \\ & \\ -\frac{\partial h^*}{\partial t} - v.\operatorname{grad} h^* - w.\operatorname{grad} h^* - g \operatorname{div} v^* = 0, & \text{dans } Q, \\ h^* = 0, & \text{sur } \Sigma^+, \\ h^*(t = T) = 0, & \text{dans } \Omega. \end{array}\right. \quad (3.3)$$

où $\big(v(x^j, t; w) + w(x_j) - u_{d_j}(t)\big) \otimes \delta(x - x^j)$ est la distribution définie par

$$\psi \longrightarrow \int_0^T \big(v(x^j, t; w) + w(x_j) - u_{d_j}(t)\big)\psi(x^j, t)dt, \quad \psi \in \mathcal{D}(Q),$$

qui est une forme linéaire continue sur $L^2\Big(0, T; \big(H^2(\Omega)^2\big)'\Big)$.

Dans le cas où l'on considère le problème $(\mathcal{P})$ linéarisé, les équations adjointes s'écrivent

$$(\mathcal{P}^*)_\ell \left\{\begin{array}{ll} -\frac{\partial v^*}{\partial t} - A\Delta v^* - \underline{h} \operatorname{grad} h^* - \omega\alpha(v^*) + Dv^* = & \\ \sum_{j=1}^m \big(v(x^j, t; w) + w(x_j) - u_{d_j}(t)\big) \otimes \underset{.}{e}lta(x - x^j) & \text{dans } Q, \\ v^*.n = 0, & \text{sur } \Sigma, \\ \operatorname{rot} v^* = 0, & \text{sur } \Sigma, \\ v^*(t = T) = 0, & \text{dans } \Omega. \\ & \\ -\frac{\partial h^*}{\partial t} - g \operatorname{div} v^* = 0, & \text{dans } Q, \\ h^*(t = T) = 0, & \text{dans } \Omega. \end{array}\right. \quad (3.4)$$

[1]Obtenues en différenciant les équations (2.4) par rapport à la variable de contrôle.

Les propriétés de l'opérateur adjoint nous permettent alors d'établir l'équation vérifiée par le minimum local de la fonctionnelle J, qui s'écrit[2], $\forall\theta \in L^2(0,T;W)$

$$\Big\langle \frac{\partial v^*}{\partial t} + v\,\mathrm{div}\ v^* + w\,\mathrm{div}\ v^* + \mathrm{rot}\ v\alpha(v^*) - Dv^* + \omega\alpha(v^*)$$

$$- h^*\ \mathrm{grad}\ h + C^*(C(v+w) - u_d) + \epsilon\Lambda_W w, \theta \Big\rangle_{L^2(0,T;W'),L^2(0,T;W)} = 0 \tag{3.5}$$

dans le cas non linéaire, et

$$\Big\langle \frac{\partial v^*}{\partial t} - Dv^* + \omega\alpha(v^*)$$

$$+ C^*(C(v+w) - u_d) + \epsilon\Lambda_W w, \theta \Big\rangle_{L^2(0,T;W'),L^2(0,T;W)} = 0 \tag{3.6}$$

dans le cas linéaire.

4. Conditions d'existence et d'unicité du contrôle optimal dans le cas linéaire

Avec les notations introduites au chapitre précédent, on a,

Théorème 4.1 - *On suppose que $J(w)$ est donnée par (3.1) et que $G \in H^1\big(0,T;H^{\frac{3}{2}}(\gamma)\big)$. Le contrôle optimal $w \in L^2(0,T;W)$ est caractérisé de manière unique par les problèmes $(\mathcal{P})_\ell$, $(\mathcal{P}^*)_\ell$ et l'équation (3.6), avec*

$$v \in \Big(L^2\big(0,T;V\big) \cap H^1\big(0,T;L^2(\Omega)^2\big)\Big),$$
$$h \in L^\infty\big(0,T;H^1(\Omega)\big),$$
$$v^* \in L^2\big(0,T;L^2(\Omega)^2\big),$$
$$h^* \in L^2\big(0,T;L^2(\Omega)\big).$$

Démonstration. Les résultats d'existence et d'unicité du problème $(\mathcal{P})_\ell$ ont été démontrés dans Orenga (1991). Pour démontrer que le problème adjoint admet une solution unique, on opère par transposition (Lions–Magenes, 1972).

[2] Où Λ_W représente l'isomorphisme canonique de $L^2(0,T;W)$ sur $L^2(0,T;W')$.

Pour simplifier les notations, on considère, dans les équations (3.4), les constantes du problème égales à un. Le système se réécrit alors

$$-\frac{\partial v^*}{\partial t} - \Delta v^* - \nabla h^* = \sum_{j=1}^{m} \big(v(x^j, t; w) + w(x^j) - u_{d_j}(t)\big) \otimes \delta(x - x^j), \quad (4.1)$$

$$-\frac{\partial h^*}{\partial t} - \operatorname{div} v^* = 0, \quad (4.2)$$

$$v^*.n = 0, \ \operatorname{rot} v^* = 0 \text{ sur } \gamma, \quad (4.3)$$

$$v^*(t = T) = 0, \ h^*(t = T) = 0. \quad (4.4)$$

Introduisons l'espace

$$\Xi = \big\{\psi \in L^2\big(0, T; H^2(\Omega)^2\big) \cap H^1\big(0, T; L^2(\Omega)^2\big); \\ \psi.n = 0, \ \operatorname{rot} \psi = 0 \text{ sur } \gamma, \ \psi(0) = 0\big\},$$

et le problème

$$\frac{\partial \psi}{\partial t} - \Delta\psi + \nabla\varphi = F, \quad (4.5)$$

$$\frac{\partial \varphi}{\partial t} + \operatorname{div} \psi = 0, \quad (4.6)$$

$$\psi.n = 0, \operatorname{rot} \psi = 0 \text{ sur } \gamma, \quad (4.7)$$

$$\psi(0) = 0, \ \varphi(0) = 0. \quad (4.8)$$

On a le résultat suivant:

Si $F \in L^2\big(0, T; L^2(\Omega)^2\big)$ alors le problème (4.5)–(4.8) admet une solution unique dans Ξ.

En effet, l'application T définie par

$$\psi \longmapsto \frac{\partial \psi}{\partial t} - \Delta\psi + \nabla\varphi,$$

où φ vérifie:

$$\frac{\partial \varphi}{\partial t} + \operatorname{div} \psi = 0,$$

est un isomorphisme de Ξ dans $L^2\big(0, T; L^2(\Omega)^2\big)$.

Par transposition, on en déduit qu'il existe $v^* \in L^2\big(0, T; L^2(\Omega)^2\big)$ et $h^* \in L^2\big(0, T; L^2(\Omega)\big)$ uniques, vérifiant

$$\int_Q v^* \left(\frac{\partial \psi}{\partial t} - \Delta\psi + \nabla\varphi\right) dQ = M(\psi), \ \frac{\partial h^*}{\partial t} + \operatorname{div} v^* = 0,$$

où:

$$M(\psi) = \sum_{j=1}^{m} \int_0^T \big(v(x^j, t; w) + w(x^j) - u_{d_j}(t)\big)\psi(x^j, t)\, dt,$$

définit une forme linéaire continue sur Ξ' et où Ξ' est le dual de Ξ.

Si $\epsilon > 0$, il y a existence et unicité du contrôle optimal w (Lions, 1969), caractérisé par

$$\left\langle \frac{\partial v^*}{\partial t} - Dv^* + \omega\alpha(v^*) + C^*\big(C(v+w) - u_d\big) + \epsilon\Lambda_W w, \theta \right\rangle_{L^2(0,T;W'),L^2(0,T;W)} = 0,$$

for all $\theta \in L^2(0,T;W)$. ∎

5. Principe de la méthode de résolution

5.1. Résolution des équations du modèle et des équations adjointes

La méthode utilisée pour la résolution des équations du modèle et des équations adjointes est basée sur l'utilisation de la méthode de Galerkin. On transforme, par troncature d'une base de l'espace considéré, le système d'équations aux dérivées partielles par un système d'équations différentielles ordinaires dont les inconnues sont les projections de la solution du problème approché sur la base.

Dans le cadre de ce travail, nous avons utilisé la base spéciale dont les propriétés ont été mises en évidence dans Orenga (1992). En particulier, on a le résultat suivant:

Théorème - *Soient*

$$V = \left\{u \in L^2(\Omega)^2, \text{div } u \in L^2(\Omega), \text{rot } u \in L^2(\Omega); u.n = 0 \text{ sur } \gamma\right\},$$

$$H_0(\text{div } 0, \text{rot } 0) = \left\{u \in L^2(\Omega)^2, \text{div } u = 0, \text{rot } u = 0; u.n = 0 \text{ sur } \gamma\right\},$$

$|.|, (.)$ *la norme et le produit scalaire dans* $L^2(\Omega)$ *ou* $L^2(\Omega)^2$.
On considère les problèmes

$$(\mathcal{P}_1)\begin{cases} -\Delta u = \lambda u & \text{dans } \Omega \\ u.n = 0 & \text{sur } \gamma \\ \text{rot } u = 0 & \text{sur } \gamma \end{cases}$$

$$(\mathcal{P}_2)\begin{cases} -\Delta p = \lambda p & \text{dans } \Omega \\ \text{grad } p.n = 0 & \text{sur } \gamma \end{cases} \qquad (\mathcal{P}_3)\begin{cases} -\Delta q = \mu q & \text{dans } \Omega \\ q = 0 & \text{sur } \gamma \end{cases}$$

et les p problèmes

$$(\mathcal{P}_{4_i})\begin{cases}-\Delta r=0 & \text{dans } \Omega\\ r=1 & \text{sur } \gamma_i\\ r=0 & \text{sur } \gamma_j\end{cases}$$

avec $j \neq i$ avec $i = 1, \ldots, p$ et $j = 1, \ldots, p$.

On a alors

- *Si* (λ, p) *est solution de* $(\mathcal{P}_2)$*, alors* $(\lambda, \text{grad } p)$ *est solution de* $(\mathcal{P}_1)$.
- *Si* (μ, q) *est solution de* $(\mathcal{P}_3)$*, alors* $(\mu, \text{Rot } q)$ *est solution de* $(\mathcal{P}_1)$. *Cette propriété montre en particulier l'existence de solutions de* $(\mathcal{P}_1)$ *à divergence nulle.*
- *Si* r_i *est solution de* $(\mathcal{P}_{4_i})$*, alors* $(0, \text{Rot } r_i)$ *est solution de* $(\mathcal{P}_1)$.
- *Si* Ω *est simplement connexe, alors* 0 *n'est pas valeur propre de* $(\mathcal{P}_1)$*, sinon l'espace propre associé à la valeur propre* 0 *est l'espace*

  $$H_0(\text{div } 0, \text{rot } 0),$$

  de dimension p, engendré par les p solutions des problèmes $(\mathcal{P}_{4_i})$.
- *Soit* $\{p_i, i \in \mathbb{N}\}$*, un ensemble de solutions de* $(\mathcal{P}_2)$ *formant une base orthogonale de* $L^2(\Omega)$*, soit* $\{q_j, j \in \mathbb{N}\}$*, un ensemble de solutions de* $(\mathcal{P}_3)$ *formant une base orthogonale de* $L^2(\Omega)$ *et soit* $\{r_1, \ldots, r_p\}$ *les p solutions indépendantes des problèmes* $(\mathcal{P}_{4_i})$*. Alors l'ensemble des* grad p_i, Rot q_j *et* Rot r_k *forme une base orthogonale de* $L^2(\Omega)^2$ *et de* V.

5.2. Approximation du gradient de la fonctionnelle

Nous considérons ici le cas d'observations ponctuelles, où les observations sont connues en tout temps mais en un nombre restreint de points du domaine.

Rappelons que l'espace des contrôles est donné par $L^2(0, T; W)$, la vitesse désirée est telle que:

$$u_d \in \left(L^2(\mathbb{R})\right)^m,$$

l'opérateur d'observabilité est noté par:

$$C : L^2(0, T; V \cup W) \longrightarrow \left(L^2(\mathbb{R})\right)^m.$$

On introduit l'espace:

$$\mathcal{H} = \left\{ w \in L^2(\Omega)^2; w = \nabla p,\ \Delta p = 0,\ \frac{\partial p}{\partial n} = 0 \text{ sur } \gamma_f \right\}.$$

Soit $\mathcal{H}'$, le dual de $\mathcal{H}$. On identifie $\mathcal{H}$ et $\mathcal{H}'$, et on a

$$W \hookrightarrow \mathcal{H} = \mathcal{H}' \hookrightarrow W'.$$

On va déterminer le minimum de $J(w)$ par une méthode itérative. On écrit l'équation vérifiée par le minimum de la façon suivante:

$$\begin{aligned}\epsilon\Big(w,\theta\Big)_{L^2(0,T;W)} + \Big(Cw, C\theta\Big)_{\left(L^2(\mathbb{R})\right)^m} &= -\Big(Cv - u_d, C\theta\Big)_{\left(L^2(\mathbb{R})\right)^m} \\ &\quad - \left\langle \frac{\partial v^*}{\partial t} - Dv^* + \omega\alpha(v^*), \theta \right\rangle_{L^2(0,T;W'),L^2(0,T;W)} \end{aligned} \tag{5.1}$$

La méthode consiste à considérer le second membre de l'équation (5.1) fonction de w à l'itération précédente. Ainsi on va calculer w^n à la $n^{\text{ième}}$ itération par

$$\begin{aligned}\epsilon\Big(w^n,\theta\Big)_{L^2(0,T;W)} + \Big(Cw^n, C\theta\Big)_{\left(L^2(\mathbb{R})\right)^m} &= -\Big(Cv^{n-1} - u_d, C\theta\Big)_{\left(L^2(\mathbb{R})\right)^m} \\ &\quad - \left\langle \frac{\partial}{\partial t}v^{*n-1} - Dv^{*n-1} + \omega\alpha(v^{*n-1}), \theta \right\rangle_{L^2(0,T;W'),L^2(0,T;W)} .\end{aligned}$$

Dans le cas non linéaire, on résout de la même manière

$$\begin{aligned}\epsilon\Big(w,\theta\Big)_{L^2(0,T;W)} &+ \Big\langle w \operatorname{div} v^*, \theta\Big\rangle_{L^2(0,T;W'),L^2(0,T;W)} + \Big(Cw, C\theta\Big)_{\left(L^2(\mathbb{R})\right)^m} \\ &- \Big(Cv - u_d, C\theta\Big)_{\left(L^2(\mathbb{R})\right)^m} - \left\langle \frac{\partial v^*}{\partial t} + v \operatorname{div} v^* + \operatorname{rot} v\alpha(v^*) \right. \\ &\left. - Dv^* + \omega\alpha(v^*) - h^* \operatorname{grad} h, \theta \right\rangle_{L^2(0,T;W'),L^2(0,T;W)} .\end{aligned} \tag{5.2}$$

Or, la résolution de (5.1) ou de (5.2) nécessite l'introduction de l'opérateur C^*, opérateur adjoint de l'opérateur C. En effet, on doit écrire

$$\Big(Cw, C\theta\Big)_{\left(L^2(\mathbb{R})\right)^m} = \Big\langle C^*Cw, \theta\Big\rangle_{L^2(0,T;W'),L^2(0,T;W)}$$

et

$$\Big(Cv - u_d, C\theta\Big)_{\left(L^2(\mathbb{R})\right)^m} = \Big\langle C^*(Cv - u_d), \theta\Big\rangle_{L^2(0,T;W'),L^2(0,T;W)}$$

La difficulté d'obtenir une caractérisation numérique de l'opérateur C^*, nous a conduit à nous orienter vers une méthode de résolution plus appropriée.

Nous proposons donc de résoudre les équations (5.1) ou (5.2) par une méthode d'approximation variationnelle, en utilisant une base de fonctions de l'espace W.

Choix de la base de W

Nous devons tout d'abord définir un produit scalaire sur W; pour cela, on utilise l'équivalence des normes suivantes (Dautray–Lions, 1988):

$$\|w\|_{H^{k+1}(\Omega)^2} \quad \text{et} \quad \|\|w\|\|$$

où

$$\|\|w\|\| = \left(\|w\|^2_{L^2(\Omega)^2} + \|\text{div } w\|^2_{H^k(\Omega)^2} + \|\text{rot } w\|^2_{H^k(\Omega)^2} + \|w.n\|^2_{H^{k+\frac{1}{2}}(\gamma)}\right)^{\frac{1}{2}}.$$

Ce résultat nous permet de munir W du produit scalaire

$$\left(\left(w, \varphi\right)\right)_W = \left(w, \varphi\right)_{L^2(\Omega)^2} + \left(w.n, \varphi.n\right)_{H^{\frac{3}{2}}(\gamma)} \quad , \tag{5.3}$$

puisque, d'après les propriétés de l'espace W,

$$\text{div } w = \text{rot } w = 0.$$

En outre, si $w = \nabla p$ est une fonction de W, alors p vérifie le problème

$$\begin{cases} \Delta p = 0 & \text{dans } \Omega \\ p = \mu & \text{sur } \gamma_e \\ \dfrac{\partial p}{\partial n} = 0 & \text{sur } \gamma_f \end{cases} \tag{5.4}$$

où $\mu \in H^{\frac{5}{2}}(\gamma_e)$.

Considérons à présent les solutions ξ_i des problèmes de Neumann suivants:

$$\begin{cases} \Delta \xi_i = 0 & \text{dans } \Omega \\ \xi_i = \psi_i & \text{sur } \gamma_e \\ \dfrac{\partial \xi_i}{\partial n} = 0 & \text{sur } \gamma_f \end{cases} \tag{5.5}$$

dans lesquels $\left\{\psi_i\right\}_{i=1}^{\infty}$ est une base de $H^{\frac{5}{2}}(\gamma_e)$ et vérifie:

$$\int_\gamma \psi_i = 0, \qquad \forall i \in \mathbb{N}.$$

On définit alors l'application

$$\mathcal{R} : H^{\frac{5}{2}}(\gamma) \longrightarrow H^3(\Omega)$$
$$\psi_i \longmapsto \xi_i$$

qui est linéaire et continue (Girault–Raviart, 1979).

Avec ces notations, on a le

Lemme 5.1 - $\{\nabla\xi_i\}_{i=1}^{\infty}$ *forme une base de* W.

Démonstration. Soit

$$\zeta \in \left\{p \in H^3(\Omega); \Delta p = 0, \ \frac{\partial p}{\partial n} = 0 \text{ sur } \gamma_f\right\}.$$

Il existe $\mu \in H^{\frac{5}{2}}(\gamma_e)$, tel que

$$\zeta = \mu \text{ sur } \gamma_e.$$

$\{\psi_i\}_{i=1}^{\infty}$ étant une base de $H^{\frac{5}{2}}(\gamma_e)$, il existe μ_n,vérifiant

$$\mu_n = \sum_{i=1}^{n} a_i\psi_i \xrightarrow{H^{\frac{5}{2}}(\gamma)} \mu,$$

et d'après la continuité de l'application $\mathcal{R}$:

$$\mathcal{R}\mu_n = \zeta_n = \sum_{i=1}^{n} a_i\xi_i \xrightarrow{H^3(\Omega)} \Theta, \qquad \text{où } \xi_i = \mathcal{R}\psi_i$$

et Θ vérifie

$$\begin{cases} \Theta = \mu & \text{sur } \gamma_e \\ \dfrac{\partial\Theta}{\partial n} = 0 & \text{sur } \gamma_f \end{cases}$$

De plus

$$0 = \Delta\zeta_n \longrightarrow \Delta\Theta = 0.$$

D'après le résultat d'unicité de la solution du problème (5.5), on en déduit que $\Theta = \zeta$ et donc que $\{\xi_i\}_{i=1}^{\infty}$ est une base de

$$\left\{p \in H^3(\Omega); \Delta p = 0, \ \frac{\partial p}{\partial n} = 0 \text{ sur } \gamma_f\right\}.$$

En conséquence, $\{\nabla\xi_i\}_{i=1}^{\infty}$ est une base de W. ■

Démontrons alors le

Lemme 5.2 - $\{\nabla\xi_i\}_{i=1}^{\infty}$ *forme une base de* $\mathcal{H}$.

Démonstration. Soit $f = \nabla\eta \in \mathcal{H}$, telle que

$$\left(f, \nabla\xi_i\right)_{\mathcal{H}} = 0 \qquad \forall i \in \mathbb{N}$$

On a

$$\left(f, \nabla\xi_i\right)_{\mathcal{H}} = \int_\Omega \nabla\eta.\nabla\xi_i$$

Or

$$\int_\Omega \nabla\eta.\nabla\xi_i = -\int_\Omega \Delta\eta\,\xi_i + \int_\gamma \frac{\partial\eta}{\partial n}\,\xi_i = \int_{\gamma_e} \frac{\partial\eta}{\partial n}\,\psi_i$$

soit

$$\left(f, \nabla\xi_i\right)_{\mathcal{H}} = \int_{\gamma_e} \frac{\partial\eta}{\partial n}\,\psi_i = 0 \qquad \forall i \in \mathbb{N}$$

$\left\{\psi_i\right\}_{i=1}^{\infty}$ étant une base de $H^{\frac{5}{2}}(\gamma_e)$ donc de $L^2(\gamma)$, on en déduit que $\frac{\partial\eta}{\partial n} = 0$ sur γ_e. Finalement, η vérifie $\Delta\eta = 0$ et $\frac{\partial\eta}{\partial n} = 0$ sur γ, d'où on déduit que $\eta = 0$ et $f = 0$, ce qui démontre que $\left\{\nabla\xi_i\right\}_{i=1}^{\infty}$ est une base de $\mathcal{H}$. ∎

On va maintenant approcher w par $w_r \in L^2(0,T;W_r)$, où W_r est le sous-espace engendré par les r fonctions $\left\{\theta_i = \nabla\xi_i\right\}_{i=1}^{r}$; on raisonne dans le cas linéaire, le cas non linéaire ne posant pas plus de difficultés. On note

$$f(v^*, h^*) = \frac{\partial v^*}{\partial t} - Dv^* + \omega\alpha(v^*), \tag{5.6}$$

$$w_r = \sum_{k=1}^{r} \mu_k\,\theta_k,$$

et avec $i = 1, \ldots, r$, l'équation (5.1) s'écrit

$$\sum_{k=1}^{r} \mu_k \left\{\epsilon\left(\theta_k, \theta_i\right)_{L^2(0,T;W)} + \sum_{j=1}^{m} \theta_k(x_j).\theta_i(x_j)\right\} =$$
$$-\sum_{j=1}^{m}\left(v - u_d\right)(x_j).\theta_i(x_j) - \left(f(v^*, h^*), \theta_i\right)_{L^2(Q)^2}\,. \tag{5.7}$$

6. Mise en œuvre numérique

6.1. Notations

On note $\mathcal{M}$ le maillage du domaine, M_i, $i = 1 \ldots N_1$, les points de l'intérieur de Ω de coordonnées (X_i, Y_i) et F_j, $j = 1 \ldots N_2$, les points frontière, de coordonnées (x_j, y_j).

$$\mathcal{M} = \left\{M_i\right\}_{i=1}^{N_1} \cup \left\{F_j\right\}_{j=1}^{N_2}.$$

6.2. Calcul des vecteurs de la base propre et de la base de W

Nous avons vu à la section 5.1 que les éléments de la base propre sont obtenus par résolution des problèmes $(\mathcal{P}_2)$, $(\mathcal{P}_3)$, $(\mathcal{P}_{4_i})$. Dans le cas d'un carré de côté unitaire, les solutions de ces problèmes sont connues de manière analytique. Dans le cas d'un domaine quelconque, les solutions sont calculées à l'aide du logiciel Modulef, par la méthode des éléments finis. Les éléments utilisés sont de type hermite à trois degrés de liberté, ce qui nous permet d'avoir accès, en tous les points du maillage, à la valeur numérique de la fonction solution ainsi que de ses dérivées premières en x et en y (Vidrascu–Georges, 1990).

Pour construire numériquement la base de W_r, on résout les r problèmes scalaires suivants:

$$\begin{cases} \Delta\xi_\alpha = 0 \\ \xi_\alpha(x_j, y_j) = \begin{cases} 1 & \text{si } j = \alpha, \text{ où } (x_j, y_j) \in \gamma_e \\ 0 & \text{si } j \neq \alpha, \end{cases} \\ \frac{\partial \xi_\alpha}{\partial n}(x_j, y_j) = 0 \quad \text{si } (x_j, y_j) \in \gamma_f \end{cases}$$

La résolution numérique de ces problèmes est, comme précédemment, effectuée par la méthode des éléments finis à l'aide du logiciel Modulef, en utilisant les mêmes éléments.

6.3. Résolution des équations du modèle et des équations adjointes

Pour résoudre les problèmes (2.4) et (3.4), nous utilisons la méthode de Galerkin, associéeà la base propre définie à la section 5.1.

On transforme ainsi les systèmes d'équations aux dérivées partielles en des systèmes d'équations différentielles ordinaires. Ces systèmes sont alors résolus par la méthode d'Adams implicite, initialisée par la méthode d'Euler implicite.

6.4. Méthode de calcul du minimum

La principale difficulté pour le calcul du minimum provient de la présence du terme $(w, \theta)_{L^2(0,T;W)}$ dans les équations (5.1) et (5.2). Or, nous avons vu en (5.3) que l'on pouvait munir W du produit scalaire

$$(\theta_k, \theta_l)_W = (\theta_k, \theta_l)_{L^2(\Omega)^2} + (\theta_k.n, \theta_l.n)_{H^{\frac{3}{2}}(\gamma)}.$$

Pour calculer la valeur du produit scalaire de $H^{\frac{3}{2}}(\gamma)$, on utilise alors le résultat suivant (Dautray–Lions, 1988):

Soit f, une fonction périodique définie sur $]0,a[$*; on définit la norme de l'espace de Sobolev* $H_p^s(0,a)$

$$\|f\|_s = \Big(\sum_{j\in\mathbb{Z}} (1+j^2)^s |C_j igr|^2\Big)^{\frac{1}{2}}, \tag{6.1}$$

où

$$C_j = \int_0^a f(x).e^{-i\,\frac{2\pi}{a}\,jx}dx.$$

■

On associe alors à (6.1) le produit scalaire suivant:

$$\big(f,g\big)_s = \Big(\sum_{j\in\mathbb{Z}} (1+j^2)^s \, C_j(f).\overline{C_j}(g)\Big), \tag{6.2}$$

où f et g sont deux fonctions périodiques définies sur $]0,a[$.

Dans notre cas, où la frontière γ_e est constituée par la réunion de q segments de longueur a_i, les fonctions $(\theta_k.n)$ sont définies sur $]0,a_i[$ et sont nulles au bord. On a

$$\begin{aligned} C_j(\theta_k.n) &= \sum_{i=1}^{q} \int_0^{a_i} (\theta_k.n)(x).e^{-i\,\frac{2\pi}{a_i}\,jx}dx \\ &= -\sum_{i=1}^{q} i \int_0^{a_i} (\theta_k.n)(x).\sin\Big(\frac{2\pi}{a_i}\,jx\Big)dx, \end{aligned}$$

et

$$\begin{aligned} \overline{C_j}(\theta_l.n) &= \sum_{i=1}^{q} \int_0^{a_i} (\theta_l.n)(x).e^{i\,\frac{2\pi}{a_i}\,jx}dx \\ &= \sum_{i=1}^{q} i \int_0^{a_i} (\theta_l.n)(x).\sin\Big(\frac{2\pi}{a_i}\,jx\Big)dx. \end{aligned}$$

Au total, le produit scalaire s'exprime donc sous la forme

$$\begin{aligned} \Big(\theta_k.n,\theta_l.n\Big)_{H^{\frac{3}{2}}(\gamma)} = \sum_{j\in\mathbb{Z}} \sum_{i=1}^{q} \Big(1+j^2\Big)^{\frac{3}{2}} \Bigg[\Big(\int_0^{a_i} (\theta_k.n)(x).\sin\Big(\frac{2\pi}{a_i}\,jx\Big)dx\Big) \\ .\Big(\int_0^{a_i} (\theta_l.n)(x).\sin\Big(\frac{2\pi}{a_i}\,jx\Big)dx\Big)\Bigg]. \end{aligned}$$

L'équation (5.7) se présente alors comme un système algébrique de r équations, où les r inconnues sont les μ_k et m représente le nombre de points d'observation, $\{x_j,\ j=1,\ldots,m\}$.

Ce système matriciel, du type

$$AX = B,$$

d'inconnues

$$X = \begin{pmatrix} \mu_1 \\ \mu_2 \\ \vdots \\ \mu_r \end{pmatrix},$$

est alors résolu par la méthode du pivot de Gauss.

Une fois connues les valeurs des coefficients μ_k $(k = 1, \ldots, r)$, on reconstruit la solution approchée

$$w_r = \sum_{k=1}^{r} \mu_k \, \theta_k,$$

qui correspond à la valeur approchée de w pour l'itération suivante de l'algorithme de contrôle.

En effet, on doit donc, à ce niveau, calculer les nouvelles valeurs de v, h, v^*, h^* correspondant à cette nouvelle valeur de w_r et ainsi de suite jusqu'à convergence de la méthode (Fig. 6.1).

6.5. Convergence de l'algorithme de calcul du minimum

Au chapitre 3., nous avons donné l'expression de la différentielle de la fonctionnelle, qui s'écrit dans le cas linéaire

$$\frac{1}{2}\langle J'(w), \theta\rangle = \langle f(v^*, h^*) + C^*(Cv - u_d) + C^*Cw \\ + \epsilon\Lambda_W w, \theta\rangle_{L^2(0,T;W'),L^2(0,T;W)} \quad (6.3)$$

où Λ_W est l'isomorphisme canonique de $L^2(0, T; W)$ sur $L^2(0, T; W')$ et $f(v^*, h^*)$ est définie comme en (5.6).

Pour démontrer, dans le cas linéaire la convergence de l'algorithme de calcul du minimum, on raisonne par récurrence[3].

- On se fixe w^0 quelconque et on calcule (v^0, h^0) et (v^{*0}, h^{*0}) en résolvant les problèmes $(\mathcal{P})_\ell$ et $(\mathcal{P}^*)_\ell$.

[3]Nous donnons un résultat de convergence uniquement dans le cas linéaire, car dans le cas non linéaire, nous n'avons pas de résultats concernant la convexité de la fonctionnelle.

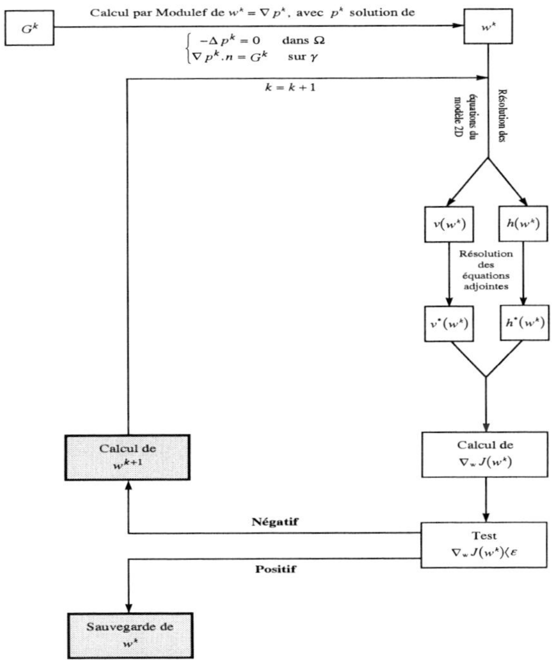

Fig. 6.1: *Schéma général de la résolution*

1) Si w^0 vérifie

$$\frac{1}{2}\langle J'(w^0), \theta\rangle = 0, \qquad \forall \theta \in L^2(0,T;W), \tag{6.4}$$

alors, d'après le théorème 4.1, w^0 est le contrôle optimal.

2) Si w^0 ne vérifie pas (6.4), on détermine w^1 par

$$(C^*C + \epsilon\Lambda_W)w^1 = -\Big(f\big(v^{*0}, h^{*0}\big) + C^*\big(Cv^0 - u_d\big)\Big).$$

L'opérateur

$$\Upsilon = \big(C^*C + \epsilon\Lambda_W\big) : W \longrightarrow W',$$

étant continu et elliptique, on peut écrire

$$w^1 = -\Upsilon^{-1}\Big(f\big(v^{*0}, h^{*0}\big) + C^*\big(Cv^0 - u_d\big)\Big).$$

Utilisons (6.3) pour donner une expression de $\langle J'(w^0), w^1 - w^0\rangle$; il vient

$\frac{1}{2}\langle J'(w^0), w^1 - w^0\rangle$
$= \langle f\big(v^{*0}, h^{*0}\big) + C^*\big(Cv^0 - u_d\big) + \Upsilon w^0, w^1 - w^0\rangle_{L^2(0,T;W'),L^2(0,T;W)}$
$= \Big\langle f\big(v^{*0}, h^{*0}\big) + C^*\big(Cv^0 - u_d\big) + \Upsilon w^0,$
$\quad -\Upsilon^{-1}\Big(f\big(v^{*0}, h^{*0}\big) + C^*\big(Cv^0 - u_d\big)\Big) - w^0\Big\rangle_{L^2(0,T;W'),L^2(0,T;W)}$
$= \Big\langle f\big(v^{*0}, h^{*0}\big) + C^*\big(Cv^0 - u_d\big) + \Upsilon w^0,$
$\quad -\Upsilon^{-1}\Big(f\big(v^{*0}, h^{*0}\big) + C^*\big(Cv^0 - u_d\big) + \Upsilon w^0\Big)\Big\rangle_{L^2(0,T;W'),L^2(0,T;W)}$
$\le -K\Big\|f\big(v^{*0}, h^{*0}\big) + C^*\big(Cv^0 - u_d\big) + \Upsilon w^0\Big\|^2_{L^2(0,T;W')}$
(où K est une constante positive)
$< 0.$

Dans le cas linéaire, la fonctionnelle étant convexe, on en déduit que

$$J(w^1) < J(w^0),$$

et on calcule (v^1, h^1) et (v^{*1}, h^{*1}) en résolvant les problèmes $(\mathcal{P})_\ell$ et $(\mathcal{P}^*)_\ell$.

- 1) Si w^{n-1} vérifie (6.4), alors d'après le théorème 4.1 w^{n-1} est le contrôle optimal.

  2) Si w^{n-1} ne vérifie pas (6.4), on détermine w^n par:

$$w^n = -\Upsilon^{-1}\Big(f\big(v^{*n-1}, h^{*n-1}\big) + C^*\big(Cv^{n-1} - u_d\big)\Big).$$

Utilisons comme précédemment l'équation (6.3) pour donner une expression de $\langle J'(w^{n-1}), w^n - w^{n-1}\rangle$; il vient

$$\frac{1}{2}\langle J'(w^{n-1}), w^n - w^{n-1}\rangle$$
$$= bigl < f\left(v^{*n-1}, h^{*n-1}\right) + C^*\left(Cv^{n-1} - u_d\right) + \Upsilon w^{n-1},$$
$$w^n - w^{n-1}\rangle_{L^2(0,T;W'),L^2(0,T;W)}$$
$$= \langle f\left(v^{*n-1}, h^{*n-1}\right) + C^*\left(Cv^{n-1} - u_d\right) + \Upsilon w^{n-1},$$
$$-\Upsilon^{-1}\Big(f\left(v^{*n-1}, h^{*n-1}\right) + C^*\left(Cv^{n-1} - u_d\right) + \Upsilon w^{n-1}\Big)\rangle_{L^2(0,T;W'),L^2(0,T;W)}$$
$$\leq -K\left\|f\left(v^{*n-1}, h^{*n-1}\right) + C^*\left(Cv^{n-1} - u_d\right) + \Upsilon w^{n-1}\right\|^2_{L^2(0,T;W')}$$

(où K est une constante positive)
< 0.

Pour la même raison que précédemment, on en déduit que

$$J(w^n) < J(w^{n-1}),$$

et on calcule (v^n, h^n) et (v^{*n}, h^{*n}) en résolvant les problèmes $(\mathcal{P})_\ell$ et $(\mathcal{P}^*)_\ell$.

On obtient ainsi une suite réelle $\{J(w^n)\}_n$, positive et décroissante; elle converge donc vers $x \in \mathbb{R}$, où x vérifie

$$x = \inf_{w^n} J(w^n).$$

Mais J étant convexe, il existe w unique qui réalise le minimum, soit

$$x = J(w).$$

On en déduit que

$$w^n \longrightarrow w, \qquad \text{faiblement dans } L^2(0,T;W).$$

7. Résultats numériques

7.1. Cas analytique

Pour valider la méthode numérique implémentée, nous avons effectué des tests sur une géométrie simplifiée, constituée par un carré de côté unitaire. Dans ce cas, le champ de vitesse désirée n'est pas constitué par un ensemble de mesures, mais est issu de la résolution des équations du problème de *shallow water* en considérant un relèvement donné, w_d. Ceci nous a permis de comparer non seulement le champ de vitesse calculé à la solution choisie comme référence (u_d), mais également le relèvement calculé par la méthode de contrôle au champ w_d.

Le sous domaine à l'intérieur duquel la vitesse u_d est supposée connue a été défini de deux manières différentes:

- En considérant des points dispersés de manière régulière à l'intérieur du domaine mais avec un pas d'espace beaucoup plus grand, ce qui correspond au cas où les données sont issues de résultats de modèles de plus grande emprise.

- En considérant des points à l'intérieur d'un rectangle inclus dans le domaine et qui correspond physiquement à un ensemble de mesures effectuées dans une zone particulière.

Dans le premier cas, nous avons pu étudier l'influence du nombre de données disponibles sur la "qualité" de la solution obtenue. Ceci montre qu'un nombre minimum de points de mesure est nécessaire pour reconstituer un champ de vitesse en accord avec la réalité physique.

Le deuxième cas nous a permis d'étudier l'influence de la localisation des points de mesure à l'intérieur du domaine. En particulier, il nous a ainsi été possible de vérifier que les meilleurs résultats sont obtenus quand les observations sont situées au centre du domaine et réparties sur un axe orthogonal au courant principal.

Les tests numériques effectués sur le carré ne sont pas présentés ici où nous donnons uniquement quelques résultats significatifs obtenus sur un cas réel.

7.2. Cas réel

Une deuxième série de tests a ensuite été effectuée sur une géométrie plus complexe, représentant la baie de Calvi. Le choix de ce site a été motivé de par la proximité de la station de recherches océanographiques STARESO, ce qui a permis la réalisation de nombreuses campagnes de mesures; en parallèle, des simulations numériques ont également été effectuées par Norro (1995). Nous avons donc utilisé les résultats de ces simulations pour reconstruire les conditions aux limites sur les frontières ouvertes.

Dans ce cas, où le champ de vitesse est connu dans tout le domaine, le sous domaine dans lequel la vitesse désirée est supposée connue a été défini de la même manière que dans le cas du carré. En outre, la méthode utilisée étant de type itératif, nous avons dû nous fixer une valeur de démarrage; pour les tests que nous présentons ici, nous sommes partis de conditions aux limites nulles aux bords.

La figure 7.1 représente un champ de vitesse établi à partir d'une simulation effectuée par Norro (1995); il s'agit du champ que nous avons utilisé comme référence.

Dans un premier temps, nous avons supposé ce champ connu sur la moitié des points du maillage. Les résultats obtenus après 20 itérations de la méthode de contrôle sont présentés aux figures 7.2. La figure 7.2_b, représentant la vitesse correspondant au relèvement obtenu par le code d'optimisation, montre une bonne adéquation de la solution avec le champ de référence, ce qui est confirmé à la figure 7.2_c, où est représentée l'évolution de l'erreur relative aux points de mesure:

$$\eta_1 = \frac{\left\| u(x^j) - u_{d_j} \right\|^2_{\left(L^2(\mathbb{R})\right)^m}}{\left\| u_{d_j} \right\|^2_{\left(L^2(\mathbb{R})\right)^m}}.$$

Les figures 7.3 représentent quant à elles les résultats obtenus en considérant la vitesse désirée connue seulement en un point sur vingt du maillage[4]. On s'aperçoit alors sur la figure 7.3_b que la vitesse calculée dans ce cas ne correspond plus au champ de référence de la figure 7.1, ce qui est vérifié sur la figure 7.3_c en représentant l'évolution de l'erreur η_1 en fonction du nombre d'itérations.

Ces remarques, ainsi que la figure 7.4 où est représentée l'évolution de l'erreur C^0:

$$\eta_2 = \sup \frac{\left| u(x^j) - u_{d_j} \right|}{\left| u_{d_j} \right|},$$

en fonction du nombre d'itérations, confirment les résultats préalablement obtenus sur le carré que nous rappelions dans la section 7.1, à savoir qu'en dessous d'un nombre minimum de points de mesure, la solution calculée n'est plus en accord avec le champ de référence.

Enfin, nous présentons aux figures 7.5, les résultats numériques obtenus dans le cas non linéaire. Dans les mêmes conditions que pour le cas de la figure 7.2 (solution désirée connue en un point sur deux du domaine), nous remarquons que les résultats sont semblables à ceux obtenus dans le cas linéaire, ce qui n'est pas étonnant car dans le cas des écoulements géophysiques, les termes de Coriolis, de gradient de pression et de viscosité sont prépondérants.

Conclusion

Le but de notre travail est la mise en œuvre d'une méthode numérique adaptée à la détermination de conditions aux limites en mer ouverte d'un problème de *shallow water*.

Les premiers tests effectués sur le carré se sont révélés encourageants. En particulier, dans ce cas simplifié, il nous a été possible de reconstituer

[4]Sur la figure, les points du maillage où la vitesse est connue sont représentés par des croix.

des conditions aux limites assez complexes avec une précision tout à fait convenable.

En outre, la méthode de Galerkin s'est avérée bien adaptée à la résolution des équations adjointes ainsi qu'au calcul du minimum de la fonctionnelle et nous a permis de conserver des temps de calcul raisonnables.

Dans un deuxième temps, nous avons appliqué le code numérique à un domaine plus proche de la réalité représentant la baie de Calvi, qui correspond bien aux objectifs que nous nous sommes fixés, de par sa taille et sa situation géographique. Les résultats obtenus dans ce cas se sont également montrés encourageants.

Actuellement, nous travaillons à une évolution du code numérique en vue d'effectuer un contrôle simultané sur la vitesse et la hauteur. Cette étude nous paraît être justifiée par la nécessité (Chatelon–Orenga, 1997) de se fixer la valeur de la hauteur sur la partie de la frontière où le fluide est entrant. En outre, l'évolution récente des techniques de mesure, en particulier l'utilisation du satellite, devrait nous permettre d'obtenir des mesures ponctuelles d'élévation dans le domaine que nous utiliserons pour reconstituer la hauteur sur la frontière en utilisant le code d'optimisation. Enfin, dans le cas non linéaire, il serait intéressant d'étudier le cas de petits domaines où les termes de Coriolis, de gradient de pression et de viscosité sont moins importants.

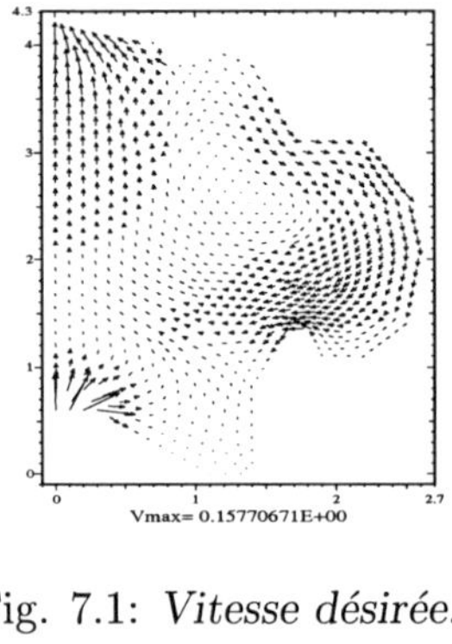

Fig. 7.1: *Vitesse désirée.*

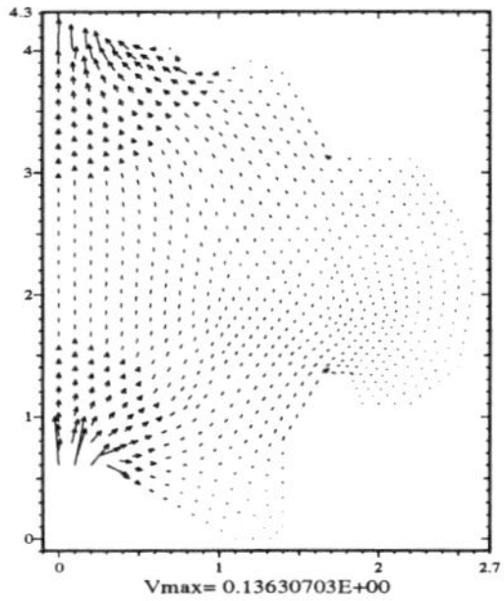

Fig. 7.2a: *Relèvement calculé en considérant la vitesse connue en un point sur deux.*

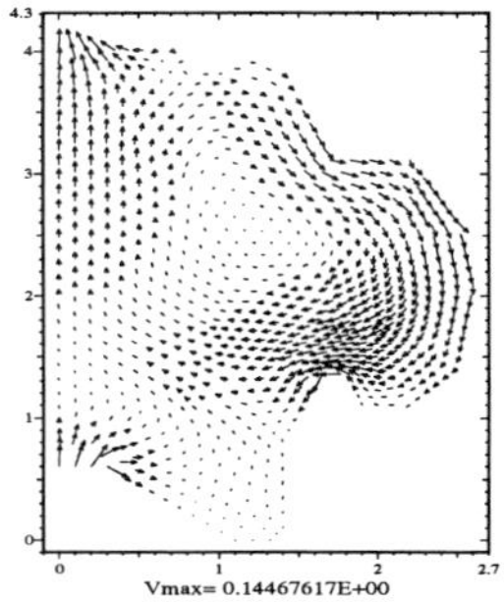

Fig. 7.2b: *Vitesse correspondante.*

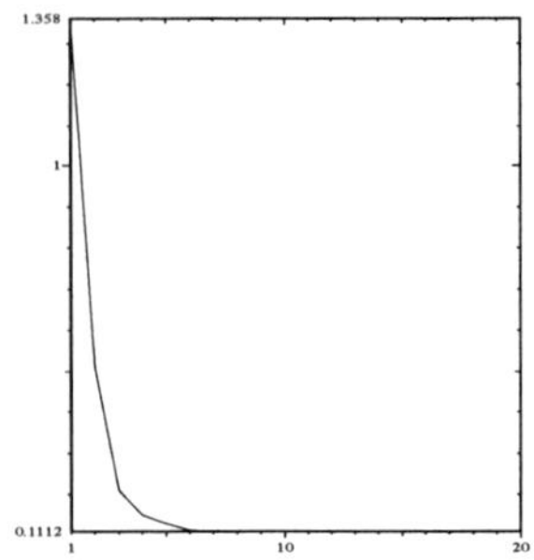

Fig.7.2c: *Evolution de l'erreur* $\eta_1 = \dfrac{\left\|u(x^j)-u_{d_j}\right\|^2_{\left(L^2(\mathbb{R})\right)^m}}{\|u_{d_j}\|^2_{\left(L^2(\mathbb{R})\right)^m}}$

en fonction du nombre d'itérations.

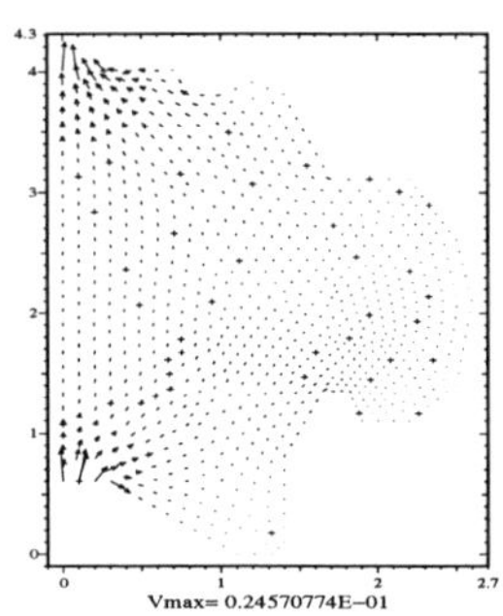

Fig.7.3a: *Evolution Relèvement calculé*
en considérant la vitesse connue en un point sur vingt.

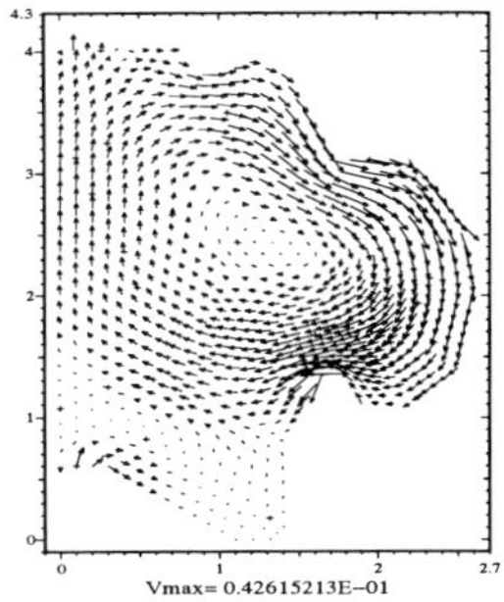

Fig.7.3b: *Vitesse correspondante.*

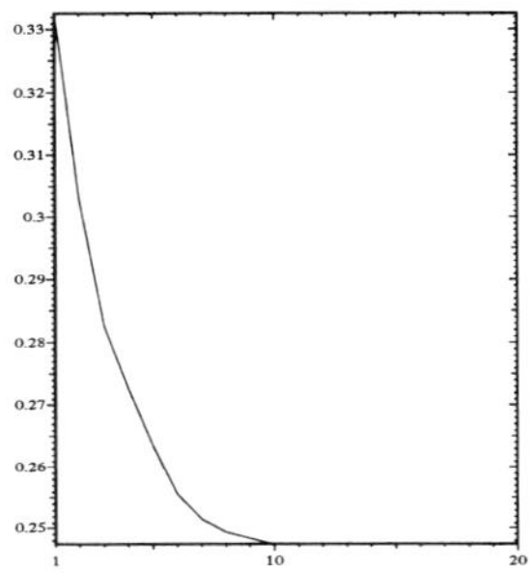

Fig.7.3c: *Evolution de l'erreur* $\eta_1 = \dfrac{\left\| u(x^j) - u_{d_j} \right\|^2_{\left(L^2(\mathbb{R})\right)^m}}{\| u_{d_j} \|^2_{\left(L^2(\mathbb{R})\right)^m}}$

en fonction du nombre d'itérations.

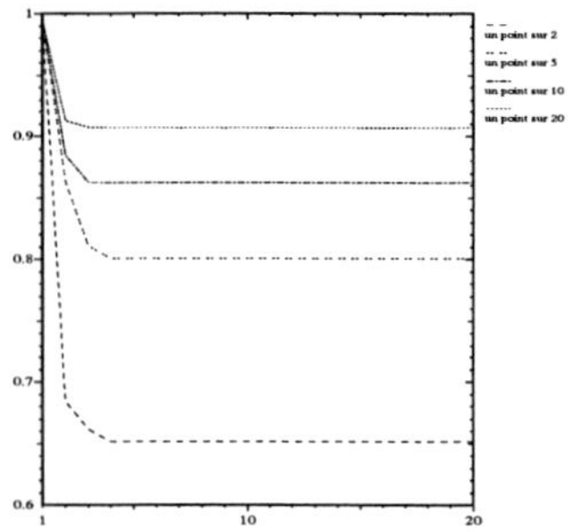

Fig.7.4: *Evolution de l'erreur* $\eta_2 = \sup \frac{|u(x^j) - u_{d_j}|}{|u_{d_j}|}$ *en fonction du nombre d'itérations.*

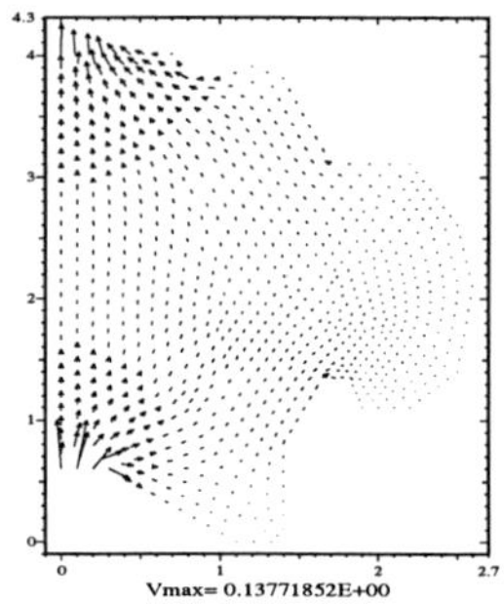

Fig.7.5a: *Relèvement calculé en considérant la vitesse connue en un point sur deux.*

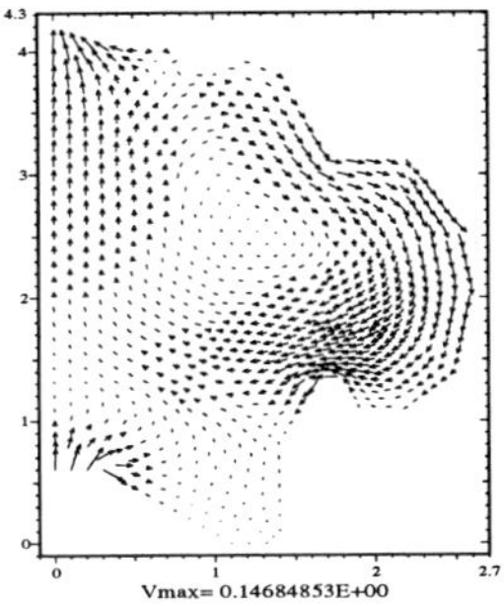

Fig.7.5b: *Vitesse correspondante.*

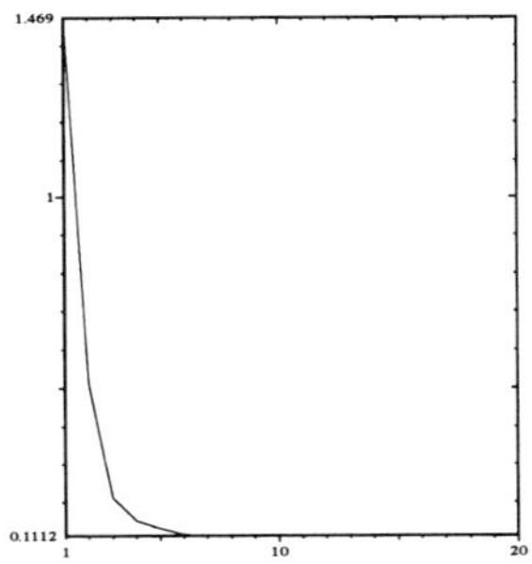

Fig.7.5c: *Evolution de l'erreur* $\eta_1 = \dfrac{\left\| u(x^j) - u_{d_j} \right\|^2_{\left(L^2(\mathbb{R})\right)^m}}{\| u_{d_j} \|^2_{\left(L^2(\mathbb{R})\right)^m}}$
en fonction du nombre d'itérations.

References

Bisgambiglia P., Traitement numérique et informatique de la modélisation spectrale. Thèse de Doctorat, Université de Corse 1989.

Chatelon F.J., Orenga P., On a non-homogeneous shallow water problem, Math. Modelling and Num. Anal.,31, 1 (1997), 27–55.

Dautray R., Lions J.L., Analyse mathématique et calcul numérique pour les sciences et les techniques. Masson, Paris 1988.

Girault V., Raviart P.A., Finite elements approximation of the Navier-Stockes equations. Springer-Verlag 1979.

Lions J.L., Contrôle optimal de systèmes gouvernés par des équations aux dérivées partielles. Dunod 1968.

Lions J.L., Magenes E., Problèmes aux limites non homogènes et applications. Dunod 1972.

Nihoul J.C.J., Modèles mathématiques et Dynamique de l'environnement. Elsevier Publ., Liège 1977.

Norro A., Etude pluridisciplinaire d'un milieu côtier. Approches expérimentale et de modélisation de la baie de Calvi (Corse). Thèse de Doctorat, Université de Liège 1995.

Orenga P., Analyse de quelques problèmes en océanographie physique. Thèse d'Habilitation, Université de Corse 1991.

Orenga P., Construction d'une base spéciale pour la résolution de quelques problèmes d'océanographie physique en dimension deux, CRAS, 314 (1) (1992), 587–590.

Orenga P., Un théorème d'existence de solutions d'un problème de shallow water, Arch. Rational Mech. Anal., Springer-Verlag, 130 (1995), 183–204.

Talagrand O., Courtier P., Variational assimilation of meteorological observations with the adjoint vorticity equation. I: Theory, Q.J.R. Meteorol. Soc., 113 (1987), 1311–1328.

Vidrascu M., Georges P.L., Guide d'utilisation et normes de programmation Modulef, I.N.R.I.A, Le Chesnay 1990.

Frédéric Bosseur et Pierre Orenga
Université de Corse Pascal Paoli
SDEM–URA CNRS 2053
Campus Grossetti
20250 Corte
France

E-mail : bosseur@univ-corse.fr, orenga@univ-corse.fr

Studies in Mathematics and its Applications, Vol. 31
D. Cioranescu and J.L. Lions (Editors)

Chapter 7

EFFECTIVE DIFFUSION IN VANISHING VISCOSITY

F. CAMPILLO AND A. PIATNITSKI

1. Introduction

In the present article we consider the asymptotics of the effective diffusion for elliptic operators with vanishing diffusion and with potential first order terms, the potential being a statistically homogeneous field.

The homogenization problems for singular perturbed operators have many important applications, among them fluid dynamics in porous media Bear [5] or groundwater pollution Fried [11]. In the recent years, various such questions were considered in detail for the operator with divergence–free vector fields. Many interesting asymptotics were constructed for the periodic case (Bensoussan–Lions–Papanicolaou [7], Fannjiang–Papanicolaou [9]) — and then in the random case (Avellaneda–Majda [1, 2, 4], Carmona–Xu [8], Fannjiang–Papanicolaou [10]).

In contrast to divergence–free case, where the effective diffusion is usually much greater than the initial one (see, for instance, Fannjiang–Papanicolaou [9]), we typically have in case of potential vector field the exponential decay of the effective coefficients.

For operators with periodic coefficients this phenomenon was investigated in Kozlov [14] and Kozlov–Piatnitski [15] where the logarithmic asymptotics of effective coefficient was found in terms of "Morse properties" of the potential on the torus of periodicity.

The operators whose first order terms are not potential but show in a way similar behavior, were considered in Kozlov–Piatnitski [16], where the logarithmic asymptotics of the effective diffusion was established.

In the present work, we study a particular case of operators with random potential first order terms. Namely, we assume that the potential is a random perturbation of a given periodic function. Considering this random perturbation, we assume that it does not change essentially the topological structure of the initial potential. This allows us to use the results from the percolation theory and to find the required asymptotics in terms of the proper percolation levels.

All the exact assumptions are provided in Section 1. Then, in Section 2, we prove the general result on asymptotic behavior of homogenized coefficients. One of the key condition of this statement is non-explicit. In Section 3 we present a couple of sufficient conditions expressed in explicit terms.

2. The setup

Let us consider a potential on $\mathbb{R}^2$ (with orthonormal basis $\{e_1, e_2\}$) of the form $U = U_0 + U_1$ where U_0 is a deterministic smooth potential which is supposed to be periodic with with period 1 in each coordinate directions. We denote the cell of periodicity $[0,1]^2$ by $\square$ and identify it with the 2D torus $\mathcal{T}$; U_1 is an isotropic random field, it represents a small random perturbation of U_0.

If S denotes the rotation matrix of angle $\pi/2$, we suppose that:

(*i*) $U_0(S\,x) = U_0(x)$ for all $x \in \mathcal{T}$,

(*ii*) the distribution of U_1 is invariant with respect to any integer shift of $\mathbb{R}^2$ and to S: $\text{law}(U_1(S\,x)) = \text{law}(U_1(x))$ for all $x \in \mathcal{T}$,

(*iii*) there exists $\gamma_0 > 0$ such that $|U_1(x)| \le \gamma_0$, for all $x \in \mathbb{R}^2$, a.s.

(*iv*) there is $\rho > 0$ such that any $\sigma(U_1(x), x \in G_1)$-measurable random variable η_1 and $\sigma(U_1(x), x \in G_2)$-measurable η_2 are independet whenever $\text{dist}(G_1, G_2) > \rho$; here G^1 and G^2 are arbitrary subsets of $\mathbb{R}^2$.

The two first conditions ensure the isotropy of the effective media.

Under condition (*i*), the potential U_0 has a specific structure: in the simplest case — other cases rely on the same arguments — the minimum number of degenerate points that U_0 could admit on $\mathcal{T}$, i.e. points x such that $\nabla U_0(x) = 0$, is four: one minimum point x_{min}, one maximum point x_{max}, and two saddle points x_{s}.

In $\mathbb{R}^2$ minimum points, maximum points, and saddle points will be denoted x_{min}, x_{max}, and x_{s} respectively.

Without loss of generality we may assume that in $\mathbb{R}^2$, the set of maximum points is $\text{X}_{\text{max}} = \mathbb{Z}^2$, then the set of minimum points should coincide with $\text{X}_{\text{min}} = \mathbb{Z}^2 + (\frac{1}{2}, \frac{1}{2})$, and the set of saddle points with $\text{X}_{\text{s}} = \mathbb{Z}^2 + (0, \frac{1}{2}) \cup \mathbb{Z}^2 + (\frac{1}{2}, 0)$, (see Figure 1).

The case of a more general potential $U_0(x)$, having more singular points, including minimum points, could be treated with the same reasoning.

We make the following hypothesis:

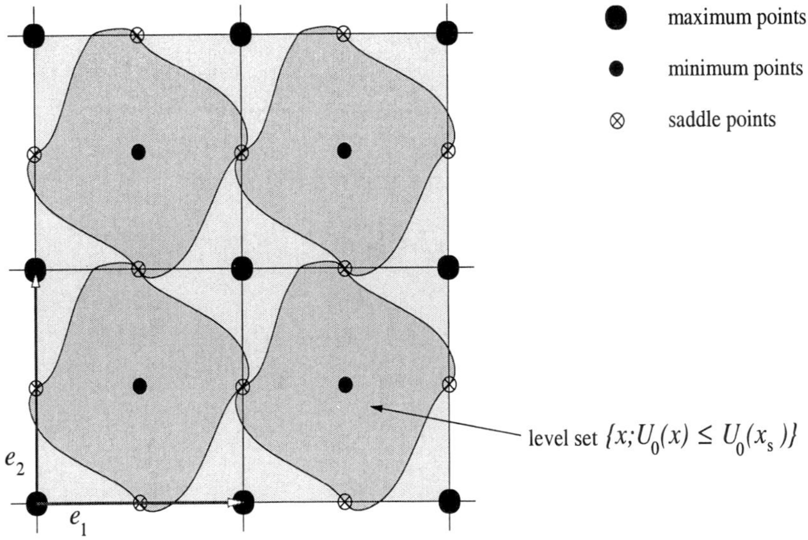

Figure 1: Example of structure for the potential U_0.

Hypothesis 1 - *For all minimum point $x_{\min}$ (the set of minimum points is $\mathrm{X}_{\min}$), let us consider:*

$$\alpha(x_{\min}, e_i) = \inf_{X(\cdot)\in\mathcal{X}(x_{\min},e_i)} \sup_{0\le t\le 1} U(X(t))\,,\ i=1,2\,, \tag{1}$$

where $\mathcal{X}(x_{\min}, e_i)$ is the set of functions $[0,1] \ni t \mapsto X(t) \in \mathbb{R}^2$ such that $X(0) = x_{\min}$, $X(1) = x_{\min} + e_i$, and which are homothetic to $[0,1] \ni t \mapsto X_0(t) \ni \mathbb{R}^2 \setminus \mathrm{X}_{\max}$ defined as follows: $X_0(t) = x_{\min} + t\, e_i$.

We suppose that $\{\alpha(x_{\min}, e_i)\,;\, x_{\min} \in \mathrm{X}_{\min},\ i = 1, 2\}$ is a family of independent random variables.

This last assumption is rather non explicit, a couple of sufficient conditions that ensure the above independence are supplied in Section 4.

Consider the following homogenization problem:

$$\left\{ \begin{array}{l} \mu\,\Delta u^\varepsilon(x) + \frac{1}{\varepsilon}\ \nabla_z[-U(z)]\big|_{z=\frac{x}{\varepsilon}} \cdot \nabla_x u^\varepsilon(x) = f(x)\,,\ x \in Q\,, \\ u^\varepsilon \in H_0^1(Q) \end{array} \right. \tag{2}$$

for some bounded domain Q in $\mathbb{R}^2$.

We assume first that the viscosity parameter μ is small and fixed, and pass to the limit as $\varepsilon \downarrow 0$. Then we study the asymptotic behavior of the effective (scalar) diffusivity $\sigma(\mu)$ as $\mu \downarrow 0$.

3. Effective diffusion

In this section we show that, under Hypothesis 1, the logarithmic limit of the effective diffusivity $\sigma(\mu)$ can be found in terms of a proper critical percolation level of the potential U.

After standard transformations, Equation (2) reads:

$$\mu\, e^{U(\frac{x}{\varepsilon})/\mu} \sum_{i=1,2} \frac{\partial}{\partial x_i}\left(e^{-U(\frac{x}{\varepsilon})/\mu}\,\frac{\partial}{\partial x_i} u^{\varepsilon}(x)\right) = f(x)$$

(from now on we will always suppose that $u^{\varepsilon} \in H_0^1(Q)$).

Multiply each term of this last equation by $e^{-U(\frac{x}{\varepsilon})/\mu}$ so that:

$$\mu \sum_{i=1,2} \frac{\partial}{\partial x_i}\left(e^{U(\frac{x}{\varepsilon})/\mu}\,\frac{\partial}{\partial x_i} u^{\varepsilon}(x)\right) = e^{-U(\frac{x}{\varepsilon})/\mu}\, f(x)\,. \tag{3}$$

Without loss of generality, we can assume that:

$$\operatorname{ess}\inf_{x\in\mathbb{R}^2} U(x) = 0\,. \tag{4}$$

Then, for any $\mu > 0$, $e^{-U(\frac{\cdot}{\varepsilon})/\mu} f \rightharpoonup \beta(\mu)\, f$ weakly in $L^2(Q)$ as $\varepsilon \downarrow 0$, where:

$$\beta(\mu) \triangleq \int_{\Box} e^{-U(\frac{x}{\varepsilon})/\mu}\, dx\,.$$

Moreover $\mu \log \beta(\mu) \to 0$ as $\mu \downarrow 0$.

For each μ, the family of operators appearing in (3) is coercive, uniformly in ε. Thus, it suffices to homogenize the following PDE:

$$\mu \sum_{i=1,2} \frac{\partial}{\partial x_i}\left(e^{U(\frac{x}{\varepsilon})/\mu}\,\frac{\partial}{\partial x_i} v^{\varepsilon}(x)\right) = \beta(\mu)\, f(x)\,. \tag{5}$$

Then, $v^{\varepsilon} \sim u^{\varepsilon}$ as $\varepsilon \downarrow 0$ (in the sense that these functions have the same limit in $H_0^1(Q)$ as $\varepsilon \downarrow 0$). Clearly, we can omit both factors μ and $\beta(\mu)$ for a while; we end up with the equation:

$$\sum_{i=1,2} \frac{\partial}{\partial x_i}\left(e^{U(\frac{x}{\varepsilon})/\mu}\,\frac{\partial}{\partial x_i} v^{\varepsilon}(x)\right) = f(x)\,. \tag{6}$$

According to Jikov et al [12], under above conditions, Equation (6) admits the effective diffusion matrix $\Sigma(\mu)$ which is isotropic: $\Sigma(\mu) = \sigma(\mu)\, I$ and the scalar effective diffusion coefficient is supplied by the following variational problem:

$$\sigma(\mu) = \liminf_{\varepsilon \downarrow 0} \inf_{\substack{v \in H^1(\Box) \\ v(0,\cdot) \equiv 0 \\ v(1,\cdot) \equiv 1}} \int_\Box e^{U(\frac{x}{\varepsilon})/\mu} |\nabla v(x)|^2 \, dx \,. \tag{7}$$

3.1. Lower bound

Let $\Box_1$ and $\Box_2$ be two neighbor cells, let say that $\Box_2 = \Box_1 + e_1$, and $x^1_{\min} \in \Box_1$, $x^2_{\min} \in \Box_2$ the corresponding minimum points of U_0. We introduce the following random open set:

$$G^-_\eta \triangleq \{x \in \mathbb{R}^2 \; : \; U(x) < \eta\}$$

and the events:

- A^0_η : the set of ω such that there is a path connecting $x^1_{\min}$ and $x^2_{\min}$ which belongs to $\mathcal{X}(x^1_{\min}, e_1)$ and which is included in $G^-_\eta(\omega)$.
- A^n_η : the set of ω such that there is a smooth curve from $\mathcal{X}(x^1_{\min}, e_1)$ of length not greater than n such that its $\frac{1}{n}$–neighborhood is included in $G^-_\eta(\omega)$.

In our case, A^0_η could be also defined as the event: the set of ω such that $G^-_\eta(\omega) \cap (\Box_1 \cup \Box_2)$ contains $x^1_{\min}$ and $x^2_{\min}$.

Clearly $\cup_{n>0} A^n_\eta = A^0_\eta$ and, hence:

$$\lim_{n \uparrow \infty} P(A^n_\eta) = P(A^0_\eta) \,. \tag{8}$$

It is also obvious that under our assumptions on the structure of U, the above events are independent for different pairs $(x^1_{\min}, x^2_{\min})$ of neighbor minimum points.

Consider standard bond percolation model using minimum points of U_0 as sites, and let p_c be the critical probability of the appearance of the infinite cluster: $p_c = \frac{1}{2}$. We define the *critical value* η_c as follows:

$$P(A^0_{\eta_c}) = \frac{1}{2} \,,$$

or, if such a η_c does not exist:

$$\eta_c = \inf\{\eta \,;\, P(A^0_\eta) < \tfrac{1}{2}\} = \sup\{\eta \,;\, P(A^0_\eta) > \tfrac{1}{2}\} \,. \tag{9}$$

This last equality is, in fact, an additional assumption which is supposed to be fulfilled later on.

For all $\gamma > 0$ small enough, $P(A^0_{\eta_c+\gamma}) > \frac{1}{2}$. Thus, using (8), $P(A^n_{\eta_c+\gamma}) > \frac{1}{2}$ for sufficiently large n. We fix such a n and denote it by n^0 ; we also denote

$$p^0 = P(A^{n^0}_{\eta_c+\gamma}).$$

We say that a bond $(x^1_{\min}, x^2_{\min})$ is open if the corresponding ω belongs to the set $A^{n^0}_{\eta_c+\gamma}(x^1_{\min}, x^2_{\min})$.

As proved in Kesten [13], for almost all realizations and for all sufficiently large N, the square $[0,N]^2$ contains at least $c(p^0)\,N$ mutually non intersecting channels connecting left and right sides of the square. Finally, we arrive at the following conclusion:

Conclusion. For sufficiently large N, $[0,N]^2$ contains at least $c(p^0)\,N$ mutually non intersecting smooth $\frac{1}{n^0}$–pipes connecting left and right sides of the square such that along each of these pipes:

$$U(y) < \eta_c + \gamma\,. \tag{10}$$

Denote the above pipes by $Q_1, \dots, Q_{k(N)}$, $k(N) \geq c(p^0)\,N$. Without loss of generality we assume that for any function $x \mapsto u(x)$ such that $u(0,x_2) \equiv 0$, $u(N,x_2) \equiv 1$ we have:

$$\int_{Q_m} \frac{\partial u(y)}{\partial \ell}\, dy \geq \frac{1}{2}\,\frac{1}{n^0}$$

(here ℓ is a variable directed along the pipe after rescaling). Indeed, taking a smooth pipe included in Q_m and choosing, if necessary, a larger value of n_0, one can achieve the above lower bound.

After rescaling $x = \varepsilon\, y$, $\varepsilon = 1/N$, we find:

$$\int_{Q^\varepsilon_m} \frac{\partial u(x)}{\partial \ell}\, dx \geq \frac{1}{2}\,\frac{1}{n^0}\,\varepsilon \qquad \text{with } Q^\varepsilon_m = \varepsilon\, Q_m\,.$$

By the Shwartz inequality:

$$\varepsilon^2\,\frac{1}{(n^0)^2}\,\frac{1}{4} \leq \left[\int_{Q^\varepsilon_m} \frac{\partial u(x)}{\partial \ell}\, dx\right]^2 \leq |Q^\varepsilon_m| \int_{Q^\varepsilon_m} |\nabla u(x)|^2\, dx\,.$$

Thus,

$$\int_{Q^\varepsilon_m} |\nabla u(x)|^2\, dx \geq \frac{\varepsilon}{4}\,\frac{1}{(n^0)^2}\,\frac{1}{c_1(p^0)}\,.$$

Summing up over m leads to:

$$\sum_{m=1}^{k(N)} \int_{Q_m^\varepsilon} |\nabla u(x)|^2 \, dx \geq c(p^0) \, \varepsilon \, \frac{\varepsilon}{4} \, \frac{1}{(n^0)^2} \, \frac{1}{c_1(p^0)} = \frac{c(p^0)}{c_1(p^0)} \, \frac{1}{(2\,n^0)^2} \, .$$

From (10), we have:

$$\begin{aligned} \int_\square e^{-U(\frac{x}{\varepsilon})/\mu} \, |\nabla u(x)|^2 \, dx &\geq \sum_{m=1}^{k(N)} \int_{Q_m^\varepsilon} e^{-U(\frac{x}{\varepsilon})/\mu} \, |\nabla u(x)|^2 \, dx \\ &\geq e^{-(\eta_c+\gamma)/\mu} \sum_{m=1}^{k(N)} \int_{Q_m^\varepsilon} |\nabla u(x)|^2 \, dx \\ &\geq e^{-(\eta_c+\gamma)/\mu} \, \frac{c(p^0)}{c_1(p^0)} \, \frac{1}{(2\,n^0)^2} \, . \end{aligned}$$

Using Definition (7) of $\sigma(\mu)$, and taking into account the fact that γ is an arbitrary positive number, we obtain:

$$\liminf_{\mu \downarrow 0} \mu \, \log \sigma(\mu) \geq -\eta_c \, .$$

3.2. Upper bound

Let $\square_1$ and $\square_2$ be two neighbor cells, let say that $\square_2 = \square_1 + e_1$, and $x^1_{\max} \in \square_1$, $x^2_{\max} \in \square_2$ the corresponding maximum points of U_0.

We introduce the random set:

$$G_\eta^+ \triangleq \{ x \in \mathbb{R}^2 \, ; \, U(x) > \eta \}$$

and the events:

- B_η^0: the set of ω such that there is a path connecting $x^1_{\max}$ and $x^2_{\max}$ which belongs to $\mathcal{X}(x^1_{\max}, e_1)$ and which is included in $G_\eta^+(\omega)$.
- B_η^n : the set of ω such that there is a smooth curve of length not greater than n such that its $\frac{1}{n}$–neighborhood is included in $G_\eta^+(\omega)$.

Comparing this setting with the one used for the proof of the lower bound, one can easily see that:

$$\eta_c = \max\{\eta \, ; \, P(B_\eta^0) < \tfrac{1}{2}\} = \min\{\eta \, ; \, P(B_\eta^0) > \tfrac{1}{2}\} \, .$$

Thus, for any small positive γ we have:

$$P(B^0_{\eta_c-\gamma}) > \frac{1}{2}\,.$$

This implies the existence of $n_0 = n_0(\gamma) > 0$ such that:

$$P(B^{n_0}_{\eta_c-\gamma}) > \frac{1}{2}\,.$$

We use the notation $p^0 = P(B^{n_0}_{\eta_c-\gamma})$.

In the same way as above one can assert that for sufficiently large N, the square $[0,N]^2$ contains at least $c(p^0)\,N$ mutually non intersecting smooth $\frac{1}{n_0}$ pipes connecting bottom and top sides of the square such that along each of these pipes:

$$U(x) > \eta_c - \gamma\,.$$

We consider a specific test function $\bar{v}$ such that:

(*i*) $\bar{v}(0,x_2) \equiv 0$ and $\bar{v}(1,x_2) \equiv 1$, $\bar{v}$ is continuous,

(*ii*) $\bar{v}$ is constant between any pair of channels (pipes), and also between $\{x\,;\,x_1 = 0\}$ and the first pipe from one side, and between the last pipe and $\{x\,;\,x_1 = 1\}$ from the other side,

(*iii*) crossing each channel (pipe), $\bar{v}$ makes a jump of amplitude $1/(c(p^0)\,N)$; inside a channel $\bar{v}$ is linear in the direction orthogonal to the curve that forms the channels.

Hence $|\nabla\bar{v}| \le n_0/c(p^0)$, and letting $\varepsilon = 1/N$, we get:

$$\begin{aligned}
\inf_{\substack{v\,\in\,H^1(\square)\\ v(0,\cdot)\,\equiv\,0\\ v(1,\cdot)\,\equiv\,1}} \int_\square e^{U(\frac{x}{\varepsilon})/\mu}\,|\nabla v(x)|^2\,dx \;&\le\; \int_\square e^{U(\frac{x}{\varepsilon})/\mu}\,|\nabla\bar{v}(x)|^2\,dx\\
&=\; \int_{\text{channels}} e^{-U(\frac{x}{\varepsilon})/\mu}\,|\nabla\bar{v}(x)|^2\,dx\\
&\le\; \frac{n_0^2}{c^2(p^0)}\,e^{-(\eta_c-\gamma)/\mu}\\
&=\; C(\gamma)\,e^{-(\eta_c-\gamma)/\mu}
\end{aligned}$$

here we also used the fact that $|\nabla\bar{v}| \equiv 0$ outside the channels. Back to Definition (7) of $\sigma(\mu)$, we get $\sigma(\mu) \le C(\gamma)\,e^{-(\eta_c-\gamma)/\mu}$. Taking the lim–sup as $\mu\downarrow 0$, we find:

$$\limsup_{\mu\downarrow 0}\,\mu\,\log\sigma(\mu) \le -\eta_c + \gamma\,.$$

Since γ is an arbitrary positive number, this relation implies:

$$\limsup_{\mu\downarrow 0} \mu \log \sigma(\mu) \leq -\eta_c \,.$$

3.3. Main result

Theorem 2 - *Under above assumptions, in particular Hypothesis 1, the log–behavior of the effective diffusion $\sigma(\mu)$ in the small viscosity case is given by:*

$$\lim_{\mu\downarrow 0} \mu \log \sigma(\mu) = -\eta_c$$

where η_c is the critical value given by (9).

4. Hypothesis 1: Sufficient conditions

In this section we provide two different sufficient conditions for validity of Hypothesis 1.

Lemma 3. - *Let the random field U_1 be equal to 0 everywhere in the vicinity of the level set $\mathcal{L}_0 = \{x : U_0(x) = U_0(x_{\rm s})\}$ except for some neighborhoods of the saddle points. Then, under the assumptions (i)-(iv) of Section 2, for sufficiently small γ_0 and ρ the random variables $\alpha(x_{\min}, e_i)$, $x_{\min} \in \mathrm{X}_{\min}$, $i = 1, 2$, are independent.*

Proof. Each periodic cell has two saddle points $x_{\rm s}^1$ and $x_{\rm s}^2$, for each one of these saddle points $x_{\rm s}^i$ (see Figure 2) we denote by $x_{\min}^{i,-} = x_{\min}$ and $x_{\min}^{i,+} = x_{\min} + e_i$ the two neighbor minimum points, symbol + corresponds to the greater value of one of the coordinates. Similarly, by $x_{\max}^{i,+} = x_{\max}$ and $x_{\max}^{i,-} = x_{\max} - e_2$ (if $i = 1$), $= x_{\max} - e_1$ (if $i = 2$) we denote the neighbor maximum points.

We begin by constructing a periodic family of sufficiently small neighborhoods $Q(x_{\rm s})$ of saddle points $x_{\rm s} \in \mathrm{X}_{\rm s}$ that possesses the following properties (see Figure 2 and Figure 3):

(i) for all $x_{\rm s} \in \mathrm{X}_{\rm s}$, $Q(x_{\rm s})$ is a smooth domain that contains no singular points except $x_{\rm s}$;

(ii) the σ–algebras generated by $\{U_1(x),\ x \in Q(x_{\rm s})\}$, $x_{\rm s} \in \mathrm{X}_{\rm s}$, are independent;

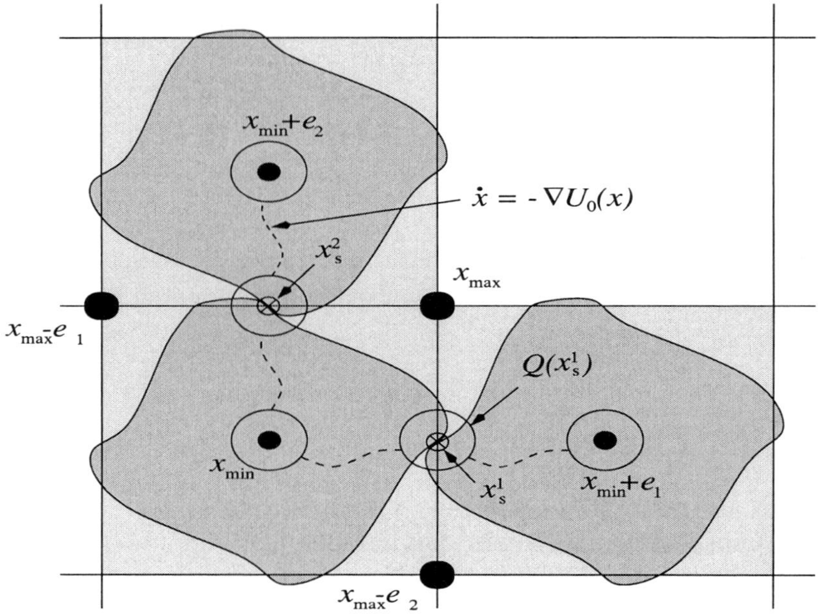

Figure 2: Sufficient condition, Lemma 3. (example of Figure 1).

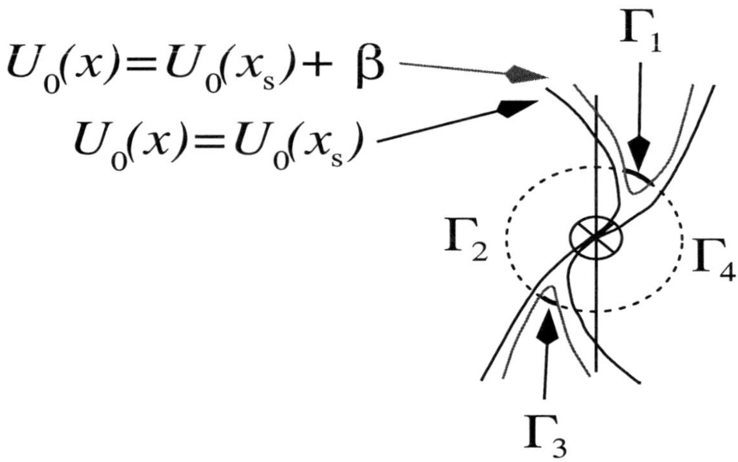

Figure 3: Zoom on point x_s^1 in Figure 2 with level lines $\{x\,;\,U_0(x)=U_0(x_s^1)\}$ and $\{x\,;\,U_0(x)=U_0(x_s^1)+\beta\}$, and the decomposition $\Gamma_1\cup\Gamma_2\cup\Gamma_3\cup\Gamma_4$ of $\partial Q(x_s^1)$.

(*iii*) there exists $\beta > 0$ such that, for all $x_s \in \mathrm{X}_s$, the following decomposition is valid: $\partial Q(x_s) = \cup_{j=1}^{4} \Gamma_j$, where Γ_j are connected components of $\partial Q(x_s)$ such that $U_0(x) > U_0(x_s) + \beta$ if $x \in \Gamma_1 \cup \Gamma_3$, and

$$\min_{x\in\Gamma_2} U_0(x) < U_0(x_s) - \beta, \quad \min_{x\in\Gamma_4} U_0(x) < U_0(x_s) - \beta,$$

$$\max_{x\in\Gamma_2\cup\Gamma_4} U_0(x) \le U_0(x_s) + \beta\,;$$

(*iv*) for all $x_s \in \mathrm{X}_s$: if $x \in \partial Q(x_s)$ and $U_0(x_s) - \beta \le U_0(x) \le U_0(x_s) + \beta$ then $U_1(x) \equiv 0$;

(*v*) all the trajectories of the equation $\dot{x} = -\nabla U_0(x)$ starting at Γ_2, are attracted with $x_{\min}^{i,-} = x_{\min}$ while the trajectories starting at Γ_4 are attracted with $x_{\min}^{i,+}$.

Under the above assumptions on U_0 and U_1 the said neighborhoods do evidently exist if β is small enough.

We are going to show now that for $\gamma_0 < \beta/2$ the random variables $\alpha(x_{\min}, e_i)$ are independent. To this end we consider arbitrary two neighbor minimum points $x_{\min}$ and $x_{\min} + e_i$ and a minimizing sequence of curves $\{\varphi^\delta(\cdot)\}$ such that $\varphi^\delta(0) = x_{\min}$, $\varphi^\delta(1) = x_{\min} + e_i$, $\varphi^\delta \in \mathcal{X}(x_{\min}, e_i)$ and

$$\max_{0\le t\le 1} U(\varphi(t)) \le \alpha(x_{\min}, e_i) + \delta\,.$$

Due to the structure of U_0 and the choice of $Q(x_s^i)$, the intersection of $\varphi(\cdot)$ with $Q(x_s^i)$ is nontrivial for all sufficiently small δ. It is also clear that φ^δ only intersects $\partial Q(x_s^i)$ at the points located at $\Gamma_2 \cup \Gamma_4$. Denote

$$\tau_1 = \max\{t\,;\,\varphi^\delta(t) \in \Gamma_2\}\,, \quad \tau_2 = \min\{t > \tau_1\,;\,\varphi^\delta(t) \in \Gamma_4\}\,.$$

Now one can replace the segments $\{\varphi(t)\,;\,0 \le t \le \tau_1\}$ and $\{\varphi(t)\,;\,\tau_2 \le t \le 1\}$ by the new ones in such a way that the curve $\tilde{\varphi}(\cdot)$ obtained is continuous, still belongs to $\mathcal{X}(x_{\min}, e_i)$ and satisfies the estimates:

$$U(\tilde{\varphi}(t)) < \alpha(x_{\min}, e_i)\,, \quad \text{for all } t < \tau_1 \text{ and } t > \tau_2\,.$$

Thus $\alpha(x_{\min}, e_i)$ only depends on $\{U_1(x)\,;\,x \in Q(x_s^i)\}$, and the statement of the lemma follows. □

The proof of the next assertion is similar to that of the preceding lemma and will be omitted.

Lemma 4 - *Let $U_1(x)$ be statistically homogeneous field (whose distributions are invariant w.r.t. any shifts) supported by Lipschitz functions, and suppose that*

$$|U_1(x)| \leq \gamma_0 \,, \ |U_1(x^1) - U_1(x^2)| \leq \gamma_1 |x^1 - x^2| \,, \ x, x^1, x^2 \in \mathbb{R}^2,$$

and that $\sigma\{U_1(x)\,;\, x \in G^1\}$ and $\sigma\{U_1(x)\,;\, x \in G^2\}$ are independent whenever

$$\text{dist}(G^1, G^2) > \rho.$$

Then for sufficiently small γ_0, γ_1 and ρ the random variables $\alpha(x_{\min}, e_i)$ are independent.

References

[1] M. Avellaneda and A.J. Majda, Mathematical models with exact renormalization for turbulent transport. Communications in Mathematical Physics, 131 (1990), 381–429.

[2] M. Avellaneda and A.J. Majda, Superdiffusion in nearly stratified flows. J. Stat. Phys., 69(3–4) (1992), 689–729.

[3] M. Avellaneda and A.J. Majda, Application of an approximate R–N–G theory, to a model for turbulent transport, with exact renormalization. In Turbulence in fluid flows. A dynamical systems approach, The IMA volumes in Mathematics and its Applications, G.R. Sell et al., editors, 1–31, Springer Verlag New York, 1993.

[4] M. Avellaneda and A.J. Majda, Simple examples with features of renormalization for turbulent transport. Phil. Trans. R. Soc. Lond. A, 346 (1994), 205–233.

[5] J.J. Bear, Dynamics of Fluids in Porous Media. Elsevier, New York, 1972.

[6] A.Yu. Belyaev and Ya.R. Efendiev, Homogenization of the stokes equations with a random potential. Math. Notes, 59, 4 (1996), 361–372.

[7] A. Bensoussan, J.L. Lions, and G. Papanicolaou, Asymptotic Analysis for Periodic Structures, volume 5 of Studies in Mathematics and its Applications. North–Holland, 1978.

[8] R.A. Carmona and L. Xu, Homogenization for time dependent 2–D incompressible Gaussian flows. Preprint.

[9] A. Fannjiang and G. Papanicolaou, Convection enhanced diffusion for periodic flows. SIAM Journal on Applied Mathematics, 54:333–408, .

[10] A. Fannjiang and G. Papanicolaou, Diffusion in turbulence. Probability Theory and Related Fields, 105 (1994), 279–334.

[11] J. Fried, Groundwater Pollution. Elsevier, Amsterdam, 1975.

[12] V. V. Jikov, S. M. Kozlov, and O. A. Oleinik, Homogenization of Differential Operators and Integral Functionals. Springer Verlag, 1994.

[13] H. Kesten, Percolation Theory for Mathematicians, volume 2 of Progress in Probability. Birkhäuser, Boston, 1982.

[14] S. Kozlov, Geometric aspects of homogenization, Russian Mathematical Surveys, 44, 2 (1989), 91–144.

[15] S.M. Kozlov and A.L. Pyatnitskii, Averaging on a background of vanishing viscosity. Math. USSR Sbornik, 70, 1 (1991), 241–261.

[16] A.L. Pyatnitski and S.M. Kozlov, Homogenization and vanishing viscosity. In B. Grigelionis, editor, Probability Theory and Mathematical Statistics. Proc. Fifth Vilnius Conference, 1989, 330–339. VSP/Mokslas, 1990.

Fabien Campillo
SYSDYS, INRIA/LATP, CMI
39 rue F.Joliot-Curie
13453 Marseille Cedex
France

E-mail: campillo@sophia.inria.fr

Andrey Piatnitski
Narvik University College HiN
Department of Mathematics
P.O. Box 385
8505 Narvik
Norway

and

Lebedev Physical Institute
Russian Academy of Science
Leninski Prospect 53
Moscow 117333
Russia

E-mail: andrey@sci.lebedev.ru

Studies in Mathematics and its Applications, Vol. 31
D. Cioranescu and J.L. Lions (Editors)

Chapter 8

VIBRATION OF A THIN PLATE WITH A "ROUGH" SURFACE

G. CHECHKIN AND D. CIORANESCU

1. Introduction

Rough surface problems have attracted much attention in the context of wave propagation and scattering (see, for example, [1], [2], [5] and [19]). The frictional behavior of deformable bodies depends explicitly on the structure of contact surfaces. Micro-characteristics such as the roughness of the contact surface or material properties near the surface, influence the large scale behavior.

The asymptotic analysis of these problems was treated for instance, in [4] and [31], while different homogenization problems in domains with rapidly oscillating boundary were considered in for instance in [6], [17], [18], [19], [29] and [30].

Based on classical Kirchhoff and Reissner plates theories, new models for bending of plates with rapidly varying thickness was proposed in [20]–[23] (see also [7], [16], [24], [25] and [33]). In these papers, motivated by the development of structural optimization (see for instance, [26]) the authors have studied symmetric, linearly elastic plates with thickness of order ε varying on a length scale of order ε^η. They restricted their attention to locally periodic plates, and to loads transverse to their midplanes. The transverse loads and the symmetry allow, in particular, to reduce the limit problem to one fourth-order equation. There are three different regimes, depending on whether $\eta < 1$ (the case of relatively slow variation of the thickness), $\eta = 1$ (when the variation is on the same scale as the mean thickness), or $\eta > 1$ (the case of relatively fast variation of the thickness). For each case, an effective rigidity tensor $M_{\alpha\gamma}^{\beta\delta}$ relating the bending moment to midplane curvature of the limit plate obtained as $\varepsilon \to 0$, was determined. In the limit problem, the vertical displacement of the midplane solves a fourth-order equation of the form

$$\sum_{\alpha,\beta,\gamma,\delta=1}^{2} \frac{\partial^2}{\partial x_\alpha\, \partial x_\beta} \left(M_{\alpha\gamma}^{\beta\delta} \frac{\partial^2 \mathcal{W}}{\partial x_\gamma\, \partial x_\delta} \right) = \varphi.$$

The methods developed in [20]–[23] relied essentially on the symmetry of plates. In the present paper we drop out the symmetry assumption, and consider a nonsymmetric locally periodic plate with oscillating boundary. In contrast to the case of plates with symmetric geometry, the limit equations in the nonsymmetric case cannot be decomposed into equations describing the vertical displcement and equations describing the horizontal ones. Due to the absence of a symmetry in the initial problem, a nontrivial interrelation between the leading terms of the asymptotic expansion do appear. In particular, the limit problem is characterized by the coupling of vertical and horizontal displacements.

In Section 1 we introduce necessary notation, define a class of thin domains and formulate the problem. In Section 2 we prove a priori estimates for the solution and investigate the asymptotic behavior of the moments and means of stresses. Then we introduce a family of auxiliary problems and obtain by a formal asymptotic expansion method the homogenized system. In the last section we prove a convergence theorem which justifies the results from Section 2.

1. Statement of the problem

Let $\Omega \subset \mathbb{R}^3$ be a domain of thickness e,

$$\Omega = \{x \mid 0 < x_1 < a,\ 0 < x_2 < b,\ 0 < x_3 < e\},$$

with $e = k\varepsilon$, where ε is a small parameter.

Suppose we are given a smooth nonpositive function $\varphi = \varphi(\widehat{x}, \widehat{\xi})$, with $\widehat{\xi} = \{\xi_1, \xi_2\}$ and $\widehat{x} = \{x_1, x_2\}$, φ being 1-periodic in $\widehat{\xi}$. We introduce the domain Π_ε by setting

$$\Pi_\varepsilon = \Big\{x \mid 0 < x_1 < a,\ 0 < x_2 < b,\ \varepsilon\varphi\big(\widehat{x}, \frac{\widehat{x}}{\varepsilon}\big) < x_3 < 0\Big\}.$$

The domain Ω_ε (the plate with a "rough" surface) is defined as follows:

$$\Omega_\varepsilon = \Omega \cup \Pi_\varepsilon \cup \{x_3 = 0\} = \Big\{x \mid 0 < x_1 < a,\ 0 < x_2 < b,\ \varepsilon\varphi\big(\widehat{x}, \frac{\widehat{x}}{\varepsilon}\big) < x_3 < e\Big\}.$$

Introduce now the surfaces

$$\omega^\varepsilon = \Big\{x \mid x \in \Omega_\varepsilon,\ x_3 = \varepsilon\varphi\Big(\widehat{x}, \frac{\widehat{x}}{\varepsilon}\Big)\Big\},$$
$$\omega\ = \{x \mid x \in \Omega_\varepsilon,\ x_3 = 0\} = \{x \mid 0 < x_1 < a,\ 0 < x_2 < b,\ x_3 = 0\},$$
$$\omega_e = \{x \mid x \in \Omega_\varepsilon,\ x_3 = e\} = \{x \mid 0 < x_1 < a,\ 0 < x_2 < b,\ x_3 = e\},$$
$$S_\varepsilon = \{x \mid x \in \partial\Omega_\varepsilon \backslash (\omega^\varepsilon \cup \omega_e)\},$$

and denote by $\nu^\varepsilon = (\nu_1^\varepsilon, \nu_2^\varepsilon, \nu_3^\varepsilon)$ the unit outward normal to ω^ε, by $n = (n_1, n_2, n_3) = (0, 0, 1)$ the unit outward normal vector to ω_e and by $n^\varepsilon = (n_1^\varepsilon, n_2^\varepsilon, 0)$ the unit outward normal to S_ε.

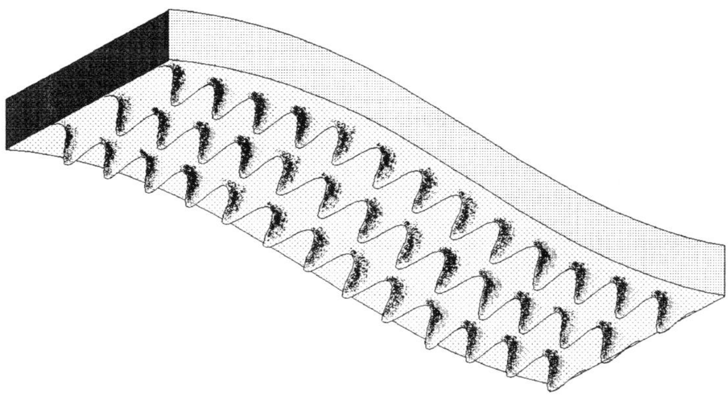

Figure 1. The plate Ω_ε

By construction, the surface ω^ε is the lower face of the plate occupying the domain Ω_ε, while ω_e is its upper face. The surface S_ε is the the lateral boundary of the plate. Observe that ω^ε is the "rough" oscillating part of the boundary $\partial\Omega_\varepsilon$ of the plate. The surface ω introduced above, is the lower face of the domain Ω; as a matter of fact, it is the mid surface of the thin domain Ω_1.

For simplicity, we will suppose that φ is compactly supported in ω, uniformly in $\widehat{\xi}$ and such that $|\varphi| < k$ for any $\widehat{x}$ and $\widehat{\xi}$.

Notation. In the sequel, the Einstein summation convention on repeated indices is adopted. As a rule, throughout the paper, Greek indices α, β, γ, δ, ρ take values in $\{1, 2\}$ while Latin indices i, j, k, l, h, p, q take values in $\{1, 2, 3\}$. The current point in Ω_ε is $x = (x_1, x_2, x_3)$, while the current point on the surface ω^ε is denoted $\widehat{x} = (\widehat{x}_1, \widehat{x}_2, \widehat{x}_3)$.

We consider the following boundary-value problem in linearized elasticity:

$$\begin{cases} L[u^\varepsilon] \equiv \dfrac{\partial}{\partial x_k}\Big(A^{kl}\dfrac{\partial u^\varepsilon}{\partial x_l}\Big) = -f^\varepsilon \quad \text{in } \Omega_\varepsilon, \\ u^\varepsilon = 0 \quad \text{on } S_\varepsilon, \\ A^{3l}\,\dfrac{\partial u^\varepsilon}{\partial x_l}\, n_3 = 0 \quad \text{on } \omega_e, \\ A^{kl}\,\dfrac{\partial u^\varepsilon}{\partial x_l}\,\nu_k^\varepsilon = g^\varepsilon\big(\widehat{\cdot, \tfrac{\cdot}{\varepsilon}}\big) \quad \text{on } \omega^\varepsilon, \end{cases} \tag{1}$$

where the unknown is the displacement $u^\varepsilon = (u_1^\varepsilon, u_2^\varepsilon, u_3^\varepsilon)$.

In system (1), $A^{kl} = A^{kl}(x)$ are $(3\times 3)-$matrices, whose elements $a_{ij}^{kl} = a_{ij}^{kl}(x)$ are bounded measurable functions satisfying the usual symmetries in elasticity

$$a_{ij}^{kl}(x) = a_{ji}^{lk}(x) = a_{il}^{kj}(x) \quad \text{a.e. in } \Omega_\varepsilon. \tag{2}$$

We suppose also that there exist $\lambda_1,\ \lambda_2 \in \mathbb{R}$, $0 < \lambda_1 < \lambda_2$ such that

$$\lambda_1\ \eta_{ik}\ \eta_{ik} \le a_{ij}^{kl}(x)\ \eta_{ik}\ \eta_{jl} \le \lambda_2\ \eta_{ik}\ \eta_{ik} \quad \text{a.e. in } \Omega_\varepsilon. \tag{3}$$

for any symmetric matrix $(\eta)_{i,j}$ with real elements.

The first boundary condition in (1) means that the plate is clamped on the lateral boundary S_ε.

Suppose we are given $f^\varepsilon = f^\varepsilon(x)$, $f^\varepsilon = (f_1^\varepsilon, f_2^\varepsilon, f_3^\varepsilon)$, the volume density of applied body forces and $g^\varepsilon = g^\varepsilon\big(\widehat{\widehat{x}, \tfrac{x}{\varepsilon}}\big)$ where $g^\varepsilon = (g_1^\varepsilon, g_2^\varepsilon, g_3^\varepsilon)$, the density of surface forces, satisfying the following assumption:

$$\begin{cases} f_\alpha^\varepsilon(x) = O(\varepsilon), \quad f_3^\varepsilon(x) = O(\varepsilon^2) \quad \text{for } x \in \Omega_\varepsilon, \\ g_\alpha^\varepsilon\big(\widehat{\widehat{x}, \dfrac{x}{\varepsilon}}\big) = O(\varepsilon^2), \quad g_3^\varepsilon\big(\widehat{\widehat{x}, \dfrac{x}{\varepsilon}}\big) = O(\varepsilon^3) \quad \text{for } x \in \omega^\varepsilon. \end{cases} \tag{4}$$

Now, let us introduce the strain tensor

$$e_{ik}(u^\varepsilon) \equiv \frac{1}{2}\left(\frac{\partial u_i^\varepsilon}{\partial x_k} + \frac{\partial u_k^\varepsilon}{\partial x_i}\right), \tag{5}$$

and the stress tensor

$$\sigma_{jl}(u^\varepsilon) \equiv a_{ij}^{kl} e_{ik}(u^\varepsilon).$$

With this notation, one can rewrite (1) in the equivalent form

$$\begin{cases} -\dfrac{\partial \sigma_{ij}}{\partial x_j}(u^\varepsilon) = f_i^\varepsilon \quad \text{in } \Omega_\varepsilon, \\ u^\varepsilon = 0 \quad \text{on } S_\varepsilon, \\ \sigma_{i3}(u^\varepsilon)\, n_3 = 0 \quad \text{on } \omega_e, \\ \sigma_{ij}(u^\varepsilon)\,\nu_j^\varepsilon = g_i^\varepsilon\big(\widehat{\cdot, \tfrac{\cdot}{\varepsilon}}\big) \quad \text{on } \omega^\varepsilon. \end{cases} \tag{6}$$

Definition 1– *The vector function $u^\varepsilon \in [H^1(\Omega_\varepsilon, S_\varepsilon)]^3$ is a weak solution of the problem (1) if the integral identity*

$$\int_{\Omega_\varepsilon} A^{kl}(x)\frac{\partial u^\varepsilon(x)}{\partial x_l}\,\frac{\partial v(x)}{\partial x_k}\,dx = \int_{\omega^\varepsilon}\Big(g^\varepsilon(\widehat{x},\frac{\widehat{x}}{\varepsilon}),v(x)\Big)\,ds + \int_{\Omega_\varepsilon}\big(f^\varepsilon(x),v(x)\big)\,dx, \tag{7}$$

holds for any $v \in [H^1(\Omega_\varepsilon, S_\varepsilon)]^3$.

Here, as usual, $[H^1(\Omega_\varepsilon, S_\varepsilon)]^3$ is the the closure with respect to the norm of the Sobolev space $[H^1(\Omega_\varepsilon)]^3$, of the set of $[C^\infty(\overline{\Omega}_\varepsilon)]^3$– vector functions, vanishing in a neighbourhood of S_ε (see for instance, [32]):

$$[H^1(\Omega_\varepsilon, S_\varepsilon)]^3 = \big\{u \in [H^1(\Omega_\varepsilon)]^3 \;\big|\; u = 0 \quad \text{on } S_\varepsilon\big\},$$

endowed with the usual gradient norm

$$\|u\|_{[H^1(\Omega_\varepsilon,S_\varepsilon)]^3} = \|\nabla u\|_{[L_2(\Omega_\varepsilon)]^{3\times 3}}.$$

In problem (5) we also used the notation

$$(u,v) = \sum_{i=1}^{3} u_i\, v_i, \quad \text{for } u,v \in [L(\Omega_\varepsilon)]^3.$$

Using definition (5) of the strain tensor, the left-hand side of (7) can be rewritten in the form

$$\int_{\Omega_\varepsilon} A^{kl}(x)\frac{\partial u^\varepsilon}{\partial x_l}\,\frac{\partial v}{\partial x_k}\,dx = \int_{\Omega_\varepsilon} a_{ij}^{kl}(x)e_{ik}(u^\varepsilon)\; e_{jl}(v)\,dx.$$

Using Korn's inequality (see [10], [27] and [28]) and the Lax-Milgram theorem, one can prove the existence and the uniqueness of a weak solution of problem (1) (or (6) or (7)). Moreover the following estimate takes place:

$$\begin{aligned}\|u^\varepsilon\|_{[H^1(\Omega_\varepsilon,S_\varepsilon)]^3} &\le K_1\|e(u^\varepsilon)\|_{[L_2(\Omega_\varepsilon)]^{3\times 3}}\\ &\le K_2\left(\|f^\varepsilon\|_{[L_2(\Omega_\varepsilon)]^3} + \|g^\varepsilon\|_{[L_2(\omega^\varepsilon)]^3}\right),\end{aligned} \tag{8}$$

where K_1 and K_2 are independent of ε (see, for example [27] or [30]).

Let us transform the domain Ω_ε by means of the dilatation in the x_3-direction

$$\widehat{z} := \widehat{x}, \qquad z_3 := \frac{x_3}{\varepsilon}.$$

Denote by $\widetilde{\Omega}_\varepsilon$ the transformed domain. Under this dilatation, any vector-function is changed as follows: we introduce a new vector-function $\Psi(z_1, z_2, z_3)$ instead of vector-function $\psi(x_1, x_2, x_3)$ by

$$\Psi_\alpha(z_1, z_2, z_3) = \frac{1}{\varepsilon}\psi_\alpha(z_1, z_2, \varepsilon z_3),$$

$$\Psi_3(z_1, z_2, z_3) = \psi_3(z_1, z_2, \varepsilon z_3).$$

We will also use the notation

$$\widetilde{\omega}^\varepsilon = \Big\{z \in \partial\widetilde{\Omega}_\varepsilon \mid z_3 = \varphi\Big(\widehat{z}, \frac{\widehat{z}}{\varepsilon}\Big)\Big\},$$

$$\widetilde{\omega}_e = \Big\{z \in \partial\widetilde{\Omega}_\varepsilon \mid z_3 = k\Big\},$$

$$S = \Big\{z \mid z \in \partial\widetilde{\Omega}_\varepsilon \setminus (\widetilde{\omega}^\varepsilon \cup \widetilde{\omega}_e)\Big\},$$

that are the transformed of respectively, ω^ε, ω_e and S_ε.

Let us denote by U^ε, F^ε and G^ε the transformed of u^ε, f^ε and g^ε respectively. Then, in the new coordinates problem (6) becomes

$$\begin{cases} \varepsilon\dfrac{\partial\sigma_{\alpha\beta}(U^\varepsilon)}{\partial z_\beta} + \dfrac{\partial\sigma_{\alpha 3}(U^\varepsilon)}{\partial z_3} = -\varepsilon F^\varepsilon_\alpha \quad \text{in } \widetilde{\Omega}_\varepsilon, \\ \varepsilon\dfrac{\partial\sigma_{3\beta}(U^\varepsilon)}{\partial z_\beta} + \dfrac{\partial\sigma_{33}(U^\varepsilon)}{\partial z_3} = -F^\varepsilon_3 \quad \text{in } \widetilde{\Omega}_\varepsilon, \\ U^\varepsilon = 0 \quad \text{on } S, \\ \sigma_{i3}(U^\varepsilon)\, n_3 = 0 \quad \text{on } \widetilde{\omega}_e, \\ \varepsilon\sigma_{\alpha\beta}(U^\varepsilon)\, \widetilde{\nu}^\varepsilon_\beta + \sigma_{\alpha 3}(U^\varepsilon)\, \widetilde{\nu}^\varepsilon_3 = \sqrt{\dfrac{1+\varepsilon^2|\nabla_z\varphi|^2}{1+|\nabla_z\varphi|^2}}\, G^\varepsilon_\alpha\Big(\widehat{z}, \dfrac{\widehat{z}}{\varepsilon}\Big) \quad \text{on } \widetilde{\omega}^\varepsilon, \\ \varepsilon\sigma_{3\beta}(U^\varepsilon)\, \widetilde{\nu}^\varepsilon_\beta + \sigma_{33}(U^\varepsilon)\, \widetilde{\nu}^\varepsilon_3 = \dfrac{1}{\varepsilon}\sqrt{\dfrac{1+\varepsilon^2|\nabla_z\varphi|^2}{1+|\nabla_z\varphi|^2}}\, G^\varepsilon_3\Big(\widehat{z}, \dfrac{\widehat{z}}{\varepsilon}\Big) \quad \text{on } \widetilde{\omega}^\varepsilon, \end{cases} \tag{9}$$

where

$$\begin{cases} \sigma_{ij}(U^\varepsilon) = a^{\beta j}_{\alpha i}(z) e_{\alpha\beta}(U^\varepsilon) + \dfrac{2}{\varepsilon} a^{3j}_{\alpha i}(z) e_{\alpha 3}(U^\varepsilon) + \dfrac{1}{\varepsilon^2} a^{3j}_{3i}(z) e_{33}(U^\varepsilon), \\ a^{kl}_{ij}(z) = a^{kl}_{ij}(z_1, z_2, \varepsilon z_3). \end{cases} \tag{10}$$

Assumption (4) reads

$$\begin{cases} F^\varepsilon_\alpha(z) = O(1), \quad F^\varepsilon_3(z) = O(\varepsilon^2), \\ G^\varepsilon_\alpha\Big(\widehat{z}, \dfrac{\widehat{z}}{\varepsilon}\Big) = O(\varepsilon), \quad G^\varepsilon_3\Big(\widehat{z}, \dfrac{\widehat{z}}{\varepsilon}\Big) = O(\varepsilon^3). \end{cases} \tag{11}$$

Denote

$$F_\alpha(z) := F^\varepsilon_\alpha(z), \quad F_3(z) := \frac{1}{\varepsilon^2}\, F^\varepsilon_3(z),$$

$$G_\alpha(\widehat{z}, \widehat{\xi}) := \frac{1}{\varepsilon}\, G^\varepsilon_\alpha(\widehat{z}, \widehat{\xi}), \quad G_3(\widehat{z}, \widehat{\xi}) := \frac{1}{\varepsilon^3}\, G^\varepsilon_3(\widehat{z}, \widehat{\xi}).$$

Definition 2 – *The vector-function $U^\varepsilon \in [H^1(\widetilde{\Omega}_\varepsilon, S)]^3$ is a weak solution of problem (9) if the integral identity*

$$\begin{aligned} \int\limits_{\widetilde{\Omega}_\varepsilon} \Big(\varepsilon \sigma_{i\beta}(U^\varepsilon)\, \frac{\partial v_i(z)}{\partial z_\beta} + \sigma_{i3}(U^\varepsilon)\, \frac{\partial v_i(z)}{\partial z_3}\Big)\, dz \\ = \int\limits_{\widetilde{\omega}^\varepsilon} \sqrt{\frac{1 + \varepsilon^2 |\nabla_z \varphi|^2}{1 + |\nabla_z \varphi|^2}}\, G^\varepsilon_\alpha\Big(\widehat{z}, \frac{\widehat{z}}{\varepsilon}\Big)\, v_\alpha(z) dz \\ + \int\limits_{\widetilde{\omega}^\varepsilon} \frac{1}{\varepsilon} \sqrt{\frac{1 + \varepsilon^2 |\nabla_z \varphi|^2}{1 + |\nabla_z \varphi|^2}}\, G^\varepsilon_3\Big(\widehat{z}, \frac{\widehat{z}}{\varepsilon}\Big)\, v_3(z)\, dS \\ + \int\limits_{\widetilde{\Omega}_\varepsilon} \big[\varepsilon\, F^\varepsilon_\alpha(z)\, v_\alpha(z) + F^\varepsilon_3(z)\, v_3(z)\big]\, dz \end{aligned} \tag{12}$$

holds for any $v \in [H^1(\widetilde{\Omega}_\varepsilon, S)]^3$.

2. Preliminary results

2.1. A priori estimates

From variational formulation (7), due to the coerciveness condition (3), to Friedrichs' inequality and keeping in mind assumptions (4), we derive the following a priori estimate:

$$\int\limits_{\Omega_\varepsilon} \Big(|\nabla u^\varepsilon_1(x)|^2 + |\nabla u^\varepsilon_2(x)|^2 + |\nabla u^\varepsilon_3(x)|^2\Big)\, dx \le K_3\, \varepsilon^2,$$

whence, in the new variables,

$$\int_{\widetilde{\Omega}_\varepsilon} \Big[\varepsilon^2\Big(\frac{\partial U_1^\varepsilon(z)}{\partial z_1}\Big)^2 + \varepsilon^2\Big(\frac{\partial U_1^\varepsilon(z)}{\partial z_2}\Big)^2 + \Big(\frac{\partial U_1^\varepsilon(z)}{\partial z_3}\Big)^2\Big]\, dz$$

$$+ \int_{\widetilde{\Omega}_\varepsilon} \Big[\varepsilon^2\Big(\frac{\partial U_2^\varepsilon(z)}{\partial z_1}\Big)^2 + \varepsilon^2\Big(\frac{\partial U_2^\varepsilon(z)}{\partial z_2}\Big)^2 + \Big(\frac{\partial U_2^\varepsilon(z)}{\partial z_3}\Big)^2\Big]\, dz$$

$$+ \int_{\widetilde{\Omega}_\varepsilon} \Big[\Big(\frac{\partial U_3^\varepsilon(z)}{\partial z_1}\Big)^2 + \Big(\frac{\partial U_3^\varepsilon(z)}{\partial z_2}\Big)^2 + \frac{1}{\varepsilon^2}\Big(\frac{\partial U_3^\varepsilon(z)}{\partial z_3}\Big)^2\Big]\, dz \le K_3\,\varepsilon^2,$$

from which, the following a priori estimates are straightforward:

$$\|e_{\alpha\beta}(U^\varepsilon)\|_{L_2(\widetilde{\Omega}_\varepsilon)} \le K_4, \tag{13}$$

$$\|e_{\alpha 3}(U^\varepsilon)\|_{L_2(\widetilde{\Omega}_\varepsilon)} \le \varepsilon\, K_4, \tag{14}$$

$$\|e_{33}(U^\varepsilon)\|_{L_2(\widetilde{\Omega}_\varepsilon)} \le \varepsilon^2\, K_4. \tag{15}$$

Then, by using (2) and (3), one immediately has

$$\|\sigma_{ij}(U^\varepsilon)\|_{L_2(\widetilde{\Omega}_\varepsilon)} \le K_4. \tag{16}$$

Now, let $\Omega_\pm$ be the domain defined by

$$\Omega_\pm = \{z \mid 0 < z_1 < a;\ 0 < z_2 < b;\ -k < z_3 < k\}.$$

Let us extend $\sigma_{ij}(U^\varepsilon)$ by zero in $\Omega_\pm$ and still denote $\sigma_{ij}(U^\varepsilon)$ this extension. Then, from estimate (16) and Rellich theorem it follows that there exists a subsequence $\{\varepsilon'\}$ such that, as $\varepsilon' \to 0$,

$$\sigma_{ij}(U^{\varepsilon'}) \rightharpoonup \sigma_{ij}^0 \quad \text{weakly in } L_2(\Omega_\pm). \tag{17}$$

2.2. Means and moments

To obtain the limit equations as $\varepsilon \to 0$, we introduce the moments and the means of stresses.

Definition 3– *The means of stresses and of its moments are defined respectively, by*

$$M_{\alpha\beta}^\varepsilon(\widehat{z}) = \int_{\varphi(\widehat{z},\frac{\widehat{z}}{\varepsilon})}^{k} z_3\,\sigma_{\alpha\beta}\, dz_3, \tag{18}$$

$$N^{\varepsilon}_{\alpha\beta}(\widehat{z}) = \int\limits_{\widehat{\varphi(\widehat{z},\frac{z}{\varepsilon})}}^{k} \sigma_{\alpha\beta}\, dz_3, \quad Q^{\varepsilon}_{\alpha}(\widehat{z}) = \frac{1}{\varepsilon} \int\limits_{\widehat{\varphi(\widehat{z},\frac{z}{\varepsilon})}}^{k} \sigma_{\alpha 3}\, dz_3. \tag{19}$$

Remark. Note that, due to (16), $M^{\varepsilon}_{\alpha\beta}, N^{\varepsilon}_{\alpha\beta}$ and Q^{ε}_{α} belong to $L_2(\omega)$.

Let $v(\widehat{z}) \in [C_0^{\infty}(\omega)]^3$ and take $V(z) = \int\limits_0^{z_3} v(\widehat{z})\, dz_3$, as test function in (12) to obtain

$$\begin{aligned}\int\limits_{\widetilde{\Omega}_\varepsilon} \Big(\varepsilon \sigma_{i\alpha}(U^{\varepsilon}) \frac{\partial V_i(z)}{\partial z_\alpha} + \sigma_{i3}(U^{\varepsilon}) v_i(\widehat{z})\Big)\, dz \\ = \int\limits_{\widetilde{\omega}^{\varepsilon}} \sqrt{\frac{1+\varepsilon^2|\nabla_z \varphi|^2}{1+|\nabla_z\varphi|^2}}\, G^{\varepsilon}_{\alpha}\Big(\widehat{z}, \frac{\widehat{z}}{\varepsilon}\Big)\, V_\alpha(z)\, dz \\ + \frac{1}{\varepsilon} \int\limits_{\widetilde{\omega}^{\varepsilon}} \sqrt{\frac{1+\varepsilon^2|\nabla_z \varphi|^2}{1+|\nabla_z\varphi|^2}}\, G^{\varepsilon}_{3}\Big(\widehat{z}, \frac{\widehat{z}}{\varepsilon}\Big)\, V_3(z)\, dS \\ + \int\limits_{\widetilde{\Omega}_\varepsilon} \Big[\varepsilon\, F_\alpha(z)\, V_\alpha(z) + F_3(z)\, V_3(z)\Big]\, dz.\end{aligned}$$

Keeping in mind (11) and (16), and using the fact that

$$\int\limits_{\widetilde{\omega}^{\varepsilon}} \sqrt{\frac{1+\varepsilon^2|\nabla_z \varphi|^2}{1+|\nabla_z\varphi|^2}}\, h(\widehat{z}, \varphi)\, dS = \int\limits_{\omega} \sqrt{1+\varepsilon^2\big|\nabla_z\varphi\big|^2}\, h(\widehat{z},\varphi)\, d\widehat{z},$$

we have that, as $\varepsilon \to 0$,

$$\int\limits_{\widetilde{\Omega}_\varepsilon} \sigma_{i3}(U^{\varepsilon})\, v_i\, dz \to 0,$$

whence

$$\sigma^0_{i3} = 0. \tag{20}$$

Taking $(0, 0, v)$ with $v \in H^1_0(\omega)$, as a test function in (12), and recalling definition (19), we get

$$\int_\omega Q^\varepsilon_\beta(\widehat{z})\,\frac{\partial v(\widehat{z})}{\partial z_\beta}\,d\widehat{z} = \frac{1}{\varepsilon^3}\int_\omega \sqrt{1+\varepsilon^2\left|\nabla_z\varphi\right|^2}\;G^\varepsilon_3\Big(\widehat{z},\frac{\widehat{z}}{\varepsilon}\Big)\,v(\widehat{z})\,d\widehat{z} \\ + \frac{1}{\varepsilon^2}\int_\omega\Big(\int_{\varphi(\widehat{z},\frac{\widehat{z}}{\varepsilon})}^{k} F^\varepsilon_3(z)\,dz_3\Big)\,v(\widehat{z})\,d\widehat{z}. \tag{21}$$

Due to hypotheses (11) and using Lemma 5 from [9] (see also [3]), one obtains the estimate

$$\left|\int_\omega\Big[\frac{1}{\varepsilon^3}\sqrt{1+\varepsilon^2\left|\nabla_z\varphi\right|^2}\;G^\varepsilon_3\Big(\widehat{z},\frac{\widehat{z}}{\varepsilon}\Big) - \mathcal{G}_3(\widehat{z})\Big]v(\widehat{z})\,d\widehat{z}\right| \le C_1\,\varepsilon\,\|v\|_{H^1(\omega)},$$

where

$$\mathcal{G}_3(\widehat{z}) = \int_T \sqrt{1+\left|\nabla_\xi\varphi(\widehat{z},\widehat{\xi})\right|^2}\;G_3(\widehat{z},\widehat{\xi})\,d\widehat{\xi},$$

and

$$T = \{\xi \;:\; 0<\xi_\alpha<1\}. \tag{22}$$

Now, observe that the second integral in the right-hand side of (21) can be rewritten as follows:

$$\frac{1}{\varepsilon^2}\int_\omega\Big(\int_{\varphi(\widehat{z},\frac{\widehat{z}}{\varepsilon})}^{k} F^\varepsilon_3(z)\,dz_3\Big)\,v(\widehat{z})\,d\widehat{z} = \int_\omega\Big(\int_{-k}^{k}\chi_{[\varphi(\widehat{z},\frac{\widehat{z}}{\varepsilon}),k]}(z_3)\,F_3(z)\,dz_3\Big)\,v(\widehat{z})\,d\widehat{z},$$

where $\chi_{[\varphi(\widehat{z},\frac{\widehat{z}}{\varepsilon}),k]}(z_3)$ is a characteristic function of $\widetilde{\Omega}_\varepsilon$, i.e.,

$$\chi_{[\varphi(\widehat{z},\frac{\widehat{z}}{\varepsilon}),k]}(z_3) = \begin{cases} 1 & \text{if } \varphi\Big(\widehat{z},\dfrac{\widehat{z}}{\varepsilon}\Big)\le z_3\le k, \quad \widehat{z}\in\omega,\\[2mm] 0 & \text{if } -k\le z_3<\varphi\Big(\widehat{z},\dfrac{\widehat{z}}{\varepsilon}\Big), \quad \widehat{z}\in\omega,\\[2mm] 0 & \text{if } \widehat{z}\notin\omega. \end{cases} \tag{23}$$

By Lemma 5 from [9], we have the estimate

$$\left|\int_\omega\Big[\int_{-k}^{k}\Big(\chi_{[\varphi(\widehat{z},\frac{\widehat{z}}{\varepsilon}),k]}(z_3) - X(z)\Big)\,F_3(z)dz_3\Big]\,v(\widehat{z})\,d\widehat{z}\right| \le C_2\,\varepsilon\,\|v\|_{H^1(\omega)}$$

where

$$X(z) = \int_T \chi_{[\varphi(\widehat{z},\widehat{\xi}),k]}(z_3)\, d\widehat{\xi}. \tag{24}$$

Consequently, when passing to the limit in (21), and recalling (20), we obtain the limit system

$$\begin{cases} -\dfrac{\partial}{\partial z_\beta} Q^0_\beta(\widehat{z}) = \mathcal{G}_3(\widehat{z}) + \displaystyle\int_{-k}^{k} X(z)\, F_3(z)\, dz_3 \quad \text{for } \widehat{z} \in \omega, \\ Q^0_\beta n^\varepsilon_\beta = 0 \quad \text{on } \partial\omega. \end{cases} \tag{25}$$

Here $Q^0_\alpha(\widehat{z})$ denotes the weak $L_2(\omega)$–limit of $Q^\varepsilon_\alpha(\widehat{z})$ as $\varepsilon \to 0$.

Let now take $(v_1, v_2, 0)$ with $v_\alpha \in H^1_0(\omega)$ as a test function in (12). We get (see definition (19)),

$$\begin{aligned} \int_\omega N^\varepsilon_{\alpha\beta}(\widehat{z})\, \frac{\partial v_\alpha(\widehat{z})}{\partial z_\beta}\, d\widehat{z} &= \frac{1}{\varepsilon} \int_\omega \sqrt{1 + \varepsilon^2 |\nabla_z \varphi|^2}\;\, G^\varepsilon_\alpha\Big(\widehat{z}, \frac{\widehat{z}}{\varepsilon}\Big)\, v_\alpha(\widehat{z})\, d\widehat{z} \\ &\quad + \int_\omega \Big(\int_{\varphi(\widehat{z},\widehat{\frac{z}{\varepsilon}})}^{k} F^\varepsilon_\alpha(z)\, dz_3 \Big)\, v_\alpha(\widehat{z})\, d\widehat{z}. \end{aligned}$$

This, together with (25), leads to the system

$$\begin{cases} -\dfrac{\partial}{\partial z_\beta} N^0_{\alpha\beta}(\widehat{z}) = \mathcal{G}_\alpha(\widehat{z}) + \displaystyle\int_{-k}^{k} X(z)\, F_\alpha(z)\, dz_3 \quad \text{for } \widehat{z} \in \omega, \\ N^0_{\alpha\beta} n^\varepsilon_\beta = 0 \quad \text{on } \partial\omega, \end{cases} \tag{26}$$

where $N^0_{\alpha\beta}(\widehat{z})$ is the weak $L_2(\omega)$– limit of $N^\varepsilon_{\alpha\beta}(\widehat{z})$ as $\varepsilon \to 0$, and

$$\mathcal{G}_\alpha(\widehat{z}) = \int_T \sqrt{1 + |\nabla_\xi \varphi(\widehat{z}, \widehat{\xi})|^2}\; G_\alpha(\widehat{z}, \widehat{\xi})\, d\widehat{\xi}.$$

Finally, let $v_\alpha \in H^1_0(\omega)$ and take $(z_3 v_1, z_3 v_2, 0)$ as test function in (12).

Using defitions (18) and (19), it follows that

$$\int_\omega \Big(M^\varepsilon_{\alpha\beta}(\widehat{z}) \, \frac{\partial v_\alpha(\widehat{z})}{\partial z_\beta} + Q^\varepsilon_\alpha(\widehat{z}) \, v_\alpha(\widehat{z}) \Big) \, d\widehat{z}$$

$$= \int_\omega \Big(\int_{\varphi(\widehat{z},\frac{\widehat{z}}{\varepsilon})}^{k} z_3 \, F^\varepsilon_\alpha(z) \, dz_3 \Big) \, v_\alpha(\widehat{z}) \, d\widehat{z} \tag{27}$$

$$+ \frac{1}{\varepsilon} \int_\omega \sqrt{1 + \varepsilon^2 |\nabla_z \varphi|^2} \; G^\varepsilon_\alpha\Big(\widehat{z}, \frac{\widehat{z}}{\varepsilon}\Big) \, \varphi\Big(\widehat{z}, \frac{\widehat{z}}{\varepsilon}\Big) \, v_\alpha(\widehat{z}) \, d\widehat{z}.$$

As above, bearing in mind hypotheses (11) and using Lemma 5 from [9], we have

$$\Big| \int_\omega \Big[\frac{1}{\varepsilon} \sqrt{1 + \varepsilon^2 |\nabla_z \varphi|^2} \, G^\varepsilon_\alpha\Big(\widehat{z}, \frac{\widehat{z}}{\varepsilon}\Big) \, \varphi\Big(\widehat{z}, \frac{\widehat{z}}{\varepsilon}\Big) - \mathcal{G}^\varphi_\alpha(\widehat{z}) \Big] v_\alpha(\widehat{z}) \, d\widehat{z} \Big| \le C_3 \, \varepsilon \, \|v\|_{H^1(\omega)},$$

where

$$\mathcal{G}^\varphi_\alpha(\widehat{z}) = \int_T \sqrt{1 + |\nabla_\xi \varphi(\widehat{z}, \widehat{\xi})|^2} \; G_\alpha(\widehat{z}, \widehat{\xi}) \, \varphi(\widehat{z}, \widehat{\xi}) \, d\widehat{\xi}.$$

Recalling the definition of the set T (see (22)), the first integral in the right-hand side of (27) can be rewritten in the form

$$\int_\omega \Big[\int_{\varphi(\widehat{z},\frac{\widehat{z}}{\varepsilon})}^{k} z_3 \; F^\varepsilon_\alpha(z) \, dz_3 \Big] v_\alpha(\widehat{z}) \, d\widehat{z} = \int_\omega \Big[\int_{-k}^{k} z_3 \, \chi_{[\varphi(\widehat{z},\frac{\widehat{z}}{\varepsilon}),k]}(z_3) \; F_\alpha(z) \, dz_3 \Big] \, v_\alpha(\widehat{z}) \, d\widehat{z},$$

where $\chi_{[\varphi(\widehat{z},\frac{\widehat{z}}{\varepsilon}),k]}$ was defined in (23). Then, again by Lemma 5 from [9], we have the estimate

$$\Big| \int_\omega \Big[\int_{-k}^{k} z_3 \Big(\chi_{[\varphi(\widehat{z},\frac{\widehat{z}}{\varepsilon}),k]}(z_3) - X(z) \Big) F_\alpha(z) \, dz_3 \Big] \, v_\alpha(\widehat{z}) \, d\widehat{z} \le C_4 \, \varepsilon \, \|v\|_{H^1(\omega)},$$

where X was defined by (24). Passing to the limit as $\varepsilon \to 0$ in (27), we obtain

$$\begin{cases} -\dfrac{\partial}{\partial z_\beta} M^0_{\alpha\beta}(\widehat{z}) + Q^0_\alpha(\widehat{z}) = \mathcal{G}^\varphi_\alpha(\widehat{z}) + \displaystyle\int_{-k}^{k} z_3 \, X(z) \, F_\alpha(z) \, dz_3 \quad \text{in } \omega, \\ M^0_{\alpha\beta} n^\varepsilon_\beta = 0 \quad \text{on } \partial\omega, \end{cases} \tag{28}$$

where $M^0_{\alpha\beta}$ is the weak $L_2(\omega)-$ limit of the moment $M^\varepsilon_{\alpha\beta}$, as $\varepsilon \to 0$.

To finish this subsection, observe that the functions $N^\varepsilon_{\alpha\beta}$ and $M^\varepsilon_{\alpha\beta}$ can be rewritten in the form

$$N^\varepsilon_{\alpha\beta}(\widehat{z}) = \int_{\varphi(\widehat{z},\widehat{\frac{z}{\varepsilon}})}^{k} \sigma_{\alpha\beta}(U^\varepsilon)\, dz_3 = \int_{-k}^{k} \chi_{[\varphi(\widehat{z},\widehat{\frac{z}{\varepsilon}}),k]}(z_3)\, \sigma_{\alpha\beta}(U^\varepsilon)\, dz_3, \tag{29}$$

respectively,

$$M^\varepsilon_{\alpha\beta}(\widehat{z}) = \int_{\varphi(\widehat{z},\widehat{\frac{z}{\varepsilon}})}^{k} z_3\, \sigma_{\alpha\beta}(U^\varepsilon)\, dz_3 = \int_{-k}^{k} z_3\, \chi_{[\varphi(\widehat{z},\widehat{\frac{z}{\varepsilon}}),k]}(z_3)\, \sigma_{\alpha\beta}(U^\varepsilon)\, dz_3. \tag{30}$$

2.3. Auxiliary problems

To understand what $M^0_{\alpha\beta}$ and $N^0_{\alpha\beta}$ are, we have to know the limit of $\sigma_{\alpha\beta}(U^\varepsilon)$. To do so, we will introduce a formal asymptotic expansion of the function U^ε of the form

$$U^\varepsilon(z) = U^0(z) + \varepsilon U^1(z, \frac{\widehat{z}}{\varepsilon}) + \varepsilon^2 U^2(z, \frac{\widehat{z}}{\varepsilon}) + \varepsilon^3 U^3(z, \frac{\widehat{z}}{\varepsilon}) + \ldots \tag{31}$$

We substitute (31) for U^ε into all the equations in (9) and equate the power-like terms of ε. We do not detail here the computations since they are obvious (and rather technical).

From the first equations in (9), taking into account that U^1, U^2 and U^3 are independent of ξ_3, we obtain an infinite chain of equations. When looking for the coefficients of ε^2 in (9), we obtain that U^1_3 satisfies

$$a^{\gamma\alpha}_{3\beta}(z)\frac{\partial^2 U^1_3}{\partial\xi_\gamma\,\partial\xi_\beta} + a^{3\alpha}_{3\beta}(z)\frac{\partial^2 U^1_3}{\partial z_3\,\partial\xi_\beta} + a^{\gamma\alpha}_{33}(z)\frac{\partial^2 U^1_3}{\partial\xi_\gamma\,\partial z_3} + a^{3\alpha}_{33}(z)\frac{\partial^2 U^1_3}{\partial z_3\,\partial z_3} = 0, \tag{32}$$

and

$$a^{\gamma 3}_{3\beta}(z)\frac{\partial^2 U^1_3}{\partial\xi_\gamma\,\partial\xi_\beta} + a^{33}_{3\beta}(z)\frac{\partial^2 U^1_3}{\partial z_3\,\partial\xi_\beta} + a^{\gamma 3}_{33}(z)\frac{\partial^2 U^1_3}{\partial\xi_\gamma\,\partial z_3} + a^{33}_{33}(z)\frac{\partial^2 U^1_3}{\partial z_3\,\partial z_3} = 0. \tag{33}$$

The next expressions (coefficients of ε^1 in (8)) are

$$
\begin{aligned}
& a_{3\beta}^{\gamma\alpha}(z)\frac{\partial^2 U_3^1}{\partial \xi_\gamma\,\partial z_\beta} + a_{3\beta}^{3\alpha}(z)\frac{\partial^2 U_3^1}{\partial z_3\,\partial z_\beta} + \frac{a_{\delta\beta}^{\gamma\alpha}(z)}{2}\left(\frac{\partial^2 U_\gamma^1}{\partial \xi_\delta\,\partial \xi_\beta} + \frac{\partial^2 U_\delta^1}{\partial \xi_\gamma\,\partial \xi_\beta}\right) \\
&+ a_{3\beta}^{\gamma\alpha}(z)\left(\frac{\partial^2 U_\gamma^1}{\partial z_3\,\partial \xi_\beta} + \frac{\partial^2 U_3^1}{\partial z_\gamma\,\partial \xi_\beta}\right) + a_{3\beta}^{\gamma\alpha}(z)\frac{\partial^2 U_3^2}{\partial \xi_\gamma\,\partial \xi_\beta} + a_{3\beta}^{3\alpha}(z)\frac{\partial^2 U_3^2}{\partial z_3\,\partial \xi_\beta} \\
&+ \frac{a_{\delta 3}^{\gamma\alpha}(z)}{2}\left(\frac{\partial^2 U_\gamma^0}{\partial z_\delta\,\partial z_3} + \frac{\partial^2 U_\delta^0}{\partial z_\gamma\,\partial z_3}\right) + \frac{a_{\delta 3}^{\gamma\alpha}(z)}{2}\left(\frac{\partial^2 U_\gamma^1}{\partial \xi_\delta\,\partial z_3} + \frac{\partial^2 U_\delta^1}{\partial \xi_\gamma\,\partial z_3}\right) \qquad (34) \\
&+ a_{33}^{\gamma\alpha}(z)\left(\frac{\partial^2 U_\gamma^1}{\partial z_3\,\partial z_3} + \frac{\partial^2 U_3^1}{\partial z_\gamma\,\partial z_3}\right) + a_{33}^{\gamma\alpha}(z)\frac{\partial^2 U_3^2}{\partial \xi_\gamma\,\partial z_3} + a_{33}^{3\alpha}(z)\frac{\partial^2 U_3^2}{\partial z_3\,\partial z_3} \\
&= 0,
\end{aligned}
$$

and

$$
\begin{aligned}
& a_{3\beta}^{\gamma 3}(z)\frac{\partial^2 U_3^1}{\partial \xi_\gamma\,\partial z_\beta} + a_{3\beta}^{33}(z)\frac{\partial^2 U_3^1}{\partial z_3\,\partial z_\beta} + \frac{a_{\delta\beta}^{\gamma 3}(z)}{2}\left(\frac{\partial^2 U_\gamma^1}{\partial \xi_\delta\,\partial \xi_\beta} + \frac{\partial^2 U_\delta^1}{\partial \xi_\gamma\,\partial \xi_\beta}\right) \\
&+ a_{3\beta}^{\gamma 3}(z)\left(\frac{\partial^2 U_\gamma^1}{\partial z_3\,\partial \xi_\beta} + \frac{\partial^2 U_3^1}{\partial z_\gamma\,\partial \xi_\beta}\right) + a_{3\beta}^{\gamma 3}(z)\frac{\partial^2 U_3^2}{\partial \xi_\gamma\,\partial \xi_\beta} + a_{3\beta}^{33}(z)\frac{\partial^2 U_3^2}{\partial z_3\,\partial \xi_\beta} \\
&+ \frac{a_{\delta 3}^{\gamma 3}(z)}{2}\left(\frac{\partial^2 U_\gamma^0}{\partial z_\delta\,\partial z_3} + \frac{\partial^2 U_\delta^0}{\partial z_\gamma\,\partial z_3}\right) + \frac{a_{\delta 3}^{\gamma 3}(z)}{2}\left(\frac{\partial^2 U_\gamma^1}{\partial \xi_\delta\,\partial z_3} + \frac{\partial^2 U_\delta^1}{\partial \xi_\gamma\,\partial z_3}\right) \qquad (35) \\
&+ a_{33}^{\gamma 3}(z)\left(\frac{\partial^2 U_\gamma^1}{\partial z_3\,\partial z_3} + \frac{\partial^2 U_3^1}{\partial z_\gamma\,\partial z_3}\right) + a_{33}^{\gamma 3}(z)\frac{\partial^2 U_3^2}{\partial \xi_\gamma\,\partial z_3} + a_{33}^{33}(z)\frac{\partial^2 U_3^2}{\partial z_3\,\partial z_3} \\
&= 0.
\end{aligned}
$$

The treatement of boundary conditions in (9) is more delicate since they contain the terms

$$\varepsilon\sqrt{\frac{1+\varepsilon^2|\nabla_z\varphi|^2}{1+|\nabla_z\varphi|^2}}\,G_\alpha\Big(\widehat{z},\frac{\widehat{z}}{\varepsilon}\Big) \quad \text{and} \quad \varepsilon^2\sqrt{\frac{1+\varepsilon^2|\nabla_z\varphi|^2}{1+|\nabla_z\varphi|^2}}\,G_3\Big(\widehat{z},\frac{\widehat{z}}{\varepsilon}\Big).$$

It is easy to see that, if $|\nabla_\xi\varphi(\widehat{z},\widehat{\xi})| \neq 0$, then

$$
\begin{aligned}
\sqrt{\frac{1+\varepsilon^2\big|\nabla_z\varphi(\widehat{z},\frac{\widehat{z}}{\varepsilon})\big|^2}{1+\big|\nabla_z\varphi(\widehat{z},\frac{\widehat{z}}{\varepsilon})\big|^2}} &= \varepsilon\,\frac{\sqrt{1+\big|\nabla_\xi\varphi(\widehat{z},\widehat{\xi})\big|^2}}{\big|\nabla_\xi\varphi(\widehat{z},\widehat{\xi})\big|}\Bigg|_{\widehat{\xi}=\frac{\widehat{z}}{\varepsilon}} \\
&\quad -\varepsilon^2\,\frac{\big(\nabla_z\varphi(\widehat{z},\widehat{\xi}),\nabla_\xi\varphi(\widehat{z},\widehat{\xi})\big)}{\big|\nabla_\xi\varphi(\widehat{z},\widehat{\xi})\big|^3\sqrt{1+\big|\nabla_\xi\varphi(\widehat{z},\widehat{\xi})\big|^2}}\Bigg|_{\widehat{\xi}=\frac{\widehat{z}}{\varepsilon}} + O(\varepsilon^3).
\end{aligned}
$$

At the points where $|\nabla_\xi \varphi(\widehat{z}, \widehat{\xi})| = 0$, one has

$$\sqrt{\frac{1+\varepsilon^2\big|\nabla_z\varphi(\widehat{z},\dfrac{\widehat{z}}{\varepsilon})\big|^2}{1+\big|\nabla_z\varphi(\widehat{z},\dfrac{\widehat{z}}{\varepsilon})\big|^2}} = O(1).$$

This allows us to equate the power-like terms of ε for the terms coming from the boundary $\tilde{\omega}^\varepsilon$ in (9). Remembering the transformation of the normal to $\widetilde{\omega}^\varepsilon$ by means of the change of variables $(z_1, z_2, z_3) \to (\xi_1, \xi_2, z_3)$, we end up with the following expressions:

$$\Big[a_{3\beta}^{\gamma\alpha}\frac{\partial U_3^1}{\partial \xi_\gamma} + a_{3\beta}^{3\alpha}\frac{\partial U_3^1}{\partial z_3}\Big]\widetilde{\nu}_\beta^\varepsilon + a_{33}^{\gamma\alpha}\frac{\partial U_3^1}{\partial \xi_\gamma}\widetilde{\nu}_3^\varepsilon = 0 \quad \text{on } \tilde{\omega}^\varepsilon, \tag{36}$$

$$\Big[a_{3\beta}^{\gamma 3}\frac{\partial U_3^1}{\partial \xi_\gamma} + a_{3\beta}^{33}\frac{\partial U_3^1}{\partial z_3}\Big]\widetilde{\nu}_\beta^\varepsilon + a_{33}^{\gamma 3}\frac{\partial U_3^1}{\partial \xi_\gamma}\widetilde{\nu}_3^\varepsilon = 0 \quad \text{on } \tilde{\omega}^\varepsilon, \tag{37}$$

and also

$$\begin{aligned}
&\Big[\frac{a_{\delta\beta}^{\gamma\alpha}}{2}\Big(\frac{\partial U_\gamma^0}{\partial z_\delta} + \frac{\partial U_\delta^0}{\partial z_\gamma}\Big) + \frac{a_{\delta\beta}^{\gamma\alpha}}{2}\Big(\frac{\partial U_\gamma^1}{\partial \xi_\delta} + \frac{\partial U_\delta^1}{\partial \xi_\gamma}\Big) + a_{3\beta}^{\gamma\alpha}\Big(\frac{\partial U_\gamma^1}{\partial z_3} + \frac{\partial U_3^1}{\partial z_\gamma}\Big) \\
&\quad + a_{3\beta}^{\gamma\alpha}\frac{\partial U_3^2}{\partial \xi_\gamma} + a_{3\beta}^{3\alpha}\frac{\partial U_3^2}{\partial z_3}\Big]\widetilde{\nu}_\beta^\varepsilon + \Big[a_{33}^{3\alpha}\frac{\partial U_3^1}{\partial z_3} + \frac{a_{\delta 3}^{\gamma\alpha}}{2}\Big(\frac{\partial U_\gamma^0}{\partial z_\delta} + \frac{\partial U_\delta^0}{\partial z_\gamma}\Big) \\
&\quad + \frac{a_{\delta 3}^{\gamma\alpha}}{2}\Big(\frac{\partial U_\gamma^1}{\partial \xi_\delta} + \frac{\partial U_\delta^1}{\partial \xi_\gamma}\Big) + a_{33}^{\gamma\alpha}\Big(\frac{\partial U_\gamma^1}{\partial z_3} + \frac{\partial U_3^1}{\partial z_\gamma}\Big) \\
&\quad + a_{33}^{\gamma\alpha}\frac{\partial U_3^2}{\partial \xi_\gamma} + a_{33}^{3\alpha}\frac{\partial U_3^2}{\partial z_3}\Big]\widetilde{\nu}_3^\varepsilon = 0,
\end{aligned} \tag{38}$$

and

$$\begin{aligned}
&\Big[\frac{a_{\delta\beta}^{\gamma 3}}{2}\Big(\frac{\partial U_\gamma^0}{\partial z_\delta} + \frac{\partial U_\delta^0}{\partial z_\gamma}\Big) + \frac{a_{\delta\beta}^{\gamma 3}}{2}\Big(\frac{\partial U_\gamma^1}{\partial \xi_\delta} + \frac{\partial U_\delta^1}{\partial \xi_\gamma}\Big) + a_{3\beta}^{\gamma 3}\Big(\frac{\partial U_\gamma^1}{\partial z_3} + \frac{\partial U_3^1}{\partial z_\gamma}\Big) \\
&\quad + a_{3\beta}^{\gamma 3}\frac{\partial U_3^2}{\partial \xi_\gamma} + a_{3\beta}^{33}(z)\frac{\partial U_3^2}{\partial z_3}\Big]\widetilde{\nu}_\beta^\varepsilon + \Big[a_{33}^{33}(z)\frac{\partial U_3^1}{\partial z_3} + \frac{a_{\delta 3}^{\gamma 3}}{2}\Big(\frac{\partial U_\gamma^0}{\partial z_\delta} + \frac{\partial U_\delta^0}{\partial z_\gamma}\Big) \\
&\quad + a_{\delta 3}^{\gamma 3}\frac{1}{2}\Big(\frac{\partial U_\gamma^1}{\partial \xi_\delta} + \frac{\partial U_\delta^1}{\partial \xi_\gamma}\Big) + a_{33}^{\gamma 3}\Big(\frac{\partial U_\gamma^1}{\partial z_3} + \frac{\partial U_3^1}{\partial z_\gamma}\Big) \\
&\quad + a_{33}^{\gamma 3}(z)\frac{\partial U_3^2}{\partial \xi_\gamma} + a_{33}^{33}\frac{\partial U_3^2}{\partial z_3}\Big]\widetilde{\nu}_3^\varepsilon = 0,
\end{aligned} \tag{39}$$

Lemma 1 – *The vector-function $U^0 = U^0(z)$ has the form*

$$U^0(z) = \Big(-z_3\frac{\partial U(\widehat{z})}{\partial z_1} + \mathcal{U}_1(\widehat{z}),\, -z_3\frac{\partial U(\widehat{z})}{\partial z_2} + \mathcal{U}_2(\widehat{z}),\, U(\widehat{z})\Big),$$

where $U \in H_0^2(\omega)$ and $\mathcal{U} \in H_0^1(\omega)]^2$.

Proof. Estimate (16) implies

$$\Big\|\frac{\partial U_3^\varepsilon}{\partial z_3}\Big\|_{L_2(\widetilde{\Omega}_\varepsilon)} \leq K_4\varepsilon^2. \tag{40}$$

Substituting expansion (31) of U_3^ε into (40), and passing afterwards to the limit as $\varepsilon \to 0$, one obtains, due to (17),

$$\frac{\partial U_3^0}{\partial z_3} = 0.$$

This means that

$$U_3^0 = U,$$

where

$$U = U(\widehat{z}) \qquad \text{with } U \in H_0^1(\omega).$$

Similarly, estimates (13)–(15) yield

$$\frac{\partial U_\alpha^0}{\partial z_3} + \frac{\partial U_3^0}{\partial z_\alpha} = 0,$$

and hence,

$$U_\alpha^0 = -z_3\frac{\partial U(\widehat{z})}{\partial z_\alpha} + \mathcal{U}_\alpha(\widehat{z}), \qquad \text{with } \mathcal{U}_\alpha \in H_0^1(\omega).$$

Following [11] (see also [12], [13], [14] and [15]) one gets successively

$$U \in H^2(\omega), \qquad U|_{\partial\omega} = \frac{\partial U}{\partial n^\varepsilon}\Big|_{\partial\omega} = 0,$$

and this ends the proof of the lemma. ■

From equations (32)–(35) and boundary conditions (36)–(39), we are led to look for $U^1 = (U_1^1, U_2^1, U_3^1)$ in the form

$$\left\{\begin{aligned} U_1^1\big(z,\frac{\widehat{z}}{\varepsilon}\big) &= \Big[V_1^{\alpha\beta}\big(z,\frac{\widehat{z}}{\varepsilon}\big) + \frac{1}{2}\, z_3^2\, \big[a_{pq}^{33}(\widehat{z},0)\big]_{j1}^{-1}\; a_{j\alpha}^{3\beta}(\widehat{z},0)\Big] \frac{\partial^2 U(\widehat{z})}{\partial z_\alpha\, \partial z_\beta} \\ &\quad + \Big[W_1^{\alpha\beta}\big(z,\frac{\widehat{z}}{\varepsilon}\big) - z_3\, \big[a_{pq}^{33}(\widehat{z},0)\big]_{j1}^{-1}\; a_{j\alpha}^{3\beta}(\widehat{z},0)\Big] e_{\alpha\beta}(\mathcal{U}), \\ U_2^1\big(z,\frac{\widehat{z}}{\varepsilon}\big) &= \Big[V_2^{\alpha\beta}\big(z,\frac{\widehat{z}}{\varepsilon}\big) + \frac{1}{2}\, z_3^2\, \big[a_{pq}^{33}(\widehat{z},0)\big]_{j2}^{-1}\; a_{j\alpha}^{3\beta}(\widehat{z},0)\Big] \frac{\partial^2 U(\widehat{z})}{\partial z_\alpha\, \partial z_\beta} \\ &\quad + \Big[W_2^{\alpha\beta}\big(z,\frac{\widehat{z}}{\varepsilon}\big) - z_3\, \big[(a_{pq}^{33}(\widehat{z},0)\big]_{j2}^{-1}\; a_{j\alpha}^{3\beta}(\widehat{z},0)\Big] e_{\alpha\beta}(\mathcal{U}), \\ U_3^1\big(z,\frac{\widehat{z}}{\varepsilon}\big) &= 0. \end{aligned}\right. \tag{41}$$

and for $U^2 = (U_1^2, U_2^2, U_3^2)$ in the form

$$\left\{\begin{aligned} U_1^2\big(z,\frac{\widehat{z}}{\varepsilon}\big) &= 0, \\ U_2^2\big(z,\frac{\widehat{z}}{\varepsilon}\big) &= 0, \\ U_3^2\big(z,\frac{\widehat{z}}{\varepsilon}\big) &= \Big[V_3^{\alpha\beta}\big(z,\frac{\widehat{z}}{\varepsilon}\big) + \frac{1}{2}\, z_3^2\, \big[a_{pq}^{33}(\widehat{z},0)\big]_{j3}^{-1}\; a_{j\alpha}^{3\beta}(\widehat{z},0)\Big] \frac{\partial^2 U(\widehat{z})}{\partial z_\alpha\, \partial z_\beta} \\ &\quad + \Big[W_3^{\alpha\beta}\big(z,\frac{\widehat{z}}{\varepsilon}\big) - z_3\, \big[a_{pq}^{33}(\widehat{z},0)\big]_{j3}^{-1}\; a_{j\alpha}^{3\beta}(\widehat{z},0)\Big] e_{\alpha\beta}(\mathcal{U}). \end{aligned}\right. \tag{42}$$

In these formulas, for each α and ρ, the vector-function $V^{\alpha\rho} = V^{\alpha\rho}(z, \widehat{z}/\varepsilon)$ satisfies in the domain $\omega_T = \omega \times T$ (see(22)), the following equation:

$$\begin{aligned} &\frac{a_{\delta\beta}^{\gamma i}}{2}(\widehat{z},0)\left(\frac{\partial^2 V_\gamma^{\alpha\rho}}{\partial\xi_\delta\,\partial\xi_\beta} + \frac{\partial^2 V_\delta^{\alpha\rho}}{\partial\xi_\gamma\,\partial\xi_\beta}\right) + a_{3\beta}^{\gamma i}(\widehat{z},0)\frac{\partial^2 V_\gamma^{\alpha\rho}}{\partial z_3\,\partial\xi_\beta} \\ &\quad + \frac{a_{\delta 3}^{\gamma i}}{2}(\widehat{z},0)\left(\frac{\partial^2 V_\gamma^{\alpha\rho}}{\partial\xi_\delta\,\partial z_3} + \frac{\partial^2 V_\delta^{\alpha\rho}}{\partial\xi_\gamma\,\partial z_3}\right) + a_{33}^{\gamma i}(\widehat{z},0)\frac{\partial^2 V_\gamma^{\alpha\rho}}{\partial z_3\,\partial z_3} \\ &\quad + a_{3\beta}^{\gamma i}(\widehat{z},0)\frac{\partial^2 V_3^{\alpha\rho}}{\partial\xi_\gamma\,\partial\xi_\beta} + a_{3\beta}^{3i}(\widehat{z},0)\frac{\partial^2 V_3^{\alpha\rho}}{\partial z_3\,\partial\xi_\beta} \\ &\quad + a_{33}^{\gamma i}(\widehat{z},0)\frac{\partial^2 V_3^{\alpha\rho}}{\partial\xi_\gamma\,\partial z_3} + a_{33}^{3i}(\widehat{z},0)\frac{\partial^2 V_3^{\alpha\rho}}{\partial z_3\,\partial z_3} = b_{\alpha 3}^{\rho i}(\widehat{z},0) \end{aligned} \tag{43}$$

with the boundary conditions on $\partial\omega_T$

$$\begin{aligned} &\Big[\frac{a_{\delta k}^{\gamma i}}{2}(\widehat{z},0)\Big(\frac{\partial V_\gamma^{\alpha\rho}}{\partial\xi_\delta} + \frac{\partial V_\delta^{\alpha\rho}}{\partial\xi_\gamma}\Big) + a_{3k}^{\gamma i}(\widehat{z},0)\Big(\frac{\partial V_\gamma^{\alpha\rho}}{\partial z_3} + \frac{\partial V_3^{\alpha\rho}}{\partial\xi_\gamma}\Big) \\ &\qquad + a_{3k}^{3i}(\widehat{z},0)\frac{\partial V_3^{\alpha\rho}}{\partial z_3}\Big]\nu_k = z_3\, b_{\alpha k}^{\rho i}(\widehat{z},0)\,\nu_k. \end{aligned} \tag{44}$$

Also, for each α and ρ, the vector-function $W^{\alpha\rho} = W^{\alpha\rho}(z, \widehat{z}/\varepsilon)$ from formulas (41) and (42), satisfies in ω_T the equation

$$\begin{aligned}
&\frac{a^{\gamma i}_{\delta\beta}}{2}(\widehat{z},0)\left(\frac{\partial^2 W^{\alpha\rho}_{\gamma}}{\partial\xi_\delta\,\partial\xi_\beta}+\frac{\partial^2 W^{\alpha\rho}_{\delta}}{\partial\xi_\gamma\,\partial\xi_\beta}\right)+a^{\gamma i}_{3\beta}(\widehat{z},0)\frac{\partial^2 W^{\alpha\rho}_{\gamma}}{\partial z_3\,\partial\xi_\beta}\\
&+\frac{a^{\gamma i}_{\delta 3}}{2}(\widehat{z},0)\left(\frac{\partial^2 W^{\alpha\rho}_{\gamma}}{\partial\xi_\delta\,\partial z_3}+\frac{\partial^2 W^{\alpha\rho}_{\delta}}{\partial\xi_\gamma\,\partial z_3}\right)+a^{\gamma i}_{33}(\widehat{z},0)\frac{\partial^2 W^{\alpha\rho}_{\gamma}}{\partial z_3\,\partial z_3}\\
&+a^{\gamma i}_{3\beta}(\widehat{z},0)\frac{\partial^2 W^{\alpha\rho}_{3}}{\partial\xi_\gamma\,\partial\xi_\beta}+a^{3i}_{3\beta}(\widehat{z},0)\frac{\partial^2 W^{\alpha\rho}_{3}}{\partial z_3\,\partial\xi_\beta}\\
&+a^{\gamma i}_{33}(\widehat{z},0)\frac{\partial^2 W^{\alpha\rho}_{3}}{\partial\xi_\gamma\,\partial z_3}+a^{3i}_{33}(\widehat{z},0)\frac{\partial^2 W^{\alpha\rho}_{3}}{\partial z_3\,\partial z_3}=0
\end{aligned}\tag{45}$$

with the boundary conditions

$$\begin{aligned}
\Big[\frac{a^{\gamma i}_{\delta k}}{2}(\widehat{z},0)\Big(\frac{\partial W^{\alpha\rho}_{\gamma}}{\partial\xi_\delta}+\frac{\partial W^{\alpha\rho}_{\delta}}{\partial\xi_\gamma}\Big)+a^{\gamma i}_{3k}(\widehat{z},0)\Big(\frac{\partial W^{\alpha\rho}_{\gamma}}{\partial z_3}+\frac{\partial W^{\alpha\rho}_{3}}{\partial\xi_\gamma}\Big)\\
+a^{3i}_{3k}(\widehat{z},0)\frac{\partial W^{\alpha\rho}_{3}}{\partial z_3}\Big]\nu_k=-b^{\rho i}_{\alpha k}(\widehat{z},0)\,\nu_k.
\end{aligned}\tag{46}$$

Remark. Observe that with systems (43)–(44) and (45)–(46), we have determined the first three terms of expansion (31). The complete characterization of its first term, the "homogenized solution" will be given in the next section.

3. Limit problems

The aim of this section is to state the "homogenization" result showing that the limit function U^0 is a good approximation of the "physical" displacement U^ε, solution of system (9). Actually, one has the following result:

Theorem 1 – *Let $U^\varepsilon \in [H^1(\widetilde{\Omega}_\varepsilon, S)]^3$ be the solution of system (9) and assume that hypothesis (11) holds true. Then as $\varepsilon \to 0$, one has the convergence*

$$U^\varepsilon \rightharpoonup U^0 \quad \textit{weakly in } [H^1(\widetilde{\Omega}_\varepsilon, S)]^3.$$

The limit function $U^0 = (U^0_1, U^0_2, U^0_3)$ has the form

$$U^0_3(z) = U(\widehat{z}), \qquad U^0_\alpha(z) = -z_3\frac{\partial U(\widehat{z})}{\partial z_\alpha} + \mathcal{U}_\alpha(\widehat{z}),$$

with $U \in H_0^2(\omega)$ and $\mathcal{U} \in [H_0^1(\omega)]^2$ and satisfying the limit problems

$$
\begin{aligned}
&\int\limits_{\omega} b_{\alpha\gamma}^{\beta\delta}(\widehat{z},0)\Big[\mathcal{M}_X^2(\widehat{z})\frac{\partial^2 U}{\partial z_\gamma\,\partial z_\delta} - e_{\gamma\delta}(\mathcal{U})\,\mathcal{M}_X^1(\widehat{z})\Big]\frac{\partial^2 v}{\partial z_\alpha\,\partial z_\beta}\,d\widehat{z} \\
&\qquad - \int\limits_{\omega}\Big[\mathcal{M}_V^1(\widehat{z})\,\frac{\partial^2 U}{\partial z_\gamma\,\partial z_\delta} - \mathcal{M}_W^1(\widehat{z})\,e_{\gamma\delta}(\mathcal{U})\Big]\frac{\partial^2 v}{\partial z_\alpha\,\partial z_\beta}\,d\widehat{z} \\
&\quad = \int\limits_{\omega}\Big[\int\limits_{-k}^{k} F_3^0 X(z)\,dz_3\Big]\,v\,d\widehat{z} + \int\limits_{\omega}\mathcal{G}_3^0\,v\,d\widehat{z} \\
&\qquad - \int\limits_{\omega}\Big[\int\limits_{-k}^{k} z_3 F_\alpha^0\,X(z)\,dz_3\Big]\frac{\partial v}{\partial z_\alpha}\,d\widehat{z} - \int\limits_{\omega}\mathcal{G}_\alpha^0\,\frac{\partial v}{\partial z_\alpha}\,d\widehat{z},\ \forall v \in H_0^2(\omega)
\end{aligned}
\tag{47}
$$

and

$$
\begin{aligned}
&\int\limits_{\omega} b_{\alpha\gamma}^{\beta\delta}(\widehat{z},0)\Big[\mathcal{M}_X^0(\widehat{z})\,e_{\gamma\delta}(\mathcal{U}) - \mathcal{M}_X^1(\widehat{z})\,\frac{\partial^2 U}{\partial z_\gamma\,\partial z_\delta}\Big]\,e_{\alpha\beta}(w)\,d\widehat{z} \\
&\qquad + \int\limits_{\omega}\Big[\mathcal{M}_W^0(\widehat{z})\,\frac{\partial^2 U}{\partial z_\gamma\,\partial z_\delta} - \mathcal{M}_V^0(\widehat{z})\,e_{\gamma\delta}(\mathcal{U})\Big]\,e_{\alpha\beta}(w)\,d\widehat{z} \\
&\quad = \int\limits_{\omega}\Big[\int\limits_{-k}^{k} F_\alpha^0\,X(z)\,dz_3\Big]\,w_\alpha\,d\widehat{z} + \int\limits_{\omega}\mathcal{G}_\alpha^0 w_\alpha\,d\widehat{z}, \quad \forall w \in [H_0^1(\omega)]^2.
\end{aligned}
\tag{48}
$$

where

$$
b_{\alpha\gamma}^{\beta\delta}(\widehat{z},z_3) = a_{\alpha\gamma}^{\beta\delta}(\widehat{z},z_3) - a_{\alpha j}^{\beta 3}(\widehat{z},z_3)\left(a_{pq}^{33}(\widehat{z},z_3)\right)_{ji}^{-1} a_{i\gamma}^{3\delta}(\widehat{z},z_3),
$$

$$
\mathcal{M}_X^i(\widehat{z}) = \int\limits_{-k}^{k} z_3^i\,X(z)\,dz_3, \quad \text{for } i = 0,\,1,\,2
$$

$$
\mathcal{M}_V^0(\widehat{z}) = \int\limits_{-k}^{k}\int\limits_{T} z_3^i\,\chi_{[\varphi(\widehat{z},\widehat{\xi}),k]}(z_3)\,\sigma_{\alpha\beta}(V^{\gamma\delta})\,d\widehat{\xi}\,dz_3
$$

$$
\mathcal{M}_W^0(\widehat{z}) = \int\limits_{-k}^{k}\int\limits_{T} z_3^i\,\chi_{[\varphi(\widehat{z},\widehat{\xi}),k]}(z_3)\,\sigma_{\alpha\beta}(V^{\gamma\delta})\,d\widehat{\xi}\,dz_3.
$$

Proof. Substituting (31) into definition (10) leads to

$$\begin{aligned}\sigma_{ij}(U^\varepsilon) = b^{\rho j}_{\alpha i}(\widehat{z},\varepsilon z_3)\, e_{\alpha\rho}(\mathcal{U}) - b^{\rho j}_{\alpha i}(\widehat{z},\varepsilon z_3)\, z_3\, \frac{\partial^2 U}{\partial z_\alpha\, \partial z_\rho} \\ + \sigma_{ij}(V^{\alpha\rho})\, \frac{\partial^2 U}{\partial z_\alpha\, \partial z_\rho} + \sigma_{ij}(W^{\alpha\rho})\, e_{\alpha\rho}(\mathcal{U}) + O(\varepsilon),\end{aligned} \tag{49}$$

where $\sigma_{ij}(V^{\alpha\rho})$ and $\sigma_{ij}(W^{\alpha\rho})$ are defined respectively, by

$$\begin{aligned}\sigma_{ij}(V^{\alpha\rho}) = \frac{a^{\gamma j}_{\delta i}}{2}(\widehat{z},\varepsilon z_3)\Big(\frac{\partial V^{\alpha\rho}_\gamma}{\partial \xi_\delta} + \frac{\partial V^{\alpha\rho}_\delta}{\partial \xi_\gamma}\Big) + a^{\gamma j}_{3i}(\widehat{z},\varepsilon z_3)\Big(\frac{\partial V^{\alpha\rho}_\gamma}{\partial z_3} + \frac{\partial V^{\alpha\rho}_3}{\partial \xi_\gamma}\Big) \\ + a^{3j}_{3i}(\widehat{z},\varepsilon z_3)\frac{\partial V^{\alpha\rho}_3}{\partial z_3}\end{aligned}$$

$$\begin{aligned}\sigma_{ij}(W^{\alpha\rho}) = \frac{a^{\gamma j}_{\delta i}}{2}(\widehat{z},\varepsilon z_3)\Big(\frac{\partial W^{\alpha\rho}_\gamma}{\partial \xi_\delta} + \frac{\partial W^{\alpha\rho}_\delta}{\partial \xi_\gamma}\Big) + a^{\gamma j}_{3i}(\widehat{z},\varepsilon z_3)\Big(\frac{\partial W^{\alpha\rho}_\gamma}{\partial z_3} + \frac{\partial W^{\alpha\rho}_3}{\partial \xi_\gamma}\Big) \\ + a^{3j}_{3i}(\widehat{z},\varepsilon z_3)\frac{\partial W^{\alpha\rho}_3}{\partial z_3}.\end{aligned}$$

Using formula (49), we can write the function $\sigma^0_{\alpha\beta}$ (see also the technique introduced in [7] and [8]) in the form

$$\sigma^0_{\alpha\beta} = b^{\beta\delta}_{\alpha\gamma}(\widehat{z},0)\; e_{\gamma\delta}(U^0) + \big\langle\sigma_{\alpha\beta}(V^{\gamma\delta})\big\rangle\; \frac{\partial^2 U}{\partial z_\gamma\, \partial z_\delta} + \big\langle\sigma_{\alpha\beta}(W^{\gamma\delta})\big\rangle\; e_{\alpha\rho}(\mathcal{U}). \tag{50}$$

Recall now formulas (29) and (30), satisfied respectively by the moments $N^\varepsilon_{\alpha\beta}$ and $M^\varepsilon_{\alpha\beta}$. Using (50) in these formulas and passing to the limit, we get

$$\begin{cases} N^0_{\alpha\beta} = A^{\beta\delta}_{\alpha\gamma}\; e_{\gamma\delta}(\mathcal{U}) - B^{\beta\delta}_{\alpha\gamma}\dfrac{\partial^2 U}{\partial z_\gamma\, \partial z_\delta}, \\ M^0_{\alpha\beta} = C^{\beta\delta}_{\alpha\gamma}\; e_{\gamma\delta}(\mathcal{U}) - D^{\beta\delta}_{\alpha\gamma}\; \dfrac{\partial^2 U}{\partial z_\gamma\, \partial z_\delta}, \end{cases} \tag{51}$$

where $b^{\beta\delta}_{\alpha\gamma}$ is the following matrix:

$$b^{\beta\delta}_{\alpha\gamma} = a^{\beta\delta}_{\alpha\gamma} - a^{\beta 3}_{\alpha j}\,\big(a^{33}_{pq}\big)^{-1}_{ji}\, a^{3\delta}_{i\gamma},$$

and

$$A^{\beta\delta}_{\alpha\gamma}(\widehat{z}) = b^{\beta\delta}_{\alpha\gamma}(\widehat{z},0)\int_{-k}^{k} X(z)\, dz_3 + \int_{-k}^{k}\int_{T} \sigma_{\alpha\beta}(W^{\gamma\delta})\,\chi_{[\varphi(\widehat{z},\widehat{\xi}),k]}(z_3)d\widehat{\xi}\, dz_3,$$

$$B_{\alpha\gamma}^{\beta\delta}(\widehat{z}) = b_{\alpha\gamma}^{\beta\delta}(\widehat{z},0) \int_{-k}^{k} z_3\, X(z)\, dz_3 - \int_{-k}^{k} \int_{T} \sigma_{\alpha\beta}(V^{\gamma\delta})\, \chi_{[\varphi(\widehat{z},\widehat{\xi}),k]}(z_3) d\widehat{\xi}\, dz_3,$$

$$C_{\alpha\gamma}^{\beta\delta}(\widehat{z}) = b_{\alpha\gamma}^{\beta\delta}(\widehat{z},0) \int_{-k}^{k} z_3\, X(z)\, dz_3 + \int_{-k}^{k} \int_{T} z_3\, \sigma_{\alpha\beta}(W^{\gamma\delta})\, \chi_{[\varphi(\widehat{z},\widehat{\xi}),k]}(z_3) d\widehat{\xi}\, dz_3,$$

$$D_{\alpha\gamma}^{\beta\delta}(\widehat{z}) = b_{\alpha\gamma}^{\beta\delta}(\widehat{z},0) \int_{-k}^{k} z_3^2\, X(z)\, dz_3 - \int_{-k}^{k} \int_{T} z_3\, \sigma_{\alpha\beta}(V^{\gamma\delta})\, \chi_{[\varphi(\widehat{z},\widehat{\xi}),k]}(z_3) d\widehat{\xi}\, dz_3.$$

One easily can check, that the matrix $b_{\alpha\gamma}^{\beta\delta}$ is positive definite.

To prove the assertions of the theorem, we follow along the lines of the proof from [7] and [8] (see also [14] and [15]). Let $v \in H_0^2(\omega)$ and multiply (28) by $\partial v/\partial z_\alpha$ and (25) by v. Passing to the limit and using (51), we obtain (47). Similarly, multiplying (26) by $w \in [H_0^1(\omega)]^2$, we deduce (48) and this ends the proof of the theorem.

References

[1] Beckmann P., Spizzichino A., The scattering of electromagnetic waves from rough surfaces, New York, Pergamon Press 1963.

[2] Belyaev A.G., Mikheev A.G., Shamaev A.S., . Plane wave diffraction by a rapidly oscillating surface, Zh. Vychisl. Mat. i Mat. Fiz., 32, 8 (1992), 1258–1272. (English translation Comput. Math. Math. Phys., 32, 8 (1992), 1121–1133.

[3] Belyaev, A. G., Piatnitski, A.L., and Chechkin, G.A., Asymptotic behavior of solution for boundary-value problem in a perforated domain with oscillating boundary, Siberian Math. Journal, 39, 4 (1998), 621–644.

[4] Bouchitte G., Lidouh A. and Suquet P., Homogénéisation de frontière pour la modélisation du contact entre un corps déformable non linéaire et un corps rigide. C. R. Acad. Sci. Paris, Ser. I, 313 (1991), 967–972.

[5] Brizzi R., Chalot J.P., Homogénéisation de frontière, Ricerche di Mat. 46, 2 (1997), 341–387.

[6] Butazzo G. and Kohn R., 1986. Reinforcement by a thin layer with oscillating thickness, Appl. Math. and Optimiz., 16, 3 (1987), 247–261.

[7] Caillerie D., The effect of a thin inclusion of high rigidity in an elastic body, Math. Methods Applied Sci. , 2 (1980), 251–270

[8] Caillerie D., Thin elastic and periodic plates, Math. Methods Appl. Sci., 6 (1984), 159–191

[9] Chechkin G. A., Friedman A. and Piatnitski A. L., The boundary value problem in domains with very rapidly oscillating boundary, Journal of Math. Anal. and Applic., 231, 1 (1999), 213–234.

[10] Chechkin G.A.and Pichugina E.A., Weighted Korn's inequality for a thin plate with a rough surface. Russian Journal of Math. Physics, 7, 3 (2000), 279–287.

[11] Ciarlet P. G.and Destuynder P., A justification of the two-dimensional linear plate model, J. Mécanique , 18 (1979), 315–344

[12] Ciarlet P.G., Plates and junctions in elastic multi–structures: an asymptotic analysis. R.M.A., Vol. 14., Masson and Springer–Verlag 1990.

[13] Ciarlet, P.G., Mathematical elasticity. Volume II. Theory of plates., Studies in Mathematics and its Applications, Vol. 27 , Elsevier 1997.

[14] Cioranescu D. and Saint Jean Paulin J., Asymptotic analysis of elastic wireworks, Preprint of Université Pierre et Marie Curie, R 89008, Paris (1989).

[15] Cioranescu D. and Saint Jean Paulin J., Homogenization of reticulated structures, Applied Mathematical Sciences 136, Springer–Verlag New York 1999.

[16] Constanda C., Lobo M. and Peréz E., On the bending of plates with transverse shear deformation and mixed periodic boundary conditions, Math. Methods Appl. Sci., 18, 5 (1995), 337–344.

[17] Damlamian A. and Vogelius M., Homogenization limits of the equations of elasticity in thin domains, IMA Preprint Series 170. Minneapolis: Institute for Mathematics and its Applications, 1985.

[18] Gaudiello, A., Homogenization of an elliptic transmission problem, Advances in Math. Sci. and Appl., Gakkotosho, 5, 2 (1995), 639–657.

[19] Kohler W., Papanicolaou G., Varadhan S., Boundary and interface problems in regions with very rough boundaries, in Multiple scattering and waves in random media, Chow P.L., Kohler W.E., Papanicolaou G.C. eds., Amsterdam, North-Holland (1981), 165–197.

[20] Kohn R. and Vogelius M., A new model for thin plates with rapidly varying thickness, J. Solid Struct., 20, 4 (1984), 333–350

[21] Kohn R. and Vogelius M., 1985. A new model for thin plates with rapidly varying thickness. II. A convergence proof. *Quart. of Appl. Math.*, XLIII, 1 (1985), 1–22

[22] Kohn R. and Vogelius M., 1986. A new model for thin plates with rapidly varying thickness. III. Comparison of different scalings, Quart. Appl. Math., XLIV, 1 (1986), 35–48

[23] Kohn R. and Vogelius M., 1986. Thin plates with rapidly varying thickness, and their relation to structural optimization, in Homogenization and effective moduli of materials and media, J. L. Ericksen, D. Kinderlehrer, R. Kohn and J.-L. Lions eds., Springer–Verlag (1986), 126–149

[24] Kolpakov A.G., Thin elastic plates with periodic structure and internal unilateral contact conditions, Prikl. Mekh. Tekh. Fizika, 5 (1991), 136–142.

[25] Lewiński T. and Telega J.J., Plates, laminates and shells. Asymptotic analysis and homogenization, World Scientific 2000.

[26] Lurie K.A., Cherkaev A.V. and Fedorov A.V., Regularization of optimal design problems for bars and plates, I and II, J. Opt. Th. Appl., 37 (1982), 499–543.

[27] Nazarov S.A., On the accuracy of asymptotic approximations for longitudinal deformations of a thin plate, Modélisation Math. et Analyse Numérique, M^2AN, 2 (1996), 185–213.

[28] Nazarov S.A., Korn's inequalities for junctions of spatial bodies and thin rods, Math. Methods Appl. Sci. 20, 3 (1997), 219–243.

[29] Nevard J. and Keller J.B., 1997. Homogenization of rough boundaries and interfaces, SIAM J. Appl. Math. 57, 6 (1997), 1660–1686.

[30] Oleinik O. A., Shamaev A. S. and Yosifian G. A., Mathematical problems in elasticity and homogenization, Amsterdam, North-Holland 1992.

[31] Sanchez-Palencia E. and Suquet P., Friction and homogenization of a boundary, in Free Boundary Problems: Theory and applications, A. Fasano and M. Primicerio eds., London, Pitman (1983), 561–571.

[32] Sobolev S.L., Some applications of functional analysis in mathematical physics. Third Edition. Translations of Mathematical Monographs Serie, Volume 90, AMS Press 1991.

[33] Vlasov B.F., On equations of plates bending, Izv. Akad. Nauk. SSSR O. M. N., 12 (1957), 57–60.

Gregory A. Chechkin
Department of Differential Equations
Faculty of Mechanics and Mathematics
Moscow State University
Moscow 119899
Russia

E-mail:

Doina Cioranescu
Laboratoire Jacques-Louis Lions
C.N.R.S and Université Pierre et Marie Curie
Boite postale 187
4 place Jussieu
75252 Paris Cedex 05
France

E-mail: cioran@ann.jussieu.fr

Studies in Mathematics and its Applications, Vol. 31
D. Cioranescu and J.L. Lions (Editors)

Chapter 9

ANISOTROPY AND DISPERSION IN ROTATING FLUIDS

J.-Y. Chemin, B. Desjardins, I. Gallagher and E. Grenier

1. Introduction

In this paper, our aim is to study dispersion phenomena occuring in singular perturbation models of fluid dynamics in $\mathbb{R}^3$. We focus on the small Rossby number limit of solutions to the incompressible Navier–Stokes equations, and are going to study dispersion phenomena in the following system:

$$(RF^\varepsilon)\left\{\begin{aligned} \partial_t u^\varepsilon + u^\varepsilon\cdot\nabla u^\varepsilon - \nu\Delta u^\varepsilon + \frac{u^\varepsilon\times e_3}{\varepsilon} &= -\nabla p^\varepsilon \quad \text{in} \quad \mathbb{R}^3\\ \operatorname{div} u^\varepsilon &= 0 \quad \text{in} \quad \mathbb{R}^3\\ u^\varepsilon_{|t=0} &= u_0,\end{aligned}\right.$$

where $e_3 = (0,0,1)$, $\varepsilon > 0$ denotes the Rossby number and $\nu > 0$ is the viscosity. We refer to [12] and [24] for a physical justification of the model.

It is well known that dispersive effects are of great importance in the study of non linear partial differential equations. Historically, the use of dispersive effects in partial differential equations appeared in the study of the wave equation with the proof of the Strichartz estimates. The idea is that despite the fact that the wave equation is time reversible, it induces a time–decay in L^p norms, of course for p greater than 2. But those decaying properties also provide smoothing effects; it has been the beginning of a long series of works (see for instance [4], [14], [17]–[22]) where those smoothing effects are used to improve, in non linear wave equations, the classical results of wellposedness (by classical results, we mean results proved with energy estimates). The same has been used in the framework of non linear Schrödinger equations. Let us have a quick look at the way Strichartz estimates are proved in the case of the wave equation. The wave equation can be reduced to the following system:

$$\partial_t u_\pm \pm i|D|u_\pm = 0 \quad \text{in} \quad \mathbb{R}\times\mathbb{R}^d, \quad \text{with} \quad |D|a = \mathcal{F}^{-1}(|\xi|\widehat{a}(\xi)). \tag{1.1}$$

So the solution is of the form

$$u(t) = \mathcal{F}^{-1}\left(e^{it|\xi|}\gamma^+(\xi) + e^{-it|\xi|}\gamma^-(\xi)\right), \tag{1.2}$$

where $\mathcal{F}$ denotes the Fourier transform on $\mathbb{R}^d$. Let us suppose that the support of the Fourier transform of the initial data γ^+ and γ^- is included in a fixed ring $\mathcal{C}$ of $\mathbb{R}^d$. The Strichartz estimate is based on the so–called "dispersive estimate"

$$\|u(t)\|_{L^\infty(\mathbb{R}^d)} \leq \frac{C}{t^{\frac{d-1}{2}}}(\|\gamma^+\|_{L^1(\mathbb{R}^d)} + \|\gamma^-\|_{L^1(\mathbb{R}^d)}). \tag{1.3}$$

Then functional analysis arguments (see for instance [11]) imply that

$$\|u\|_{L^p(\mathbb{R},L^q(\mathbb{R}^d))} \leq C(\|\gamma^+\|_{L^2(\mathbb{R}^d)} + \|\gamma^-\|_{L^2(\mathbb{R}^d)})$$

for suitable p and q, which is clearly a decay property. Using scaling arguments and Sobolev embeddings, it is easily proved that, for s strictly greater than $(d+1)/2$,

$$\|\nabla u\|_{L^2([0,T];L^\infty(\mathbb{R}^d))} \leq C_T(\|\gamma^+\|_{H^s(\mathbb{R}^d)} + \|\gamma^-\|_{H^s(\mathbb{R}^d)}).$$

Now let us return to our problem. A number of studies exist concerning systems of the type (RF^ε) in the viscous case as well as in the inviscid case ($\nu = 0$). We refer, among others, to [1]-[3], [7], [10], [13], [15], [25]. Let us introduce some notation: let P be the Leray projector onto divergence free vector fields; the fact that the operator

$$L_{RF}u^\varepsilon \stackrel{\text{def}}{=} P(u^\varepsilon \times e_3)$$

is skew–symmetric implies that if the initial data u_0 is an element of the space $L^2(\mathbb{R}^3)$, then we obtain a sequence u^ε of Leray's weak solutions, uniformly bounded in the space $L^\infty(\mathbb{R}^+;L^2(\mathbb{R}^3)) \cap L^2(\mathbb{R}^+;\dot{H}^1(\mathbb{R}^3))$. We have noted $\dot{H}^s$ the homogeneous Sobolev space of order s, defined as the closure of $\mathcal{D}(\mathbb{R}^3)$ for $\|\cdot\|_{\dot{H}^s}$, where $\|u\|_{\dot{H}^s} \stackrel{\text{def}}{=} \|\,|\xi|^s\widehat{u}(\xi)\,\|_{L^2(\mathbb{R}^3)}$, and $\widehat{u}$ is the Fourier transform of u.

The aim of this study is to analyze that system in the limit when ε goes to zero. Formally, one expects the solutions to converge to elements of the kernel of the penalization operator. This leads us to the well–known bidimensional Navier–Stokes system

$$(NS2D)\begin{cases} \partial_t\bar{u} + \bar{u}'\cdot\nabla_h\bar{u} - \nu\Delta_h\bar{u} &= -(\nabla_h p, 0) \quad \text{in} \quad \mathbb{R}^2 \\ \operatorname{div}\bar{u} &= 0 \quad \text{in} \quad \mathbb{R}^2 \\ \bar{u}_{|t=0} &= \bar{u}_0, \end{cases}$$

where $\Delta_h = \partial_1^2 + \partial_2^2$, $\nabla_h = (\partial_1, \partial_2)$, $u' = (u_1, u_2)$ denotes the first two components of u, and $\bar{u}_0$ is a bidimensional, divergence free vector field.

Finally, it will be useful in the following to define the solutions of the free linear equations associated with (RF^ε):

$$(FRF^\varepsilon) \begin{cases} \partial_t w^\varepsilon_{_F} - \nu\Delta w^\varepsilon_{_F} + \dfrac{1}{\varepsilon} L_{_{RF}} w^\varepsilon_{_F} &= f \quad \text{in} \quad \mathbb{R}^3 \\ w^\varepsilon_{_F\,|t=0} &= w_0, \end{cases}$$

where w_0 is a divergence free, tridimensional vector field.

In the spirit of the work of S. Ukai (see [26]) and of two of the authors (see [8]), we shall use dispersive effects in the system (FRF_ε) to prove strong convergence results.

Theorem 1 - *The set u^ε converges to 0 in $L^2_{loc}(\mathbb{R}^+, L^q(\mathbb{R}^3))$, for any $q \in]2, 6[$.*

Remark. A similar theorem can be stated for the compressible Navier-Stokes equations for isentropic gases in the low Mach number limit (see [8]).

Since the limit of the system (RF^ε) is the bidimensional Navier–Stokes equation $(NS2D)$, it is natural to consider initial data of the type $u_0 = \bar{u}_0 + w_0$ where $\bar{u}_0$ depends only on two variables.

Theorem 2 - *Let u_0 be a tridimensional divergence free vector field, such that*

$$u_0 = \bar{u}_0 + w_0, \quad \textit{with} \quad \bar{u}_0 \in \left(L^2(\mathbb{R}^2)\right)^3 \quad \textit{and} \quad w_0 \in \left(\dot{H}^{\frac{1}{2}}(\mathbb{R}^3)\right)^3,$$

with $\bar{u}_0$ and w_0 divergence free. Then there exists $\varepsilon_0 > 0$ such that for all $\varepsilon \leq \varepsilon_0$, there is a unique global solution u^ε to the system (RF^ε), which satisfies

$$u^\varepsilon - \bar{u} \in C^0_b(\mathbb{R}^+, \dot{H}^{\frac{1}{2}}(\mathbb{R}^3)) \cap L^2(\mathbb{R}^+, \dot{H}^{\frac{3}{2}}(\mathbb{R}^3)),$$

$$u^\varepsilon - \bar{u} - w^\varepsilon_{_F} \to 0 \quad \textit{in} \quad L^\infty(\mathbb{R}^+, \dot{H}^{\frac{1}{2}}(\mathbb{R}^3)) \cap L^2(\mathbb{R}^+, \dot{H}^{\frac{3}{2}}(\mathbb{R}^3)),$$

where $\bar{u}$ and $w^\varepsilon_{_F}$ are the respectively the solutions of $(NS2D)$ and (FRF^ε) associated respectively with $\bar{u}_0$ and w_0.

The statement of Theorem 2 shows clearly that anisotropic phenomena are involved in the study of (RF^ε). The solutions of the linearized equation

$$\partial_t v + L_{_{RF}} v - \nu\varepsilon\Delta v = 0 \quad \text{in} \quad \mathbb{R}^3 \tag{1.4}$$

are given in a similar way to (1.2): we have

$$v(t) = \mathcal{F}^{-1}\left(e^{\pm it\frac{\xi_3}{|\xi|} - \nu t\varepsilon|\xi|^2}\gamma_\pm\right),$$

and it is proved in this paper that operators of the following type, where ξ_3 is seen as a parameter,

$$e^{\pm it\frac{\xi_3}{(\xi_3^2-\Delta_h)^{1/2}}-\nu t\varepsilon(\xi_3^2-\Delta_h)}\gamma_\pm(\cdot,\xi_3)$$

satisfy (anisotropic) Strichartz–type estimates.

The structure of the paper is as follows. In Section 2, we prove some basic Strichartz estimates (see (2.6)) for systems of the type (1.4); they will be used to prove Theorem 1. Although these Strichartz estimates are not explicitly anisotropic, their proof requires an anisotropic decomposition of the frequency space, with respect to the vertical frequency variable ξ_3. That is due to the fact that the phase $\dfrac{\xi_3}{|\xi|}$ is almost constant when ξ_3 is almost equal to 0, or when $|\xi_3|$ is much larger than $|\xi_h|$, so in particular is almost stationary. So, there is no hope of getting a dispersive effect in that case.

In Section 3, we prove anisotropic Strichartz–type estimates, which take into account the viscosity effect. These estimates are of the type

$$\|u(t)\|_{L^\infty(\mathbb{R}^2_{x_h},L^2(\mathbb{R}_{x_3}))} \leq \frac{C}{t^{\frac{1}{4}}}2^{3j}\|u_0\|_{L^1(\mathbb{R}^2_{x_h},L^2(\mathbb{R}_{x_3}))},$$

for any function whose Fourier transform is supported in the set $2^j\mathcal{C}\cap(\mathbb{R}^2\times 2^k[a,b])$, with $a<\frac{1}{2}$ and $b>2$, and $(j,k)\in\mathbb{Z}^2$.

In Section 4, we prove Theorem 2. The method used is classical in the study of incompressible Navier–Stokes equations, once the dispersive estimates have been obtained.

2. Convergence of Leray's solutions

The aim of this part is to prove Theorem 1. We first derive Strichartz type estimates for Coriolis force and then apply it to (RF^ε).

2.1. Strichartz type estimates for Coriolis force

In order to prove Theorem 1, we need to derive dispersion estimates for wave equations related to the rotating Navier-Stokes model (RF^ε). The proof follows closely the lines of [11] and [19].

The linearized version of (RF^ε) reads

$$\left\{\begin{array}{rcl} \partial_t v+\frac{1}{\varepsilon}v\times e_3-\nu\Delta v+\nabla p=0,\ \operatorname{div} v &=& 0,\quad \text{in } \mathbb{R}_t\times\mathbb{R}^3_x,\\ v_{|t=0} &=& v_0 \ \text{ in } \mathbb{R}^3_x, \end{array}\right. \tag{2.1}$$

which yields in Fourier variables $\xi \in \mathbb{R}^3$

$$\partial_t \widehat{v} + \nu |\xi|^2 \widehat{v} - \frac{\xi_3 \xi \times \widehat{v}}{\varepsilon |\xi|^2} = 0 \text{ in } \mathbb{R}_t \times \mathbb{R}^3_\xi, \quad \text{and} \quad \widehat{v}_{|t=0} = \widehat{v}_0. \tag{2.2}$$

Hence we are led to studying

$$\begin{aligned} \mathcal{G}^\varepsilon_\nu(t) : f \mapsto & \int_{\mathbb{R}^3_\xi} \widehat{f}(\xi) e^{\pm it\frac{\xi_3}{|\xi|} - \nu t \varepsilon |\xi|^2 + ix\cdot\xi} \, d\xi \\ & = \int_{\mathbb{R}^3_\xi \times \mathbb{R}^3_y} f(y) e^{\pm it\frac{\xi_3}{|\xi|} - \nu t\varepsilon|\xi|^2 + i(x-y)\cdot\xi} \, d\xi dy, \end{aligned} \tag{2.3}$$

first considering the case when $\widehat{f}$ is supported in $\mathcal{C}_{r,R}$ for some $r < R$, defined by

$$\mathcal{C}_{r,R} = \{\xi \in \mathbb{R}^3 \; / \; |\xi_3| \geq r \text{ and } |\xi| \leq R\}. \tag{2.4}$$

Let us introduce

$$K(t,\tau,z) \stackrel{\text{def}}{=} \int_{\mathbb{R}^3_\xi} \psi(\xi) e^{ita(\xi,z) + iz\cdot\xi - \tau|\xi|^2} d\xi, \quad \text{where} \quad a(\xi,z) \stackrel{\text{def}}{=} \frac{\xi_3}{|\xi|},$$

and ψ is a function of $\mathcal{D}(\mathbb{R}^3 \setminus \{0\})$, which of course depends on r and R, which is radial with respect to the horizontal variable $\xi_h = (\xi_1, \xi_2)$ and whose value is 1 near $\mathcal{C}_{r,R}$.

Lemma 1 - *For any (r,R) such that $0 < r < R$, there exists a constant $C_{r,R}$ such that*

$$\|K(t,\tau,.)\|_{L^\infty(\mathbb{R}^3)} \leq C_{r,R} \min\{1, t^{-\frac{1}{2}}\} e^{-r^2\tau}.$$

Proof. The proof follows the lines of a stationary phase method, in a very simple way. First, using the rotation invariance in (ξ_1, ξ_2), we restrict to the case when $z_2 = 0$. Next, denoting $\alpha(\xi) = -\partial_{\xi_2} a(\xi) = \xi_2 \xi_3 / |\xi|^3$, we introduce the following differential operator:

$$\mathcal{L} \stackrel{\text{def}}{=} \frac{1}{1 + t\alpha^2(\xi)} \left(1 + i\alpha(\xi)\partial_{\xi_2}\right)$$

which acts on the ξ_2 variable, and satisfies $\mathcal{L}(e^{ita}) = e^{ita}$. Integrating by parts, we obtain

$$K(t,\tau,z) = \int_{\mathbb{R}_{\xi_2}} e^{ita(\xi,z) + iz\cdot\xi} \left({}^t\mathcal{L}(\psi(\xi) e^{-\nu\tau|\xi|^2})\right) \, d\xi_2.$$

Easy computations then yield

$$
\begin{aligned}
{}^t\mathcal{L}\left(\psi(\xi)e^{-\tau|\xi|^2}\right) &= \left(\frac{1}{1+t\alpha^2} - i(\partial_{\xi_2}\alpha)\frac{1-t\alpha^2}{(1+t\alpha^2)^2}\right)\psi(\xi)e^{-\tau|\xi|^2} \\
&\quad -\frac{i\alpha}{1+t\alpha^2}\partial_{\xi_2}\left(e^{-\tau|\xi|^2}\psi(\xi)\right), \qquad (2.5)
\end{aligned}
$$

so using the fact that ξ is in a fixed ring of $\mathbb{R}^3$, and $\psi \in \mathcal{D}(\mathbb{R}^3)$, we get

$$
\|K(t,\tau,\cdot)\|_{L^\infty} \leq C_{r,R}e^{-\tau|\xi|^2}\int_{\mathbb{R}} \frac{d\xi_2}{1+t\xi_2^2},
$$

which proves Lemma 1. ■

Remark. The above lemma can be extended to more general functions a. Indeed, in the case when the phase a is non stationary on $\operatorname{supp}\psi \times \{z \,/\, |z| \leq \rho\}$, i.e. $\nabla_\xi a \neq 0$, multiple integrations by parts yield arbitrary polynomial time decay, whereas if ψ is supported in a neighborhood of points $\xi_0(z)$ such that $\nabla a(\xi_0(z), z) = 0$ and $D^2_\xi a(\xi_0(z), z)$ has j non–zero eigenvalues, then such a stationary phase argument yields a time decay of order $t^{-j/2}$. For instance, in the case of acoustic waves occuring in the low Mach number limit of compressible flows, we obtain a decay in $t^{-\frac{d-1}{2}}$, which is closely related to Strichartz' estimate for the wave equation (see [11]).

Lemma 1 yields the following corollary in a standard way (see [11]).

Corollary 1 - *For any constants r and R, let $\mathcal{C}_{r,R}$ be the domain defined in (2.4). Then a constant $C_{r,R}$ exists such that if w_0 and $f(t,\cdot)$ are two vector fields such that* Supp $\widehat{w}_0$ $\cup$ Supp $\widehat{f}$ $\subset$ $\mathcal{C}_{r,R}$, *and if w^ε_F is the solution of the linear equation (FRF^ε) with forcing term f and initial data w_0, then*

$$
\|w^\varepsilon_F\|_{L^4(\mathbb{R}^+,L^\infty)} \leq C_{r,R}\,\varepsilon^{\frac{1}{4}}\left(\|w_0\|_{L^2} + \|f\|_{L^1(\mathbb{R}^+,L^2)}\right). \qquad (2.6)
$$

2.2. Proof of Theorem 1

We are now able to prove Theorem 1. Let us define the function $u^\varepsilon_R \overset{\text{def}}{=} \chi\Big(\frac{|D|}{R}\Big)u^\varepsilon$, where $|D| = (-\Delta)^{\frac{1}{2}}$, and χ is a function in $\mathcal{D}(\mathbb{R})$, whose value is 1 near the origin. By Sobolev embeddings and stating $\alpha \overset{\text{def}}{=} \frac{3}{q} - \frac{1}{2}$, we can write

$$
\begin{aligned}
\|u^\varepsilon - u^\varepsilon_R\|_{L^2(\mathbb{R}^+,L^q(\mathbb{R}^3))} &\leq C\|u^\varepsilon - u^\varepsilon_R\|_{L^2(\mathbb{R}^+,H^{1-\alpha}(\mathbb{R}^3))} \\
&\leq CR^{-\alpha}\|u^\varepsilon\|_{L^2(\mathbb{R}^+,H^1(\mathbb{R}^3))} \\
&\leq \frac{C}{\sqrt{\nu}}R^{-\alpha}\|u_0\|_{L^2(\mathbb{R}^3)}. \qquad (2.7)
\end{aligned}
$$

Now let us estimate u_R^ε. We have

$$\partial_t u_R^\varepsilon - \nu \Delta u_R^\varepsilon + \frac{1}{\varepsilon} P(u_R^\varepsilon \times e_3) = f_R^\varepsilon,$$

with

$$f_R^\varepsilon(t) \stackrel{\text{def}}{=} -\chi\Big(\frac{|D|}{R}\Big) P \operatorname{div}(u^\varepsilon \otimes u^\varepsilon)(t),$$

which implies, by Duhamel's formula, that

$$u_R^\varepsilon = e^{t(\nu\Delta + \sigma(D))} u_{0,R} + \int_0^t e^{(t-t')(\nu\Delta+\sigma(D))} f_R^\varepsilon(t')\, dt',$$

where $\sigma(\xi)v \stackrel{\text{def}}{=} \dfrac{\xi_3}{\varepsilon|\xi|^2}\xi \times v$. Now, using the Berstein inequalities, we have, for any $r > 0$,

$$\|e^{t(\nu\Delta+\sigma(D))} u_{0,R}\|_{L^\infty(\mathbb{R}^3)} \leq C(R^2 r)^{\frac{1}{2}} \|e^{t(\nu\Delta+\sigma(D))} u_{0,R}\|_{L^2(\mathbb{R}^3)},$$

where $D_3 = -i\partial_3$; so with Corollary 1, we have

$$\begin{aligned}
\Big\|e^{t(\nu\Delta+\sigma(D))} u_{0,R}\Big\|_{L_T^4(L^\infty(\mathbb{R}^3))} &\leq \Big\|e^{t(\nu\Delta+\sigma(D))} \chi\Big(\frac{D_3}{r}\Big) u_{0,R}\Big\|_{L_T^4(L^\infty(\mathbb{R}^3))} \\
&\quad + C\left\|e^{t(\nu\Delta+\sigma(D))} \Big(\operatorname{Id} - \chi\Big(\frac{D_3}{r}\Big)\Big) u_{0,R}\right\|_{L_T^4(L^\infty(\mathbb{R}^3))} \\
&\leq CT^{\frac{1}{4}} (R^2 r)^{\frac{1}{2}} \|u_0\|_{L^2} + \varepsilon^{\frac{1}{4}} C_{r,R} \|u_0\|_{L^2}, \qquad (2.8)
\end{aligned}$$

where to simplify, we have noted $L_T^p(L^q(\mathbb{R}^d)) = L^p([0,T]; L^q(\mathbb{R}^d))$. So, $e^{t(\nu\Delta+\sigma(D))} u_{0,R}$ goes to zero in $L_T^4(L^\infty(\mathbb{R}^3))$.

Moreover, we have

$$\begin{aligned}
\Big\|\chi\Big(\frac{D_3}{r}\Big) f_R^\varepsilon\Big\|_{L_T^\infty(L^2(\mathbb{R}^3))} &\leq R\, \|\chi\Big(\frac{|D|}{R}\Big)\chi\Big(\frac{D_3}{r}\Big) u^\varepsilon \otimes u^\varepsilon\|_{L_T^\infty(L^2(\mathbb{R}^3))} \\
&\leq R(rR^2)^{\frac{1}{2}} \|u_0\|_{L^2(\mathbb{R}^3)}^2.
\end{aligned}$$

Then, Corollary 1 yields, as for (2.8),

$$\begin{aligned}
&\Big\|\int_0^t e^{(t-t')(\nu\Delta+\sigma(D))} f_R^\varepsilon(t')\, dt'\Big\|_{L_T^4(L^\infty(\mathbb{R}^3))} \\
&\qquad \leq CR(rR^2)^{\frac{1}{2}} T^{\frac{5}{4}} \|u_0\|_{L^2(\mathbb{R}^3)}^2 + \varepsilon^{\frac{1}{4}} C_{r,R} T \|u_0\|_{L^2}^2.
\end{aligned}$$

Using (2.8), we get

$$\begin{aligned}
\|u_R^\varepsilon\|_{L_T^4(L^\infty(\mathbb{R}^3))} &\leq CT^{\frac{1}{4}} (R^2 r)^{\frac{1}{2}} \|u_0\|_{L^2} + \varepsilon^{\frac{1}{4}} C_{r,R} \|u_0\|_{L^2} \\
&\quad + CR(rR^2)^{\frac{1}{2}} T^{\frac{5}{4}} \|u_0\|_{L^2(\mathbb{R}^3)}^2 + \varepsilon^{\frac{1}{4}} C_{r,R} T \|u_0\|_{L^2}^2.
\end{aligned}$$

¿From this, we deduce that for a fixed R,

$$\lim_{\varepsilon\to 0} u_R^\varepsilon = 0 \quad \text{in} \quad L^2_{loc}(\mathbb{R}^+; L^\infty(\mathbb{R}^3)),$$

which, with (2.7), finally implies that

$$u^\varepsilon \to 0 \quad \text{in} \quad L^2_{loc}(\mathbb{R}^+; L^q(\mathbb{R}^3)) \quad \forall q \in]2,6[,$$

and this proves Theorem 1.

3. Anisotropic Strichartz estimates

In this section, we are going to prove anisotropic, and viscous, Strichartz estimates, which will enable us in the following section to prove Theorem 2.

Let us first introduce some notation: for all $x \in \mathbb{R}^3$, we set $x = (x_h, x_3)$, with $x_h = (x_1, x_2)$, and denote $L^{p,q}_{x_h,x_3}$, for $1 \le p, q \le \infty$, the space defined by the norm

$$\|f\|_{L^{p,q}_{x_h,x_3}} \stackrel{\text{def}}{=} \Big\| \|f\|_{L^q_{x_3}} \Big\|_{L^p_{x_h}}.$$

Let us also recall some definitions of Littlewood–Paley theory (we refer for instance to [6] for details): Δ_j (resp. S_j) denote the Littlewood–Paley spectral localization operators, defined in the following way:

$$\forall j \in \mathbb{Z}, \quad \Delta_j \stackrel{\text{def}}{=} \varphi(2^{-j}|D|) \quad \text{and} \quad S_j = \sum_{j' \le j-1} \Delta_{j'}, \tag{3.1}$$

where $\varphi \in \mathcal{D}(\mathbb{R})$, is such $\varphi(s) = 0$ near 0.

$$\varphi(s) = 1 \text{ if } \frac{1}{2} \le |s| \le 1; \qquad \forall t \in \mathbb{R} \setminus \{0\}, \sum_{j \in \mathbf{Z}} \varphi(2^{-j} t) = 1.$$

In the rest of the paper, $N_0 \in \mathbb{N}$ is a fixed constant, depending only on φ, such that if $|j - j'| \ge N_0$, then $\operatorname{supp} \varphi(2^{-j}\cdot) \cap \operatorname{supp} \varphi(2^{-j'}\cdot) = \emptyset$. As we are interested in proving anisotropic estimates, we are also led to introduce an anisotropic Littlewood–Paley decomposition, in the spirit of [16]. To do so, let us define the operators Δ_k^v and S_k^v by

$$\forall k \in \mathbb{Z}, \quad \Delta_k^v \stackrel{\text{def}}{=} \varphi(2^{-k} D_3) \quad \text{and} \quad S_k^v = \sum_{k' \le k-1} \Delta_{k'}^v.$$

Theorem 3 - *Let $\mathcal{G}_\nu^\varepsilon$ be the operator defined in Equation (2.3); for any $p \in [1, \infty]$ and for any $\alpha > 0$,there exists a constant C such that, for any vector field f, we have*

$$\left\| \Delta_j \mathcal{G}_\nu^\varepsilon \left(\frac{\cdot}{\varepsilon} \right) f \right\|_{L_T^p(L^\infty(\mathbb{R}^3))} \leq C 2^{j(\frac{3}{2} - \frac{2}{p})} (\varepsilon 2^{2j})^{\frac{1}{4p(1+\alpha)}} \|\Delta_j f\|_{L^2(\mathbb{R}^3)}, \quad (3.2)$$

$$\left\| \Delta_j \Delta_k^v \mathcal{G}_\nu^\varepsilon \left(\frac{\cdot}{\varepsilon} \right) f \right\|_{L_T^p(L^{\infty,2}_{x_h,x_3})} \leq C 2^{j(1-\frac{2}{p})} \min \left\{ 1, (\varepsilon 2^{2j})^{\frac{1}{4p}} 2^{\frac{1}{2p}(j-k)} \right\} \times \|\Delta_j \Delta_k^v f\|_{L^2(\mathbb{R}^3)}. \quad (3.3)$$

Proof. Like all Strichartz–type inequalities, (3.2) and (3.3) are consequences of the following dispersive estimates:

Lemma 2 - *For any function f, we have*

$$\|\Delta_j \Delta_k^v \mathcal{G}(\tau, \theta) f\|_{L^\infty(\mathbb{R}^3)} \leq C \min \left\{ 2^{k-j}, \tau^{-\frac{1}{2}} \right\} 2^{3j} e^{-C\theta 2^{2j}} \|f\|_{L^1(\mathbb{R}^3)}, \quad (3.4)$$

and

$$\|\Delta_j \Delta_k^v \mathcal{G}(\tau, \theta) f\|_{L^{\infty,2}_{x_h,x_3}} \leq C \min \left\{ 1, \tau^{-\frac{1}{2}} 2^{(j-k)} \right\} 2^{2j} e^{-C\theta 2^{2j}} \|f\|_{L^{1,2}_{x_h,x_3}}, \quad (3.5)$$

where

$$\mathcal{G}(\tau, \theta) f \stackrel{\text{def}}{=} \mathcal{F}^{-1} \left(e^{i\tau \frac{\xi_3}{|\xi|} - \theta |\xi|^2} \widehat{f}(\xi) \right).$$

First, let us remark that inequality (3.4) is nothing but Lemma 1 with a precise control of the constant.

Proof of Lemma 2. We start by the proof of (3.4), and (3.5) will be obtained by very similar methods.

We can write

$$\Delta_j \Delta_k^v \mathcal{G}(\tau, \theta) f(x) = K_{j,k}(\tau, \theta) * f(x),$$

with

$$K_{j,k}(\tau, \theta, x) \stackrel{\text{def}}{=} \int_{\mathbb{R}^3} e^{ix\cdot\xi + i\tau \frac{\xi_3}{|\xi|} - \theta |\xi|^2} \varphi(2^{-j}|\xi|) \varphi(2^{-k}\xi_3) \, d\xi, \quad (3.6)$$

hence all we have to prove is that

$$\|K_{j,k}(\tau, \theta, \cdot)\|_{L^\infty(\mathbb{R}^3)} \leq \frac{C}{\tau^{\frac{1}{2}}} e^{-C\theta 2^{2j}} 2^{3j}. \quad (3.7)$$

Indeed, (3.7) implies that

$$\|\Delta_j\Delta_k^v\mathcal{G}(\tau,\theta)f\|_{L^\infty(\mathbb{R}^3)}\leq\frac{C}{\tau^{\frac{1}{2}}}e^{-C\theta 2^{2j}}2^{3j}\|f\|_{L^1(\mathbb{R}^3)},\tag{3.8}$$

and we also have

$$\|\Delta_j\Delta_k^v\mathcal{G}(\tau,\theta)f\|_{L^2(\mathbb{R}^3)}\leq Ce^{-C\theta 2^{2j}}\|\Delta_j\Delta_k^v f\|_{L^2(\mathbb{R}^3)}.\tag{3.9}$$

Moreover, Bernstein's inequality yields

$$\|\Delta_j\Delta_k^v\mathcal{G}(\tau,\theta)f\|_{L^\infty(\mathbb{R}^3)}\leq C2^{\frac{3j}{2}}2^{\frac{k-j}{2}}\|\Delta_j\Delta_k^v\mathcal{G}(\tau,\theta)f\|_{L^2(\mathbb{R}^3)},\tag{3.10}$$

so with (3.9), we have

$$\begin{aligned}\|\Delta_j\Delta_k^v\mathcal{G}(\tau,\theta)f\|_{L^\infty(\mathbb{R}^3)} &\leq C2^{\frac{3j}{2}}2^{\frac{k-j}{2}}e^{-C\theta 2^{2j}}\|\Delta_j\Delta_k^v f\|_{L^2(\mathbb{R}^3)}\\ &\leq Ce^{-C\theta 2^{2j}}2^{3j}2^{k-j}\|f\|_{L^1(\mathbb{R}^3)}.\end{aligned}\tag{3.11}$$

Inequality (3.11) associated with (3.8) implies that (3.4) is proved, except for the fact that we still have to prove (3.7). Using the rotation invariance of $K_{j,k}$, we have

$$K_{j,k}(\tau,\theta,x)=\int_{\mathcal{B}(0,C2^j)}e^{ix_1\xi_1}\left(\int_{\mathbb{R}^2}e^{i\tau\frac{\xi_3}{|\xi|}-\theta|\xi|^2}\varphi(2^{-j}|\xi|)\varphi(2^{-k}\xi_3)\,d\xi_2d\xi_3\right)d\xi_1.$$

Then we write $K_{j,k}(\tau,\theta,x)=2^{3j}\widetilde{K}_{j,k}(\tau,2^{2j}\theta,2^jx)$, with

$$\widetilde{K}_{j,k}(\tau,\eta,y)\stackrel{\text{def}}{=}\int_{\mathcal{B}(0,C)}e^{iy_1\xi_1}\left(\int_{\mathbb{R}^2}e^{i\tau\frac{\xi_3}{|\xi|}-\eta|\xi|^2}\varphi(|\xi|)\varphi(2^{j-k}\xi_3)\,d\xi_2d\xi_3\right)d\xi_1.$$

As in the proof of Lemma 1 in Section 2, we shall integrate by parts in the following integral:

$$\Phi(\xi_1,\xi_3)\stackrel{\text{def}}{=}\int_{\mathbb{R}}e^{i\tau\frac{\xi_3}{|\xi|}-\eta|\xi|^2}\varphi(|\xi|)\,d\xi_2.$$

To do so, let us introduce the operator $\mathcal{L}$ defined as

$$\forall a,\ \mathcal{L}a\stackrel{\text{def}}{=}\frac{1}{1+\tau\alpha^2(\xi)}(a+i\alpha(\xi)\partial_{\xi_2}a)\quad\text{with}\quad\alpha(\xi)=-\partial_{\xi_2}a(\xi)=\xi_2\xi_3/|\xi|^3.$$

An easy computation yields

$$\mathcal{L}(e^{i\tau\frac{\xi_3}{|\xi|}})=e^{i\tau\frac{\xi_3}{|\xi|}},$$

which implies immediately that

$$\Phi(\xi_1,\xi_3) = \int_{\mathbb{R}} e^{i\tau\frac{\xi_3}{|\xi|}}\ {}^t\mathcal{L}\left(e^{-\eta|\xi|^2}\varphi(|\xi|)\right) d\xi_2.$$

But we can compute, for any function Ψ, as in (2.5),

$${}^t\mathcal{L}\left(\Psi(\xi)\right) = \left(\frac{1}{1+\alpha^2\tau} - i(\partial_{\xi_2}\alpha)\frac{1-\tau\alpha^2}{(1+\tau\alpha^2)^2}\right)\Psi(\xi) - \frac{i\alpha}{1+\tau\alpha^2}\partial_{\xi_2}\Psi(\xi),$$

so, using the fact that ξ is in a fixed ring of $\mathbb{R}^3$, that $\varphi \in \mathcal{D}(\mathbb{R})$, we get, after elementary computations,

$$\left|{}^t\mathcal{L}\left(e^{-\eta|\xi|^2}\varphi(|\xi|)\right)\right| \leq \frac{C}{1+\tau\xi_2^2\xi_3^2}e^{-\eta|\xi|^2}.$$

Finally, we have

$$\begin{aligned}\|\widetilde{K}_{j,k}(\tau,\eta,\cdot)\|_{L^\infty(\mathbb{R}^3)} &\leq Ce^{-C\eta}\int_{c\leq|\xi_2|\leq C}\int_{c2^{k-j}\leq|\xi_3|\leq C2^{k-j}}\frac{d\xi_2 d\xi_3}{1+\tau\xi_2^2\xi_3^2}\\ &\leq Ce^{-C\eta}\tau^{-\frac{1}{2}}\int_{\mathbb{R}}\frac{d\zeta_2}{1+\zeta_2^2}. \qquad (3.12)\end{aligned}$$

So the lemma is proved in the case (3.4).

The proof of (3.5) is exactly along the same lines, only we keep the variable ξ_3 as a parameter in the computations above. In other words, let us define

$$I_{j,k}(\tau,\theta,x_h,\xi_3) \stackrel{\text{def}}{=} \int_{\mathbb{R}^2} e^{ix_h\cdot\xi_h+i\tau\frac{\xi_3}{|\xi|}-\theta|\xi|^2}\varphi(2^{-j}|\xi|)\varphi(2^{-k}\xi_3)\, d\xi_h.$$

Then we have

$$\|\Delta_j\Delta_k^v\mathcal{G}(\tau,\theta)f\|_{L^{\infty,2}_{x_h,x_3}} \leq C\|I_{j,k}(\tau,\theta,\cdot,\cdot)\|_{L^{\infty,\infty}_{x_h,\xi_3}}\|f\|_{L^{1,2}_{x_h,x_3}},$$

where we have used a Young inequality in the horizontal variables and Hölder's inequality in the vertical one, associated with Plancherel's formula.

Now, all we need is an estimate of $I_{j,k}(\tau,\theta,\cdot,\cdot)$ in $L^{\infty,\infty}_{x_h,\xi_3}$, which is obtained exactly as in the isotropic case above: we have

$$I_{j,k}(\tau,\theta,x_h,\xi_3) = 2^{2j}\widetilde{I}_{j,k}(\tau,2^{2j}\theta,2^jx_h,2^{-j}\xi_3)$$

with

$$\widetilde{I}_{j,k}(\tau,\eta,y_h,\zeta_3) \stackrel{\text{def}}{=} \varphi(2^{j-k}\zeta_3)\int_{\mathbb{R}^2} e^{iy_h\cdot\zeta_h+i\tau\frac{\zeta_3}{|\zeta|}-\eta|\zeta|^2}\varphi(|\zeta|)\, d\zeta_h.$$

An identical computation as for (3.12) yields

$$\begin{aligned}\|\widetilde{I}_{j,k}(\tau,\eta,y_h,\zeta_3)\|_{L^{\infty,\infty}_{y_h,\zeta_3}} &\leq Ce^{-C\eta}\Big\|\varphi(2^{j-k}\zeta_3)\int_{c\leq|\xi_2|\leq C}\frac{d\zeta_2}{1+\tau\zeta_2^2\zeta_3^2}\Big\|_{L^\infty_{\zeta_3}}\\ &\leq Ce^{-C\eta}\tau^{-\frac{1}{2}}\|\varphi(2^{j-k}\zeta_3)\zeta_3^{-1}\|_{L^\infty_{\zeta_3}}\\ &\leq Ce^{-C\eta}2^{(j-k)}\tau^{-\frac{1}{2}}.\end{aligned}$$

So finally we have obtained that

$$\|\Delta_j\Delta_k^v\mathcal{G}(\tau,\theta)f\|_{L^{\infty,2}_{x_h,x_3}}\leq Ce^{-C\theta 2^{2j}}2^{2j}2^{(j-k)}\tau^{-\frac{1}{2}}\|f\|_{L^{1,2}_{x_h,x_3}}. \tag{3.13}$$

Now to conclude to (3.5), we write similar estimates as (3.10),(3.11): we have, by Sobolev embeddings,

$$\begin{aligned}\|\Delta_j\Delta_k^v\mathcal{G}(\tau,\theta)f\|_{L^{\infty,2}_{x_h,x_3}} &\leq C2^j\|\Delta_j\Delta_k^v\mathcal{G}(\tau,\theta)f\|_{L^2(\mathbb{R}^3)}\\ &\leq C2^je^{-C\theta 2^{2j}}\|\Delta_j\Delta_k^vf\|_{L^2(\mathbb{R}^3)}\\ &\leq C2^{2j}e^{-C\theta 2^{2j}}\|\Delta_j\Delta_k^vf\|_{L^{1,2}_{x_h,x_3}},\end{aligned}$$

which with (3.13) implies that

$$\|\Delta_j\Delta_k^v\mathcal{G}(\tau,\theta)f\|_{L^{\infty,2}_{x_h,x_3}}\leq Ce^{-C\theta 2^{2j}}2^{2j}\min\left\{\frac{1}{\tau^{\frac{1}{2}}}2^{(j-k)},1\right\}\|f\|_{L^{1,2}_{x_h,x_3}}.$$

The lemma is proved. ■

End of the proof of Theorem 3. To prove the result, we are going to use duality arguments, based on the isotropic estimate (3.2). In the following, we shall call

$$\mathcal{B}\stackrel{\text{def}}{=}\left\{\Psi\in\mathcal{D}(\mathbb{R}^3),\|\Psi\|_{L^\infty_T(L^1(\mathbb{R}^3))}\leq 1\right\}. \tag{3.14}$$

The following set of equalities is standard:

$$\begin{aligned}\|\Delta_j\mathcal{G}_\nu^\varepsilon f\|_{L^1_T(L^\infty(\mathbb{R}^3))} &= \sup_{\Psi\in\mathcal{B}}\int_{\mathbb{R}^+}<\Delta_j\mathcal{G}_\nu^\varepsilon(t)f,\Psi(t)>\,dt\\ &= \sup_{\Psi\in\mathcal{B}}(2\pi)^{-3}\int_{\mathbb{R}^+\times\mathbb{R}^3}\varphi(2^{-j}|\xi|)e^{-i\frac{t}{\varepsilon}\frac{\xi_3}{|\xi|}-\nu t|\xi|^2}\widehat{f}(\xi)\widehat{\Psi}(t,\xi)\,dtd\xi\\ &= \sup_{\Psi\in\mathcal{B}}(2\pi)^{-3}\sum_{k\leq j}\int_{\mathbb{R}^+\times\mathbb{R}^3}\widehat{f}_{j,k}(\xi)\widehat{\Psi}_{j,k}(t,\xi)e^{-i\frac{t}{\varepsilon}\frac{\xi_3}{|\xi|}-\nu t|\xi|^2}\,dtd\xi,\end{aligned}$$

where we have defined $f_{j,k}=\Delta_j\Delta_k^vf$ and $\Psi_{j,k}=\Delta_j\Delta_k^v\Psi$. So it follows that

$$\begin{aligned}\|\Delta_j\mathcal{G}_\nu^\varepsilon f\|_{L^1_T(L^\infty(\mathbb{R}^3))} \leq{}& \sup_{\Psi\in\mathcal{B}}(2\pi)^{-3}\sum_{k\leq j}\|f_{j,k}\|_{L^2(\mathbb{R}^3)}\\ &\times\left\|\int_{\mathbb{R}^+}e^{-i\frac{t}{\varepsilon}\frac{\xi_3}{|\xi|}-\nu t|\xi|^2}\widehat{\Psi}_{j,k}(t,\xi)\,dt\right\|_{L^2(\mathbb{R}^3)}.\end{aligned} \tag{3.15}$$

But we have

$$\left\| \int_{\mathbb{R}^+} e^{-i\frac{t}{\varepsilon}\frac{\xi_3}{|\xi|}-\nu t|\xi|^2}\widehat{\Psi}_{j,k}(t,\xi)\,dt \right\|^2_{L^2(\mathbb{R}^3)}$$

$$= \int_{(\mathbb{R}^+)^2\times\mathbb{R}^3} e^{-i\frac{t}{\varepsilon}\frac{\xi_3}{|\xi|}-\nu t|\xi|^2}\widehat{\Psi}_{j,k}(t,\xi)e^{i\frac{s}{\varepsilon}\frac{\xi_3}{|\xi|}-\nu s|\xi|^2}\overline{\widehat{\Psi}}_{j,k}(s,\xi)\,dtdsd\xi$$

$$\leq \int_{(\mathbb{R}^+)^2} \|\Psi_{j,k}(s,\cdot)\|_{L^1(\mathbb{R}^3)}\|\Psi_{j,k}(t,\cdot)\|_{L^1(\mathbb{R}^3)} \times \min\left\{2^{k-j}, \frac{\varepsilon^{\frac{1}{2}}}{|t-s|^{\frac{1}{2}}}\right\} 2^{3j}e^{-\nu(t+s)2^{2j}}\,dtds, \tag{3.16}$$

where we have used (3.4). So

$$\|\Delta_j\mathcal{G}^\varepsilon_\nu f\|^2_{L^1_T(L^\infty(\mathbb{R}^3))} \leq C\sum_{k\leq j} 2^{3j}\|f_{j,k}\|^2_{L^2(\mathbb{R}^3)} \times \int_{(\mathbb{R}^+)^2} \min\left\{2^{k-j}, \frac{\varepsilon^{\frac{1}{2}}}{|t-s|^{\frac{1}{2}}}\right\} e^{-\nu(t+s)2^{2j}}\,dtds. \tag{3.17}$$

On the one hand, the upper bound (3.17) can be estimated directly in the following way:

$$2^{3j}\int_{\mathbb{R}^+\times\mathbb{R}^+} \min\left\{2^{k-j}, \frac{\varepsilon^{\frac{1}{2}}}{|t-s|^{\frac{1}{2}}}\right\} e^{-\nu(t+s)2^{2j}}\,dtds \leq C2^{k-j}2^{-j}. \tag{3.18}$$

On the other hand, we can write

$$2^{3j}\int_{\mathbb{R}^+\times\mathbb{R}^+} \frac{1}{|t-s|^{\frac{1}{2}}}e^{-\nu(t+s)2^{2j}}\,dtds = \int_{\mathbb{R}\times\mathbb{R}} \frac{1}{|t'-s'|^{\frac{1}{2}}}e^{-\nu(t'+s')}\,dt'ds', \tag{3.19}$$

which finally implies

$$\left\| \int_{\mathbb{R}^+} e^{-i\frac{t}{\varepsilon}\frac{\xi_3}{|\xi|}-\nu t|\xi|^2}\widehat{\Psi}_{j,k}(t,\xi)\,dt \right\|^2_{L^2(\mathbb{R}^3)} \leq C2^{k-j}2^{-j}\min\left\{1, (\varepsilon 2^{2j})^{\frac{1}{2}}2^{(j-k)}\right\}.$$

Using the fact that

$$\sum_k \|\Delta_j\Delta^v_k f\|^2_{L^2(\mathbb{R}^3)} = \|\Delta_j f\|^2_{L^2(\mathbb{R}^3)},$$

we have

$$\|\Delta_j\mathcal{G}^\varepsilon_\nu f\|_{L^1_T(L^\infty(\mathbb{R}^3))} \leq C2^{-\frac{j}{2}}\|\Delta_j f\|_{L^2(\mathbb{R}^3)}\sum_{k\leq j} 2^{\frac{k-j}{2}}\min\left\{1, (\varepsilon 2^{2j})^{\frac{1}{4}}2^{\frac{j-k}{2}}\right\}.$$

Finally we find, for any $r > 1$,

$$\|\Delta_j \mathcal{G}_\nu^\varepsilon f\|_{L^1_T(L^\infty(\mathbb{R}^3))} \leq C 2^{-\frac{j}{2}} \|\Delta_j f\|_{L^2(\mathbb{R}^3)} \sum_{k \leq j} 2^{\frac{k-j}{2}(1-\frac{1}{r})} (\varepsilon 2^{2j})^{\frac{1}{4r}},$$

which means in other words that for any $\alpha > 0$, choosing $r = 1 + \alpha$,

$$\|\Delta_j \mathcal{G}_\nu^\varepsilon f\|_{L^1_T(L^\infty(\mathbb{R}^3))} \leq C_\alpha 2^{-\frac{j}{2}} \|\Delta_j f\|_{L^2(\mathbb{R}^3)} (\varepsilon 2^{2j})^{\frac{1}{4(1+\alpha)}}. \tag{3.20}$$

Now, to get the L^p_T estimate, we just have to interpolate (3.20) with

$$\|\Delta_j \mathcal{G}_\nu^\varepsilon f\|_{L^\infty_T(L^2(\mathbb{R}^3))} \leq C \|\Delta_j f\|_{L^2(\mathbb{R}^3)}, \tag{3.21}$$

which, for all $p \in [1, \infty]$ and for all $\alpha > 0$, and with $1/q = 1/2 - 1/2p$ yields

$$\|\Delta_j \mathcal{G}_\nu^\varepsilon f\|_{L^p_T(L^q(\mathbb{R}^3))} \leq C_\alpha 2^{-\frac{j}{2p}} (\varepsilon 2^{2j})^{\frac{1}{4p(1+\alpha)}} \|\Delta_j f\|_{L^2(\mathbb{R}^3)}.$$

Then Bernstein's inequality implies the result in the isotropic case.

We are finally left with the anisotropic case. To prove estimate (3.3), the computation is identical to the one leading to estimate (3.2): we just have to change the set $\mathcal{B}$ defined in (3.14) into

$$\mathcal{B}' \stackrel{\text{def}}{=} \left\{ \Psi \in \mathcal{D}(\mathbb{R}^3), \|\Psi\|_{L^\infty_T(L^{1,2}_{x_h,x_3})} \leq 1 \right\}.$$

Then estimate (3.16) becomes, with (3.5),

$$\left\| \int_{\mathbb{R}^+} e^{-i\frac{t}{\varepsilon}\frac{\xi_3}{|\xi|} - \nu t |\xi|^2} \widehat{\Psi}_{j,k}(t,\xi)\, dt \right\|^2_{L^2(\mathbb{R}^3)}$$

$$\leq \int_{(\mathbb{R}^+)^2} \|\Psi_{j,k}(s,\cdot)\|_{L^{1,2}_{x_h,x_3}} \|\Psi_{j,k}(t,\cdot)\|_{L^{1,2}_{x_h,x_3}}$$

$$\times \min\left\{1, \frac{\varepsilon^{\frac{1}{2}} 2^{(j-k)}}{|t-s|^{\frac{1}{2}}}\right\} 2^{2j} e^{-\nu(t+s)2^{2j}}\, dt ds.$$

So finally, using (3.18) and (3.19) we have

$$\|\Delta_j \Delta_k^v \mathcal{G}_\nu^\varepsilon f\|_{L^1_T(L^{\infty,2}_{x_h,x_3})} \leq C \min\left\{1, (\varepsilon 2^{2j})^{\frac{1}{4}} 2^{\frac{j-k}{2}}\right\} 2^{-j} \|\Delta_j \Delta_k^v f\|_{L^2(\mathbb{R}^3)}.$$

The end of the proof is then exactly that of the isotropic case, and we obtain estimate (3.3). So Theorem 3 is proved. ∎

4. Global existence and convergence of smooth solutions

Let us recall that we have noted w^ε_F for the solution of the linear equation

$$\left\{\begin{aligned} \partial_t w^\varepsilon_F - \nu\Delta w^\varepsilon_F + \frac{1}{\varepsilon} L_{RF} w^\varepsilon_F &= 0 \\ w^\varepsilon_{F|t=0} &= w_0 . \end{aligned}\right.$$

Classical results on the heat equation imply that

$$\forall t \in \mathbb{R}^+, \quad \|w^\varepsilon_F(t)\|^2_{\dot H^{\frac12}(\mathbb{R}^3)} + 2\nu \int_0^t \|w^\varepsilon_F(s)\|^2_{\dot H^{\frac32}(\mathbb{R}^3)}\, ds \le \|w_0\|^2_{\dot H^{\frac12}(\mathbb{R}^3)}.$$

Moreover, it is obvious that

$$\lim_{\substack{N\to+\infty\\ N_3\to-\infty}} (\mathrm{Id} - S^v_{N_3}) S_N w_0 = w_0 \quad \text{in} \quad \dot H^{\frac12}(\mathbb{R}^3).$$

Consequently, for any $\eta > 0$, there exist two integers N and N_3 such that if $w^\varepsilon_{Fm} \stackrel{\text{def}}{=} (\mathrm{Id} - S^v_{N_3}) S_N w^\varepsilon_F$, then, for all $t \in \mathbb{R}^+$,

$$\|w^\varepsilon_F(t) - w^\varepsilon_{Fm}(t)\|^2_{\dot H^{\frac12}(\mathbb{R}^3)} + 2\nu \int_0^t \|w^\varepsilon_F(s) - w^\varepsilon_{Fm}(s)\|^2_{\dot H^{\frac32}(\mathbb{R}^3)}\, ds \le \eta. \quad (4.1)$$

Now let us define $w^\varepsilon \stackrel{\text{def}}{=} u^\varepsilon - \bar u - w^\varepsilon_{Fm}$, where $\bar u$ is the solution of the bidimensional Navier–Stokes equation. Then the function w^ε satisfies

$$\left\{\begin{aligned} \partial_t w^\varepsilon - \nu\Delta w^\varepsilon + \frac{1}{\varepsilon} L_{RF} w^\varepsilon &= Q(w^\varepsilon, w^\varepsilon + 2\bar u + 2w^\varepsilon_{Fm}) \\ &\quad + Q(w^\varepsilon_{Fm}, w^\varepsilon_{Fm}) + 2\,Q(\bar u, w^\varepsilon_{Fm}) \\ w^\varepsilon_{|t=0} &= (\mathrm{Id} - S_N(\mathrm{Id} - S^v_{N_3})) w_0 , \end{aligned}\right. \quad (4.2)$$

where $Q(a,b) = A_{\ell,m}(D) a^m b^\ell$, $A_{\ell,m}(D)$ is a Fourier multiplier of order 1.

Classical results on the incompressible bidimensional Navier–Stokes equations (see [23] for instance) imply that

$$\forall t \in \mathbb{R}^+, \quad \|\bar u(t)\|^2_{L^2(\mathbb{R}^2)} + 2\nu \int_0^t \|\nabla \bar u(s)\|^2_{L^2(\mathbb{R}^2)}\, ds = \|\bar u_0\|^2_{L^2(\mathbb{R}^2)}. \quad (4.3)$$

It follows from (4.1) that Theorem 2 is proved. if we show that (4.2) is well posed, for ε small enough, in the space $C^0_b(\mathbb{R}^+, \dot H^{\frac12}(\mathbb{R}^3)) \cap L^2(\mathbb{R}^+, \dot H^{\frac32}(\mathbb{R}^3))$, and also that

$$w^\varepsilon \to 0 \quad \text{in} \quad L^\infty(\mathbb{R}^+, \dot H^{\frac12}(\mathbb{R}^3)) \cap L^2(\mathbb{R}^+, \dot H^{\frac32}(\mathbb{R}^3)). \quad (4.4)$$

Let us now apply the operators Δ_j defined earlier (see (3.1)), to equation (4.2), and write an energy estimate in $L^2(\mathbb{R}^3)$. In order to estimate the terms in (4.2) containing the bidimensional function $\bar{u}$, which is less smooth than w^ε, we use Lemma 5.1 of [9]: it states that the product of any tridimensional function by $\bar{u}$ is half a derivative smoother than if $\bar{u}$ were tridimensional; the setting in [9] is that of the torus, but the results extend unchanged to the whole space.

Let us recall the rules of product in Sobolev spaces, which will be used extensively throughout the proofs:

$$\forall s,t<\frac{d}{2},\ s+t>0, \quad \|ab\|_{\dot{H}^{s+t-\frac{d}{2}}(\mathbb{R}^d)} \leq C_{s,t}\|a\|_{\dot{H}^s(\mathbb{R}^d)}\|b\|_{\dot{H}^t(\mathbb{R}^d)},$$

$$\forall s,t<1,\ s+t>0, \quad \|ab\|_{\dot{H}^{s+t-1}(\mathbb{R}^3)} \leq C_{s,t}\|a\|_{\dot{H}^s(\mathbb{R}^2)}\|b\|_{\dot{H}^t(\mathbb{R}^3)}.$$

We shall also use the classical fact (see [6]) that

$$\forall d\geq 1, \quad \|v\|_{\dot{H}^s(\mathbb{R}^d)} \sim \|2^{js}\|\Delta_j v\|_{L^2(\mathbb{R}^d)}\|_{\ell^2(\mathbf{Z})},$$

$$\|\nabla\Delta_j v\|_{L^2(\mathbb{R}^d)} \geq c2^j\|\Delta_j v\|_{L^2(\mathbb{R}^d)}.$$

Now let us apply operator Δ_j to equation (4.2) satisfied by w^ε. We then take the scalar product in $L^2(\mathbb{R}^3)$ with $\Delta_j w^\varepsilon$, and obtain

$$\begin{aligned}
&\frac{1}{2}\frac{d}{dt}\|\Delta_j w^\varepsilon(t)\|^2_{L^2(\mathbb{R}^3)} + c\nu 2^{2j}\|\Delta_j w^\varepsilon(t)\|^2_{L^2(\mathbb{R}^3)}\\
&\qquad\leq C\, c_j^\varepsilon(t)2^{\frac{j}{2}}\|F^\varepsilon(t)\|_{\dot{H}^{-\frac{1}{2}}(\mathbb{R}^3)}\|\Delta_j w^\varepsilon(t)\|_{L^2(\mathbb{R}^3)}\\
&\qquad\quad + C\, c_j^\varepsilon(t)2^j\|\bar{F}^\varepsilon(t)\|_{\dot{H}^{-1}(\mathbb{R}^3)}\|\Delta_j w^\varepsilon(t)\|_{L^2(\mathbb{R}^3)}\\
&+Cc_j^\varepsilon(t)2^{\frac{j}{2}}\|w^\varepsilon(t)\|_{\dot{H}^1(\mathbb{R}^3)}\left(\|w^\varepsilon(t)\|_{\dot{H}^1(\mathbb{R}^3)}+\|\bar{u}(t)\|_{\dot{H}^{\frac{1}{2}}(\mathbb{R}^2)}\right)\|\Delta_j w^\varepsilon(t)\|_{L^2(\mathbb{R}^3)},
\end{aligned} \tag{4.5}$$

where we have written

$$F^\varepsilon \stackrel{\text{def}}{=} 2\,Q(w^\varepsilon, w^\varepsilon_{Fm}) + Q(w^\varepsilon_{Fm}, w^\varepsilon_{Fm}), \quad \text{and} \quad \bar{F}^\varepsilon \stackrel{\text{def}}{=} 2\,Q(\bar{u}, w^\varepsilon_{Fm}).$$

Here like throughout the paper, $c_j^\varepsilon(t)$ is a sequence in $\ell^2(\mathbb{Z})$, of norm 1, and C denotes any "universal" constant.

To simplify the notation of the space–time norms, we shall write

$$L^p_T(\dot{H}^s(\mathbb{R}^d)) = L^p([0,T];\dot{H}^s(\mathbb{R}^d)) \text{ and } L^p(\dot{H}^s(\mathbb{R}^d)) = L^p(\mathbb{R}^+;\dot{H}^s(\mathbb{R}^d)).$$

Observe that estimate (4.5) has exactly the same form as estimate (B.8) of [10] (proof of Lemma B.1), with the following correspondance of notation:

$$\sigma=\frac{1}{2}, \quad q=j, \quad A=w^\varepsilon, \quad |B(t)|_{\sigma+\frac{1}{2}}=\|\bar{u}(t)\|_{\dot{H}^{\frac{1}{2}}(\mathbb{R}^2)}, \quad \text{and} \quad W=F^\varepsilon.$$

The only difference with estimate (B.8) is the presence of an extra forcing term, $\bar{F}^\varepsilon$, but it is an easy adaptation of the computations of [10] to see that as in Lemma B.1 of [10], we have

$$X_4^\varepsilon(t)e^{-C\bar{U}_4^4(t)} \leq C\|F^\varepsilon\|_{L_t^2(\dot{H}^{-\frac{1}{2}}(\mathbb{R}^3))} + C\|\bar{F}^\varepsilon\|_{L_t^4(\dot{H}^{-1}(\mathbb{R}^3))} + C(X_4^\varepsilon)^2(t), \tag{4.6}$$

and

$$X_2^\varepsilon(t) + X_\infty^\varepsilon(t) \leq C\|F^\varepsilon\|_{L_t^2(\dot{H}^{-\frac{1}{2}}(\mathbb{R}^3))} + C\|\bar{F}^\varepsilon\|_{L_t^4(\dot{H}^{-1}(\mathbb{R}^3))} + CX_4^\varepsilon(t)(X_4^\varepsilon(t) + \bar{U}_4(t)), \tag{4.7}$$

where we have noted

$$X_4^\varepsilon(t) = \|w^\varepsilon\|_{L_t^4(\dot{H}^1(\mathbb{R}^3))}, \quad X_2^\varepsilon(t) = \|w^\varepsilon\|_{L_t^2(\dot{H}^{\frac{3}{2}}(\mathbb{R}^3))},$$

$$X_\infty^\varepsilon(t) = \|w^\varepsilon\|_{L_t^\infty(\dot{H}^{\frac{1}{2}}(\mathbb{R}^3))}, \quad \text{and} \quad \bar{U}_4(t) = \|\bar{u}\|_{L_t^4(\dot{H}^{\frac{1}{2}}(\mathbb{R}^2))}.$$

So to conclude, all we need is an estimate on F^ε in $L^2(\dot{H}^{-\frac{1}{2}}(\mathbb{R}^3))$ and on $\bar{F}^\varepsilon$ in $L^4(\dot{H}^{-1}(\mathbb{R}^3))$. These estimates are given in the following proposition, the proof of which is postponed for a while.

Proposition 1 - *Under the assumptions of Theorem 2, we have*

$$\|F^\varepsilon\|_{L_T^2(\dot{H}^{-\frac{1}{2}}(\mathbb{R}^3))} \leq \eta_1(\varepsilon)X_4^\varepsilon(t) + \eta_2(\varepsilon), \quad \textit{and} \quad \|\bar{F}^\varepsilon\|_{L_T^4(\dot{H}^{-1}(\mathbb{R}^3))} \leq \eta_3(\varepsilon),$$

with $\lim_{\varepsilon\to 0} \eta_i = 0$, *for* $i \in \{1,2,3\}$.

Let us now conclude the proof of the theorem. Estimate (4.6), along with Proposition 1, implies that

$$X_4^\varepsilon(t)\Big(e^{-C\bar{U}_4^4(t)} - \eta_1(\varepsilon)\Big) \leq C(X_4^\varepsilon)^2(t) + \eta(\varepsilon),$$

with $\eta(\varepsilon) \stackrel{\text{def}}{=} \eta_2(\varepsilon) + \eta_3(\varepsilon)$. But the energy equality in $(NS2D)$ recalled in (4.3), implies that

$$\forall t \geq 0, \quad \bar{U}_4^4(t) \leq C\nu^{-2}\|\bar{u}_0\|_{L^2(\mathbb{R}^2)}^4,$$

so let us choose $\varepsilon_0 > 0$ such that

$$\forall \varepsilon \leq \varepsilon_0, \quad \eta_1(\varepsilon) \leq \frac{1}{4}\exp\Big(-C\nu^{-2}\|\bar{u}_0\|_{L^2(\mathbb{R}^2)}^4\Big).$$

We get, for any $\varepsilon \leq \varepsilon_0$

$$X_4^\varepsilon(t) \leq Ce^{C\nu^{-2}\|\bar{u}_0\|^4_{L^2(\mathbb{R}^2)}}\left((X_4^\varepsilon)^2(t) + \eta(\varepsilon)\right).$$

So we have, for any $\varepsilon \leq \varepsilon_0$,

$$X_4^\varepsilon(t) \leq Ce^{C\nu^{-2}\|\bar{u}_0\|^4_{L^2(\mathbb{R}^2)}}\eta(\varepsilon).$$

Finally, from inequalities (4.6) and (4.7), we obtain

$$\lim_{\varepsilon\to 0} w^\varepsilon = 0 \quad \text{in the space} \quad L^\infty(\dot{H}^{\frac{1}{2}}(\mathbb{R}^3)) \cap L^2(\dot{H}^{\frac{3}{2}}(\mathbb{R}^3)).$$

The global well–posedness of (4.2) for ε small enough follows by classical arguments; this concludes the proof of the theorem. ■

Proof of Proposition 1. Let us recall that

$$F^\varepsilon = 2\,Q(w^\varepsilon, w^\varepsilon_{_{Fm}}) + Q(w^\varepsilon_{_{Fm}}, w^\varepsilon_{_{Fm}}) \quad \text{and} \quad \bar{F}^\varepsilon = 2\,Q(\bar{u}, w^\varepsilon_{_{Fm}}).$$

It is here that we are going to use the Strichartz estimates proved in Section 3, namely estimates (3.2) and (3.3); this is the reason why we got rid of the very low and very high frequencies of the linear solution above: the function $w^\varepsilon_{_{Fm}}$ has frequencies such that

$$2^{N_3} \leq |\xi| \leq 2^N \quad \text{and} \quad 2^{N_3} \leq |\xi_3|,$$

so estimates (3.2) and (3.3) simply become respectively

$$\|\Delta_j w^\varepsilon_{_{Fm}}\|_{L^p_T(L^\infty(\mathbb{R}^3))} \leq C(\varepsilon 2^{2N})^{1/(4p\alpha)}\|\Delta_j w_{0,m}\|_{L^2(\mathbb{R}^3)}, \quad \forall \alpha > 1, \quad (4.8)$$

and

$$\|w^\varepsilon_{_{Fm}}\|_{L^p_T(L^{\infty,2}_{x_h,x_3})} \leq C(\varepsilon 2^{2N})^{1/4p}\|w_{0,m}\|_{L^2(\mathbb{R}^3)}, \quad (4.9)$$

where we have omitted the dependence of the constants on N and N_3, which are fixed once and for all.

Let us start by estimating $Q(w^\varepsilon_{_{Fm}}, w^\varepsilon_{_{Fm}} + w^\varepsilon)$ in $L^2_T(H^{-\frac{1}{2}}(\mathbb{R}^3))$. We are going to use the paraproduct algorithm introduced by J.-M. Bony in [5]: we have, with the notation introduced in (3.1), and with the usual summation convention,

$$Q(c,b) = A_{\ell,m}(D)(T'_{b^\ell}b^m + T_{b^m}c^\ell), \quad (4.10)$$

with

$$T'_c b = \sum_j S_{j+2}c\,\Delta_j b \quad \text{and} \quad T_b c = \sum_j S_{j-1}b\,\Delta_j c.$$

We can write, for any $\alpha > 1$ and according to (4.8),

$$2^{-\frac{j}{2}} \|\Delta_j A_{\ell,m}(D) T'_{w^{\varepsilon,}_{Fm}\, ell} b^m\|_{L^2_T(L^2(\mathbb{R}^3))}$$

$$\leq C \sum_{j' \geq j - N_0} \|S_{j'} w^\varepsilon_{Fm}\|_{L^4_T(L^\infty(\mathbb{R}^3))} \|\Delta_{j'} b\|_{L^4_T(L^2(\mathbb{R}^3))} 2^{\frac{j}{2}}$$

$$\begin{aligned}
&\leq C_\varepsilon \sum_{\substack{j' \geq j - N_0 \\ N_3 \leq j'' \leq \min\{j'-1, N\}}} \|\Delta_{j''} w_0\|_{L^2(\mathbb{R}^3)} 2^{\frac{j}{2}} \|\Delta_{j'} b\|_{L^4_T(L^2(\mathbb{R}^3))} \\
&\leq C_\varepsilon \|w_{0,m}\|_{L^2(\mathbb{R}^3)} \sum_{\substack{j' \geq j - N_0 \\ N_3 \leq j'' \leq \min\{j'-1, N\}}} 2^{\frac{j}{2}} 2^{\frac{j''-N}{2}} \|\Delta_{j'} b\|_{L^4_T(L^2(\mathbb{R}^3))} \\
&\leq C_\varepsilon \|w_{0,m}\|_{L^2(\mathbb{R}^3)} \sum_{j' \geq j - N_0} 2^{-N/2} 2^{j'} \|\Delta_{j'} b\|_{L^4_T(L^2(\mathbb{R}^3))} 2^{\frac{j-j'}{2}},
\end{aligned}$$

where C_ε denotes a constant multiplied by any positive power of $\varepsilon 2^{2N}$. So we have obtained, using Young's inequality,

$$\begin{aligned}
&\|A_{\ell,m}(D) T'_{w^{\varepsilon,\ell}_{Fm}} (w^\varepsilon + w^\varepsilon_{Fm})^m\|^2_{L^2_T(\dot{H}^{-\frac{1}{2}}(\mathbb{R}^3))} \\
&\quad \leq C^2_\varepsilon \|w_{0,m}\|^2_{L^2(\mathbb{R}^3)} \times \left(\|w_{0,m}\|^2_{L^2(\mathbb{R}^3)} + \|w^\varepsilon\|^2_{L^4_T(\dot{H}^1(\mathbb{R}^3))} \right). \qquad (4.11)
\end{aligned}$$

As to the term $T_{b^m} w^{\varepsilon,\ell}_{Fm}$, let us define

$$\mathcal{C}_j \stackrel{\text{def}}{=} 2^{-\frac{j}{2}} \|\Delta_j A_{\ell,m}(D) T_{b^m} w^{\varepsilon,\ell}_{Fm}\|_{L^2_T(L^2(\mathbb{R}^3))}.$$

We have

$$\begin{aligned}
\mathcal{C}_j &\leq C \sum_{\substack{|j-j'| \leq N_0 \\ N_3 \leq j' \leq N}} 2^{\frac{j}{2}} \|S_{j'} b\|_{L^4_T(L^2(\mathbb{R}^3))} \|\Delta_{j'} w^\varepsilon_{Fm}\|_{L^4_T(L^\infty(\mathbb{R}^3))} \\
&\leq C_\varepsilon \sum_{\substack{|j-j'| \leq N_0 \\ N_3 \leq j' \leq N}} 2^{\frac{j}{2}} \|\Delta_{j''} w_0\|_{L^2(\mathbb{R}^3)} \|S_{j'} b\|_{L^4_T(L^2(\mathbb{R}^3))} \\
&\leq C_\varepsilon \|w_{0,m}\|_{L^2(\mathbb{R}^3)} \|b\|_{L^4_T(\dot{H}^1(\mathbb{R}^3))}.
\end{aligned}$$

So with (4.11), we obtain finally

$$\begin{aligned}
&\|Q(w^\varepsilon_{Fm}, w^\varepsilon + w^\varepsilon_{Fm})\|^2_{L^2_T(\ psob-\frac{1}{2}(\mathbb{R}^3))} \\
&\quad \leq C^2_\varepsilon \|w_{0,m}\|^2_{L^2(\mathbb{R}^3)} \left(\|w_{0,m}\|^2_{L^2(\mathbb{R}^3)} + \|w^\varepsilon\|^2_{L^4_T(\dot{H}^1(\mathbb{R}^3))} \right). \qquad (4.12)
\end{aligned}$$

Now, to end the proof, we need to estimate the term $Q(w^{\varepsilon}_{Fm}, \bar{u})$; it is with this estimate in mind that we proved an anisotropic Strichartz estimate in Section 3. We have simply

$$\begin{aligned}
\|A_{\ell,m}(D)w^{\varepsilon,\ell}_{Fm}\,\bar{u}^m\|_{L^4_T(\dot{H}^{-1}(\mathbb{R}^3))} &\leq C\|w^{\varepsilon,\ell}_{Fm}\,\bar{u}^m\|_{L^4_T(L^2(\mathbb{R}^3))}\\
&\leq C\|\bar{u}\|_{L^\infty_T(L^2(\mathbb{R}^2))}\|w^{\varepsilon}_{Fm}\|_{L^4_T(L^{\infty,2}_{x_h,x_3})}\\
&\leq C_\varepsilon\|\bar{u}_0\|_{L^2(\mathbb{R}^2)}\|w_{0,m}\|_{L^2(\mathbb{R}^3)},
\end{aligned}$$

according to (4.9). This estimate, along with (4.12), proves Proposition 1.

References

[1] A. Babin, A. Mahalov, and B. Nicolaenko, Global splitting, integrability and regularity of 3D Euler and Navier–Stokes equations for uniformly rotating fluids, European J. of Mechanics, 15 (1996), 291–300.

[2] A. Babin, A. Mahalov, and B. Nicolaenko, Resonances and regularity for Boussinesq equations, Russ. J. of Math. Physics, 4 (1996), 417–428.

[3] A. Babin, A. Mahalov, B. Nicolaenko, and Y. Zhou, On the asymptotic regimes and the strongly stratified limit of rotating Boussinesq equations, Journal of Theoretical and Comp. Fluid Dynamics, 9 (1997), 223–251.

[4] H. Bahouri and J.-Y. Chemin, Equations d'ondes quasilinéaires et estimation de Strichartz, Prépublication du Laboratoire d'Analyse Numérique, to appear in American Journal of Mathematics, 121 (1999).

[5] J.-M. Bony, Calcul symbolique et propagation des singularités pour les équations aux dérivées partielles non linéaires, Annales de l'École Normale Supérieure, 14 (1981), 209–246.

[6] J.-Y. Chemin, Fluides parfaits incompressibles, Astérisque, 230 (1995).

[7] J.-Y. Chemin, A propos d'un problème de pénalisation de type antisymétrique, Journal Math. Pures Appl., 76 (1997), 739–755.

[8] B. Desjardins and E. Grenier, Low Mach number limit of compressible flows in the whole space, Proc. Royal Society of London, 455 (1999), 2271–2279.

[9] I. Gallagher, The Tridimensional Navier–Stokes equations with almost bidimensional data: stability, uniqueness and life span, International Math. Research Notices, 18 (1997), 919–935.

[10] I. Gallagher, Applications of Schochet's methods to parabolic equations, Journal Math. Pures Appl., 77 (1998), 989–1054.

[11] J. Ginibre and G. Velo, Generalized Strichartz inequalities for the wave equation, Journal of Functional Anal., 133 (1995), 50–68.

[12] H.P. Greenspan, The theory of rotating fluids, Cambridge monographs on mechanics and applied mathematics (1969).

[13] E. Grenier, Oscillatory perturbations of the Navier Stokes equations, Journal Math. Pures Appl., 76 (1997), 477–498.

[14] L. Hörmander, Lectures on nonlinear hyperbolic differential equations, Math. and Applications, 26, Springer (1996).

[15] D. Iftimie, La résolution des équations de Navier–Stokes dans des domaines minces et la limite quasigéostrophique, Thèse de l'Université Paris 6 (1997).

[16] D. Iftimie, The resolution of the Navier–Stokes equations in anisotropic spaces, Revista Matematica Ibero–Americana, 15 (1999), 1–36.

[17] F. John, Existence for large times of strict solutions of non linear wave equations in three space dimensions, Comm. Pure and Appl. Math., 40 (1987), 79–109.

[18] L. Kapitanski, Some generalization of the Strichartz-Brenner inequality, Leningrad Math. Journal, 1 (1990), 693–721.

[19] M. Keel and T. Tao, Endpoint Strichartz estimates, American Journal of Mathematics, 120 (1998), 955–980.

[20] S. Klainerman, Global existence for nonlinear wave equations, Comm. Pure and Appl. Math., 33 (1980), 43–101.

[21] S. Klainerman, Uniform decay estimates and the Lorentz invariance of the classical wave equation Comm. Pure and Appl. Math., 38 (1985), 321–332.

[22] S. Klainerman and M. Machedon, Remark on Strichartz type inequalites (with an appendix of J. Bourgain and D. Tataru), International Math. Research Notices, 5 (1996), 201–220.

[23] P.-L. Lions, Mathematical topics in fluid dynamics, Vol. 1, Incompressible models, Oxford University Press (1998).

[24] J. Pedlosky, Geophysical fluid dynamics, Springer (1979).

[25] S. Schochet, Fast singular limits of hyperbolic PDEs, J. Diff. Equ. 114 (1994), 476–512.

[26] S. Ukai, The incompressible limit and the initial layer of the compressible Euler equation, J. Math. Kyoto Univ. 26, no. 2 (1986), 323–331.

Jean-Yves Chemin
Centre de Mathématiques
Ecole Polytechnique
91128 -Palaiseau
France
E-mail: chemin@math.polytechnique.fr

Benoit Desjardins
DMA - ENS, 45 rue d'Ulm
75230 Paris Cedex 05

and

CEA/DIF/DCSA
B.P. 12
91680 Bruyères le Châtel
France
E-mail: desjardins@na-net.ornl.gov

Isabelle Gallagher
Département de Mathématiques
Université de Paris-Sud
91405 Orsay Cedex
and
Centre de Mathématiques
École Polytechnique
91128 Palaiseau Cedex, France

Emmanuel Grenier
UMPA (CNRS UMR 5669
E.N.S. Lyon
46 allée d'Italie
69364 Lyon Cedex 07
France
E-mail: egrenier@umpa.ens-lyon.fr

Studies in Mathematics and its Applications, Vol. 31
D. Cioranescu and J.L. Lions (Editors)

Chapter 10

INTEGRAL EQUATIONS AND SADDLE POINT FORMULATION FOR SCATTERING PROBLEMS

F. COLLINO AND B. DESPRES

1. Introduction

In this work we deal with integral equations for solving obstacle scattering by time-harmonic electromagnetic waves. We show that it is possible to design a system of integral equations with a positive spectrum. This kind of system has been introduced in [8], [9], in the 2D acoustic case and in [7] for Maxwell's equations. The system of equations may be viewed as coming from the minimization of a positive quadratic functional. This approach seems to be new. It can be either derived from some more or less simple manipulations of classical integral operators.

One interesting feature of the new system of equations, comparing to the classical theory, is that the functional framework is L^2 based. It comes directly from the construction of the quadratic functional. A natural penalisation procedure may be used to get more coercivity on the multiplier, even if the inf-sup condition is already true for the weak formulation.

The strong coercivity properties of this system allow to get a global and coherent framework for the introduction of numerical iterative algorithms for the numerical computation of the discrete solutions, as it is the case for many coercive problems : Jacobi algorithms, conjugate gradient algorithms, and so on, may be introduced with comprehensive proof of convergence [10].

The outline of the paper is the following. After some definitions we derive the system of equations by minimization of a natural quadratic functional under constraint. It explains the presence of a Lagrange multiplier in the unknowns. After that, we discuss well posedness of the system, introduce penalisation, and propose various algorithms to solve the system. Finally some properties of the integral operators are proved, based on the analysis of Calderon projectors.

2. Some definitions

In this work the x and y notations may denote either the space variable needed to define the integral operators $x \in R^3$ and $y \in R^3$, or the pair unknown-multiplier $(x, y) \in X \times Y$ where X and Y or the appropriate functional spaces. Natural interpretation of these notations is enough, and no confusion is possible.

Let $D^- \subset R^3$ a bounded domain with regular boundary Γ, $n(x)$ the unit normal vector to Γ directed into $D^+ \subset R^3$, the exterior domain of D^-. The problem we address is the determination of the out-coming electromagnetic field scattered by Γ. Let $k > 0$ be the wavenumber and $Z_0 > 0$ the impedance of the vacuum. The unknown field is a solution to (1), (2) where

$$\begin{cases} \nabla \wedge E^+ - ikZ_0 H^+ = 0, & \text{in } D^+ \\ \nabla \wedge H^+ + ikZ_0^{-1} E^+ = 0, & \text{in } D^+, \end{cases} \tag{1}$$

is the Maxwell's system and

$$\lim_{|x| \to \infty} |x| \left(Z_0 H^+ \wedge \frac{x}{|x|} - E^+ \right) = 0, \tag{2}$$

is the Sommerfeld radiation condition at infinity. System (1-2) must be supplemented by a boundary condition on Γ. Let us assume for a while that the boundary condition is an absorbing boundary condition

$$n(x) \wedge (E^+_{/\Gamma}(x) \wedge n(x)) + Z_0(H^+_{/\Gamma}(x) \wedge n(x)) = G^{in}. \tag{3}$$

Other boundary conditions will been introduced in section . Following [5] We define the classical integral operators $\tilde{T}$ and $\tilde{K}$ defined from Γ to D^+

$$\begin{cases} \tilde{T}J(x) & = & \dfrac{1}{k} \nabla_x \wedge \left(\nabla_x \wedge \displaystyle\int_\Gamma G(x,y) J(y) \, d\Gamma(y) \right) \\ \tilde{K}J(x) & = & -\nabla_x \wedge \displaystyle\int_\Gamma G(x,y) J(y) \, d\Gamma(y), \end{cases} \tag{4}$$

or, in another form,

$$\begin{cases} \tilde{T}J(x) & = & k \displaystyle\int_\Gamma \left(G(x,y) J(y) + \frac{1}{k^2} \nabla_x G(x,y) \nabla^t \cdot J(y) \right) \, d\Gamma(y) \\ \tilde{K}J(x) & = & \displaystyle\int_\Gamma \nabla_y G(x,y) \wedge J(y) \, d\Gamma(y). \end{cases} \tag{5}$$

Here $\nabla^t \cdot J(y)$ denotes the surface divergence of J. The kernel $G(x,y)$ is the radiating Green function for the Helmholtz equation

$$G(x,y) = \frac{\exp^{ik|x-y|}}{4\pi|x-y|}. \tag{6}$$

Passing to the limit $x \to \Gamma$, $x \in D^-$ we define the classical surface to surface integral operators

$$\begin{cases} TJ(x) = \lim_{y\to x} n(x) \wedge (\tilde{T}J(y)) \wedge n(x)) \\ KJ(x) = \lim_{\epsilon\to 0} n(x) \wedge \left(\left(\int_{\Gamma,\ |x-y|>\epsilon} \nabla_y G(x,y) \wedge J(y)\, d\Gamma(y) \right) \wedge n(x) \right). \end{cases} \tag{7}$$

Now we define the 2×2 integral operator $\mathbf{S}$ by

$$\mathbf{S} = \begin{bmatrix} T & K - \frac{1}{2}n\wedge \\ K - \frac{1}{2}n\wedge & T \end{bmatrix}, \tag{8}$$

To go further we introduce the decomposition in real and imaginary part of the Green function $G(x,y)$

$$G(x,y) = \frac{\cos(k|x-y|)}{4\pi|x-y|} + i\frac{\sin(k|x-y|)}{4\pi|x-y|}. \tag{9}$$

We obtain a similar decomposition for $\mathbf{S}$ in real and imaginary parts

$$\mathbf{S} = \mathbf{T} + \mathbf{iR}, \tag{10}$$

with

$$\mathbf{T} = \begin{bmatrix} T_r & K_r - \frac{1}{2}n\wedge \\ K_r - \frac{1}{2}n\wedge & T_r \end{bmatrix}, \quad \mathbf{R} = \begin{bmatrix} T_i & K_i \\ K_i & T_i \end{bmatrix}. \tag{11}$$

From a mathematical point of view T_i, T_r, K_r, K_i are symmetric, $n(x)\wedge$ is antisymmetric and $G_i(x,y)$ is a regular kernel. So $\mathbf{R}$ appears as symmetric and regularizing while $\mathbf{T}^*$ the adjoint of $\mathbf{T}$ satisfies

$$\mathbf{T} - \mathbf{T}^* = \mathbf{\Pi} = \begin{bmatrix} 0 & -n(x)\wedge \\ -n(x)\wedge & 0 \end{bmatrix}. \tag{12}$$

Next we derive the following decomposition of $\mathbf{R}$. If $\hat{d}$ is a direction of the unit sphere S^2 and $(J(x), M(x))$ are two fields on Γ we define the far field operator

$$\mathbf{a}^\infty \mathbf{J}(\hat{\mathbf{d}}) = \frac{\mathbf{k}}{4\pi} \int_{\mathbf{\Gamma}} \hat{\mathbf{d}} \wedge (\mathbf{J}(\mathbf{x}) \wedge \hat{\mathbf{d}}) \mathbf{e}^{-\mathbf{ikx}\cdot\hat{\mathbf{d}}} \mathbf{d\Gamma}(\mathbf{x}). \tag{13}$$

We combine this into a new far field operator

$$\mathbf{A}^{\infty} : \begin{bmatrix} J(x) \\ M(x) \end{bmatrix} \quad \longrightarrow \quad \mathbf{A}^{\infty} \begin{bmatrix} J \\ M \end{bmatrix} (\hat{\mathbf{d}}) = \mathbf{a}^{\infty}\mathbf{J}(\hat{\mathbf{d}}) - \mathbf{i}\hat{\mathbf{d}} \wedge \mathbf{a}^{\infty}\mathbf{M}(\hat{\mathbf{d}}). \quad (14)$$

It may be proved that

$$\mathbf{R} = (\mathbf{A}^{\infty})^{*} \, \mathbf{A}^{\infty}. \quad (15)$$

3. Derivation of the system

The idea is to consider a set of solutions of the Maxwell's system

$$\begin{cases} \nabla \wedge E - ikZ_0 H = 0, & \text{in } D^+ \\ \nabla \wedge H + ikZ_0^{-1} E = 0, & \text{in } D^+, \end{cases} \quad (16)$$

with the following expansion at infinity

$$\begin{aligned} E(x) \quad &\to \quad \frac{e^{ik|x|}}{|x|} a_{\infty}^{out}(\hat{x}) + \frac{e^{-ik|x|}}{|x|} a_{\infty}^{in}(\hat{x}), \ |x| \to \infty, \quad (\hat{x} = \frac{x}{|x|}) \\ Z_0 H(x) \quad &\to \quad \frac{e^{ik|x|}}{|x|} (\hat{x} \wedge a_{\infty}^{out}(\hat{x})) - \frac{e^{-ik|x|}}{|x|} (\hat{x} \wedge a_{\infty}^{in}(\hat{x})), \ |x| \to \infty, \end{aligned} \quad (17)$$

where $a_{\infty}^{out}(E,H)$ and $a_{\infty}^{in}(E,H)$ are some tangential fields on the unit sphere, and the convergence $e \to e^{\infty}$ holds in the sense

$$\lim_{R \to \infty} \frac{1}{R} \int_{R \le |x| \le 2R} |e(x) - e^{\infty}(x)|^2 \, dx = 0. \quad (18)$$

In other words, we consider all the in and out-coming electromagnetic fields on D^+. In general, we are only interested in out-coming fields but nevertheless we begin by embedding the problem in this larger set of solutions.

For technical reasons, we will assume that all these fields possess tangential traces on Γ of square integrable modulus. Indeed, we will prove further that such electromagnetic fields satisfying in addition to

$$\lim_{R \to \infty} \frac{1}{R} \int_{B_R} |E|^2 + Z_0 |H|^2 < \infty, \quad (19)$$

where $B_R = \{|x| \le R, x \in D^+\}$, have the asymptotic behavior (17) with a_{∞}^{out}, a_{∞}^{in} in $L^2(TS^2)$. Let

$$\begin{cases} G_{\Gamma}^{in} = G_{\Gamma}^{in}(E,H) = n(x) \wedge (E_{/\Gamma} \wedge n(x)) + Z_0 H_{/\Gamma} \wedge n(x) \\ G_{\Gamma}^{out} = G_{\Gamma}^{out}(E,H) = -n(x) \wedge (E_{/\Gamma} \wedge n(x)) + Z_0 H_{/\Gamma} \wedge n(x), \end{cases} \quad (20)$$

and G^{in} some given tangential field of $L^2(T\Gamma)$. We consider the following functional

$$\left\{\begin{array}{c} I(E,H)=\frac{1}{4}\|G^{in}_\Gamma(E,H)\|^2_X+\frac{1}{4}\|G^{out}_\Gamma(E,H)\|^2_X \\ +\|a^{out}_\infty(E,H)\|^2_Z+\|a^{in}_\infty(E,H)\|^2_Z-\Re e\left(G^{in}_\Gamma(E,H),G^{=EEn}\right), \end{array}\right. \tag{21}$$

where $(.,.)$ and $\|.\|$ denotes respectively the inner product and norm in the following spaces

Definition 1 – *Let $L^2(T\Gamma)$ be the space of tangent fields*

$$L^2(T\Gamma)=\{\varphi\in L^2(\Gamma)^3,\ \varphi.n=0)\}\subset L^2(\Gamma)^3.$$

We define the functional spaces

$$X=L^2(T\Gamma)\times L^2(T\Gamma)$$

and (with same definition for the tangent space)

$$Z=L^2(TS^2).$$

This functional has the following feature.

Theorem 1 – *The minimum of $I(E,H)$ in the space of all out-coming and in-coming electro-magnetic fields E,H, with tangential traces on Γ of square integrable modulus and growth behavior (19), is reached by the solution of the following problem*

$$\left\{\begin{array}{lr} \nabla\wedge E^+-ikZ_0H^+=0 & in\ \ D^+, \\ \nabla\wedge H^++ikZ_0^{-1}E^+=0 & in\ \ D^+, \\ n(x)\wedge(E^+_{/\Gamma}(x)\wedge n(x))+Z_0(H^+_{/\Gamma}(x)\wedge n(x))=G^{in} & on\ \ \Gamma, \\ a^{in}_\infty(E^+,H^+)=0 & at\ infinity. \end{array}\right. \tag{22}$$

This result means that it is possible to relax both the condition at infinity and the boundary condition, and to recover them through the minimization process. The condition at infinity is treated exactly like the boundary condition on Γ.

The key point of the proof rests on an isometry lemma, which is exactly equivalent to the unitarity of the scattering matrix in scattering theory [11].

Lemma 1 (Isometry Lemma) – *Let (E,H) some electromagnetic field satisfying (17) with tangential traces on Γ of square integrable modulus. Then, the following equality holds*

$$\left\{\begin{array}{ll} & \frac{1}{4}\|G^{in}_\Gamma(E,H)\|^2+\|a^{in}_\infty(E,H)\|^2_Z \\ = & \frac{1}{4}\|G^{out}_\Gamma(E,H)\|^2+\|a^{out}_\infty\|^2_Z. \end{array}\right. \tag{23}$$

To prove this lemma, we introduce $B_r = \{|x| \leq r, x \in D^+\}$ for some large r. The outward normal is denoted as ν. So $\nu = -n(x)$ on the interior boundary of B_r that coincides with Γ. We begin with

$$\left\{\begin{array}{l} \displaystyle\int_{\partial B_r} \left|\nu(x) \wedge (E_{/\Gamma}(x) \wedge \nu(x)) + Z_0(H_{/\Gamma}(x) \wedge \nu(x))\right|^2 - \\ \displaystyle\int_{\partial B_r} \left|-\nu(x) \wedge (E_{/\Gamma}(x) \wedge \nu(x)) + Z_0(H_{/\Gamma}(x) \wedge \nu(x))\right|^2 \\ \displaystyle = 4\Re e \int_{\partial B_r} \nu(x) \wedge (E_{/\Gamma}(x) \wedge \nu(x)) \cdot \overline{H_{/\Gamma}(x) \wedge \nu(x)}. \end{array}\right. \tag{24}$$

Let us recall the integration by part formula

$$\int_\Theta (\nabla \wedge U \cdot V - \nabla \wedge V \cdot U) = \int_{\partial\Theta} (\nu \wedge (U_{/\partial\Theta} \wedge \nu)) \cdot (V_{/\partial\Theta} \wedge \nu) d\sigma, \tag{25}$$

for all bounded regular domain Θ. Equations (16) imply that

$$0 = \int_{B_r} (\nabla \wedge E - ikZ_0 H) \cdot \overline{H} - \overline{(\nabla \wedge H + ikZ_0^{-1} E)} \cdot E,$$

that is

$$\int_{\partial B_r} (\nu \wedge (E \wedge \nu)) \cdot \overline{H} \wedge \nu = ik \int_{B_r} (Z_0 |H|^2 + Z_0^{-1} |E|^2)$$

So

$$\Re e \int_{\partial B_r} (\nu \wedge (E \wedge \nu)) \cdot (\overline{H} \wedge \nu) = 0,$$

what implies

$$\int_{\partial B_r} |-(\nu \wedge (E \wedge \nu)) + Z_0(H \wedge \nu)|^2 = \int_{\partial B_r} |+(\nu \wedge (E \wedge \nu)) + Z_0(H \wedge \nu)|^2$$

and

$$\int_R^{2R} \frac{1}{R} \left(\int_{\partial B_r} |-(\nu \wedge (E \wedge \nu)) + Z_0(H \wedge \nu)|^2 \right) dr = \int_R^{2R} \frac{1}{R} \left(\int_{\partial B_r} |+(\nu \wedge (E \wedge \nu)) + Z_0(H \wedge \nu)|^2 \right) dr.$$

We then let R goes to infinity. With the help of Hypothesis (17), we get

$$\begin{cases} \lim_{R\to\infty} \frac{1}{R}\int_R^{2R}\int_{\partial B_r} |-(\nu\wedge(E\wedge\nu))+Z_0(H\wedge\nu)|^2 dr = \\ = \int_\Gamma |G_\Gamma^{in}(x)|^2 d\Gamma(x) + 4\int_{S^2} |a^{in}(\hat{d}))|^2 d\sigma(\hat{d}). \end{cases} \tag{26}$$

and, in a same way,

$$\begin{cases} \lim_{R\to\infty} \frac{1}{R}\int_R^{2R}\int_{\partial B_r} |+(\nu\wedge(E\wedge\nu))+Z_0(H\wedge\nu)|^2 dr = \\ = \int_\Gamma |G_\Gamma^{out}(x)|^2 d\Gamma(x) + 4\int_{S^2} |a^{out}(\hat{d}))|^2 d\sigma(\hat{d}). \end{cases} \tag{27}$$

It proves (23).

Once the isometry Lemma has been obtained, the minimization of $I(E,H)$ becomes obvious. Indeed, definition (21) and equation (23) gives

$$\begin{cases} I(E,H) & = & \frac{1}{2}\|G_\Gamma^{in}(E,H)\|_X^2 + 2\|a_\infty^{in}(E,H)\|_Z^2 - \Re e\left(G_\Gamma^{in}(E,H), G^{in}\right)_X \\ & = & \frac{1}{2}\|G_\Gamma^{in}(E,H) - G^{in}\|_X^2 + 2\|a_\infty^{in}(E,H)\|_Z^2 - \frac{1}{2}\|G^{in}\|_X^2. \end{cases} \tag{28}$$

It is then clear that the minimum is reached exactly for the electromagnetic field such that both the condition at infinity and the boundary condition are satisfied

$$a_\infty^{in}(E,H) = 0, \qquad G_\Gamma^{in}(E,H) = G^{in}. \tag{29}$$

Now, all the remaining difficulty is to choose an appropriate -and useful for practical computations- parameterization of our set of in and out-coming electromagnetic fields to derive the expressions of the related quantities G_Γ^{in}, G_Γ^{out}, a_∞^{in} and a_∞^{out}. For doing that, we consider $(\tilde{E},\tilde{H})$ the extension by zero of (E,H) in D^-. It is classical to show that $(\tilde{E},\tilde{H})$ satisfies in the sense of distributions of $\mathbf{R}^3$

$$\begin{cases} k^2\tilde{E} + \vec{\Delta}\tilde{E} = -ikZ_0\left(J\delta_\Gamma + \frac{1}{k^2}\vec{\nabla}\vec{\nabla}\cdot(J\delta_\Gamma)\right) + \vec{\nabla}\wedge(M\delta_\Gamma) \\ k^2\tilde{H} + \vec{\Delta}\tilde{H} = -\frac{ik}{Z_0}\left(M\delta_\Gamma + \frac{1}{k^2}\vec{\nabla}\vec{\nabla}\cdot(M\delta_\Gamma))\right) - \vec{\nabla}\wedge(J\delta_\Gamma) \\ \vec{\nabla}\cdot\tilde{E} = \frac{Z_0}{ik}\vec{\nabla}\cdot(J\delta_\Gamma), \quad \vec{\nabla}\cdot\tilde{H} = +\frac{1}{Z_0 ik}\vec{\nabla}\cdot(M\delta_\Gamma), \end{cases} \tag{30}$$

with

$$J = n \wedge H_{/\Gamma}, \quad M = -n \wedge E_{/\Gamma}. \tag{31}$$

The general solution of system (30) is

$$\begin{cases} \tilde{E}(x) &= iZ_0 \tilde{T}_r J(x) + \tilde{K}_r M(x) + E^{Her}(x) \\ \tilde{H}(x) &= -\tilde{K} J(x) + iZ_0^{-1} \tilde{T}_r M(x) + H^{Her}(x), \end{cases} \tag{32}$$

where we choose $\tilde{T}_r$, $\tilde{K}_r$ identical to $\tilde{T}$, $\tilde{K}$ given in (4) and (5) except that $G(x,y)$ is replaced by $\Re e G(x,y)$ while $(E^{Her}(x), H^{Her}(x))$ is some free electromagnetic field

$$\begin{cases} k^2 E^{Her} + \vec{\Delta} E^{Her} = 0, \quad k^2 H^{Her} + \vec{\Delta} H^{Her} = 0 \\ \vec{\nabla} \cdot E^{Her} = 0, \quad \vec{\nabla} \cdot H^{Her} = 0. \end{cases} \tag{33}$$

Note that the choice of the kernel $\Re e G(x,y)$ for the particular solution of system (30) is completely arbitrary at this stage, but it will be convenient for our purpose. Now, we use Theorem 6.30 of [5] which asserts that every entire solution of the Maxwell's system with the growth property (19) is an Herglotz pair. It means that there exists some tangential field γ in $L^2(S^2)$ such that

$$\begin{cases} E^{Her}(x) = \dfrac{ik\sqrt{iZ_0}}{4\pi} \displaystyle\int_{S^2} \gamma(\hat{d}) e^{ik\hat{d}x} d\sigma(\hat{d}) \\ H^{Her}(x) = \dfrac{ik}{4\pi\sqrt{iZ_0}} \displaystyle\int_{S^2} (i\hat{d} \wedge \gamma(\hat{d})) e^{ik\hat{d}x} d\sigma(\hat{d}). \end{cases} \tag{34}$$

Once more time, the normalization constant $\frac{ik\sqrt{iZ_0}}{4\pi}$ is here for convenience. Thus, a possible parameterization of in and out-coming fields might be (x, γ) with

$$x = \begin{bmatrix} J_1 \\ M_1 \end{bmatrix} = \begin{bmatrix} \sqrt{iZ_0} & J(x) \\ \sqrt{iZ_0}^{-1} & M(x) \end{bmatrix} \tag{35}$$

Reciprocally, every (x, γ) is associated to an electromagnetic field through (32)-(34)-(35). But it remains to verify that the fields vanish in D^-, the open complement of D^+ (if not, equality (31) would not hold anymore). For that, it is enough to ensure that both the tangential interior traces on Γ are zero. Using once more time the classical jump conditions for the potentials, we get

$$\begin{aligned} 0 &= n(x) \wedge (\tilde{E}^-_{/\Gamma}(x) \wedge n(x)) \\ &= iZ_0 (T_r J)(x) + (K_r M)(x) - \frac{1}{2} n(x) \wedge M(x) + ie^{Her}(x), \end{aligned} \tag{36}$$

$$
\begin{aligned}
&0 = n(x) \wedge (H^{+}_{/\Gamma}(x) \wedge n(x)) = \\
&-(K_r J)(x) + iZ_0^{-1}(T_r M)(x) + \frac{1}{2} n(x) \wedge J(x) + ih^{Her}(x),
\end{aligned}
\tag{37}
$$

with

$$
\left\{
\begin{aligned}
e^{Her}(x) &= n(x) \wedge \left(\frac{k\sqrt{iZ_0}}{4\pi} \int_{S^2} \gamma(\hat{d}) e^{ik\hat{d}x} d\sigma(\hat{d}) \wedge n(x) \right) \\
h^{Her}(x) &= n(x) \wedge \left(\frac{k}{4\pi\sqrt{iZ_0}} \int_{S^2} (i\hat{d} \wedge \gamma(\hat{d})) e^{ik\hat{d}x} d\sigma(\hat{d}) \wedge n(x) \right).
\end{aligned}
\right.
\tag{38}
$$

Let us normalize (36) by $\sqrt{iZ_0}^{-1}$ and (37) by $-\sqrt{iZ_0}$. A look at (11) and (14) gives us immediately after transposition

$$
\mathbf{Tx} + \mathbf{i}(\mathbf{A}^{\infty})^{*}\gamma = \mathbf{0}. \tag{39}
$$

Equation (39) defines a closed linear sub-manifold $\mathcal{M}$ of $X \times Z$ in which lives the pair (x, γ).

Definition 2 – *Let $\mathcal{M}$ be the closed linear sub-manifold of $X \times Z$ defined by*

$$
\mathcal{M} = \{(x, \gamma) \in X \times Z, \;\; \mathbf{Tx} + \mathbf{i}(\mathbf{A}^{\infty})^{*}\gamma = \mathbf{0}\} \subset \mathbf{X} \times \mathbf{Z}.
$$

To every (x, γ) is associated an electromagnetic field through (32)-(34)-(35), vanishing in D^- and with tangential traces satisfying to (31).

Once this parameterization has been constructed, it only remains to rewrite the functional $I(E, H)$. At first, we have

$$
\left\{
\begin{aligned}
&\frac{1}{4}\|G_\Gamma^{in}(E,H)\|_X^2 + \frac{1}{4}\|G_\Gamma^{out}(E,H)\|_X^2 = \frac{1}{2}\|E \wedge n\|_X^2 + \frac{Z_0^2}{2}\|H \wedge n\|_X^2 \\
&\qquad = \frac{Z_0}{2}\left(\|J_1\|_X^2 + \|M_1\|_X^2\right) = \frac{Z_0}{2}\|x\|_X^2,
\end{aligned}
\right.
\tag{40}
$$

$$
(G_\Gamma^{in}(E,H), G^{in})_X = Z_0 \int_\Gamma (J_1 - in \wedge M_1) \cdot \overline{\frac{G^{in}}{i\sqrt{iZ_0}}} = Z_0\,(x, g)_X\,, \tag{41}
$$

where g is defined in (79).

To get the expressions of the far fields, we let x goes to infinity. For the potential, the calculations are classical (see [5]) page 157). From,

$$\begin{cases} \nabla_x \wedge G(x,y)a(y) = \dfrac{e^{ik|x|}}{4\pi|x|}\left(+ik\hat{x}\wedge a(y)e^{-ik\hat{x}y} + O\left(\dfrac{|a|_Z}{|x|}\right)\right), \\ \nabla_x \wedge (\nabla_x \wedge G(x,y)a(y)) = \\ \qquad \dfrac{k^2 e^{ik|x|}}{4\pi|x|}\left(\hat{x}\wedge(a(y)\wedge\hat{x})e^{-ik\hat{x}y} + O\left(\dfrac{|a|_Z}{|x|}\right)\right), \end{cases} \tag{42}$$

as $|x|$ goes to infinity, uniformly for all y in Γ, we get

$$\begin{aligned} \frac{E(x) - E^{Her}(x)}{\sqrt{iZ_0}} &= \frac{1}{2}\mathbf{A}^\infty(\mathbf{J_1},\mathbf{M_1};\hat{\mathbf{x}})\frac{\mathbf{e}^{\mathbf{ik|x|}}}{|\mathbf{x}|} \\ &+\frac{1}{2}\mathbf{A}^\infty(\mathbf{J_1},\mathbf{M_1};-\hat{\mathbf{x}})\frac{\mathbf{e}^{-\mathbf{ik|x|}}}{|\mathbf{x}|} + \mathbf{O}\left(\frac{\mathbf{1}}{|\mathbf{x}|^\mathbf{2}}\right), \\ (H(x) - H^{Her}(x))\sqrt{iZ_0} &= \frac{1}{2}(-i\hat{x}\wedge\mathbf{A}^\infty(\mathbf{J_1},\mathbf{M_1};\hat{\mathbf{x}}))\frac{\mathbf{e}^{\mathbf{ik|x|}}}{|\mathbf{x}|} \\ &+\frac{1}{2}(+i\hat{x}\wedge\mathbf{A}^\infty(\mathbf{J_1},\mathbf{M_1};-\hat{\mathbf{x}}))\frac{\mathbf{e}^{-\mathbf{ik|x|}}}{|\mathbf{x}|} + \mathbf{O}\left(\frac{\mathbf{1}}{|\mathbf{x}|^\mathbf{2}}\right). \end{aligned} \tag{43}$$

The asymptotic behavior of the Herglotz wave is obtained thanks to the stationary phase theorem for the sphere. If $a \in C^1(S^2)$ we have

$$\int_{S^2} a(\hat{d})e^{ik\hat{d}x}d\sigma(\hat{d}) \to \frac{2\pi}{i}\left(a(\hat{x})\,\frac{e^{ik|x|}}{k|x|} - a(-\hat{x})\,\frac{e^{-ik|x|}}{k|x|}\right), \tag{44}$$

as $|x|$ goes to infinity. It is also true in our context for $a \in L^2(S^2)$ with convergence in the sense (18). We deduce

$$\begin{cases} \dfrac{E^{Her}(x)}{\sqrt{iZ_0}} \to \dfrac{1}{2}\gamma(\hat{x})\,\dfrac{e^{ik|x|}}{k|x|} - \dfrac{1}{2}\gamma(-\hat{x})\,\dfrac{e^{-ik|x|}}{k|x|}, \quad |x| \to \infty \\ H^{Her}(x)\sqrt{iZ_0} \to \dfrac{1}{2}(-i\hat{x}\wedge\gamma(\hat{x}))\,\dfrac{e^{ik|x|}}{k|x|} - \dfrac{1}{2}(i\hat{x}\wedge\gamma(-\hat{x}))\,\dfrac{e^{-ik|x|}}{k|x|}. \end{cases} \tag{45}$$

Finally, gathering the expressions, we recover the asymptotic behavior (17) with

$$\begin{cases} a_\infty^{out}(E,H) = \dfrac{\sqrt{iZ_0}}{2}\left(\mathbf{A}^\infty(\mathbf{J_1},\mathbf{M_1};\hat{\mathbf{x}})) + \gamma(\hat{\mathbf{x}})\right) \\ a_\infty^{in}(E,H) = \dfrac{\sqrt{iZ_0}}{2}\left(\mathbf{A}^\infty(\mathbf{J_1},\mathbf{M_1};-\hat{\mathbf{x}})) - \gamma(-\hat{\mathbf{x}})\right), \end{cases} \tag{46}$$

and, in particular,

$$\|a_\infty^{out}(E,H)\|_Z^2 + \|a_\infty^{in}(E,H)\|_Z^2 = \frac{Z_0}{2}\|\gamma\|_Z^2 + \frac{Z_0}{2}\|\mathbf{A}^\infty(\mathbf{J_1},\mathbf{M_1})\|_{\mathbf{Z}}^{\mathbf{2}}. \quad (47)$$

After all those tedious calculations, we get the expression for the functional

$$I(E,H) = Z_0 J(x,\gamma) = Z_0\left(\frac{1}{2}\|x\|_X^2 + \frac{1}{2}\|\gamma\|_Z^2 + \frac{1}{2}\|\mathbf{A}^\infty\mathbf{x}\|_{\mathbf{Z}}^{\mathbf{2}} - \Re\mathbf{e}\,(\mathbf{x},\mathbf{g})_{\mathbf{X}}\right). \quad (48)$$

Let us adopt the definition

Definition 3 – *For all* $(x,\gamma) \in X \times Z$ *we define*

$$J(x,\gamma) = \frac{1}{2}\|x\|_X^2 + \frac{1}{2}\|\gamma\|_Z^2 + \frac{1}{2}\|\mathbf{A}^\infty\mathbf{x}\|_{\mathbf{Z}}^{\mathbf{2}} - \Re\mathbf{e}\,(\mathbf{x},\mathbf{g})_{\mathbf{X}}\,.$$

Theorem 2 – *Let* $X = \left(L^2(T\Gamma)\right)^2$. *The minimum of the functional* $J(x,\gamma)$ *given in (48), over all the pairs* $(x,\gamma) \in \mathcal{M}$

$$\mathbf{Tx} + \mathbf{i}(\mathbf{A}^\infty)^*\gamma = \mathbf{0}, \quad (49)$$

is reached at

$$x = \begin{bmatrix} \sqrt{iZ_0}\; n \wedge H^+_{/\Gamma} \\ -\sqrt{iZ_0}^{\,-1}\; n \wedge E^+_{/\Gamma} \end{bmatrix}, \quad \gamma = \frac{a_\infty^{out}(E^+,H^+)}{\sqrt{iZ_0}}, \quad (50)$$

where (E^+,H^+) *is the solution of the Maxwell system (22).*

Definition 4 – *We define the Hilbertian space*

$$Y = \{y \in X,\ \mathbf{T^*y} \in \mathbf{X}\}, \quad (51)$$

equipped with the following norm,

$$\|y\|_Y = \|y\|_X + \|\mathbf{T^*y}\|_{\mathbf{X}}. \quad (52)$$

If Y' is the dual space of Y, the constraint (49) can be viewed as an equality in Y'

$$\forall y \in Y, \quad (x,\mathbf{T^*y})_{\mathbf{X}} + \mathbf{i}(\gamma,(\mathbf{A}^\infty)^*\mathbf{y})_{\mathbf{Z}} = \mathbf{0}. \quad (53)$$

At this point it is classical to dualize the constraint, introducing the Lagrangian

Definition 5 – *For all* $(x,\gamma,y) \in X \times Z \times Y$ *we define*

$$\mathcal{L}(x,\gamma,y) = J(x,\gamma) - \Re e\left((x,\mathbf{T^*y})_{\mathbf{X}} - \mathbf{i}(\gamma,(\mathbf{A}^\infty)\mathbf{y})_{\mathbf{Z}}\right). \quad (54)$$

It is well known (cf. [2]) that if the Lagrangian admits a saddle point, then its first argument is the minimum argument of J in $\mathcal{M}$

$$\mathcal{L}(x,\gamma,y) = \inf_{(x^*,\gamma^*)\in X\times Z} \sup_{y^*\in W} \mathcal{L}(x^*,\gamma^*,y^*) = \min_{(x^*,\gamma^*)\in\mathcal{M}} J(x^*,\gamma^*), \tag{55}$$

Furthermore, as J is quadratic, such a saddle point exists if and only if $D_x\mathcal{L}$ $(x,\gamma,y) = 0$, $D_\gamma\mathcal{L}(x,\gamma,y) = 0$, $D_y\mathcal{L}(x,\gamma,y) = 0$, i.e.

$$\begin{cases} x + (\mathbf{A}^\infty)^* \, \mathbf{A}^\infty \mathbf{x} - \mathbf{T}^*\mathbf{y} = \mathbf{g} \\ \gamma + i\,\mathbf{A}^\infty \mathbf{y} = \mathbf{0} \\ -\mathbf{T}\mathbf{x} - \mathbf{i}(\mathbf{A}^\infty)^* \, \gamma = \mathbf{0}. \end{cases} \tag{56}$$

Discarding γ, it gives

$$\begin{cases} x + (\mathbf{A}^\infty)^* \, \mathbf{A}^\infty \mathbf{x} - \mathbf{T}^*\mathbf{y} = \mathbf{g} \\ -\mathbf{T}\mathbf{x} - (\mathbf{A}^\infty)^* \, \mathbf{A}^\infty \mathbf{y} = \mathbf{0}. \end{cases} \tag{57}$$

This is the basic integral system we discuss in this work. The interesting and new feature of this system is that it is is clearly derived from the minimization of a natural quadratic functional under constraint, and that the functional is clearly related to the isometry Lemma (equivalent to the unitarity of the scattering matrix). The additional unknown is interpreted as the Lagrange multiplier of our constrained minimization problem. It is possible to re-interpret the multiplier and prove that $y = ix$ (up to a term in $Ker\mathbf{T}^*$ actually equal to $Ker\mathbf{A}^\infty$, see Appendix).

To our knowledge such a system has not been discovered beforehand, even if many researchers have stressed the interest of these kind of structure for practical computations. Nevertheless it is in some sense natural, since based on the isometry result which is equivalent to the unitarity of the scattering matrix, what has been well known for a very long time, [11]. To every unitary operator $\mathbf{U}\mathbf{U}^* = \mathbf{I}$, is associated at least formally a hermitian operator $\mathbf{H} = \frac{\mathbf{I}-\mathbf{i}\mathbf{U}}{\mathbf{I}+\mathbf{i}\mathbf{U}}$ with inverse formula $\mathbf{U} = \frac{\mathbf{I}+\mathbf{i}\mathbf{H}}{\mathbf{I}-\mathbf{i}\mathbf{H}}$. So in some sense this mixed integral system corresponds to a "scattering" like integral theory.

4. Variational formulation and well-posedness. The penalized systems

Variational formulation are useful for minimization problems. It provides a good framework for the study of uniqueness and existence, and also for discretization and convergence of the discrete solution. The natural variational formulation of our problem is as follows

$$\left\{\begin{array}{ll} x \in V_x, \quad y \in V_y, & \\ (x,\tilde{x}) + (\mathbf{A}_\infty \mathbf{x}, \mathbf{A}_\infty \tilde{\mathbf{x}}) - (\mathbf{T}^* \mathbf{y}, \tilde{\mathbf{x}}) = (\mathbf{g}, \tilde{\mathbf{x}}), & \forall \tilde{x} \in V_x, \\ (x, \mathbf{T}^* \tilde{\mathbf{y}}) + (\mathbf{A}_\infty \mathbf{y}, \mathbf{A}_\infty \tilde{\mathbf{y}}) = \mathbf{0}, & \forall \tilde{y} \in V_y. \end{array}\right. \tag{58}$$

It remains to define the functional spaces V_x and V_y. Let us remark that both smooth analytic operators $\mathbf{A}_\infty \mathbf{y}, \mathbf{A}_\infty$ are not involved in the coerciveness of the system. Due to the L^2 coerciveness of the formulation it is clear that the only choice for V_x is

$$V_x = X = L^2(T\Gamma) \times L^2(T\Gamma).$$

It may seem at first sight impossible to define a compatible space V_y such that the inf-sup condition of Babuska-Brezzi holds, [2],

$$\max \frac{(x, \mathbf{T}^* \mathbf{y})}{||x||_{L^2}} \geq k||y||_{W_y}, \; k > 0, \quad \text{with } W_y = \frac{V_y}{Ker\mathbf{T}^*}, \tag{59}$$

just because standard functional spaces in which $\mathbf{T}$ is continuous are known to be based on

$$H^{-\frac{1}{2}}(div, \Gamma) \text{ and } H^{-\frac{1}{2}}(curl, \Gamma).$$

Nevertheless it is at least possible to provide an abstract framework in which the inf-sup condition holds. Let us take $V_x = X$, $V_y = Y$, defined in (51).

Definition 6 – *We define the quotient Hilbertian space*

$$W = \frac{Y}{Ker\mathbf{T}^*}$$

equipped with the norm

$$\inf_{y_0 \in Ker\mathbf{T}^*} \|y - y_0\|_Y = \|\mathbf{T}^* \mathbf{y}\|_\mathbf{X} + \inf_{\mathbf{y_0} \in \mathbf{Ker}\mathbf{T}^*} \|\mathbf{y} - \mathbf{y_0}\|_\mathbf{X}.$$

System (57) is well-posed as soon as the inf-sup condition

$$\sup_{x \in X} \frac{(x, \mathbf{T}^* \mathbf{y})_\mathbf{X}}{\|x\|_X} \geq C\|y\|_W \tag{60}$$

holds for some strictly positive constant $C > 0$. This inequality can be derived as follows. Picking $x = \mathbf{T}^* \mathbf{y}$ in (60), we find

$$\sup_{x \in X} \frac{(x, \mathbf{T}^* \mathbf{y})_\mathbf{X}}{\|x\|_X} \geq \|\mathbf{T}^* \mathbf{y}\|_\mathbf{X}. \tag{61}$$

Let $y \in Y$. In appendix it is proved that $-\mathbf{\Pi T}$ with $\mathbf{\Pi}$ defined in (12), is a projector (it is a Calderon Projector):

$$\mathbf{T\Pi T} = -\mathbf{T}. \tag{62}$$

Let $y_1 = -\mathbf{\Pi^* T^* y}$. We have that $\mathbf{T^*(y - y_1) = T^*y + T^*\Pi^*T^*y = 0}$. So $y_0 = y - y_1 \in Ker\mathbf{T}^*$ hence

$$\inf_{y_0 \in Ker\mathbf{T}^*} \|y - y_0\|_X \leq \|y_1\|_X = \|\mathbf{\Pi^* T^* y}\|_{\mathbf{X}} = \|\mathbf{T^* y}\|_{\mathbf{X}}, \tag{63}$$

where we have used the isometric property $\mathbf{\Pi^* \Pi = I}$. So we have

$$\sup_{x \in X} \frac{(x, \mathbf{T^* y})_{\mathbf{X}}}{\|x\|_X} \geq \frac{1}{2}\|\mathbf{T^* y}\|_{\mathbf{X}} + \frac{\mathbf{1}}{\mathbf{2}} \inf_{\mathbf{y_0 \in KerT^*}} \|\mathbf{y - y_0}\|_{\mathbf{X}} = \frac{\mathbf{1}}{\mathbf{2}}\|\mathbf{y}\|_{\mathbf{W}}. \tag{64}$$

Thus it gives

Lemma 2 – *The inf-sup condition (60) in spaces* $= X \times W$ *is true with* $C = \frac{1}{2}$.

Since the natural continuity $\mathbf{T^* : Y \to X}$ holds and one has the bound (98) $\|\mathbf{A}_\infty^* \mathbf{A}_\infty \mathbf{y}\|_{\mathbf{X}} \leq \mathbf{C'}\|\mathbf{y}\|_{\mathbf{W}}$ for some $C' > 0$, it proves following [2]

Theorem 3 – *The variational system (58) is well posed, that is, for every pair* $(g, g_y) \in (V_x', V_y') = (V_x, V_y')$ *there exists a unique* $(x, y) \in V_x \times W$ *such that*

$$\left\{ \begin{array}{ll} (x, \tilde{x}) + (\mathbf{A}_\infty \mathbf{x}, \mathbf{A}_\infty \tilde{\mathbf{x}}) - (\mathbf{T^* y}, \tilde{\mathbf{x}}) = (\mathbf{g}, \tilde{\mathbf{x}}), & \forall \tilde{x} \in V_x, \\ (x, \mathbf{T}^* \tilde{\mathbf{y}}) + (\mathbf{A}_\infty \mathbf{y}, \mathbf{A}_\infty \tilde{\mathbf{y}}) =_{\mathbf{W}'} \langle \mathbf{g_y}, \tilde{\mathbf{y}} \rangle_{\mathbf{W}}, & \forall \tilde{y} \in V_y. \end{array} \right. \tag{65}$$

However when discretization is considered a difficulty arises with the use of $(X, Y) = (V_x, V_y)$ spaces. The reason is that we want to avoid the construction of some discrete space compatible with the L^2 based space (X, Y). We would like to take a classical integral code based on the duality $H^{-\frac{1}{2}}(div, \Gamma)$ and $H^{-\frac{1}{2}}(curl, \Gamma)$, rearrange all the routines, and use the iterative algorithms described further in order to solve our new discrete integral system. Then the question of the convergence of the discrete solution to the exact solution arises. All our efforts in order to prove the convergence using this strategy failed. The reason seems to be that the classical discretization of integral operator is based on $H^{-\frac{1}{2}}(div, \Gamma)$ and $H^{-\frac{1}{2}}(curl, \Gamma)$, and not on (X, Y), [4]. Moreover some numerical results in 2D for the Helmholtz equation, [1], show that this problem may be a real one. It seems that there are cases where the discrete solution obtained through the strategy described above does not converge to the exact solution, even in some very simple and regular cases. Of course this conclusion has to be re-evaluated if the discretization of the integral operators are compatible with (X, Y). Note also that the kernel $Ker\mathbf{T}^*$ is a large space of infinite dimension. Some algorithms are very sensitive to the dimension of this kernel. It is our purpose now to modify the system and to present what we will call the penalized problem, with much stronger coercivity.

Let β some positive penalisation parameter (for instance $\beta = 1$). Remembering that $y = ix$, we modify system (15) to obtain the penalized system

$$\begin{cases} (1+\beta)x + (\mathbf{A}^\infty)^* \, \mathbf{A}^\infty \mathbf{x} - \mathbf{T}^*\mathbf{y} + \mathbf{i}\beta\mathbf{y} = \mathbf{g} \\ +\mathbf{T}\mathbf{x} - \mathbf{i}\beta\mathbf{x} + \beta\mathbf{y} + (\mathbf{A}^\infty)^* \, \mathbf{A}^\infty\mathbf{y} = \mathbf{0}. \end{cases} \tag{66}$$

which appears to be a system of the form

$$\mathcal{A}_\beta \begin{bmatrix} x \\ y \end{bmatrix} = \begin{bmatrix} g \\ 0 \end{bmatrix}. \tag{67}$$

A simple calculation shows that

$$\begin{cases} \Re e \left(\mathcal{A}_\beta \begin{bmatrix} x \\ y \end{bmatrix}, \begin{bmatrix} x \\ y \end{bmatrix} \right)_{X \times X} \\ = \|x\|_X^2 + \beta\|x+iy\|_X^2 + \|\mathbf{A}^\infty\mathbf{y}\|_\mathbf{Z}^2 + \|\mathbf{A}^\infty\mathbf{x}\|_\mathbf{Z}^2 \\ \geq \min(\frac{\beta}{2}, \frac{1}{3}) \left(\|x\|_X^2 + \|y\|_X^2 \right), \end{cases} \tag{68}$$

and the system appears now as coercive in the x variable and also in the y variable.

Another possibility is to modify system (15) according to

$$\begin{cases} (1-\beta)x + (\mathbf{A}^\infty)^* \, \mathbf{A}^\infty\mathbf{x} - (\mathbf{T}^*\mathbf{y} + \mathbf{i}\beta\mathbf{y}) = \mathbf{g} \\ (\mathbf{T}\mathbf{x} - \mathbf{i}\beta\mathbf{x}) + \beta\mathbf{y} + (\mathbf{A}^\infty)^* \, \mathbf{A}^\infty\mathbf{y} = \mathbf{0}, \end{cases} \tag{69}$$

where β is now some positive number less than 1 (let $\beta = \frac{1}{2}$). The interest of (69) is that it corresponds to a saddle point for the Lagrangian

$$\begin{aligned} \mathcal{L}_\beta(x,\gamma,y) &= (1-\beta)\frac{1}{2}\|x\|_X^2 - \beta\frac{1}{2}\|y\|_X^2 - \Re e(g,x)_X \\ +\frac{1}{2}\|\gamma\|_Z^2 + \frac{1}{2}\|\mathbf{A}^\infty\mathbf{x}\|_\mathbf{Z}^2 &- \Re\mathbf{e}\left((\mathbf{x}, \mathbf{T}^*\mathbf{y} + \mathbf{i}\beta\mathbf{y})_\mathbf{X} - \mathbf{i}(\gamma, (\mathbf{A}^\infty)\mathbf{y})_\mathbf{Z}\right). \end{aligned} \tag{70}$$

The problem appears as a penalized saddle point problem.

5. System for general impedance boundary conditions

We turn now to the case of a general boundary condition. We assume that the electro-magnetic field is determined by some impedance boundary condition of the type

$$n(x) \wedge (E^+_{/\Gamma}(x) \wedge n(x)) + Z_0 Z_r (H^+_{/\Gamma}(x) \wedge n(x)) = F, \tag{71}$$

where Z_r is some impedance operator that we assume symmetric with a positive real part, i.e.

$$(\Re e Z_r J, J) \geq 0, \quad \forall J \in D(Z_r). \tag{72}$$

We associate to it its reflection coefficient operator

$$R = (Id - Z_r)(Id + Z_r)^{-1}, \tag{73}$$

which, thanks to (72), satisfies

$$\|R\| \leq 1 \Leftrightarrow \|RJ\|^2_{L^2(T\Gamma)} \leq \|J\|^2_{L^2(T\Gamma)}. \tag{74}$$

We begin by rewriting the boundary condition with the only help of R. We have

$$G^{in} = -RG^{out} + (Id + R)F, \tag{75}$$

where $G^{in} = n(x) \wedge (E^+_{/\Gamma}(x) \wedge n(x)) + Z_0(H^+_{/\Gamma}(x) \wedge n(x))$ while

$$G^{out} = n(x) \wedge (E^+_{/\Gamma}(x) \wedge n(x)) - Z_0(H^+_{/\Gamma}(x) \wedge n(x)). \tag{76}$$

Note that this notation is compatible with (3) with $Z_r = 1$. Setting

$$F_0 = \frac{1}{\sqrt{iZ_0}}(Id + R)F, \tag{77}$$

$$f = \begin{bmatrix} -iF_0 \\ -n(x) \wedge F_0 \end{bmatrix}, \tag{78}$$

and using

$$g = \begin{bmatrix} -i \frac{1}{\sqrt{iZ_0}} G^{in}(x) \\ -n(x) \wedge \frac{1}{\sqrt{iZ_0}} G^{in}(x) \end{bmatrix}, \tag{79}$$

for g with (75) for G^{in} in system (57), we have

$$\begin{cases} x + (\mathbf{A}^\infty)^* \mathbf{A}^\infty \mathbf{x} - \mathbf{T}^* \mathbf{y} = \mathbf{f} - \mathbf{N_R}\, \mathbf{x} \\ -\mathbf{T}\mathbf{x} - (\mathbf{A}^\infty)^* \mathbf{A}^\infty \mathbf{y} = \mathbf{0}, \end{cases} \tag{80}$$

where

$$\mathbf{N_R}\,\mathbf{x} = \mathbf{N_R}\begin{bmatrix} J_1(x) \\ M_1(x) \end{bmatrix} = \begin{bmatrix} -iR\,\frac{G^{out}(x)}{\sqrt{iZ_0}} \\ -n(x)\wedge R\,\frac{G^{out}(x)}{\sqrt{iZ_0}} \end{bmatrix}, \tag{81}$$

or

$$\mathbf{N_R}\,\mathbf{x} = \begin{bmatrix} -iR\,(n\wedge M_1 - iJ_1) \\ -n\wedge R\,(n\wedge M_1 - iJ_1) \end{bmatrix}. \tag{82}$$

Let β some positive parameter (for instance $\beta = 1$), remembering that $y = ix$, we finally modify the system to obtain

$$\begin{cases} (1+\beta)x + (\mathbf{A}^\infty)^*\,\mathbf{A}^\infty\mathbf{x} - \mathbf{T}^*\mathbf{y} + \mathbf{N_R}\,\mathbf{x} + \mathbf{i}\beta\mathbf{y} = \mathbf{f} \\ \mathbf{T}\mathbf{x} - \mathbf{i}\beta\mathbf{x} + \beta\mathbf{y} + (\mathbf{A}^\infty)^*\,\mathbf{A}^\infty\mathbf{y} = \mathbf{0}. \end{cases} \tag{83}$$

6. Calderon projectors

Our aim in this section is to make the link between the operators we have defined with the Calderon projectors, [4] page 87, [6] page 93.

Let $(\hat{J}_1, \hat{M}_1)$ two fields given on Γ, (not necessarily corresponding to the traces of some exterior electro-magnetic fields). We can associate to them the two fields in Ω^+

$$\begin{cases} \dfrac{E^+(x)}{\sqrt{iZ_0}} & = \tilde{T}\hat{J}_1(x) + \tilde{K}\hat{M}_1(x) \\ -\sqrt{iZ_0}H^+(x) & = \tilde{K}\hat{J}_1(x) + \tilde{T}\hat{M}_1(x), \end{cases} \tag{84}$$

where $\tilde{T}J, \tilde{K}M$ are given in (5). If x goes to some points on Γ, we have via the jump conditions

$$\begin{cases} n(x)\wedge(\dfrac{1}{\sqrt{iZ_0}}\,E^+_{/\Gamma}(x)\wedge n(x)) = \\ \qquad +T\hat{J}_1(x) + K\hat{M}_1(x) + \dfrac{1}{2}n(x)\wedge\hat{M}_1(x), \end{cases} \tag{85}$$

$$\begin{cases} n(x)\wedge(-\sqrt{iZ_0}H^+_{/\Gamma}(x)\wedge n(x)) = \\ \qquad T\hat{M}_1(x) + K\hat{J}_1(x) + \dfrac{1}{2}n(x)\wedge\hat{J}_1(x). \end{cases} \tag{86}$$

Now, we can proceed exactly as in the first section: from the exterior traces, we construct the fields (J_1, M_1) and x by

$$x = \begin{bmatrix} J_1 \\ M_1 \end{bmatrix} = \begin{bmatrix} \sqrt{iZ_0}J \\ \sqrt{iZ_0}^{-1}M \end{bmatrix}, \tag{87}$$

and we have

$$\mathbf{Sx} = \mathbf{S} \begin{bmatrix} J_1 \\ M_1 \end{bmatrix} = \mathbf{0}. \tag{88}$$

But, it is easy to see that (86)- (85) reads (see definition (12) and (11))

$$-\mathbf{\Pi} \begin{bmatrix} J_1 \\ M_1 \end{bmatrix} = (\mathbf{S} - \mathbf{\Pi}) \begin{bmatrix} \hat{J}_1 \\ \hat{M}_1 \end{bmatrix}. \tag{89}$$

Multiplying by $\mathbf{S\Pi}$ and using both $-\mathbf{\Pi^2} = \mathbf{Id}$ and (88) we get

$$0 = \mathbf{S} \begin{bmatrix} J_1 \\ M_1 \end{bmatrix} = (\mathbf{S\Pi S} + \mathbf{S}) \begin{bmatrix} \hat{J}_1 \\ \hat{M}_1 \end{bmatrix}. \tag{90}$$

In other words,

$$-\mathbf{\Pi S} = (-\mathbf{\Pi S})^{\mathbf{2}}, \tag{91}$$

appears as a projector: it is one of the Calderon projector. With our decomposition in real and imaginary part $\mathbf{S} = \mathbf{T} + \mathbf{iR}$, we get

$$\begin{cases} \mathbf{T\Pi T} - \mathbf{R\Pi R} = -\mathbf{T} \\ \mathbf{T\Pi R} + \mathbf{R\Pi T} = -\mathbf{R}. \end{cases} \tag{92}$$

Up to now, what we have obtained corresponds to the Green function (6), but we can operate exactly in the same way with

$$G(x, y) = \frac{\exp^{-ik|x-y|}}{4\pi|x-y|}. \tag{93}$$

the only modification being the radiation condition at infinity. The $\frac{1}{|x-y|}$ singularity of the kernel remains , so we get the same relation (91) also for $\tilde{\mathbf{S}} = \mathbf{T} - \mathbf{iR}$. It expands into

$$\begin{cases} \mathbf{T\Pi T} + \mathbf{R\Pi R} = -\mathbf{T} \\ \mathbf{T\Pi R} + \mathbf{R\Pi T} = -\mathbf{R}. \end{cases} \tag{94}$$

By comparison we get

$$\mathbf{T\Pi T} = -\mathbf{T} \tag{95}$$

and

$$\mathbf{R\Pi R} = \mathbf{0}. \tag{96}$$

It remains to prove that

$$\mathbf{R\Pi T}^* = \mathbf{R}. \tag{97}$$

This is a simple consequence of the fact that free fields are in the kernel of "exterior" integral operators [5]. To see this let us consider the free field (that is a continuous field in the whole space, with continuous traces on Γ) $\mathbf{Ry}$ for some arbitrary smooth y. Let $(J, M) = \mathbf{\Pi Ry}$. By integration by parts we get

$$\begin{cases} 0 &= iZ_0(\tilde{T}J)(\hat{x}) + (\tilde{K}M)(\hat{x}) \\ 0 &= -(\tilde{K}J)(\hat{x}) + iZ_0^{-1}(\tilde{T}M)(\hat{x}), \end{cases}$$

for $\hat{x}$ in D^+. Passing to the limit $\hat{x} \to \Gamma$ we get that

$$(\mathbf{T} + \mathbf{\Pi} + \mathbf{iR})\mathbf{\Pi R y} = \mathbf{0}.$$

It means that $\mathbf{T\Pi R} - \mathbf{R} = \mathbf{0}$ which implies by transposition the desired relation (97). A consequence of the boundedness of $\mathbf{R\Pi}$ is the continuity bound

$$\|\mathbf{Ry}\|_{\mathbf{X}} \leq \|\mathbf{R\Pi}\|_{\mathcal{L}(\mathbf{X})} \|\mathbf{T}^*\mathbf{y}\|_{\mathbf{X}} \leq \mathbf{C}' \|\mathbf{y}\|_{\mathbf{W}}.. \tag{98}$$

Finally note that the identity (95) also proves by a density argument that $\mathbf{\Pi Ty} \in \mathbf{Y}$, with the notations of Definition (4). In other words $-\mathbf{\Pi T}$ is a well-defined projector in Y.

References

[1] N. Bartoli, F. Collino, and B. Desprès, Integral Equations via Saddle Point Problems for Acoustic Problems. in preparation.

[2] F. Brezzi and M. Fortin, Mixed and Hybrid Finite Element Methods, Number 15 in Springer Series in Computational Mathematics. Springer-Verlag, 1991.

[3] F. Collino and B. Despres, Integral equations via saddle point problems for time-harmonic maxwell's equations, in preparation.

[4] M. Cessenat, Mathematical Methods in Electromagnetism, Number 41 in Series on advances in Mathematics for Applied Sciences. World-Scientific, 1996.

[5] D. Colton and R. Kress, Inverse Acoustic and electromagnetic scattering theory, Applied Mathematical Sciences 93. Springer-Verlag, 1992.

[6] G. Chen and J. Zhou, Boundary Element Methods, Academic Press, 1992.

[7] B. Despres, Fonctionnelle quadratique et équations intégrales pour les équations de maxwell en domain extérieur, Comptes Rendus de l'Académie des Sciences, Paris, Série I, 323(1996), 547–552.

[8] B. Despres, Fonctionnelle quadratique et équations integrales pour les problèmes d'onde harmonique en domaine extérieur, M2AN, 31(1997), 679–732.

[9] B. Despres, Quadratic functional and integral equations for harmonic wave equations, In Mathematical and numerical aspects of wave propagation, 56–66, 1998.

[10] P. Lascaux and R. Theodor, Analyse numérique matricielle appliquée à l'art de l'ingénieur, Masson, 1987.

[11] M. Reed and B. Simon, Scattering Theory, Academic-Press, New York, 1979.

Francis Collino
CERFACS
42 avenue Coriolis
31057 Toulouse
France

E-mail:@inria.fr

Bruno Despres
Laboratoire d'Analyse Numérique
Université Paris VI - CNRS
175 rue du Chevaleret
75013 Paris
France

E-mail: despres@ann.jussieu.fr

Studies in Mathematics and its Applications, Vol. 31
D. Cioranescu and J.L. Lions (Editors)

Chapter 11

EXISTENCE AND UNIQUENESS OF A STRONG SOLUTION FOR NONHOMOGENEOUS MICROPOLAR FLUIDS

C. CONCA, R. GORMAZ, E. ORTEGA AND M. ROJAS

1. Introduction

In this paper, we study the equations of a nonhomogeneous viscous incompressible micropolar fluid. These equations are studied in a bounded domain $\Omega \subset \mathbb{R}^3$, with boundary Γ, in a time interval $[0, T]$. Let $u(x,t) \in \mathbb{R}^3$, $w(x,t) \in \mathbb{R}^3$, $\rho(x,t) \in \mathbb{R}$ and $p(x,t) \in \mathbb{R}$, denote respectively, the velocity, the angular velocity of internal rotation, the density and the pressure at a point $x \in \Omega$ and at time $t \in [0, T]$. The governing equations for this fluid are

$$\begin{aligned}
&\rho\frac{\partial u}{\partial t} + \rho(u \cdot \nabla)u - (\mu + \mu_r)\Delta u + \nabla p = 2\mu_r \mathrm{rot}\, w + \rho f,\\
&\mathrm{div}\ u = 0,\\
&\rho\frac{\partial w}{\partial t} + \rho(u \cdot \nabla)w - (c_a + c_d)\Delta w - (c_0 + c_d - c_a)\nabla\ \mathrm{div}\ w\\
&\qquad\qquad +4\mu_r w = 2\mu_r \mathrm{rot}\, u + \rho g,\\
&\frac{\partial \rho}{\partial t} + (u \cdot \nabla)\rho = 0,
\end{aligned} \tag{1.1}$$

in $Q_T := \Omega \times (0, T)$, with the following boundary and initial conditions

$$\begin{aligned}
&u(x,t) = 0,\ w(x,t) = 0 \ \text{ on } \ \Gamma \times (0,T),\\
&u(x,0) = u_0(x),\ w(x,0) = w_0(x) \ \text{ in } \ \Omega,\\
&\rho(x,0) = \rho_0(x) \ \text{ in } \ \Omega.
\end{aligned} \tag{1.2}$$

Here $f(x,t)$ and $g(x,t)$ are densities of the external forces field and the external torque field respectively. The positive constants μ, μ_r, c_0, c_a, c_d characterize isotropic properties of the fluid; μ is the usual Newtonian dynamic viscosity; μ_r represent the dynamic microrotation viscosity; c_0, c_a

and c_d are called coefficients of angular viscosities. These new viscosities are related to the asymmetry of the stress tensor and in consequence related to the appearance of the field of internal rotation w. Furthermore, these positive constants satisfy $c_0 + c_d > c_a$. The expressions ∇, Δ, div and rot denote, the gradient, Laplacian, divergence and rotational operators respectively. We denote $\partial u/\partial t$ by u_t and the ith component of $(u\cdot\nabla)v$ in cartesian coordinates is given by

$$[(u\cdot\nabla)v]_i = \sum_{j=1}^{3} u_j \frac{\partial v_i}{\partial x_j}; \quad u\cdot\nabla\rho = \sum_{j=1}^{3} u_j \frac{\partial \rho}{\partial x_j}.$$

For the derivation and physical discussion of equations (1.1)–(1.2), see D. Condiff & J. Dahler (1964), L. Petrosyan (1984) and G. Lukaszewicz (1999). We observe that this model of fluids includes as a particular case the classical Navier-Stokes equations, which has been thoroughly studied (see for instance the classical books of O. Ladyzhenskaya (1969), J.L. Lions (1969), R. Temam (1979) and the references there in). It also includes the reduced model of the nonhomogeneous Navier-Stokes equations, which has been less studied than the previous case (see for instance S. Antoncev et al. (1990), J. Simon (1990), J. Kim (1987), O. Ladyzhenskaya & V. Solonnikov (1975), R. Salvi (1991), and J.L. Boldrini & M. Rojas (1992), (1997)).

Concerning the model of a micropolar fluid that we consider in this paper, G. Lukaszewicz (1990) established the existence of local weak solutions for (1.1)–(1.2), using linearization and an almost fixed point theorem. In this paper Lukaszewicz remarked the possibility of proving the existence of a strong solution, under the hypothesis that the initial density is separated from zero, by the techniques used in G. Lukaszewicz (1988), (1989), that is, linearization and fixed point theorems; notice that Lukaszewicz assume constant density.

The first result on the existence and uniqueness of a strong solution (local and global) for problem (1.1)–(1.2) was proved in J.L. Boldrini & M. Rojas (1998) using the spectral semi-Galerkin method and compactness arguments. The convergence rate of this method was established in J.L. Boldrini & M. Rojas (1996).

In this work, we use a different approach in order to obtain the existence and uniqueness of a strong solution. We propose an iterative process, which solves at each iteration a linear problem. For each linear problem it is easy to prove the existence and uniqueness of a strong solution. We obtain a priori estimates for the sequence generated by the iterative process. In the next step we show that this sequence of solutions of linear problems is a Cauchy sequence in an appropriate Banach space, and consequently, converges. From this convergence, the strong solution for the full original nonlinear problem is obtained. As by-product we obtain bounds for the convergence rate of the method. We think that the techniques developed here could be adapted to a numerical scheme that solves these equations.

We remark that, from the technical point of view, the hyperbolic character of the transport equation in (1.1) and the extra nonlinearities of the problem make the technical arguments more elaborated than those used in the case of constant density (compare with E. Ortega & M. Rojas (1997)).

This paper is organized as follows: In Section 2, we describe the iterative method. In Section 3, we derive some important a priori estimates. In Section 4, we prove the Cauchy character of the sequence of solutions of the linearized problems and we give an error bound between an element of the sequence and the unique solution of the problem. We give also an error estimate for the pressure.

Finally, a word about the notation adopted in this work. The constants appearing in the various estimates depending exclusively on Ω, are generically denoted by C_Ω, except for the smallest eigenvalue of the Laplace operator $B = -\Delta$ in Ω with a Dirichlet boundary condition, which will be denoted by λ.

2. Preliminaries

Let Ω be a bounded domain in $\mathbb{R}^3$ with a smooth boundary $\partial\Omega$ and $T > 0$ a positive real number. Denote by D either Ω or $\Omega \times (0,T)$. For $n \in \mathbb{N}$ and $1 \leq q < \infty$, we consider the usual Sobolev space

$$W^{m,q}(D) = \{\ f\ \in L^q(D) \,|\, \partial^\alpha f \in L^q(D),\ |\alpha| \leq m\ \},$$

endowed with the usual $W^{m,q}$-norm. If $q = 2$ we denote $H^m(D) := W^{m,2}(D)$ and $H_0^m(D) :=$ the closure of $\mathcal{D}(D)$ in $H^m(D)$. We put

$$\begin{aligned} \mathcal{V}(\Omega) &= \{v \in \mathcal{D}(\Omega)^3 \,|\, \operatorname{div} v = 0 \quad \text{in} \quad \Omega\} \\ H &= \quad \text{closure of} \quad \mathcal{V}(\Omega) \quad \text{in} \quad L^2(\Omega)^3, \\ V &= \quad \text{closure of} \quad \mathcal{V}(\Omega) \quad \text{in} \quad H^1(\Omega)^3. \end{aligned}$$

Since Ω has a smooth boundary, it is well known that

$$V = \{v\ \in H_0^1(\Omega)^3 \,|\, \operatorname{div} v = 0 \ \ \text{in}\ \Omega\}.$$

We recall the Helmholtz decomposition of vector fields $L^2(\Omega) = H \oplus G$, where $G = \{\phi \,|\, \phi = \nabla p,\ p \in H^1(\Omega)\}$.

Throughout this paper P will denote the orthogonal projection from $L^2(\Omega)$ onto H. The Stokes operator $A : D(A) \hookrightarrow H \longrightarrow H$ is given by

$A = -P\Delta$ with domain $D(A) = V \cap H^2(\Omega)$. It is well known that A is a positive definite, self-adjoint operator and it is characterized by the relation

$$(Aw, v) = (\nabla w, \nabla v), \quad \forall w \in D(A), \quad v \in V.$$

If Ω is of class $C^{1,1}$, the norms $\|u\|_{H^2}$ and $\|Au\|$ are equivalent in $D(A)$ (see C. Amrouche & V. Girault (1991)). Similarly, we introduce the Laplacian operator $B = -\Delta$ with domain $D(B) = H_0^1(\Omega) \cap H^2(\Omega)$.

By means of P, we can reformulate problem (1.1)–(1.2) as follows: Find u, w, ρ (in spaces which will be defined later on), satisfying

$$P(\rho u_t) + (\mu + \mu_r)Au + P(\rho u \cdot \nabla u) = 2\mu_r P(\operatorname{rot} w) + P(\rho f), \quad (2.1)$$
$$\rho w_t + (c_a + c_d)Bw + \rho u \cdot \nabla w - (c_0 + c_d - c_a)\nabla \operatorname{div} w + 4\mu_r w = 2\mu_r \operatorname{rot} u + \rho g, \quad (2.2)$$
$$\rho_t + u \cdot \nabla \rho = 0, \quad (2.3)$$
$$u(x,0) = u_0(x), \; w(x,0) = w_0(x), \; \rho(x,0) = \rho_0(x) \text{ in } \Omega. \quad (2.4)$$

We propose the following iterative process for approximating the solution of problem (2.1)-(2.4). For $n = 1$, we define

$$u^1(t) = u_0, \quad w^1(t) = w_0, \text{ and } \rho^1(x,t) = \rho_0(x).$$

Assuming that u^n, w^n and ρ^n have already been computed, we define u^{n+1}, w^{n+1} and ρ^{n+1} as the unique solution of the following system of linear equations:

$$P(\rho^n u_t^{n+1}) + (\mu + \mu_r)Au^{n+1} + P(\rho^n u^n \cdot \nabla u^{n+1}) = 2\mu_r P(\operatorname{rot} w^n) + P(\rho^n f), \quad (2.5)$$
$$\rho^n w_t^{n+1} + (c_a + c_d)Bw^{n+1} + \rho^n u^n \cdot \nabla w^{n+1} - (c_o + c_d - c_a)\nabla \operatorname{div} w^{n+1} + 4\mu_r w^{n+1} = 2\mu_r \operatorname{rot} u^n + \rho^n g, \quad (2.6)$$
$$\rho_t^{n+1} + u^{n+1} \cdot \nabla \rho^{n+1} = 0, \quad (2.7)$$
$$u^{n+1}(x,0) = u_0(x), \quad w^{n+1}(x,0) = w_0(x), \quad \rho^{n+1}(x,0) = \rho_0(x) \text{ in } \Omega. \quad (2.8)$$

For simplicity, from now on we consider $u_0(x) = 0$ and $w_0(x) = 0$. In this case, it is clear that the first iterate is $(u^1, w^1, \rho^1) = (0, 0, \rho_0)$. In this way, we have reduced a nonlinear coupled system into a sequence of linear, weakly coupled systems. Our goal is to prove that the resulting solutions (u^n, w^n, ρ^n) define a Cauchy sequence in a suitable Banach space, which converges to a strong solution of our problem. More precisely, in this paper we prove:

Theorem 2.1 - *Assuming that the hypotheses (3.1), (3.12) and (3.28) hold (see Section 3 below), then there exists a unique solution (u, w, ρ) of problem (2.1)-(2.4). Moreover, the sequences u^n, w^n and ρ^n satisfy*

$$\begin{aligned}
u^n &\longrightarrow u \quad \textit{strongly in} \quad L^\infty(0,T;V) \cap L^2(0,T;H^2(\Omega) \cap V),\\
w^n &\longrightarrow w \quad \textit{strongly in} \quad L^\infty(0,T;H_0^1(\Omega)) \cap L^2(0,T;H^2(\Omega) \cap H_0^1(\Omega)) \textit{ and}\\
\rho^n &\longrightarrow \rho \quad \textit{strongly in} \quad L^\infty(0,T;L^\infty(\Omega)).
\end{aligned}$$

3. A priori estimates

Throughout this paper the external fields f and g are assumed to be $L^2(Q_T)$ functions, small enough with respect to the viscosities coefficients of the model (a precise formulation of this hypothesis will be given later on; see Section 3.2) μ, μ_r, c_a and c_d. Concerning the initial density ρ_0, we assume that it is a continuously differentiable function ($\rho_0 \in \mathcal{C}^1$), and that there exist α, β such that

$$0 < \alpha \leq \rho_0(x) \leq \beta \quad \forall x \in \bar{\Omega}. \tag{3.1}$$

The Galerkin method used, resolves at each step a linear system. As a result, J.L. Boldrini & M. Rojas (1998) proved that the solutions (u^n, w^n, ρ^n) enjoy the following conditions concerning their regularity:

$$\begin{aligned}
&u^n \in L^\infty(0,T;V) \cap L^2(0,T;D(A)),\\
&u_t^n \in L^2(0,T;H),\\
&Au^n \in L^2(0,T;L^2(\Omega)),\\
&w^n \in L^\infty(0,T;H_0^1(\Omega)) \cap L^2(0,T;D(B)),\\
&w_t^n \in L^2(0,T;L^2(\Omega)),\\
&Bw^n \in L^2(0,T;L^2(\Omega)).
\end{aligned}$$

We are going to prove on the one hand that these sequences are furthermore uniformly bounded in the corresponding spaces. On the other hand, applying the method of characteristics to the continuity equation (2.7), it follows immediately that whenever ρ^n exists, it satisfies $0 < \alpha \leq \rho^n \leq \beta$. In particular we have that

$$\{\rho^n\} \text{ is uniformly bounded in } L^\infty(0,T;L^\infty(\Omega)). \tag{3.2}$$

Furthermore, the hypothesis on ρ_0 make possible to apply the results of O. Ladyzhenskaya & V. Solonnikov (1978) (see Lemma 1.3, p. 705). In our case, uniform bounds for $\nabla \rho^n$ and ρ_t^n in $L^\infty(0,T;L^\infty(\Omega))$ are also obtained.

3.1. The L^2 estimates of u^n and w^n

Multiplying the equation (2.7) by v, and taking scalar product of the resulting identity also by v we get $(\rho_t^n v, v) = -(\mathrm{div}\,(\rho^n u^n)v, v) = 2(\rho^n u^n \cdot \nabla v, v)$ and consequently, for all $v \in H_0^1(\Omega)$ such that $v_t \in L^2(\Omega)$,

$$\frac{1}{2}\frac{d}{dt}\|\sqrt{\rho^n}v\|^2 = \frac{1}{2}(\rho_t^n v, v) + (\rho^n v_t, v) = (\rho^n u^n \cdot \nabla v, v) + (\rho^n v_t, v).$$

Let us keep this identity in mind. Multiplying (2.5) by u^{n+1} and (2.6) by w^{n+1} we obtain

$$\frac{1}{2}\frac{d}{dt}\|\sqrt{\rho^n}u^{n+1}\|^2 + (\mu+\mu_r)\|\nabla u^{n+1}\|^2 = 2\mu_r(\mathrm{rot}\, w^n, u^{n+1}) + (\rho^n f, u^{n+1}), \quad (3.3)$$

$$\frac{1}{2}\frac{d}{dt}\|\sqrt{\rho^n}w^{n+1}\|^2 + (c_a+c_d)\|\nabla w^{n+1}\|^2 + (c_0+c_d-c_a)\|\mathrm{div}\ w^{n+1}\|^2 + 4\mu_r\|w^{n+1}\|^2 = 2\mu_r(\mathrm{rot}\, u^n, w^{n+1}) + (\rho^n g, w^{n+1}). \quad (3.4)$$

It is well known that for all $u \in H_0^1(\Omega)$ the following classical inequalities hold

$$\begin{cases} \|\mathrm{rot}\, u\| \le \|\nabla u\|, & \|u\|_{L^4} \le \sqrt{2}\,\|u\|^{1/4}\|\nabla u\|^{3/4} \\ \|u\|^2 \le \lambda^{-1}\|\nabla u\|^2, \end{cases} \quad (3.5)$$

where, recall that λ denotes the smallest eigenvalue of the Laplace operator.

Using these inequalities, classical estimates for the right hand side of (3.3) and (3.4) lead to the following differential inequalities

$$\frac{d}{dt}\|\sqrt{\rho^n}u^{n+1}\|^2 + (\mu+\mu_r)\|\nabla u^{n+1}\|^2 \le \frac{8\mu_r^2}{\mu+\mu_r}\|w^n\|^2 + \frac{2\beta^2\lambda^{-1}}{\mu+\mu_r}\|f\|^2,$$

$$\frac{d}{dt}\|\sqrt{\rho^n}w^{n+1}\|^2 + (c_a+c_d)\|\nabla w^{n+1}\|^2 + 2(c_0+c_d-c_a)\|\mathrm{div}\ w^{n+1}\|^2 \le \frac{4\mu_r^2}{c_a+c_d}\|u^n\|^2 + \frac{\beta^2}{8\mu_r}\|g\|^2.$$

Therefore, adding up both inequalities and integrating from 0 to t, we have

$$\alpha(\|u^{n+1}(t)\|^2 + \|w^{n+1}(t)\|^2) + (\mu+\mu_r)\int_0^t \|\nabla u^{n+1}(\tau)\|^2 d\tau + (c_a+c_d)\int_0^t \|\nabla w^{n+1}(\tau)\|^2 d\tau + 2(c_0+c_d-c_a)\int_0^t \|\mathrm{div}\ w^{n+1}(\tau)\|^2 d\tau$$
$$\le \frac{8\mu_r^2}{\mu+\mu_r}\int_0^t \|w^n(\tau)\|^2 d\tau + \frac{4\mu_r^2}{c_a+c_d}\int_0^t \|u^n(\tau)\|^2 d\tau$$

$$+\frac{2\beta^2\,\lambda^{-1}}{\mu+\mu_r}\|f\|^2_{L^2(Q_T)}+\frac{\beta^2}{8\mu_r}\|g\|^2_{L^2(Q_T)}$$

(recall that $u_0 = w_0 = 0$). This means that we can choose constants M and C such that

$$\begin{aligned}\|u^{n+1}(t)\|^2+\|w^{n+1}(t)\|^2+\frac{\mu+\mu_r}{\alpha}\int_0^t\|\nabla u^{n+1}(\tau)\|^2d\tau\\ +\frac{c_a+c_d}{\alpha}\int_0^t\|\nabla w^{n+1}(\tau)\|^2d\tau\\ \le C\int_0^t(\|u^n(\tau)\|^2+\|w^n(\tau)\|^2)d\tau+M.\end{aligned}\tag{3.6}$$

Effectively, the following choice provides the desired inequality

$$C = \max\{\frac{8\mu_r^2}{\alpha(\mu+\mu_r)},\frac{4\mu_r^2}{\alpha(c_a+c_d)}\},\tag{3.7}$$

$$M = \frac{2\beta^2\,\lambda^{-1}}{\alpha(\mu+\mu_r)}\|f\|^2_{L^2(Q_T)}+\frac{\beta^2}{8\alpha\,\mu_r}\|g\|^2_{L^2(Q_T)}.\tag{3.8}$$

Thus, setting

$$\varphi_n(t)=\|u^n(t)\|^2+\|w^n(t)\|^2,$$

(3.6) reduces in particular to

$$\varphi_{n+1}(t)\le M+C\int_0^t\varphi_n(\tau)d\tau.$$

Observing that $\varphi_1(t) = 0$, a straightforward induction argument shows that for $n \ge 2$,

$$\varphi_n(t)\le M\sum_{k=0}^{n-2}\frac{(Ct)^k}{k!}\le M\,e^{Ct},\tag{3.9}$$

and hence for all n we have

$$\sup_{t\in[0,T]}(\,\|u^n(t)\|^2+\|w^n(t)\|^2\,)\le\sup_{t\in[0,T]}M\,e^{Ct}=M\,e^{CT}.\tag{3.10}$$

Combining (3.6) and (3.9), we get the following result:

Lemma 3.1 - *The sequences u^n and w^n satisfy*

$$\|u^n\|^2_{L^2(0,T;V)}\le\frac{\alpha Me^{CT}}{\mu+\mu_r}\ \text{and}\ \|w^n\|^2_{L^2(0,T;H_0^1(\Omega))}\le\frac{\alpha Me^{CT}}{c_a+c_d},\tag{3.11}$$

i.e., u^n remains bounded in $L^2(0,T;V)$ and w^n in $L^2(0,T;H_0^1(\Omega))$, uniformly as $n \longrightarrow \infty$.

3.2. The L^∞ estimates of u^n and w^n

These bounds are harder to obtain. Notice that inequality (3.9) already gives $L^\infty(0,T;L^2(\Omega))$ estimates. For bounds in $L^\infty(0,T;H^1(\Omega))$ it is necessary to introduce two parameters, δ and η, and prove that there exist at least one choice of these parameters for which a certain key inequality (see (3.20)) holds with positive left hand side terms. On the other hand, we will need that the following technical condition

$$\left(\|f\|^2_{L^2(Q_T)} + \|g\|^2_{L^2(Q_T)}\right)\left(\frac{\beta}{\mu+\mu_r}\right)^3\left(\frac{1}{2} + \left(\frac{4}{\lambda}+\frac{1}{4}\right)\frac{\mu_r}{c_a+c_d}e^{CT}\right) \\ \leq \frac{\lambda^{1/4}}{160C_\Omega^2}\min\{\frac{\alpha}{4\beta}, \frac{\alpha^5}{\beta^5}\} \tag{3.12}$$

holds, where C is defined by (3.7) and C_Ω is introduced in (3.17). Notice that this hypothesis is fulfilled if f and g are small enough with respect to the viscosities μ and μ_r.

Multiplying (2.5) firstly by $\delta A u^{n+1}$ and secondly by u_t^{n+1}, an integration in Ω yields

$$\delta(\mu+\mu_r)\|Au^{n+1}\|^2 = -\delta\,(\rho^n u_t^{n+1}, Au^{n+1}) + 2\mu_r\,\delta\,(\operatorname{rot} w^n, Au^{n+1}) \\ +\delta\,(\rho^n f, Au^{n+1}) - \delta(\rho^n u^n.\nabla u^{n+1}, Au^{n+1}) \tag{3.13}$$

and

$$\|\sqrt{\rho^n}u_t^{n+1}\|^2 + \frac{\mu+\mu_r}{2}\frac{d}{dt}\|\nabla u^{n+1}\|^2 = 2\mu_r(\operatorname{rot} w^n, u_t^{n+1}) \\ +(\rho^n f, u_t^{n+1}) - (\rho^n u^n.\nabla u^{n+1}, u_t^{n+1}), \tag{3.14}$$

respectively. Thus, using the fact that $\alpha \leq \rho^n \leq \beta$, we have

$$\alpha\|u_t^{n+1}\|^2 + \frac{\mu+\mu_r}{2}\frac{d}{dt}\|\nabla u^{n+1}\|^2 + \delta(\mu+\mu_r)\|Au^{n+1}\|^2 \\ \leq |\delta\,(\rho^n u_t^{n+1}, Au^{n+1})| + |2\mu_r\,\delta\,(\operatorname{rot} w^n, Au^{n+1})| \\ + |2\mu_r(\operatorname{rot} w^n, u_t^{n+1})| + |(\rho^n f, u_t^{n+1})| \\ + |\delta\,(\rho^n f, Au^{n+1})| + |\delta(\rho^n u^n.\nabla u^{n+1}, Au^{n+1})| \\ + |(\rho^n u^n.\nabla u^{n+1}, u_t^{n+1})|. \tag{3.15}$$

Using the classical estimates (3.5), Hölder and Young's inequalities, the sixth term in the right hand side of (3.15) can be estimated as follows

$$|\delta(\rho^n u^n \cdot \nabla u^{n+1}, Au^{n+1})| \leq \delta\,\beta\|u^n\|_{L^4}\|\nabla u^{n+1}\|_{L^4}\|Au^{n+1}\|$$

$$\leq \delta\,\beta\,\sqrt{2}\,\|u^n\|^{1/4}\|\nabla u^n\|^{3/4}\|\nabla u^{n+1}\|_{L^4}\|Au^{n+1}\|$$
$$\leq \delta\,\beta\,\sqrt{2}\,\lambda^{-1/8}\|\nabla u^n\|\|\nabla u^{n+1}\|_{L^4}\|Au^{n+1}\|. \qquad (3.16)$$

Since the embedding $H^2(\Omega) \hookrightarrow W^{1,4}(\Omega)$ is continuous, for all $u \in D(A)$ we have

$$\|\nabla u\|_{L^4} \leq \|u\|_{W^{1,4}} \leq C_\Omega\|u\|_{H^2} \leq C_\Omega\|Au\|. \qquad (3.17)$$

Thus, from (3.16) and (3.17), we obtain

$$|\delta(\rho^n u^n \cdot \nabla u^{n+1}, Au^{n+1})| \leq \delta\,\beta\,\sqrt{2}\,\lambda^{-1/8}C_\Omega\|\nabla u^n\|\|Au^{n+1}\|^2. \qquad (3.18)$$

Similarly, we can estimate the seventh term in the right hand side of (3.15) as follows

$$\begin{aligned}|(\rho^n u^n.\nabla u^{n+1}, u_t^{n+1})| &\leq \beta\,\|u^n\|_{L^4}\|\nabla u^{n+1}\|_{L^4}\|u_t^{n+1}\| \\ &\leq \delta\,\beta\,\sqrt{2}\;\lambda^{-1/8}C_\Omega\|\nabla u^n\|\|Au^{n+1}\|^2 \\ &\quad + \frac{\beta\,\sqrt{2}\;\lambda^{-1/8}C_\Omega}{4\delta}\|\nabla u^n\|\|u_t^{n+1}\|^2. \qquad (3.19)\end{aligned}$$

Therefore, using classical estimates for the remaining terms of the right hand side of (3.15), and by virtue of Young's inequality we obtain,

$$\begin{aligned}&(\mu+\mu_r)\frac{d}{dt}\|\nabla u^{n+1}\|^2 + 2\left(\alpha - 3\eta - \frac{\beta\,\sqrt{2}\,\lambda^{-1/8}C_\Omega}{4\delta}\|\nabla u^n\|\right)\|u_t^{n+1}\|^2 \\ &+2\delta\left((\mu+\mu_r) - \frac{3\delta\,\beta^2}{4\eta} - \beta\,\sqrt{8}\,\lambda^{-1/8}C_\Omega\|\nabla u^n\|\right)\|Au^{n+1}\|^2 \\ &\leq \left(\frac{8\mu_r^2\,\eta}{\beta^2} + \frac{2\mu_r^2}{\eta}\right)\|\nabla w^n\|^2 + \left(2\eta + \frac{\beta^2}{2\eta}\right)\|f\|^2,\end{aligned}$$

where $\eta > 0$ is any positive parameter. Integrating from 0 to t this last inequality, we obtain

$$\begin{aligned}&(\mu+\mu_r)\|\nabla u^{n+1}(t)\|^2 \\ &+2\int_0^t\left(\alpha - 3\eta - \frac{\beta\,\sqrt{2}\,\lambda^{-1/8}C_\Omega}{4\delta}\|\nabla u^n(\tau)\|\right)\|u_t^{n+1}(\tau)\|^2 d\tau \\ &+2\delta\int_0^t\left((\mu+\mu_r) - \frac{3\delta\,\beta^2}{4\eta} - \beta\,\sqrt{8}\,\lambda^{-1/8}C_\Omega\|\nabla u^n(\tau)\|\right)\|Au^{n+1}(\tau)\|^2 d\tau \\ &\leq \left(\frac{8\mu_r^2\,\eta}{\beta^2} + \frac{2\mu_r^2}{\eta}\right)\int_0^t\|\nabla w^n(\tau)\|^2 d\tau + (2\eta + \frac{\beta^2}{2\eta})\|f\|^2_{L^2(Q_T)} \\ &\leq \left(\frac{4\eta}{\beta^2} + \frac{1}{\eta}\right)\frac{2\mu_r^2\alpha M e^{CT}}{c_a + c_d} + \left(2\eta + \frac{\beta^2}{2\eta}\right)\|f\|^2_{L^2(Q_T)}. \qquad (3.20)\end{aligned}$$

Here, we have to choose carefully the values of the two parameters η and δ in order to guarantee positiveness of the left-hand-side terms. Choosing

$$\eta = \frac{\alpha}{4}; \quad \delta = \frac{\alpha(\mu+\mu_r)}{4\,\beta^2},$$

we have

$$\begin{aligned}
&(\mu+\mu_r)\|\nabla u^{n+1}(t)\|^2 \\
&+2\int_0^t \left(\frac{\alpha}{4} - \frac{\beta^3\sqrt{2}\,\lambda^{-1/8}C_\Omega}{\alpha(\mu+\mu_r)}\|\nabla u^n(\tau)\|\right)\|u_t^{n+1}(\tau)\|^2 d\tau \\
&+\frac{\alpha(\mu+\mu_r)}{2\,\beta^2}\int_0^t \left(\frac{\mu+\mu_r}{4} - \beta\sqrt{8}\,\lambda^{-1/8}C_\Omega\|\nabla u^n(\tau)\|\right)\|Au^{n+1}(\tau)\|^2 d\tau \\
&\le 2(\alpha^2+4\beta^2)\frac{\mu_r^2}{\beta^2(c_a+c_d)}Me^{CT} + \frac{1}{2\alpha}(\alpha^2+4\beta^2)\|f\|^2_{L^2(Q_T)} \\
&\le \frac{5\beta^2}{\alpha}\left(\|f\|^2_{L^2(Q_T)} + \|g\|^2_{L^2(Q_T)}\right)\left(\frac{1}{2} + \frac{\mu_r e^{CT}}{c_a+c_d}\left(\frac{1}{4}+\frac{4}{\lambda}\right)\right) \qquad (3.21)
\end{aligned}$$

where we used in this latter expression the definition of M (see (3.8)) and other standard estimates.

Let ε be any positive number. Choosing carefully the size of the data f, g and the viscosities μ and μ_r, it is possible to obtain

$$\begin{aligned}
&(\mu+\mu_r)\|\nabla u^{n+1}(t)\|^2 + \\
&+2\int_0^t \left(\frac{\alpha}{4} - \frac{\beta^3\sqrt{2}\,\lambda^{-1/8}C_\Omega}{\alpha(\mu+\mu_r)}\|\nabla u^n(\tau)\|\right)\|u_t^{n+1}(\tau)\|^2 d\tau \\
&+\frac{\alpha(\mu+\mu_r)}{2\,\beta^2}\int_0^t \left(\frac{\mu+\mu_r}{4} - \beta\sqrt{8}\,\lambda^{-1/8}C_\Omega\|\nabla u^n(\tau)\|\right)\|Au^{n+1}(\tau)\|^2 d\tau \\
&\le \varepsilon^2 \qquad (3.22)
\end{aligned}$$

Setting $n=1$ in (3.22), we get

$$\begin{aligned}
(\mu+\mu_r)\|\nabla u^2(t)\|^2 + 2\int_0^t \frac{\alpha}{4}\|u_t^2(\tau)\|^2 d\tau + \\
+\frac{\alpha(\mu+\mu_r)}{2\,\beta^2}\int_0^t \frac{\mu+\mu_r}{4}\|Au^2(\tau)\|^2 d\tau \le \varepsilon^2,
\end{aligned}$$

for all $t \in [0,T]$ since $u^1 = 0$. Thus

$$\|\nabla u^2(t)\| \le \frac{\varepsilon}{(\mu+\mu_r)^{1/2}}. \qquad (3.23)$$

Using classical induction on n, that is, assuming that

$$\|\nabla u^n(t)\| \leq \frac{\varepsilon}{(\mu+\mu_r)^{1/2}},$$

we prove that for ε sufficiently small, a similar inequality holds for u^{n+1}. Effectively, it is clear that for ε sufficiently small we have

$$\frac{\alpha}{4} - \frac{\beta^3\sqrt{2}\,\lambda^{-1/8}C_\Omega\,\varepsilon}{\alpha(\mu+\mu_r)^{3/2}} > 0 \quad \text{and} \quad \frac{\mu+\mu_r}{4} - \frac{\beta\sqrt{8}\,\lambda^{-1/8}C_\Omega\,\varepsilon}{(\mu+\mu_r)^{1/2}} > 0.$$

This condition, stated in terms of ε^2, is written as

$$\varepsilon^2 < \frac{(\mu+\mu_r)^3\lambda^{1/4}}{32\beta^2C_\Omega^2}\min\{\frac{\alpha^4}{\beta^4},\frac{1}{4}\}. \tag{3.24}$$

Therefore, from (3.22) it follows the desired estimate for u^{n+1}. Thus, identifying ε^2 with the right hand side of inequality (3.21), we have proved:

Lemma 3.2 - *Assume that hypothesis* (3.12) *is fulfilled. Then there exists* ε *which depends on the smallness of* f *and* g *with respect to the viscosities* μ *and* μ_r*, such that*

$$\sup_{t\in[0,T]}\|\nabla u^n(t)\| \leq \frac{\varepsilon}{(\mu+\mu_r)^{1/2}}, \tag{3.25}$$

i.e., u^n *remains uniformly bounded in* $L^\infty(0,T;V)$ *as* $n \longrightarrow \infty$.

Similar arguments leads us to an analogous result for w^n.

3.3. The L^2 estimates of the time derivatives

Throughout this section we assume that hypothesis (3.12) is fulfilled. By virtue of Lemma 3.2, inequality (3.22) reduces to

$$(\mu+\mu_r)\|\nabla u^{n+1}(t)\|^2 + 2\int_0^t\left(\frac{\alpha}{4} - \frac{\beta^3\sqrt{2}\,\lambda^{-1/8}C_\Omega\,\varepsilon}{\alpha(\mu+\mu_r)^{3/2}}\right)\|u_t^{n+1}(\tau)\|^2d\tau$$
$$+\frac{\alpha(\mu+\mu_r)}{2\,\beta^2}\int_0^t\left(\frac{\mu+\mu_r}{4} - \frac{\beta\sqrt{8}\,\lambda^{-1/8}C_\Omega\,\varepsilon}{(\mu+\mu_r)^{1/2}}\right)\|Au^{n+1}(\tau)\|^2d\tau \leq \varepsilon^2.$$

Therefore, we conclude that there exists a constant C_ε, independent of n, such that

$$\|\nabla u^{n+1}(t)\|^2 + \int_0^t\|u_t^{n+1}(\tau)\|^2d\tau + \int_0^t\|Au^{n+1}(\tau)\|^2d\tau \leq C_\varepsilon, \tag{3.26}$$

which implies u_t^n and Au^n remain uniformly bounded in $L^2(0,T;H)$, as $n \longrightarrow \infty$.

Similarly, for all n, we can prove that there exists a second constant C'_ε, independent of n such that

$$\|\nabla w^{n+1}(t)\|^2 + \int_0^t \|Bw^{n+1}(\tau)\|^2 d\tau + \int_0^t \|w_t^{n+1}(\tau)\|^2 d\tau \leq C'_\varepsilon, \qquad (3.27)$$

and analogous bounds for w^n hold.

3.4. The L^2 estimates of the time derivatives

In this section we assume the following additional hypotheses on the data f and g:

$$\begin{aligned} \nabla f,\, \nabla g \;&\in\; L^2(Q_T)^3 \text{ and} \\ f_t,\, g_t \;&\in\; L^2(Q_T). \end{aligned} \qquad (3.28)$$

Differentiating (2.5) with respect to t, we obtain

$$\begin{aligned} &P(\rho_t^n u_t^{n+1}) + P(\rho^n u_{tt}^{n+1}) + (\mu + \mu_r)Au_t^{n+1} \\ &\quad = 2\,\mu_r\, P(\operatorname{rot} w_t^n) + P(\rho_t^n f) + P((\rho^n f_t) - P(\rho_t^n u^n \cdot \nabla u^{n+1}) \\ &\qquad -P(\rho^n u_t^n \cdot \nabla u^{n+1}) - P(\rho^n u^n \cdot \nabla u_t^{n+1}). \end{aligned} \qquad (3.29)$$

Multiplication of (3.29) by u_t^{n+1} yields

$$\begin{aligned} &\frac{1}{2}\frac{d}{dt}\|\sqrt{\rho^n} u_t^{n+1}\|^2 + (\mu+\mu_r)\|\nabla u_t^{n+1}\|^2 = -\frac{1}{2}(\rho_t^n u_t^{n+1}, u_t^{n+1}) \\ &\quad +2\,\mu_r\,(\operatorname{rot} w_t^n, u_t^{n+1}) + (\rho_t^n f, u_t^{n+1}) + (\rho^n f_t, u_t^{n+1})) \\ &\quad -(\rho_t^n u^n \cdot \nabla u^{n+1}, u_t^{n+1}) - (\rho^n u_t^n . \nabla u^{n+1}, u_t^{n+1}) \\ &\quad -(\rho^n u^n \cdot \nabla u_t^{n+1}, u_t^{n+1})) \\ &= \frac{1}{2}(\operatorname{div}(\rho^n u^n) u_t^{n+1}, u_t^{n+1}) + 2\,\mu_r\,(w_t^n, \operatorname{rot} u_t^{n+1}) \\ &\quad -(\operatorname{div}(\rho^n u^n) f, u_t^{n+1})) + (\rho^n f_t, u_t^{n+1}) \\ &\quad +(\operatorname{div}(\rho^n u^n) u^n \cdot \nabla u^{n+1}, u_t^{n+1}) - (\rho^n u_t^n \cdot \nabla u^{n+1}, u_t^{n+1})) \\ &\quad -(\rho^n u^n \cdot \nabla u_t^{n+1}, u_t^{n+1}) \end{aligned} \qquad (3.30)$$

since from (2.3), $\rho_t^n = -\operatorname{div}(\rho^n u^n)$.

Using classical estimates, Hölder and Young's inequalities, each of the seven right hand side terms in (3.30) can be estimated. A straightforward but tedious estimate yields

$$\begin{aligned}&\frac{1}{2}\frac{d}{dt}\|\sqrt{\rho^n}u_t^{n+1}\|^2+(\mu+\mu_r)\|\nabla u_t^{n+1}\|^2\\&\quad\le C_\eta\|u_t^{n+1}\|^2+C_\eta\|w_t^n\|^2+C_\eta\|f\|_{H^1}^2+C_\eta\|f_t\|^2+C_\eta\|Au^n\|^2+\\&\quad+C_\eta\|Au^{n+1}\|^2+C_\eta\|u_t^n\|^2\|Au^{n+1}\|^2+C_\eta\|u_t^{n+1}\|^2+8\eta\,\|\nabla u_t^{n+1}\|^2.\end{aligned}$$

where η is any positive parameter, and C_η is a constant independent of n. Choosing

$$\eta=\frac{\mu+\mu_r}{16},$$

and denoting C the corresponding constant C_η we get

$$\begin{aligned}\frac{d}{dt}\|\sqrt{\rho^n}u_t^{n+1}\|^2+(\mu+\mu_r)\|\nabla u_t^{n+1}\|^2&\le C\,\|u_t^{n+1}\|^2+C\,\|w_t^n\|^2\\&+C\,\|f\|_{H^1}^2+C\,\|Au^n\|^2+C\,\|Au^{n+1}\|^2\\&+C\,\|f_t\|^2+C\,\|u_t^n\|^2\|Au^{n+1}\|^2+C\,.\end{aligned}\tag{3.31}$$

Let us now estimate $\|Au^n\|^2$ in terms of $\|u_t^n\|^2$. In order to do this, we multiply (2.5) by Au^{n+1} and obtain

$$\begin{aligned}(\mu+\mu_r)\|Au^{n+1}\|^2&=-(\rho^n u_t^{n+1},Au^{n+1})+2\,\mu_r\,(\operatorname{rot}w^n,Au^{n+1})\\&\quad+(\rho^n f,Au^{n+1})-(\rho^n u^n\cdot\nabla u^{n+1},Au^{n+1}).\end{aligned}$$

Since

$$\begin{aligned}|(\rho^n u^n.\nabla u^{n+1},Au^{n+1})|&\le\beta\,\|u^n\|_{L^4}\|\nabla u^{n+1}\|_{L^4}\|Au^{n+1}\|\\&\le C\,\|\nabla u^{n+1}\|^{1/4}\|Au^{n+1}\|^{7/4}\\&\le C_\delta\|\nabla u^{n+1}\|^2+\delta\|Au^{n+1}\|^2,\end{aligned}$$

where δ is any positive parameter used in this last application of Young's inequality, and using classical estimations for the remaining terms, we get

$$\begin{aligned}(\mu+\mu_r)\|Au^{n+1}\|^2&\le C_\delta\|u_t^{n+1}\|^2+C_\delta\|\nabla w^n\|^2+C_\delta\|f\|^2+\\&\quad+C_\delta\|\nabla u^{n+1}\|^2+4\,\delta\,\|Au^{n+1}\|^2.\end{aligned}$$

Then, choosing $\delta>0$ sufficiently small, from the last inequality, we obtain as needed

$$\|Au^{n+1}\|^2\le C\,\|u_t^{n+1}\|^2+C.\tag{3.32}$$

Thus, (3.31) can be reduced to

$$\begin{aligned}&\frac{d}{dt}\|\sqrt{\rho^n}u_t^{n+1}\|^2+(\mu+\mu_r)\|\nabla u_t^{n+1}\|^2\\&\qquad\le C\,\|u_t^{n+1}\|^2+C\,\|w_t^n\|^2+C\,\|f\|_{H^1}^2+C\,\|f_t\|^2\\&\qquad+C\,\|u_t^n\|^2\|u_t^{n+1}\|^2+C\,\|u_t^n\|^2+C.\end{aligned}$$

Integrating this inequality from 0 to t, we have

$$\begin{aligned}
&\alpha\|u_t^{n+1}(t)\|^2 + (\mu+\mu_r)\int_0^t \|\nabla u_t^{n+1}(\tau)\|^2 d\tau \\
&\quad \le C\int_0^t (\|u_t^{n+1}(\tau)\|^2 + \|u_t^n(\tau)\|^2 + \|w_t^n(\tau)\|^2) d\tau \\
&\quad\quad + C\int_0^t \|f(\tau)\|_{H^1}^2 + \|f_t(\tau)\|^2) d\tau \\
&\quad\quad + C\int_0^t \|u_t^n(\tau)\|^2\, \|u_t^{n+1}(\tau)\|^2 d\tau + \alpha\|u_t^{n+1}(0)\|^2 + Ct.
\end{aligned}$$

From equation (3.14), we can easily bound the rightmost term, $\|u_t^{n+1}(0)\|^2$. This is due to the fact that $(d/dt)\|\nabla u^{n+1}(t)\|^2$ is non negative at $t=0$ since $\nabla u^{n+1}(0) = 0$. Applying the estimates for $\int_0^t \|u_t^{n+1}(\tau)\|^2 d\tau$ and $\int_0^t \|w_t^{n+1}(\tau)\|^2 d\tau$ and hypothesis (3.28), we get

$$\|u_t^{n+1}(t)\|^2 + \int_0^t \|\nabla u_t^{n+1}(\tau)\|^2 d\tau \le C + C\int_0^t \|u_t^n(\tau)\|^2 \|u_t^{n+1}(\tau)\|^2 d\tau.$$

Let us denote $\varphi(t) = \|u_t^{n+1}(t)\|^2$. The above inequality can be written as

$$\varphi(t) \le C + C\int_0^t \|u_t^n(\tau)\|^2 \varphi(t)\, d\tau$$

which, by Gronwall's lemma, implies

$$\|u_t^{n+1}(t)\|^2 = \varphi(t) \le C\exp(C\int_0^t \|u_t^n(\tau)\|^2\, d\tau) \le C\exp(C\,C_\varepsilon),$$

where C_ε comes from one more application of (3.26). We conclude finally that there exists a constant C independent of n such that

$$\|u_t^{n+1}(t)\|^2 + \int_0^t \|\nabla u_t^{n+1}(\tau)\|^2 d\tau \le C, \tag{3.33}$$

which means that $u_t^n \in L^\infty(0,T;H) \cap L^2(0,T;V)$. Moreover, from (3.32) we have for all n

$$\sup_t \|Au^{n+1}(t)\|^2 \le C, \tag{3.34}$$

that is, $u^n \in L^\infty(0,T,D(A))$.

Similarly, it can also be proven that there exists a constant C, independent of n, such that

$$\left\{ \begin{array}{l} \|w_t^{n+1}(t)\|^2 + \int_0^t \|\nabla w_t^{n+1}(\tau)\|^2 d\tau \leq C \quad \text{and} \\ \sup_t \|Bw^{n+1}(t)\|^2 \leq C. \end{array} \right. \tag{3.35}$$

3.5. An $L^2(0,T;W^{1,\infty})$ estimate of u^n

Let us write (2.5) as

$$(\mu + \mu_r)Au^{n+1} = P(F), \tag{3.36}$$

where

$$F = 2\mu_r \text{rot}\, w^n + \rho^n f - \rho^n u_t^{n+1} - \rho^n u^n \cdot \nabla u^{n+1}.$$

Making use of all the previous estimates, it is possible to prove that $F \in L^2(0,T;L^6(\Omega))$. As a consequence of the Amrouche-Girault's results (1991), we conclude that u^n remains uniformly bounded in $L^2(0,T;W^{2,6}(\Omega))$. This, using the Sobolev embedding, implies that u^n remains uniformly bounded in $L^2(0,T;W^{1,\infty}(\Omega))$.

3.6. Three key estimates of the second order derivatives

Without any extra regularity assumed for the solution at $t = 0$, we can still obtain bounds for ∇u_t^n in $L^\infty(\varepsilon,T;L^2(\Omega))$ and for u_{tt}^n and Au_t^n in $L^2(\varepsilon,T;L^2(\Omega))$ for any $\varepsilon > 0$. Analogously for w^n.

Let first multiply (3.29) by u_{tt}^{n+1}. Using (3.2), (3.25), previously obtained estimates, and Hölder and Young's inequalities, we have

$$\begin{aligned} \alpha \|u_{tt}^{n+1}\|^2 + \frac{\mu+\mu_r}{2}\frac{d}{dt}\|\nabla u_t^{n+1}\|^2 &\leq C_\eta \|u_t^{n+1}\|^2 + C_\eta \|\nabla w_t^n\|^2 + \\ &+ C_\eta \|f\|^2 + C_\eta \|f_t\|^2 + C_\eta \|Au^{n+1}\|^2 + C_\eta \|\nabla u_t^n\|^2 \|Au^{n+1}\|^2 \\ &+ C_\eta \|\nabla u_t^{n+1}\|^2 + 7\eta \|u_{tt}^{n+1}\|^2, \end{aligned}$$

where η is any positive parameter, and C_η is a constant independent of n. Choosing $\eta = \dfrac{\alpha}{14}$ and taking into account (3.33)-(3.35), we have

$$\begin{aligned} \alpha \|u_{tt}^{n+1}\|^2 + (\mu+\mu_r)\frac{d}{dt}\|\nabla u_t^{n+1}\|^2 \quad &\leq \quad C \|\nabla w_t^n\|^2 + C \|f\|^2 + C \|f_t\|^2 \\ &+ C \|\nabla u_t^n\|^2 + C \|\nabla u_t^{n+1}\|^2 + C. \end{aligned}$$

It is possible to overcome the lack of boundedness for $\|\nabla u_t^{n+1}\|^2$ at $t = 0$. In fact, multiplying this latter inequality by $\sigma(t) = \min\{1, t\}$, we get

$$\begin{aligned} &\alpha\,\sigma(t)\|u_{tt}^{n+1}\|^2 + (\mu+\mu_r)\frac{d}{dt}(\sigma(t)\|\nabla u_t^{n+1}\|^2) \\ &\quad \leq (\mu+\mu_r)\sigma'(t)\|\nabla u_t^{n+1}\|^2 + C\,\sigma(t)(\|\nabla u_t^n\|^2 + \|\nabla w_t^n\|^2) \\ &\qquad + C\,\sigma(t)(\|f\|^2 + \|f_t\|^2) + C\,\sigma(t)(\|\nabla u_t^{n+1}\|^2 + 1). \end{aligned} \tag{3.37}$$

Observe that as a consequence of (3.33), it exists a sequence $\varepsilon_k \longrightarrow 0$, such that $\varepsilon_k\,\|\nabla u_t^{n+1}(\varepsilon_k)\|^2 \leq C$. Therefore, since $\sigma(t) \leq 1$ and $\sigma'(t) \leq 1$ a.e. in $[0,T]$, applying (3.33)-(3.35) and integrating (3.37) from ε_k to t, we obtain

$$\begin{aligned} &\alpha \int_{\varepsilon_k}^t \sigma(\tau)\|u_{tt}^{n+1}(\tau)\|^2 d\tau + (\mu+\mu_r)\sigma(t)\|\nabla u_t^{n+1}(t)\|^2 \\ &\qquad \leq (\mu+\mu_r)\sigma(\varepsilon_k)\,\|\nabla u_t^{n+1}(\varepsilon_k)\|^2 + C. \end{aligned}$$

Taking limit as $\varepsilon_k \longrightarrow 0$ reduces the previous inequality to

$$\int_0^t \sigma(\tau)\|u_{tt}^{n+1}(\tau)\|^2 d\tau + \sigma(t)\|\nabla u_t^{n+1}(t)\|^2 \leq C,$$

which provides, for any $\varepsilon > 0$, uniform estimates of u_{tt}^n and ∇u_t^n in the space $L^2(\varepsilon, T; L^2(\Omega))$ as $n \longrightarrow \infty$.

Analogously, for all n, it can be proved that

$$\int_0^t \sigma(\tau)\|w_{tt}^{n+1}(\tau)\|^2 d\tau + \sigma(t)\|\nabla w_t^{n+1}(t)\|^2 \leq C$$

and similar uniform bounds hold for w^n.

Finally, from (3.29) we obtain

$$(\mu+\mu_r)\int_0^t \sigma(\tau)\|Au_t^{n+1}(\tau)\|^2 d\tau \leq \int_0^t \sigma(\tau)\|G^n(\tau)\|^2 d\tau$$

where

$$\begin{aligned} G^n \;=\;& 2\mu_r \mathrm{rot}\, w_t^n + \rho_t^n f + \rho^n f_t - \rho_t^n u_t^{n+1} - \rho^n u_{tt}^{n+1} - \rho_t^n u^n \cdot \nabla u^{n+1} \\ & -\rho^n u_t^n \cdot \nabla u^{n+1} - \rho^n u^n \cdot \nabla u_t^{n+1}. \end{aligned}$$

Gathering together all the estimates already derived, we conclude that $\sigma^{1/2}(t)G^n$ remains uniformly bounded in $L^2(0,T; L^2(\Omega))$ and therefore, for any $\varepsilon > 0$, Au_t^n is uniformly bounded in $L^2(\varepsilon, T; L^2(\Omega))$. A similar conclusion hold for w^n.

Remark. Using classical compactness arguments, we conclude that, up to a subsequence, the approximate solutions (u^n, w^n, ρ^n) converges to a strong solution of problem (1.1)-(1.2). Alternatively, this can be proved using a different approach which we develop in section 4.

4. The convergence of the sequence

This section is devoted to prove that u^n, w^n and ρ^n are Cauchy sequences.

Let us introduce the following notation for the difference of two terms of a sequence. For $n,\ s \geq 1$,

$$u^{n,s}(t) = u^{n+s}(t) - u^n(t),\ w^{n,s}(t) = w^{n+s}(t) - w^n(t),$$

and

$$\rho^{n,s}(t) = \rho^{n+s}(t) - \rho^n(t).$$

It is clear that $u^{n,s}$, $w^{n,s}$ and $\rho^{n,s}$ satisfy the following equations

$$\begin{aligned} P(\rho^{n-1+s}u_t^{n,s}) + (\mu+\mu_r)Au^{n,s} &= 2\mu_r P(\text{rot}\, w^{n-1,s}) + P(\rho^{n-1,s}f) \\ &\quad -P(\rho^{n-1,s}u_t^n) - P(\rho^{n-1+s}u^{n-1+s}\cdot\nabla u^{n,s}) \\ &\quad -P(\rho^{n-1+s}u^{n-1,s}\cdot\nabla u^n) - P(\rho^{n-1,s}u^{n-1}\cdot\nabla u^n) \quad (4.1) \\ \rho^{n-1+s}w_t^{n,s} + (c_a+c_d)Bw^{n,s} &- (c_0+c_d-c_a)\nabla\ \text{div}\ w^{n,s} \\ &\quad +4\mu_r w^{n,s} = 2\mu_r(\text{rot}\, u^{n-1,s}) + \rho^{n-1,s}g - \rho^{n-1,s}w_t^n \\ &\quad -\rho^{n-1+s}u^{n-1+s}\cdot\nabla w^{n,s} - \rho^{n-1+s}u^{n-1,s}\cdot\nabla w^n \\ &\quad -\rho^{n-1,s}u^{n-1}\cdot\nabla w^n \quad (4.2) \\ \rho_t^{n,s} + u^{n,s}\cdot\nabla\rho^{n+s} + u^n\cdot\nabla\rho^{n,s} &= 0. \quad (4.3) \end{aligned}$$

The following lemma which can be easily proven, is fundamental in order to obtain error estimates.

Lemma 4.1 - *Let* $0 \leq \phi_1(t) \leq M$ *for all* $t \in [0,T]$ *and assume that there exist* $C > 0$ *such that for all* $n \geq 2$, *we have*

$$0 \leq \phi_n(t) \leq C\int_0^t \phi_{n-1}(\tau)d\tau.$$

Then

$$\phi_n(t) \leq M\frac{(Ct)^{n-1}}{(n-1)!}$$

for all $t \in [0,T]$, *and* $n \geq 2$.

The next lemma, proven in W. Varhorn (1994, Lemma 3.10 p. 122) is a variant of Gronwall's lemma and it will be also needed.

Lemma 4.2 - *Let $\gamma \in \mathbb{R}$ ($\gamma > 0$) and let Φ, f, $g \in C([0,T])$ real functions ($f \geq 0$, $g \geq 0$ on $[0,T]$) satisfying for all $t \in [0,T]$ the inequality*

$$\Phi(t) + \int_0^t f(\tau)\, d\tau \leq \int_0^t g(\tau)\, d\tau + \gamma \int_0^t \Phi(\tau)\, d\tau.$$

Then for all $t \in [0,T]$ we obtain the estimate

$$\Phi(t) + \int_0^t f(\tau)\, d\tau \leq \left(\int_0^t g(\tau)\, d\tau \right) e^{\gamma t}.$$

4.1. Bounding the error of the density sequence

The density sequence can be bounded in term of the velocity sequence. For this purpose, let us multiply (4.3) by $(\rho^{n,s})^5$ and integrate in Ω. We get

$$\begin{aligned}
\frac{1}{6}\frac{d}{dt}\int_\Omega |\rho^{n,s}|^6 dx &= -\int_\Omega u^{n,s} \cdot \nabla \rho^{n+s} (\rho^{n,s})^5 dx - \frac{1}{6}\int_\Omega u^n \cdot \nabla (\rho^{n,s})^6 dx \\
&\leq \int_\Omega |u^{n,s}||\nabla \rho^{n+s}||\rho^{n,s}|^5 dx + \frac{1}{6}\int_\Omega \text{div } u^n (\rho^{n,s})^6 dx \\
&\leq \|\nabla \rho^{n+s}\|_{L^\infty(0,T;L^\infty(\Omega))} \int_\Omega |u^{n,s}||\rho^{n,s}|^5 dx \\
&\leq C \left(\int_\Omega |u^{n,s}|^6 dx \right)^{1/6} \left(\int_\Omega |\rho^{n,s}|^6 dx \right)^{5/6},
\end{aligned}$$

that is,

$$\frac{1}{6}\frac{d}{dt}\|\rho^{n,s}\|_{L^6}^6 \leq C\, \|u^{n,s}\|_{L^6} \|\rho^{n,s}\|_{L^6}^5.$$

Since on one hand $\frac{1}{6}(d/dt)\|\rho^{n,s}\|_{L^6}^6 = \|\rho^{n,s}\|_{L^6}^5 (d/dt)\|\rho^{n,s}\|_{L^6}$, and on the other hand the embedding $H^1(\Omega) \hookrightarrow L^6(\Omega)$ is continuous,

$$\frac{d}{dt}\|\rho^{n,s}\|_{L^6} \leq C\, \|\nabla u^{n,s}\|.$$

Integrating from 0 to t and applying the Cauchy-Schwarz inequality, we get

$$\|\rho^{n,s}(t)\|_{L^6} \leq C \int_0^t \|\nabla u^{n,s}(\tau)\| d\tau \leq CT \left(\int_0^t \|\nabla u^{n,s}(\tau)\|^2 d\tau \right)^{1/2}. \quad (4.4)$$

The latter inequality means that ρ^n is a Cauchy sequence in $L^\infty(0,T;L^6(\Omega))$ whenever u^n is a Cauchy sequence in $L^2(0,T;V)$. Furthermore, from (4.3) we have

$$\begin{aligned} \rho_t^{n,s} + u^n.\nabla\rho^{n,s} + u^{n,s}\cdot\nabla\rho^{n+s} &= 0 \\ \rho^{n,s}(0) &= 0. \end{aligned}$$

which, using the characteristic method (see O. Ladyzhenskaya & V. Solonnikov (1978), p. 730), lead us to

$$\|\rho^{n,s}(t)\|_{L^\infty} \leq \|\nabla\rho^{n+s}\|_{L^\infty(0,T;L^\infty(\Omega))}\int_0^t \|u^{n,s}(\tau)\|_{L^\infty}d\tau. \tag{4.5}$$

4.2. Convergence of u^n and w^n in L^∞

In this paragraph we prove that u^n and w^n are Cauchy sequences in the space $L^\infty(0,T;H^1(\Omega))$. Multiplying (4.1) by $\delta Au^{n,s}$, where δ is any positive parameter, an integration over Ω and some straightforward estimations yield,

$$\begin{aligned} \delta(\mu+\mu_r)\|Au^{n,s}\|^2 \quad \leq \quad & \eta\|u_t^{n,s}\|^2 + \eta\|\nabla w^{n-1,s}\|^2 + \eta\|\rho^{n-1,s}\|_{L^6}^2\|f\|_{L^3}^2 \\ & +\eta\|\rho^{n-1,s}\|_{L^6}^2\|\nabla u_t^n\|^2 + \eta\|\nabla u^{n,s}\|^2 \\ & +\eta\|\nabla u^{n-1,s}\|^2 + \eta\|\rho^{n-1,s}\|_{L^6}^2 \\ & +\frac{1}{4\eta}\delta^2 C\|Au^{n,s}\|^2. \end{aligned} \tag{4.6}$$

where η is any positive parameter and C is a constant independent of n. Analogously, multiplying (4.1) by $u_t^{n,s}$, we get

$$\begin{aligned} & \alpha\|u_t^{n,s}\|^2 + \frac{\mu+\mu_r}{2}\frac{d}{dt}\|\nabla u^{n,s}\|^2 \\ & \leq C_\eta\|\nabla w^{n-1,s}\|^2 + C_\eta\|\rho^{n-1,s}\|_{L^6}^2\|f\|_{L^3}^2 \\ & \quad +C_\eta\|\rho^{n-1,s}\|_{L^6}^2\|\nabla u_t^n\|^2 + C_\eta\|\nabla u^{n,s}\|^2 + C_\eta\|\nabla u^{n-1,s}\|^2 \\ & \quad +C_\eta\|\rho^{n-1,s}\|_{L^6}^2 + 6\eta\|u_t^{n,s}\|^2, \end{aligned} \tag{4.7}$$

where C_η is a constant independent of n. Adding up (4.6) and (4.7) we obtain

$$\begin{aligned} & \alpha\|u_t^{n,s}\|^2 + \frac{\mu+\mu_r}{2}\frac{d}{dt}\|\nabla u^{n,s}\|^2 + \delta(\mu+\mu_r)\|Au^{n,s}\|^2 \\ & \leq C_\eta + \eta)\left\{\|\nabla w^{n-1,s}\|^2 + \|\rho^{n-1,s}\|_{L^6}^2\|f\|_{L^3}^2 + \|\rho^{n-1,s}\|_{L^6}^2\|\nabla u_t^n\|^2\right. \\ & \quad \left.+\|\nabla u^{n,s}\|^2 + \|\nabla u^{n-1,s}\|^2 + \|\rho^{n-1,s}\|_{L^6}^2\right\} + 7\eta\|u_t^{n,s}\|^2 \\ & \quad +\frac{C\delta^2}{4\eta}\|Au^{n,s}\|^2. \end{aligned}$$

We choose $\eta = \frac{\alpha}{14}$, and $\delta > 0$ such that $(\mu + \mu_r)\delta - \frac{C\delta^2}{4\eta} \equiv C_1 > 0$. This reduces the previous inequality to

$$\begin{aligned}
&\alpha\|u_t^{n,s}\|^2 + (\mu + \mu_r)\frac{d}{dt}\|\nabla u^{n,s}\|^2 + C_1\|Au^{n,s}\|^2 \\
&\quad \leq C\|\nabla w^{n-1,s}\|^2 + C\|\rho^{n-1,s}\|_{L^6}^2\|f\|_{L^3}^2 + C\|\rho^{n-1,s}\|_{L^6}^2\|\nabla u_t^n\|^2 \\
&\qquad +C\|\nabla u^{n,s}\|^2 + C\|\nabla u^{n-1,s}\|^2 + C\|\rho^{n-1,s}\|_{L^6}^2 \qquad (4.8)
\end{aligned}$$

with positive constants C_1, C independent of n.

From (4.2), we have

$$\begin{aligned}
&\rho^{n-1+s}w_t^{n,s} + Lw^{n,s} + 4\mu_r w^{n,s} \\
&\quad = 2\mu_r(\mathrm{rot}\, u^{n-1,s}) + \rho^{n-1,s}g - \rho^{n-1,s}w_t^n - \rho^{n-1+s}u^{n-1+s}\cdot\nabla w^{n,s} \\
&\qquad -\rho^{n-1+s}u^{n-1,s}\cdot\nabla w^n - \rho^{n-1,s}u^{n-1}\cdot\nabla w^n \qquad (4.9)
\end{aligned}$$

where $Lw^{n,s} = (c_a + c_d)Bw^{n,s} - (c_0 + c_d - c_a)\nabla\, \mathrm{div}\, w^{n,s}$. Since L is a strongly elliptic operator (see O. Ladyzhenskaya et al. (1968) p. 70), there exists a positive constant N_0 depending exclusively on $c_a + c_d$, $c_0 + c_d - c_a$ and Γ, such that

$$(Lw^{n,s}, Bw^{n,s}) \geq (c_a + c_d)\|Bw^{n,s}\|^2 - N_0\|\nabla w^{n,s}\|^2. \qquad (4.10)$$

Multiplying (4.9) by $\theta Bw^{n,s}$, where θ is any positive parameter, the use of (4.10) and some classical estimations yield

$$\begin{aligned}
\theta(c_a + c_d)\|Bw^{n,s}\|^2 \leq\; & \theta C_\zeta\|\nabla w^{n,s}\|^2 + \zeta\|w_t^{n,s}\|^2 + \zeta\|\nabla u^{n-1,s}\|^2 \\
& +\zeta\|\rho^{n-1,s}\|_{L^6}^2\|g\|_{L^3}^2 + \zeta\|\rho^{n-1,s}\|_{L^6}^2\|\nabla w_t^n\|^2 \\
& +\zeta\|\nabla w^{n,s}\| + \zeta\|\nabla u^{n-1,s}\| \\
& +\zeta\|\rho^{n-1,s}\|_{L^6}^2 + \frac{\theta^2 C}{4\zeta}\|Bw^{n,s}\|^2, \qquad (4.11)
\end{aligned}$$

where ζ is any positive parameter and C is a constant independent of n. Multiplying (4.2) by $w_t^{n,s}$, classical estimations lead us to

$$\begin{aligned}
&\alpha\|w_t^{n,s}\|^2 + \frac{c_a + c_d}{2}\frac{d}{dt}\|\nabla w^{n,s}\|^2 + \frac{c_0 + c_d - c_a}{2}\frac{d}{dt}\|\mathrm{div}\; w^{n,s}\|^2 \\
&\qquad +2\mu_r\frac{d}{dt}\|w^{n,s}\|^2 \leq C_\zeta\|\nabla u^{n-1,s}\|^2 + C_\zeta\|\rho^{n-1,s}\|_{L^6}^2\|g\|_{L^3}^2 \\
&\qquad\qquad +C_\zeta\|\rho^{n-1,s}\|_{L^6}^2\|\nabla w_t^n\|^2 + C_\zeta\|\nabla w^{n,s}\|^2 \\
&\qquad\qquad +C_\zeta\|\nabla u^{n-1,s}\|^2 + C_\zeta\|\rho^{n-1,s}\|_{L^6}^2 \\
&\qquad\qquad +6\zeta\|w_t^{n,s}\|^2. \qquad (4.12)
\end{aligned}$$

Adding up (4.11) and (4.12), and proceeding as before, but choosing this time

$$\zeta = \frac{\alpha}{12}; \quad \theta = \frac{(c_a + c_d)\alpha}{25\beta^2}$$

and denoting

$$C_2 = \frac{(c_a + c_d)^2\alpha}{(25\beta)^2},$$

we have

$$\begin{aligned}
&\alpha\|w_t^{n,s}\|^2 + (c_a + c_d)\frac{d}{dt}\|\nabla w^{n,s}\|^2 + C_2\|Bw^{n,s}\|^2 \\
&\quad +(c_0 + c_d - c_a)\frac{d}{dt}\|\text{div } w^{n,s}\|^2 + 4\mu_r\frac{d}{dt}\|w^{n,s}\|^2 \\
&\quad \leq C\|\nabla u^{n-1,s}\|^2 + C\|\rho^{n-1,s}\|_{L^6}^2\|g\|_{L^3}^2 \\
&\qquad +C\|\rho^{n-1,s}\|_{L^6}^2\|\nabla w_t^n\|^2 + C\|\nabla w^{n,s}\|^2 \\
&\qquad +C\|\rho^{n-1,s}\|_{L^6}^2,
\end{aligned} \tag{4.13}$$

where C is again a constant independent of n. Adding up (4.8) and (4.13) and integrating from 0 to t, we obtain

$$\begin{aligned}
&(\mu + \mu_r)\|\nabla u^{n,s}(t)\|^2 + (c_a + c_d)\|\nabla w^{n,s}(t)\|^2 + \alpha\int_0^t (\|u_t^{n,s}(\tau)\|^2 \\
&\quad +\|w_t^{n,s}(\tau)\|^2)d\tau + C_1\int_0^t \|Au^{n,s}(\tau)\|^2 d\tau + C_2\int_0^t \|Bw^{n,s}(\tau)\|^2 d\tau \\
&\quad +(c_0 + c_d - c_a)\|\text{div } w^{n,s}(t)\|^2 \\
&\leq C\int_0^t (\|\nabla u^{n-1,s}(\tau)\|^2 + \|\nabla w^{n-1,s}(\tau)\|^2)d\tau \\
&\quad +C\int_0^t \|\rho^{n-1,s}(\tau)\|_{L^6}^2(\|f(\tau)\|_{L^3}^2 + \|g(\tau)\|_{L^3}^2)d\tau \\
&\quad +C\int_0^t \|\rho^{n-1,s}(\tau)\|_{L^6}^2(\|\nabla u_t^n(\tau)\|^2 + \|\nabla w_t^n(\tau)\|^2)d\tau \\
&\quad +C\int_0^t (\|\nabla u^{n,s}(\tau)\|^2 + \|\nabla w^{n,s}(\tau)\|^2)d\tau \\
&\quad +C\int_0^t \|\rho^{n-1,s}(\tau)\|_{L^6}^2 d\tau.
\end{aligned} \tag{4.14}$$

Bounding the right hand side term $\|\rho^{n-1,s}(t)\|_{L^6}^2$ in (4.14) with estimate (4.4), we obtain

$$
\begin{aligned}
&C_3(\|\nabla u^{n,s}(t)\|^2 + \|\nabla w^{n,s}(t)\|^2) + C_3 \int_0^t (\|u_t^{n,s}(\tau)\|^2 + \|w_t^{n,s}(\tau)\|^2) d\tau \\
&\quad + C_3 \int_0^t (\|Au^{n,s}(\tau)\|^2 + \|Bw^{n,s}(\tau)\|^2) d\tau + C_3 \|\text{div } w^{n,s}(t)\|^2 \\
&\leq C \int_0^t (\|\nabla u^{n-1,s}(\tau)\|^2 + \|\nabla w^{n-1,s}(\tau)\|^2) d\tau \\
&\quad + C \int_0^t \|\nabla u^{n-1,s}(t_1)\|^2 dt_1 \int_0^t (\|f(\tau)\|_{L^3}^2 + \|g(\tau)\|_{L^3}^2) d\tau \\
&\quad + C \int_0^t \|\nabla u^{n-1,s}(t_1)\|^2 dt_1 \int_0^t (\|\nabla u_t^n(\tau)\|^2 + \|\nabla w_t^n(\tau)\|^2) d\tau \\
&\quad + C \int_0^t (\|\nabla u^{n,s}(\tau)\|^2 + \|\nabla w^{n,s}(\tau)\|^2) d\tau + CT \int_0^t \|\nabla u^{n-1,s}(\tau)\|^2 d\tau,
\end{aligned}
$$

where $C_3 = \min\{\ \mu + \mu_r,\ c_a + c_d,\ \alpha,\ C_1,\ C_2\ \}$. Then

$$
\begin{aligned}
&\|\nabla u^{n,s}(t)\|^2 + \|\nabla w^{n,s}(t)\|^2 + \int_0^t (\|u_t^{n,s}(\tau)\|^2 + \|w_t^{n,s}(\tau)\|^2) d\tau \\
&\quad + \int_0^t \|Au^{n,s}(\tau)\|^2 d\tau + \int_0^t \|Bw^{n,s}(\tau)\|^2 d\tau \\
&\leq C \int_0^t (\|\nabla u^{n-1,s}(\tau)\|^2 + \|\nabla w^{n-1,s}(\tau)\|^2) d\tau \\
&\quad + C \int_0^t (\|\nabla u^{n,s}(\tau)\|^2 + \|\nabla w^{n,s}(\tau)\|^2) d\tau.
\end{aligned}
$$

Applying Lemma 4.2, we get

$$
\begin{aligned}
&\|\nabla u^{n,s}(t)\|^2 + \|\nabla w^{n,s}(t)\|^2 + \int_0^t (\|u_t^{n,s}(\tau)\|^2 + \|w_t^{n,s}(\tau)\|^2) d\tau \\
&\quad + \int_0^t \|Au^{n,s}(\tau)\|^2 d\tau + \int_0^t \|Bw^{n,s}(\tau)\|^2 d\tau \\
&\leq M_1 \int_0^t (\|\nabla u^{n-1,s}(\tau)\|^2 + \|\nabla w^{n-1,s}(\tau)\|^2) d\tau \qquad (4.15)
\end{aligned}
$$

where M_1 is a positive constant, independent of n. Thus, we have

$$
\begin{aligned}
&\|\nabla u^{n,s}(t)\|^2 + \|\nabla w^{n,s}(t)\|^2 \\
&\quad \leq M_1 \int_0^t (\|\nabla u^{n-1,s}(\tau)\|^2 + \|\nabla w^{n-1,s}(\tau)\|^2) d\tau.
\end{aligned}
$$

Since the application of (3.26) and (3.27) gives $\|\nabla u^{1,s}(t)\|^2 + \|\nabla w^{1,s}(t)\|^2 \leq M_2$, $\forall\, s$ and $t \in [0,T]$ for some positive constant M_2, using the Lemma 4.1, we obtain

$$\|\nabla u^{n,s}(t)\|^2 + \|\nabla w^{n,s}(t)\|^2 \leq M_2 \frac{(M_1 t)^{n-1}}{(n-1)!} \leq \frac{(MT)^{n-1}}{(n-1)!}, \tag{4.16}$$

where M is again a constant independent of n. This inequality proves that u^n and w^n are Cauchy sequences in $L^\infty(0,T;H^1(\Omega))$, and then ρ^n is a Cauchy sequence in $L^\infty(0,T;L^\infty(\Omega))$ in view of (4.4)-(4.5).

Finally, since

$$\begin{aligned} M_1 \int_0^t (\|\nabla u^{n-1,s}(\tau)\|^2 + \|\nabla w^{n-1,s}(\tau)\|^2) d\tau \\ \leq M_1 \int_0^t \frac{(M\tau)^{n-2}}{(n-2)!} d\tau \leq M_1 \frac{(Mt)^{n-1}}{(n-1)!} \end{aligned} \tag{4.17}$$

from (4.15) we have

$$\int_0^t (\|u_t^{n,s}(\tau)\|^2 + \|w_t^{n,s}(\tau)\|^2) d\tau \leq M \frac{(M_1 t)^{n-1}}{(n-1)!} \leq M \frac{(M_1 T)^{n-1}}{(n-1)!}. \tag{4.18}$$

$$\int_0^t (\|Au^{n,s}(\tau)\|^2 + \|Bw^{n,s}(\tau)\|^2) d\tau \leq M \frac{(M_1 t)^{n-1}}{(n-1)!} \leq M \frac{(M_1 T)^{n-1}}{(n-1)!}. \tag{4.19}$$

Let now summarize the results already obtained in the following lemma:

Lemma 4.3 - *We assume that hypotheses (3.1), (3.12) and (3.28) are fulfilled. Then there exist* $u \in L^\infty(0,T;V) \cap L^2(0,T;H^2(\Omega) \cap V)$, $w \in L^\infty(0,T;H_0^1(\Omega)) \cap L^2(0,T;H^2(\Omega) \cap H_0^1(\Omega))$ *and* $\rho \in L^\infty(0,T;L^\infty(\Omega))$ *such that*

$$u^n \longrightarrow u \text{ strongly in } L^\infty(0,T;V) \cap L^2(0,T;H^2(\Omega) \cap V), \tag{4.20}$$

$$\begin{aligned} w^n \longrightarrow w \text{ strongly in } L^\infty(0,T;H_0^1(\Omega)) \cap L^2(0,T;H^2(\Omega)) \\ \text{and strongly in } L^2(0,T;H_0^1(\Omega)) \end{aligned} \tag{4.21}$$

$$\rho^n \longrightarrow \rho \text{ strongly in } L^\infty(0,T;L^\infty(\Omega)). \tag{4.22}$$

Following a similar pattern to the proof of Lemma 4.3 and multiplying the resulting inequalities by $\sigma(t) = \min\{1,t\}$ (as in Paragraph 3.6), the additional convergences are proven in Conca et al. (2002):

Lemma 4.4 *Under the same hypotheses of Lemma 4.3, we have the following convergences:*

$$
\begin{aligned}
u_t^n &\longrightarrow u_t \text{ strongly in } L^\infty(\epsilon,T;H)\cap L^2(\epsilon,T;V),\\
u_{tt}^n &\longrightarrow u_{tt} \text{ strongly in } L^2(\epsilon,T;H),\\
w_t^n &\longrightarrow w_t \text{ strongly in } L^\infty(\epsilon,T;L^2(\Omega))\cap L^2(\epsilon,T;H_0^1(\Omega)),\\
w_{tt}^n &\longrightarrow w_{tt} \text{ strongly in } L^2(\epsilon,T;L^2(\Omega)),\\
u^n &\longrightarrow u \text{ strongly in } L^\infty(\epsilon,T;H^2(\Omega)\cap V)\cap L^2(\epsilon,T;W^{1,\infty}(\Omega)\cap V),\\
w^n &\longrightarrow w \text{ strongly in } L^\infty(\epsilon,T;H^2(\Omega)\cap H_0^1(\Omega))\cap L^2(\epsilon,T;W^{1,\infty}(\Omega))\\
&\qquad\qquad \text{and strongly in } L^2(\epsilon,T;H_0^1(\Omega)),
\end{aligned}
$$

where ε is any small positive constant.

4.3. Passage to the limit

Applying Lemma 4.3, as a standard procedure we obtain

$$
\int_0^T \langle \rho u_t + \rho u\cdot\nabla u - \rho f - 2\,\mu_r \mathrm{rot}\, w - (\mu+\mu_r)\Delta u, v\,\rangle \phi(t)\,dt = 0,
$$

$$
\int_0^T \langle \rho w_t + \rho u\cdot\nabla w - \rho g - 2\,\mu_r \mathrm{rot}\, u + 4\,\mu_r\, w - (c_a+c_d)\Delta w
$$
$$
-(c_0+c_d-c_a)\nabla \mathrm{div}\, w, z\,\rangle \psi(t)\,dt = 0,
$$

for all $z,\, v\in L^2(\Omega)$ and $\phi,\, \psi\in L^\infty(0,T)$. This clearly implies

$$
\begin{aligned}
&\langle \rho u_t + \rho u\cdot\nabla u - \rho f - 2\,\mu_r \mathrm{rot}\, w - (\mu+\mu_r)\Delta u, v\,\rangle = 0,\\
&\langle \rho w_t + \rho u\cdot\nabla w - \rho g - 2\,\mu_r \mathrm{rot}\, u + 4\,\mu_r\, w - (c_a+c_d)\Delta w\\
&\qquad\qquad -(c_0+c_d-c_a)\nabla \mathrm{div}\, w, z\,\rangle = 0
\end{aligned}
$$

a.e. in $[0,T]$, for every $v\in H,\ z\in L^2(\Omega)$.

From these equalities we have

$$
P(\rho u_t + \rho u\cdot\nabla u - \rho f - 2\,\mu_r \mathrm{rot}\, w - (\mu+\mu_r)\Delta u\,) = 0
$$

and

$$
\begin{aligned}
&\rho w_t + \rho u\cdot\nabla w - \rho g - 2\,\mu_r \mathrm{rot}\, u + 4\,\mu_r\, w\\
&\qquad\qquad -(c_a+c_d)\Delta w - (c_0+c_d-c_a)\nabla \mathrm{div}\, w = 0.
\end{aligned}
$$

In the case of the density, we observe that

$$
\begin{aligned}
&u^n \longrightarrow u \ \text{ strongly in } \ L^2(0,T;L^2(\Omega)),\\
&\rho_t^n \longrightarrow \rho_t, \ \text{ and } \ \nabla\rho^n \longrightarrow \nabla\rho \ \text{ weakly in } \ L^2(0,T;L^2(\Omega)),
\end{aligned}
$$

Thus, when $n \to \infty$, the approximated continuity equation gives

$$\rho_t + u \cdot \nabla \rho = 0 \text{ in the } L^2(0,T;L^2(\Omega)) - \text{sense.}$$

Let us prove the continuity of the solution (u, w, ρ) at $t = 0$. Since $u \in L^\infty(0,T;D(A))$ and $u_t \in L^2(\varepsilon,T;D(A))$, then by classical interpolation theory (see R. Temam (1979), p. 260) u is a.e. equal to a continuous function from $[\varepsilon, T]$ into $D(A)$, i.e.,

$$u \in C([\varepsilon,T];D(A)) \;\; \forall\, \varepsilon > 0.$$

On the other hand, since $u_t \in L^2(\varepsilon,T;D(A))$, $u_{tt} \in L^2(\varepsilon,T;H)$ and using again interpolation theory we have

$$u_t \in C([\varepsilon,T];V) \;\; \forall\, \varepsilon > 0.$$

Therefore,

$$u \in C^1([\varepsilon,T];V) \cap C([\varepsilon,T];D(A)) \;\; \forall\, \varepsilon > 0.$$

Analogously, we prove that

$$w \in C^1([\varepsilon,T];H_0^1(\Omega)) \cap C([\varepsilon,T];D(B)) \;\; \forall\, \varepsilon > 0.$$

To prove continuity at $t = 0$, we already have

$$\lim_{t \to 0^+} \|u(t) - u(0)\| = 0, \quad \lim_{t \to 0^+} \|\nabla u(t) - \nabla u(0)\| = 0.$$

Also, we can prove indeed that

$$\lim_{t \to 0^+} \|Au(t) - Au(0)\| = 0.$$

In fact, it is sufficient to show that

$$\lim_{t \to 0^+} \sup \|Au(t)\| \le \|Au_0\|$$

since we already know that $u(t) \longrightarrow u_0$ in $H^1(\Omega)$.

Multiplying (2.5) by Au_t^{n+1} and integrating in Ω, we obtain

$$\begin{aligned} \frac{\mu + \mu_r}{2} \frac{d}{dt} \|Au^{n+1}\|^2 + \|\sqrt{\rho^n} \nabla u_t^{n+1}\|^2 = \\ = -(\rho^n u^n \cdot \nabla u^{n+1}, Au_t^{n+1}) + 2\mu_r (\operatorname{rot} w^n, Au_t^{n+1}) \\ +(\rho^n f, Au_t^{n+1}) - (\nabla \rho^n \cdot \nabla u_t^{n+1}, u_t^{n+1}), \end{aligned}$$

which, integrating from 0 to t, implies

$$\begin{aligned}\|Au^{n+1}(t)\|^2 \le\ & \|Au_0\|^2 + \frac{2}{\mu+\mu_r}[(-\rho^n(t)u^n(t)\cdot\nabla u^{n+1}(t) + 2\mu_r \mathrm{rot}\, w^n(t)\\ &+\rho^n(t)f(t), Au^{n+1}(t)) - (-\rho_0^n u^n(0)\cdot\nabla u_0^{n+1} + 2\mu_r\mathrm{rot}\, w_0^n\\ &+\rho_0^n f(0), Au_0^{n+1})] + \frac{2}{\mu+\mu_r}N(t)\end{aligned}$$

where

$$\begin{aligned}N(t) = &\int_0^t |(\rho_t^n u^n\cdot\nabla u^{n+1} + \rho^n u_t^n\cdot\nabla u^{n+1} + \rho^n u^n\cdot\nabla u_t^{n+1}\\ &-2\mu_r\mathrm{rot}\, w_t^n - \rho_t^n f - \rho^n f_t, Au^{n+1})|\,d\tau + \int_0^t |(\nabla\rho^n\cdot\nabla u_t^{n+1}, u_t^{n+1})|\,d\tau\\ &\le c\int_0^t (\|\nabla u^{n+1}\| + \|\nabla u_t^n\| + \|\nabla u_t^{n+1}\| + \|\nabla w_t^n\|\\ &+\|f\| + \|f_t\|)\,d\tau \le c\,t^{1/2},\end{aligned}$$

by virtue of Hölder's inequality and the estimates given in Lemmata 4.3–4.4. From this, we conclude

$$\begin{aligned}\|Au(t)\|^2 \le \|Au_0\|^2 + c\,[&(-\rho(t)u(t)\cdot\nabla u(t) + 2\mu_r\mathrm{rot}\, w(t)\\ &+\rho(t)f(t), Au(t)) - (-\rho_0 u(0)\cdot\nabla u_0 + 2\mu_r\mathrm{rot}\, w_0\\ &+\rho_0 f(0), Au_0)] + c\,t^{1/2}.\end{aligned}$$

Since $\rho(t)u(t)\cdot\nabla u(t) \to \rho_0 u_0\cdot\nabla u_0$, $\rho(t)f(t) \to \rho_0 f(0)$, $\mathrm{rot}\, w(t) \to \mathrm{rot}\, w_0$ in $L^2(\Omega)$ and $Au(t) \to Au_0$ weakly in $L^2(\Omega)$ as $t \to 0^+$, we obtain the desired result. It is now clear that

$$\lim_{t\to 0^+} \|u_t(t) - u_t(0)\| = 0.$$

The continuity of $u(t)$ is therefore proved. The results for w are proved in a similar way. This completes the proof of existence in Theorem 2.1.

4.4. Uniqueness

Suppose that there is another solution (u_1, w_1, ρ_1) of (2.1)-(2.4). Define

$$U = u_1 - u, \quad W = w_1 - w \quad \text{and} \quad R = \rho_1 - \rho.$$

These auxiliary functions satisfy a set of equations similar to (4.1)–(4.3). If we multiply the first equation by U, the second by W and the third by

R and repeat arguments similar to those given in Paragraph 4.2, we obtain for $\varphi(t) = \|U(t)\|^2 + \|W(t)\|^2 + \|R(t)\|^2$ an inequality of the type:

$$\varphi(t) \leq C \int_0^t \varphi(\tau)\, d\tau$$

which, by Gronwall's inequality, is equivalent to assert $U = 0$, $W = 0$ and $R = 0$.

4.5. Results on the pressure

To conclude, let us refer the interested reader to Conca et al. (2002), where the following result concerning the pressure is proven.

Lemma 4.5 - *Assume that the hypotheses of Lemma 4.3 are fulfilled. Then, for each n, there exists $p^n \in L^2(0,T;H^1(\Omega)/\mathbb{R})$ such that (u^n, w^n, ρ^n, p^n) satisfy the first equation in (1.1). The approximate pressure p^n converges to a limiting element p in $L^2(0,T;H^1(\Omega)/\mathbb{R})$ and (u, w, ρ, p) is the unique solution of (1.1)-(1.2), where (u, w, ρ) is the unique solution of problem (2.1)-(2.4). Moreover, we have the error estimate*

$$\int_0^T \|p^n(\tau) - p(\tau)\|^2_{H^1(\Omega)/\mathbb{R}} d\tau \leq \frac{(MT)^{n-1}}{(n-1)!},$$

where again M is a positive constant.

Also, p^n converge to p in $L^\infty(\varepsilon, T; H^1(\Omega)/\mathbb{R})$, for all $\varepsilon > 0$ and we have the following error estimate:

$$\sup_t \sigma(t) \|p^n(t) - p(t)\|^2_{H^1(\Omega)/\mathbb{R}} \leq \frac{(MT)^{n-2}}{(n-2)!}.$$

Acknowledgments. This work was supported by the Fondap Programme in Mathematical Mechanics. Elva Ortega was supported by CNPq-Brazil and Marko Rojas was partially supported by CNPq-Brazil, Grant 300116/93-4(RN) and FAPESP-Brazil, Grant 1997/3711-0.

References

[1] Amrouche C. and Girault V., On the existence and regularity of the solutions of the Stokes problem in arbitrary dimensions. Proc. Japan Acad. Ser. A Math. Sci. 67 (1991), 171–175.

[2] Antoncev S.N., Kazhikov A.V. and Monakhov V.N., Boundary Value Problems in Mechanics of Nonhomogeneous Fluids, North-Holland, Amsterdam 1990.

[3] Boldrini J.L. and Rojas-Medar M.A., Global solutions to the equations for the motion of stratified incompressible fluids, Mat. Contemp. 3 (1992), 1–8.

[4] Boldrini J.L. and Rojas-Medar M.A., On the convergence rate of spectral approximation for the equations for nonhomogeneous asymmetric fluids, Math. Model. Numer. Anal. 30(2) (1996), 123–155.

[5] Boldrini J.L. and Rojas-Medar M.A., Global strong solutions of the equations for the motion of nonhomogeneous incompressible fluids, in Numerical Methods in Mechanics, C. Conca et al. eds., 35–45, Pitman Research Notes in Mathematics Series 371, Longman, Harlow (1997).

[6] Boldrini J.L. and Rojas-Medar M.A., Strong solutions of the equations for nonhomogeneous asymmetric fluids, Technical Report N<u>o</u> 21 (1998), IMECC-UNICAMP, Brazil.

[7] Conca C., Gormaz R., Ortega-Torres E.E. and Rojas-Medar M.A., The equations of nonhomogeneous asymmetric fluids: An iterative approach, Math. Methods Appl. Sci. (2002), in press.

[8] Condiff D.W. and Dahler J.S., Fluid mechanical aspects of antisymmetric stress, Phys. Fluids 7(6) (1964), 842–854.

[9] Kim J.U., Weak solutions of an initial boundary value problem for an incompressible viscous fluid with nonnegative density, Siam J. Math. Anal. 18(1987), 89–96.

[10] Ladyzhenskaya O.A., Solonnikov V.A. and Uraltseva N.N., Linear and Quasilinear Equations of Parabolic Type, American Mathematical Society 23, revised edition, (1988).

[11] Ladyzhenskaya O.A., The Mathematical Theory of Viscous Incompressible Flow, Gordon and Breach, New York (1969).

[12] Ladyzhenskaya O.A. and Solonnikov V.A., Unique solvability of an initial and boundary value problem for viscous nonhomogeneous incompressible fluids, Zap. Nauch. Sem. Leningrad Otdel Math. Inst. Steklov 52 (1975), 52–109; Engl. Transl. (1978), J. Soviet Math. 9, 697–749.

[13] Lions J.L., Quelques Méthodes de Résolution des Problèmes aux Limites non Linéaires, Dunod/Gauthier-Villars, Paris (1969).

[14] Lukaszewicz G., On nonstationary flows of asymmetric fluids, Rend. Accad. Naz. Sci. XL Mem. Mat. XII (1988), 35–44.

[15] Lukaszewicz G., On the existence, uniqueness and asymptotic properties for solutions of flows of asymmetric fluids, Rend. Accad. Naz. Sci. XL Mem. Mat. XIII (1989), 105–120.

[16] Lukaszewicz G., On nonstationary flows of incompressible asymmetric fluids, Math. Methods Appl. Sci. 13(3) (1990), 219–232.

[17] Lukaszewicz G., Micropolar Fluids: Theory and Applications, Birkhäuser, Berlin (1999).

[18] Ortega-Torres E.E. and Rojas-Medar M.A., The equations of a viscous asymmetric fluid: An iterational approach, Technical Report N$^{\underline{o}}$ 42 (1997), IMECC-UNICAMP, Brazil.
[19] Petrosyan L.G., Some Problems of Mechanics of Fluids with Antisymmetric Stress Tensor, Erevan (in Russian) 1984.
[20] Salvi R., The equations of viscous incompressible nonhomogeneous fluids: On the existence and regularity, J. Austral. Math. Soc. Ser. B, 33(1) (1991), 94–110.
[21] Simon J., Nonhomogeneous viscous incompressible fluids: Existence of velocity, density and pressure, Siam J. Math. Anal. 21 (1990), 1093–1117.
[22] Temam R., Navier-Stokes Equations, revised ed., North-Holland, Amsterdam 1979.
[23] Varnhorn W., The Stokes Equations, Mathematical Research 76, Akademie-Verlag, Berlin (1994).

Carlos Conca, Raúl Gormaz
Departamento de Ingenieria Matemática
and CMM, UMR 2071 CNRS-UCHILE
Universidad de Chile
Casilla 170/3-Correo 3
Santiago
CHILE
E-mail: cconca@dim.uchile.cl, E-mail: rgormaz@dim.uchile.cl

Elva Ortega
Departamento de Matemática
Universidad de Antofagasta
Casilla 170
Antofagasta
CHILE
E-mail: eortega@uantof.cl

Marko Rojas
Departamento de Matemática Aplicada
IMECC-UNICAMP C.P. 6065
13081-970 Campinas-SP
BRAZIL
E-mail: marko@ime.unicamp.br

Studies in Mathematics and its Applications, Vol. 31
D. Cioranescu and J.L. Lions (Editors)

Chapter 12

HOMOGENIZATION OF DIRICHLET MINIMUM PROBLEMS WITH CONDUCTOR TYPE PERIODICALLY DISTRIBUTED CONSTRAINTS

R. DE ARCANGELIS

0. Introduction

The study of the homogenization of variational problems subject to pointwise oscillating constraints on the gradient was proposed in the 1978 book [3] by A. Bensoussan, J.L. Lions and G. Papanicolaou (cf. also [28], [1], [2], [14], [24], [4], [10] besides [3], for general references on homogenization theory), where also a general conjecture on the limit problems was formulated (cf. §17 of Chapter 1 in [3]). It originated from some problems from elastic-plastic torsion theory for media with periodic structures (cf. for example [18], [6], [29], [22], [25], [19]) that involved integral energies defined on sets of functions u subject to constraints of the type

$$|\nabla u(x)| \leq \varphi(hx) \text{ for a.e. } x, \tag{0.1}$$

where φ is an essentially bounded and $]0,1[^n$-periodic function, and h ($\in \mathbb{N}$) is the homogenization parameter (cf. [10] for a bibliography on the subject).

Analogously, some models in electrostatics and elasticity (cf. [27]) suggested the treatment of homogenization problems in presence of periodically distributed conductors, or of rigid inclusions in the case of elasticity. Such models lead again to the consideration of gradient constraints as in (0.1), but with φ taking only the values 0 and $+\infty$ (cf. [13], [30], [5], [15], [7]), or, in order to consider also "thin" (i.e. with null Lebesgue measure) conductors, to constraints as (cf. [11])),

$$u \text{ is constant in each element of a }]0,1/h[^n\text{-periodic family} \tag{0.2}$$

$$\text{of subsets of } \mathbb{R}^n.$$

In all these papers the homogenization of some classes of Dirichlet minimum problems with null boundary data, or sometimes of Neumann minimum problems, for variational integrals has been performed.

For example, by using the results of [11], the following homogenization result in BV spaces can be proved (cf. [11], [10], and Theorem 5.2 below).

Let $Y =]0,1[^n$, $\mathcal{C}_Y$ be a collection of subsets of Y, and let $\mathcal{C}$ be the set of the periodically distributed conductors defined as

$$\mathcal{C} = \{(i_1,\ldots,i_n) + C : (i_1,\ldots,i_n) \in \mathbb{Z}^n,\ C \in \mathcal{C}_Y\}. \tag{0.3}$$

Let f be a function satisfying

$$\begin{cases} f\colon (x,z) \in \mathbb{R}^n \times \mathbb{R}^n \mapsto f(x,z) \in [0,+\infty[\\ f(\cdot,z) \text{ measurable, } Y\text{-periodic and in } L^1(Y) \text{ for every } z \in \mathbb{R}^n \\ f(x,\cdot) \text{ convex for a.e. } x \in \mathbb{R}^n, \end{cases} \tag{0.4}$$

and let f_{hom} be defined by

$$\begin{aligned} f_{\text{hom}}\colon z \in \mathbb{R}^n \mapsto \inf\Big\{ \int_Y f(y, z+\nabla v)dy : v \in W^{1,\infty}_{\text{loc}}(\mathbb{R}^n), \\ v\ Y\text{-periodic},\ u_z + v \text{ constant in } S \text{ for every } S \in \mathcal{C}\Big\}, \end{aligned} \tag{0.5}$$

where, for every $z \in \mathbb{R}^n$, we denote by u_z the linear function having the gradient z.

Then, by assuming that

$$|z| \le f(x,z) \text{ for a.e. } x \in \mathbb{R}^n \text{ and every } z \in \mathbb{R}^n, \tag{0.6}$$

$$\text{int}(\{z \in \mathbb{R}^n : f_{\text{hom}}(z) < +\infty\}) \ne \emptyset, \tag{0.7}$$

and

$$S \text{ is connected for every } S \in \mathcal{C}, \tag{0.8}$$

it results that f_{hom} is convex and finite in the whole $\mathbb{R}^n$, and that

$$|z| \le f_{\text{hom}}(z) \text{ for every } z \in \mathbb{R}^n. \tag{0.9}$$

Moreover, for every bounded open subset Ω of $\mathbb{R}^n$ with Lipschitz boundary, $\lambda \in]0,+\infty[$, $p \in]1, \frac{n}{n-1}[$, $\beta \in L^\infty(\Omega)$, the values

$$\begin{aligned} i_h(\Omega,0,\beta) = \inf\Big\{ \int_\Omega f(hx,\nabla u)dx + \int_\Omega \beta u dx + \lambda \int_\Omega |u|^p dx : \\ u \in W^{1,\infty}_0(\Omega),\ u \text{ constant in } \Omega \cap \frac{1}{h} S \text{ for every } S \in \mathcal{C}\Big\} \end{aligned} \tag{0.10}$$

converge, as h tends to $+\infty$, to the finite value

$$i_\infty(\Omega,0,\beta)=\min\Big\{\int_\Omega f_{\rm hom}(\nabla u)dx+\int_\Omega f_{\rm hom}^\infty\Big(\frac{dD^su}{d|D^su|}\Big)d|D^su|+$$

$$+\int_{\partial\Omega} f_{\rm hom}^\infty(-u\mathbf{n}_\Omega)d\mathcal{H}^{n-1}+\int_\Omega \beta u dx+\lambda\int_\Omega |u|^p dx : u\in BV(\Omega)\Big\},$$

where $f_{\rm hom}^\infty$ is the recession function of $f_{\rm hom}$, $\mathbf{n}_\Omega$ is the outward unit vector normal to $\partial\Omega$, $\mathcal{H}^{n-1}$ is the $(n-1)$-dimensional Hausdorff measure, $BV(\Omega)$ is the space of the functions in $L^1(\Omega)$ whose distributional partial derivatives are Borel measures with finite total variation in Ω, and D^su, $|D^su|$, $\frac{dD^su}{d|D^su|}$ are defined in §1.

Finally, convergence results in $L^1(\Omega)$ for minimizing sequences of the problems in (0.10) to minimizers of $i_\infty(\Omega,0,\beta)$ too hold.

In the above results the choice of the boundary datum ($=0$) has been forced by the shape of the constraints, and in [12] it has been observed by means of an example in one space dimension that, at least under constraints as in (0.1) with φ in $L^\infty(Y)$, homogenization result of the type just described can be no more true if non homogeneous boundary data are taken into account.

Analogously, in the present paper and in the context of homogenization under constraints as in (0.2), we produce an example showing that, for $n=2$, $f(x,z)=|z|^2$ for a.e. $x\in\mathbb{R}^2$ and every $z\in\mathbb{R}^2$, for a suitable $\mathcal{C}_Y$ verifying (0.7) and (0.8), $\Omega=Y$, $\beta\in L^\infty(\Omega)$, $\lambda\in\mathbb{R}$, $p\in\,]1,3/2[$, and a nice boundary datum u_0, infima in (0.10), but taken in $u_0+W_0^{1,\infty}(\Omega)$ in place of $W_0^{1,\infty}(\Omega)$, can be $+\infty$ for every $h\in\mathbb{N}$, whilst the minimum

$$\min\Big\{\int_\Omega f_{\rm hom}(\nabla u)dx+\int_\Omega \beta u dx+\lambda\int_\Omega |u|^p dx : u\in u_0+W_0^{1,2}(\Omega)\Big\}$$

can be finite (cf. Example 6.1).

These features are essentially due to the fact that in general, for non zero boundary data, for every $h\in\mathbb{N}$ there can be no admissible function verifying the corresponding constancy constraint. Therefore it seems more natural to consider, in place of a fixed boundary datum, a sequence of boundary data, each one fulfilling the constraint, and converging to a limit boundary datum.

This has already been done in [12] in the setting of constraints as in (0.1) with φ in $L^\infty(Y)$.

In the present paper we want to treat the same homogenization problem, but in the setting of constraints as in (0.2).

To do this, we consider a collection $\mathcal{C}_Y$ of subsets of Y, $\mathcal{C}$ given by (0.3), f by (0.4), $f_{\hom}$ by (0.5), a bounded open set Ω with Lipschitz boundary, a sequence of boundary data $\{u_{0,h}\} \subseteq W^{1,\infty}_{\mathrm{loc}}(\mathbb{R}^n)$, and prove, for $\lambda \in [0, +\infty[$, $p \in [1, \frac{n}{n-1}[$, $\beta \in L^\infty(\Omega)$, some homogenization results for the problems

$$i_h(\Omega, u_{0,h}, \beta) = \inf\left\{ \int_\Omega f(hx, \nabla u)dx + \int_\Omega \beta u dx + \lambda \int_\Omega |u|^p dx : \right. \tag{0.11}$$

$$\left. u \in u_{0,h} + W^{1,\infty}_0(\Omega),\ u \text{ constant in } \Omega \cap \frac{1}{h}S \text{ for every } S \in \mathcal{C} \right\},$$

under different sets of assumptions depending on the coerciveness properties of f.

A first results deals with the case of superlinear coerciveness, and is the following.

Let $\phi: \mathbb{R}^n \to [0, +\infty[$ be convex and superlinear, in the sense that

$$\lim_{z \to \infty} \frac{\phi(z)}{|z|} = +\infty. \tag{0.12}$$

Let us assume that

$$\phi(z) \leq f(x, z) \text{ for a.e. } x \in \mathbb{R}^n \text{ and every } z \in \mathbb{R}^n, \tag{0.13}$$

and that (0.7) and the following conditions hold

$$u_{0,h} \text{ is constant in } \Omega \cap \frac{1}{h}S \text{ for every } h \in \mathbb{N} \text{ and } S \in \mathcal{C}, \tag{0.14}$$

$$\text{the integrals } \int_{\cdot} f(hx, \nabla u_{0,h})dx \text{ are equi-absolutely} \tag{0.15}$$

$$\text{continuous in } \Omega \text{ when } h \in \mathbb{N},$$

$$\text{there exist } \sigma > 1,\ M \geq 0 \text{ such that} \tag{0.16}$$

$$\int_\Omega f(hx, \sigma \nabla u_{0,h})dx \leq M \text{ for every } h \in \mathbb{N}.$$

Moreover, let us assume that there exists $u_0 \in W^{1,1}_{\mathrm{loc}}(\mathbb{R}^n)$ such that

$$u_{0,h} \to u_0 \text{ in } L^1(\Omega), \tag{0.17}$$

$$f_{\hom}(\sigma \nabla u_0) \in L^1(\Omega) \text{ for every } \sigma \in \mathbb{R}. \tag{0.18}$$

Then $f_{\rm hom}$ is convex and finite in the whole $\mathbb{R}^n$,

$$\phi(z) \leq f_{\rm hom}(z) \text{ for every } z \in \mathbb{R}^n, \tag{0.19}$$

and for every $\lambda \in [0, +\infty[$, $p \in [1, \frac{n}{n-1}[$, $\beta \in L^\infty(\Omega)$ the values $i_h(\Omega, u_{0,h}, \beta)$ in (0.11) converge, as h tends to $+\infty$, to the finite value

$$i_\infty(\Omega, u_0, \beta) = \min\Big\{ \int_\Omega f_{\rm hom}(\nabla u)dx + \int_\Omega \beta u dx + \lambda \int_\Omega |u|^p dx : \quad u \in u_0 + W_0^{1,1}(\Omega)\Big\} \tag{0.20}$$

(cf. Theorem 5.1). We emphasize that in this case we do not need to assume (0.8).

When (0.13) is replaced by the linear coerciveness assumption (0.6), the homogenization process must be carried out in the context of BV spaces, but to do this some of the above conditions must be strengthened. Thus we assume that (0.6)$\div$(0.8) and (0.16) are fulfilled, and tat, if Ω' is an open set such that $\overline{\Omega} \subseteq \Omega'$, the following conditions hold

$$u_{0,h} \text{ is constant in } \Omega' \cap \frac{1}{h}S \text{ for every } h \in \mathbb{N} \text{ and } S \in \mathcal{C}, \tag{0.21}$$

$$\text{the integrals } \int_\cdot f(hx, \nabla u_{0,h})dx \text{ are equi-absolutely continuous in } \Omega' \text{ when } h \in \mathbb{N}, \tag{0.22}$$

$$u_{0,h} \to u_0 \text{ in } L^1(\Omega'), \tag{0.23}$$

$$f_{\rm hom}(\sigma \nabla u_0) \in L^1(\Omega') \text{ for every } \sigma \in \mathbb{R}. \tag{0.24}$$

Then, we prove that $f_{\rm hom}$ is convex and finite in the whole $\mathbb{R}^n$, that (0.9) holds, and that for every $\lambda \in]0, +\infty[$, $p \in]1, \frac{n}{n-1}[$, $\beta \in L^\infty(\Omega)$ the values $i_h(\Omega, u_{0,h}, \beta)$ in (0.11) converge, as h tends to $+\infty$, to the finite value

$$i_\infty(\Omega, u_0, \beta) = \min\Big\{ \int_\Omega f_{\rm hom}(\nabla u)dx + \int_\Omega f^\infty_{\rm hom}\Big(\frac{dD^s u}{d|D^s u|}\Big) d|D^s u| + \int_{\partial\Omega} f^\infty_{\rm hom}((u_0 - u)\mathbf{n}_\Omega) d\mathcal{H}^{n-1} + \int_\Omega \beta u dx + \lambda \int_\Omega |u|^p dx : u \in BV(\Omega)\Big\} \tag{0.25}$$

(cf. Theorem 5.2).

Of course, problem in (0.25) reduces to the one in (0.20) when (0.13) is fulfilled. In particular, when $u_{0,h} \equiv 0$ for every $h \in \mathbb{N}$, the above described convergence result for the problems in (0.10) follows as corollary.

In both cases, convergence results in $L^1(\Omega)$ for the minimizing sequences of the problems in (0.11) to minimizers of $i_\infty(\Omega, u_0, \beta)$ are also proved.

In particular, the above results continue to hold if $a \in L^1_{\text{loc}}(\mathbb{R}^n)$ is Y-periodic, $q \in [1, +\infty[$, f satisfies

$$f(x,z) \leq a(x) + |z|^q \text{ for a.e. } x \in \mathbb{R}^n \text{ and every } z \in \mathbb{R}^n, \tag{0.26}$$

and if assumptions (0.15), (0.16), (0.18), respectively (0.22), (0.16), (0.24), are replaced by

$$\text{the integrals } \int_\cdot |\nabla u_{0,h}|^q dx \text{ are equi-absolutely} \qquad (0.27)$$

$$\text{continuous in } \Omega \text{ when } h \in \mathbb{N},$$

respectively by

$$\text{the integrals } \int_\cdot |\nabla u_{0,h}|^q dx \text{ are equi-absolutely} \qquad (0.28)$$

$$\text{continuous in } \Omega' \text{ when } h \in \mathbb{N}$$

(cf. corollaries 4.3 and 4.4).

Our results are obtained by exploiting De Giorgi's Γ-convergence theory, together with some recent results and techniques introduced in [11], [9], and [16].

1. Notations and preliminary results

We first recall some properties of BV spaces. We refer to [21] and [31] for a complete exposition on the matter.

Let Ω be an open set. For every $u \in BV(\Omega)$ we denote by $|Du|$ the total variation of the $\mathbb{R}^n$-valued measure Du. Moreover, according to Lebesgue Decomposition Theorem, we have

$$Du(E) = \int_E \nabla u dx + D^s u(E) \text{ for every Borel subset } E \text{ of } \Omega,$$

where we have denoted by ∇u the density of the absolutely continuous part of Du, and with $D^s u$ the singular part of Du, both with respect to Lebesgue measure.

We recall that $BV(\Omega)$ is a Banach space with norm

$$\|\cdot\|_{BV(\Omega)} \colon u \in BV(\Omega) \mapsto \int_\Omega |u| dx + |Du|(\Omega).$$

If Ω has Lipschitz boundary, $BV(\Omega)$ continuously embeds in $L^{\frac{n}{n-1}}(\Omega)$ and compactly in $L^p(\Omega)$ for every $p \in [1, \frac{n}{n-1}[$.

If Ω has Lipschitz boundary and $u \in BV(\Omega)$, then the null extension u_0 of u to $\mathbb{R}^n$ is in $BV(\mathbb{R}^n)$, and there exists a function in $L^1(\partial\Omega)$ (endowed with the $\mathcal{H}^{n-1}$ measure), called the trace of u and again denoted by u, such that $Du_0 = -u\mathbf{n}\mathcal{H}^{n-1}$ in $\partial\Omega$. As consequence we have that if Ω' is an open set such that $\overline{\Omega} \subseteq \Omega'$, $u \in BV(\Omega)$, and $v \in BV(\Omega' \setminus \overline{\Omega})$, then the function

$$w = \begin{cases} u & \text{in } \Omega \\ v & \text{in } \Omega' \setminus \overline{\Omega} \end{cases}$$

is in $BV(\Omega')$, $v - u \in L^1(\partial\Omega)$, and $Dw = (v-u)\mathbf{n}_\Omega\mathcal{H}^{n-1}$ in $\partial\Omega$.

For every Lebesgue measurable set E we denote by $|E|$ its measure.

For every $f: \mathbb{R}^n \to [0, +\infty[$ convex, we define the recession function f^∞ of f by

$$f^\infty: z \in \mathbb{R}^n \mapsto \lim_{t \to +\infty} \frac{1}{t} f(tz).$$

It is well known that f^∞ is convex, lower semicontinuous, and positively 1-homogeneous.

We now introduce the Γ-convergence theory. We refer to [17] and [14] for a complete exposition on the subject.

Let (U, τ) be a topological space satisfying the first countability axiom.

Definition 1.1. - *For every $h \in \mathbb{N}$ let $F_h: U \to [-\infty, +\infty]$. We define the $\Gamma^-(\tau)$-lower limit and the $\Gamma^-(\tau)$-upper limit of $\{F_h\}$ as*

$$\Gamma^-(\tau) \liminf_{h \to +\infty} F_h: u \in U \mapsto \inf\left\{ \liminf_{h \to +\infty} F_h(u_h) : u_h \to u \text{ in } \tau \right\},$$

and

$$\Gamma^-(\tau) \limsup_{h \to +\infty} F_h: u \in U \mapsto \inf\left\{ \limsup_{h \to +\infty} F_h(u_h) : u_h \to u \text{ in } \tau \right\}.$$

If in u one has

$$\Gamma^-(\tau) \liminf_{h \to +\infty} F_h(u) = \Gamma^-(\tau) \limsup_{h \to +\infty} F_h(u),$$

we say that in u there exists the $\Gamma^-(\tau)$-limit of $\{F_h\}$, and we define it as

$$\Gamma^-(\tau) \lim_{h \to +\infty} F_h(u) = \Gamma^-(\tau) \liminf_{h \to +\infty} F_h(u) = \Gamma^-(\tau) \limsup_{h \to +\infty} F_h(u).$$

We recall that

$$\Gamma^-(\tau)\liminf_{h\to+\infty} F_h(u) \le \Gamma^-(\tau)\liminf_{k\to+\infty} F_{h_k}(u) \le \tag{1.1}$$

$$\le \Gamma^-(\tau)\limsup_{k\to+\infty} F_{h_k}(u) \le \Gamma^-(\tau)\limsup_{h\to+\infty} F_h(u)$$

for every $\{h_k\} \subseteq \mathbb{N}$ strictly increasing, $u \in U$,

$$\Gamma^-(\tau)\liminf_{h\to+\infty} F_h \text{ and } \Gamma^-(\tau)\limsup_{h\to+\infty} F_h \tag{1.2}$$

are τ-lower semicontinuous in U.

For every $h \in \mathbb{N}$, let $F_h: U \to [-\infty, +\infty]$. We say that F_h are equicoercive if for every $c \in \mathbb{R}$ there exists a compact set $K_c \subseteq U$ such that $\cup_{h\in\mathbb{N}}\{u \in U : F_h(u) \le c\} \subseteq K_c$.

Theorem 1.2. - *For every $h \in \mathbb{N}$ let $F_h: U \to [-\infty, +\infty]$. Assume that the functionals F_h are equicoercive, and that for every $u \in U$ the limit $\Gamma^-(\tau)\lim_{h\to+\infty} F_h(u)$ exists. Then $\Gamma^-(\tau)\lim_{h\to+\infty} F_h$ attains its minimum in U, and*

$$\min\left\{\Gamma^-(\tau)\lim_{h\to+\infty} F_h(u) : u \in U\right\} = \lim_{h\to+\infty} \inf\{F_h(u) : u \in U\}.$$

Moreover, if $\lim_{h\to+\infty}\inf\{F_h(u) : u \in U\} < +\infty$, and if $\{u_h\} \subseteq U$ is such that $\lim_{h\to+\infty}(F_h(u_h) - \inf\{F_h(u) : u \in U\}) = 0$, then $\{u_h\}$ is τ-compact, and its converging subsequences converge to solutions of $\min\{\Gamma^-(\tau)\lim_{h\to+\infty} F_h(u) : u \in U\}$.

We now come to the homogenization of integral functionals.

Let $\mathcal{C}_Y$ be a collection of subsets of Y, $\mathcal{C}$ be given by (0.3), f be as in (0.4), $\{u_{0,h}\} \subseteq W^{1,\infty}_{\text{loc}}(\mathbb{R}^n)$, and let Ω be a bounded open set. Then, by (0.4), it soon follows that $\int_\Omega f(hx, \nabla u)dx < +\infty$ for every $h \in \mathbb{N}$, $u \in W^{1,\infty}_{\text{loc}}(\mathbb{R}^n)$, and let us define

$$F_h(\Omega, \cdot): u \in L^1(\Omega) \mapsto \begin{cases} \int_\Omega f(hx, \nabla u)dx & \text{if } u \in W^{1,\infty}_{\text{loc}}(\mathbb{R}^n), \\ & u \text{ constant in } \Omega \cap \frac{1}{h}S \text{ for every } S \in \mathcal{C} \\ +\infty & \text{otherwise,} \end{cases} \tag{1.3}$$

and

$$F_{0,h}(\Omega, \cdot): u \in L^1(\Omega) \mapsto \begin{cases} \int_\Omega f(hx, \nabla u)dx & \text{if } u \in u_{0,h} + W^{1,\infty}_0(\Omega), \\ & u \text{ constant in } \Omega \cap \frac{1}{h}S \text{ for every } S \in \mathcal{C} \\ +\infty & \text{otherwise.} \end{cases} \tag{1.4}$$

It is clear that

$$\Gamma^-(L^1(\Omega))\liminf_{h\to+\infty} F_h(\Omega,u)\leq \Gamma^-(L^1(\Omega))\liminf_{h\tau o+\infty} F_{0,h}(\Omega,u), \tag{1.5}$$

$$\Gamma^-(L^1(\Omega))\limsup_{h\to+\infty} F_h(\Omega,u)\leq \Gamma^-(L^1(\Omega))\limsup_{h\tau o+\infty} F_{0,h}(\Omega,u)$$

$$\text{for every bounded open set } \Omega,\ u\in L^1(\Omega).$$

The following results are proved in [11].

Lemma 1.3. - *Let f be as in (0.4), and $f_{\rm hom}$ be defined by (0.5). Assume that (0.7) holds. Then there exists $L\in[0,+\infty[$, and, for every bounded open set Ω and each compact subset K of Ω, there exist $\{\psi_h\}\subseteq W_0^{1,\infty}(\Omega)$, and $h_{\Omega,K}\in\mathbb{N}$ such that*

$$\psi_h \text{ is constant in } \frac{1}{h}S \text{ for every } S\in\mathcal{C} \text{ and } h\geq h_{\Omega,K},$$

and

$$0\leq\psi_h\leq 1 \text{ in } \Omega,\quad \psi_h=1 \text{ in } K,\quad |||\nabla\psi_h|||_{L^\infty(\Omega)}\leq\frac{L}{\operatorname{dist}(K,\partial\Omega)}$$

$$\text{for every } h\in\mathbb{N}.$$

Lemma 1.4. - *Let f be as (0.4), E be a measurable set with $|E|<+\infty$, $r>0$, and $\{m_h\}\subseteq(L^1(E))^n$ be such that $\|m_h\|_{(L^\infty(E))^n}\leq r$ for every $h\in\mathbb{N}$, and $m_h\to 0$ in $(L^1(E))^n$. Then*

$$\limsup_{h\to+\infty}\int_E f(hx,m_h)dx\leq 2^n|E|\int_Y f(y,0)dy.$$

Theorem 1.5. - *Let $\mathcal{C}_Y$ be a collection of subsets of Y, $\mathcal{C}$ be given by (0.3), f be as in (0.4), $f_{\rm hom}$ be given by (0.5), and let, for every $h\in\mathbb{N}$, F_h be defined by (1.3). Assume that (0.7) holds. Then $f_{\rm hom}$ is convex and finite in $\mathbb{R}^n$, moreover for every bounded open set Ω with Lipschitz boundary, $u\in BV(\Omega)$ the limit $\Gamma^-(L^1(\Omega))\lim_{h\to+\infty}F_h(\Omega,u)$ exists and*

$$\Gamma^-(L^1(\Omega))\lim_{h\to+\infty}F_h(\Omega,u)=\int_\Omega f_{\rm hom}(\nabla u)dx+\int_\Omega f^\infty_{\rm hom}\Big(\frac{dD^su}{d|D^su|}\Big)d|D^su|.$$

The following approximation result easily comes from Proposition 2.6 in [20].

Proposition 1.6. - *Let $g\colon \mathbb{R}^n \to [0,+\infty[$ be convex. Then for every bounded open set Ω with Lipschitz boundary, and $v \in W_0^{1,1}(\Omega)$ there exists $\{v_k\} \subseteq C_0^\infty(\Omega)$ such that $v_k \to v$ in $L^1(\Omega)$, $\nabla v_k \to \nabla v$ in $(L^1(\Omega))^n$, and*

$$\lim_{k\to+\infty} \int_\Omega g(\nabla v_k)dx = \int_\Omega g(\nabla v)dx.$$

Finally, we recall from [16] the following relaxation result in BV spaces.

Theorem 1.7. - *Let $g\colon \mathbb{R}^n \to [0,+\infty[$ be convex, $u_0 \in W_{\mathrm{loc}}^{1,1}(\mathbb{R}^n)$, Ω be a bounded open set with Lipschitz boundary, and Ω' an open set such that $\overline{\Omega} \subseteq \Omega'$. Assume that*

$$g(\sigma\nabla u_0) \in L^1(\Omega) \text{ for every } \sigma \in \mathbb{R}.$$

Then

$$\inf\Big\{ \liminf_{h\to+\infty} \int_\Omega g(\nabla u_h)dx : \{u_h\} \subseteq u_0 + C_0^1(\Omega),\ u_h \to u \text{ in } L^1(\Omega)\Big\} =$$

$$= \int_\Omega g(\nabla u)dx + \int_\Omega g^\infty\Big(\frac{dD^s u}{d|D^s u|}\Big)d|D^s u| + \int_{\partial\Omega} g^\infty((u_0 - u)\mathbf{n}_\Omega)d\mathcal{H}^{n-1}$$

$$\text{for every } u \in BV(\Omega).$$

Remark 1.8. We observe that in [16] a slightly different version of Theorem 1.7 above is proved. In fact in [16], without introducing any open set Ω', it is assumed that

$$g(\sigma\nabla u_0) \in L^1_{\mathrm{loc}}(\mathbb{R}^n) \text{ for every } \sigma \in \mathbb{R},$$

and the same representation formula of Theorem 1.7 is deduced not only for $u \in BV(\Omega)$, but for every bounded open set Ω with Lipschitz boundary, and $u \in BV(\Omega)$.

On the contrary, when the open set Ω is fixed a priori, an analysis of the results of [16] proves that the assumptions of Theorem 1.7 are sufficient in order to obtain the representation formula only for every $u \in BV(\Omega)$.

2. Estimates from above

Lemma 2.1. - *Let $\mathcal{C}_Y$ be a collection of subsets of Y, $\mathcal{C}$ be given by (0.3), f be as in (0.4), f_{hom} be given by (0.5), $\{u_{0,h}\} \subseteq W_{\mathrm{loc}}^{1,\infty}(\mathbb{R}^n)$, and $u_0 \in$*

$W^{1,1}_{\rm loc}(\mathbb{R}^n)$. For every $h \in \mathbb{N}$ let F_h and $F_{0,h}$ be defined by (1.3) and (1.4). Let Ω be a bounded open set with Lipschitz boundary, and assume that (0.7), and (0.14)÷(0.17) hold. Then

$$\Gamma^-(L^1(\Omega)) \limsup_{h\to+\infty} F_{0,h}(\Omega, u) \leq \Gamma^-(L^1(\Omega)) \limsup_{h\to+\infty} F_h(\Omega, u)$$

for every $u \in L^1(\Omega)$ such that $\operatorname{spt}(u - u_0) \subseteq \Omega$.

Proof. Let $u \in L^1(\Omega)$ be such that $\operatorname{spt}(u - u_0) \subseteq \Omega$. Clearly we can assume that $\Gamma^-(L^1(\Omega)) \limsup_{h\to+\infty} F_h(\Omega, u) < +\infty$, so that there exists $\{u_h\} \subseteq W^{1,\infty}_{\rm loc}(\mathbb{R}^n)$ such that $u_h \to u$ in $L^1(\Omega)$, u_h is constant in $\Omega \cap \frac{1}{h}S$ for every $S \in \mathcal{C}$ and $h \in \mathbb{N}$, and

$$\limsup_{h\to+\infty} \int_\Omega f(hx, \nabla u_h)dx = \Gamma^-(L^1(\Omega)) \limsup_{h\to+\infty} F_h(\Omega, u). \tag{2.1}$$

For every $k \in \mathbb{N}$, let $\chi_k : \mathbb{R} \to \mathbb{R}$ be a smooth function such that $\chi_k(t) = -(k+1)$ if $t \leq -(k+2)$, $\chi_k(t) = t$ if $-k \leq t \leq k$, $\chi_k(t) = k+1$ if $t \geq k+2$, $0 \leq \chi'_k \leq 1$, and set

$$\hat{u}_{k,h} = u_{0,h} + \chi_k(u_h - u_{0,h}), \quad \hat{u}_k = u_0 + \chi_k(u - u_0).$$

Let us fix $\varepsilon > 0$, then, since for every $k \in \mathbb{N}$ (0.17) yields

$$\limsup_{h\to+\infty} |\{x \in \Omega : |u_h(x) - u_{0,h}(x)| > k\}| \leq$$

$$\leq \frac{1}{k} \limsup_{h\to+\infty} \int_\Omega |u_h - u_{0,h}|dx = \frac{1}{k} \int_\Omega |u - u_0|dx,$$

by (0.15) we deduce the existence of $k_\varepsilon \in \mathbb{N}$ such that

$$\limsup_{h\to+\infty} \int_{\{x\in\Omega:|u_h(x)-u_{0,h}(x)|>k\}} f(hx, \nabla u_{0,h})dx \leq \varepsilon \text{ for every } k \geq k_\varepsilon. \tag{2.2}$$

Moreover, again by (0.15), let A be an open set such that $\overline{A} \subseteq \Omega$, $\operatorname{spt}(u - u_0) \subseteq A$, and

$$\limsup_{h\to+\infty} \int_{\Omega\setminus A} f(hx, \nabla u_{0,h})dx \leq \varepsilon, \tag{2.3}$$

and let $\{\psi_h\} \subseteq W^{1,\infty}_0(\Omega)$, $h_{\Omega,A} \in \mathbb{N}$ be given by Lemma 1.3 with $\psi_h = 1$ in A.

For every h, $k \in \mathbb{N}$, $t \in]0,1[$ let us now define

$$w_{t,k,h} = t^2(2-t)(\psi_h \hat{u}_{k,h} + (1-\psi_h)u_{0,h}) + (1-t)(1+t-t^2)u_{0,h},$$

$$w_{t,k} = t^2(2-t)\hat{u}_k + (1-t)(1+t-t^2)u_0,$$

$$w_t = t^2(2-t)u + (1-t)(1+t-t^2)u_0.$$

Then, by using also (0.14), it follows that, for every h, $k \in \mathbb{N}$, $t \in]0,1[$, $w_{t,k,h} \in u_{0,h} + W_0^{1,1}(\Omega)$, and that $w_{t,k,h}$ is constant in $\Omega \cap \frac{1}{h}S$ for every $S \in \mathcal{C}$ provided $h \geq h_{\Omega,A}$.

By the convexity properties of f we get

$$\int_\Omega f(hx, \nabla w_{t,k,h})dx \leq \tag{2.4}$$

$$\leq t\int_\Omega f(hx, t(2-t)(\psi_h \nabla \hat{u}_{k,h} + (1-\psi_h)\nabla u_{0,h} + (\hat{u}_{k,h} - u_{0,h})\nabla\psi_h))dx +$$

$$+(1-t)\int_\Omega f(hx, (1+t-t^2)\nabla u_{0,h})dx \leq$$

$$\leq t^2(2-t)\int_\Omega f(hx, \psi_h \nabla \hat{u}_{k,h} + (1-\psi_h)\nabla u_{0,h})dx +$$

$$+t(1-t(2-t))\int_\Omega f\Big(hx, \frac{t(2-t)}{1-t(2-t)}(\hat{u}_{k,h} - u_{0,h})\nabla\psi_h\Big)dx +$$

$$+(1-t)\int_\Omega f(hx, (1+t-t^2)\nabla u_{0,h})dx \text{ for every } h,\ k \in \mathbb{N},\ t \in]0,1[.$$

We estimate the first term in the right-hand side of (2.4). Again by the convexity of f we have successively

$$\int_\Omega f(hx, \psi_h \nabla \hat{u}_{k,h} + (1-\psi_h)\nabla u_{0,h})dx = \tag{2.5}$$

$$= \int_A f(hx, \nabla \hat{u}_{k,h})dx + \int_{\Omega\setminus A} f(hx, \psi_h \nabla \hat{u}_{k,h} + (1-\psi_h)\nabla u_{0,h})dx \leq$$

$$\leq \int_A f(hx, \nabla \hat{u}_{k,h})dx + \int_{\Omega\setminus A} \psi_h f(hx, \nabla \hat{u}_{k,h})dx +$$

$$+ \int_{\Omega\setminus A} (1-\psi_h) f(hx, \nabla u_{0,h})dx \leq$$

$$\leq \int_\Omega f(hx, \nabla \hat{u}_{k,h})dx + \int_{\Omega\setminus A} f(hx, \nabla u_{0,h})dx =$$

$$= \int_{\{x\in\Omega:|u_h(x)-u_{0,h}(x)|\geq k+2\}} f(hx,\nabla u_{0,h})dx+$$

$$+\int_{\{x\in\Omega:|u_h(x)-u_{0,h}(x)|\leq k\}} f(hx,\nabla u_h)dx+$$

$$+\int_{\{x\in\Omega:k<|u_h(x)-u_{0,h}(x)|<k+2\}} f(hx,\chi_k'(u_h-u_{0,h})\nabla u_h+(1-\chi_k'(u_h-u_{0,h}))\nabla u_{0,h})dx+$$

$$+\int_{\Omega\setminus A} f(hx,\nabla u_{0,h})dx \leq$$

$$\leq \int_{\{x\in\Omega:|u_h(x)-u_{0,h}(x)|\geq k+2\}} f(hx,\nabla u_{0,h})dx+$$

$$+\int_{\{x\in\Omega:|u_h(x)-u_{0,h}(x)|\leq k\}} f(hx,\nabla u_h)dx+$$

$$+\int_{\{x\in\Omega:k<|u_h(x)-u_{0,h}(x)|<k+2\}} \chi_k'(u_h-u_{0,h})f(hx,\nabla u_h)dx+$$

$$+\int_{\{x\in\Omega:k<|u_h(x)-u_{0,h}(x)|<k+2\}} (1-\chi_k'(u_h-u_{0,h}))f(hx,\nabla u_{0,h})dx+$$

$$+\int_{\Omega\setminus A} f(hx,\nabla u_{0,h})dx \leq \int_\Omega f(hx,\nabla u_h)dx+$$

$$+\int_{\{x\in\Omega:|u_h(x)-u_{0,h}(x)|>k\}} f(hx,\nabla u_{0,h})dx + \int_{\Omega\setminus A} f(hx,\nabla u_{0,h})dx$$

$$\text{for every } h,\ k\in\mathbb{N},\ t\in\,]0,1[.$$

Therefore, by (2.5), (2.1), (2.2), and (2.3) we obtain

$$\limsup_{h\to+\infty}\int_\Omega f(hx,\psi_h\nabla\hat{u}_{k,h}+(1-\psi_h)\nabla u_{0,h})dx \leq \tag{2.6}$$

$$\leq \Gamma^-(L^1(\Omega))\limsup_{h\to+\infty} F_h(\Omega,u)+2\varepsilon \text{ for every } k\geq k_\varepsilon,\ t\in\,]0,1[.$$

We now fix $k\geq k_\varepsilon$, $t\in\,]0,1[$, and observe that $\|\hat{u}_{k,h}-u_{0,h}\|_{L^\infty(\Omega)}\leq k+2$ for every $h\in\mathbb{N}$. Then, since $\operatorname{spt}(u-u_0)\subseteq A$, by Lemma 1.3 we obtain

$$\limsup_{h\to+\infty}\int_\Omega |(\hat{u}_{k,h}-u_{0,h})\nabla\psi_h|dx \leq \frac{L}{\operatorname{dist}(A,\partial\Omega)}\limsup_{h\to+\infty}\int_{\Omega\setminus A}|\hat{u}_{k,h}-u_{0,h}|dx =$$

$$= \frac{L}{\text{dist}(A, \partial\Omega)} \int_{\Omega\setminus A} |\hat{u}_k - u_0| dx = \frac{L}{\text{dist}(A, \partial\Omega)} \int_{\Omega\setminus A} |\chi_k(u - u_0)| dx = 0.$$

Therefore, by Lemma 1.4 applied with

$$E = \Omega,\ r = \frac{t(2-t)}{1-t(2-t)}(k+2),\ \{m_h\} = \Big\{\frac{t(2-t)}{1-t(2-t)}(\hat{u}_{k,h} - u_{0,h})\nabla\psi_h\Big\},$$

we get

$$\limsup_{h\to+\infty} \int_\Omega f\Big(hx, \frac{t(2-t)}{1-t(2-t)}(\hat{u}_{k,h} - u_{0,h})\nabla\psi_h\Big) dx \leq \tag{2.7}$$

$$\leq 2^n|\Omega| \int_Y f(y,0)dy \text{ for every } k \geq k_\varepsilon,\ t \in\,]0,1[.$$

Finally, again by the convexity of f, and (0.16) we infer

$$\limsup_{h\to+\infty} \int_\Omega f(hx, (1+t-t^2)\nabla u_{0,h}) dx \leq \tag{2.8}$$

$$\leq \limsup_{h\to+\infty} \Big\{\frac{1+t-t^2}{\sigma} \int_\Omega f(hx, \sigma\nabla u_{0,h})dx +$$

$$+\Big(1 - \frac{1+t-t^2}{\sigma}\Big) \int_\Omega f(hx,0)dx\Big\} \leq$$

$$\leq \frac{1+t-t^2}{\sigma} M + \Big(1 - \frac{1+t-t^2}{\sigma}\Big)|\Omega| \int_Y f(y,0)dy$$

$$\text{for every } t \in\,]0,1[\text{ such that } 1+t-t^2 \leq \sigma.$$

So, once we observe that for every $k \in \mathbb{N}$ and $t \in\,]0,1[$, $w_{t,k,h} \to w_{t,k}$ in $L^1(\Omega)$ as h diverges, by (2.4), (2.6), (2.7), and (2.8), we obtain

$$\Gamma^-(L^1(\Omega)) \limsup_{h\to+\infty} F_{0,h}(\Omega, w_{t,k}) \leq \limsup_{h\to+\infty} \int_\Omega f(hx, \nabla w_{t,k,h}) dx \leq \tag{2.9}$$

$$\leq t^2(2-t)\Big\{\Gamma^-(L^1(\Omega)) \limsup_{h\to+\infty} F_h(\Omega, u) + 2\varepsilon\Big\} +$$

$$+t(1-t(2-t))2^n|\Omega| \int_Y f(y,0)dy +$$

$$+(1-t)\Big\{\frac{1+t-t^2}{\sigma} M + \Big(1 - \frac{1+t-t^2}{\sigma}\Big)|\Omega| \int_Y f(y,0)dy\Big\}$$

$$\text{for every } k \geq k_\varepsilon,\ t \in\,]0,1[\text{ such that } 1+t-t^2 \leq \sigma.$$

We now remark that for every $t \in]0,1[$, $w_{t,k} \to w_t$ in $L^1(\Omega)$ as k tends to $+\infty$, and that $w_t \to u$ in $L^1(\Omega)$ as t increases to 1. Therefore by (2.9), (1.2), and (0.4) the lemma follows letting first k go to $+\infty$, then t go to 1, and finally ε decrease to 0. ■

By using Lemma 2.1 we can prove the estimate from above for the $\Gamma^-(L^1)$-upper limit of $\{F_{0,h}\}$. We start with the case dealing with Sobolev functions.

Proposition 2.2. - *Let $\mathcal{C}_Y$ be a collection of subsets of Y, $\mathcal{C}$ be given by (0.3), f be as in (0.4), f_{hom} be given by (0.5), $\{u_{0,h}\} \subseteq W^{1,\infty}_{\text{loc}}(\mathbb{R}^n)$, and $u_0 \in W^{1,1}_{\text{loc}}(\mathbb{R}^n)$. For every $h \in \mathbb{N}$ let $F_{0,h}$ be defined by (1.4). Let Ω be a bounded open set with Lipschitz boundary, and assume that (0.7), and (0.14)÷(0.18) hold. Then*

$$\Gamma^-(L^1(\Omega)) \limsup_{h\to+\infty} F_{0,h}(\Omega,u) \leq \int_\Omega f_{\text{hom}}(\nabla u)dx \text{ for every } u \in u_0 + W^{1,1}_0(\Omega).$$

Proof. Let $u \in u_0 + W^{1,1}_0(\Omega)$, and let us assume that $\int_\Omega f_{\text{hom}}(\nabla u)dx < +\infty$. Let us set $u = u_0 + v$ with $v \in W^{1,1}_0(\Omega)$, and take $t \in [0,1[$. Then, by the convexity of f_{hom}, and Proposition 1.6, there exists $\{v_{t,k}\} \subseteq C^\infty_0(\Omega)$ such that $v_{t,k} \to tv$ in $L^1(\Omega)$, $\nabla v_{t,k} \to t\nabla v$ in $(L^1(\Omega))^n$, and

$$\lim_{k\to+\infty} \int_\Omega f_{\text{hom}}(\nabla v_{t,k})dx = \int_\Omega f_{\text{hom}}(t\nabla v)dx. \tag{2.10}$$

Let us observe now that, by (0.18) and the convexity of f_{hom}, it follows that

$$\int_\Omega f_{\text{hom}}(t\nabla v)dx \leq t\int_\Omega f_{\text{hom}}(\nabla u)dx + (1-t)\int_\Omega f_{\text{hom}}\Big(\frac{t}{t-1}\nabla u_0\Big)dx < +\infty,$$

from which, together with (2.10), it is not difficult to deduce that

$$\begin{aligned}&\text{for every } t \in [0,1[\text{ the integrals } \int_\cdot f_{\text{hom}}(\nabla v_{t,k})dx \\ &\text{are equi-absolutely continuous in } \Omega \text{ when } k \in \mathbb{N}.\end{aligned} \tag{2.11}$$

For every $k \in \mathbb{N}$ we now set $u_{t,k} = u_0 + tv_{t,k}$. Then $u_{t,k} \in u_0 + C^\infty_0(\Omega)$ for every $k \in \mathbb{N}$, and $u_{t,k} \to u_0 + t^2 v$. Moreover, the convexity of f_{hom} yields

$$\begin{aligned}&f_{\text{hom}}(\nabla u_{t,k}(x)) \leq tf_{\text{hom}}(\nabla v_{t,k}(x)) + (1-t)f_{\text{hom}}\Big(\frac{1}{1-t}\nabla u_0(x)\Big) \\ &\text{for a.e. } x \in \Omega, \text{ and every } k \in \mathbb{N},\end{aligned}$$

by (2.11), Vitali-Lebesgue Theorem, and (0.18) we conclude that

$$\lim_{k\to+\infty}\int_\Omega f_{\rm hom}(\nabla u_{t,k})dx = \int_\Omega f_{\rm hom}(\nabla u_0 + t^2\nabla v)dx \qquad (2.12)$$

$$\text{for every } t \in\,]0,1[.$$

Finally, by the convexity inequality

$$\int_\Omega f_{\rm hom}(\nabla u_0 + t^2\nabla v)dx \le t^2\int_\Omega f_{\rm hom}(\nabla u)dx + (1-t^2)\int_\Omega f_{\rm hom}(\nabla u_0)dx$$

$$\text{for every } t \in\,]0,1[,$$

(0.18), and (2.12) we conclude that there exists $\{u_k\} \subseteq u_0 + C_0^\infty(\Omega)$ such that $u_k \to u$ in $L^1(\Omega)$, $\nabla u_k \to \nabla u$ in $(L^1(\Omega))^n$, and

$$\lim_{k\to+\infty}\int_\Omega f_{\rm hom}(\nabla u_k)dx = \int_\Omega f_{\rm hom}(\nabla u)dx. \qquad (2.13)$$

In conclusion, by (1.2), Lemma 2.1, Theorem 1.5, and (2.13) we obtain

$$\Gamma^-(L^1(\Omega))\limsup_{h\to+\infty} F_{0,h}(\Omega,u) \le \liminf_{k\to+\infty}\Gamma^-(L^1(\Omega))\limsup_{h\to+\infty} F_{0,h}(\Omega,u_k) \le$$

$$\le \int_\Omega f_{\rm hom}(\nabla u)dx,$$

that is the thesis. ■

Finally, we treat the case of BV functions.

Proposition 2.3. - *Let $\mathcal{C}_Y$ be a collection of subsets of Y, $\mathcal{C}$ be given by (0.3), f be as in (0.4), $f_{\rm hom}$ be given by (0.5), $\{u_{0,h}\} \subseteq W^{1,\infty}_{\rm loc}(\mathbb{R}^n)$, and $u_0 \in W^{1,1}_{\rm loc}(\mathbb{R}^n)$. For every $h \in \mathbb{N}$ let $F_{0,h}$ be defined by (1.4). Let Ω be a bounded open set with Lipschitz boundary, Ω' be an open set such that $\overline{\Omega} \subseteq \Omega'$, and assume that (0.7), (0.14)÷(0.17), and (0.24) hold. Then*

$$\Gamma^-(L^1(\Omega))\limsup_{h\to+\infty} F_{0,h}(\Omega,u) \le \qquad (2.14)$$

$$\le \int_\Omega f_{\rm hom}(\nabla u)dx + \int_\Omega f^\infty_{\rm hom}\Big(\frac{dD^s u}{d|D^s u|}\Big)d|D^s u| + \int_{\partial\Omega} f^\infty_{\rm hom}((u_0-u)\mathbf{n}_\Omega)d\mathcal{H}^{n-1}$$

$$\text{for every } u \in BV(\Omega).$$

Proof. Let $u \in BV(\Omega)$, and assume that the right-hand side of (2.14) is finite. Then, by (0.24) and Theorem 1.7 there exists $\{u_h\} \subseteq u_0 + C_0^1(\Omega)$ such that $u_h \to u$ in $L^1(\Omega)$, and

$$\lim_{h\to+\infty} \int_\Omega f_{\rm hom}(\nabla u)dx = \tag{2.15}$$

$$= \int_\Omega f_{\rm hom}(\nabla u)dx + \int_\Omega f_{\rm hom}^\infty\Big(\frac{dD^s u}{d|D^s u|}\Big)d|D^s u| + \int_{\partial\Omega} f_{\rm hom}^\infty((u_0 - u)\mathbf{n}_\Omega)d\mathcal{H}^{n-1}.$$

Consequently, by (1.2), Lemma 2.1, Theorem 1.5, and (2.15), the proposition follows as in Proposition 2.2. ∎

3. Estimates from below

Lemma 3.1. - *Let $\mathcal{C}_Y$ be a collection of subsets of Y, $\mathcal{C}$ be given by (0.3), f be as in (0.4), $f_{\rm hom}$ be given by (0.5), and $\{u_{0,h}\} \subseteq W^{1,\infty}_{\rm loc}(\mathbb{R}^n)$. For every $h \in \mathbb{N}$ let $F_{0,h}$ be defined by (1.4). Let Ω be a bounded open set with Lipschitz boundary, and assume that (0.7) holds. Then*

$$\Gamma^-(L^1(\Omega)) \liminf_{h\to+\infty} F_{0,h}(\Omega, u) \geq \int_\Omega f_{\rm hom}(\nabla u)dx + \int_\Omega f_{\rm hom}^\infty\Big(\frac{dD^s u}{d|D^s u|}\Big)d|D^s u|$$

for every $u \in BV(\Omega)$.

Proof. Follows from (1.5), and Theorem 1.5. ∎

It is clear that Lemma 3.1 does not fit well with the corresponding estimate from above given by Proposition 2.3. To do this, we need to assume (0.8) and, for a given bounded open set Ω, to consider another open set Ω' such that $\overline{\Omega} \subseteq \Omega'$, for which the following conditions hold

$$u_{0,h} \text{ is constant in } (\Omega' \setminus \overline{\Omega}) \cap \frac{1}{h}S \text{ for every } h \in \mathbb{N} \text{ and } S \in \mathcal{C}, \tag{3.1}$$

$$\text{the integrals } \int_\cdot f(hx, \nabla u_{0,h})dx \text{ are equi-absolutely} \tag{3.2}$$

$$\text{continuous in } \Omega' \setminus \overline{\Omega} \text{ when } h \in \mathbb{N},$$

$$u_{0,h} \to u_0 \text{ in } L^1(\Omega' \setminus \overline{\Omega}). \tag{3.3}$$

Proposition 3.2. - *Let $\mathcal{C}_Y$ be a collection of subsets of Y, $\mathcal{C}$ be given by (0.3), f be as in (0.4), $f_{\rm hom}$ be given by (0.5), $\{u_{0,h}\} \subseteq W^{1,\infty}_{\rm loc}(\mathbb{R}^n)$, and*

$u_0 \in W^{1,1}_{\mathrm{loc}}(\mathbb{R}^n)$. For every $h \in \mathbb{N}$ let $F_{0,h}$ be defined by (1.4). Let Ω be a bounded open set with Lipschitz boundary, Ω' be an open set such that $\overline{\Omega} \subseteq \Omega'$, and assume that (0.7), (0.8), and (3.1)÷(3.3) hold. Then

$$\Gamma^-(L^1(\Omega)) \liminf_{h\to+\infty} F_{0,h}(\Omega,u) \geq$$

$$\geq \int_\Omega f_{\mathrm{hom}}(\nabla u)dx + \int_\Omega f^\infty_{\mathrm{hom}}\Big(\frac{dD^s u}{d|D^s u|}\Big)d|D^s u| + \int_{\partial\Omega} f^\infty_{\mathrm{hom}}((u_0-u)\mathbf{n}_\Omega)d\mathcal{H}^{n-1}$$

for every $u \in BV(\Omega)$.

Proof. Let $u \in BV(\Omega)$, and assume that $\Gamma^-(L^1(\Omega)) \liminf_{h\to+\infty} F_{0,h}(\Omega,u) < +\infty$, then there exist $\{u_h\} \subseteq W^{1,\infty}_{\mathrm{loc}}(\mathbb{R}^n)$, and $\{h_k\} \subseteq \mathbb{N}$ strictly increasing such that $u_h \in u_{0,h} + W^{1,\infty}_0(\Omega)$ for every $h \in \mathbb{N}$, $u_h \to u$ in $L^1(\Omega)$, u_{h_k} is constant in $\Omega \cap \frac{1}{h_k}S$ for every $S \in \mathcal{C}$ and $k \in \mathbb{N}$, and

$$\Gamma^-(L^1(\Omega)) \liminf_{h\to+\infty} F_{0,h}(\Omega,u) = \lim_{k\to+\infty} \int_\Omega f(h_k x, \nabla u_{h_k})dx. \tag{3.4}$$

Let $\varepsilon > 0$, and let, by (3.2), $A \subseteq \Omega'$ be a bounded open set with Lipschitz boundary such that $\overline{\Omega} \subseteq A$, and

$$\limsup_{h\to+\infty} \int_{A\setminus\overline{\Omega}} f(hx, \nabla u_{0,h})dx < \varepsilon. \tag{3.5}$$

Then, by (3.3), (3.1), and (1.1), it is clear that

$$u_h \to w = \begin{cases} u & \text{in } \Omega \\ u_0 & \text{in } A\setminus\Omega \end{cases} \text{ in } L^1(A),$$

and that u_{h_k} is constant in $A \cap \frac{1}{h_k}S$ for every $S \in \mathcal{C}$ and $k \in \mathbb{N}$. In fact, given $S \in \mathcal{C}$ and $k \in \mathbb{N}$, this comes trivially if $\frac{1}{h_k}S \subseteq \Omega$, whilst, if $\frac{1}{h_k}S \cap (\mathbb{R}^n \setminus \Omega) \neq \emptyset$, it follows once we observe that (0.8) implies the coincidence of the constant value taken by u_{h_k} in $\Omega \cap \frac{1}{h_k}S$ with the one taken by u_{0,h_k} in $\frac{1}{h_k}S$. Moreover,

$$\liminf_{k\to+\infty} \int_A f(h_k x, \nabla u_{h_k})dx \geq \Gamma^-(L^1(A)) \liminf_{k\to+\infty} F_{h_k}(A,w) \geq \tag{3.6}$$

$$\geq \Gamma^-(L^1(A)) \liminf_{h\to+\infty} F_h(A,w).$$

By combining (3.4) with (3.6), (3.5), and Lemma 3.1, it results

$$\Gamma^-(L^1(\Omega))\liminf_{h\to+\infty} F_{0,h}(\Omega,u) = \tag{3.7}$$

$$= \liminf_{k\to+\infty}\int_A f(h_k x, \nabla u_{h_k})dx - \limsup_{k\to+\infty}\int_{A\setminus\Omega} f(h_k x, \nabla u_{0,h_k})dx \geq$$

$$\geq \int_A f_{\mathrm{hom}}(\nabla w)dx + \int_A f^\infty_{\mathrm{hom}}\Big(\frac{dD^s w}{d|D^s w|}\Big)d|D^s w| - \varepsilon \geq$$

$$\geq \int_\Omega f_{\mathrm{hom}}(\nabla u)dx + \int_\Omega f^\infty_{\mathrm{hom}}\Big(\frac{dD^s u}{d|D^s u|}\Big)d|D^s u| +$$

$$+ \int_{\partial\Omega} f^\infty_{\mathrm{hom}}\Big(\frac{dD^s w}{d|D^s w|}\Big)d|D^s w| - \varepsilon.$$

By (3.7) the proposition follows as ε tends to 0, once we recall that $Dw = (u_0 - u)\mathcal{H}^{n-1}$ in $\partial\Omega$. ■

4. Representation results for homogenized functionals

In this section we collect the previously obtained estimates to get some integral representation results, in Sobolev and BV spaces, for the $\Gamma^-(L^1)$-limit of the functionals in (1.4).

Theorem 4.1. - *Let $\mathcal{C}_Y$ be a collection of subsets of Y, $\mathcal{C}$ be given by (0.3), f be as in (0.4), f_{hom} be given by (0.5), $\{u_{0,h}\} \subseteq W^{1,\infty}_{\mathrm{loc}}(\mathbb{R}^n)$, and $u_0 \in W^{1,1}_{\mathrm{loc}}(\mathbb{R}^n)$. For every $h \in \mathbb{N}$ let $F_{0,h}$ be defined by (1.4). Let Ω be a bounded open set with Lipschitz boundary, and assume that (0.7), and (0.14)÷(0.18) hold. Then f_{hom} is convex and finite in $\mathbb{R}^n$, for every $u \in u_0 + W^{1,1}_0(\Omega)$ the limit $\Gamma^-(L^1(\Omega))\lim_{h\to+\infty} F_{0,h}(\Omega,u)$ exists and*

$$\Gamma^-(L^1(\Omega))\lim_{h\to+\infty} F_{0,h}(\Omega,u) = \int_\Omega f_{\mathrm{hom}}(\nabla u)dx.$$

Proof. The properties of f_{hom} follow from Theorem 1.5, whilst the remaining part of the theorem by Proposition 2.2, and Lemma 3.1. ■

Theorem 4.2. - *Let $\mathcal{C}_Y$ be a collection of subsets of Y, $\mathcal{C}$ be given by (0.3), f be as in (0.4), f_{hom} be given by (0.5), $\{u_{0,h}\} \subseteq W^{1,\infty}_{\mathrm{loc}}(\mathbb{R}^n)$, and $u_0 \in W^{1,1}_{\mathrm{loc}}(\mathbb{R}^n)$. For every $h \in \mathbb{N}$ let $F_{0,h}$ be defined by (1.4). Let Ω be a bounded open set with Lipschitz boundary, Ω' be an open set such that $\overline{\Omega} \subseteq \Omega'$, and assume that (0.7), (0.8), (0.16), and (0.21)÷(0.24) hold.*

Then f_{hom} is convex and finite in $\mathbb{R}^n$, for every $u \in BV(\Omega)$ the limit $\Gamma^-(L^1(\Omega)) \lim_{h\to+\infty} F_{0,h}(\Omega, u)$ exists and

$$\Gamma^-(L^1(\Omega)) \lim_{h\to+\infty} F_{0,h}(\Omega, u) =$$

$$= \int_\Omega f_{\text{hom}}(\nabla u)dx + \int_\Omega f_{\text{hom}}^\infty\Big(\frac{dD^s u}{d|D^s u|}\Big)d|D^s u| + \int_{\partial\Omega} f_{\text{hom}}^\infty((u_0 - u)\mathbf{n}_\Omega)d\mathcal{H}^{n-1}.$$

Proof. The properties of f_{hom} follow from Theorem 1.5, whilst the remaining part of the theorem from Proposition 2.3, and Proposition 3.2. ■

In particular, by the above results we deduce the following corollaries.

Corollary 4.3. - *Let $\mathcal{C}_Y$ be a collection of subsets of Y, $\mathcal{C}$ be given by (0.3), f be as in the first two lines of (0.4), f_{hom} be given by (0.5), $\{u_{0,h}\} \subseteq W^{1,\infty}_{\text{loc}}(\mathbb{R}^n)$, $u_0 \in W^{1,1}_{\text{loc}}(\mathbb{R}^n)$, $a \in L^1_{\text{loc}}(\mathbb{R}^n)$ be Y-periodic, and $q \in [1, +\infty[$. For every $h \in \mathbb{N}$ let $F_{0,h}$ be defined by (1.4). Let Ω be a bounded open set with Lipschitz boundary, and assume that (0.7), (0.14), (0.17), (0.26), and (0.27) hold. Then f_{hom} is convex and finite in $\mathbb{R}^n$, for every $u \in u_0 + W^{1,1}_0(\Omega)$ the limit $\Gamma^-(L^1(\Omega)) \lim_{h\to+\infty} F_{0,h}(\Omega, u)$ exists and*

$$\Gamma^-(L^1(\Omega)) \lim_{h\to+\infty} F_{0,h}(\Omega, u) = \int_\Omega f_{\text{hom}}(\nabla u)dx.$$

Proof. We prove that the assumptions of Theorem 4.1 are fulfilled. It is clear that (0.15) and (0.16) follow trivially from (0.26) and (0.27).

Let us observe now that (0.17), (0.27), and the $L^1(\Omega)$-lower semicontinuity of the functional $v \in W^{1,q}(\Omega) \mapsto \int_\Omega |\nabla v|^q dx$ imply that $u_0 \in W^{1,q}(\Omega)$. By virtue of this, since by (0.26) and Jensen's inequality it results

$$f_{\text{hom}}(z) \leq \int_Y a dy + \inf\Big\{ \int_Y |z + \nabla v|^q dy : v \in W^{1,\infty}_{\text{loc}}(\mathbb{R}^n),$$

$$v \ Y\text{-periodic}, \ u_z + v \text{ constant in } S \text{ for every } S \in \mathcal{C}\Big\} = \int_Y a dy + |z|^q$$

$$\text{for every } z \in \mathbb{R}^n,$$

condition (0.18) too follows, in fact

$$f_{\text{hom}}(\sigma\nabla u_0) \leq \int_Y a dy + \sigma^q |\nabla u_0|^q \in L^1(\Omega) \text{ for every } \sigma \in \mathbb{R}.$$

In conclusion, the corollary follows from Theorem 4.1. ■

Corollary 4.4. - *Let $\mathcal{C}_Y$ be a collection of subsets of Y, $\mathcal{C}$ be given by (0.3), f be as in the first two lines of (0.4), f_{hom} be given by (0.5), $\{u_{0,h}\} \subseteq W^{1,\infty}_{\text{loc}}(\mathbb{R}^n)$, $u_0 \in W^{1,1}_{\text{loc}}(\mathbb{R}^n)$, $a \in L^1_{\text{loc}}(\mathbb{R}^n)$ be Y-periodic, and $q \in [1, +\infty[$. For every $h \in \mathbb{N}$ let $F_{0,h}$ be defined by (1.4). Let Ω be a bounded open set with Lipschitz boundary, Ω' be an open set such that $\overline{\Omega} \subseteq \Omega'$, and assume that (0.7), (0.8), (0.21), (0.23), (0.26), and (0.28) hold. Then f_{hom} is convex and finite in $\mathbb{R}^n$, for every $u \in BV(\Omega)$ the limit $\Gamma^-(L^1(\Omega)) \lim_{h\to+\infty} F_{0,h}(\Omega, u)$ exists and*

$$\Gamma^-(L^1(\Omega)) \lim_{h\to+\infty} F_{0,h}(\Omega, u) =$$

$$= \int_\Omega f_{\text{hom}}(\nabla u)dx + \int_\Omega f^\infty_{\text{hom}}\Big(\frac{dD^s u}{d|D^s u|}\Big) d|D^s u| + \int_{\partial\Omega} f^\infty_{\text{hom}}((u_0 - u)\mathbf{n}_\Omega) d\mathcal{H}^{n-1}.$$

Proof. Follows the outlines of the proof of Corollary 4.3, but by exploiting Theorem 4.2 in place of Theorem 4.1. ■

5. The convergence of infima and of minimizing sequences

In the present section we derive, from the results of §4, the statements on the convergence of infima and of minimizing sequences.

Theorem 5.1. - *Let $\mathcal{C}_Y$ be a collection of subsets of Y, $\mathcal{C}$ be given by (0.3), f be as in (0.4), f_{hom} be given by (0.5), $\phi: \mathbb{R}^n \to [0, +\infty[$ be convex and verifying (0.12), $\{u_{0,h}\} \subseteq W^{1,\infty}_{\text{loc}}(\mathbb{R}^n)$, and $u_0 \in W^{1,1}_{\text{loc}}(\mathbb{R}^n)$. Let Ω be a bounded open set with Lipschitz boundary, and assume that (0.7), and (0.13)÷(0.18) hold. Then f_{hom} is convex and finite in $\mathbb{R}^n$, and (0.19) holds.*

For every $\lambda \in [0, +\infty[$, $p \in [1, \frac{n}{n-1}[$, and $\beta \in L^\infty(\Omega)$ let $i_h(\Omega, u_{0,h}, \beta)$ be given by (0.11), and $i_\infty(\Omega, u_0, \beta)$ by (0.20). Then the following facts hold:
a) $i_\infty(\Omega, u_0, \beta)$ has at least one solution, and

$$i_\infty(\Omega, u_0, \beta) = \lim_{h\to+\infty} i_h(\Omega, u_{0,h}, \beta) < +\infty;$$

b) if for every $h \in \mathbb{N}$ $u_h \in u_{0,h} + W^{1,\infty}_0(\Omega)$ is constant in $\Omega \cap \frac{1}{h}S$ for every $S \in \mathcal{C}$, and

$$\lim_{h\to+\infty} \Big\{ \int_\Omega f(hx, \nabla u_h)dx + \int_\Omega \beta u_h dx + \lambda \int_\Omega |u_h|^p dx - i_h(\Omega, u_{0,h}, \beta) \Big\} = 0,$$

then $\{u_h\}$ is compact in $L^1(\Omega)$, and its converging subsequences converge to solutions of (0.20).

Proof. The properties of f_{hom} come by Theorem 1.5.

For every $h \in \mathbb{N}$ let $F_{0,h}(\Omega, \cdot)$ be given by (1.4), and let λ, p and β be as above. Then, by (0.13), and Theorem 4.1, it soon follows that for every $u \in L^1(\Omega)$ the limit $\Gamma^-(L^1(\Omega)) \lim_{h\to+\infty}\{F_{0,h}(\Omega,u)+\int_\Omega \beta u dx+\lambda \int_\Omega |u|^p dx\}$ exists and

$$\Gamma^-(L^1(\Omega)) \lim_{h\to+\infty} \left\{F_{0,h}(\Omega,u) + \int_\Omega \beta u dx + \lambda \int_\Omega |u|^p dx\right\} = \tag{5.1}$$

$$= \begin{cases} \int_\Omega f_{\mathrm{hom}}(\nabla u)dx + \int_\Omega \beta u dx + \lambda \int_\Omega |u|^p dx & \text{if } u \in u_0 + W_0^{1,1}(\Omega) \\ +\infty & \text{if } u \in L^1(\Omega) \setminus (u_0 + W_0^{1,1}(\Omega)). \end{cases}$$

In fact, if $u \in u_0 + W_0^{1,1}(\Omega)$ is such that $\int_\Omega f_{\mathrm{hom}}(\nabla u)dx + \int_\Omega \beta u dx + \lambda \int_\Omega |u|^p dx < +\infty$, then, Theorem 4.1 provides $\{u_h\}$ such that $\{u_h - u_{0,h}\} \subseteq W_0^{1,\infty}(\Omega)$, $u_h \to u$ in $L^1(\Omega)$, and

$$\int_\Omega f_{\mathrm{hom}}(\nabla u)dx = \limsup_{h\to+\infty} \int_\Omega f(hx, \nabla u_h)dx.$$

Now, by making use of (0.13), $\{u_h\}$ turns out to be bounded in $W^{1,1}(\Omega)$. Consequently, the Rellich-Kondrachov Compactness Theorem implies that $u_h \to u$ in $L^p(\Omega)$. Therefore

$$\Gamma^-(L^1(\Omega)) \limsup_{h\to+\infty} \left\{F_{0,h}(\Omega,u) + \int_\Omega \beta u dx + \lambda \int_\Omega |u|^p dx\right\} \leq \tag{5.2}$$

$$\leq \limsup_{h\to+\infty} \left\{ \int_\Omega f(hx, \nabla u_h)dx + \int_\Omega \beta u_h dx + \lambda \int_\Omega |u_h|^p dx\right\} =$$

$$= \int_\Omega f_{\mathrm{hom}}(\nabla u)dx + \int_\Omega \beta u dx + \lambda \int_\Omega |u|^p dx \text{ for every } u \in u_0 + W_0^{1,1}(\Omega).$$

Conversely, if $u \in L^1(\Omega)$ satisfies $\Gamma^-(L^1(\Omega)) \liminf_{h\to+\infty}\{F_{0,h}(\Omega,u) + \int_\Omega \beta u dx + \lambda \int_\Omega |u|^p dx\} < +\infty$, there exists $\{u_h\} \subseteq L^1(\Omega)$ and $\{h_k\} \subseteq \mathbb{N}$ strictly increasing, such that $u_h \to u$ in $L^1(\Omega)$, $\{u_{h_k} - u_{0,h_k}\} \subseteq W_0^{1,\infty}(\Omega)$, and

$$\Gamma^-(L^1(\Omega)) \liminf_{h\to+\infty} \left\{F_{0,h}(\Omega,u) + \int_\Omega \beta u dx + \lambda \int_\Omega |u|^p dx\right\} = \tag{5.3}$$

$$= \liminf_{k\to+\infty} \left\{ \int_\Omega f(h_k x, \nabla u_{h_k})dx + \int_\Omega \beta u_{h_k} dx + \lambda \int_\Omega |u_{h_k}|^p dx\right\}.$$

Then, (0.13) and the de la Vallée Poussin Compactness Theorem provide that $u \in W^{1,1}(\Omega)$ and that $u_{h_k} \to u$ in weak-$W^{1,1}(\Omega)$. Moreover the weak-$W^{1,1}(\Omega)$-closedness of $W_0^{1,1}(\Omega)$ also implies that actually $u \in u_0 + W_0^{1,1}(\Omega)$.

Consequently, (5.3), again the Rellich-Kondrachov Compactness Theorem, and Theorem 4.1 imply that

$$\Gamma^-(L^1(\Omega)) \liminf_{h\to+\infty} \left\{ F_{0,h}(\Omega, u) + \int_\Omega \beta u dx + \lambda \int_\Omega |u|^p dx \right\} \geq \tag{5.4}$$

$$\geq \liminf_{k\to+\infty} \int_\Omega f(h_k x, \nabla u_{h_k}) dx + \int_\Omega \beta u dx + \lambda \int_\Omega |u|^p dx \geq$$

$$\geq \begin{cases} \int_\Omega f_{\mathrm{hom}}(\nabla u) dx + \int_\Omega \beta u dx + \lambda \int_\Omega |u|^p dx & \text{if } u \in u_0 + W_0^{1,1}(\Omega) \\ +\infty & \text{if } u \in L^1(\Omega) \setminus (u_0 + W_0^{1,1}(\Omega)). \end{cases}$$

By (5.2) and (5.4), equality (5.1) follows.

Finally, again by (0.13), it soon follows that the functionals $u \in L^1(\Omega) \mapsto F_{0,h}(\Omega, u) + \int_\Omega \beta u dx + \lambda \int_\Omega |u|^p dx$ are equicoercive in $L^1(\Omega)$.

In conclusion, the theorem follows from (5.1), and Theorem 1.2, once we observe that, by (0.22), (0.23), (0.13), and the Sobolev Imbedding Theorem

$$\lim_{h\to+\infty} i_h(\Omega, u_{0,h}, \beta) \leq$$

$$\leq \limsup_{h\to+\infty} \int_\Omega f(hx, \nabla u_{0,h}) dx + \int_\Omega \beta u_{0,h} dx + \lambda \int_\Omega |u_{0,h}|^p dx < +\infty.$$

■

Theorem 5.2. - *Let $\mathcal{C}_Y$ be a collection of subsets of Y, $\mathcal{C}$ be given by (0.3), f be as in (0.4), f_{hom} be given by (0.5), $\{u_{0,h}\} \subseteq W^{1,\infty}_{\mathrm{loc}}(\mathbb{R}^n)$, and $u_0 \in W^{1,1}_{\mathrm{loc}}(\mathbb{R}^n)$. Let Ω be a bounded open set with Lipschitz boundary, Ω' be an open set such that $\overline{\Omega} \subseteq \Omega'$, and assume that (0.7), (0.8), (0.6), (0.16), and (0.21)÷(0.24) hold. Then f_{hom} is convex and finite in $\mathbb{R}^n$, and (0.9) holds.*

For every $\lambda \in]0, +\infty[$, $p \in]1, \frac{n}{n-1}[$, and $\beta \in L^\infty(\Omega)$ let $i_h(\Omega, u_{0,h}, \beta)$ be given by (0.11), and $i_\infty(\Omega, u_0, \beta)$ by (0.25). Then the following facts hold:
a) $i_\infty(\Omega, u_0, \beta)$ has at least one solution, and

$$i_\infty(\Omega, u_0, \beta) = \lim_{h\to+\infty} i_h(\Omega, u_{0,h}, \beta) < +\infty;$$

b) if for every $h \in \mathbb{N}$ $u_h \in u_{0,h} + W_0^{1,\infty}(\Omega)$ $\{u_h\}$ is constant in $\Omega \cap \frac{1}{h} S$ for every $S \in \mathcal{C}$, and

$$\lim_{h\to+\infty} \left\{ \int_\Omega f(hx, \nabla u_h) dx + \int_\Omega \beta u_h dx + \lambda \int_\Omega |u_h|^p dx - i_h(\Omega, u_{0,h}, \beta) \right\} = 0,$$

then $\{u_h\}$ is compact in $L^1(\Omega)$, and its converging subsequences converge to solutions of (0.25).

Proof. The proof is similar to the one of Theorem 5.1. We develop it by emphasizing the main differences.

The properties of $f_{\hom}$ come by Theorem 1.5.

For every $h \in \mathbb{N}$ let $F_{0,h}(\Omega, \cdot)$ be given by (1.4), and let λ, p and β be as above. Then, by (0.6), and Theorem 4.2, it soon follows that for every $u \in L^1(\Omega)$ the limit $\Gamma^-(L^1(\Omega)) \lim_{h\to+\infty}\{F_{0,h}(\Omega, u)+\int_\Omega \beta u dx+\lambda \int_\Omega |u|^p dx\}$ exists and

$$\Gamma^-(L^1(\Omega)) \lim_{h\to+\infty} \left\{F_{0,h}(\Omega, u) + \int_\Omega \beta u dx + \lambda \int_\Omega |u|^p dx\right\} = \tag{5.5}$$

$$= \begin{cases} \int_\Omega f_{\hom}(\nabla u)dx + \int_\Omega f^\infty_{\hom}\left(\frac{dD^s u}{d|D^s u|}\right)d|D^s u| + \\ \int_{\partial\Omega} f^\infty_{\hom}((u_0 - u)\mathbf{n}_\Omega)d\mathcal{H}^{n-1} + \int_\Omega \beta u dx + \lambda \int_\Omega |u|^p dx \quad \text{if } u \in BV(\Omega) \\ +\infty \quad \text{if } u \in L^1(\Omega) \setminus BV(\Omega). \end{cases}$$

In fact, if $u \in BV(\Omega)$ satisfies $\int_\Omega f_{\hom}(\nabla u)dx+\int_\Omega f^\infty_{\hom}\left(\frac{dD^s u}{d|D^s u|}\right)d|D^s u|+\int_{\partial\Omega} f^\infty_{\hom}((u_0-u)\mathbf{n}_\Omega)d\mathcal{H}^{n-1}+\int_\Omega \beta u dx+\lambda \int_\Omega |u|^p dx < +\infty$, then, Theorem 4.2 provides $\{u_h\}$ such that $\{u_h - u_{0,h}\} \subseteq W^{1,\infty}_0(\Omega)$, $u_h \to u$ in $L^1(\Omega)$, and

$$\int_\Omega f_{\hom}(\nabla u)dx + \int_\Omega f^\infty_{\hom}\left(\frac{dD^s u}{d|D^s u|}\right)d|D^s u| + \int_{\partial\Omega} f^\infty_{\hom}((u_0-u)\mathbf{n}_\Omega)d\mathcal{H}^{n-1} =$$

$$= \limsup_{h\to+\infty} \int_\Omega f(hx, \nabla u_h)dx.$$

Now, since $\lambda > 0$ and $p > 1$, by (0.6) it follows that $\{u_h\}$ is bounded in $BV(\Omega)$. Consequently, the compact embedding of $BV(\Omega)$ in $L^p(\Omega)$, implies that $u_h \to u$ in $L^p(\Omega)$. Therefore

$$\Gamma^-(L^1(\Omega)) \limsup_{h\to+\infty} \left\{F_{0,h}(\Omega, u) + \int_\Omega \beta u dx + \lambda \int_\Omega |u|^p dx\right\} \leq \tag{5.6}$$

$$\leq \limsup_{h\to+\infty} \left\{ \int_\Omega f(hx, \nabla u_h)dx + \int_\Omega \beta u_h dx + \lambda \int_\Omega |u_h|^p dx\right\} =$$

$$= \int_\Omega f_{\hom}(\nabla u)dx + \int_\Omega f^\infty_{\hom}\left(\frac{dD^s u}{d|D^s u|}\right)d|D^s u| + \int_{\partial\Omega} f^\infty_{\hom}((u_0-u)\mathbf{n}_\Omega)d\mathcal{H}^{n-1} +$$

$$+ \int_\Omega \beta u dx + \lambda \int_\Omega |u|^p dx \text{ for every } u \in BV(\Omega).$$

Conversely, if $u \in L^1(\Omega)$ satisfies $\Gamma^-(L^1(\Omega)) \liminf_{h\to+\infty}\{F_{0,h}(\Omega,u) + \int_\Omega \beta u dx + \lambda \int_\Omega |u|^p dx\} < +\infty$, there exists $\{u_h\} \subseteq L^1(\Omega)$ and $\{h_k\} \subseteq \mathbb{N}$ strictly increasing, such that $u_h \to u$ in $L^1(\Omega)$, $\{u_{h_k} - u_{0,h_k}\} \subseteq W_0^{1,\infty}(\Omega)$, and

$$\Gamma^-(L^1(\Omega)) \liminf_{h\to+\infty} \left\{F_{0,h}(\Omega,u) + \int_\Omega \beta u dx + \lambda \int_\Omega |u|^p dx\right\} = \tag{5.7}$$

$$= \liminf_{k\to+\infty} \left\{ \int_\Omega f(h_k x, \nabla u_{h_k}) dx + \int_\Omega \beta u_{h_k} dx + \lambda \int_\Omega |u_{h_k}|^p dx\right\}.$$

Then, since $\lambda > 0$ and $p > 1$, (0.6) and again the compact embedding of $BV(\Omega)$ in $L^p(\Omega)$ provide that $u_{h_k} \to u$ in $L^p(\Omega)$, and that $u \in BV(\Omega)$. Consequently, (5.7) and Theorem 4.2 imply that

$$\Gamma^-(L^1(\Omega)) \liminf_{h\to+\infty} \left\{F_{0,h}(\Omega,u) + \int_\Omega \beta u dx + \lambda \int_\Omega |u|^p dx\right\} \geq \tag{5.8}$$

$$\geq \liminf_{k\to+\infty} \int_\Omega f(h_k x, \nabla u_{h_k}) dx + \int_\Omega \beta u dx + \lambda \int_\Omega |u|^p dx \geq$$

$$\geq \begin{cases} \int_\Omega f_{\mathrm{hom}}(\nabla u) dx + \int_\Omega f^\infty_{\mathrm{hom}}\left(\frac{dD^s u}{d|D^s u|}\right) d|D^s u| + \\ \int_{\partial\Omega} f^\infty_{\mathrm{hom}}((u_0 - u)\mathbf{n}_\Omega) d\mathcal{H}^{n-1} + \int_\Omega \beta u dx + \lambda \int_\Omega |u|^p dx & \text{if } u \in BV(\Omega) \\ +\infty & \text{if } u \in L^1(\Omega) \setminus BV(\Omega). \end{cases}$$

By (5.6) and (5.8), equality (5.5) follows.

Finally, since $\lambda > 0$ and $p > 1$, again by (0.6) it soon follows that the functionals $u \in L^1(\Omega) \mapsto F_{0,h}(\Omega,u) + \int_\Omega \beta u dx + \lambda \int_\Omega |u|^p dx$ are equicoercive in $L^1(\Omega)$.

In conclusion, the theorem follows from (5.5), and Theorem 1.2, once we observe that, by (0.22), (0.23), (0.6), and Sobolev Imbedding Theorem

$$\lim_{h\to+\infty} i_h(\Omega, u_{0,h}, \beta) \leq$$

$$\leq \limsup_{h\to+\infty} \int_\Omega f(hx, \nabla u_{0,h}) dx + \int_\Omega \beta u_{0,h} dx + \lambda \int_\Omega |u_{0,h}|^p dx < +\infty.$$

■

6. An example

This section provides an example proving that the previous homogenization results can be no more true when a single Dirichlet boundary datum is taken into account.

Example 6.1. Let $n = 2$, $\mathcal{C}_Y = \{]0,1/2[^2\}$, $f\colon (x,z) \in \mathbb{R}^n \times \mathbb{R}^n \mapsto |z|^2$, and f_{hom} be given by (0.5).

Let us first prove that

$$\frac{4}{3}|z|^2 \leq f_{\text{hom}}(z) \leq 2|z|^2 \text{ for every } z \in \mathbb{R}^n, \tag{6.1}$$

from which (0.7) follows trivially.

To do this, we observe that for every $v \in W^{1,\infty}_{\text{loc}}(\mathbb{R}^n)$ such that v is Y-periodic and $u_z + v$ is constant in S for every $S \in \mathcal{C}$, by Jensen's inequality, Gauss-Green Theorem, and the periodicity of v, it results

$$\int_Y |z+\nabla v|^2 dy = \int_{Y\setminus]0,1/2[^2} |z+\nabla v|^2 dy \geq$$

$$\geq \frac{4}{3}\left|\int_{Y\setminus]0,1/2[^2}(z+\nabla v)dy\right|^2 = \frac{4}{3}\left|\int_{\partial(Y\setminus]0,1/2[^2)}(u_z+v)\mathbf{n}_{Y\setminus]0,1/2[^2}d\mathcal{H}^{n-1}\right|^2 =$$

$$= \frac{4}{3}|z|^2 \text{ for every } z \in \mathbb{R}^n,$$

from which the left-hand side of (6.1) follows. In order to prove the remaining inequality, we take $z \in \mathbb{R}^n$ and define $v^{(z)}$ as the Y-periodic extension of

$$v^{(z)}\colon (x_1,x_2) \in Y \mapsto \begin{cases} -x_1z_1 - x_2z_2 & \text{if } (x_1,x_2) \in]0,1/2[\times]0,1/2[\\ -(1-x_1)z_1 - x_2z_2 & \text{if } (x_1,x_2) \in]1/2,1[\times]0,1/2[\\ -(1-x_1)z_1 - (1-x_2)z_2 & \\ & \text{if } (x_1,x_2) \in]1/2,1[\times]1/2,1[\\ -x_1z_1 - (1-x_2)z_2 & \text{if } (x_1,x_2) \in]0,1/2[\times]1/2,1[, \end{cases}$$

then $v^{(z)} \in W^{1,\infty}_{\text{loc}}(\mathbb{R}^n)$, $u_z + v^{(z)}$ is constant in S for every $S \in \mathcal{C}$, and

$$f_{\text{hom}}(z) \leq \int_Y |z+\nabla v^{(z)}|^2 dy = 2|z|^2.$$

Let us now take $\Omega = Y$, and $u_0\colon x \in \mathbb{R}^n \mapsto x_1$, then it is clear that all the assumptions of Theorem 5.1, except (0.14), are fulfilled with $u_{0,h} = u_0$ for every $h \in \mathbb{N}$, and that for every $h \in \mathbb{N}$ there cannot be functions in $u_0 + W^{1,\infty}_0(Y)$ that are constant in each set of the type $(\frac{i}{h},0)+]0,\frac{1}{2h}[^2$, or $(0,\frac{i}{h})+]\frac{1}{2h},0[^2$, for some $i \in \{0,\ldots,h-1\}$. Consequently, for every $\beta \in L^\infty(Y)$, and $\lambda \in \mathbb{R}$, it results

$$\inf\left\{\int_Y f(hx,\nabla u)dx + \int_Y \beta u dx + \lambda\int_Y |u|dx :\right.$$

$$u \in u_0 + W_0^{1,\infty}(\Omega),\ u \text{ constant in } \Omega \cap \frac{1}{h}S \text{ for every } S \in \mathcal{C}\Big\} = +\infty,$$

whilst, by (6.1),

$$\min\Big\{\int_\Omega f_{\text{hom}}(\nabla u)dx + \int_\Omega \beta u dx + \lambda \int_\Omega |u|dx : u \in u_0 + W_0^{1,1}(\Omega)\Big\} < +\infty.$$

References

[1] Attouch H., Variational Convergence for Functions and Operators, Pitman (1984).

[2] Bakhvalov N.S., Panasenko G.P., Homogenization: Averaging Processes in Periodic Media. Math. Appl. (Soviet Ser.) 36, Kluwer Academic Publishers (1989).

[3] Bensoussan A., Lions J.L., Papanicolaou G., Asymptotic Analysis for Periodic Structures. Stud. Math. Appl. 5, North Holland (1978).

[4] Braides A., Defranceschi A., Homogenization of Multiple Integrals. Oxford Lecture Ser. Math. Appl. 12, Oxford University Press (1998).

[5] Braides A., Garroni A., Homogenization of periodic nonlinear media with stiff and soft inclusions, Math. Models Methods Appl. Sci. 5, (1995), 543-564.

[6] Brezis H., Sibony M., Équivalence de deux inéquations variationnelles et applications, Arch. Rational Mech. Anal. 41, (1971), 254-265.

[7] Briane M., Homogenization in some weakly connected domains, Ricerche Mat. 47, (1998), 51-94.

[8] Buttazzo G., Semicontinuity, Relaxation and Integral Representation in the Calculus of Variations. Pitman Res. Notes Math. Ser. 207, Longman Scientific & Technical (1989).

[9] Carbone L., De Arcangelis R., On the relaxation of Dirichlet minimum problems for some classes of unbounded integral functionals, Ricerche Mat. 48-Suppl., (1999), 347-372; special issue in memory of Ennio De Giorgi.

[10] Carbone L., De Arcangelis R., Unbounded Functionals in the Calculus of Variations. Chapman & Hall/CRC Monogr. Surv. Pure Appl. Math. 125, Chapman & Hall/CRC (2001).

[11] Carbone L., De Arcangelis R., De Maio U., Homogenization of media with periodically distributed conductors, Asymptotic Anal. 23, (2000), 157-194.

[12] Carbone L., Salerno S., Some remarks on a problem of homogenization with fixed traces, Appl. Anal. 22, (1986), 71-86.

[13] Cioranescu D., Saint Jean Paulin J., Homogenization in open sets with holes, J. Math. Anal. Appl. 71, (1979), 590-607.

[14] Dal Maso G., An Introduction to Γ-Convergence. Progr. Nonlinear Differential Equations Appl. 8, Birkhäuser-Verlag (1993).

[15] De Arcangelis R., Gaudiello A., Paderni G., Some cases of homogenization of linearly coercive gradient constrained variational problems, Math. Models Methods Appl. Sci 6, (1996), 901-940.

[16] De Arcangelis R., Trombetti C., On the relaxation of some classes of Dirichlet minimum problems, Comm. Partial Differential Equations 24, (1999), 975-1006.

[17] De Giorgi E., Franzoni T., Su un tipo di convergenza variazionale, Atti Accad. Naz. Lincei Rend. Cl. Sci. Fis. Mat. Natur. (8), 58, (1975), 842-850.

[18] Duvaut G., Lanchon H., Sur la solution du problème de torsion élastoplastique d'une barre cylindrique de section quelconque, C. R. Acad. Sci. Paris Sér. I Math. 264, (1967), 520-523.

[19] Duvaut G., Lions J.L., Inequalities in Mechanics and Physics. Grundlehren Math. Wiss. 219, Springer-Verlag (1976).

[20] Ekeland I., Temam R., Convex Analysis and Variational Problems. Stud. Math. Appl. 1, North-Holland American Elsevier (1976).

[21] Giusti E., Minimal Surfaces and Functions of Bounded Variations. Monogr. Math. 80, Birkhäuser-Verlag (1984).

[22] Glowinski R., Lanchon H., Torsion élastoplastique d'une barre cylindrique de section multiconnexe, J. Mécanique 12, (1973), 151-171.

[23] Goffman C., Serrin J., Sublinear functions of measures and variational integrals, Duke Math. J. 31, (1964), 159-178.

[24] Jikov V.V., Kozlov S.M., Oleinik O.A., Homogenization of Differential Operators and Integral Functionals. Springer-Verlag (1994).

[25] Lanchon H., Torsion élastoplastique d'une barre cylindrique de section simplement ou multiplement connexe, J. Mécanique 13, (1974), 267-320.

[26] Murat F., Tartar L., H-convergence, in "Topics in the Mathematical Modelling of Composite Materials", A. Cherkaev and R. Kohn editors, Progr. Nonlinear Differential Equations Appl. 31, Birkäuser-Verlag (1997), 21-44.

[27] Rauch J., Taylor M., Electrostatic screening, J. Math. Phys. 16, (1975), 284-288.

[28] Sanchez-Palencia E., Nonhomogeneous Media and Vibration Theory. Lecture Notes in Phys. 127, Springer-Verlag (1980).

[29] Ting T.W., Elastic-plastic torsion of simply connected cylindrical bars, Indiana Univ. Math. J. 20, (1971), 1047-1076.

[30] Zhikov V.V., Averaging of functionals of the calculus of variations and elasticity theory, Math. USSR Izv. 29, (1987), 33-66.
[31] Ziemer W.P., Weakly Differentiable Functions. Grad. Texts in Math. 120, Springer-Verlag, (1989).

Riccardo De Arcangelis
Università di Napoli "Federico II"
Dipartimento di Matematica e Applicazioni "Renato Caccioppoli"
via Cintia, Complesso Monte S. Angelo
80126 Napoli
Italy
E-mail: dearcang@unina.it

Studies in Mathematics and its Applications, Vol. 31
D. Cioranescu and J.L. Lions (Editors)

Chapter 13

TRANSPORT OF TRAPPED PARTICLES IN A SURFACE POTENTIAL

P. DEGOND

1. Introduction

The purpose of this paper is the rigorous derivation of a diffusion model for trapped particles in a surface potential subject to collisions with the surface. This physical situation frequently occurs in plasma as well as semiconductor physics. For instance, surface discharges on satellite solar generators induce severe damage when metallic parts are biased (by the plasma environment) at a lower potential than the neighbouring dielectric parts. In this configuration, the electrons emitted from the metallic parts are confined along the surface of the dielectric parts by the surrounding potential and undergo many collisions with the dielectric surface (see e.g. [12] for more details and references). A similar situation occurs for electrons moving inside the inversion layer of an MOS transistor: they are confined in the vicinity of the insulator surface by the potential profile of the MOS structure and undergo many collisions with the insulator surface in their way from the source to the drain. It has been estimated that the electron mobility in an MOS inversion layer is strongly reduced (by a factor 3 at least) compared with that of a bulk semiconductor because of collisions with the oxide surface (see e.g. [10]).

Particles colliding with a surface may suffer a large number of physical processes, like specular reflection, attachment to the wall, secondary emission, etc (see e.g. [25]). Our assumption is that particles are reemitted with their incident energy and random directions (in other words, the interaction with the surface is supposed elastic). The resulting macroscopic dynamics is a diffusion process in an extended space consisting of the position and energy coordinates of the particles. The associated model is often referred to in the literature as the "SHE" model (for Spherical Harmonics Expansion, a terminology arising from its early derivation by physicists [26]). It consists of a diffusion equation for the energy distribution function $F(x, \varepsilon)$ (i.e. the number density of the particles at position x and energy ε). Although seemingly too simple, the assumption of elastic collisions often leads to a satisfactory description of the macroscopic dynamics. For instance, the diffusion of particles between two plates subject to a large magnetic field has been correctly described by such an approach [14], [15], both qualitatively and quantitatively.

The diffusion dynamics induced by particle-surface interaction has first been investigated in the case of inelastic interactions in [1], [2] (see also [11] for references) where the collisionless motion of particles in between two plates is considered. In the case of inelastic collisions, the resulting diffusion process takes place in the usual position space. However, for this problem, [8] shows that the diffusivity is actually infinite and that a time rescaling is necessary to recover a finite diffusivity. This is the so-called 'anomalous diffusion' phenomenon. An alternate proof is given in [21]. In our framework of elastic collisions and confinement by a surface potential, anomalous diffusion can occur if the potential is too flat in the vicinity of the surface. An example of such a situation will be given.

The SHE model naturally appears in semiconductor modeling [26], [20] and gas discharge modeling [25], when the collisions of the particles against the medium (phonons in semiconductors, atoms in gas discharges) are assumed elastic. In the physics literature, the SHE model is heuristically derived from a truncated expansion of the Boltzmann equation in Spherical Harmonics [26], [20], [25]. In [18] (in the case of relaxation operators) and in [5] (in the general case), the SHE model is shown to derive from a diffusion approximation of the Boltzmann equation (see also [13] for a review and references). The derivation of the SHE model in connection with surface collision mechanisms is reported in [14], [15] to describe the confinement of particles between two plates subject to a large transverse magnetic field. The SHE model also appears in [6] and [17] in the context of scattering by plane interfaces (a situation which bears some similarities with surface scattering), with applications to the modeling of semiconductor superlattices.

The diffusion approximation (or diffusion limit) is the singular limit of the Boltzmann equation when the ratio α of the collision mean free path to the characteristic length scale is small, and when simultaneously, the ratio of the mean time between collisions to the characteristic time scale is of order α^2. This limit has been extensively studied in the literature, in various contexts (see e.g. [23], [7], [4] for neutron transport, [3] for radiative transfer and in [24], [22] for semiconductors). In the present work, we investigate the diffusion limit for the collisionless Boltzmann equation describing trapped particles in a surface potential subject to elastic collisions with the surface. We shall focus on a *formal* result, leaving the rigorous proof of convergence to future work. We refer to [16] and [6] for rigourous convergence results in related frameworks.

The summary of the paper is as follows. In Section 2, the starting point of our analysis i.e. the kinetic equation describing trapped particles in the vicinity of a solid wall is established. The main result of the paper i.e. the formal convergence of the kinetic model to the SHE model as the small parameter α tends to 0 is also stated. Section 3 is devoted to important properties of the boundary collision operator. Then, the main theorem is proved in Section 4. Explicit computations of the diffusion coefficients are possible in a certain number of examples listed in Section 5. Finally, in Section 6, we investigate various extensions of the model.

2. A kinetic model for the transport of trapped particles in a surface potential

We consider the domain $\Omega = \mathbb{R}^2 \times (-\infty, 0]$ and denote the position vector by $x = (x_1, x_2, z) \in \Omega$. A solid wall is supposed to be located along the plane $\Gamma = \partial\Omega = \{z = 0\} \cong \mathbb{R}^2$. Our aim is to describe the motion of the particles subject to a force field in Ω and to collisions with the solid boundary at Γ. We decompose x into its perpendicular and parallel components $\underline{x}$ and z relative to the boundary Γ: $\underline{x} = (x_1, x_2) \in \mathbb{R}^2$ and $z \in (-\infty, 0]$. Similarly, the velocity vector of the particles $v = (v_1, v_2, v_z) \in \mathbb{R}^3$ is decomposed into $v = (\underline{v}, v_z)$ with $\underline{v} = (v_1, v_2) \in \mathbb{R}^2$ and $v_z \in \mathbb{R}$. We suppose that the particles are subject to a potential force field $F(x) = (F_1, F_2, F_z)$, where $F(x) = -\nabla_x \phi(x)$ and ϕ is the potential. Again, denoting by $\underline{F} = (F_1, F_2)$, we have $\underline{F} = -\nabla_{\underline{x}} \phi$, where $\nabla_{\underline{x}}$ denotes the 2-dimensional gradient with respect to $\underline{x}$.

The set of particles is described by its number density in phase space (x, v) (or distribution function). It is assumed to evolve without collisions in the domain Ω. Therefore, the distribution function $f(x, v, t)$ satisfies the collisionless Boltzmann equation for $(x, v) \in \Omega \times \mathbb{R}^3$:

$$\partial_t f + v \cdot \nabla_x f + F \cdot \nabla_v f = 0\,, \tag{2.1}$$

where the mass is set equal to 1. Expression (2.1) simply expresses the conservation of the number density in phase space along the classical trajectories of the particles. We now need to specify the boundary condition at the boundary $\Gamma \times \mathbb{R}^3$ of the phase space. We introduce the traces (or boundary values) of f on $\Gamma \times \mathbb{R}^3$ according to:

$$\gamma(f) = f|_{\{z=0\}}\,, \quad \gamma^{\pm}(f) = f|_{\{z=0,\, \pm v_z > 0\}}\,.$$

$\gamma^+(f)$ is the distribution function of the particles exiting the domain Ω at the boundary Γ, while $\gamma^-(f)$ is that of the incoming particles. A well-posed kinetic problem requires the prescription of the incoming trace. Here, we suppose that it is a function of the outgoing trace through an operator $\mathcal{B}$ which expresses the interaction of the particles with the solid boundary:

$$\gamma^-(f) = \mathcal{B}(\gamma^+(f)) \,. \tag{2.2}$$

We shall consider an expression of $\mathcal{B}$ as follows:

$$\begin{aligned} \mathcal{B}\phi(v) &= \int_{\omega' \in \mathbb{S}^2_+} K(\underline{x}, |v|^2/2; \omega' \to \omega)\phi(|v|\omega')|\omega'_z| d\omega', \\ & \quad \forall v \in \mathbb{R}^3 \text{ s.t. } v_z < 0 \,, \end{aligned} \tag{2.3}$$

where $v = |v|\omega$, $|\omega| = 1$ is the decomposition of v into spherical coordinates, $\mathbb{S}^2 = \{\omega \in \mathbb{R}^3, |\omega| = 1\}$ is the unit shere and $\mathbb{S}^2_\pm = \{\omega \in \mathbb{S}^2, \pm\omega_z > 0\}$ are the two hemispheres. Expression (2.3) models an elastic bounce against the solid wall, with a random deflection of the velocity direction. K is an integral kernel which describes the reflection law of the particles. Indeed, $K(\underline{x}, |v|^2/2, \omega' \to \omega)|\omega_z| d\omega$ is the probability for a particle hitting the wall at point $\underline{x}$ with velocity $v' = |v|\omega'$ to be reflected with the same $|v|$ and velocity direction ω in the solid angle $d\omega$. We shall comment on the physical relevence of this model in Section3. Note that $\mathcal{B} = \mathcal{B}(\underline{x}, |v|^2/2)$ operates on the angular variable ω only while $\underline{x}$ and $|v|$ are mere parameters.

We are interested in the situation where the potential ϕ confines the particles close to the boundary Γ. Therefore, we assume

Hypothesis 2.1 - *For any given fixed $\underline{x} \in \Gamma$, the function $z \in (-\infty, 0] \to \phi(\underline{x}, z)$ is decreasing and*

$$\lim_{z \to -\infty} \phi(\underline{x}, z) = +\infty.$$

This hypothesis could somehow be relaxed but we shall leave this point to future work. We define:

$$\phi_0(\underline{x}) := \phi(\underline{x}, z = 0) \,, \quad \psi(\underline{x}, z) = \phi(\underline{x}, z) - \phi_0(\underline{x}) \,. \tag{2.4}$$

We note that $\psi(\underline{x}, 0) = 0$.

Under hypothesis 2.1, the motion of the particles can easily be pictured. After some excursion in the domain of negative z, the particles are attracted back to the wall by the potential field. As they hit the wall, they are reflected elastically into the domain and a new ballistic loop begins. As a result of this succession of bounces, the large scale dynamics in the direction parallel to the boundary should resemble a diffusion process. Therefore, we are led to rescale the longitudinal space coordinate $\underline{x}$ and the time t according to $\underline{x}' = \alpha \underline{x}$, $t' = \alpha^2 t$, where $\alpha \ll 1$ is the (supposedly small) ratio of the size of a typical ballistic loop to the longitudinal length scale under consideration (e.g. the size of the device). The time rescaling like α^2 is typical of a diffusion process. After rescaling, the problem reads (dropping the primes for clarity):

$$\begin{aligned} &\alpha \partial_t f^\alpha + \left(\underline{v} \cdot \nabla_{\underline{x}} - \nabla_{\underline{x}} \phi \cdot \nabla_{\underline{v}}\right) f^\alpha \\ &\qquad + \frac{1}{\alpha}\left(v_z \frac{\partial}{\partial z} - \frac{\partial \phi}{\partial z}\frac{\partial}{\partial v_z}\right) f^\alpha = 0 \,, \qquad (2.5) \\ &\gamma^-(f^\alpha) = \mathcal{B}(\gamma^+(f^\alpha)) \,. \qquad (2.6) \end{aligned}$$

In (2.5), it is implicitely assumed that the potential has only macroscopic scale variations in the direction parallel to the boundary Γ. We supplement the problem with an initial condition

$$f^\alpha(x, v, t = 0) = f_I(x, v) \,, \quad \forall (x, v) \in \Omega \times \mathbb{R}^3 \,. \qquad (2.7)$$

The goal of this work is to investigate the limit $\alpha \to 0$ in problem (2.5), (2.6), (2.7). Defining the measure

$$\delta_\varepsilon(x, v) = \delta\left(\frac{1}{2}|v|^2 + \psi(\underline{x}, z) - \varepsilon\right) \,, \qquad (2.8)$$

where δ denotes the delta measure, we aim at the following *formal* result:

Theorem 2.1 - *Under hypotheses 2.1 and 3.2, the solution f^α of problem (2.5), (2.6), (2.7) formally converges as $\alpha \to 0$ to a function*

$$F(\underline{x}, \frac{1}{2}|v|^2 + \psi(\underline{x}, z), t),$$

where the function $F(\underline{x}, \varepsilon, t)$ satisfies the following diffusion ('SHE') problem:

$$\begin{aligned} &N(\underline{x}, \varepsilon)\partial_t F + \left(\nabla_{\underline{x}} - \nabla_{\underline{x}}\phi_0 \, \partial_\varepsilon\right) \cdot \underline{J} = 0 \,, \qquad (2.9) \\ &\underline{J}(\underline{x}, \varepsilon, t) = -D(\underline{x}, \varepsilon) \left(\nabla_{\underline{x}} - \nabla_{\underline{x}}\phi_0 \, \partial_\varepsilon\right) F \,, \qquad (2.10) \\ &F(\underline{x}, \varepsilon, t = 0) = F_I(\underline{x}, \varepsilon) = \frac{1}{N(\underline{x}, \varepsilon)} \int f_I \, \delta_\varepsilon(x, v) \, dv dz, \qquad (2.11) \\ &\underline{J}(\underline{x}, \varepsilon = 0, t) = 0 \,. \qquad (2.12) \end{aligned}$$

The coefficients N (the 'density-of-states') and D (the 'diffusivity tensor') are given by:

$$N(\underline{x}, \varepsilon) = \int \delta_\varepsilon(x, v) \, dv \, dz \,, \quad D(\underline{x}, \varepsilon) = \int \underline{\chi} \otimes \underline{v} \, \delta_\varepsilon(x, v) \, dv \, dz \,. \qquad (2.13)$$

The vector function $\underline{\chi} = \underline{\chi}(x, v) = (\chi_1, \chi_2)$ *is the unique solution of the 'auxiliary problem'*

$$\left(-v_z \frac{\partial}{\partial z} + \frac{\partial \psi}{\partial z} \frac{\partial}{\partial v_z}\right) \underline{\chi} = \underline{v}, \quad \gamma^+(\underline{\chi}) = \mathcal{B}^*(\gamma^-(\underline{\chi})), \tag{2.14}$$

satisfying $\int \underline{\chi} \delta_\varepsilon(x, v) dv\, dz = 0$, *for all* $(\underline{x}, \varepsilon)$, *where* $\mathcal{B}^*$ *is the operator adjoint of* $\mathcal{B}$. *Equation (2.14) must be understood componentwise (i.e.* χ_i *is the solution with* v_i *at the right-hand-side of (2.14) for* $i = 1, 2$*).*

This theorem describes the large scale dynamics of the particles in terms of a distribution function which only depends on time, on the position coordinate along the surface $\underline{x}$ and on the 'energy variable' $\varepsilon = \frac{1}{2}|v|^2 + \psi(\underline{x}, z)$. ε is the kinetic energy of a particle when it hits the wall (since then, $z = 0$ and $\psi(\underline{x}, 0) = 0$). When $z \neq 0$, ε is not the kinetic energy of the particle, but rather, its total energy in the potential ψ. Note that it is not the actual total energy in the 'true' potential ϕ.

The quantity $N(\underline{x}, \varepsilon) F(\underline{x}, \varepsilon, t) d\underline{x} d\varepsilon$ is the number of particles in a certain elementary volume of phase space. This volume is the cartesian product of the interval of width $d\varepsilon$ about the energy ε and the infinite cylindrical volume consisting of all half lines parallel to z originating from the surface element $d\underline{x}$. The quantity $N(\underline{x}, \varepsilon) d\underline{x} d\varepsilon$ is the geometrical volume of this region of phase space. It bear similarities with the 'density-of-states' in solid-state physics (hence justifying our terminology). Similarly, $\underline{J}(\underline{x}, \varepsilon, t)$ is the parallel component to the wall of the total flux carried by the particles belonging to the same elementary volume. In view of this interpretation, the prescription of the initial datum by (2.11) is natural.

System (2.9), (2.10) is a diffusion equation in the extended space $(\underline{x}, \varepsilon) \in \mathbb{R}^2 \times [0, \infty)$ known as the 'SHE' model (see bibliography in Section 1). in the present case, the drift is due to the gradient of the potential ϕ_0 at the wall, which results in an average force $F_0 = \nabla_{\underline{x}} \phi_0$ in the direction parallel to the wall. At variance, both the density-of-states and the diffusivity only depend on the transverse profile of the potential described by ψ. Therefore, the longitudinal and transverse variations of the confining potential ϕ play different roles in the limit model. This decoupling is due to our assumption on the different scales of the longitudinal and transverse motions.

Formula (2.13) can be understood by means of the coarea formula [19]. For a function $\varphi(z, v)$, we have, omitting the dependence with respect to $\underline{x}$:

$$\int \varphi(z, v)\, \delta_\varepsilon(x, v)\, dv\, dz$$
$$= \int_{Z(\varepsilon)}^{0} \left(\int_{\mathbb{S}^2} \varphi(z, \sqrt{2(\varepsilon - \psi(z))}\, \omega)\, d\omega \right) \sqrt{2(\varepsilon - \psi(z))}\, dz, \tag{2.15}$$

where $Z(\varepsilon) = Z(\underline{x}, \varepsilon)$ is the unique root (by virtue of hypothesis 2.1) of the equation $\psi(z) = \varepsilon$. $Z(\varepsilon)$ is also the turning point of a particle starting from the wall with directional energy $\frac{1}{2}|v_z|^2 = \varepsilon$. In particular, formula (2.15) applied to (2.9) gives

$$N(\underline{x}, \varepsilon) = 4\pi \int_{Z(\varepsilon)}^{0} \sqrt{2(\varepsilon - \psi(z))}\, dz < \infty\,. \tag{2.16}$$

The auxiliary function $\underline{\chi}$ provides the response of the microscopic scale to gradients of the macroscopic variables. System (2.14) is constructed by means of the operator adjoint of the leading order term (in factor of $1/\alpha$) of the original equation (2.5). This system has multiple solutions and the integral constraint is used to single out one particular solution. It is easy to see (cf (2.13)) that the limit model does not depend on this choice.

3. The boundary collision operator

We denote by $L^2(\mathbb{S}^2_\pm)$ the space of square integrable functions φ from $\mathbb{S}^2_\pm$ into $\mathbb{R}$ for the measure $|\omega_z| d\omega$. We suppose that $\mathcal{B}(\underline{x}, \varepsilon)$ is a bounded operator from $L^2(\mathbb{S}^2_+)$ onto $L^2(\mathbb{S}^2_-)$. Furthermore, we suppose, for all $\underline{x} \in \mathbb{R}^2$, $\varepsilon > 0$:

Hypothesis 3.2 - *(i) flux conservation:*

$$\int_{\omega \in \mathbb{S}^2_-} K(\underline{x}, \varepsilon; \omega' \to \omega) |\omega_z|\, d\omega = 1\,, \quad \forall \omega' \in \mathbb{S}^2_+\,, \tag{3.1}$$

(ii) Reciprocity:

$$K(\underline{x}, \varepsilon; \omega' \to \omega) = K(\underline{x}, \varepsilon; -\omega \to -\omega')\,, \quad \forall \omega \in \mathbb{S}^2_-,\ \omega' \in \mathbb{S}^2_+\,. \tag{3.2}$$

(iii) Positivity: $K(\underline{x}, \varepsilon; \omega' \to \omega) > 0$, for all $\omega \in \mathbb{S}^2_-$, $\omega' \in \mathbb{S}^2_+$.
(iv) $\mathcal{B}(\underline{x}, \varepsilon)$ is a compact operator from $L^2(\mathbb{S}^2_+)$ onto $L^2(\mathbb{S}^2_-)$.

Relation (3.1) expresses the conservation of the normal flux of particles of given energy ε at the point $\underline{x}$ of the boundary. Indeed, the magnitude of the incoming normal flux J_z^- is :

$$J_z^-(\underline{x}, \varepsilon, t) = \int_{\omega \in \mathbb{S}^2_-} f(\underline{x}, z = 0, |v|\omega)\, |v| |\omega_z|\, d\omega\,.$$

Using boundary conditions (2.6), we have:

$$\begin{aligned} J_z^- &= \int_{\omega \in \mathbb{S}_-^2} \left(\int_{\omega' \in \mathbb{S}_+^2} K(\omega' \to \omega) f(|v|\omega') \, |\omega_z'| d\omega' \right) |v||\omega_z| d\omega \\ &= \int_{\omega' \in \mathbb{S}_+^2} f(|v|\omega') \, |v||\omega_z'| d\omega' \; = \; J_z^+ \, , \end{aligned} \tag{3.3}$$

where the integrations with respect to ω and ω' have been exchanged, hypothesis (3.1) has been used and J_z^+ denotes the outgoing normal flux.

This hypothesis is crucial for the validity of theorem 2.1. In practice, this assumption is not rigorously satisfied because surface interactions, like secondary emission [25] can induce jumps of the electron energy. However, over a wide energy range, these jumps are small and the interaction process can be approximated by the elastic process (2.3) with good accuracy. Even if the discrepancy between the two processes is large, the limit SHE model (2.9)-(2.10) can still give a good qualitative picture of the phenomena (see an example in a slightly different context in [15]).

The reciprocity relation (3.2) is a macroscopic effect of the time reversibility of elementary particle-surface interactions. It may not always be true (see [11] for references about the validity of the reciprocity relation), but it considerably simplifies the analysis and we shall take it for granted. By the change ω ito $-\omega$ in (3.1) and the use of (3.2), we easily deduce the following 'normalization' relation:

$$\int_{\omega \in \mathbb{S}_+^2} K(\underline{x}, \varepsilon; \omega' \to \omega) |\omega_z'| \, d\omega' = 1 \, , \quad \forall \omega \in \mathbb{S}_-^2 \, , \tag{3.4}$$

The normalization relation, together with the positivity of K has important consequences. First, from (2.3), $\mathcal{B}\varphi(\omega)$ for $\omega \in \mathbb{S}_-^2$ appears as a convex mean value of $\varphi(\omega')$ over $\mathbb{S}_+^2$. From Jensen's inequality, we deduce the following inequality, which bears some similarities with the Darrozès-Guiraud inequality in gas dynamics [11]:

$$\int_{\mathbb{S}_-^2} |\mathcal{B}\varphi(\omega)|^2 \, |\omega_z| d\omega \leq \int_{\mathbb{S}_+^2} |\varphi(\omega)|^2 \, |\omega_z| d\omega \, , \tag{3.5}$$

Therefore, the operator norm of $\mathcal{B}$ in the space $\mathcal{L}(L^2(\mathbb{S}_+^2), L^2(\mathbb{S}_-^2))$ of bounded operators from $L^2(\mathbb{S}_+^2)$ to $L^2(\mathbb{S}_-^2)$ is less or equal one: $||\mathcal{B}|| \leq 1$. Furthermore, from (3.4), if φ is constant over $\mathbb{S}_+^2$, $\mathcal{B}\varphi$ is constant over $\mathbb{S}_-^2$. Therefore, $||\mathcal{B}|| = 1$.

Hypothesis (iv) is a regularity hypothesis on the integral kernel K. Its consequences are best expressed if we introduce the specular reflection operator $\mathcal{J}$ which operates from $L^2(\mathbb{S}_+^2)$ to $L^2(\mathbb{S}_-^2)$ according to $\mathcal{J}\varphi(\omega) = \varphi(\omega^*)$, where $\omega = (\underline{\omega}, \omega_z)$ and $\omega^* = (\underline{\omega}, -\omega_z)$. Its adjoint $\mathcal{J}^*$ is the specular reflection operator from $L^2(\mathbb{S}^2_-)$ to $L^2(\mathbb{S}^2_+)$. Then, the operator $\mathcal{B}\mathcal{J}^*$ and its adjoint $\mathcal{J}\mathcal{B}^*$ operate on $L^2(\mathbb{S}^2_-)$ while $\mathcal{B}^*\mathcal{J}$ and $\mathcal{J}^*\mathcal{B}$ operate on $L^2(\mathbb{S}^2_+)$. By hypothesis (iv), the operators $\mathrm{I} - \mathcal{B}\mathcal{J}^*$, $\mathrm{I} - \mathcal{B}^*\mathcal{J}$, etc, are Fredholm operators [9]. By using Fredholm theory and the Krein-Rutman theorem [9], it is possible to prove the following (see [16], [6]):

Lemma 3.2 - *(i) The Null-Spaces $N(\mathrm{I} - \mathcal{B}^*\mathcal{J})$ and $N(\mathrm{I} - \mathcal{J}^*\mathcal{B})$ are spanned by the constant functions on $\mathbb{S}^2_-$ and $\mathbb{S}^2_+$ respectively.*
(ii) The range $R(\mathrm{I} - \mathcal{B}^*\mathcal{J})$ is such that $R(\mathrm{I} - \mathcal{B}^*\mathcal{J}) = N(\mathrm{I} - \mathcal{J}^*\mathcal{B})^\perp$. Equivalently, the equation $(\mathrm{I} - \mathcal{B}^*\mathcal{J})f = g$ has a solution f if and only if $\int_{\mathbb{S}^2_+} g(\omega)|\omega_z|d\omega = 0$. Then, the solution f is unique under the additional constraint $\int_{\mathbb{S}^2_+} f(\omega)|\omega_z|d\omega = 0$.

4. The macroscopic limit

In this section, we give a proof of theorem 2.1. The proof is divided in four steps. First, we show that the limit is of the form $F(\underline{x}, \frac{1}{2}|v|^2 + \psi(\underline{x}, z), t)$. Second, we prove the continuity equation (2.9). Third, we show the existence of the auxiliary function $\underline{\chi}$. Finally, we prove the current equation (2.10).

4.1. The limit is a function of the energy

We suppose that $f^\alpha \to f$ as $\alpha \to 0$ in adequate functional spaces which will not be precised any further in the present work. Formally, from (2.5), (2.6), f is a solution of the problem

$$\left(v_z \frac{\partial}{\partial z} - \frac{\partial \phi}{\partial z} \frac{\partial}{\partial v_z} \right) f = 0 \,, \quad \gamma^-(f) = \mathcal{B}(\gamma^+(f)) \,. \tag{4.1}$$

We show the

Lemma 4.3 - *The solutions of (4.1) are of the form*

$$f(x, v, t) = F(\underline{x}, \frac{1}{2}|v|^2 + \psi(\underline{x}, z), t) \,. \tag{4.2}$$

Proof. Obviously, $\underline{x}$ and t are mere parameters in problem (4.1) and will be omitted in the remainder of the discussion. For $(z, \underline{v})$ given, we introduce the change of variables

$$u_z = \mathrm{sgn}(v_z)\sqrt{v_z^2 + 2\psi(z)} \,, \quad v_z = \mathrm{sgn}(u_z)\sqrt{u_z^2 - 2\psi(z)} \,, \tag{4.3}$$

between $v_z \in \mathbb{R}$ and

$$u_z \in D_z = (-\infty, -\sqrt{2\psi(z)}] \cup [\sqrt{2\psi(z)}, \infty).$$

We note $u = (\underline{v}, u_z)$ and $f(z,v) = \bar{f}(z,u)$. Notice that $u_z = v_z$ for $z = 0$. Then, $\bar{f}$ satisfies the problem:

$$v_z(z,u_z)\frac{\partial}{\partial z}\bar{f} = 0\,, \quad \gamma^-(\bar{f}) = \mathcal{B}(\gamma^+(\bar{f}))\,. \tag{4.4}$$

Let $u \in \mathbb{R}^3$ be given and z be such that $u_z \in D_z$, i.e. $z \in [Z(u_z^2/2), 0]$, where $Z(\varepsilon)$ is the turning point as defined in section 2.1. We shall denote it simply by Z when the context is clear. Integrating the first equation (4.4) with respect to z between Z and 0, we obtain for $u_z > 0$ that $\bar{f}(0,\underline{v},u_z) = \bar{f}(Z,\underline{v},u_z)$ and $\bar{f}(0,\underline{v},-u_z) = \bar{f}(Z,\underline{v},-u_z)$. But $\bar{f}(Z,\underline{v},u_z) = \bar{f}(Z,\underline{v},-u_z) = f(Z,\underline{v},0)$ since Z is the turning point. So, $f(0,\underline{v},u_z) = f(0,\underline{v},-u_z)$, or equivalently: $\gamma^+(f) = \mathcal{J}^*\gamma^-(f)$.

Inserting this equation into the second equation (4.1), we deduce that $(\mathrm{I} - \mathcal{B}\mathcal{J}^*)(\gamma^-(f)) = 0$, which, by virtue of lemma 3.2, implies that $\gamma^-(f) = F(|v|^2/2)$. Then, going back to (4.4) and integrating it in z on the interval $[z,0]$ easily leads to (4.2) and concludes the proof. ■

4.2. The continuity equation

For finite $\alpha > 0$, we define the average density F^α and current J^α by

$$\begin{aligned} F^\alpha(\underline{x},\varepsilon,t) &= \frac{1}{N(\underline{x},\varepsilon)}\int f^\alpha\,\delta_\varepsilon(x,v)\,dvdz, \\ J^\alpha(\underline{x},\varepsilon,t) &= \frac{1}{\alpha}\int \underline{v} f^\alpha\,\delta_\varepsilon(x,v)\,dvdz\,. \end{aligned} \tag{4.5}$$

We prove:

Lemma 4.4 - *F^α and J^α satisfy the continuity equation (2.9).*

Proof. We multply (2.5) by $\delta_\varepsilon(x,v)$ and integrate it with respect to v and z. First, using Green's formula, we have, recalling (2.8) and omitting the arguments in the delta measures:

$$\begin{aligned} &\int \left(v_z\frac{\partial}{\partial z} - \frac{\partial \phi}{\partial z}\frac{\partial}{\partial v_z}\right) f^\alpha\,\delta\left(\frac{1}{2}|v|^2 + \psi(\underline{x},z) - \varepsilon\right)\,dzdv = \\ &\quad = \int v_z f^\alpha \delta dv|_{z=0} - \int v_z f^\alpha \delta' \frac{\partial \psi}{\partial z} dvdz + \int v_z f^\alpha \delta' \frac{\partial \phi}{\partial z} dvdz \\ &\quad = \int v_z f^\alpha|_{z=0}\,\delta(\frac{1}{2}|v|^2 - \varepsilon)dv = 0\,, \end{aligned} \tag{4.6}$$

because

$$\frac{\partial \psi}{\partial z} = \frac{\partial \phi}{\partial z}$$

and by virtue of the flux conservation relation (3.3). δ' denotes the derivative of the delta measure.

Then, we compute :

$$\begin{aligned}
\int (\underline{v} \cdot \nabla_{\underline{x}} - \nabla_{\underline{x}}\phi \cdot \nabla_{\underline{v}}) f^\alpha \, \delta \left(\frac{1}{2}|v|^2 + \psi(\underline{x}, z) - \varepsilon \right) dz dv = \\
&= \nabla_{\underline{x}} \cdot \int \underline{v} f^\alpha \delta dv dz - \int \underline{v} f^\alpha \delta' \cdot \nabla_{\underline{x}} \psi dv dz + \int \underline{v} f^\alpha \delta' \cdot \nabla_{\underline{x}} \phi dv dz \\
&= \alpha \nabla_{\underline{x}} \cdot J^\alpha + \nabla_{\underline{x}} \phi_0 \cdot \int \underline{v} f^\alpha \delta' dv dz \\
&= \alpha \left(\nabla_{\underline{x}} \cdot J^\alpha - \nabla_{\underline{x}} \phi_0 \cdot \frac{\partial J^\alpha}{\partial \varepsilon} \right),
\end{aligned} \tag{4.7}$$

because $\int \underline{v} f^\alpha \delta' dv dz = -(\partial/\partial\varepsilon)(\int \underline{v} f^\alpha \delta dv dz)$.

Finally, collecting (4.6), (4.7) and (4.5) leads to the continuity equation (2.9) for F^α and J^α. ∎

Obviously, $F^\alpha \to F$ as $\alpha \to 0$. Thus, the continuity equation (2.9) for F and J will be proved as soon as we know that $J^\alpha \to J$. This is the aim of the next two sections. We start with proving that the auxiliary equation has a non empty set of solutions.

4.3. The auxiliary equation

In this section, we consider the most general problem of which (2.14) is a particular case. Indeed, let $g(z, v)$ be given and let us consider the problem of finding χ such that:

$$\left(-v_z \frac{\partial}{\partial z} + \frac{\partial \psi}{\partial z} \frac{\partial}{\partial v_z} \right) \chi = g \, , \quad \gamma^+(\chi) = \mathcal{B}^*(\gamma^-(\chi)) \, . \tag{4.8}$$

Again, we omit the dependence upon the $\underline{x}$ variable in the forthcoming discussion. We prove:

Lemma 4.5 - *Problem (4.8) has a solution if and only if g satisfies*

$$\int g \delta_\varepsilon(x, v) dz dv = 0.$$

Furthermore, if this condition is satisfied, the solution χ is unique under the condition $\int \chi \delta_\varepsilon(x, v) dz dv = 0$ and the set of solutions is the one-dimensional linear manifold $\{\chi + F(|v|^2/2 + \psi)$ with $F(\varepsilon)$ arbitrary.

Proof. We use the change of coordinates (4.3). With the same notations as in section 4.1 the problem is then written:

$$-v_z(z,u_z)\frac{\partial}{\partial z}\bar{\chi} = \bar{g}\,, \quad \gamma^+(\bar{\chi}) = \mathcal{B}^*(\gamma^-(\bar{\chi}))\,. \tag{4.9}$$

Let u be fixed. We integrate the first equation (4.9) with respect to z between the turning point $Z(u_z^2/2)$ and 0 and obtain for $u_z > 0$:

$$\bar{\chi}(0,\underline{v},u_z) \quad - \quad \bar{\chi}(Z,\underline{v},u_z) = -\int_Z^0 \bar{g}(z,\underline{v},u_z)|v_z(z,u_z)|^{-1}dz, \tag{4.10}$$

$$\bar{\chi}(0,\underline{v},-u_z) \quad - \quad \bar{\chi}(Z,\underline{v},-u_z) = \int_Z^0 \bar{g}(z,\underline{v},-u_z)|v_z(z,u_z)|^{-1}dz, \tag{4.11}$$

Since $\bar{\chi}(Z,\underline{v},u_z) = \bar{\chi}(Z,\underline{v},-u_z) = \chi(Z,\underline{v},0)$, we deduce that $\bar{\chi}(0,\underline{v},-u_z) - \bar{\chi}(0,\underline{v},u_z) = G(\underline{v},u_z)$ with

$$G(\underline{v},u_z) = \int_Z^0 (\bar{g}(z,\underline{v},u_z) + \bar{g}(z,\underline{v},-u_z))\,|v_z(z,u_z)|^{-1}dz\,. \tag{4.12}$$

This is equivalently written

$$\gamma^-(\bar{\chi}) = \mathcal{J}\gamma^+(\bar{\chi}) + G\,. \tag{4.13}$$

We insert this relation into the last equation (4.9) and obtain

$$(\,\mathrm{I}\, - \mathcal{B}^*\mathcal{J})\gamma^+(\bar{\chi}) = \mathcal{B}^*G\,. \tag{4.14}$$

By lemma 3.2 (ii), a solution $\gamma^+(\bar{\chi})$ exists if and only if

$$\int_{\mathsf{S}_+^2} (\mathcal{B}^*G)(|u|\omega)|\omega_z|d\omega = 0,$$

which, in view of the flux conservation relation for $\mathcal{B}^*$ (which results from (3.4)), is equivalent to

$$\int_{\mathsf{S}_-^2} G(|u|\omega)|\omega_z|d\omega = 0.$$

This relation can equivalently be written, for all $\varepsilon = |u|^2/2$:

$$\begin{aligned}
0 \quad &= \quad \int_{\{u\in\mathbf{R}^3,u_z<0\}} G(u)\delta\left(\frac{1}{2}|u|^2 - \varepsilon\right)|u_z|du \\
&= \quad \int_{\mathbf{R}^3}\left(\int_{Z(u_z^2/2)}^0 \bar{g}(z,\underline{v},u_z)|v_z(z,u_z)|^{-1}dz\right)\delta\left(\frac{1}{2}|u|^2-\varepsilon\right)|u_z|du \\
&= \quad \int_{-\infty}^0\left(\int_{D_z}\bar{g}(z,\underline{v},u_z)\delta\left(\frac{1}{2}|u|^2-\varepsilon\right)|v_z(z,u_z)|^{-1}|u_z|du\right)dz\,.
\end{aligned} \tag{4.15}$$

Using the change of variables (4.3) to change to the variable v in the integral (4.15), we find:

$$0 = \int_{-\infty}^{0} \left(\int_{\mathbf{R}^3} g(z, \underline{v}, v_z) \delta \left(\frac{1}{2}|v|^2 + \psi(z) - \varepsilon \right) dv \right) dz \,, \tag{4.16}$$

which is the condition of lemma 4.5.

Therefore, under this condition, there exists a solution $\gamma^+(\bar{\chi})$ of (4.14), which determines the outgoing trace of $\bar{\chi}$. Then, the incoming trace $\gamma^-(\bar{\chi})$ is determined by (4.13). Once the traces are known, $\bar{\chi}(z, u)$ for all possible values of z are obtained by integrating (4.9) between z and 0. Two solutions of (4.14) for $\gamma^+(\bar{\chi})$ differ by a constant function of ω, i.e. a function of $|u|^2$ only. Then, the associated $\gamma^-(\bar{\chi})$ differ by the same function of $|u|^2$ and so do the associated $\bar{\chi}$. Back to the v variable, we deduce that two solutions of (4.8) differ by a function of $|v|^2 + \psi(z)$. A unique solution can clearly be singled out by imposing that

$$\int \chi \delta_\varepsilon(x, v) dz dv = 0.$$

This ends the proof. ■

The function $g(z, v) = \underline{v}$ obviously satisfies the solvability condition of lemma 4.5 by oddness. Therefore, the auxiliary function $\underline{\chi}$ defined by (2.14) exists and is unique under the constraint $\int \chi \delta_\varepsilon(x, v) dz d\underline{v} = 0$. We are now ready to derive the current equation.

4.4. The current equation

Lemma 4.6 - *The current equation (2.10) is satisfied.*

Proof. We multiply equation (2.5) by $\underline{\chi}\delta_\varepsilon(x, v)$. Using Green's formula and the fact that

$$\frac{\partial \phi}{\partial z} = \frac{\partial \psi}{\partial z},$$

we compute, omitting the arguments in the delta measures:

$$\begin{aligned} K^\alpha \quad &:= \int \left(v_z \frac{\partial}{\partial z} - \frac{\partial \phi}{\partial z} \frac{\partial}{\partial v_z} \right) f^\alpha \, \underline{\chi} \, \delta \left(\frac{1}{2}|v|^2 + \psi(\underline{x}, z) - \varepsilon \right) dv \, dz \\ &= \left(\int v_z f^\alpha \underline{\chi} \delta dv \right) \Big|_{z=0} + \int f^\alpha \left(-v_z \frac{\partial}{\partial z} + \frac{\partial \phi}{\partial z} \frac{\partial}{\partial v_z} \right) \underline{\chi} \, \delta \, dv dz. \end{aligned} \tag{4.17}$$

Now, we have, using (2.6) and the second equation (2.14):

$$\left(\int v_z f^\alpha \underline{\chi} \delta dv \right) \Big|_{z=0}$$

$$= |v|^2 \left(\int_{\mathbb{S}^2_+} \gamma^+(f)\gamma^+(\underline{\chi})|\omega_z|d\omega - \int_{\mathbb{S}^2_-} \gamma^-(f)\gamma^-(\underline{\chi})|\omega_z|d\omega \right)$$

$$= |v|^2 \left(\int_{\mathbb{S}^2_+} \gamma^+(f)\gamma^+(\underline{\chi})|\omega_z|d\omega - \int_{\mathbb{S}^2_-} \mathcal{B}(\gamma^+(f))\gamma^-(\underline{\chi})|\omega_z|d\omega \right)$$

$$= |v|^2 \int_{\mathbb{S}^2_+} \gamma^+(f) \left(\gamma^+(\underline{\chi}) - \mathcal{B}^*(\gamma^-(\underline{\chi})\right) |\omega_z|d\omega = 0\,, \tag{4.18}$$

and with the first equation (2.14):

$$\int f^\alpha \left(-v_z \frac{\partial}{\partial z} + \frac{\partial \phi}{\partial z}\frac{\partial}{\partial v_z} \right) \underline{\chi}\,\delta\, dvdz = \int f^\alpha \underline{v}\,\delta\, dvdz = \alpha J^\alpha\,. \tag{4.19}$$

Therefore, inserting (4.18) and (4.19) into (4.17), we deduce that $K^\alpha = \alpha J^\alpha$. This a posteriori justifies the definition of $\underline{\chi}$.

¿From (2.5) and (4.17), we deduce:

$$J^\alpha = -\alpha \int \partial_t f^\alpha\, \underline{\chi}\, \delta_\varepsilon(x,v)\, dvdz - \int \left(\underline{v}\cdot\nabla_{\underline{x}} - \nabla_{\underline{x}}\phi\cdot\nabla_{\underline{v}}\right) f^\alpha\, \underline{\chi}\, \delta_\varepsilon(x,v)\, dvdz\,.$$

Taking the limit $\alpha \to 0$ and using (4.2), we obtain:

$$J = -\int \left(\underline{v}\cdot\nabla_{\underline{x}} - \nabla_{\underline{x}}\phi\cdot\nabla_{\underline{v}}\right) F\, \underline{\chi}\, \delta_\varepsilon(x,v)\, dvdz\,.$$

But

$$\left(\underline{v}\cdot\nabla_{\underline{x}} - \nabla_{\underline{x}}\phi\cdot\nabla_{\underline{v}}\right) F = \underline{v}\cdot \left(\nabla_{\underline{x}} - \nabla_{\underline{x}}\phi_0 \frac{\partial}{\partial \varepsilon}\right) F\,.$$

So, we get

$$J = -\left(\int \underline{\chi}\otimes\underline{v}\, \delta_\varepsilon(x,v)\, dvdz \right) \left(\nabla_{\underline{x}} - \nabla_{\underline{x}}\phi_0 \frac{\partial}{\partial \varepsilon}\right) F\,,$$

which leads to equ. (2.10) and (2.13). This ends the proof. ■

5. Examples

In this section, we consider particular examples of boundary collision operators $\mathcal{B}$. A rotationally invariant $\mathcal{B}$ means that K only depends on ω_z, ω'_z and the angle between $\underline{\omega}$ and $\underline{\omega}'$, i.e. $K(\omega' \to \omega) = K(\omega_z, \omega'_z, u_{\underline{\omega}} \cdot u_{\underline{\omega}'})$, where $u_{\underline{\omega}} = \underline{\omega}/|\underline{\omega}| \in \mathbb{S}^1$ and $\mathbb{S}^1$ is the unit circle. If in addition, K does not depend on $u_{\underline{\omega}} \cdot u_{\underline{\omega}'}$, $\mathcal{B}$ is said to be isotropic. We show that in the case of an isotropic $\mathcal{B}$, the density-of-states and diffusivity can be expressed in terms of the bounce period of a particle, given by

$$T(\underline{x}, u_z^2/2) = 2 \int_{Z(\underline{x}, u_z^2/2)}^{0} |v_z(z, u_z)|^{-1} dz \,, \tag{5.1}$$

where u_z is the transverse velocity at the origin $z = 0$.

Lemma 5.7 - *If $\mathcal{B}$ is isotropic, we have*

$$\begin{aligned} N(\underline{x}, \varepsilon) &= 2\pi\varepsilon \int_0^1 T(\underline{x}, \varepsilon\rho) d\rho, \\ D(\underline{x}, \varepsilon) &= \pi\varepsilon^2 \int_0^1 T^2(\underline{x}, \varepsilon\rho)(1-\rho) d\rho \; I \,, \end{aligned} \tag{5.2}$$

where I is the 2×2 identity matrix.

Proof. We first begin with N. Using the change of variables (4.3), and the coarea formula, we have, ignoring the $\underline{x}$-dependence:

$$\begin{aligned} N(\varepsilon) &= \int \delta\left(\frac{1}{2}|u|^2 - \varepsilon\right) |v_z(z, u_z)|^{-1} |u_z| du dz \\ &= \int_{\mathbf{R}^3} \left(\int_{Z(u_z^2/2)}^{0} |v_z(z, u_z)|^{-1} dz \right) \delta\left(\frac{1}{2}|u|^2 - \varepsilon\right) |u_z| du \\ &= \frac{1}{2} \int_{\mathbf{R}^3} T(u_z^2/2) \delta\left(\frac{1}{2}|u|^2 - \varepsilon\right) |u_z| du = \varepsilon \int_{\mathbb{S}^2} T(\varepsilon|\omega_z|^2) |\omega_z| d\omega \end{aligned}$$

Using $\underline{\omega} \in B(0,1)$ as parametrization of $\mathbb{S}^2_{\pm}$, where $B(0,1)$ is the unit ball in $\mathbb{R}^2$ and noting that $|\omega_z| d\omega = d\underline{\omega}$, we obtain:

$$N(\varepsilon) = 2\varepsilon \int_{B(0,1)} T(\varepsilon(1 - |\underline{\omega}|^2)) d\underline{\omega} \,.$$

Changing to cylindrical coordinates in $B(0,1)$, we finally get:

$$N(\varepsilon) = 4\pi\varepsilon \int_0^1 T(\varepsilon(1-\rho^2)) \rho d\rho = 2\pi\varepsilon \int_0^1 T(\varepsilon\rho) d\rho \,,$$

which is the first formula (5.2).

Now, we turn to D. We use the notations of section 4.3. First, we note that with $g = v_i$, $(i = 1, 2)$, equ. (4.12) gives $G(\underline{v}, u_z) = v_i T(u_z^2/2)$. Therefore, G is an odd function of v_i and since $\mathcal{B}$ is isotropic, $\mathcal{B}G = 0$. It follows from (4.14) and lemma 3.2 (i) that $\gamma^+(\chi_i)$ does not depend on the angular variable ω. In fact, we can take $\gamma^+(\chi_i) = 0$ since any other choice will simply add a function of $|v|^2/2 + \psi$ to χ_i, which will not modify the value of D. From (4.13), it follows that $\gamma^-(\chi_i) = G$. Then, integrating (4.9) between z and 0, we deduce, for $u_z > 0$:

$$\begin{aligned} \bar{\chi}_i(z, \underline{v}, u_z) &= v_i\, \tau(z, u_z^2/2), \\ \bar{\chi}_i(z, \underline{v}, -u_z) &= v_i\, \left(T(u_z^2/2) - \tau(z, u_z^2/2)\right), \end{aligned} \tag{5.3}$$

where

$$\tau(z, u_z^2/2) = \int_z^0 |v_z(\zeta, u_z)|^{-1} d\zeta \,.$$

Now, it is clear that the vector $\underline{\chi}$ is proportional to $\underline{v}$, so that the diffusivity $D = d\,\mathrm{I}$ is a scalar with

$$d = \int \chi_1 v_1 \delta_\varepsilon(x, v) dz dv = \int \bar{\chi}_1 v_1 \delta\left(\frac{1}{2}|u|^2 - \varepsilon\right) |v_z(z, u_z)|^{-1} |u_z| du dz \,.$$

Additionally, the term proportional to τ in χ_1 is odd with respect to u_z and has a vanishing contribution to d. Thus, using the coarea formula and with the same computations as for N, we get:

$$\begin{aligned} d &= \int_{u_z<0} v_1^2 T(u_z^2/2) \delta\left(\frac{1}{2}|u|^2 - \varepsilon\right) |v_z(z, u_z)|^{-1} |u_z| du dz \\ &= \frac{1}{2} \int v_1^2 T(u_z^2/2) \left(\int_{Z(u_z^2/2)}^0 |v_z(z, u_z)|^{-1} dz\right) \delta\left(\frac{1}{2}|u|^2 - \varepsilon\right) |u_z| du \\ &= \frac{1}{4} \int v_1^2 T^2(u_z^2/2) \delta\left(\frac{1}{2}|u|^2 - \varepsilon\right) |u_z| du = \frac{1}{2}\varepsilon^2 \int_{\mathbb{S}^2} |\underline{\omega}|^2 T^2(\varepsilon\omega_z^2) |\omega_z| d\omega \\ &= \varepsilon^2 \int_{B(0,1)} |\underline{\omega}|^2 T^2(\varepsilon(1 - |\underline{\omega}|^2)) d\underline{\omega} = 2\pi\varepsilon^2 \int_0^1 \rho^2 T^2(\varepsilon(1 - \rho^2)) \rho d\rho \\ &= \pi\varepsilon^2 \int_0^1 (1 - \rho) T^2(\varepsilon\rho) d\rho \,, \end{aligned}$$

which was the formula to be proved. ■

We now review a certain number of transverse potential profiles ψ:
(i) Linear potential: $\psi(z) = -Fz$, where F is a constant force:

$$T = 2\sqrt{2}\frac{\sqrt{\varepsilon}}{F}, \quad N = \frac{8\sqrt{2}\pi}{3}\frac{\varepsilon^{3/2}}{F}, \quad D = \frac{4\pi}{3}\frac{\varepsilon^3}{F^2}. \tag{5.4}$$

(ii) Harmonic potential: $\psi(z) = \frac{1}{2}\omega^2 z^2$ where ω is the angular frequency of the harmonic oscillator:

$$T = \frac{\pi}{\omega}, \quad N = 2\pi^2 \frac{\varepsilon}{\omega}, \quad D = \pi^3 \frac{\varepsilon^2}{2\omega^2}. \tag{5.5}$$

(iii) Infinite wall at distance L: $\psi(z) = 0$ for $z \in [-L, 0]$; $\psi(z) = +\infty$ for $z < -L$: Then :

$$T = \sqrt{2} L \frac{1}{\sqrt{\varepsilon}}, \quad N = 4\sqrt{2}\pi L \sqrt{\varepsilon}, \quad D = +\infty . \tag{5.6}$$

The divergence of the diffusivity is related to anomalous diffusion (see introduction).

(iv) Exponential potential: $\psi(z) = \psi_\infty(1 - \exp(z/\lambda))$. This the most realistic potential profile for plasmas or semiconductor modeling. ψ_∞ is the potential at infinity and λ is the screening (or Debye) length. It does not satisfy hypothesis 2.1 because ψ has a finite limit as $z \to -\infty$, but the theory is applicable for energies $\varepsilon \le \psi_\infty$. After some computations, we get, denoting $e = (\varepsilon/\psi_\infty)^{1/2}$:

$$T = \frac{2\sqrt{2}\,\lambda}{\sqrt{\psi_\infty}} \frac{\arcsin e}{\sqrt{1-e^2}}, \quad N = 8\sqrt{2}\pi\lambda\sqrt{\psi_\infty}\nu(e), \quad D = 16\pi\lambda^2\psi_\infty\delta(e) \tag{5.7}$$

with

$$\nu(e) = \int_0^e \arcsin u \, \frac{u\,du}{\sqrt{1-u^2}}, \quad \delta(e) = \int_0^e \arcsin^2 u \, \frac{e^2 - u^2}{1-u^2} \, u\,du .$$

ν can also be expressed according to $\nu(e) = e - \sqrt{1-e^2}\arcsin e$. As expected, N and D are only defined for $\varepsilon \le \psi_\infty$. For $\varepsilon \approx 0$, formulae (5.7) reduce to those of the linear potential case (5.4) with $F = \psi_\infty/\lambda$. For $\varepsilon \to \psi_\infty$, we have

$$N \to 3N_{\text{lin}}(\psi_\infty), \quad D \to \frac{3}{2}(\frac{\pi^2}{2} - 1)D_{\text{lin}}(\psi_\infty),$$

where $N_{\text{lin}}(\psi_\infty)$ and $D_{\text{lin}}(\psi_\infty)$ are the value of N and D at $\varepsilon = \psi_\infty$ in the linear potential case. Since

$$\frac{3}{2}(\frac{\pi^2}{2} - 1) \approx 6,$$

we see that the linear and exponential cases give rise to densities-of-state and diffusivities of the same order of magnitude through the entire range of admissible energies $\varepsilon \in [0, \psi_\infty]$. Therefore, it is reasonable to think that, in most physical situations, the linear potental case gives acceptable results, within the error bounds on the physical data. We also notice that the diffusivity in the exponential case is finite even at the boundary of the domain of admissible energies. However, its derivative with respect to the energy is infinite at this point, as it should be.

6. Extensions

In this section, we review a certain number of extensions of the model. So far, only elastic surface interactions have been considered in the model. However, as already mentioned, particle-surface interactions can be quite complex. In particular, they certainly involve inelastic mechanisms such as attachment, secondary emission, etc.

Aditionally, the medium corresponding to Ω may be something else than pure vacuum (like, for instance, a gas, a plasma or a crystal). In this case, the particles can suffer collisions with the medium during their excursion into Ω (e.g. collisions with neutral molecules or ions in a gas or a plasma, collisions with phonons and impurities in a crystal). These so-called 'bulk' collisions can be elastic or inelastic. In this section, we shall first discuss bulk and surface inelastic collisions. Then, we shall consider bulk elastic collisions.

6.1. The current equation

To account for inelastic interactions, we need to suppose that they are a perturbation of order α^2 of the elastic interactions. More precisely, we start from the following kinetic model

$$\alpha\partial_t f^\alpha + \left(\underline{v}\cdot\nabla_{\underline{x}} - \nabla_{\underline{x}}\phi\cdot\nabla_{\underline{v}}\right) f^\alpha + \frac{1}{\alpha}\left(v_z\frac{\partial}{\partial z} - \frac{\partial\phi}{\partial z}\frac{\partial}{\partial v_z}\right) f^\alpha = \alpha Q(f^\alpha)\,, \tag{6.1}$$

$$\gamma^-(f^\alpha) = (1-\alpha^2)\mathcal{B}(\gamma^+(f^\alpha)) + \alpha^2\mathcal{I}(\gamma^+(f^\alpha))\,. \tag{6.2}$$

where $Q(f)$ is the bulk inelastic collision operator while $\mathcal{I}(\gamma^+(f))$ is the inelastic surface interaction operator. We shall leave the expressions of Q and $\mathcal{I}$ unspecified since they strongly depend on the precise physics involved. We see that the bulk inelastic operator in (6.1) appears mutiplied by a factor α, to be compared with the factor α^{-1} of the transport operator modeling the confinment in the z direction. Similarly, the boundary condition (6.2) expresses that elastic collisions appear with probability of order $(1-\alpha^2)$ while inelastic ones appear only with probability α^2. For this model, the following modification of Theorem 2.1 can be proved:

Theorem 6.8 - *Under the same hypotheses as in theorem 2.1, the same conclusion holds true for system (6.1), (6.2), except for a modification of the continuity equation (2.9), now written:*

$$N(\underline{x},\varepsilon)\partial_t F + \left(\nabla_{\underline{x}} - \nabla_{\underline{x}}\phi_0\,\partial_\varepsilon\right)\cdot\underline{J} = I(F)\,, \tag{6.3}$$

with $I(F)$ given by $I(F) = \bar{\mathcal{I}}(F) + \bar{Q}(F)$ and

$$\bar{Q}(F) = \int Q(F)\, dzdv \tag{6.4}$$

$$\begin{aligned}\bar{\mathcal{I}}(F) &= -F \int_{v_z<0} \delta\left(\frac{1}{2}|v|^2 - \varepsilon\right) |v_z| dv \\ &\quad + \int_{v_z<0} \mathcal{I}(F)\, \delta\left(\frac{1}{2}|v|^2 - \varepsilon\right) |v_z| dv\,. \end{aligned} \tag{6.5}$$

The "collision operator" $I(F)$ of the SHE model collects the contributions of the bulk inelastic collision operator $\bar{Q}$ and of the surface inelastic operator $\bar{\mathcal{I}}$. The structure of $\bar{\mathcal{I}}$ shows a loss term (with a minus sign) corresponding to the absorption of particles by the boundary and a gain term (with a plus sign) corresponding to the inelastic reemission. The loss term can be written

$$\bar{\mathcal{I}}^-(F) = -\nu_b N F,$$

where $\nu_b = \nu_b(\varepsilon)$ is a collision frequency with the surface, given by

$$\nu_b = N^{-1} \int_{v_z<0} \delta(|v|^2/2 - \varepsilon)|v_z| dv = 2\pi\varepsilon N^{-1}.$$

Proof. Only the proof of the continuity equ. (2.9) is modified, and more precisely, the computation of (4.6). Indeed, we get, using (6.2) and the flux conservation relation (3.3):

$$\begin{aligned}\alpha^{-2} \int \left(v_z \frac{\partial}{\partial z} - \frac{\partial \phi}{\partial z}\frac{\partial}{\partial v_z}\right) f^\alpha\, \delta\left(\frac{1}{2}|v|^2 + \psi(\underline{x}, z) - \varepsilon\right) dzdv = \\ = \alpha^{-2} \int_{v_z<0} \left(J(\gamma^+(f^\alpha)) - \gamma^-(f^\alpha)\right) \delta(\frac{1}{2}|v|^2 - \varepsilon)\, |v_z| dv \\ = \int_{v_z<0} \left(J(\gamma^+(f^\alpha)) - \mathcal{J}(\gamma^+(f^\alpha))\right) \delta(\frac{1}{2}|v|^2 - \varepsilon)\, |v_z| dv\,, \end{aligned}$$

which, in the limit $\alpha \to 0$ leads to (6.5). The obtention of (6.4) is straightforward. ■

6.2. Bulk elastic collisions

Contrary to inelastic collisions, bulk elastic collisions must be of order 1 (i.e. of the same order as surface elastic collisions) to yield a modification of the limit SHE model. Thus, we consider the following kinetic model

$$\alpha\partial_t f^\alpha + \left(\underline{v}\cdot\nabla_{\underline{x}} - \nabla_{\underline{x}}\phi\cdot\nabla_{\underline{v}}\right) f^\alpha + \frac{1}{\alpha}\left(v_z\frac{\partial}{\partial z} - \frac{\partial\phi}{\partial z}\frac{\partial}{\partial v_z}\right) f^\alpha = \frac{1}{\alpha}\mathcal{E}(f^\alpha), \tag{6.6}$$

$$\gamma^-(f^\alpha) = \mathcal{B}(\gamma^+(f^\alpha)). \tag{6.7}$$

where $\mathcal{E}(f)$ is the bulk elastic collision operator. We shall restrict to relaxation operators of the form

$$\mathcal{E}(f) = \nu_e(x,e)\left(-f + \frac{1}{4\pi}\int_{\mathbb{S}^2} f(v(e)\omega)d\omega\right), \tag{6.8}$$

where $e = |v|^2/2$ is the kinetic energy of the particle, $v(e) = (2e)^{1/2}$ and ν_e is the elastic collision frequency. The treatment of more general elastic operators is left to future work. The presence of the elastic operator at the right-hand-side of (6.6) results in a modification of the diffusivity. More precisely:

Theorem 6.9 - *Under the same hypotheses as in theorem 2.1 and the additional isotropy assumption on $\mathcal{B}$ (see Section 5), the same conclusion holds true for system (6.6), (6.7), except for a modified auxiliary function, now a solution of:*

$$\left(-v_z\frac{\partial}{\partial z} + \frac{\partial\psi}{\partial z}\frac{\partial}{\partial v_z} - \mathcal{E}\right)\underline{\chi} = \underline{v}, \quad \gamma^+(\underline{\chi}) = \mathcal{B}^*(\gamma^-(\underline{\chi})). \tag{6.9}$$

Proof. The proof of theorem 6.9 consists of the same four points as for theorem 2.1 (see Section 4). However, we shall solely concentrate on the existence of a solution of the auxiliary problem (6.9) since the other steps can be proved by using the elementary properties of $\mathcal{E}$ (such as particle conservation, entropy, etc, see e.g. [5], [13]). To solve (6.9), we first solve:

$$\left(-v_z\frac{\partial}{\partial z} + \frac{\partial\psi}{\partial z}\frac{\partial}{\partial v_z} + \nu_e\right)\chi_i = v_i, \quad \gamma^+(\underline{\chi}) = \mathcal{B}^*(\gamma^-(\underline{\chi})). \tag{6.10}$$

for $i = 1, 2$, and prove that χ_i is odd with respect to v_i. This implies that

$$\int_{\mathbb{S}^2} \chi_i(z, v(e)\omega)d\omega = 0,$$

for all $z \leq 0$ which shows that χ_i is also a solution of (6.9). Of course, this method does not gives the uniqueness of the solution; this question is left to future work. As usual, we forget $\underline{x}$ in the following discusssion.

Using the same method and notations as in section 4.3, we obtain, by changing to the variable u:

$$-v_z(z,u_z)\frac{\partial\bar{\chi}_i}{\partial z}+\bar{\nu}_e\bar{\chi}_i=v_i\,, \tag{6.11}$$

with

$$\bar{\nu}_e(z,|u|^2/2)=\nu_e(z,|u|^2/2-\psi(z)).$$

Integrating the first equation (6.11) between $Z(u_z^2/2)$ and 0, we obtain

$$\gamma^-(\bar{\chi}_i)=\alpha(0,Z)^2\mathcal{J}(\gamma^+(\bar{\chi}_i))+G,$$

with

$$\alpha(z_1,z_2)=\exp\left(\int_{z_1}^{z_2}\bar{\nu}_e(\zeta,|u|^2/2)|v_z(\zeta,u_z)|^{-1}\,d\zeta\right),$$

$$G=v_iT_e\,,\quad T_e=\alpha(0,Z)\int_Z^0(\alpha(Z,z)+\alpha(z,Z))|v_z(z,u_z)|^{-1}dz\,.$$

Inserting it into the second equation (6.11), we obtain

$$(\mathrm{I}-\mathcal{B}^*\alpha(0,Z)^2\mathcal{J})(\gamma^+(\bar{\chi}_i))=\mathcal{B}^*G.$$

Since G is odd with respect to v_i, the isotropy assumption on $\mathcal{B}$ implies that $\mathcal{B}^*G=0$. Now, because $Z<0$, we have $\alpha(0,Z)<1$ and the operator norm of $\mathcal{B}^*\alpha(0,Z)^2\mathcal{J}$ is strictly less than 1. It tollows that $(\mathrm{I}-\mathcal{B}^*\alpha(0,Z)^2\mathcal{J})$ is invertible and therefore $\gamma^+(\bar{\chi}_i)=0$. Then $\gamma^-(\bar{\chi}_i)=G$. Now, integrating (6.11) between z and 0, we deduce, for $u_z>0$

$$\begin{aligned}
&\bar{\chi}_i(z,\underline{v},u_z)=v_i\,\tau^+(z),\\
&\bar{\chi}_i(z,\underline{v},-u_z)=v_i\left(\alpha(z,0)T_e-\tau^-(z)\right), \qquad (6.12)\\
&\tau^+(z)=\int_z^0\alpha(\zeta,z)|v_z(\zeta,u_z)|^{-1}d\zeta,\\
&\tau^-(z)=\int_z^0\alpha(z,\zeta)|v_z(\zeta,u_z)|^{-1}d\zeta. \qquad (6.13)
\end{aligned}$$

From (6.12), (6.13), it is clear that $\bar{\chi}_i$ is odd with respect to v_i, proving that it is also a solution of (6.9). This concludes the proof. ■

The computation can be pursued to obtain an almost explicit formula for the diffusivity:

$$D=2\pi\varepsilon^2\int_0^1 I(\varepsilon,\rho)(1-\rho)d\rho,$$

$$I\left(\frac{|u|^2}{2},\frac{|u_z^2|}{|u|^2}\right)=\int_Z^0(\tau^+-\tau^-+\alpha(z,0)T_e)\frac{dz}{|v_z(z,u_z)|}.$$

The computation can be pursued even further in the case of a constant ν_e, then leading to

$$I(\varepsilon,\rho)=\nu_e^{-2}\left[\kappa(\varepsilon\rho)-\left(1-e^{-\kappa(\varepsilon\rho)}\right)\right],$$

where $\kappa(\varepsilon)=\nu_e T(\varepsilon)$, T being the bounce period in the absence of collisions (5.1). The quantity κ measures the number of collisions in a bounce period. Finally, in the case of a linear confining potential $\psi(z)=-Fz$, we obtain

$$D(\varepsilon) = 4\pi\varepsilon^2\nu_e^{-2}\delta(\kappa(\varepsilon)) \tag{6.14}$$

$$\delta(\kappa) = \frac{2}{15}\kappa-\frac{1}{4}+\frac{1}{\kappa^2}-\frac{6}{\kappa^4}+e^{-\kappa}\left(\frac{2}{\kappa^2}+\frac{6}{\kappa^3}+\frac{6}{\kappa^4}\right). \tag{6.15}$$

It is easy to check that in the limit $\nu_e\to 0$, expression (6.14) yields (5.4). It is remarkable that, in spite of the complexity of the situation, an analytic formula for the diffusivity can be obtained.

Acknowledgements. The author would express his gratitude to J. P. Boeuf for having suggested this problem and provided encouragements and references. This work has been supported by the TMR network No. ERB FMBX CT97 0157 on "Asymptotic methods in kinetic theory" of the European Community.

References

[1] H. Babovsky, On the Knudsen flows within thin tubes, J. Statist. Phys., 44(1986), 865–878.

[2] H. Babovsky, C. Bardos and T. Platkowski, Diffusion approximation for a Knudsen gas in a thin domain with accommodation on the boundary, Asymptotic Analysis, 3(1991), 265–289.

[3] C. Bardos, F. Golse and B. Perthame, The Rosseland approximation for the radiative transfer equations, Comm. Pure Appl. Math. 40(1987), 691–721 and 42(1989), 891–894.

[4] C. Bardos, R. Santos and R. Sentis, Diffusion approximation and computation of the critical size, Trans. A. M. S., 284(1984), 617–649.

[5] N. Ben Abdallah, P. Degond, On a hierarchy of macroscopic models for semiconductors, J. Math. Phys. 37(1996), 3306–3333.

[6] N. Ben Abdallah, P. Degond, A. Mellet and F. Poupaud, Electron transport in semiconductor superlattices, preprint

[7] A. Bensoussan, J. L. Lions and G. Papanicolaou, Boundary layers and homogenization of transport processes, Publ. RIMS Kyoto Univ. 15(1979), 53–157.

[8] C. Börgers, C. Greengard et E. Thomann, The diffusion limit of free molecular flow in thin plane channels, SIAM J. Appl. Math., Vol. 52, No. 4, (1992), 1057–1075.

[9] H. Brezis, Analyse Fonctionnelle, théorie et applications, Masson, Paris, 1983.

[10] R. Castagné, Dispositifs semiconducteurs et structures élémentaires des circuits intégrés, Vol. 3, Cours de DEA, University Paris Sud.

[11] C. Cercignani, The Boltzmann equation and its applications, Springer, New-York, 1998.

[12] M. Cho and D. E. Hastings, Dielectric charging processes and arcing rates of high voltage solar arrays, J. Spacecraft and Rockets 28(1991), 698–706.

[13] P. Degond, Mathematical modeling of microelectronics semiconductor devices Proceedings, Morningside Institute, Beijing, to appear.

[14] P. Degond, A model of near-wall conductivity and its application to plasma thrusters, SIAM J. Appl. Math., 58(1998), 1138–1162.

[15] P. Degond, V. Latocha, L. Garrigues, J. P. Boeuf, Electron transport in stationary plasma thrusters, Transp. Th. Stat. Phys. 27(1998), 203–221.

[16] P. Degond and S. Mancini, Diffusion driven by collisions with the boundary, preprint, 1999.

[17] P. Degond, K. Zhang, Diffusion approximation of a scattering matrix model of semiconductor superlattices, preprint, 1999.

[18] P. Dmitruk, A. Saul and L. Reyna, High electric field approximation in semiconductor devices, Appl. Math. Letters 5(1992), 99–102.

[19] H. Federer,Geometric Measure Theory, Springer, Berlin, (1969).

[20] N. Goldsman, L. Henrickson, J. Frey, A physics based analytical numerical solution to the Boltzmann transport equation for use in device simulation, Solid State Electron. 34(1991), 389–396.

F. Golse, Anomalous diffusion limit for the Knudsen gas, Asymptotic Analysis, (1998).

[22] F. Golse and F. Poupaud, Limite fluide des équations de Boltzmann des semiconducteurs pour une statistique de Fermi-Dirac, Asymptotic Analysis 6(1992), 135–160.

[23] E. W. Larsen and J. B. Keller, Asymptotic solution of neutron transport problems for small mean free paths, J. Math. Phys. 15(1974), 75–81.

[24] F. Poupaud, Diffusion approximation of the linear semiconductor equation: analysis of boundary layers, Asymptotic Analysis 4(1991), 293–317.

[25] Yu. P. Raizer, Gas discharge Physics, Springer, Berlin, 1997.

[26] D. Ventura, A. Gnudi, G. Baccarani and F. Odeh, Multidimensional spherical harmonics expansion of Boltzmann equation for transport in semiconductors, Appl. Math. Letters 5 (1992), 85–90.

P. Degond
MIP, UMR CNRS 5640
Université Paul Sabatier
118, route de Narbonne
31062 Toulouse Cedex
France

E-mail: degond@mip.ups-tlse.fr

Studies in Mathematics and its Applications, Vol. 31
D. Cioranescu and J.L. Lions (Editors)

Chapter 14

DIFFUSIVE ENERGY BALANCE MODELS IN CLIMATOLOGY

J.I. Díaz

1. Introduction

This paper contains an expanded and updated version of my lecture at the Collège de France, on May 1997, where I collected several results on a diffusive energy balance model given by a nonlinear parabolic problem formulated in the following terms:

$$(P)\begin{cases} u_t - div(|\nabla u|^{p-2}\nabla u) \in QS(x)\beta(u) - \mathcal{G}(u) + f(x,t) & \text{in } (0,T)\times\mathcal{M}, \\ u(x,0) = u_0(x) & \text{in } \mathcal{M}, \end{cases}$$

where $\mathcal{M}$ is a C^∞ two-dimensional compact connectedoriented Riemannian manifold without boundary. We assume $T > 0$ arbitrarily fixed, $Q > 0$, $S \in L^\infty(\mathcal{M})$ and $p \geq 2$. The function $\mathcal{G}$ is increasing and β represents a bounded maximal monotone graph in $\mathbb{R}^2$ (of Heaviside type). We also consider the associate stationary problem

$$(P_{Q,f}) \qquad -div(|\nabla u|^{p-2}\nabla u) + \mathcal{G}(u) \in QS(x)\beta(u) + f_\infty(x) \text{ on } \mathcal{M}.$$

Through the paper we shall use the notation $div(|\nabla u|^{p-2}\nabla u) = \Delta_p u$. Problem (P) arises in the modeling of some problems in Climatology: the so-called *Energy Balance Models* introduced independently, in 1969 by M.I. Budyko [15] and W.D. Sellers [64]. The models have a diagnostic character and intended to understand the evolution of the global climate on a long time scale. Their main characteristic is the high sensitivity to the variation ofsolar and terrestrial parameters. This kind of models has been used inthe study of the Milankovitch theory of the ice-ages (see, e.g. North, Mengel and Short [60]).

The model is obtained from the thermodynamics equation of the *atmosphere primitive equations* via averaging process (see, e.g. Lions, Temam and Wang [53] for a mathematical study of those equations, Kiehl [50] for the application of averaging processes and Remark 1 for some nonlocal variants of (P)). More simply, the model can be formulated by using the energy balance on the Earth's surface: internal energy flux variation $= R_a - R_e + D$, were R_a and R_e represent the absorbed solar and the emitted terrestrial energy flux, respectively and D is the horizontal heat diffusion.

Let us express the components of the above balance in mathematical terms. The distribution of temperature $u(x,t)$ is expressed pointwise after standard average process, where the spatial variable x is in the Earth's surface which may be identified with a compact Riemannian manifold without boundary $\mathcal{M}$ (for instance, the two-sphere S^2), and t is the time variable. The time scale is considered relatively long. Nevertheless, in the so called *seasonal models* a smaller scale of time is introduced in order to analyze the effect of the seasonal cycles in the climate and in particular in the ice caps formation (see Remark 2 for the connection with the associate time periodical problem).

To simplify the presentation we assume that the internal energy flux variation is simply given as the product of the heat capacity c (a given constant which can be assumed equal to one by rescaling) and the partial derivative of the temperature u with respect to the time. For a more general modeling see Remark 1.

The *absorbed energy* R_a depends on the planetary *coalbedo* β. The coalbedo function represents the fraction of the incoming radiation flux which is absorbed by the surface. In ice-covered zones, reflection is greater than over oceans, therefore, the coalbedo is smaller. One observes that there is a sharp transition between zones of high and low coalbedo. In the energy balance climate models, a main change of the coalbedo occurs in a neighborhood of a critical temperature for which ice becomes white, usually taken as $u = -10^0C$.

The different coalbedo is modelled as a discontinuous function of the temperature in the *Budyko model*. Here it will be treated as a maximal monotone graph in $\mathbb{R}^2$

$$\beta(u) = \begin{cases} m & u < -10 \\ [m, M] & u = -10 \\ M & u > -10, \end{cases} \tag{1}$$

where $m = \beta_i$ and $M = \beta_w$ represent the coalbedo in the ice-covered zone and the free-ice zone, respectively and $0 < \beta_i < \beta_w < 1$ (the value of these constants has been estimated by observation from satellites). In the *Sellers model*, β is assumed to be a more regular function (at least Lipschitz continuous), as for instance

$$\beta(u) = \begin{cases} m & u < u_i, \\ m + (\frac{u-u_i}{u_w-u_i})(M-m) & u_i \le u \le u_w, \\ M & u > u_w, \end{cases}$$

where u_i and u_w are fixed temperatures closed to -10^0C.In both models, the absorbed energy is given by $R_a = QS(x)\beta(u)$where $S(x)$ is the *insolation function* and Q is the so-called *solar constant.*

The Earth's surface and atmosphere, warmed by the Sun, reemit part of the absorbed solar flux as an infrared long-wave radiation. This energy R_e is represented, in the Budyko model, according to the Newton cooling law, that is,

$$R_e = Bu + C. \tag{2}$$

Here, B and C are positive parameters, which are obtained by observation, and can depend on the *greenhouse effect.* However, in the Sellers model, R_e is expressed according to the Stefan - Boltzman law

$$R_e \;=\; \sigma u^4, \tag{3}$$

where σ is called *emissivity constant* and now u is in Kelvin degrees.

The *heat diffusion* D is given by the divergence of the conduction heat flux F_c and the advection heat flux F_a. Fourier's law expresses $F_c = k_c\nabla u$ where k_c is the *conduction coefficient.* The advection heat flux is given by $F_a = \mathbf{v}\cdot\nabla u$ and it is known (see e.g. Ghil and Childress [35]) that, to the level of the planetary scale, it can be modeled in terms of $k_a\nabla u$ for a suitable diffusion coefficient k_a. So, $D = \div(k\nabla u)$ with $k = k_c+k_a$. In the pioneering models, the diffusion coefficient k was considered as a positive constant. Nevertheless, in 1972, P.H. Stone [68] proposed a coefficient $k = |\nabla u|$, in order to consider negative feedback in the eddy fluxes. So, in that case the heat diffusion is represented by the quasilinear operator $D = div(|\nabla u|\nabla u)$. Our formulation (P) takes into account such a case which corresponds to the special choice $p = 3$ (notice that the case $p = 2$ leads to the linear diffusion). These physical laws lead to problem (P) with $R_e(u) = \mathcal{G}(u) - f$.

In Section 2 we start by presenting some results on the existence and uniqueness of solutions which generalize some previous results in the literature for a one-dimensional simplified formulation. Such simplification considers the averaged temperature over each parallel as the unknown. So, the two-dimensional model (P) is reduced in a one-dimensional model when $\mathcal{M}$ is the two dimensional sphere and considering the spherical coordinates. Therefore, the model becomes

$$(P^1)\begin{cases} u_t - (\rho(x)|u_x|^{p-2}u_x)_x \in QS(x)\beta(u) - R_e(u) & \text{in } (-1,1)\times(0,T), \\ u(x,0) = u_0(x) & \text{in } (-1,1), \end{cases}$$

with $\rho(x) = (1-x^2)^{\frac{p}{2}}$ where $x = \sin\theta$ and θ is the latitude. Notice that again there is no boundary condition since the meridional heat flux $(1-x^2)^{\frac{p}{2}}|u_x|^{p-2}u_x$ vanishes at the poles $x = \pm 1$. We also include in this section some comments on the *free boundaries* associated to the Budyko type model (the curves separating the regions $\{x : u(x,t) < -10\}$ and $\{x : u(x,t) > -10\}$). We end the section with a result on the stabilization of solutions as $t \to \infty$. Some references on the question of the approximate controllability for the transient model are given in Remark 4.

Section 3 is devoted to the study of the number of stationary solutions according to the parameter Q, when β is not necessarily Lipschitz continuous and $p \geq 2$. We start by estimating an interval of values for Q where there exist at least *three* stationary solutions and other complementary intervals for Q where the stationary solution is unique. A more precise study of the bifurcation diagram of solutions for different positive values of Q is available once we specialize $f_\infty(x) \equiv C$ with C a prescribed constant. Then problem $(P_{Q,f})$ becomes

$$(P_{Q,C}) \qquad -\operatorname{div}(|\nabla u|^{p-2}\nabla u) + \mathcal{G}(u) + C \ \in \ QS(x)\beta(u) \text{ on } \mathcal{M}.$$

We denote by Σ the set of pairs $(Q,u) \in \mathbb{R}^+ \times V$, where u satisfies the equation $(P_{Q,C})$. We show that, under suitable conditions, Σ contains an unbounded connected component which is S -shaped containing $(0, \mathcal{G}^{-1}(-C))$ with at least one turning point to the right (and so at least another one to the left). We end Section 3 with a remark on a simplified version of problem $(P_{Q,C})$ for which it is still possible to find more precise answers: if $Q_1 < Q < Q_2$, for some suitable positive constants $Q_1 < Q_2$, then we have infinitely many solutions. More precisely, there exists $k_0 \in \mathbb{N}$ such that for every $k \in \mathbb{N}$, $k \geq k_0 \in \mathbb{N}$ there exists at least a solution u_k which crosses the level $u_k = -10$, exactly k times.

2. The transient model

2.1. On the existence of solutions

Motivated by the model background described in the Introduction, we introduce the following structure hypotheses: $p \geq 2$, $Q > 0$,
– $(\mathrm{H}_\mathcal{M})$ $\mathcal{M}$ is a C^∞ two-dimensional compact connected oriented Riemannian manifold of $\mathbb{R}^3$ without boundary,
– (H_β) β is a bounded maximal monotone graph in $\mathbb{R}^2$, i.e. $m \leq z \leq M$, $\forall z \in \beta(s)$, $\forall s \in \mathbb{R}$,
– $(\mathrm{H}_\mathcal{G})$ $\mathcal{G} : \mathbb{R} \to \mathbb{R}$ is a continuous strictly increasing function such that $\mathcal{G}(0) = 0$, and $|\mathcal{G}(\sigma)| \geq C|\sigma|^r$ for some $r \geq 1$,

– (H$_s$) $S : \mathcal{M} \to \mathbb{R}$, $S \in L^\infty(\mathcal{M})$, $S_1 \geq S(x) \geq S_0 > 0$ a.e.$x \in \mathcal{M}$,
– (H$_f$) $f \in L^\infty(\mathcal{M} \times (0,T))$, (resp.– (H$_f^\infty$) $f \in L^\infty(\mathcal{M} \times (0,\infty))$),
– (H$_0$) $u_0 \in L^\infty(\mathcal{M})$.

The possible discontinuity in the coalbedo function causes that (P) does not have classical solutions in general, even if the data u_0 and f are smooth. Therefore, we must introduce the notion of weak solution. The natural "energy space" associated to (P) is the one given by

$$V := \{u : \mathcal{M} \to R,\ u \in L^2(\mathcal{M}),\ \nabla_{\mathcal{M}} u \in L^p(T\mathcal{M})\},$$

which is a reflexive Banach space if $1 < p < \infty$. Here $T\mathcal{M}$ denotes the tangent bundle and any differential operator must be understood in terms of the *Riemannian metric* $\mathbf{g}$ given on $\mathcal{M}$ (see, e.g. Aubin [8] and Díaz and Tello [26]).

Definition 1 – *We say that $u : \mathcal{M} \to \mathbb{R}$ is a bounded weak solution of (P) if i) $u \in C([0,T]; L^2(\mathcal{M})) \cap L^p(0,T;V) \cap L^\infty(\mathcal{M} \times (0,T))$ and ii) there exists $z \in L^\infty(\mathcal{M} \times (0,T))$ with $z(x,t) \in \beta(u(x,t))$ a.e. $(x,t) \in \mathcal{M}\times(0,T)$ such that*

$$\int_{\mathcal{M}} u(x,T)v(x,T)dA - \int_0^T < v_t(x,t), u(x,t) >_{V' \times V} dt +$$

$$+ \int_0^T \int_{\mathcal{M}} < |\nabla u|^{p-2} \nabla u,\ \nabla v > dAdt + \int_0^T \int_{\mathcal{M}} \mathcal{G}(u) v\, dAdt =$$

$$= \int_0^T \int_{\mathcal{M}} QS(x)z(x,t)\ v dAdt + \int_0^T \int_{\mathcal{M}} f\, v\, dAdt + \int_{\mathcal{M}} u_0(x)v(x,0)\, dA$$

$$\forall v \in L^p(0,T;V) \cap L^\infty(\mathcal{M}\times(0,T)) \text{ such that } v_t \in L^{p'}(0,T;V'),$$

where $<,>_{V' \times V}$ denotes the duality product in $V' \times V$.

We have

Theorem 1 – *There exists at least a bounded weak solution of (P). Moreover, if $T = +\infty$ and f verifies (H_f^∞), the solution u of (P) can be extended to $[0,\infty) \times \mathcal{M}$ in such a way that $u \in C([0,\infty), L^2(\mathcal{M})) \cap L^\infty(\mathcal{M} \times (0,\infty))$ $\cap\ L^p_{loc}(0,\infty;V)$.*

The above result can be proved in different ways. As in the case of the one-dimensional model (Diaz [19]) we can apply the techniques of Diaz and Vrabie [30] based on fixed point arguments which are useful for multivalued non monotone equations. We start by defining the operator $A{:}D(A) \subset$

$L^2(\mathcal{M}) \longrightarrow L^2(\mathcal{M}), A(u) = -\Delta_p u + \mathcal{G}(u)$ if $u \in D(A) = \{u \in L^2(\mathcal{M}) : -\Delta_p u + \mathcal{G}(u) \in L^2(\mathcal{M})\}$. The Cauchy associated problem

$$(P_h) \begin{cases} \dfrac{du}{dt}(t) + Au(t) \ni h(t) & t \in (0,T), \text{ in } X = L^2(\mathcal{M}) \\ u(0) = u_0, & u_0 \in L^2(\mathcal{M}), \end{cases}$$

is well posed (it has a unique mild solution in $C([0,T]; L^2(\mathcal{M}))$ for every $h \in L^2(0,T; L^2(\mathcal{M}))$ by the abstract results of Brezis [14]) since we have

Proposition 1 – *Let*

$$\phi(u) = \begin{cases} \dfrac{1}{p}\displaystyle\int_{\mathcal{M}} |\nabla u|^p dA + \int_{\mathcal{M}} G(u)dA & u \in D(\phi) \\ +\infty & u \notin D(\phi) \end{cases} \tag{4}$$

where $G(u) = \displaystyle\int_0^u \mathcal{G}(\sigma)d\sigma$ *with* $D(\phi) := \{u \in L^2(\mathcal{M}),\ \nabla u \in L^p(T\mathcal{M})$ *and* $\int_{\mathcal{M}} G(u)dA < +\infty\}$. *Then i) ϕ is proper, convex and lower semicontinuous in $L^2(\mathcal{M})$. ii)$A = \partial\phi$, and $\overline{D(A)} = L^2(\mathcal{M})$, and iii) A generates a compact semigroup of contractions $S(t)$ on $L^2(\mathcal{M})$.*

Besides, from Brezis [14] we know that u, solution of(P_h), verifies that $u \in L^p(0,T;V)$, $\sqrt{t}u_t \in L^2(0,T;L^2(\mathcal{M}))$, $u \in W^{1,2}(\delta,T;L^2(\mathcal{M}))$, $0 < \delta < T$. Let us prove the existence of solutions for the problem (P) via a fixed point for a certain operator $\mathcal{L}$. Let $Y = L^p(0,T;L^2(\mathcal{M}))$ and define $\mathcal{L} : K \to 2^{L^p(0,T;L^2(\mathcal{M}))}$ by the following process: Let us define

$$K = \{z \in L^p(0,T;L^\infty(\mathcal{M})) \ : \ \|z(t)\|_{L^\infty(\mathcal{M})} \le C_0 \text{ a.e.} t \in (0,T)\}$$

with $C_0 = QS_1M + \|f\|_{L^\infty(0,T;L^\infty(\mathcal{M}))}$.

Now, we fix $u_0 \in L^2(\mathcal{M})$ and define the *solution operator* (or generalized Green operator) $I_0 : K \to C([0,T];L^2(\mathcal{M}))$ by $I_0(z) = v$ solution of (P_h) associated to $h \equiv z$. Given $f \in L^p((0,T);L^2(\mathcal{M}))$, we define the *superposition operator* $\mathcal{F}$:$L^p((0,T);L^2(\mathcal{M})) \to 2^{L^p((0,T);L^2(\mathcal{M}))}$ by $\mathcal{F}(v(t)) = \{h : h(x,t) \in QS(x)\beta(v(x,t)) + f(x,t) \text{ a.e. } x \in \mathcal{M}\}$.

Finally, we define

$$\mathcal{L}(z) = \{h \in L^p(0,T;L^2(\mathcal{M})) \text{: } h(t) \in \mathcal{F}(I_0(z)(t)) \text{ in } L^2(\mathcal{M}) \text{ a.e. } t \in (0,T)\}.$$

It is not difficult to check (see Diaz and Tello [26]) that using the results by Vrabie [71] and Diaz and Vrabie [30], $\mathcal{L}$ verifies the assumptions required to apply a version of the Schauder-Tychonoff Theorem due to Arino et al. [7] and hence the existence of a local solution to (P) is proved.

In order to complete the proof of Theorem 1, we are going to show that the solution u can be continued up to $t = \infty$, when (H_f^∞) is fulfilled. Taking u as test function we get

$$\int_{\mathcal{M}} u_t u dA + \int_{\mathcal{M}} |\nabla u|^p dA + \int_{\mathcal{M}} \mathcal{G}(u) u dA$$
$$= \int_{\mathcal{M}} QSzudA + \int_{\mathcal{M}} fudA, \quad z \in \beta(u). \tag{5}$$

Then

$$\frac{1}{2}\frac{d}{dt}\int_{\mathcal{M}} |u|^2 dA + \int_{\mathcal{M}} |\nabla u|^p dA + c\int_{\mathcal{M}} |u|^2 dA \leq C + \epsilon \|u\|^2_{L^2(\mathcal{M})}$$

and by Gronwall's inequality $\|u(t)\|_{L^2(\mathcal{M})} \leq k \quad \forall t > 0$ with k independent of t. By a well known result (see e.g. Theorem 4.3.4 of Cazenave and Haraux [16]) u can be extended to $(0, \infty)$ and so $u \in C([0,\infty); L^2(\mathcal{M}))$. To see that $u \in L^\infty(\mathcal{M} \times (0,\infty))$ we introduce $\overline{u}(x,t)$ and $\underline{u}(x,t)$, given as the unique solutions of the problem

$$\begin{cases} \overline{u}_t - \Delta_p \overline{u} + \mathcal{G}(\overline{u}) = MQS(x) + f^+(x,t) & \text{on } \mathcal{M} \times (0,\infty) \\ \overline{u}(0,x) = u_0^+(x) = \max\{0, u_0(x)\} & \text{on} \mathcal{M}, \end{cases}$$

and

$$\begin{cases} \underline{u}_t - \Delta_p \underline{u} + \mathcal{G}(\underline{u}) = mQS(x) + f^-(x,t) \\ \underline{u}(0,x) = u_0^-(x) = \min\{0, u_0(x)\}, \end{cases}$$

respectively. Since the operator $-\Delta_p u + \mathcal{G}(u)$ is T-accretive in $L^2(\mathcal{M})$, it is easy to see that $\underline{u}(x,t) \leq u(x,t) \leq \overline{u}(x,t)$ which proves the assertion once that

$$\|\overline{u}\|_{L^\infty((0,\infty)\times\mathcal{M})} \leq \max\{\|u_0^+\|_\infty \,,\, \mathcal{G}^{-1}(\|MQS\|_\infty + \|f^+\|_\infty)\}$$

$$\|\underline{u}\|_{L^\infty((0,\infty)\times\mathcal{M})} \leq \max\{\|u_0^-\|_\infty \,,\, \mathcal{G}^{-1}(\|mQS\|_\infty + \|f^-\|_\infty)\}.$$

■

Remark 1. More realistic energy balance models are formulated in terms of functional equations adding some non local terms to problem (P) . So, for instance, in linking the albedo to the surface temperature u alone, one neglects the very important long response times the cryosphere exhibits. E.g. the expansion or the retreat of the huge continental ices heets occurs with response times of thousands of years, a feature which Bhattacharya, Ghil and Vulis [13] proposed to incorporate by substituting u by a long term average of u, e.g. $w(x,t) := \int_{-T}^{0} \gamma(x,s) u(x,t+s) ds$ with $T \approx 10^4$ years with

$\gamma(.,-T) \equiv 0$, $\gamma(.,s) > 0$ for $s \in (-T,0]$ and $\int_{-T}^{0} \gamma(.,s)ds = 1$. Of course, one can refine this procedure by having independent ice- and snow-lines. In that case, one understands ice-lines as the boundaries of regions that are covered by continental ice-sheets or huge glaciers (slow response times in comparison with the ten-year mean), whereas snow-lines refer to boundaries of regions where the variations in ice- or snow cover occur on the time-scale of u. This approach was chosen in Diaz and Hetzer [25] (see also Hetzer and Schmidt [48] and Hetzer [45],[46], [44]).

On the other hand, if we disregard the latent energy stored in continental ice sheets and glaciers, the internal energy flux is given by $e(x,t) = c(x)u(x,t)$ with c the heat capacity which varies considerably with x due to the land-water distribution. However, a more accurate modeling suggests to set $e(x,t) = c(x)u(x,t) + h(w(x,t))$ where c denotes the thermal inertia and $h(w(x,t))$ stands for the latent energy density due to huge ice accumulations. This approach is closely related to the one for the Stefan problem (cf., e.g., Meirmanov [55])with the obvious change that h should depend on the long-term temperature mean w rather than on u in view of the time scales relevant for the latent fluxes. Here h is a nonnegative bounded decreasing function with derivative ξ having compact support. Using that $e_t = [cu + h(w)]_t = cu_t + \xi(w)w_t$ and observing that $w_t = \int_{-T}^{0} \gamma(s,.)u_t(.,t+s)ds$ in case that u is sufficiently smooth, one obtains $cu_t + \xi(w)\gamma(.,0)u - \xi(w)\int_{-T}^{0} \gamma_s(.,s)u(.,t+s)ds$ via integration by parts for e_t. Collecting all terms one is led to a non linear and nonlocal quasilinear parabolic problem. Some results on the existence of solutions for such a model can be found in Diaz and Hetzer [25]. We also mention the treatment made in Bermejo, Diaz and Tello [12] for the study of the general case $c = c(x)$ (but without any nonlocal terms) and study of the multi-layer model made in Hetzer and Tello [49].

Remark 2. A more realistic description of the *incoming solar radiation flux* $QS(x)$ is obtained by replacing it by a time depending function $Q(x,t)$ under the general assumption $Q(x,t)$ is $T-$periodical in time and $Q(x,t) \geq 0$. This last inequality allows to consider the polar night phenomena (time where $Q(x,t) = 0$ for some subsets of the manifold $\mathcal{M}$). The consideration of the periodicity of the forcing term is motivated by the seasonal variation of the incoming solar radiation flux during one natural year. The existence of periodical solutions for the associated model was the object of the paper Badii and Diaz [9]. The existence of periodic solutions for the Sellers type model was obtained by considering the Poincaré map F associated to the Cauchy problem (P), i.e. the operator assigning, to every initial data, the solution of (P) evaluated after T-period. We prove that F is a continuous, compact and pointwise increasing map and so, the Schauder fixed point theorem can be applied. The existence of a periodic solution for the Budyko type model needs some different arguments. The Poincaré map can not be well defined as a univalued operator and so we apply a variant of the Schauder-Tychonoff fixed point theorem for a suitable multivalued operator.

Remark 3. For different purposes it is useful to get existence results via regularization of the multivalued term $\beta(u)$. See, e.g., Xu [72] and Feireisl and Norbury [33] for some special formulations when $p = 2$. In our case it can be obtained as an easy adaptation of the results of Section 3. We also mention some results on the numerical approach due to Lin and North [51], Hetzer, Jarausch and Mackens [42], Bermejo [11] and Diaz, Bermejo and Tello [12].

2.2. On the uniqueness of solutions

The question of uniqueness has different answers for the different coalbedo functions under consideration depending on whether the coalbedo is supposed to be discontinuous or not. For the Sellers model (β locally Lipschitz), the uniqueness is obtained by standard methods (see e.g. Diaz [19]). Nevertheless, in the Budyko model (β multivalued), there are cases of nonuniqueness (in spite of the parabolic nature of(P)). The first nonuniqueness result in this context seems to be the one given in Diaz [19] where infinitely many solutions are found for the one-dimensional model (P^1) for any initial condition u_0 satisfying

$$\left.\begin{array}{l} u_0 \in C^\infty(I),\ u_0(x) = u_0(-x)\ \forall x \in [0,1], \\ u_0(0) = -10,\ u_0^{(k)}(0) = 0,\ k = 1,2 \\ u_0'(1) = 0,\ u_0'(x) < 0,\ x \in (0,1). \end{array}\right\} \tag{6}$$

Notice that these initial data u_0 are very "flat" at the level-10. A similar non uniqueness result for the Budyko model with a suitable initial datum carries over to the two-dimensional model when $\mathcal{M} = S^2$. Each solution $u_1(x,t)$ of (P^1) generates a solution $u_2(x,y,t)$ of 2D model by rotation about the axis through the poles (notice that the initial datum $u_2(x,y,0)$ is independent of the longitude), i.e. $u_2(x,y,t) = u_1(\sin\theta, t)$ where $(x,y) \in S^2$ with latitude θ. It is not difficult to prove that u_2 is asolution of (P) for the initial datum $u_1 \sin\theta, 0)$. Other non uniqueness results can be found by using selfsimilar special solutions as in Gianni and Hulshof [37].

Since non uniqueness of solutions may arise, it is useful to know (see Diaz and Tello [26]) that in any case problem (P) has a maximal solution u^* and a minimal solution u_* (i.e. u^* and u_* are solutions of (P) such that everysolution u of (P) verifies that $u_* \leq u \leq u^*$ in$(0,T) \times \mathcal{M})$. In order to obtain a criterion for the uniqueness of solutions for Budyko type models we introduce the notion of *nondegeneracy property* for functions defined on $\mathcal{M}$.

Definition 2 – *Let $w \in L^\infty(\mathcal{M})$. We say that w satisfies the strong nondegeneracy property (resp. weak) if there exist $C > 0$ and $\epsilon_0 > 0$ such that for any $\epsilon \in (0, \epsilon_0)$, $|\{x \in \mathcal{M} : |w(x) + 10| \leq \epsilon\}| \leq C\epsilon^{p-1}$ (resp. $|\{x \in \mathcal{M} : 0 < |w(x) + 10| \leq \epsilon\}| \leq C\epsilon^{p-1}$), where $|E|$ denotes the Lebesgue measure on the manifold $\mathcal{M}$ for all $E \subset \mathcal{M}$.*

Theorem 2 – *i) Assume that there exists a solution u of (P) suchthat $u(.,t)$ verifies the strong nondegeneracy property for any $t \in [0,T]$. Then u is the unique bounded weak solution of (P). ii) There exists at most one solution of (P) verifying the weak nondegeneracy property.*

The proof is based in the fact (adapted from Feireisl and Norbury [33]) that β generates a continuous operator from $L^\infty(\mathcal{M})$ to $L^q(\mathcal{M})$, $\forall q \in [1, \infty)$, although β is discontinuous, when the domain of such operator is the set of functions verifying the strong nondegeneracy property. More precisely, we have (see Diaz [19], Diaz and Tello [26])

Lemma 1 – *(i) Let $w, \hat{w} \in L^\infty(\mathcal{M})$ and assume that w satisfies the strong nondegeneracy property. Then for any $q \in [1, \infty)$, there exists $\tilde{C} > 0$ such that for any $z, \hat{z} \in L^\infty(\mathcal{M})$ with $z(x) \in \beta(w(x))$ and $\hat{z}(x) \in \beta(\hat{w}(x))$ a.e. $x \in \mathcal{M}$, we have*

$$\| z - \hat{z} \|_{L^q(\mathcal{M})} \leq (b_w - b_i) \min\{\tilde{C} \| w - \hat{w} \|_{L^\infty(\mathcal{M})}^{(p-1)/q} , |\mathcal{M}|^{1/q}\}. \tag{7}$$

(ii) If $w, \hat{w} \in L^\infty(\mathcal{M})$ and satisfy the weak nondegeneracy property then

$$\int_{\mathcal{M}} (z(x) - \hat{z}(x))(w(x) - \hat{w}(x))dA \leq (b_w - b_i)C \| w - \hat{w} \|_{L^\infty(\mathcal{M})}^p . \tag{8}$$

Idea of the proof of Theorem 2. Assume that there exist two bounded weak solutions u and $\hat{u}$ of (P), where u verifies the strong nondegeneracy property, i.e. $u_t - \Delta_p u + \mathcal{G}(u) = QSz + f$ and $\hat{u}_t - \Delta_p \hat{u} + \mathcal{G}(\hat{u}) = QS\hat{z} + f$, in $(0,T) \times \mathcal{M}$, for some $z \in \beta(u)$ and $\hat{z} \in \beta(\hat{u})$. Taking $(u - \hat{u})$ as the test function we get

$$\frac{1}{2}\frac{d}{dt}\int_{\mathcal{M}} |u(t) - \hat{u}(t)|^2 dA + \int_{\mathcal{M}} (\mathcal{G}(u) - \mathcal{G}(\hat{u}))(u - \hat{u})dA +$$

$$\int_{\mathcal{M}} < |\nabla u(t)|^{p-2}\nabla u(t) - |\nabla \hat{u}(t)|^{p-2}\nabla \hat{u}(t),\ \nabla u(t) - \nabla \hat{u}(t) > dA =$$

$$= Q\int_{\mathcal{M}} S(x)(z(x,t) - \hat{z}(x,t))(u(x,t) - \hat{u}(x,t))dA. \tag{9}$$

By using the embedding $V \hookrightarrow L^\infty(\mathcal{M})$ if $p > 2$ and $V \subset L^\sigma(\mathcal{M})$ for all $\sigma \in [1, \infty)$ if $p = 2$ (recall that $\mathcal{M}$ is a two-dimensional compact Riemannian manifold: see, e.g. Aubin [8]) we arrive at

$$\begin{aligned}\frac{1}{2}\frac{d}{dt} \parallel u - \hat{u} \parallel^2_{L^2(\mathcal{M})} &\leq (C_l Q \parallel S \parallel_{L^\infty(\mathcal{M})} - \frac{C_0}{C_{1,p,\infty}}) \parallel u - \hat{u} \parallel^p_{L^\infty(\mathcal{M})} + \\ &+ \tilde{C}_0 \parallel u - \hat{u} \parallel^2_{L^2(\mathcal{M})}, \end{aligned} \quad (10)$$

in the case $p > 2$ and

$$\begin{aligned}\frac{1}{2}\frac{d}{dt} \parallel u - \hat{u} \parallel^2_{L^2(\mathcal{M})} &\leq (C_l Q \parallel S \parallel_{L^\infty(\mathcal{M})} - \frac{|\mathcal{M}|^{\frac{2}{\sigma}}}{C_{1,2,\sigma}}) \parallel u - \hat{u} \parallel^2_{L^\infty(\mathcal{M})} + \\ &+ \parallel u - \hat{u} \parallel^2_{L^2(\mathcal{M})} + \frac{\epsilon}{C_{1,2,\sigma}}, \end{aligned} \quad (11)$$

for the case $p = 2$ where ϵ and $\sigma = \sigma(\epsilon)$ .Now, we distinguish two cases: CASE 1: if $C_l Q \parallel S \parallel_\infty - \dfrac{C_0}{C_{1,p,\infty}} \leq 0$ and $p > 2$, then

$$\frac{1}{2}\frac{d}{dt} \parallel u - \hat{u} \parallel^2_{L^2(\mathcal{M})} \leq \tilde{C}_0 \parallel u - \hat{u} \parallel^2_{L^2(\mathcal{M})} .$$

and the result holds by Gronwall's Lemma. If $p = 2$ and $C_l Q \parallel S \parallel_\infty - \dfrac{1}{C_{1,2,\sigma}} \leq 0$

$$\begin{aligned}\parallel u - \hat{u} \parallel^2_{L^2(\mathcal{M})} &\leq e^{2t} \parallel u_0 - \hat{u}_0 \parallel^2_{L^2(\mathcal{M})} - \frac{2\epsilon}{C_{1,2,\sigma}}(e^{2t} - 1) \leq \\ &\leq -2\epsilon C_l Q \parallel S \parallel_\infty (e^{2t} - 1)\end{aligned}$$

and since the above inequality is true for all ϵ, we conclude the uniqueness.

CASE 2: if $C_l Q \parallel S \parallel_\infty - \frac{1}{C_{1,p,\infty}} > 0$, we consider a suitable rescaling $(\mathcal{M} \mapsto \mathcal{M}_\delta)$ given by the dilatation D of magnitude $\delta > 0$ on the manifold $(\mathcal{M}, \mathbf{g})$, $D : \mathcal{M} \subset \mathbb{R}^3 \to \mathbb{R}^3$, $D(x) = \tilde{x} = \delta x$. So, if u is any function defined on $\mathcal{M}$, its local representation in the new coordinates is $\tilde{u} : \mathcal{M}_\delta \to \mathbb{R}$, $\tilde{u}(\tilde{x}) = u(\frac{\tilde{x}}{\delta})$ and we have

$$\frac{\partial \tilde{u}}{\partial \tilde{x}_i}(\tilde{x}) = \frac{1}{\delta}\frac{\partial u}{\partial x_i}(\frac{\tilde{x}}{\delta}) \qquad i = 1, 2, 3.$$

So problem (P) in the new coordinates becomes

$$(P_\delta) \begin{cases} \tilde{u}_t - \delta^p \mathrm{div}_{\mathcal{M}_\delta} \left(|\nabla_{\mathcal{M}_\delta} \tilde{u}|^{p-2} \nabla_{\mathcal{M}_\delta} \tilde{u}\right) + \mathcal{G}(\tilde{u}) \in QS\beta(\tilde{u}) + f & \text{in } (0,T) \times \mathcal{M}_\delta \\ \tilde{u}(0, \tilde{x}) = u_0(\frac{\tilde{x}}{\delta}) & \text{on } \mathcal{M}_\delta. \end{cases}$$

Clearly, if $\tilde{u}$ is a solution of (P_δ) then $\tilde{u}(\delta x, t)$ is a solution of (P). Moreover, the uniqueness of (P_δ) implies the uniqueness of (P), and conversely.

Let us see that there exists $\delta > 0$ such that the solution of (P_δ) is unique. Let u_δ and $\tilde{u}_\delta$ be two solutions of (P_δ) with u_δ verifying the strong nondegeneracy property. Arguing as before we arrive at

$$\frac{1}{2}\frac{d}{dt}\int_{\mathcal{M}_\delta} |u_\delta(t) - \hat{u}_\delta(t)|^2 dA_\delta$$

$$+\delta^p \int_{\mathcal{M}_\delta} < |\nabla u_\delta|^{p-2}\nabla u_\delta - |\nabla \hat{u}_\delta|^{p-2}\nabla \hat{u}_\delta, \nabla u_\delta - \nabla \hat{u}_\delta > dA_\delta$$

$$+\int_{\mathcal{M}_\delta} (\mathcal{G}(u_\delta) - \mathcal{G}(\hat{u}_\delta))(u_\delta - \hat{u}_\delta) dA_\delta = Q \int_{\mathcal{M}_\delta} S_\delta (z_\delta - \hat{z}_\delta)(u_\delta - \hat{u}_\delta) dA_\delta$$

for some $z_\delta \in \beta(u_\delta)$ and $\hat{z}_\delta \in \beta(\hat{u}_\delta)$. Here, S_δ is defined by $S_\delta : \mathcal{M}_\delta \to \mathbb{R}$, $S_\delta(\tilde{x}) = S(\frac{\tilde{x}}{\delta})$. (10) and (11) allow us toestimate $u_\delta - \hat{u}_\delta$ for u_δ and $\hat{u}_\delta$ solutions of (P_δ)

$$\begin{aligned}
&\frac{1}{2}\frac{d}{dt} \| u_\delta - \hat{u}_\delta \|^2_{L^2(\mathcal{M}_\delta)} \\
&\quad \le (C_{l,\delta}\, Q \| S_\delta \|_{L^\infty(\mathcal{M}_\delta)} - \frac{\delta^2 |\mathcal{M}_\delta|^{\frac{1}{\sigma}}}{C_{1,2,\sigma,\delta}}) \| u_\delta - \hat{u}_\delta \|^2_{L^\infty(\mathcal{M}_\delta)} \\
&\quad + \| u_\delta - \hat{u}_\delta \|^2_{L^2(\mathcal{M}_\delta)} + \frac{\epsilon}{C_{1,2,\sigma,\delta}}.
\end{aligned} \tag{12}$$

A careful study of the dependence on δ of the involved constants (see Diaz and Tello [26]) allows us to see that if we define the constant

$$K_{p,\delta} = C_{l,\delta} Q \; \| S_\delta \|_{L^\infty(\mathcal{M}_\delta)} - \frac{\delta^2 |\mathcal{M}|^{\frac{2}{\sigma}}}{C_{1,2,\sigma,\delta}},$$

we have that $\lim_{\delta \to 0} K_{p,\delta} = 0$. This reduces the proof to Case 1 and the proof of (i) follows. For the proof of (ii) we use the second part of Lemma 1 and so

$$\frac{1}{2}\frac{d}{dt} \| u - \hat{u} \|_2^2 \le (C_d Q \| S \|_{L^\infty(\mathcal{M})} - \frac{C_0}{C_{1,p,q}}) \| u - \hat{u} \|^p_\infty + \tilde{C}_0 \| u - \hat{u} \|_2^2$$

where C_d is the constant of the weak nondegeneracy property (Lemma1). The uniqueness follows as in (i), by studying the sign of the constant $C_d Q \| S \|_{L^\infty(\mathcal{M})} - \frac{1}{C_{1,p,q}}$ and by rescaling when it is negative. ■

Remark 4. It is possible to give several sufficient criteria for the nondegeneracy property. For instance, in the one-dimensional case and $p = 2$, if $u_0 \in C^1((-1,1))$ is such that there exists $\epsilon_0 > 0$ such that the set $\{x \in (-1,1) : |u_0(x) + 10| \leq \epsilon_0\}$ has a finite number of connected components I_j with $j = 1,..,N$ and for any j there exists $x_j \in I_j$ such that $u_0(x_j) = -10$,and $|u_{0x}(x)| \geq \delta_0$ for some $\delta_0 > 0$ and any $x \in I_j$ close to x_j, then there exists a solution $u(x,t)$ satisfying the strong nondegeneracy property on $(0, T^*)$ for some T^* (see Diaz and Tello [26]). Some results on solutions with $|\nabla u| \neq 0$ on the level where β becomes multivalued for a similar bidimensional problem are given in Gianni [36].

2.3. On the free boundary for Budyko type models

The discontinuity of the albedo function assumed in the Budyko model (β multivalued) generates a natural *free boundary* or interface $\zeta(t)$ between the ice-covered area ($\{x \in \mathcal{M} : u(x,t) < -10\}$) and the ice-free area ($\{x \in \mathcal{M} : u(x,t) > -10\}$). The free boundary is then given as $\zeta(t) = \{x \in \mathcal{M} : u(x,t) = -10\}$. In Xu [72], the Budyko model for $p = 2$ is considered in the one-dimensional case. He shows that if the initial datum u_0 satisfies

$$\begin{aligned}&u_0(x) = u_0(-x),\ u_0 \in C^3([-1,1]),\ u_0'(x) < 0 \text{ for any } x \in (0,1)\\ &\text{and there exists } \zeta(0) \in (0,1) \text{ such that } (u_0(x) + 10)(x - \zeta(0)) < 0\\ &\text{for any } x \in [0, \zeta(0)) \cup (\zeta(0), 1],\end{aligned}$$

then there exists a bounded weak solution u of (P) for which one has $\zeta(t) = \{\zeta_+(t)\} \cup \{\zeta_-(t)\}$ with $x = \zeta_+(t)$ a smooth curve, $\zeta_-(t) = \zeta_+(t)$ and $\zeta_+(\cdot) \in C^\infty([0, T^*))$, where T^* is characterized as the first time t for which $\zeta_+(t) = 1$. He also gives an expression for the derivative $\zeta_+'(t)$ (some related results for a model corresponding to $\rho(x) = 1$ can be found in Feireisl and Norbury [33], Gianni and Hulshof [37] and Stakgold [67]). We point out that the uniqueness result can be applied for such an initial datum. For the study of the free boundary in the bidimensional case see Diaz [22] and Gianni [36].

The interpretation of the size of the separating zone $\zeta(t)$ for other models is in fact a controversial question. So, some satellite pictures (Image of the Weddell sea taken by the satellite Spot on December 10, 1987) show that the separating region between the ice-free and the ice-covered zones is not a simple line on the Earth but a narrow zone where ice and water are mixed. Mathematically it could correspond to say that the set

$$M(t) = \{x \in \mathcal{M} \colon\ u(x,t) = -10\}$$

is a positively measured set. In the following we shall denote this set as the *mushy region* (since it plays the same role than in changing phase problems, see e.g. Diaz , Fasano and Meirmanov [23]).

Using the strong maximum principle, it is possible to show that if $p = 2$ the interior set of the mushy region $M(t)$ is empty even if the interior of

$M(0)$ is a nonempty open set (see Gianni and Hulshof [37]). As we shall see, this is not the case when $p > 2$ (recall that $p = 3$ in Stone [68]). A necessary condition for the Budyko model (with $R_e = Bu + C$) for $\overset{\circ}{M}(t) \neq \emptyset$ is that

$$C - 10B \in [\beta_i QS(x), \beta_w QS(x)] \quad \text{for a.e. } x \in \mathcal{M}. \tag{13}$$

It is possible to show that if $p > 2$, this condition is also sufficient. Here we merely present a result for the one-dimensional case (see Diaz [22] for the bidimensional case):

Theorem 3 – *Let $p > 2$. Assume (13) and $u_0 \in L^\infty(I)$ such that there exist $x_0 \in I$ and $R_0 > 0$ satisfying*

$$M(0) = \{x \in I : u_0(x) = -10\} \supset B(x_0, R_0)(= \{x \in I : |x - x_0| < R_0\}).$$

If u is the bounded weak solution of (P) satisfying the weak nondegeneracy property, then there exist $T^* \in (0, T]$ and a nonincreasing function $R(t)$ with $R(0) = R_0$ such that

$$M(t) = \{x \in I : u(x,t) = -10\} \supset B(x_0, R(t))$$

for any $t \in [0, T^*)$.

Proof. We shall use an energy method as developed in Diaz and Veron [29]. Given u bounded weak solution of (P), we define $v = u + 10$. As in Lemma 3.1 of the above reference, by multiplying the equation by v we obtain that for a.e. $R \in (0, R_0)$ and $t \in (0, T)$, we have

$$\frac{1}{2}\int_{B(x_0,R)} |v(x,t)|^2 dx + \int_0^t \int_{B(x_0,R)} \rho(x)|v_x|^p dx d\tau$$
$$+B \int_0^t \int_{B(x_0,R)} |v(x,\tau)|^2 dx d\tau \leq$$
$$\leq \int_0^t \int_{S(x_0,R)} \rho |v_x|^{p-2} v_x \cdot \vec{n} v ds d\tau + \int_0^t \int_{B(x_0,R)} \{QSz - C + 10B\} v dx d\tau$$
$$= I_1 + I_2,$$

where $\rho(x) = (1 - x^2)^{\frac{p}{2}}$, $S(x_0, R) = \partial B(x_0, R) = \{x_0 - R\} \cup \{x_0 + R\}$ and $z(x,t) \in \beta(u(x,t))$ for a.e. $x \in B(x_0, R)$ and $t \in (0, T]$. We introduce the energy functions

$$E(R,t) = \int_0^t \int_{B(x_0,R)} \rho(x)|v_x|^p dx d\tau$$
$$b(R,t) = \sup_{0 \leq \tau \leq t} \text{ess} \int_{B(x_0,R)} |v(x,\tau)|^2 dx.$$

Using Holder's inequality and the interpolation-trace Lemma of Diaz -Veron [29], we get

$$I_1 \leq \left(\frac{\partial E}{\partial R}(R,t)\right)^{(p-1)/p} \left(\int_0^t \int_{S(x_0,R)} |v|^p dx d\tau\right)^{1/p} \leq$$

$$\leq Ct^{(1-\theta)/p} \left(\frac{\partial E}{\partial R}(R,t)\right)^{(p-1)/p}$$

$$\times \left(E(R,t)^{1/p} + R^\delta t^{1/p} b(R,t)^{1/2}\right)^\theta b(R,t)^{(1-\theta)/2},$$

where

$$\theta = p/(3p-2) \text{ and } \delta = -(3p-2)/2p.$$

Using the assumption (13), we have $\hat{z}(\cdot) = [(C - 10B)/QS(\cdot)] \in \beta(-10)$. Then applying Lemma 3 we get

$$I_2 \leq (M-m)Q \parallel S \parallel_{L^\infty(I)} C' \int_0^t \parallel v(\tau) \parallel^p_{L^\infty(B(x_0,R)} d\tau.$$

From Theorem 4 of Rakotoson and Simon [62], we have the estimate

$$\parallel v \parallel_{L^\infty(J)} \leq C_1 \parallel v_x \parallel_{L^p(J:\rho)} + C_2 \parallel v \parallel_{L^1(J:\rho)}, \quad \forall v \in V, \tag{14}$$

for some positive constants independent on the interval J. Then we obtain

$$I_2 \leq (M-m)Q \parallel S \parallel_{L^\infty(I)} C'(C_1 E(R,t) + tC_3(R) b(R,t)),$$

where now

$$C_3(R) = \frac{\left(\int_{B(x_0,R)} \rho(x)^2 dx\right)^{p-2}}{C_1^p \left(\int_{B(x_0,R)} \rho(x) dx\right)^p} \parallel u + 10 \parallel^p_{L^\infty((0,T):L^2(I))}.$$

As in the proof of the uniqueness, we can assume C_1 small enough without loss of generality. Then, there exists $T^* \in (0,T]$ and $\lambda \in (0,1]$ such that

$$\lambda(E(R,t) + b(R,t)) \leq I_1$$

which implies that

$$\lambda E^\mu \leq t^{(1-\theta)/p} \frac{\partial E}{\partial R}$$

for some $\mu \in (0,1)$ and for any $t \in [0, T^*)$. The proof ends as in Diaz and Veron [29](see also and Antonsev and Diaz [4]). ■

Remark 5. The existence of the mushy region (for anyvalue of $p \in (1, \infty)$) can be proved for a different class of models by taking into account a discontinuous diffusivity (see Held, Linder and Suárez [40]). In that case the problem is a variant of the Stefan problem (see, e.g., Meirmanov [55]). It would be interesting to find sufficient conditions implying the persistence of a mushy region for any time $t > 0$. The fact that a mushy region may exist for the stationary problem can be found from the results of Diaz [?] (see Theorem 1.14).

2.4. Stabilization of solutions when $t \to +\infty$

In order to analyze the stabilization of the solutions of (P), following Diaz, Hernández and Tello [24], we assume the additional condition

- $(H_\infty)]$ $f \in L^\infty((0, \infty) \times \mathcal{M})$ and there exists $f_\infty \in V'$ such that

$$\int_{t-1}^{t+1} \|f(\tau, \cdot) - f_\infty(\cdot)\|_{V'} d\tau \to 0 \quad \text{as } t \to \infty.$$

We start by recalling a global regularity of the solutions on $(0, \infty)$.

Lemma 2 – *Assume we are given* $u_0 \in V \cap L^\infty(\mathcal{M})$, $f \in L^\infty(\mathcal{M} \times (0, \infty)) \cap W^{1,1}_{loc}((0, \infty); L^1(\mathcal{M}))$ *and* $\int_t^{t+1} \|f_t(s, \cdot)\|_{L^1(\mathcal{M})} ds \leq C_0$, $\forall t > 0$ *where* C_0*is a time independent constant. Then there exists a weak solution of* (P) *such that*

$$u \in L^\infty(0, \infty; V) \ \ \textit{and} \ \ u_t \in L^2(0, \infty; L^2(\mathcal{M})). \tag{15}$$

A key point in the proof is to check that $\varphi(t) = \frac{1}{p} \int_{\mathcal{M}} |\nabla u(x, t)|^p dA$ satisfies

$$\varphi(t + 1) \leq C[\varphi(t) - \varphi(t + 1)] + \theta(t) \qquad t > 0, \tag{16}$$

where C is a positive constant and $\theta(t) > 0$ when t is large enough with $\theta(t) = O(1)$ when $t \to \infty$. Then, thanks to a technical lemma due to Nakao [56], we conclude that $\varphi(t) = O(1)$ which is equivalent to $u \in L^\infty(0, \infty; V)$.

The following theorem proves the stabilization of the solutions u satisfying (15). As usual, given u bounded weak solution of(P), we define the ω-limit set of u by

$$\omega(u) = \{u_\infty \in V \cap L^\infty(\mathcal{M}) : \exists t_n \to +\infty \text{ such that } u(t_n, \cdot) \to u_\infty \text{ in } L^2(\mathcal{M})\}.$$

Theorem 4 – *Let* $u_0 \in L^\infty(\mathcal{M}) \cap V$ *and let* u *be any bounded weak solution satisfying (15). Then, i)* $\omega(u) \neq \emptyset$ *and if* $u_\infty \in \omega(u)$, $\exists t_n \to +\infty$ *such that* $u(\cdot, t_n + s) \to u_\infty$ *in* $L^2(-1, 1; L^2(\mathcal{M}))$ *and* $u_\infty \in V$ *is a weak solution of the stationary problem associated to* f_∞ *; ii) in fact, if* $u_\infty \in \omega(u)$, *then* $\exists \{\hat{t}_n\} \to +\infty$ *such that* $u(\cdot, \hat{t}_n) \to u_\infty$ *strongly in* V.

Proof. Let u_∞ be an element of $\omega(u)$. Then,

$$\|u(t_n+s)-u(t_n)\|^2_{L^2(\mathcal{M})} \leq 2\ \|u_t\|^2_{L^2((t_n-1,t_n+1);L^2(\mathcal{M}))}.$$

Since $u_t \in L^2(0,\infty;L^2(\mathcal{M}))$, $\|u_t\|^2_{L^2((t_n-1,t_n+1);L^2(\mathcal{M}))} \to 0$ when $t_n \to \infty$ and by the Lebesgue convergence theorem we conclude that $u(\cdot,t_n+\cdot) \to u_\infty$ in $L^2((-1,1);L^2(\mathcal{M}))$.

To prove that u_∞ is a solution of (P_Q), we consider the test functions $v(x,t) = \xi(x)\varphi(t-t_n)$ with $\xi \in V \cap L^\infty(\mathcal{M})$ and $\varphi \in \mathcal{D}(-1,1)$, $\varphi \geq 0$, $\int_{-1}^1 \varphi = 1$. Then

$$\int_{t_n-1}^{t_n+1}\int_{\mathcal{M}} u_t\xi\varphi(t-t_n) + \int_{t_n-1}^{t_n+1}\int_{\mathcal{M}} |\nabla u|^{p-2}\nabla u\cdot\nabla\xi\varphi(t-t_n)$$
$$+\int_{t_n-1}^{t_n+1}\int_{\mathcal{M}} \mathcal{G}(u)\xi\varphi(t-t_n)$$
$$= \int_{t_n-1}^{t_n+1}\int_{\mathcal{M}} Qz\xi\varphi(t-t_n) - \int_{t_n-1}^{t_n+1}\int_{\mathcal{M}} f(x,t)\xi\varphi(t-t_n) \quad z\in\beta(u(x,t)).$$

Changing variables, namely $s = t-t_n$ and defining $U_n(x,s) = u(x,t_n+s)$, we obtain that

$$\begin{array}{llll} U_n \rightharpoonup u_\infty & \text{weakly in} & L^\sigma((-1,1);V) & \forall\sigma>1 \\ |\nabla U_n|^{p-2}\nabla U_n \rightharpoonup Y & \text{weakly in} & L^\sigma((-1,1);L^p(T\mathcal{M})) & \forall\sigma>1. \end{array}$$

Applying Aubin's compactness result (see e.g. Simon [66]), a well known property of the maximal monotone graphs (see Brezis [14]) and Lebesgue's theorem, we get that $z_n \rightharpoonup z_\infty \in \beta(u_\infty)$ weakly in $L^\sigma(\mathcal{M}\times(-1,1))$ $\forall\sigma>1$ and $\mathcal{G}(U_n)\to\mathcal{G}(u_\infty)$ in $L^1(\mathcal{M}\times(-1,1))$. Letting $n\to\infty$, we arrive to

$$\int_{-1}^1\int_{\mathcal{M}} Y\cdot\nabla\xi\varphi + \int_{\mathcal{M}}\mathcal{G}(u_\infty)\xi = \int_{\mathcal{M}} QSz_\infty\xi + \int_{\mathcal{M}} f_\infty\xi \quad \forall\xi\in V\cap L^\infty(\mathcal{M}).$$

Now, the main difficulty is to prove that

$$\int_{-1}^1 Y(s,\cdot)\varphi(s) = |\nabla u_\infty|^{p-2}\nabla u_\infty.$$

Due to the coercivity of the p-Laplacian operator we obtain the following inequality :

$$\lim_{n\to\infty}\int_{-1}^1\int_{\mathcal{M}}(|\nabla U_n|^{p-2}\nabla U_n - |\nabla\chi|^{p-2}\nabla\chi)\cdot(\nabla u_\infty - \nabla\chi)\varphi(s)\,dAds \geq 0, \tag{17}$$

which holds for $\chi \in V$. We arrive to the desired convergence by taking $\chi = u_\infty + \lambda\xi$ and applying a Minty type argument to(17) as in Diaz and de Thélin [28].

The proof of (ii) uses the coerciveness of the operator and the fact that

$$\int_{-1}^{1}\int_{\mathcal{M}}(|\nabla U_n|^{p-2}\nabla U_n - |\nabla u_\infty|^{p-2}\nabla u_\infty)\cdot(\nabla U_n - \nabla u_\infty)\varphi(s)\, dAds \to 0.$$

The inequality $|\zeta - \hat{\zeta}|^p \leq (|\zeta|^{p-2}\zeta - |\hat{\zeta}|^{p-2}\hat{\zeta})\cdot(\zeta - \hat{\zeta})$, $\forall \zeta, \hat{\zeta} \in \mathbb{R}^N$ allows us to obtain

$$\lim_{n\to\infty}\int_{-1}^{1}\int_{\mathcal{M}}|\nabla U_n - \nabla u_\infty|^p\varphi(s)\, dAds \;=\; 0,\ \forall\varphi.$$

This implies that there exists a subsequence $\{s_n\}_{n\in\mathbb{N}}$, where $s_n \in (-1,1)$ such that

$$\lim_{n\to\infty}\int_{\mathcal{M}}|\nabla u(t_n + s_n, \cdot) - \nabla u_\infty|^p dA = 0$$

and so we prove the assertion. ■

Remark 6. If u_∞ is an isolated point of $\omega(u)$, it is easy to see that in fact the above convergences hold as $t \to \infty$ (and not merely for a sequence $t_n \to \infty$). The proof of this convergence is an open problem in the remaining cases. In fact, in some cases the set of stationary points is a continuum (see Remark 11) and the convergence when $t \to \infty$ is far from trivial (for some results in this direction see Feireisl and Simondon [34]).

Remark 7. A result on the convergence (in a suitable sense) of the free boundaries to the free boundary of the solution of the stationary problem is given in Diaz [22] (see also Gianni [36]).

Remark 8. The question of the approximate controllability was considered in Diaz [?] and [21]. To avoid technical difficulties, in these articles the manifold $\mathcal{M}$ is replaced by an open regular bounded set Ω of $\mathbb{R}^2$ (here $\mathbb{R}^2$ can be also substituted by $\mathbb{R}^N$ with $N \in \mathbb{N}$) and p is taken as $p = 2$. As a boundary condition on $(0,T)\times\partial\Omega$, it is chosen the one of Neumann type since it leads to a set of test functions for the weak formulation very similar to the one corresponding to the case of a Riemannian manifold without boundary. The case of an internal control is considered by taking $f(x,t)=v(x,t)\chi_\omega$ with v the control to be searched, and χ_ω, the characteristic function of ω, a given open bounded subset of Ω. Thus, the new formulation is now the following: given y_0, $y_d : \Omega \to \mathbb{R}$ and $\varepsilon > 0$, find $v_\varepsilon : \omega \times (0,T) \to \mathbb{R}$ such that $d(y(T : v_\varepsilon), y_d) \leq \varepsilon$ where, in general, $y(T : v)$ represents the solution of problem

$$(\mathcal{P}_\omega)\begin{cases} y_t - \Delta y + \mathcal{G}(y) \in QS(x)\beta(y) + v\chi_\omega & \text{in } \Omega \times (0,T) \\ \dfrac{\partial y}{\partial n} = 0 & \text{on } \partial\Omega \times (0,T) \\ y(0,\cdot) = y_0(\cdot) & \text{on } \Omega, \end{cases}$$

where n is the outer unit vector to $\partial\Omega$. It is shown that the answer to the approximate controllability question depends on the asymptotic behaviour of the nonlinearity $\mathcal{G}(y)$ of the equation. If, for instance $\mathcal{G}(y) = |y|^{p-2}y$, then the approximate controllability property holds when $p \in (0,1]$ but if $p > 1$, an *obstruction phenomenon* appears, implying the impossibility of the controllability for general data. Some results concerning a special class of data for the superlinear case $p > 1$ are presented in Diaz [21]. We point out that in 1955, John von Neumann [57] proposed to control the climate by acting on the albedo and that this still remains a mathematical open question. Finally, we mention the "rain making" (see Dennis [17]) as a practical example of the application of control problems in environment.

3. On the stationary problem

We consider the problem $(P_{Q,f})$ obtained in the last subsection. Following Diaz, Hernández and Tello [24], we made in this section the additional assumptions

$-(\mathrm{H}^*_{\mathcal{G}})$ $\mathcal{G}$ satisfies $(\mathrm{H}_{\mathcal{G}})$ and $\lim_{|s|\to\infty} |\mathcal{G}(s)| = +\infty$.

$-(\mathrm{H}_{f_\infty})$ $f_\infty \in L^\infty(\mathcal{M})$ and there exists $C_f > 0$ such that $-\|f_\infty\|_{L^\infty(\mathcal{M})} \leq f(x) \leq -C_f$ *a.e.* $x \in \mathcal{M}$

$-(\mathrm{H}^*_\beta)$ there exist two real numbers $0 < m < M$ and $\epsilon > 0$ such that $\beta(r) = \{m\}$ for any $r \in (-\infty, -10-\epsilon)$ and $\beta(r) = \{M\}$ for any $r \in (-10+\epsilon, +\infty)$.

$-(\mathrm{H}_{C_f})$ $\mathcal{G}(-10-\epsilon) + C_f > 0$ and $\dfrac{\mathcal{G}(-10+\epsilon) + \|f_\infty\|_{L^\infty(\mathcal{M})}}{\mathcal{G}(-10-\epsilon) + C_f} \leq \dfrac{S_0 M}{S_1 m}$.

A function $u \in V \cap L^\infty(\mathcal{M})$ is called a *bounded weak solution* of $(P_{Q,f})$ if there exists $z \in L^\infty(\mathcal{M})$, $z(x) \in \beta(u(x))$ a.e. $x \in \mathcal{M}$ such that

$$\int_{\mathcal{M}} (|\nabla u|^{p-2}\nabla u) \cdot \nabla v \, dA + \int_{\mathcal{M}} \mathcal{G}(u) v \, dA = \int_{\mathcal{M}} QS(x) z v \, dA + \int_{\mathcal{M}} f_\infty v \, dA,$$

for any $v \in V$.

3.1. Existence of at least three solutions for suitable Q

We start with a multiplicity result given in Diaz, Hernández and Tello [24]

Theorem 5 – *Let u_m, u_M be the (unique) solutions of the problems*

$$(P_m) \quad -\Delta_p u + \mathcal{G}(u) = QS(x)m + f_\infty(x) \text{ on } \mathcal{M},$$

$$(P_M) \quad -\Delta_p u + \mathcal{G}(u) = QS(x)M + f_\infty(x) \text{ on } \mathcal{M},$$

respectively. Then: i) for any $Q > 0$, there is a minimal solution $\underline{u}$ (resp. a maximal solution $\overline{u}$) of problem$(P_{Q,f})$. Moreover, any other solution u must satisfy

$$u_m \leq \underline{u} \leq u \leq \overline{u} \leq u_M \tag{18}$$

$$\mathcal{G}^{-1}(QS_0m - \|f_\infty\|_{L^\infty(\mathcal{M})}) \tag{19}$$
$$leq u_m \leq \mathcal{G}^{-1}(QS_1m - C_f),$$
$$\mathcal{G}^{-1}(QS_0M - \|f_\infty\|_{L^\infty(\mathcal{M})})$$
$$leq u_M \leq \mathcal{G}^{-1}(QS_1M - C_f). \tag{20}$$

ii) for any Q there is, at least, a solution u of $(P_{Q,f})$ which is a global minimum of the functional

$$J(w) = \frac{1}{p}\int_{\mathcal{M}} |\nabla w|^p dA + \int_{\mathcal{M}} G(w)dA - \int_{\mathcal{M}} f_\infty w dA - \int_{\mathcal{M}} QS(x)j(w)dA,$$

on the set $K = \{w \in V,\ G(w) \in L^1(\mathcal{M})\}$, where $\beta = \partial j$. Moreover, if (H_{C_f}) holds, then: iii) if $0 < Q < Q_1$, then $(P_{Q,f})$ has a unique solution $u = u_m$, $u < -10$, u is the minimum of J on K, and

$$\mathcal{G}^{-1}(-\|f_\infty\|_{L^\infty(\mathcal{M})}) \leq \lim_{Q \searrow 0}\ inf\, \|u\|_{L^\infty(\mathcal{M})} \leq \lim_{Q \searrow 0}\ sup\, \|u\|_{L^\infty(\mathcal{M})} u$$

$$\leq \mathcal{G}^{-1}(-C_f),$$

iv) if $Q_2 < Q < Q_3$, then $(P_{Q,f})$ has at least three solutions,u_i, $i = 1, 2, 3$ with $u_1 = u_M$, $u_1 > -10$, $u_2 = u_m$, $u_2 < -10$and $u_1 \geq u_3 \geq u_2$ on $\mathcal{M}$. Moreover, u_1 and u_2 are local minima of J on $K \cap L^\infty(\mathcal{M})$ and, if $p > 2$, and v) if $Q_4 < Q$, then $(P_{Q,f})$ has a unique solution $u = u_M$, $u > -10$, u is the minimum of J on K and $\|u\|_{L^\infty(\mathcal{M})} \to +\infty$ when $Q \to +\infty$, where

$$Q_1 = \frac{\mathcal{G}(-10-\epsilon) + C_f}{S_1 M} \qquad Q_2 = \frac{\mathcal{G}(-10+\epsilon) + \|f_\infty\|_{L^\infty(\mathcal{M})}}{S_0 M} \tag{21}$$

$$Q_3 = \frac{\mathcal{G}(-10-\epsilon) + C_f}{S_1 m} \qquad Q_4 = \frac{\mathcal{G}(-10+\epsilon) + \|f_\infty\|_{L^\infty(\mathcal{M})}}{S_0 m}. \tag{22}$$

Proof. i) It is a consequence of the fact that the comparison principle holds for problems (P_m), (P_M) and then the method of sub and supersolutions can be applied (see e.g. Amann [1]). ii) The conclusion follows from the Weierstrass argument by using Lebesgue convergence theorem. iii) From assumption (H_{C_f}) and since $0 < Q < Q_1$, we obtain that $\overline{u}_1 < -10-\epsilon$ and $overlineu_2 < -10-\epsilon$. The proof of v) is analogous to i).

The proof of iv) is divided into several steps. First, we construct two constant subsolutions V_i and two constant supersolutions U_i such that

$$V_2 \; < \; U_2 \; < \; -10-\epsilon \; < \; -10+\epsilon \; < \; V_1 \; < \; U_1, \tag{23}$$

proving the existence of, at least two solutions of $(P_{Q,f})$. Later, we introduce the approximate model (P_Q^λ)

$$(P_Q^\lambda) \qquad -\Delta_p u + \mathcal{G}(u) \;=\; QS(x)\beta_\lambda(u) + f_\infty(x) \quad \text{on } \mathcal{M},$$

where β_λ is the Lipschitz function $\beta_\lambda = \frac{1}{\lambda}(I-(I-\lambda\beta)^{-1})$, $\lambda > 0$ (the Yosida approximation of β). Since β verifies (H_β^*), we get that β_λ is a bounded and nondecreasing function $\forall\lambda > 0$, $\beta_\lambda(s) = \beta(s)$ for any $s \notin [-10-\epsilon, -10+\epsilon+\lambda M]$, $\beta_\lambda(s) \to \beta(s)$ in the sense of maximal monotone graphs when $\lambda \to 0$. If β is a Lipschitz function, we take simply $\beta_\lambda = \beta$. The existence of a solution of (P_Q^λ) is obtained by a topological fixed point argument.

Let us show the convergence of the mentioned solution of (P_Q^λ) to a third solution of $(P_{Q,f})$. For $\lambda < \lambda_0$ (a certain positive parameter) U_1, U_2 are supersolutions of(P_Q^λ) and V_1, V_2 are subsolutions of(P_Q^λ). So, arguing as in ii), we obtain two families of solutions $\{u_1^\lambda\}$ and $\{u_2^\lambda\}$ of (P_Q^λ) such that

$$\begin{aligned} -10+\epsilon+\lambda_0 M \; < \; V_1 \quad &\le \quad u_1^\lambda \le U_1 \\ V_2 \; \le u_2^\lambda \quad &\le \quad U_2 \; < \; -10-\epsilon. \end{aligned}$$

Moreover, since $\beta_\lambda(u_1^\lambda) = \beta(u_1^\lambda)$, we deduce that $u_1^\lambda = u_1$. Analogously, we conclude that $u_2^\lambda = u_2$. In order to prove that (P_Q^λ) has a third solution u_3^λ, different from u_1^λ and u_2^λ, we show the applicability of a result due to Amann [1] to the function $F(v) := (-\Delta_p + \mathcal{G})^{-1}(QS(\cdot)\beta_\lambda(v) + f_\infty(\cdot))$ on the space $E = L^\infty(\mathcal{M})$. Finally, we get the a priori estimates

$$\int_{\mathcal{M}} |\nabla u_\lambda|^p dA + \int_{\mathcal{M}} \mathcal{G}(u_\lambda)u_\lambda dA = \int_{\mathcal{M}} QS(x)\beta_\lambda(u_\lambda)u_\lambda dA + \int_{\mathcal{M}} f_\infty u_\lambda dA.$$

$$\int_{\mathcal{M}} QS(x)\beta_\lambda(u_\lambda)u_\lambda dA + \int_{\mathcal{M}} f_\infty u_\lambda dA \le C_1$$

which allows to conclude that $u_\lambda \to u$ strongly in $L^2(\mathcal{M})$, $\beta_\lambda(u_\lambda) \rightharpoonup$ weakly in $L^2(\mathcal{M})$ and that $\beta_\lambda \to \beta$ in the sense of maximal monotone graphs (so, $z \in \beta(u)$ see, e.g.,Benilan, Crandall and Saks [10]). Moreover, we get that $\lim_{\lambda\to 0} \| \nabla u_\lambda - \nabla u \|_{L^p(T\mathcal{M})} = 0$. Since $u_3^\lambda \to u_3$ uniformly and $u_1 > -10+\epsilon_0$, there exists ϵ_1 such that $\forall \epsilon < \epsilon_1$, $u_3^\epsilon > -10+\epsilon_0$, which is a contradiction (u_3 necessarily crosses the level-10). ■

Corollary 1 – *Let $R_e(u) = Bu + C$ with β given by (1), $-10B + C > 0$ and $\frac{S_1}{S_0} \leq \frac{M}{m}$. Then we have i) if $0 < Q < \frac{-10B+C}{S_1 M}$, then $(P_{Q,f})$ has a unique solution, ii) if $\frac{-10B+C}{S_0 M} < Q < \frac{-10B+C}{S_1 m}$, then $(P_{Q,f})$ has at least three solutions, iii) if $\frac{-10B+C}{S_0 m} < Q$, then $(P_{Q,f})$ has a unique solution.*

Remark 9. As pointed out in Hetzer [44], the uniqueness of solutions for Q small and Q large still holds if conditions (H_β^*) and $(H_\mathcal{G})$ are replaced by $\mathcal{G} \in C^1(\mathbb{R})$, $\beta \in C^1(\mathbb{R}-\{-10\})$, $m \leq \beta(r) \leq M$, $\forall r \in \mathbb{R}-\{-10\}$, $\inf\{\frac{\mathcal{G}'(r)}{\beta'(r)}, r \in [\underline{U}, -10-\epsilon]\} > 0$, where $\underline{U} := \mathcal{G}^{-1}(-\|f_\infty\|_{L^\infty(\mathcal{M})})$ and $\inf\{\frac{\mathcal{G}'(r)}{\beta'(r)}, r \in [-10+\epsilon, +\infty)\} > 0$. Indeed, if Q is small enough, we can construct a supersolution showing that any possible solution u satisfies that $\underline{U} \leq u \leq -10-\epsilon$ on $\mathcal{M}$. Then, any solution u must satisfy $-\Delta_p u + \mathcal{F}(x,u) = f_\infty(x)$ with $\mathcal{F}(x,u) := \mathcal{G}(u) - QS(x)\beta(u)$. Since $\mathcal{F}(x,u)$is a strictly increasing function on $[\underline{U}, -10-\epsilon]$, for a.e. $x \in \mathcal{M}$ we have the uniqueness of solutions. The assumption on $\mathcal{G}$ leads to a similar conclusion when Q is large enough.

3.2. S-shaped bifurcation diagram

As a continuation of the previous results we can improve the answer for the special formulation

$$(P_{Q,C}) \qquad -\operatorname{div}(|\nabla u|^{p-2}\nabla u) + \mathcal{G}(u) + C \in QS(x)\beta(u) \text{ on } \mathcal{M}.$$

Following Arcoya, Diaz and Tello [6], we shall describe more precisely the bifurcation diagram and in particular, we shall prove that the principal branch (emanating from $(0, \mathcal{G}^{-1}(-C)) \in \mathbb{R}^+ \times L^\infty(\mathcal{M})$) is S-shaped, i.e. it contains at least one turning point to the left and another one to the right. By a turning point to the left (respectively, to the right), we understand a point (Q^*, u^*) in the principal branch such that in a neighborhood in $\mathbb{R}^+ \times L^\infty(\mathcal{M})$ of it, the principal branch is contained in $\{(Q,u) \in \mathbb{R}^+ \times L^\infty(\mathcal{M}) \,/\, Q \leq Q^*\}$(respectively, $\{(Q,u) \in \mathbb{R}^+ \times L^\infty(\mathcal{M}) \,/\, Q \geq Q^*\}$). A previous result is due to Hetzer [43], for the special case of $p = 2$ and β a C^1 function. He proves that the principal branch of the bifurcation diagram has an even number(including zero) of turning points. Our main result already improves this information showing that indeed, this number of turning points is greater than or equal to two.

Semilinear problems with discontinuous forcing terms on an open bounded set and with Dirichlet boundary conditions have been considered in Ambrosetti [2], Ambrosetti, Calahorrano and Dobarro [3], Arcoya and Calahorrano [5] (see also Drazin and Griffel [31], North [59] and Schmidt [65] in the context of energy balance models).

We make the additional assumption

$-(\mathrm{H}_C)$ $\mathcal{G}(-10-\epsilon)+C>0$ and $\dfrac{\mathcal{G}(-10+\epsilon)+C}{\mathcal{G}(-10-\epsilon)+C} \leq \dfrac{S_2 M}{S_1 m}$.

We start by considering the problem with β a Lipschitz function (as in the Sellers model).

Theorem 6 – *Let β be a Lipschitz continuous function verifying (H_β^*). Then Σ contains an unbounded connected component which is S-shaped containing $(0, \mathcal{G}^{-1}(-C))$ with at least one turning point to the right contained in the region $(Q_1, Q_2) \times L^\infty(\mathcal{M})$, and another one to the left in $(Q_3, Q_4) \times L^\infty(\mathcal{M})$.*

Proof. Step 1. Σ has an unbounded component containing $(0, \mathcal{G}^{-1}(-C))$: We claim that the following result, due to Rabinowitz [61], can be applied to our case: "Let E a Banach space. If $F : \mathbb{R} \times E \to E$ is compact and $F(0, u) \equiv 0$, then Σ contains a pair of unbounded components C^+ and C^- in $\mathbb{R}^+ \times E$, $\mathbb{R}^- \times E$, respectively and $C^+ \cap C^- = \{(0,0)\}$". To do so, we consider the translation of u given by $v := u - \mathcal{G}^{-1}(-C)$. Obviously, v is a solution of

$$-\Delta_p v + \hat{\mathcal{G}}(v) = QS(x)\hat{\beta}(v) \qquad \text{on } \mathcal{M} \tag{24}$$

where $\hat{\mathcal{G}}(\sigma) = \mathcal{G}(\sigma + \mathcal{G}^{-1}(-C)) + C$ and $\hat{\beta}(\sigma) = \beta(\sigma + \mathcal{G}^{-1}(-C))$. We define $\hat{\Sigma}$ in an analogous way to Σ. Let $E = L^\infty(\mathcal{M})$ and define $F(Q, v) = (-\Delta_p + \hat{\mathcal{G}})^{-1}(QS(x)\hat{\beta}(v))$. Then F is the composition of a continuous operator and a compact one (recall that $p \geq 2$), so F is also compact. On the other hand, if $Q = 0$ problem (24) has a unique solution $v = 0$, so $F(0,0) = 0$. In conclusion, $\hat{\Sigma}$ contains two unbounded components $\hat{C}^+$and $\hat{C}^-$ on $\mathbb{R}^+ \times L^\infty(\mathcal{M})$ and $\mathbb{R}^- \times L^\infty(\mathcal{M})$ respectively and $\hat{C}^+ \cap \hat{C}^- = \{(0,0)\}$. Since Σ is a translation of $\hat{\Sigma}$, Σ contains two unbounded components C^+ and C^- on $\mathbb{R}^+ \times L^\infty(\mathcal{M})$ and $\mathbb{R}^- \times L^\infty(\mathcal{M})$ respectively, and that $C^+ \cap C^- = \{(0, \mathcal{G}^{-1}(-C))\}$. Since $Q \geq 0$ in the studied model, we are interested inC^+. In order to establish the behaviour of C^+, we also recall that for every $q > 0$, there exists a constant $L = L(q)$ such that, if $0 \leq Q \leq q$, then every solution u_Q of $(P_{Q,C})$ verifies $\|u_Q\|_{L^\infty(\mathcal{M})} \leq L(q)$. Since the principal component is unbounded, its projection over the Q-axis is $[0, \infty)$. On the other hand, if Q is large enough, $(P_{Q,C})$ has a unique solution u_Q, and this solution

is greater than $\mathcal{G}^{-1}(QS_0M - C)$. Since $\lim_{|s|\to\infty} |\mathcal{G}(s)| = +\infty$, then the unbounded branch containing $(0, \mathcal{G}^{-1}(-C))C^+$ should go to (∞, ∞).
Step 2. Bifurcation diagram for two auxiliary problems: We consider the auxiliary zero-dimensional models

$$\begin{array}{llcll} (P_1) & \mathcal{G}(u) + C & = & QS_1\beta(u) & u \in \mathbb{R}, \\ (P_2) & \mathcal{G}(u) + C & = & QS_2\beta(u) & u \in \mathbb{R}. \end{array}$$

The number of solutions to these problems depends clearly on the values of Q. Let us call Σ_1 and Σ_2 the bifurcation diagrams of (P_1) and (P_2), respectively. By assumptions $(H_{\mathcal{G}})$, (H_β^*) and (H_C), the principal components of Σ_1 and Σ_2 are S-shaped. We also remark that the sets

$$K_1^i := \{(Q, u_Q) \in \mathbb{R}^2 : 0 \le Q \le \frac{\mathcal{G}(-10-\epsilon)+C}{S_i m},\ u_Q = \mathcal{G}^{-1}(QS_i m - C)\},$$

$$K_2^i := \{(Q, u_Q) \in \mathbb{R}^2 : Q \ge \frac{\mathcal{G}(-10+\epsilon)+C}{S_i M},\ u_Q = \mathcal{G}^{-1}(QS_i M - C)\}$$

are contained in Σ_i, $i = 1, 2$.
Step 3. A comparison argument: If $Q < Q_3$, there exists u_Q solution of $(P_{Q,C})$ such that $u_Q < 10 - \epsilon$. Thus u_Q satisfies

$$QS_2m \le -\Delta_p u + \mathcal{G}(u) + C \le QS_1m \text{ on } \mathcal{M}.$$

Let u_Q^1 and u_Q^2 be the solutions of the problems

$$\begin{array}{lcl} \mathcal{G}(u) + C & = & QS_1m \text{ on } \mathcal{M} \\ \mathcal{G}(u) + C & = & QS_2m \text{ on } \mathcal{M}, \end{array}$$

respectively. That is, (Q, u_Q^1) and (Q, u_Q^2) live in Σ_1 and Σ_2, respectively. Now, if $Q < Q_3$,

$$-\Delta_p u_Q^2 + \mathcal{G}(u_Q^2) \le -\Delta_p u_Q + \mathcal{G}(u_Q) \le -\Delta_p u_Q^1 + \mathcal{G}(u_Q^1),$$

and so by the comparison principle for the monotone problem $-\Delta_p u + \mathcal{G}(u) = f \in L^2(\mathcal{M})$ on $\mathcal{M}$, we have that u_Q^2
leq $u_Q \le u_Q^1$. Therefore, the component of Σ starting in $(0, \mathcal{G}^{-1}(-C))$, lives between Σ_1 andΣ_2 to arrive at (Q_3, u_{Q_3}), where u_{Q_3} is the minimal solution of (P_{Q_3}). Analogously, if we denote by u_{Q_2} the maximal solution of (P_{Q_2}), we can prove that the component ofΣ which connects (Q_2, u_Q^2) with (∞, ∞) lives between Σ_1 and Σ_2. From $Q_2 < Q_3$, the branch containing $(0, \mathcal{G}^{-1}(-C))$ is un bounded and by the uniqueness of solution for $(P_{Q,C})$ when $Q > Q_4$, we get that this branch is necessarily S-shaped. ■

Our next result avoids the Lipschitz assumption made in Theorem 6.

Theorem 7 – *Let β a general maximal monotone graph satisfying (H_β^*) and assume (H_C). Then Σ has an unbounded S-shaped component containing $(0, \mathcal{G}^{-1}(-C))$ with at least one turning point to the right contained in the region $(Q_1, Q_2) \times L^\infty(\mathcal{M})$ and another one to the left in $(Q_3, Q_4) \times L^\infty(\mathcal{M})$, respectively.*

To prove Theorem 7, we approximate problem $(P_{Q,C})$ when β is not Lipschitz continuous. We only need to show the *convergence* of the principal branches C_n of these approximating problems to a S-shaped unbounded connected set C of solutions of $(P_{Q,C}\ )$. For this reason, let us recall the notions of *liminf* and *limsup* of a sequence of subsets C_n of a metric space X:

$$\begin{aligned}
\liminf_{n\to\infty}\ C_n := \quad \{p \in \quad & X:\ \text{for any neighbourhood } U(p) \text{ of } p \text{ in } X \\
& \exists n_0 \in \mathbb{N} : \mathrm{U(p)} \cap \mathrm{C_n} \neq \emptyset\ \forall \mathrm{n} \geq \mathrm{n_0}\}, \\
\limsup_{n\to\infty}\ C_n := \quad \{p \in \quad & X:\ \text{for any neighbourhood } U(p) \text{ of } p \text{ in } X \\
& U(p) \cap C_n \neq \emptyset \text{ for infinitely many } n\}.
\end{aligned}$$

A lemma due to Whyburn [70] shows that if i) $\lim_{n\to\infty}\ inf\ C_n\ \neq\ \emptyset$ and ii) $\cup_{n=1}^{\infty} C_n$ is precompact, then $\lim_{n\to\infty}\ sup\, C_n$ is a nonempty, precompact, closed and connected set. *Proof of Theorem 7.* The method of super and sub solutions proves that if $Q > Q_2$, then there exists a solution of $(P_{Q,C})$ greater than $-10+\epsilon$. Analogously, we know that if $0 \leq Q < Q_3$, then (P_Q) has a solution smaller than $-10-\epsilon$. It is clear that these functions are not the unique solutions of $(P_{Q,C})$ in those intervals and that the uniqueness holds at least in the Q-intervals $[0, Q_1)$ and (Q_4, ∞). Since we can not apply directly Rabinowitz theorem to our problem, we consider the family $\beta_n = n(I - (I - \frac{1}{n}\beta)^{-1})$, $n \in \mathbb{N}$ to approximate β in the sense of maximal monotone graphs when $n \to \infty$. Notice that since βverifies (H_β^*), then β_n is a Lipschitz bounded nondecreasing function (see Brezis [14]) and that $\beta_n(s) = \beta(s)$ for any $s \notin [-10-\epsilon, -10+\epsilon+\frac{M}{n}]$, $\forall n$.

Let u_n be the solutions of the approximated problem

$$(P_Q^n) \qquad -\Delta_p u_n + \mathcal{G}(u_n) + C \;=\; QS(x)\beta_n(u_n) \text{ on } \mathcal{M}$$

and let Σ_n the bifurcation diagrams for (P_Q^n). Let us denote by S_n the component of Σ_n containing $(0, \mathcal{G}^{-1}(-C))$. By Theorem 6, every S_n is an unbounded, connected and S-shaped set. First of all, we are going to prove that lim sup S_n is a connected and closed set of solutions to problem (P_Q). In order to apply Whyburn' result, we consider the sets C_n^j $(j > Q_4)$ defined as $S_n \cap ([0, j] \times L^\infty(\Omega))$, $\forall n \in \mathbb{N}$ containing $(0, \mathcal{G}^{-1}(-C))$. It is easy to see that these sets are connected and that i) is verified. Let us check (ii),$\cup_{n=1}^{\infty} C_n^j$ is precompact. Since X is a Banach space, it suffices to prove that every sequence $\{(Q_l, u_l)\}_{l\in\mathbb{N}} \subset \cup_{n=1}^{\infty} C_n^j$ contains a subsequence $\{(Q_{l_k}, u_{l_k})\}$ converging in X. From $Q_l \in [0, j]$, there exists $Q \in [0, j]$ and a subsequence of $\{Q_l\}$which we still call $\{Q_l\}$, such that $Q_l \to Q$. On the other hand,u_l is a solution of the problem

$$-\Delta_p u_l + \mathcal{G}(u_l) + C \;=\; QS(x)\beta_l(u_l) \text{ on } \mathcal{M}.$$

Taking u_l as a test function in this equation, we obtain the estimate

$$\int_{\mathcal{M}} |\nabla u_l|^p dA + \int_{\mathcal{M}} \frac{B}{2}|u_l|^2 dA \le \frac{(j\|S\|_\infty M + C)^2|\mathcal{M}|}{2B} \tag{25}$$

where $|\mathcal{M}|$ is the Hausdorf measure of $\mathcal{M}$. Then u_l is a bounded sequence in V. From the compact embedding $V \subset L^\infty(\mathcal{M})$ when $p > 2$, there exist $u \in L^\infty(\mathcal{M})$ and asubsequence $\{u_{l_k}\}$ of $\{u_l\}$ such that $u_{lk} \to u$in $L^\infty(\mathcal{M})$. If $p = 2$, then $\{u_l\}$ is a bounded sequence in the Sobolev space $H^2(\mathcal{M})$. From the compact embedding $H^2(\mathcal{M}) \subset C(\mathcal{M})$, we deduce the existence of a subsequence $\{u_{lk}\}$ and $u \in C(\mathcal{M})$, such that $u_{lk} \to u$ in $L^\infty(\mathcal{M})$. Thus $\cup_{n=1}^{\infty} C_n^j$ is precompact. Then by Whyburn's result $C^j \equiv \lim_{n\to\infty} sup\, C_n^j$ is a connected and compact set in X. Moreover, since every S_n is unbounded and fixed $\overline{Q}$, the solutions u_Q are uniformly bounded in $L^\infty(\mathcal{M})$, for $Q < \overline{Q}$, we have that $C_k^j \cap (\{j\} \times L^\infty(\mathcal{M})) \neq \emptyset$, for all $j \in \mathbb{N}$.

Now, we prove that the set C^j is contained in Σ. Let us see that for every $Q \in [Q_1, Q_4]$, we have that every $(Q, u) \in C^j$ verifies that u is a solution of (P_Q) (notice that it is true for every $Q \in (0, Q_1] \cup [Q_4, +\infty)$ from $C_n^j = C^j$ in these intervals). Let $(Q, u) \in C^j = \lim_{n\to\infty} \sup C_n^j$, that is, there exists a subsequence of $(Q_n, u_n) \in C_n^j$ such that $(Q_{n_k}, u_{n_k}) \to (Q, u)$ in $\mathbb{R}\times L^\infty(\mathcal{M})$. From estimate (25) and the compact embedding $H^2(\mathcal{M}) \subset L^\infty(\mathcal{M})$ (for $p = 2$) and $V \subset L^\infty(\mathcal{M})$ (for $p > 2$), we deduce the existence of $u \in L^\infty(\mathcal{M})$ and a subsequence of $\{(Q_{n_k}, u_{n_k})\}$ which we call $\{(Q_{n_k}, u_{n_k})\}$, such that

$$(Q_{n_k}, u_{n_k}) \;\to\; (Q, u) \;\text{ in } \mathbb{R} \times L^\infty(\mathcal{M}),$$

Since $\beta_n \to \beta$ in the sense of maximal monotone graphs of $\mathbb{R}^2$, we have that $\beta_{n_k}(u_{n_k}) \rightharpoonup z \in \beta(u)$ weakly in $L^2(\mathcal{M})$. Using a Minty's type argument we deduce that u is a solution of the problem $(P_{Q,C})$. Thus $(Q, u) \in \Sigma$ and $C^j \subset \Sigma$. Since for all n and j, $C_n^j \cap (\{j\} \times L^\infty(\mathcal{M})) \neq \emptyset$, there exists $\{(j, u_n)\}_{n\in\mathbb{N}}$ such that $(j, u_n) \in C_n^j$, that is,

$$-\Delta_p u_n + \mathcal{G}(u_n) = jS(x)\beta_n(u_n) - C \quad \text{in } \mathcal{M}.$$

Using that the operator $(\Delta_p + \mathcal{G})^{-1}$ is compact in $L^\infty(\mathcal{M})$, there exists a subsequence $u_{n_k} \to u$ in $L^\infty(\mathcal{M})$. Thus $(j, u) \in C^j$ and $C^j \cap (\{j\} \times$

$L^\infty(\mathcal{M})) \neq \emptyset$. Since $j > Q_4$, u_j is the unique solution of $(P_{Q,C})$. On the other hand, we know that $\Sigma \cap (j,\infty) \times L^\infty(\mathcal{M}) = \Sigma_M \cap (j,\infty) \times L^\infty(\mathcal{M})$. So, we have obtained a connected unbounded set which starts in $(0, \mathcal{G}^{-1}(-C))$. The proof ends with the argument used in the proof of Theorem 6 for $Q_2 < Q_3$. ■

Remark 10. We point out that our results remain true for the more general equation

$$-\text{div}(k(x)|\nabla u|^{p-2}\nabla u) + \mathcal{G}(u) + C \in QS(x)\beta(u) \text{ on } \mathcal{M},$$

where $k(x)$ is a given bounded function with $k(x) \geq k_0 > 0$ a.e.$x \in \mathcal{M}$, representing the *eddy diffusion coefficient.* When $\mathcal{M} = \mathcal{S}^1$, it is usually assumed that $S(x) = S(\lambda)$ and $k(x) = k(\lambda, \phi)$ with λ the latitude and ϕ the longitude. So, in that case, the corresponding solutions are not ϕ-invariant.

Remark 11. By using a shooting method, it is possible to show (see Diaz and Tello [27] that there exist infinitely many equilibrium solutions for some values of Q when we study the one-dimensional problem

$$(P_{1,Q,C}) \begin{cases} -(|u'|^{p-2}u')' + Bu + C \in Q\beta(u) & x \in (0,1), \\ u'(0) = u'(1) = 0. \end{cases}$$

If $Q_1 < Q < Q_2$ then $(P_{1,Q,C})$ has infinitely many solutions. Moreover, there exists $K_0 \in \mathbb{N}$ such that for every $K \in \mathbb{N}$, $K \geq K_0 \in \mathbb{N}$ there exists at least a solution which crosses the level $u_K = -10$, exactly K times.

Remark 12. After my lecture at the Collège de France, Professor J.L. Lions pointed out to me the reference Rahmstorf [63] where a S-shaped diagram bifurcation curve arises in the context of the Atlantic Thermohaline Circulation in reponse to changes in the hydrological cycle.

References

[1] H. Amann, Fixed point equations and nonlinear eigenvalue problems in ordered Banach spaces, SIAM Review, 18, 4 (1976), 620–709.

[2] A. Ambrosetti, Critical points and non linear variational problems. Supplement au Bulletin de la Societé Mathématique de France. Mémoire 49 (1992).

[3] A. Ambrosseti, M. Calahorrano and F.Dobarro, Global branching for discontinuous problems, Comment. Math. Univ. Carolinae, 31 (1990), 213–222.

[4] S.N. Antontsev and J.I. Diaz, New results on localization of solutions of nonlinear elliptic and parabolic equations obtained by energy methods, Soviet Math. Dokl., 38 (1989), 535–539.

[5] D. Arcoya, and M. Calahorrano, Multivalued non-positone problems, Rend. Mat. Acc. Lincei, 9 (1990), 117–123.
[6] D. Arcoya, J.I. Diaz and L. Tello, S-shaped bifurcation branch in a quasilinear multivalued model arising in climatology, Journal of Differential Equations, 150 (1998), 215–225.
[7] O. Arino, S. Gautier, and J.P. Penot, A fixed point theorem for sequentially continuous mappings with applications to ordinary differential equations, Funkcialaj Ekvacioj, 27 (1984), 273–279.
[8] T. Aubin, Nonlinear analysis on manifolds. Monge-Ampere equations. Springer-Verlag, New York, 1982.
[9] M. Badii and J.I.Diaz, Time periodic solutions for a diffusive energy balance model in climatology, J. Mathematical Analysis and Applications, 233 (1999), 713–729.
[10] Ph. Benilan, M.G. Crandall and P.Saks, Some L^1 existence and dependence results for semilinear elliptic equations under nonlinear boundary conditions, Appl. Math. Optimization, 17 (1988), 203–224.
[11] R. Bermejo, Numerical solution to a two-dimensional diffusive climate model, in Modelado de sistemas en oceanografia, climatologia y ciencias medio-ambientales: Aspectos matemáticos y numéricos, (A. Valle and C. Parés eds.), Universidad de Málaga, (1994), 15–30.
[12] R. Bermejo, J.I. Diaz and L. Tello, Article in preparation.
[13] K. Bhattacharya, M. Ghil and I.L. Vulis, Internal variability of an energy balance climate model, J. Atmosph. Sci., 39(1982), 1747–1773.
[14] H. Brezis, Opérateurs maximaux monotones et semigroupes de contractions dans les espaces de Hilbert, North Holland, Amsterdam, 1973.
[15] M.I. Budyko, The effects of solar radiation variations on the climate of the Earth, Tellus, 21 (1969), 611–619.
[16] T. Cazenave and A. Haraux, Introduction aux problèmes d'évolution semi-linéaires. Mathématiques et Applications, Ellipses, Paris, 1990.
[17] A.S. Dennis, Weather modifications by cloud seeding. Academic Press, 1980.
[18] J.I. Diaz , Nonlinear partial differential equations and free boundaries. Pitman, Londres, 1984.
[19] J.I. Diaz, Mathematical analysis of some diffusive energy balance climate models, in Mathematics, climate and environment, (J.I. Diaz and J.-L.Lions, eds.), Masson, Paris, (1993), 28–56.
[20] J.I. Diaz, On the controllability of some simple climate models, in Environment, economics and their mathematical models, (J.I.Diaz and J.-L. Lions eds.), Masson, (1994), 29–44.
[21] J.I. Diaz, On the mathematical treatment of energy balance climate models, in The mathematics of models for climatology and environment, NATO ASI Series, Serie I: Global Environmental Change, 48, (J.I. Diaz ed.), Springer, Berlin, (1996), 217–252.

[22] J.I. Diaz, On a free boundary problem arising in climatology, to appear.

[23] J.I. Diaz, A. Fasano and A.Meirmanov, On the disappearence of the mushy region in multidimensional Stefan problems, in Free boundary problems: theory and applications, Vol. VII. Pitman. London (1993), 124–129.

[24] J.I. Diaz, J. Hernandez and L. Tello, On the multiplicity of equilibrium solutions to a nonlinear diffusion equation on a manifold arising in climatology, J. Math. An. Appl., 216 (1997), 593–613.

[25] J.I. Diaz and G. Hetzer, A quasilinear functional reaction-diffusion equation arising in climatology, in Équations aux dérivées partielles et applications: Articles dédiés à Jacques-Louis Lions, Gauthier-Villars, Paris (1998), 461–480.

[26] J.I. Diaz and L. Tello, A nonlinear parabolic problem on a Riemannian manifold without boundary arising in climatology, Collect. Math., 50 (1999), 19–51.

[27] J.I. Diaz and L. Tello, Infinidad de soluciones estacionarias para un modelo climatológico simple via un método de tiro, Actas XVI CEDYA, VI CMA, (R. Montenegro, G. Montero and G. Winter eds.), Servicio de Publicaciones de la Univ. de Las Plamas, 1999, 1469–1476.

[28] J.I. Diaz and F. de Thélin, On a nonlinear parabolic problem arising in some models related to turbulent flows, SIAM J. Math. An., 25, 4 (1994), 1085–1111.

[29] J.I. Diaz and L. Veron, Local vanishing properties of solutions of elliptic and parabolic problems quasilinear equations, Trans. A.M.S., 290 (1985), 787–814.

[30] J.I. Diaz and I.I. Vrabie, Existence for reaction-diffusion systems, J. Math. Analysis and Appl., 188 (1994), 521–528.

[31] P.G. Drazin and D.H. Griffel, On the branching structure of diffusive climatological models, J. Atmos. Sci., 34 (1977), 1696–1706.

[32] E. Fereisel, A note on uniqueness for parabolic problems with discontinuous nonlinearities, Nonlinear Analysis, 16(1991), 1053–1056.

[33] E. Fereisel and J. Norbury, Some existence, uniqueness, and nonuniqueness theorems for solutions of parabolic equations with discontinuous nonlinearities, Proc. Royal Soc. Edinburgh, 119A (1991), 1–17.

[34] E. Fereisel and F. Simondon, Convergence for degenerate parabolic equations in one dimension, C.R. Acad. Sci. Paris, Sér. I, 323 (1996), 251–255.

[35] M. Ghill and S. Childress, Topics in geophysical fluid dynamics, Springer-Verlag, 1987.

[36] R. Gianni, Existence of the free boundary in a multi-dimensional combustion problem, Proc. Royal Soc. Edinburgh, 125A, (1995), 525–543.

[37] R. Gianni and J. Hulshof, Thesemilinear heat equation with a Heaviside source term. Euro. J. of Applied Mathematics, 3 (1992), 367–379.

[38] R. Gianni and P. Mannucci, Some existence theorems for an N-dimensional parabolic equation with a discontinuous source term, SIAM J. Math. Anal., (1993), 618–633.

[39] I.M. Held and M.J. Suarez, Simple albedo feedback models of the ice-caps, Tellus, 36 (1974).

[40] I.M. Held, D.I. Linder and M.J. Suarez, Albedo feedback, the meridional structure of the effective heat diffusivity, and climatic sensitivity: Results from dynamic and diffusive models, American Meteorological Society (1981), 1911–1927.

[41] K. Henderson-Sellers and McGufiie, A climate modelling primer, John Wiley and Sons, Chichester, 1987.

[42] G. Hetzer, H. Jarausch and W. Mackens, A multiparameter sensitivity analysis of a 2D diffusive climate model, Impact and computing in science and engineering, 1 (1989),327–393.

[43] G. Hetzer, The structure of the principal component for semilinear diffusion equations from energy balanceclimate models, Houston Journal of Math., 16 (1990), 203–216.

[44] G. Hetzer, S-shapedness for energy balance climate models of Sellers-type, in The Mathematics of models for climatology and environment, NATO ASI Series, Serie I: global environmental change, 48, (J.I. Diaz ed.), Springer, Berlin,(1996), 25–288.

[45] G. Hetzer, A functional reaction-diffusion equation from climate modeling: S-shapedness of the principal branch. Diff. and Integral Eq. 8 (1995), 1047–1059.

[46] G. Hetzer, Global existence, uniqueness, and continuous dependence for a reaction-diffusion equation with memory, EJDE, (1996), 1–16.

[47] G. Hetzer and P.G. Schmidt, A global attractor and stationary solutions for a reaction-diffusion system arising from climate modeling, Nonlinear analysis. TMA, 14 (1990), 915–926.

[48] G. Hetzer and P.G. Schmidt, Analysis of energy balance models, in World congress of nonlinear analysts'92, Tampa, FL, August 19-26, 1992, (V. Laksmikantham ed.), Walter de Gruyter, Berlin-New York (1996),1609–1618.

[49] G. Hetzer and L. Tello, Sobre un sistema parabólico no lineal "multicapa" en limatologia, in Actas del XV Congreso de ecuaciones diferenciales y aplicaciones, Serv. Publ. Univ. de Vigo (1998), 471–476.

[50] J. T. Kiehl, Atmoshperic general circulation modeling, In Climate system modeling, (K.E. Trenberth ed.), Cambridge University Press, (1992), 319–370.

[51] R.Q. Lin and G.R. North, A study of abrupt climate change in a simple nonlinear climate model. Climate dynamics, 4 (1990), 253–261.

[52] J.-L. Lions, El Planeta Tierra, Espasa-Calpe. Serie Instituto de España. Madrid, 1990.

[53] J.-L. Lions, R. Temam and S. Wang, New formulations of the primitive equations of atmosphere and applications, Nonlinearity, 5 (1992), 237–288.

[54] J.G. Mengel, D.A. Short and G.R. North, Seasonal snowline instability in an energy balance model, Climate Dynamics, 2 (1988), 127–131.

[55] A.M. Meirmanov, The Stefan problem, Walter de Gruyter, Berlin-New York, 1992.

[56] M. Nakao, A difference inequality and its application to nonlinear evolution equations, J. Math. Soc. Japan, 30 (1978), 747–762.

[57] J. von Neumann, Can we survive technology?, Nature, 1955. (Also in Collected works. Vol VI, Pergamon, 1966.)

[58] G.R. North, Multiple solutions in energy balance climate models, in Paleogeography, paleoclimatology, paleoecology, 82, Elsevier Science Publishers, B.V. Amsterdam (1990), 225–235.

[59] G.R. North, Introduction to simple climate model, in Mathematics, climate and environment, (J.I. Diaz and J.-L.Lions eds.), Masson, Paris (1993), 139–159.

[60] G.R. North, J.G. Mengel and D.A. Short, Simple energy balance model resolving the season and continents: Applications to astronomical theory of ice ages. J. Geophys. Res., 88 (1983), 6576–6586.

[61] P.H. Rabinowitz, A global theorem for nonlinear eigenvalue problems and applications, in Contributions to nonlinear functional analysis, (E.H. Zarantonello ed.), Academic Press, New York (1971), 11–36.

[62] J.M. Rakotoson and B. Simon, Relative rearrangement on a measure space. Application to the regularity of weighted monotone rearrangement. Part. II. Appl. Math. Lett., 6 (1993), 79–82.

[63] S. Rahmstorf, Bifurcations of the atlantic thermohaline circulation in reponse to changes in the hydrological cycle, Nature, 378 (1995), 145–149.

[64] W.D. Sellers, A global climatic model based on the energy balance of the earth-atmosphere system, J. Appl. Meteorol., 8 (1969), 392–400.

[65] B.E. Schmidt, Bifurcation of stationary solutions for Legendre-type boundary value problems arising from climate modeling. Ph. D. Thesis. Auburn Univ. 1994.

[66] J. Simon, Compact sets in the space $L^p(0,T;B)$, Annali Mat. Pura et Appl., CXLVI (1987), 65–96.

[67] I. Stakgold, Free boundary problems in climate modeling, in Mathematics, climate and environment, (J.I.Diaz and J.-L.Lions eds.), Masson (1993), 177–188.

[68] P.H. Stone, A simplified radiative-dynamical model for the static stability of rotating atmospheres, J. Atmos. Sci., 29, 3 (1972), 405–418.

[69] L. Tello, Tratamiento matemático de algunos modelos no lineales que aparecen en climatologia. Ph. D. Thesis. Univ. Complutense de Madrid, 1996.

[70] G.T. Whyburn, Topological analysis, Princeton Univ. Press, Princeton, 1955.

[71] I.I. Vrabie, Compactness methods for non linear evolutions, Pitman, London, 1987.

[72] X. Xu, Existence and regularity theorems for a free boundary problem governing a simple climate model, Aplicable Anal., 42 (1991), 33–59.

J. Ildefonso Diaz
Departamento de Matemática Aplicada
Facultad de Matemáticas
Universidad Complutense de Madrid
28040 Madrid
Spain

E-mail: jidiaz@sunma4.mat.ucm.es

Studies in Mathematics and its Applications, Vol. 31
D. Cioranescu and J.L. Lions (Editors)

Chapter 15

UNIQUENESS AND STABILITY IN THE CAUCHY PROBLEM FOR MAXWELL AND ELASTICITY SYSTEMS

M. ELLER, V. ISAKOV, G. NAKAMURA AND D. TATARU

1. Preliminaries

By $x = (x_0, x_1, ..., x_n)$ we denote the coordinates in $R \times R^n$. For the differentiation operators we set $\partial_j = \partial/\partial x_j$ and $D_j = \frac{1}{i}\partial/\partial x_j$. Sometimes we shall call x_0 the time coordinate, and use the alternate notation t for it.

Given a positive scalar function a in R^n we define the scalar associate wave operator in $R \times R^n$ by

$$\Box_a = a\partial_t^2 - \Delta.$$

By H^k we denote the classical Sobolev spaces, with the norm denoted $\|.\|_k$. Often we shall use on these spaces the weighted norms,

$$\|u\|_{k;\tau}^2 = \int \sum_{|\alpha| \leq k} \tau^{2(k-|\alpha|)} |\delta^a au|^2 dx \tag{1.1}$$

These norms can also be defined in a bounded domain Ω. To describe the regularity of boundary traces of functions on the boundary $\partial\Omega$ of a domain Ω we use a similar norm, denoted $\|.\|_{k,\tau,\partial\Omega}$.

2. Pseudoconvexity and Carleman estimates

The aim of this section is to review the concept of pseudoconvexity and state several kinds of Carleman estimates for scalar equations. The Carleman estimates have been the main tool in the study of unique continuation problems; see [5]. While we cannot transfer the unique continuation results directly from equations to systems, it turns out that sometimes we can transfer the Carleman estimates to systems; these in turn imply unique continuation results for systems.

Thus, let $P(x, D)$ be a partial differential operator with real principal part whose principal symbol we denote by $p(x, \xi)$. We will assume that the coefficients of the principal part of P are in C^1 while the other coefficients are bounded and measurable.

Let S be a smooth oriented surface in R^{n+1}. Then we can represent it as a level set $S = \{\phi = 0\}$ of a smooth function ϕ so that $\nabla\phi \neq 0$ on S. The function ϕ is uniquely determined up to multiplication by smooth positive functions.

To a smooth function ϕ we associate the operator

$$P_\phi(x, D, \tau) = P(x, D + i\tau\nabla\phi)$$

which is the conjugate of the operator P with respect to the exponential weight $e^{\tau\phi}$,

$$P_\phi(x, D, \tau) = e^{\tau\phi}P(x, D)e^{-\tau\phi}$$

The principal symbol of the conjugated operator P_ϕ is

$$p_\phi(x, \xi, \tau) = p(x, \xi + i\tau\nabla\phi)$$

The Poisson bracket of two symbols P, q is defined as

$$\{p, q\} = \nabla_\xi p \cdot \nabla_x q - \nabla_x p \cdot \nabla_\xi q$$

Let Γ be a conical subspace of the cotangent bundle $T^* R^{n+1}$. Now we are ready to define pseudoconvex functions and surfaces.

Definition 2.1 - *a) A smooth function ϕ is pseudo-convex with respect to P on Γ if*

$$\{p, \{p, \phi\}\}(x, \xi) > 0 \quad \forall\ (x, \xi) \in \Gamma, \xi \neq 0,\ p(x, \xi) = 0 \tag{2.1}$$

b) A smooth oriented surface $S = \{\phi = 0\}$ is called pseudo-convex with respect to P on Γ if

$$\{p, \{p, \phi\}\}(x, \xi) > 0 \quad \forall\ (x, \xi) \in \Gamma_S,\ \xi \neq 0,\ p(x, \xi) = \{p, \phi\}(x, \xi) = 0 \tag{2.2}$$

Definition 2.2 - *a) A smooth function ϕ is strongly pseudo-convex with respect to P on Γ if it is pseudoconvex and*

$$\frac{1}{i\tau}\{\bar{p_\phi}, p_\phi\}(x, \xi) > 0 \quad \forall\ (x, \xi) \in \Gamma,\ \xi \neq 0,\ \tau > 0,\ p_\phi(x, \xi) = 0\,. \tag{2.3}$$

b) A smooth oriented surface S is strongly pseudo-convex with respect to P on Γ if it is pseudoconvex and

$$\frac{1}{i\tau}\{\bar{p_\phi}, p_\phi\}(x, \xi) > 0\ \forall\ (x, \xi) \in \Gamma_S, \quad \begin{array}{l} \xi \neq 0,\ \tau > 0, \\ p_\phi(x, \xi) = \{p_\phi, \phi\}(x, \xi) = 0. \end{array} \tag{2.4}$$

The relation between pseudoconvex functions and surfaces is obvious,

Remark 2.3. a) An oriented surface S is (strongly) pseudoconvex with respect to P on Γ iff it is a level set of a function ϕ which is (strongly) pseudoconvex with respect to P on Γ, so that $\nabla\phi \neq> 0$ on Γ.

b) Moreover, if the oriented surface $S = \{\phi = 0\}$ is (strongly) pseudoconvex with respect to P on Γ then $e^{\lambda\phi}$ is (strongly) pseudoconvex with respect to P on Γ for large enough λ.

c) The (strong) pseudoconvexity condition for both functions and surfaces is stable with respect to small C^2 perturbations.

For second order operators the condition (2.4) is void for noncharacteristic surfaces, therefore

Proposition 2.4 - *Let P be a second order operator with real principal symbol. If a noncharacteristic oriented surface S is pseudoconvex with respect to P on Γ then it is strongly pseudoconvex with respect to P on Γ.*

Now we discuss the corresponding Carleman estimates. Since this is all we use later on, in the sequel we only refer to second order partial differential operators $P(x, D)$ with real principal symbol. The classical result (see [4]) is the following one:

Theorem 2.5 - *Let Ω be a compact subset of R^{n+1}. If ϕ is strongly pseudoconvex with respect to P in Ω then*

$$\tau\|e^{\tau\phi}u\|_{1;\tau}^2 \leq C\|e^{\tau\phi}Pu\|_0^2 \tag{2.5}$$

whenever $u \in H^1$ is supported in Ω so that the RHS is finite.

The substitution $v = e^{\tau\phi}u$ reduces the estimate to

$$\tau\|v\|_{1;\tau}^2 \leq C\|P_\phi v\|_0^2$$

which is essentially a subelliptic estimate for P_ϕ. The strong pseudoconvexity condition therefore expresses the subellipticity of P_ϕ in terms of its symbol. If instead Ω is a bounded domain with smooth boundary and u is a function in Ω, then the appropriate estimate should include the Cauchy data of u on the boundary, see Tataru [11].

Theorem 2.6 - *Let Ω be a compact subset of R^n. If ϕ is strongly pseudoconvex with respect to P in Ω then*

$$\tau\|e^{\tau\phi}u\|_{1;\tau,\Omega}^2 \leq C(\|e^{\tau\phi}Pu\|_{(0)}^2 + \tau(\|e^{\tau\phi}u\|_{1;\tau,\partial\Omega}^2 + \|e^{\tau\phi}\frac{\partial u}{\partial\nu}\|_{0,\partial\Omega}^2)) \tag{2.6}$$

whenever $u \in H^1$ is supported in Ω so that the RHS is finite.

In effect such an estimate might hold even with only one boundary data on the RHS, provided that the corresponding boundary operator satisfies a strong Lopatinskii boundary condition with respect to $d\phi$, see [11].

If the coefficients of P are analytic with respect to some variables then the unique continuation results can be improved. We state here the Carleman estimate corresponding to the simplest result of this type, namely for the wave equation with time independent coefficients. In this case, we set $\Gamma = \{\xi_0 = 0\}$, where ξ_0 is the time Fourier variable and consider functions and operators in R^{n+1}.

Theorem 2.7 - *Let P be a second order hyperbolic operator with time independent coefficients. Furthermore, let ϕ be a quadratic function in $\bar{\Omega}$ which is strongly pseudo-convex with respect to P on Γ. Then there exist $d > 0$ and $C > 0$ such that*

$$\tau\|e^{-\frac{\epsilon}{2\tau}D_0^2}e^{\tau\phi}u\|_{1;\tau}^2 \leq C\Big(\|e^{-\frac{\epsilon}{2\tau}D_0^2}e^{\tau\phi}P(x,D)u\|_2^2 + \|e^{\tau(\phi-d\epsilon)}u\|_{1;\tau}^2\Big) \quad (2.7)$$

for all $u \in H^1$, supported in Ω, provided τ is large enough and ϵ is sufficiently small.

This result was proved in [10]. In effect it holds even if we only assume that the coefficients of P are analytic in time and that ϕ is analytic. However, the proof is substantially simpler in the special case considered above. The critical point is the conjugation argument, which needs to be carried through also with respect to the Gaussian.

With the substitution $w = e^{\tau\phi}u$ the estimate (2.7) reduces to

$$\tau\|e^{-\frac{\epsilon}{2\tau}D_0^2}w\|_{1;\tau}^2 \leq C\Big(\|e^{-\frac{\epsilon}{2\tau}D_0^2}P(x,D+i\tau\nabla\phi)w\|_2^2 + \|e^{-d\epsilon\tau}w\|_{1;\tau}^2\Big) \quad (2.8)$$

Furthermore

$$e^{-\frac{\epsilon}{2\tau}D_0^2}t = (t + i\frac{\epsilon}{\tau}D_0)e^{-\frac{\epsilon}{2\tau}D_0^2}$$

Since $\nabla\phi$ is linear, this implies that

$$e^{-\frac{\epsilon}{2\tau}D_0^2}P(x,D+i\tau\nabla\phi) = P(x,D+i\tau\nabla\phi - \epsilon\partial_t(\nabla\phi)D_t)e^{-\frac{\epsilon}{2\tau}D_0^2} \quad (2.9)$$

Hence if we set $v = e^{-\frac{\epsilon}{2\tau}D_0^2}w$ then (2.8) reduces to

$$\tau\|v\|_{1;\tau}^2 \leq C\|P_{\epsilon,\phi}v\|_2^2 + \|e^{-d\epsilon\tau}w\|_{1;\tau}^2\Big)$$

with

$$P_{\epsilon,\phi} = P(x,D+i\tau\nabla\phi - \epsilon\partial_t(\nabla\phi)D_t)$$

If ϵ is small enough then $P_{\epsilon,\tau}$ is a small perturbation of P_ϕ. Then the pseudoconvexity condition for ϕ implies some subellipticity for $P_{\epsilon,\phi}$ in the region $|\xi_0| \leq c\tau$. Outside this region, the Gaussian yields exponential decay in τ and leads to the last RHS term.

3. Uniqueness and stability for principally scalar systems

Here we apply the results in the previous section to the systems of the type

$$P_j u_j + b_j(x,t;\nabla u) + c_j(x,t;u) = f_j,\ j = 1,...,m \tag{3.1}$$

Here P_j are second order operators with real principal part and C^1 coefficients and other coefficients in L^∞_{loc},b_j, c_j are linear functions of $\nabla u, u$ with L^∞_{loc}-coefficients, $u = (u_1, ..., u_m)$.

First we obtain the Carleman estimates for such systems.

Theorem 3.1 - *Let ϕ be a smooth function which is strongly pseudoconvex with respect to all P_j in a given compact set K. Then*

$$\tau \|e^{\tau\phi} u\|^2_{1,\tau} \leq c\|e^{\tau\phi} f\|^2, \qquad \tau > \tau_0 \tag{3.2}$$

whenever u is supported in K and solves (3.1).

Proof. First we apply the Carleman estimates in Theorem 2.5 to u_j with respect to P_j. This yields

$$\tau \|e^{\tau\phi} u_j\|^2_{1,\tau} \leq c\|e^{\tau\phi} P_j u_j\|^2,$$

and further

$$\tau \|e^{\tau\phi} u_j\|^2_{1,\tau} \leq c\|e^{\tau\phi} u_j\|^2_{1,\tau} + \|e^{\tau\phi} f_j\|^2_{1,\tau}$$

Summing with respect to j, the first terms on the right are absorbed on the left for sufficiently large τ and we obtain (3.2). ■

Next we consider the unique continuation problem for the corresponding homogeneous problem

$$P_j v_j + b_j(x,t;\nabla v) + c_j(x,t;v) = 0,\ j = 1,...,m \tag{3.3}$$

Theorem 3.2 - *Let $S = \{\phi = 0\}$ be an oriented surface which is strongly pseudoconvex with respect to P_j for all j. Then we have unique continuation across S for H^1_{loc} solutions u to (3.3). More precisely, given any $x_0 \in S$ and $v \in H^1_{loc}$ solving (3.3) near x_0 which is identically zero in $\{\phi > 0\}$ near S, this implies that $v = 0$ near x_0.*

Proof. Without any restriction in generality we assume that $x^0 = 0$. Since S is strongly pseudo-convex, by Remark 2.3 we can also assume that ϕ is strongly pseudoconvex with respect to P_j.

Consider the modified function $\psi(x,t) = \phi(x,t) + \epsilon(r^2 - |x|^2)$. For small ϵ, ψ is still strongly pseudoconvex with respect to all P_j near x_0. Furthermore,

$$\{\psi > 0\} \cap \{\phi < 0\} \subset B(0,r)$$

Hence, by choosing r sufficiently small we insure that (3.3) holds in $\{\psi > 0\}$.

Now let χ be a cutoff function which is 1 in $\psi > 0$ and 0 in $\psi < -\epsilon r^2$. Then the function $u = \chi v$ solves (3.1) with $f \in L^2$, supported in

$$\text{supp } \nabla\chi \subset \{\psi < 0\}.$$

Apply the Carleman estimates (3.2) to u. We obtain

$$\tau \| e^{\tau\psi} \chi f \|_{1,\tau} \leq c \| e^{\tau\psi} f \|^2$$

Since f is supported in $\{\psi < 0\}$, letting τ go to infinity yields

$$\lim_{\tau\to\infty} \| e^{\tau\psi} \chi v \| = 0$$

which implies that $v = 0$ in $\{\psi > 0\}$. Since $\psi(0) > 0$ this concludes the proof. ■

The next step is to establish the corresponding stability estimate. For this it is more interesting to work with boundary value problems. Thus, let Ω be a bounded domain in R^{n+1} with C^1 boundary $\partial\Omega$. Let S be a part of $\partial\Omega$ where one prescribes the Cauchy data for u

$$u = g_0, \quad \partial_\nu u = g_1 \quad \text{on } S \tag{3.4}$$

Given a smooth function ϕ let

$$\Omega_\epsilon = \Omega \cap \{\epsilon < \phi\}, \quad \partial\Omega_\epsilon = \partial\Omega \cap \{\epsilon < \phi\}.$$

Theorem 3.3 - *Let ϕ be strongly pseudo-convex with respect to all operators P_j on $\overline{\Omega}$ and $\partial\Omega_0 \subset S$. Let $u \in H^1(\Omega)$ solve the Cauchy problem (3.1),(3.4).*

Then with some positive constants C, $\lambda = \lambda(\epsilon) \in (0,1)$ we have

$$\|u\|_{1,\Omega_\epsilon} \leq C(\epsilon)(F + F^\lambda M^{1-\lambda})$$

where

$$F = \|g_0\|_{1,S} + \|g_1\|_{0,S} + \|f\|_{0,\Omega_0}, \quad M = \|u\|_{1,\Omega_0}.$$

Proof. We want to use the analogue of the Carleman estimate (3.2) but for boundary value problems,

$$\tau \| e^{\tau\phi} u \|^2_{1;\tau,\Omega} \leq C(\| e^{\tau\phi} f \|^2_{0,\Omega} + \tau(\| e^{\tau\phi} u \|^2_{1;\tau,\partial\Omega} + \| e^{\tau\phi} \frac{\partial u}{\partial\nu} \|^2_{0,\partial\Omega})) \tag{3.5}$$

for u solving (3.1) in Ω. This follows in the same way as (3.2) if we use (2.6) instead of (2.5).

To proceed we need to localize to Ω_0. Hence, introduce a cut-off function $\chi \in C^\infty(R^{n+1}), \chi = 1$ on $\Omega_{\epsilon/2}$ and $\chi = 0$ outside Ω_0. Then work with χu. After commuting we get

$$P_j(\chi u_j) = \chi f_j + A_{1;j} u$$

where $A_{1;j}$ is a linear differential operator of first order with measurable bounded coefficients depending on ϵ, supported in Ω_0.

Hence, applying (3.5) to χu we obtain the inequality

$$\begin{aligned} \tau \| e^{\tau\phi} \chi u \|^2_{1;\tau} &\leq C(\epsilon)(\| e^{\tau\phi} u \|^2_{1,\tau,\Omega_0} + \| e^{\tau\phi} \chi f_j \|^2_0 + \\ &+ \tau \sum_{k=0,1} \| e^{\tau\phi} \partial^k_\nu u_j \|^2_{1-j;\tau,S}) \end{aligned}$$

Since $\chi = 1$ on $\Omega_{\epsilon/2}$, this further gives

$$\begin{aligned} (\tau - C(\epsilon)) \| e^{\tau\phi} u \|^2_{1;\tau,\Omega_{\epsilon/2}} &\leq C(\epsilon) \| e^{\tau\phi} u \|^2_{1,\tau,\Omega_0 \setminus \Omega_{\epsilon/2}} + \| e^{\tau\phi} \chi f_j \|^2_0 + \\ &+ \tau \sum_{k=0,1} \| e^{\tau\phi} \partial^k_\nu u_j \|^2_{1-j;\tau,S}) \end{aligned}$$

For large enough τ this finally gives

$$\| u_j \|_{1,\tau,\Omega_\epsilon} \leq C(\epsilon)(F e^{\tau(\Phi - \epsilon/2)} + M e^{-\tau\epsilon/2})$$

where Φ is the supremum of ϕ in Ω. This inequality holds for all sufficiently large τ, say $\tau \geq \tau_0(\epsilon)$. Increasing $C(\epsilon)$ we can take $\tau_0(\epsilon) = 0$. Hence we can minimize the right hand side with respect to τ. If $M \leq F$, then Theorem 3.3 follows if we simply set $\tau = 0$. Otherwise, we choose τ so that

$$F e^{\tau(\Phi - \epsilon/2)} = M e^{-\tau\epsilon/2}$$

and then the right hand side equals

$$C(\epsilon) M^{1-\lambda} F^\lambda, \quad \lambda = \frac{\epsilon}{2\Phi}$$

and we obtain the conclusion of the theorem with $\lambda = \epsilon/(2\Phi)$. ■

If a portion of the boundary is strongly pseudoconvex, then the above result can be used locally near S. The corresponding uniqueness result follows easily from Theorem 3.2.

Corollary 3.4 - *Assume that the surface $S \subset \partial\Omega$ is strongly pseudo-convex with respect to $P_j, j = 1, ..., m$. If $u \in H^1(\Omega)$ solves the system (3.3) with*

$f = 0$ and has zero Cauchy data (3.4) on S then $u = 0$ in some neighborhood of S in Ω.

Another corollary gives an explicit description of an uniqueness domain when $S = \gamma \times (-T, T) \in C^2$ in the two following cases: 1) $\gamma = \partial\Omega'$, the space origin is in Ω' and 2) $\Omega' \subset \{-h < x_n < 0, x_1^2 + ... + x_{n-1}^2 < r^2\}, S = \partial\Omega' \cap \{x_n < 0\}$. Here we let $x' = x \in R^n$. To formulate the result we need the weight function $\psi(x,t) = x_1^2 + ... + x_{n-1}^2 + (x_n - \beta)^2 - \theta^2 t^2 - s$ to define $\Omega(\epsilon)$ as $\Omega \cap \{\psi > \epsilon\}$

Corollary 3.5 - *Assume that the coefficients $a(j), j = 1, ..., m$ of the principal part $P_j = \Box_{a(j)}$ of the system (3.1) satisfy the following conditions*

$$\theta^2 a(j)(a(j) + \partial_t a(j)/2 + a^{-1/2}(j)|t\nabla a(j)|) < a(j) + 1/2x \cdot \nabla a(j) - 1/2\beta\partial_n a(j)$$

$$\theta^2 a(j) \leq 1 \qquad on\, \bar{\Omega}\,, a(j) \in C^1(\bar{\Omega}) \tag{3.6}$$

and $\Omega' \subset B(0; \theta T), \beta = s = 0$ in case 1) and $h(h + 2\beta) < \theta^2 T^2, \beta^2 + r^2 = s$ in case 2). Assume that $u \in H^1(\Omega)$ solves the system (3.1).

Then

$$\|u\|_{1,\Omega_\epsilon} \leq C(F + M^{1-\lambda} F^\lambda)$$

where F is $\|f\|_{0,\Omega} + \|u\|_{1,S} + \|\nabla u\|_{0,S}$ and $M = \|u\|_{1,S}$. In particular, if $u = 0, \partial_\nu u = 0$ on S then $u = 0$ in Ω_0.

Proof. It is known that under the conditions (3.6) with respect to the principal coefficients the function $\phi = exp(\sigma/2\psi)$ for large σ is strictly pseudoconvex in $\bar{\Omega}$ with respect to the wave operators $\Box_{a(j)}$ [6]. Moreover the additional conditions on h, β, θ, r, T imply that $(\partial\Omega)_0 \subset S$. Applying Theorem 3.3 we complete the proof. ■

We observe that for constant $a(j)$ in case 1) the conditions (3.6) simplify to $\theta^2 a(j) < 1$ and $\Omega' \subset B(0; \theta T)$. Then Corollary 3.5 gives the sharp description of the uniqueness domain in the lateral Cauchy problem. Generally, when $\partial_n a(j) < 0$ on $\overline{\Omega}$ one can satisfy the first condition (3.6) by choosing β positive and large. The uniqueness domain Ω_0 in case 2) is discussed in [7], section 3.4.

While applicable to a wide class of systems, including semilinear ones, Theorem 3.2 can be substantially improved when the coefficients do not depend on the time variable.

Theorem 3.6 - *Let S be a C^1-smooth surface which is non- characteristic with respect to the operators $P_j, j = 1, ..., m$. Assume that P_j are hyperbolic with respect to x_0. Assume that the coefficients of the system (3.3) do not depend on x_0. Then we have unique continuation for H^1 solutions v to (3.3) across S, in the sense that if $v = 0$ on one side of S then $v = 0$ near S.*

Proof. Denote the two sides of S by S^+, respectively S^-. Let ϕ be a C^1 function so that $S = \{\phi = 0\}$. Let $y \in S$ so that $u = 0$ in S^+ near y. Then we want to show that $\phi = 0$ in S^- near y. Without any restriction in generality we can assume that $y = 0$.

To achieve this we use an argument similar to the one in the proof of Theorem 3.2. We start by constructing a quadratic function ψ with the following properties:

i) The set $\psi > 0$ cuts a small neighbourhood of 0 in S^-. More precisely

$$\{\psi > 0\} \cap S^- \cap B(0, 4r) \subset B(0, 2r)$$

ii) $\psi(0) > 0$
iii) ψ is a small C^1 perturbation of ϕ in $B(0, 4r)$.
iv) ψ is strongly pseudoconvex with respect to all P_j in $\Gamma = \{\xi_0 = 0\}$.
Here r is an arbitrarily small parameter. To achieve this set

$$\psi(x) = x\nabla\phi(0) + \frac{2\lambda\epsilon}{r}(x\nabla\phi(0))^2 - \frac{2\delta}{r}(x^2 - r^2)$$

where λ is a fixed large parameter. The first term insures that $\nabla\psi \approx \nabla\phi(0)$ in $B(0, 4r)$; the second term yields (i)-(iii); finally, the third term provides the strong pseudoconvexity condition on Γ while doing no harm to (i)-(iii).

For small positive ϵ both (ii) and (iii) are clearly fulfilled, so it remains to verify (i). It suffices to show that $\phi + \frac{4\lambda\delta}{r}\phi^2 > \psi$ in $B(0, 4r) \setminus B(0, 2r)$, or that

$$\phi(x) + \frac{4\lambda\delta}{r}\phi^2 + \delta|x| > x\nabla\phi(0) + \frac{2\lambda\delta}{r}(x\nabla\phi(0))^2 \quad x \in B(0, 4r)$$

Since

$$\phi(x) = x\nabla\phi(0) + o(|x|)|x|$$

this follows if r is sufficiently small.

It remains to verify the strong pseudoconvexity condition. We have

$$p_\psi(x, \xi, \tau) = p(x, \xi) - \tau^2 p(x, \nabla\psi) + 2i\tau p_\xi \nabla\psi$$

therefore

$$\begin{aligned}
\frac{1}{i\tau}\{\bar{p_\psi}, p_\psi\} &= p_\xi(x, \xi)(\nabla^2\psi)p_\xi(x, \xi) + \tau^2 p_\xi(x, \nabla\psi)(\nabla^2\psi)p_\xi(x, \nabla\psi) \\
&\quad + O(|\xi|^2 + \tau^2) \\
&= \frac{\lambda\delta}{r}((p_\xi\nabla\psi(0))^2 + \tau^2|p(x, \nabla\psi(0))|^2) + O(\frac{\delta}{r} + 1)(|\xi|^2 + \tau^2)
\end{aligned}$$

Since $p(x, \nabla\phi(0)) \neq 0$ and λ is large, the above expression is positive as long as τ is away from 0. But when τ is close to 0, p_ψ is close to p, which is elliptic on $\xi_0 = 0$ because of the hyperbolicity condition. This concludes the proof of (iv).

Now we can use the Carleman estimates in (2.7) with respect to the weight function ψ in $B(0,4r)$. If u solves (3.1) and is supported in $B(0,4r)$ then we can proceed as in Theorem 3.1 to write (2.7) for each u_j and sum them up. The coupling is lower order and, as in Theorem 3.1, can be absorbed into the left hand side after summation. In the end we get

$$\tau\|e^{-\frac{\epsilon}{2\tau}D_0^2}e^{\tau\phi}u\|_{1;\tau}^2 \leq C(\|e^{-\frac{\epsilon}{2\tau}D_0^2}e^{\tau\phi}f\|_2^2 + \|e^{\tau(\phi-d\epsilon)}u\|_{1;\tau}^2) \tag{3.7}$$

To use this for solutions u to (3.3) we need to truncate first. Let χ be a cut-off C^∞- function such that it is 1 on $B(0,2r)$ and 0 outside $B(0,4r)$. Then the function $u = \chi v$ is supported in $B(0,4r)$ and solves (3.1) with f in L^2, supported in $B(0,4r) \setminus B(0,2r)$.

Then (3.7) yields

$$\tau\|e^{-\frac{\epsilon}{2\tau}D_0^2}e^{\tau\phi}u\|_{1;\tau}^2 \leq ce^{\gamma\tau}, \qquad \gamma = \max\{0, \Psi - d\epsilon\} \tag{3.8}$$

where Ψ is the supremum of ψ in the support of v.

To complete the proof it suffices to make use of the following result of Tataru [10], Proposition 4.1.

Lemma 3.7 - *Let $u \in L^2$ be a function with compact support. Furthermore, let $\phi \in C^r$ and assume that*

$$\|e^{-\frac{\epsilon}{2\tau}D_0^2}e^{\tau\phi}u\|_2 \to 0 \qquad as\ \tau \to \infty$$

Then $u = 0$ in $\{x \in R^n : \phi(x) > 0\}$.

Applying this in (3.8) we get $v = 0$ in $\{\psi > \gamma\}$. Then $\Psi \geq \gamma$, therefore by the definition of Γ we get $\Psi \leq 0$, i.e. $u = v = 0$ in $\{\psi > 0\}$. The proof is complete. ∎

Corollary 3.8 - *Assume in addition that the principal part is in divergence form and the coefficients b_j are C^1. Then the same result holds for L^2 solutions u to (3.3).*

Proof. Let $\mu(t)$ be a smooth regularizing kernel supported in $[-1,1]$, with integral 1. Set

$$\mu_\delta(t) = \delta^{-1}\mu(\frac{t}{\delta})$$

and look at the mollified functions

$$u_\delta = \mu_\delta * u$$

where the convolution is only with respect to t.

Since (3.1) has time independent coefficients and v solves it, it follows that the functions v_δ solve it as well. We claim that $v_\delta \in H^1$. Clearly $\partial_t^k u_\delta \in L^2$ for all k. Using this in (3.1) we get

$$P_j^* u_j \in L_t^2(H_{loc}^{-1})$$

where P_j^* contains the part of P_j without any time derivatives. Then P_j^* is elliptic, in divergence form, with C^1 coefficients, therefore the elliptic theory yields $u_j \in L_t^2(H_{loc}^1)$, which proves our claim.

On the other hand, supp $u \cap \{\psi \geq 0\} \subset B(0,2r)$ therefore supp $u_\delta \cap \{\psi \geq 0\} \subset B(0,2r)$ if δ is sufficiently small. Then the argument in the proof of the Theorem applies to u_δ and gives $u_\delta = 0$ in $\psi > 0$. Letting $\delta \to 0$ we get the same for u, q.e.d.

It is convenient to have a global version of the previous Theorem. To formulate the next result we assume that S_θ is a continuous family of C^1 regular non-characteristic mutually disjoint surfaces, $\theta \in [0,1]$. Set $\Omega_0 = \cup_{\theta \in [0,1]} S_\theta$

Corollary 3.8 - *Under the conditions of Theorem 3.6 let u be an H^1 solution to the system (3.1) which is zero near S_1 so that supp $u \cap \Omega_0$ is compact. Then $u = 0$ in Ω_0. The same holds for L^2 solutions if the additional conditions in Corollary 3.8 are fulfilled.*

Proof. a) Let

$$s = \sup\{\theta \in [0,1];\ \text{supp } u \cap S_\theta \neq 0\}$$

Assume by contradiction that $s > 0$. Then supp $u \cap S_\theta = 0$ for $\theta > s$, while, by compactness, supp $u \cap S_s \neq 0$. If $x_0 \in$ supp $u \cap S_s$ then, by continuity, we conclude that u is 0 on one side of S_s near x. Then we can use the local argument above to conclude that $u = 0$ near x_0, which contradicts our earlier assumption. ∎

4. Applications to the Maxwell system

The results of the previous can be applied to classical systems of mathematical physics which can be reduced to the systems of the form (3.1) by relatively simple transformations. First we consider the Maxwell system

$$\begin{cases} \partial_t(\epsilon E) - \text{curl } H = j \\ \partial_t(\mu H) + \text{curl } E = 0 \\ \text{div } (\epsilon E) = 4\pi\rho \\ \text{div } (\mu H) = 0 \end{cases} \tag{4.1}$$

for electric and magnetic fields $E = (E_1, E_2, E_3), H = (H_1, H_2, H_3)$ depending on $(x,t) \in \Omega \subset R^4$. Here $\epsilon(x,t), \mu(x,t)$, denote correspondingly electric permittivity, magnetic permeability, $j(x,t)$ is the density of electrical current and $\rho(x,t)$ is the electrical charge density. We consider the isotropic case when $\epsilon, \mu \in C^2$ are scalar functions. We are interested in the Cauchy problem for (4.1) when one prescribes the Cauchy data

$$E = E_0, H = H_0 \quad \text{on } S \tag{4.2}$$

where S is a C^2 surface.

Uniqueness and stability in the Cauchy problem for the system (4.1) will follow from the results of the previous section and from the following simple lemma.

Lemma 4.1 - *For solutions to the Maxwell system (4.1) with $j = 0, \rho = 0$ we have*

$$\begin{aligned} \partial_t^2(\epsilon E) - \Delta(\mu^{-1} E) &+ \nabla(\nabla(\epsilon\mu)^{-1} \cdot \epsilon E) - \text{curl } (\nabla \mu^{-1} \times E) \\ &+ \text{curl } (\frac{\mu_t}{\mu} H) = 0. \end{aligned} \tag{4.3}$$

and

$$\begin{aligned} \partial_t^2(\mu H) - \Delta(\epsilon^{-1} H) &+ \nabla(\nabla(\epsilon\mu)^{-1} \cdot \mu H) - \text{curl } (\nabla \epsilon^{-1} \times H) \\ &- \text{curl } (\frac{\epsilon_t}{\epsilon} E) = 0. \end{aligned} \tag{4.4}$$

Note that this computation requires only C^1 regularity for E, H. If in addition both E and H are at least C^2 then one can rewrite these equations as

$$(\epsilon\mu\partial_t^2 - \Delta)E = Q_1(E, H)$$
$$(\epsilon\mu\partial_t^2 - \Delta)H = Q_2(E, H)$$

where Q_1, Q_2 are first order operators with bounded coefficients.

Proof. To reduce the system (4.1) to (4.3), (4.4) we differentiate its first equation with respect to t and use the second equation (4.1) to replace $\partial_t(\mu H)$ by curl E. We obtain

$$\partial_t^2 \epsilon E + \text{curl } (\mu^{-1} \text{curl } E) - \text{curl } (\frac{\mu_t}{\mu} H) = 0.$$

and further

$$\partial_t^2(\epsilon E) + \text{curl } (\text{curl } (\mu^{-1} E)) + \text{curl } (\nabla \mu^{-1} \times E) + \text{curl } (\frac{\mu_t}{\mu} H) = 0.$$

But

$$\text{curl curl } E = -\Delta E + \nabla \text{div } E$$

therefore

$$\partial_t^2(\epsilon E) - \Delta(\mu^{-1}E)) + \nabla \text{div } (\mu^{-1}E) - \text{curl } (\nabla\mu^{-1} \times E) + \text{curl } (\frac{\mu_t}{\mu}H) = 0.$$

Now using the third equation (4.1) we finally get

$$\partial_t^2(\epsilon E) - \Delta(\mu^{-1}E)) + \nabla(\nabla(\epsilon\mu)^{-1} \cdot \epsilon E) - \text{curl } (\nabla\mu^{-1} \times E) + \text{curl } (\frac{\mu_t}{\mu}H) = 0.$$

The equation for H can be obtained in a similar fashion.

The equations (4.3), (4.4) form a principally scalar hyperbolic system, so from Theorems 3.2, 3.3, 3.6 and their corollaries we can derive uniqueness and stability results for the Cauchy problem.

The Carleman estimates for the Maxwell's equations have the form

$$\begin{aligned} \tau(\|e^{\tau\phi}E\|_{1,\tau}^2 + \|e^{\tau\phi}H\|_{1,\tau}^2) \ &\leq \ c(\|e^{\tau\phi}(\partial_t(\epsilon E) - \text{curl } H)\|_{1,\tau}^2 \\ &+ \ \|e^{\tau\phi}(\partial_t(\mu H) - \text{curl } E)\|_{1,\tau}^2 \\ &+ \ \|\text{div } (\epsilon E)\|_{1,\tau}^2 + \|\text{div } (\mu H)\|_{1,\tau}^2). \end{aligned}$$

In the presence of the boundary one needs to add another term, namely $\tau\|e^{\tau\phi}(E,H)\|_{1,\tau,\partial\Omega}^2$; the conormal derivatives of both E and H can be derived from the equations. By Theorem 3.1, these estimates hold if the coefficients ϵ, μ are C^2, provided that ϕ is strongly pseudoconvex with respect to the wave operator $\square_{\epsilon\mu}$.

The main unique continuation result is

Theorem 4.2 - *Assume that the oriented surface $S = \{\phi = 0\}$ is strongly pseudo-convex with respect to $\square_{\epsilon\mu}$. If $(E,H) \in H^1(\Omega)$ solve the system (4.1) and are zero on S then they are zero near S in $\{\phi > 0\}$. .*

Correspondingly, from Lemma 4.1 and Theorem 3.3, we get the stability estimate

Theorem 4.3 - *Let ϕ be strongly pseudo-convex with respect to $\square_{\mu\epsilon}$ on $\overline{\Omega}$ and $(\partial\Omega)_0 \subset S$.*

Then for a solution $(E,H) \in H^1(\Omega)$ to the Cauchy Problem (4.1), (4.2) one has

$$\|E\|_{1,\Omega_\delta} + \|H\|_{1,\Omega_\delta} \leq C(F + M^{1-\lambda}F^\lambda) \tag{4.5}$$

with some positive constants $C, \lambda < 1$ depending on δ. Here $F = \|E_0\|_{1,S} + \|H_0\|_{1,S}$, $M = \|E\|_{1,\Omega_0} + \|H\|_{1,\Omega_0}$.

Now we give explicit description of uniqueness domains when $\Omega = \Omega' \times (-T,T)$ and $S = \gamma \times (-T,T)$ in the cases considered in Corollary 3.5.

Corollary 4.4 - *Assume that the coefficients $a = \epsilon\mu$ satisfy the condition (3.6) and that parameters h, β, θ, T, r also satisfy the conditions of Corollary 3.5. Then (4.5) holds.*

Finally we give a sharp description of the uniqueness domain in the Cauchy problem when ϵ, μ, σ do not depend on time; this follows from Theorem 3.6 and Corollary 3.8.

Theorem 4.5 - *Let S be a C^1-smooth surface which is non-characteristic with respect to the operator $\square_{\epsilon\mu}$. Then we have unique continuation across S for L^2 solutions E, H to (3.3) , in the sense that if $v = 0$ on one side of S then $v = 0$ near S.*

5. Applications to the classical elasticity system

The uniqueness of the continuation for the static elasticity system under C^∞-assumptions and with arbitrary first order perturbations was proven by Dehman and Robbiano [3]. Recently, Ang,Ikehata,Trong, and Yamamoto [1] reduced a static system with zero order perturbations to a principally diagonal one and obtained sharp uniqueness of the continuation results under reduced smoothness assumptions.The time dependent classical elasticity with arbitrary first order perturbations was considered by Isakov [6] by pseudo-convexity methods. Recently Yamamoto [12] applied principal diagonalization to some inverse source problems for the Maxwell system. A similar reduction is available the classical elasticity system

$$\rho\partial_t^2 u - \mu(\Delta u + \nabla \text{div } u) - \nabla(\lambda \text{div } u) - \sum_{j=1}^{3} \nabla\mu\cdot(\nabla u_j + \partial_j u)e_j = 0 \quad \text{in } \Omega \quad (5.1)$$

for the displacement vector $u = (u_1, u_2, u_3)$ depending on $(x, t) \in \Omega$. We will assume that the the density $\rho \in C^1(\overline{\Omega})$ and the Lame parameters μ, λ are functions in $C^2(\overline{\Omega})$.We are interested in the Cauchy problem for this system when one prescribes the Cauchy data

$$u = u_0, \partial_\nu u = u_1 \quad \text{on } S \quad (5.2)$$

We will extend the system (5.1) for three unknown functions u_1, u_2, u_3 to a new one for seven unknown functions by introducing two more vector functions $v = \text{div } u$ and $w = \text{curl } u$. Another, more economical reduction requires one to introduce just one auxiliary variable, namely div u. However, this requires some lengthy computations and is relegated to the appendix. For a particular case of this reduction we refer to [2].

To find differential equations for new functions we apply the divergence and the curl operators to the system (5.1).

Lemma 5.1 - *Let $v = \text{div } u, w = \text{curl } u$. If u solves (5.1) then*

$$\begin{aligned} \rho/\mu\partial_t^2 u - \Delta u + A_{1;1}(u,v) = 0 \\ \rho/(2\mu+\lambda)\partial_t^2 v - \Delta v + A_{2;1}(u,v,w) = 0 \quad in\ \Omega \\ \rho/\mu\partial_t^2 w - \Delta w + A_{3;1}(u,v,w) = 0 \end{aligned} \tag{5.3}$$

where $A_{j;1}$ are linear differential operators of first order with measurable and bounded coefficients in Ω. In addition, when the $\rho \in C^2(\overline{\Omega'})$, $\mu, \lambda \in C^3(\overline{\Omega'})$ do not depend on t, then the coefficients of $A_{j;1}$ do not depend on t and the coefficients of first order derivatives are in $C^1(\overline{\Omega'})$.

Proof. Dividing the equations (5.1) by ρ and applying the divergence we obtain

$$\partial_t^2 v - (2\mu+\lambda)/\rho\Delta v - (\nabla(\mu/\rho) + \lambda\nabla(1/\rho) + 2/(\rho)\nabla\lambda + \nabla\mu/\rho)\cdot\nabla v$$

$$-(\nabla(\mu/\rho) + \nabla\mu/\rho)\cdot u - \text{div }(1/\rho\nabla\lambda)v - \sum_{j=1}^{3}\partial_j(\nabla\mu/\rho)\cdot(\nabla u_j + \partial_j u) = 0$$

or using the identity $\Delta u = \nabla\text{div } u - \text{curl curl } u$ write it as

$$\partial_t^2 v - (2\mu+\lambda)/\rho\Delta v - 2/(\lambda+2\mu)\nabla((\lambda+2\mu)^2/\rho)\cdot\nabla v +$$

$$1/\mu\nabla(\mu^2/\rho)\cdot\text{curl } w - \sum_{j\le 3}\partial_j(\nabla\mu/\rho)\cdot(\nabla u_j + \partial_j u) - \text{div }(\nabla\lambda/\rho)v = 0$$

Similarly applying the curl and using the identity $\text{curl }(fu) = f\text{curl } u + \nabla f \times u$ leads to the equations

$$\partial_t^2 w - \mu/\rho\Delta w - (\nabla(\mu/\rho))\times(\Delta u + \nabla v) - \nabla(1/\rho)\times\nabla(\lambda v) -$$

$$\sum_{j\le 3}(((1/\rho)\partial_j\mu)\text{curl } \partial_j u + \nabla((\partial_j\mu)/\rho)\times(\partial_j u + \nabla u_j)) = 0$$

Using again the formula $\Delta u = \nabla v - \text{curl } w$ we arrive at

$$\partial_t^2 w - \mu/\rho\Delta w - \nabla(\mu/\rho)\times(\nabla v - \text{curl } w + \nabla v) -$$

$$\nabla(1/\rho)\times\nabla(\lambda v) - \sum_{j\le 3}(1/\rho\partial_j\mu\partial_j w + \nabla(\partial_j\mu/\rho)\times(\partial_j u + \nabla u_j)) = 0$$

Summing up we obtain the extended principally hyperbolic system (5.3).■

As for the Maxwell system this Lemma and our results on principally scalar systems imply the following uniqueness and stability theorems for the Cauchy problem (5.1),(5.2).

Theorem 5.2 - *Let ϕ be strongly pseudo-convex with respect to the operators $\square_{\rho/\mu}, \square_{\rho/(2\mu+\lambda)}$ on $\overline{\Omega}$ and $(\partial\Omega)_0 \subset S$.*

Then for a solution $u \in H^2(\Omega)$ to the Cauchy problem (5.1),(5.2) one has

$$\|u\|_{1,\Omega_\epsilon} + \|\text{div } u\|_{1,\Omega_\epsilon} + \|\text{curl } u\|_{1,\Omega_\epsilon} \leq C(F + M^{1-\lambda}F^\lambda) \tag{5.4}$$

with some positive constants C, λ depending on ϵ. Here

$$F = \|u_0\|_{2,S} + \|u_1\|_{1,S}, \quad M = \|u\|_{2,\Omega}.$$

In particular, when $u_0 = u_1 = 0$ on S one has $u = 0$ on Ω_0.

This result follows from Theorem 3.3. If the part S of the boundary is itself strongly pseudoconvex then we get the following local result:

Corollary 5.3 - *Assume that the surface S is strongly pseudo-convex with respect to the operators $\square_{\rho/\mu}, \square_{\rho/(2\mu+\lambda)}$ in $\overline{\Omega}$. If $u \in H^2(\Omega)$ solves the Cauchy problem (5.1),(5.2) with zero u_0, u_1 then u=0 near S on Ω.*

Now we give an explicit uniqueness domain when $\Omega = \Omega' \times (-T, T)$ and $S = \gamma \times (-T, T)$ in the cases considered in Corollary 3.5.

Corollary 5.4 - *Assume that Ω, S are the domain and the part of its boundary described in Corollary 3.5. Let the functions $a = \rho/\mu$ and $a = \rho/2(\mu + \lambda)$ and the parameters h, β, θ, T, r satisfy the conditions (4.3). Let $u \in H^2(\Omega)$ solve the Cauchy problem (5.1),(5.2). Then (5.4) holds. In particular if the Cauchy data are zero then $u = 0$ on Ω_0.*

As for the Maxwell system we can improve these results and reduce regularity assumptions about u when the elastic parameters are time independent.

Theorem 5.5 - *Assume that the coefficients $\rho \in C^2$,$\mu, \lambda \in C^3$ are time independent. Let S be a noncharacteristic surface with respect to both $\square_{\rho/\mu}, \square_{\rho/(2\mu+\lambda)}$. Then we have unique continuation across S for H^1_{loc} solutions u to (5.1).*

Appendix. A diagonalization of the classical elasticity system

Let

$$Lu = (\lambda + \mu)\nabla(\nabla \cdot u) + \mu\nabla u + (\nabla \cdot u)\nabla\lambda - \{\nabla u + (\nabla u)^T\}\nabla\mu$$

be the spatial part of the elasticity system, where $(\nabla u)^T$ denotes the transpose of ∇u. By a direct computation we obtain the following lemma:

Lemma A.1 - *Let*

$$\gamma = \mu^{-\frac{1}{2}},\ \beta = \gamma(\lambda+\mu),\ \Theta = (\lambda+\mu)\nabla\gamma + \gamma\nabla\mu$$

and

$$\begin{aligned} V &= \nabla\lambda\otimes\nabla\gamma + \nabla\gamma\otimes\nabla\mu + (\lambda+\mu)\nabla^2\gamma + (\nabla\gamma\cdot\nabla\mu + \mu\Delta\gamma)I_n, \\ W &= \gamma\{V - (\nabla\Theta)^T\} - \nabla\gamma\otimes\Theta. \end{aligned}$$

Then,

$$\gamma L(\gamma u) = \Delta u + \nabla(\gamma\beta\nabla\cdot u + \gamma\Theta\cdot u) - (\nabla\cdot u)\gamma\Theta + Wu.$$

Since

$$\nabla\cdot\{\gamma L(\gamma u)\} = \Delta\{\nabla\cdot u + \gamma\beta\nabla\cdot u + \gamma\Theta\cdot u\} - \nabla\cdot\{(\nabla\cdot u)\gamma\Theta\} + \nabla\cdot\{Wu\},$$

we have

$$\begin{aligned} &\begin{pmatrix} \nabla^T \\ I_n \end{pmatrix} \gamma L(\gamma u) \\ &= \left\{\Delta M + \begin{pmatrix} \nabla^T & 0 \\ 0 & \nabla \end{pmatrix}\begin{pmatrix} -\gamma\Theta & W \\ \gamma\beta & \gamma\Theta^T \end{pmatrix} + \begin{pmatrix} 0 & 0 \\ -\gamma\Theta & W \end{pmatrix}\right\}\begin{pmatrix} \nabla\cdot u \\ u \end{pmatrix} \\ &= \left\{\Delta I_{n+1} + \begin{pmatrix} \nabla^T & 0 \\ 0 & \nabla \end{pmatrix} V_1 + V_2\right\} M \begin{pmatrix} \nabla\cdot u \\ u \end{pmatrix} \end{aligned}$$

where I_k is the k times k identity matrix and

$$\begin{aligned} M &= \begin{pmatrix} 1+\gamma\beta & \gamma\Theta^T \\ 0 & I_n \end{pmatrix}, \\ V_1 &= \begin{pmatrix} -\dfrac{\gamma\Theta}{1+\gamma\beta} & \dfrac{\gamma^2}{1+\gamma\beta}\Theta\otimes\Theta + W \\ \dfrac{\gamma\beta}{1+\gamma\beta} & \dfrac{\gamma}{1+\gamma\beta}\Theta^T \end{pmatrix}, \\ V_2 &= \begin{pmatrix} 0 & 0 \\ -\dfrac{\gamma}{1+\gamma\beta}\Theta & \dfrac{\gamma^2}{1+\gamma\beta}\Theta\otimes\Theta + W \end{pmatrix}. \end{aligned}$$

Note that for $w = \gamma u$, we have

$$\begin{pmatrix} \nabla \cdot u \\ u \end{pmatrix} = Q \begin{pmatrix} \nabla \cdot w \\ w \end{pmatrix}$$

with

$$Q = \begin{pmatrix} \gamma^{-1} & (\nabla \gamma^{-1})^T \\ 0 & \gamma^{-1} I_n \end{pmatrix}.$$

Hence

$$\begin{pmatrix} \nabla^T \\ I_n \end{pmatrix} \gamma L w = \left\{ \Delta I_{n+1} + \begin{pmatrix} \nabla^T & 0 \\ 0 & \nabla \end{pmatrix} V_1 + V_2 \right\} M Q \begin{pmatrix} \nabla \cdot w \\ w \end{pmatrix}.$$

Moreover, from

$$\rho\gamma {\partial_t}^2 (\nabla \cdot w) + \nabla(\rho\gamma) \cdot {\partial_t}^2 w - \nabla^T(\gamma L w) = \nabla^T(\rho\gamma {\partial_t}^2 w - \gamma L w),$$

we obtain

$$\begin{pmatrix} \nabla^T \\ I_n \end{pmatrix} \gamma\{\rho {\partial_t}^2 w - L w\} = S {\partial_t}^2 \begin{pmatrix} \nabla \cdot w \\ w \end{pmatrix} - \begin{pmatrix} \nabla^T \\ I_n \end{pmatrix} \gamma L w$$

with

$$S = \begin{pmatrix} \rho\gamma & (\nabla(\rho\gamma))^T \\ 0 & (\rho\gamma) I_n \end{pmatrix}.$$

Summing up, we have proved that

$$\begin{pmatrix} \nabla^T \\ I_n \end{pmatrix} \gamma\{\rho {\partial_t}^2 w - L w\} = \qquad \text{(A.1)}$$
$$S \left[{\partial_t}^2 I_{n+1} - S^{-1} \left\{ \Delta I_{n+1} + \begin{pmatrix} \nabla^T & 0 \\ 0 & \nabla \end{pmatrix} V_1 + V_2 \right\} M Q \right] \begin{pmatrix} \nabla \cdot w \\ w \end{pmatrix}.$$

A short computation yields

$$S^{-1} M Q = K$$

with

$$K = \rho^{-1} \begin{pmatrix} \lambda + 2\mu & 2(\nabla\mu)^T - \rho^{-1}\mu(\nabla\rho)^T \\ 0 & \mu I_n \end{pmatrix}.$$

Moreover, we have

$$D := H^{-1} K H = \rho^{-1} \begin{pmatrix} \lambda + 2\mu & 0 \\ 0 & \mu I_n \end{pmatrix}$$

with

$$H = \begin{pmatrix} 1 & 2(\nabla\mu)^T - \rho^{-1}\mu(\nabla\rho)^T \\ 0 & -(\lambda+\mu)I_n \end{pmatrix}.$$

Hence, (A.1) yields the following principally diagonal conjugate to the linear elasticity operator:

Theorem A.2 - *One has*

$$\begin{pmatrix} \nabla^T \\ I_n \end{pmatrix} \gamma\{\rho\partial_t{}^2 w - Lw\} = H\{\partial_t{}^2 I_{n+1} - (D\Delta I_{n+1} + P_1)\}H^{-1} \begin{pmatrix} \nabla\cdot w \\ w \end{pmatrix}$$

where

$$\begin{aligned} P_1 &= C[\Delta I_{n+1}, H] + A[\Delta I_{n+1}, B] + A\left(\begin{pmatrix} \nabla^T & 0 \\ 0 & \nabla \end{pmatrix} R + R'\right) \\ A &= (SH)^{-1} = (A_{ij})_{1\le i,j\le 2},\ B = MQ = (B_{ij})_{1\le i,j\le 2}, \\ C &= H^{-1}K = (C_{ij})_{1\le i,j\le 2},\ R = V_1 BH = (R_{ij})_{1\le i,j\le 2}, \\ R' &= V_2 BH = (R'_{ij})_{1\le i,j\le 2}. \end{aligned}$$

The explicit forms of these are the following :

$$A_{11} = \mu^{\frac{1}{2}}\rho^{-1},$$
$$A_{12} = \tfrac{1}{2}\mu^{-\frac{1}{2}}\rho^{-1}(\lambda+\mu)^{-1}(\lambda+5\mu)(\nabla\mu)^T \\ -\mu^{\frac{1}{2}}\rho^{-2}(\lambda+\mu)^{-1}(\lambda+2\mu)(\nabla\rho)^T$$
$$A_{21} = 0,\ A_{22} = -(\lambda+\mu)^{-1}\mu^{\frac{1}{2}}\rho^{-1}I_n$$

$$B_{11} = \mu^{-\frac{1}{2}}(\lambda+2\mu),\ B_{12} = \frac{3}{2}\mu^{-\frac{1}{2}}(\nabla\mu)^T,\ B_{21} = 0,\ B_{22} = \mu^{\frac{1}{2}}I_n$$

$$C_{11} = \rho^{-1}(\lambda+2\mu),\ C_{12} = \rho^{-1}(2(\nabla\mu)^T - \rho^{-1}\mu(\nabla\rho)^T)$$
$$C_{21} = 0,\ C_{22} = -\mu(\lambda+\mu)^{-1}I_n$$

$$R_{11} = -\frac{1}{2}\mu^{-\frac{3}{2}}(\mu-\lambda)\nabla\mu$$

$$R_{12} = \frac{1}{2}\mu^{-\frac{3}{2}}(\lambda-3\mu)\nabla\mu\otimes\nabla\mu + \frac{1}{2}\rho^{-1}\mu^{-\frac{1}{2}}(\mu-\lambda)\nabla\mu\otimes\nabla\rho \\ +\mu^{-\frac{1}{2}}(\lambda+\mu)\nabla^2\mu - (\lambda+\mu)(\tfrac{1}{4}\mu^{-\frac{3}{2}}|\nabla\mu|^2 - \tfrac{1}{2}\mu^{-\frac{1}{2}}\Delta\mu)I_n$$

$$R_{21} = \mu^{-\frac{1}{2}}(\lambda+\mu),\ R_{22} = \mu^{-\frac{1}{2}}(\lambda+\mu)(\nabla\mu)^T - \rho^{-1}\mu^{\frac{1}{2}}(\lambda+\mu)(\nabla\rho)^T$$

$$R'_{11} = R'_{12} = 0,\ R'_{21} = R_{11},\ R'_{22} = R_{12}$$

References

[1] D.D. Ang, M. Ikehata, D.D. Trong, M. Yamamoto, Unique continuation for a stationary isotropic Lame system with variable coefficients Comm. Part. Diff. Equat. 23(1998), 371–385.

[2] Chelminski, K. - The Principle of Limiting Absorbtion in Elasticity, Bull. Polish Acad. Sc. Math., 41(1993), 19–30.

[3] B. Dehman, L. Robbiano, La propriété du prolongement unique pour un système elliptique, le système de Lamé, J. Math Pures Appl., 72(1993), 745–493.

[4] L. Hörmander , Linear partial differential operators, Springer-Verlag, Berlin 1966

[5] L. Hörmander, The analysis of linear partial differential operators I-IV Springer-Verlag, Berlin 1983.

[6] V. Isakov, A non-hyperbolic Cauchy problem for $\square_b\square_c$ and its applications to elasticity theory Comm. Pure Appl. Math., 39(1986), 747–767.

[7] V. Isakov, Inverse Problems for PDE, Springer-Verlag, New York 1997.

[8] R. Leis, Initial boundary value problems in mathematical physics Wiley, New York 1986.

[9] D. Tataru, A priori estimates of Carleman's type in domains with boundary, Journal des Math. Pures et Appl., 73(1994), 355–387.

[10] D. Tataru, Unique continuation for solutions to PDE's; between Hörmander's and Holmgren's theorem, Comm. PDE 20(1995), 855–884.

[11] D.Tataru, Carleman estimates and unique continuation for solutions to boundary value problems J. Math. Pures Appl. 75(1996), 367–408.

[12] M. Yamamoto, On an inverse problem of determining source terms in Maxwell's equations with a single measurement in Inverse Problems, Tomography, and Image Processing,241—256, Plenum Press, New York

Matthias Eller
Department of Mathematics
Tennessee Technological University
Cookeville, TN 38505
e-mail:ME1125@tntech.edu

Victor Isakov
Department of Mathematics and Statistics,
Wichita State University
Wichita, KS 67260-0033
e-mail: isakov@twsuvm.uc.twsu.edu

Gen Nakamura
Faculty of Engineering
Gunma University
1-5-1 Tenjin-cho, Kiryn-shi, Gunma 376
Japan
e-mail: nakamura@ns518.eg.gunma-u.ac.jp

Daniel Tataru
Dept. of Mathematics
Northwestern University
Evanston, IL 60208
http://www.math.nwu.edu/ tataru

Studies in Mathematics and its Applications, Vol. 31
D. Cioranescu and J.L. Lions (Editors)

Chapter 16

ON THE UNSTABLE SPECTRUM OF THE EULER EQUATION

S. FRIEDLANDER

Introduction

In this short paper we survey some recent results concerning the unstable spectrum of the Euler equations linearised about a steady flow of an ideal fluid. We discuss the concept of a "fluid Lyapunov exponent" which determines the essential spectral radius of the linearised Euler operator. A positive lower bound for this exponent is computable for many classes of explicit flows and hence gives an effective method for detecting instability in the essential spectrum. The problem of existence of unstable eigenvalues is more difficult and, at least at present, must be treated on a case by case basis. In the case of certain specific flows containing regions of spatial oscillation it is possible to investigate the discrete unstable spectrum using averaging techniques and to demonstrate the existence of unstable eigenvalues. In the case of plane parallel shear flow with a sinusoidal profile we construct the complete unstable spectrum. We discuss a result which proves under certain rather general conditions that linear instability implies nonlinear instability. We show how this abstract theorem can be used to conclude that some specific examples of oscillating flows with discrete unstable eigenvalues are nonlinearly unstable.

1. The Euler equations for fluid motion

The Euler equations for the motion of an invisid incompressible fluid are

$$q_t + (q \cdot \nabla)\, q = -\nabla P \tag{1.1}$$

$$\nabla \cdot q = 0\,. \tag{1.2}$$

The evolution of the velocity field $q(x,t)$ is determined by its current state, as well as by the internal pressure forces $-\nabla p(x,t)$.

Let $U_0(x)$ be a steady Euler flow. From (1.1), (1.2) the equilibrium equation for U_0 can be written in the form

$$U_0 \times \operatorname{curl} U_0 = -\nabla H \,, \ \nabla \cdot U_0 = 0 \tag{1.3}$$

where H is a scalar field known as the reduced pressure. We consider the linearised Euler equations for a small perturbation $v(x,t)$ about a steady flow U_0:

$$\left.\begin{array}{rcl} v_t & = & -(U_0 \cdot \nabla)\, v - (v \cdot \nabla)\, U_0 - \nabla p \equiv Lv \\ \nabla \cdot v & = & 0 \end{array}\right\} \tag{1.4}$$

with initial condition

$$v(x,0) = v_0(x)\,. \tag{1.5}$$

The results that we will present are proved for

$$U_0(x) \in (C^\infty(T^n))^n \ , \ \nabla \cdot U_0 = 0 \tag{1.6}$$

$$v_0(x) \in (L^2(T^n))^n \ , \ \nabla \cdot v_0 = 0 \tag{1.7}$$

where $T^n = \mathbb{R}^n/2\pi\,\mathbb{Z}^n$ is an n-dimensional torus. The applications we will discuss are examples in 2 and 3 dimensions.

In this paper we discuss the spectral problem associated with equation (1.4), in particular the nature of the stability spectrum σ. The stability spectrum of a viscous flow in a bounded domain is simple: it consists of a sequence of eigenvalues. However the more difficult mathematical problem of the spectrum of *inviscid* flows still has open questions. Unlike the viscous problem where the relevant Navier-Stokes operator is elliptic, the linearised Euler operator L is degenerate non-elliptic and a continuous spectrum is usually present. In fluid dynamics the inviscid limit is well known to be a subtle problem that is mathematically and physically challenging [1]. We will give some partial answers concerning the structure of the stability spectrum for the invisid equations (1.4).

2. The essential spectral radius

Let e^{Lt} be the evolution operator for the linearised Euler equation (1.4). We use the following classification of the elements of the spectrum σ. A point $z \in \sigma$ is called a point of the *discrete* spectrum σ_{disc} if it satisfies the following three conditions.

1. z is an isolated point in σ.
2. z has finite multiplicity.
3. The range of $z - e^{Lt}$ is closed.

If z does not satisfy $1-3$, it is called an element of the essential spectrum σ_{ess}. Thus

$$\sigma = \sigma_{\text{disc}} \cup \sigma_{\text{ess}}$$

and the union is disjoint. We denote by r_{ess} the essential spectral radius in L^2, where, following Nussbaum [2] we define

$$r_{\text{ess}}\left(e^{Lt}\right) = \sup\left\{|z| \mid z \in \sigma_{\text{ess}}\left(e^{Lt}\right)\right\}.$$

The following theorem due to Vishik [3] gives an expression for $r_{\text{ess}}\left(e^{Lt}\right)$ in terms of a geometric quantity that can be considered as a Lyapunov exponent for fluid flow.

Theorem 2.1 – *One has*

$$r_{\text{ess}}\left(e^{Lt}\right) = e^{\mu t}$$

where

$$\mu = \lim_{t\to\infty} \frac{1}{t} \log \sup_{\substack{x_0, \xi_0, b_0 \\ b_0 \cdot \xi_0 = 0 \\ |b_0| = 1,\ |\xi_0| = 1}} \left|b\left(x_0, \xi_0, b_0; t\right)\right|. \tag{2.1}$$

Here the vector b is determined by the following system of ODE's which we call the bicharacteristic amplitude equations.

$$\left.\begin{aligned} \dot{x} &= U_0(x) && (2.2.1)\\ \dot{\xi} &= -\left(\frac{\partial U_0}{\partial x}\right)^T \xi && (2.2.2)\\ \dot{b} &= -\left(\frac{\partial U_0}{\partial x}\right) b + 2\left[\left(\frac{\partial U_0}{\partial x}\right) b \cdot \xi\right] \xi/|\xi|^2 && (2.2.3) \end{aligned}\right\}$$

with $x(0) = x_0$, $\xi(0) = \xi_0$, $b(0) = b_0$.

This theorem is proved by writing the evolution operator e^{Lt} as a product of a pseudodifferential operator and a shift operator along the trajectories of the equilibrium flow U_0. This allows the growth of the evolution operator to be studied to precise exponential asymptotics. This approach refines the result of Vishik and Friedlander [4] which proves that $e^{\mu t}$ gives a lower bound for $r_{\text{ess}}(e^{Lt})$. A heuristic derivation of this result is obtained by applying a "geometric optics" treatment based on high frequency perturbations to equations (1.4). In the language of geometric optics, equation (2.2.2) is the Eikonal equation and equation (2.2.3) is the transport equation. Equation (2.2.3) is the evolution equation for the amplitude of a high frequency wavelet initially localised at x_0, with initial wave number vector ξ_0. The quantity on the RHS of (2.1) is the Lyapunov exponent corresponding to the maximal exponential growth rate of such an amplitude vector $b(t)$.

The result of Theorem 2.1 gives one piece of information concerning the stability spectrum for inviscid flows, namely the maximum growth rate of instability in the essential spectrum. Moreover it implies that any point z in the spectrum $\sigma(e^{Lt})$ such that $|z| > e^{\mu t}$ is necessarily an eigenvalue of finite multiplicity. As we will discuss later, a positive lower bound for the value of the Lyapunov exponent μ can be explicitly computed in several examples [5], [6]. Furthermore Theorem 2.1 provides an effective sufficient condition for instability of large classes of inviscid fluid flows. Since expression (2.1) involves the supremum over initial conditions (x_0, ξ_0, b_0), it is only necessary to show there exists at least one initial condition for which the solution to the system (2.2) of ODE gives

$$\lim_{t\to\infty} \frac{1}{t} \log |b| > 0$$

to conclude that $\mu > 0$, and hence the unstable essential spectrum is non-empty.

3. Examples of instability in the essential spectrum

The idea that exponential stretching of fluid particles could imply instability for the Euler equations is originally due to Arnold [7]. Friedlander and Vishik [6] use the result discussed in Section 2 to prove every flow with exponential stretching somewhere in the flow is linearly unstable.

Theorem 3.1 – *(3-dimensions). Let the flow U_0 have a positive classical Lyapunov exponent at some point x_0. Then $\mu > 0$, and hence the flow U_0 is unstable.*

The proof of this theorem follows from a result obtained from the system (2.2) of ODE, namely:

$$\frac{d}{dt}(b_1 \times b_2 \cdot \xi) = 0 \tag{3.1}$$

where b_1 and b_2 are two linearly independent solutions of (2.2.3) corresponding to a co-tangent vector ξ that satisfies (2.2.2). Since the flow is volume preserving the existence of a positive classical Lyapunov exponent (i.e. an exponentially growing tangent vector) implies the existence of an exponentially decaying co-tangent vector ξ. From (3.1) we then conclude the existence of at least one exponentially growing amplitude vector b, which implies $\mu > 0$. Therefore we note that in 3-dimensions the classical Lyapunov exponent provides a lower bound on μ.

In 2-dimensions, the system (2.2) of ODE provides an even stronger constraint on the relation between $b(t)$ and $\xi(t)$, namely

$$\frac{d}{dt}\left(|b(t)|\ |\xi(t)|\right) = 0 . \tag{3.2}$$

Hence in 2-dimensions the fluid Lyapunov exponent μ and the maximal classical Lyapunov exponent for the dynamical system $\dot{x} = U_0(x)$ are the same.

It follows from Theorem 3.1 that any flow U_0 (in 2 or 3 dimensions) with a hyperbolic stagnation point x_0 is unstable [i.e. $U(x_0) = 0$ and there exists an eigenvalue of $\left(\frac{\partial U}{\partial x}\right)_{x_0}$ with positive real part]. There are large classes of fluid flows U_0 with such stagnation points. For all such flows $r_{\text{ess}} > 1$.

A class of flows with presumably chaotic stream lines was identified by Arnold [8]. An example is the so-called ABC flow $U_0 = (\dot{x}, \dot{y}, \dot{z})$ where

$$\begin{aligned} \dot{x} &= A\sin z + C\cos y \\ \dot{y} &= B\sin x + A\cos z \\ \dot{z} &= C\sin y + B\cos x . \end{aligned} \tag{3.3}$$

For general values of the constants A, B and C numerical investigations [9], [10] indicate that ABC flows exhibit the phenomenon of Lagrangian chaos which suggests strong exponential stretching. Analytic treatment of ABC flows [11], [12], proves that for certain ranges of A, B and C there is exponential stretching either at hyperbolic points or associated with hyperbolic closed trajectories. The result of Theorem 3.1 then proves that these ABC flows are hydrodynamically unstable.

In 3-dimensions (unlike 2-dimensions), it is possible to have flows U_0 for which the classical Lyapunov exponents are all zero yet the fluid exponent μ is positive. Such an example is constructed in [6]. It is proved that the integrable flow $U_0 \times \operatorname{curl} U_0 = -\nabla H$ with $\nabla H \neq 0$ has μ positive provided that a certain geometric condition is satisfied by the stream lines. The following (non sharp) condition ensures $\mu > 0$:

$$\int_0^T \{\kappa\, \widehat{n} \cdot \nabla H - \tau_g\, U_0 \cdot \operatorname{curl} U_0 / |\nabla H|^2\}\, dt \geq 0 \tag{3.4}$$

where, for any stream line of the flow as it wraps around the toroidal surface $H = H_0$, T is the period, κ the curvature, τ_g the geodesic torsion and $\widehat{n}$ the principal unit normal to the stream line.

We have described many fluid flows where it can be shown that $\mu > 0$. In a few cases μ can be computed explicitly. For example, the 2-dimensional cellular flow

$$U_0 = (-\sin x \cos y, \cos x \sin y)\,.$$

In this case μ is given by the positive real eigenvalue of the matrix $\left(\frac{\partial U_0}{\partial x}\right)$ at a hyperbolic stagnation point. Thus $\mu = 1$ for this simple cellular flow. There are certain classes of 2-dimensional flows for which it follows from (3.2) that $\mu = 0$ [13]. In particular $\mu = 0$ (i.e. there is no unstable essential spectrum) for

(1) 2-dimensional flows with no stagnation point,

(2) 2-dimensional plane parallel shear flows.

4. Examples of instability in the discrete spectrum

We now turn to the question of existence and distribution of unstable eigenvalues in the discrete spectrum of equation (1.4). As we remarked, the linearised Euler operator is degenerate, non-elliptic and there are no general theorems that may be applied to prove the existence of unstable discrete eigenvalues. However in certain rather special examples it is possible to construct unstable eigenvalues.

The spectral problem for the linearised Euler operator is considerably simpler in 2-dimensions rather than 3-dimensions. In particular, in 2-dimensions we can define a scalar stream function to replace the divergence free velocity field. We write

$$U_0 = \widehat{k} \times \nabla\Psi(x,y)\;,\;\; v = \widehat{k} \times \nabla\phi(x,y,t)\,. \tag{4.1}$$

Hence

$$\nabla \times U_0 = \widehat{k}\;\nabla^2\,\Psi(x,y)\;,\;\; \nabla \times v = \widehat{k}\;\nabla^2\,\phi(x,y,t)\,. \tag{4.2}$$

Here $\widehat{k}$ is the unit vector perpendicular to the 2-dimensional plane with cartesian co-ordinates (x, y). The equilibrium equation (1.3) will be satisfied when Ψ satisfies an elliptic equation of the form

$$\nabla^2\,\Psi = -F(\Psi)\,. \tag{4.3}$$

Taking the curl of equation (1.4) gives the equation for the evolution of the perturbation vorticity $\omega \equiv \nabla \times v$:

$$\frac{\partial \omega}{\partial t} = \{U_0, \omega\} + \{v, \nabla \times U_0\} \tag{4.4}$$

where { , } denotes the Poisson bracket of two vector fields, i.e.

$$\{A, B\} = (B \cdot \nabla) A - (A \cdot \nabla) B \,. \tag{4.5}$$

In general the second Poisson bracket on the RHS of (4.4) is very difficult to analyze. However in 2-dimensions the problem greatly simplifies because $\widehat{k} \cdot \nabla(\cdot) \equiv 0$. We note that for 2-dimensional flows it is proved in [14] that there is a solution of the linearised Euler equation (1.4) of the form $e^{\sigma t}\, v(x)$ with $v \in L^2$, $\nabla \cdot v = 0$, if and only if there is a non-trivial solution $\omega \in H^{-1}$ of the spectral problem for the 2-dimensional reduction of equation (4.4) given by

$$\sigma\, \omega + (U_0 \cdot \nabla)\, \omega + (v \cdot \nabla)(\nabla \times U_0) = 0 \,. \tag{4.6}$$

We therefore consider the eigenvalue PDE given by subsituting (4.1) – (4.3) into (4.6), namely

$$\sigma\, \nabla^2 \phi = \left(\Psi_y \frac{\partial}{\partial x} - \Psi_x \frac{\partial}{\partial y} \right) (\nabla^2 \phi + F'(\Psi)\, \phi) \,. \tag{4.7}$$

We take the boundary conditions to be 2π periodicity in (x, y).

A simple and very classical example that has received much attention in the literature of the past 100 years is plane parallel shear flow (see, for example, [15], [16], [17]). In this case $U_0 = (U(y), 0)$ and (4.7) becomes the so-called Rayleigh equation:

$$(\sigma / ik + U(y)) \left(\frac{d^2}{dy^2} - k^2 \right) \Phi(y) - U''(y)\, \Phi(y) = 0 \tag{4.8}$$

where we have written

$$\phi(x, y, t) = \Phi(y)\, e^{ikx}\, e^{\sigma t} \,. \tag{4.9}$$

The celebrated Rayleigh stability criterion [15] says that a necessary condition for instability is the presence of an inflection point in the profile $U(y)$. As we remarked in Section 3, the concept of the "fluid Lyapunov exponent" μ given by expression (2.1), can be used to prove that equation (4.8) with periodic boundary conditions has no unstable essential spectrum for any profile $U(y)$. It remains to discuss the possibility of discrete unstable eigenvalues (i.e. σ such that $\operatorname{Re}\sigma > 0$) associated with equation (4.8) for profiles $U(y)$ that contain at least one inflection point.

Meshalkin and Sinai [18], followed by Yudovich [19] investigated the instability of a *viscous* shear flow $U(y) = \sin my$ using techniques of continued fractions. More recently Friedlander et al [13], [20], [21] showed that these techniques could be used for the invisid equation (4.8) with $U(y) = \sin my$. Eigenfunctions are constructed in terms of Fourier series that converge to C^∞-smooth functions for eigenvalues σ that satisfy the characteristic equation. We write

$$\Phi(y) = \sum_{n=-\infty}^{\infty} a_n \, e^{iny} . \tag{4.10}$$

The recurrence relation equivalent to (4.8) yields the following tridiagonal infinite algebraic system:

$$d_{n+m} = \beta_n(\sigma)\, d_n + d_{n-m} \ , \ n \in \mathbb{Z} \tag{4.11}$$

where

$$\beta_n(\sigma) = \frac{2\sigma}{k} \cdot \frac{(k^2+n^2)}{k^2+n^2-m^2} \tag{4.12}$$

and

$$d_n = a_n(k^2+n^2-m^2) . \tag{4.13}$$

The system (4.11) is treated using continued fractions to yield the characteristic equation for σ namely

$$\left(\beta_j + \frac{1}{[\beta_{j+m}, \beta_{j+2m}, \ldots]}\right)\left(\beta_{j-m} + \frac{1}{[\beta_{j-2m}, \beta_{j-3m}, \ldots]}\right) + 1 = 0 \tag{4.14}$$

for each integer $j = 0, 1, \ldots [m/2]$. Here $[\ ,\ldots]$ denotes an infinite continued fraction, i.e.

$$[\beta_{j+m}, \beta_{j+2m}, \beta_{j+3m}, \ldots] = \frac{1}{\beta_{j+m} + \frac{1}{\beta_{j+2m} + \frac{1}{\beta_{j+3m} + \cdots}}} . \tag{4.15}$$

In [20], [21] an analysis is given of all the roots σ with $\operatorname{Re}\sigma > 0$, of equation (4.14) for each fixed integer m and $0 \le j \le m/2$. It is shown that for $k^2 > m^2 - j^2$ there are no eigenvalues. For $k_c^2 < k^2 < m^2 - j^2$, for a critical wave number k_c, there are purely real roots and for $0 < k^2 < k_c^2$ there is a complex conjugate pair of roots. This describes the complete unstable spectrum associated with the flows $U = \sin my$.

The existence of unstable eigenvalues for equation (4.8) with a general rapidly oscillating profile $U(my)$, $m \gg 1$, was demonstrated in [21] using homogeneization techniques to compute the spectral asymptotics. It is proved that one root of the asymptotic characteristic equation is given by

$$\sigma/k = \pm\sqrt{\langle U^2 \rangle} \qquad \text{as } m \to \infty \tag{4.16}$$

where $\langle \cdot \rangle$ denotes the 2π-average with respect to the fast variable my. Numerical analysis is used to describe the qualative behaviour of the distribution of roots of the asymptotic characteristic equation in the parameter space of the wave numbers n and k.

The following is an example of a stream function Ψ satisfying an equation of the form (4.3) which exhibits both the features of exponential stretching at a hyperbolic stagnation point and oscillatory shear flow behaviour. We consider

$$\Psi = \frac{1}{m}\left(\cos(x + my) + a\cos(x - my)\right) \tag{4.17}$$

where a is a constant such that $0 < a < 1$. This flow has hyperbolic points at

$$x + my = 2n\pi \ , \ x - my = (2j - 1)\,\pi \, .$$

Hence by the results of Section 3 there is a non empty unstable essential spectrum associated with this flow. The fluid Lyapunov exponent μ can be calculated explicitly in this example to give

$$\mu = 2\,a^{1/2} \, . \tag{4.17a}$$

Again homogenisation techniques can be used to demonstrate the existence of unstable eigenvalues for equation (4.7) with Ψ given by (4.17) with $m \gg 1$ [22]. We introduce a change of variables $\eta = my$ and write (4.7) in the form

$$\begin{aligned} &\sigma\left(\frac{\partial^2}{\partial x^2} + m^2\,\frac{\partial^2}{\partial \eta^2}\right)\phi \\ +\ &\left[(\sin(x+\eta) - a\sin(x-\eta))\,\frac{\partial}{\partial x} - (\sin(x+\eta) + a\sin(x-\eta))\,\frac{\partial}{\partial \eta}\right] \\ \times\ &\left[\frac{\partial^2}{\partial x^2} + m^2\,\frac{\partial^2}{\partial \eta^2} + (1+m^2)\right]\phi = 0 \, . \end{aligned} \tag{4.18}$$

We seek Block eigenfunctions of the form

$$\phi = e^{ip\eta/m}\, G(x,\eta) \tag{4.19}$$

where p is an integer such that $p \ll m$, and we seek smooth functions $G(x, \eta)$ given by a regular expansion in powers of $1/m$:

$$
\begin{aligned}
&G(x,\eta) = \Phi_0(x) + \widetilde{\phi}_0(x,\eta) + \frac{1}{m}\left(\Phi_1(x) + \widetilde{\phi}_1(x,\eta)\right) + \dots \\
&\text{with } \langle\widetilde{\phi}_j\rangle = \int_0^{2\pi} \widetilde{\phi}_j \, d\eta = 0 \\
&\text{and } \sigma = \sigma_0 + \frac{1}{m}\sigma_1 + \dots .
\end{aligned}
\tag{4.20}
$$

The solvability conditions for the hierarchy of equations for $\widetilde{\phi}_j$ gives the ODE that define the averaged part of the eigenfunctions Φ_j. The equation for Φ_0 becomes

$$
\begin{aligned}
\sigma_0^2 \left(\frac{d^2}{dx^2} - p^2\right)\Phi_0 + \left(\frac{1+a^2}{2} + a\cos 2x\right)\left(\frac{d^2}{dx^2} - p^2\right)\frac{d^2\Phi_0}{dx^2} \\
-\, 4\,a\sin 2x \left(\frac{d^2}{dx^2} - p^2\right)\frac{d\Phi_0}{dx} - 4\,a\cos 2x\,\frac{d^2}{dx^2}\,\Phi_0 = 0 .
\end{aligned}
\tag{4.21}
$$

We consider the behaviour of (4.21) with the parameter a. In the limit $a = 0$ equation (4.21) reduces to the equation governing the spectral asymptotics for oscillatory shear flows. When $a = 1$ the equation is degenerate at the hyperbolic stagnation points where $\cos 2x = -1$. In this case the flow has no shear flow type regions and the degeneracy of (4.21) implies there are no discrete unstable eigenvalues with smooth eigenfunctions. For $0 < a < 1$, equation (4.21) has discrete real eigenvalues σ_0 which are the leading order part of an expansion in powers of $(1/m)$ of the eigenvalues of (4.18). For large wave number k, Weyl's formula gives the asymptotic behaviour of σ_0 as

$$
\sigma_0^2/k^2 \to (1-a^2)/2 \tag{4.22}
$$

which as $a \to 0$ is in agreement with expression (4.16) for the purely sheared flow.

As we remarked, the fully 3-dimensional vorticity equation (4.4) is very difficult to study. However some "2 1/2" dimensional examples are tractable. For example, the stability problem of a three component velocity of the form

$$
U(x,y) = \widehat{k} \times \nabla\Psi + \widehat{k}\, g(\Psi) \tag{4.23}
$$

is governed by an equation for a z-independent perturbation of the form of equation (4.7). A specific example is the "ABC" type flow with

$$
\dot{x} = \cos my \ , \ \dot{y} = \frac{1}{m}\sin mx \ , \ \dot{z} = \sin my + \frac{1}{m}\cos mx . \tag{4.24}
$$

For large m, averaging techniques again lead to an equation for the average with respect to my of the leading order eigenfunction $\Phi_0(x)$:

$$\sigma_0^2\left(\frac{d^2}{dx^2}-p^2\right)\Phi_0+\frac{1}{2}\left(\frac{d^2}{dx^2}-p^2\right)\frac{d^2\Phi_0}{dx^2}=0\,. \tag{4.25}$$

Hence

$$\sigma_0^2/k^2=1/2$$

and we conclude there exist discrete unstable eigenvalues.

The 3-dimensional general ABC flow given by (3.3) is not integrable. It has been investigated numerically by a number of authors including Henon [9] and Dombre et al [10]. It appears that the Lagrangian trajectories of the flow (3.3) densely fill some open domains of T^3. Stagnation points may occur and when they do there is numerical evidence that they are connected by a web of heteroclinic streamlines. Poincaré sections computed for more general ABC flows include patterns that are qualitively similar to the streamlines of the special flow given by (4.24). This suggests that the unstable spectrum associated with general ABC flows contains both an essential part and discrete eigenvalues.

5. Nonlinear instability

In the previous sections we have discussed the unstable spectrum of the Euler equation linearised about a given steady flow U_0. The following general theorem of Friedlander et al [13] can be applied to prove that some of the examples we considered are also nonlinearly unstable.

Consider the evolution equation

$$w_t=Lw+N(w)\ ,\ w(0)=w_0 \tag{5.1}$$

and fix a pair of Banach spaces $X\hookrightarrow Z$ with a dense embedding. Here L is the generator of a C_0-group of operators in $\mathcal{L}(Z)$, e^{Lt} leaves X invariant for $t\in\mathbb{R}$, $X\subset D(L)$ and N is a nonlinear operator $N:X\to Z$. We assume there is *local* existence theorem for (5.1) for $w_0\in X$.

Definition of nonlinear stability: The trivial solution $w=0$ of (5.1) is called nonlinearly stable in X if for all $\varepsilon>0$ there exists a $\delta>0$ such that $\|w(0)\|_X<\delta$ implies

(i) there exists a unique solution $w(t)\in L^\infty((0,\infty);X)\cap C([0,\infty),Z)$

and

(ii) $\|w(t)\|_X < \varepsilon$ for a.e. $t \in [0, \infty)$.
The trivial solution $w = 0$ is called nonlinearly unstable if it is not stable.
Note: by this definition the "blowing up" of a solution is a particular case of instability.

Theorem 5.1 – *Let (5.1) admit a local existence theorem in X. Let N and L satisfy the following conditions.*

(1) $\|N(w)\|_Z \le C_0 \|w\|_X \, \|w\|_Z$ *for* $w \in X$ *with* $\|w\|_X < \rho$ *for some* $\rho > 0$. (5.2)

(2) *A spectral "gap" condition, i.e.*

$$\sigma(e^{Lt}) = \sigma_+ \cup \sigma_- \quad \textit{with } \sigma_+ \ne \phi$$

where

$$\begin{array}{l} \sigma_+ \subset \{z \in \mathbb{C} \mid e^{Mt} < |z| < e^{\Lambda t}\} \\ \sigma_- \subset \{z \in \mathbb{C} \mid e^{\lambda t} < |z| < e^{\mu t}\} \end{array} \tag{5.3}$$

with

$$-\infty < \lambda < \mu < M < \Lambda < \infty \textit{ and } M > 0\,. \tag{5.4}$$

Then the trivial solution $w = 0$ to equation (5.1) is nonlinearly unstable.

The proof of this theorem is by contradiction and it uses the gap condition to permit a projection onto the subspace of growing modes that is guaranteed to exist by virtue of the spectral gap. We note that a stronger result, i.e. replacing condition (2) by the condition that the unstable spectrum $\sigma(e^{Lt})$ is non-empty, may in fact be true but would require a different method of proof.

We apply Theorem 5.1 to the Euler equations in the following way. We rewrite (1.1) – (1.2) with the notation

$$q = U_0 + w \tag{5.5}$$

where $U_0 \in (C^\infty(T^n))^n$ is a steady solution of (1.1), i.e.

$$w_t = -(U_0 \cdot \nabla)\, w - (w \cdot \nabla)\, U_0 - (w \cdot \nabla)\, w - \nabla P \tag{5.6}$$

$$\nabla \cdot w = 0\,. \tag{5.7}$$

Hence in this notation

$$Lw \equiv -(U_0 \cdot \nabla)\, w - (w \cdot \nabla)\, U_0 - \nabla p \tag{5.8}$$

$$N(w) \equiv -(w \cdot \nabla)\, w - \nabla q \tag{5.9}$$

and instability of the trivial solution $w = 0$ of the equation $w_t = Lw + N(w)$ corresponds to instability of the steady Euler flow U_0. We make the "natural" choices of the spaces X and Z:

$$X = X_s = \{w \in (H^s(T^n))^n \mid \nabla \cdot w = 0\}, \; s > \tfrac{n}{2} + 1 \tag{5.10}$$

$$Z = \{w \in (L^2(T^n))^n \mid \nabla \cdot w = 0\}. \tag{5.11}$$

For this choice of X local existence is a classical result [23] and condition (5.2) follows from the Sobolev embedding theorem.

The condition that is difficult to check is the spectral gap condition (5.3) - (5.4). We note that in the case of the Navier-Stokes equations in a finite domain the additional viscous term $\nu\nabla^2 q$ ensures that the equations are elliptic. Hence the spectrum is purely discrete and the existence of any unstable eigenvalue ensures the gap condition is satisfied. For the Euler equations, we return to the examples of Sections 3 and 4 where the structure of the unstable spectrum was analysed for certain choices of U_0.

In Section 4 we discussed how any plane parallel shear flow with a sufficiently oscillating profile has a non-empty discrete unstable spectrum. The results of Section 3 can be used to show that the unstable essential spectrum associated with this flow is empty. Hence the spectral gap condition required by the proof of Theorem 5.1 is satisfied and therefore all shear flows with rapidly oscillating profiles are nonlinearly unstable.

We note this result is proved with periodic boundary conditions. The cellular 2-dimensional flow given by (4.17) is also shown to have discrete unstable eigenvalues. In this case the essential unstable spectrum is non-empty. However for $a \neq 1$ the computable expressions for the essential spectral radius (see (4.17a)) and the spectral asymptotics (see (4.22)) show that there exist unstable eigenvalues outside the essential spectral radius. Hence again we can apply Theorem 5.1 to prove nonlinear instability.

References

[1] Constantine, P., A few results and open problems regarding incompressible fluids, Notices of AMS, 42, n° 6 (1995), 658-663.

[2] Nussbaum, R., The radius of the essential spectrum, Duke Math. J., 37 (1970), 473–478.

[3] Vishik, M.M., The spectrum of small oscillations of an ideal fluid and Lyapunov exponents, J. Math. Pures et Appl., 75 (1996), 531–557.

[4] Vishik, M.M. and Friedlander, S., Dynamo theory methods for hydrodynamic stability, J. Math. Pures et Appl., 72 (1993), 145–180.

[5] Friedlander, S. and Vishik, M.M., Instability criteria for the flow of an inviscid incompressible fluid, Phys. Rev. Lett., 66, n° 17 (1991), 2204–2206.

[6] Friedlander, S. and Vishik, M.M., Instability criteria for steady flows of a perfect fluid, Chaos, 2, n° 3 (1992), 455–460.

[7] Arnold, V.I., Notes on the 3-dimensional flow pattern of a perfect fluid in the presence of small perturbation of the initial velocity field, Appl. Math. Mech., 36, n° 2 (1972), 236–242.

[8] Arnold, V.I., Sur la topologie des coulements stationnaires des fluides parfaits, C.R. Acad. Sci. Paris, 261 (1965), 17–20.

[9] Henon, M., Sur la topologie des lignes de courant dans un cas particulier, C.R. Acad. Sci. Paris, 262 (1966), 312.

[10] Dombre, T., Frisch, U., Greene, J.M., Henon, M., Mehr, A. and Soward, A., Chaotic streamlines in *ABC* flows, J. Fluid Mechanics 167(1986), 353–391.

[11] Friedlander, S., Gilbert, A. and Vishik, M.M., Hydrodynamic instability for certain *ABC* flows, Geophys. Astrophys. Fluid Dynamics, 73(1993), 97–107.

[12] Chicone, C., A geometric approach to regular perturbation theory with application to hydrodynamics, Trans. AMS, 347, n° 12 (1995), 4559–4598.

[13] Friedlander, S., Strauss, W. and Vishik, M.M., Nonlinear instability in an ideal fluid, Ann. Inst. Henri Poincar, Analyse non Linéaire 14, n° 2 (1997), 187–209.

[14] Friedlander, S., Strauss, W. and Vishik, M.M., Robustness of instability for the two dimensional Euler equations, submitted for publication.

[15] Rayleigh (Lord), On the stability or instability of certain fluid motions, Proc. Lond. Math. Soc., 9 (1880), 57–70.

[16] Tollmein, W., Ein allgenienes kriterium der instabilitat laminarer geschwindigkeitsverteilungen, Nachr. Ges. Wiss. Gottingen Math. Phys., 50(1935), 79–114.

[17] Lin, C.C., The Theory of Hydrodynamic Stability, Cambridge University Press (1955).

[18] Meshalkin, L. and Sinai, Y., Investigation of the stability of a stationary solution of a system of equations for the plane movement of an incompressible viscous liquid, App. Math. and Mech., 25(1961), 1140–1143.

[19] Yudovich, V.I., An example of secondary stationary flow or periodic flow appearing while a laminar flow of a viscous incompressible fluid looses its stability, App. Math. and Mech., 29(1965), 453–467.

[20] Friedlander, S. and Howard, L., Instability in parallel flows revisited, Studies in App. Math., to appear

[21] Belenkaya, L., Friedlander, S. and Yudovich, V., The unstable spectrum of oscillating shear flows, SIAM J. App. Math., to appear

[22] Friedlander, S., Vishik, M.M. and Yudovich, V., The unstable spectrum of a 2-dimensional periodic flow of an ideal fluid, in preparation.

[23] Temam, R., Local existence of C^∞ solutions of the Euler equations of incompressible perfect fluids, Lecture Notes Math., 565 (1976), 184–194.

Susan Freidlander
Department of Mathematics (M/C 249)
University of Illinois-Chicago
Chicago, Illinois 60607
USA
E-mail: susan@math.nwu.edu

Studies in Mathematics and its Applications, Vol. 31
D. Cioranescu and J.L. Lions (Editors)

Chapter 17

DECOMPOSITION EN PROFILS DES SOLUTIONS DE L'EQUATION DES ONDES SEMI LINEAIRE CRITIQUE A L'EXTERIEUR D'UN OBSTACLE STRICTEMENT CONVEXE

I. GALLAGHER ET P. GÉRARD

Résumé. On s'intéresse à l'équation d'ondes semi linéaire critique

$$(1) \qquad \Box u_n + |u_n|^4 u_n = 0 \quad \text{dans} \quad \mathbb{R}_t \times \Omega,$$

avec conditions aux limites de Dirichlet, où Ω est l'extérieur d'un domaine strictement convexe de $\mathbb{R}^3$. En suivant la méthode introduite par H. Bahouri et P. Gérard dans [1] dans le cas où $\Omega = \mathbb{R}^3$, on démontre tout d'abord un théorème de structure pour des suites de solutions d'énergie bornée de l'équation des ondes linéaire dans Ω. La démonstration fait appel à un résultat de non concentration pour de telles suites, qui s'obtient en utilisant des mesures semi–classiques. Les estimations de Strichartz démontrées par H. Smith et C. Sogge dans [16] permettent d'en déduire une description des suites de solutions d'énergie bornée de (1), à des termes de reste près, petits en énergie et en normes de Strichartz.

Abstract. We are interested in the critical semilinear wave equation

$$(1) \qquad \Box u_n + |u_n|^4 u_n = 0 \quad \text{in} \quad \mathbb{R}_t \times \Omega,$$

with Dirichlet boundary conditions, where Ω is the outside of a strictly convex domain of $\mathbb{R}^3$. Following the method introduced by H. Bahouri and P. Gérard in [1] in the case where $\Omega = \mathbb{R}^3$, we first prove a structure theorem for sequences of solutions to the linear wave equation in Ω. The proof requires a non–concentration result for such sequences, the proof of which involves semi–classical measures. The Strichartz estimates proved by H. Smith and C. Sogge in [16] enable us to infer a description of sequences of bounded energy solutions to (1), up to small remainder terms, both in energy and in Strichartz norms.

1. Introduction

Dans [1], H. Bahouri et P. Gérard ont obtenu un développement en profils des solutions de l'équation d'ondes semi linéaire critique suivante:

$$\Box u_n + |u_n|^4 u_n = 0 \quad \text{dans} \quad \mathbb{R}_t \times \mathbb{R}^3_x, \quad (u_n, \partial_t u_n)_{|t=0} = (\varphi_n, \psi_n), \tag{1}$$

où $(\varphi_n, \psi_n)_{n\in\mathbb{N}}$ est une suite bornée de $\dot{H}^1 \times L^2(\mathbb{R}^3)$. Ce développement s'obtient à partir d'une étude sur l'équation libre

$$\Box v_n = 0 \quad \text{dans} \quad \mathbb{R}_t \times \mathbb{R}^3_x, \quad (v_n, \partial_t v_n)_{|t=0} = (\varphi_n, \psi_n),$$

en utilisant de manière cruciale des estimations de type Strichartz (voir [10] par exemple, ou l'estimation (2) ci–dessous). Notre objectif ici est de reprendre cette étude dans le cas où le système (1) est posé dans un domaine de l'espace. Le seul cas connu pour lequel on dispose d'estimations de Strichartz est celui de l'extérieur d'un obstacle strictement convexe. Ainsi dans [16], H. Smith et C. Sogge ont montré que pour tout temps $T > 0$, et pour tout couple (q, r), vérifiant $1/q + 3/r = 1/2$ et $q > 2$, il existe une constante $C_{T,r}$ telle que pour toute fonction f telle que $(f, \partial_t f) \in \mathcal{C}^0([0,T], \dot{H}^1_0(\Omega) \times L^2(\Omega))$, on a

$$\|f\|_{L^q([0,T],L^r(\Omega))} \leq C_{T,r} \left(\|\nabla f_{|t=0}\|_{L^2(\Omega)} + \|\Box f\|_{L^1([0,T],L^2(\Omega))} \right). \tag{2}$$

On a noté $\dot{H}^1_0(\Omega)$ la fermeture de $C^\infty_0(\Omega)$ pour la norme $\|\nabla \cdot\|_{L^2(\mathbb{R}^3)}$ dans $\dot{H}^1(\mathbb{R}^3)$. Dans le cas où $\Omega = \mathbb{R}^3$, un simple argument d'échelle permet de montrer que $C_{T,r}$ ne dépend pas de T. Dans notre cadre d'étude, où Ω est l'extérieur d'un obstacle strictement convexe, un article récent de H. Smith et C. Sogge (voir [17]) montre que $C_{T,r}$ ne dépend pas non plus de T.

Nous allons donc dans cette étude considérer un domaine Ω de $\mathbb{R}^3$, régulier, extérieur d'un obstacle strictement convexe, et étudier le problème suivant:

$$\left\{ \begin{array}{c} \Box u_n + |u_n|^4 u_n = 0 \quad \text{dans} \quad \mathbb{R}_t \times \Omega, \quad u_{n|\mathbb{R}_t \times \partial\Omega} = 0 \\ (u_n, \partial_t u_n)_{|t=0} = (\varphi_n, \psi_n), \end{array} \right. \tag{3}$$

où $(\varphi_n, \psi_n)_{n\in\mathbb{N}}$ est une suite bornée de $E(\Omega) \overset{\text{déf}}{=} \dot{H}^1_0(\Omega) \times L^2(\Omega)$.

Il est démontré dans [16] l'existence et l'unicité de solutions régulières, globales en temps, à (3) pour $(\varphi_n, \psi_n) \in C^\infty_0(\Omega) \times C^\infty_0(\Omega)$. Dans le cas où $(\varphi_n, \psi_n) \in E(\Omega)$, on peut montrer, en utilisant les techniques de J. Shatah et M. Struwe (voir [15]) et des estimations de [16], l'existence et l'unicité d'une solution u_n à (3), et

$$u_n \in C^0(\mathbb{R}, \dot{H}^1_0(\Omega)) \cap \dot{C}^1(\mathbb{R}, L^2(\Omega)) \cap L^5_{loc}(\mathbb{R}, L^{10}(\Omega)). \tag{4}$$

La suite (u_n) est de plus bornée dans l'espace d'énergie. L'un des résultats de cette étude est qu'elle est également bornée dans $L^5_{loc}(\mathbb{R}, L^{10}(\Omega))$. Cela permet en corollaire d'obtenir une estimation Lipschitz sur le groupe d'évolution non linéaire associant une donnée initiale (φ_n, ψ_n) dans un borné de l'espace d'énergie, à la solution u_n de (3). Nous renvoyons à [4] pour un énoncé précis.

Nous ne nous attarderons pas ici sur ce problème de résolution de (3) dans l'espace d'énergie.

Remarquons que la norme $L^5_{loc}(\mathbb{R}, L^{10}(\Omega))$ apparaissant dans (4) est naturellement associée au système semi linéaire critique (3): c'est la norme de Strichartz qui permet de considérer le terme non linéaire $|u_n|^4 u_n$ comme un terme de force dans $L^1_{loc}(\mathbb{R}, L^2(\Omega))$.

Avant d'énoncer les résultats que nous allons présenter ici, donnons quelques notations: nous appellerons v_n la solution du problème linéaire associé à (3):

$$\begin{cases} \Box v_n = 0 \quad \text{dans} \quad \mathbb{R}_t \times \Omega, \quad v_{n|\mathbb{R}_t \times \partial\Omega} = 0 \\ \qquad (v_n, \partial_t v_n)_{|t=0} = (\varphi_n, \psi_n). \end{cases} \tag{5}$$

Rappelons que l'énergie d'une fonction f est donnée par

$$E_0(f,t) \stackrel{\text{déf}}{=} \frac{1}{2} \int_\Omega \left(|\partial_t f(t,x)|^2 + |\nabla f(t,x)|^2 \right) dx,$$

et ne dépend pas de t si f vérifie (5). De même, pour une solution de (3) vérifiant (4), la quantité

$$E(f,t) \stackrel{\text{déf}}{=} \frac{1}{2} \int_\Omega \left(|\partial_t f(t,x)|^2 + |\nabla f(t,x)|^2 \right) dx + \frac{1}{6} \int_\Omega |f(t,x)|^6 \, dx$$

est une constante.

Le premier résultat de ce travail concerne les suites de solutions de (5). Commençons par donner une définition.

Définition 1.1 - *Soit $(\varphi, \psi) \in \dot{H}^1 \times L^2(\mathbb{R}^3)$, et soit $(h_n, x_n, t_n) \in (\mathbb{R}^+)^{\mathbb{N}} \times \mathbb{R}^{3\mathbb{N}} \times \mathbb{R}^{\mathbb{N}}$, tels qu'il existe $(x_\infty, t_\infty) \in \mathbb{R}^3 \times \mathbb{R}$, avec*

$$\lim_{n \to \infty} (h_n, x_n, t_n) = (0, x_\infty, t_\infty).$$

On appelle onde de concentration linéaire associée à $(\varphi, \psi, h_n, x_n, t_n)$ la solution du problème suivant:

$$\begin{cases} \Box p_n = 0 \quad \text{dans} \quad \mathbb{R}_t \times \Omega, \quad p_{n|\mathbb{R}_t \times \partial\Omega} = 0 \\ (p_n, \partial_t p_n)_{|t=t_n} = \left(\frac{1}{h_n^{1/2}} \mathcal{P}_\Omega \left(\varphi(\frac{\cdot - x_n}{h_n}) \right), \frac{1}{h_n^{3/2}} \mathbf{1}_\Omega(\cdot) \psi(\frac{\cdot - x_n}{h_n}) \right), \end{cases} \tag{6}$$

où $\mathcal{P}_\Omega$ est le projecteur orthogonal de $\dot{H}^1(\mathbb{R}^3)$ sur $\dot{H}^1_0(\Omega)$.

Remarques. On omet de spécifier, dans la définition d'une onde de concentration linéaire, le domaine dans lequel on résout l'équation d'onde linéaire (6). On supposera toujours que ce domaine est Ω. D'autre part, quitte à extraire une sous–suite et à faire une translation en temps, on peut toujours supposer que

$$\lim_{n\to\infty} \frac{t_n}{h_n} = +\infty, \quad -\infty \quad \text{ou} \quad 0.$$

Dans la suite, nous appellerons donnée concentrante toute quantité du type $(\varphi, \psi, h_n, x_n, t_n)$ de la définition 1.1. Nous dirons que (h_n, x_n, t_n) et $(\widetilde{h}_n, \widetilde{x}_n, \widetilde{t}_n)$ sont orthogonales, comme dans [1], si

$$\text{soit} \quad \lim_{n\to\infty} \left| \log\left(\frac{\widetilde{h}_n}{h_n}\right)\right| = +\infty,$$

$$\text{soit} \quad h_n = \widetilde{h}_n \quad \text{et} \quad \lim_{n\to\infty} \frac{|(x_n, t_n) - (\widetilde{x}_n, \widetilde{t}_n)|}{h_n} = +\infty.$$

D'autre part, nous dirons que deux données concentrantes sont équivalentes si la différence des ondes de concentration linéaires associées tend vers zéro en énergie, quand n tend vers l'infini.

Nous ferons dans tout ce travail l'hypothèse de compacité à l'infini suivante:

$$\lim_{R\to\infty} \limsup_{n\to\infty} \int_{|x|\geq R} \left(|\nabla \varphi_n(x)|^2 + |\psi_n(x)|^2\right) \, dx = 0. \tag{7}$$

Enfin nous supposerons pour simplifier que φ_n et ψ_n tendent faiblement vers zéro quand n tend vers l'infini.

Nous sommes à présent en mesure d'énoncer le premier résultat de cette étude.

Théorème 1 - *Soit v_n la solution de (5), où (φ_n, ψ_n) est bornée dans $E(\Omega)$ et vérifie (7). Il existe des données concentrantes $(\varphi^{(j)}, \psi^{(j)}, h_n^{(j)}, x_n^{(j)}, t_n^{(j)})$, pour $j \in \mathbb{N} \setminus \{0\}$, orthogonales deux–à–deux, telles que quitte à extraire une sous–suite, la solution v_n se décompose de la manière suivante, pour tout $\ell \in \mathbb{N} \setminus \{0\}$:*

$$v_n(t,x) = \sum_{j=1}^{\ell} p_n^{(j)}(t,x) + w_n^{(\ell)}(t,x),$$

où $p_n^{(j)}$ est l'onde de concentration associée à $(\varphi^{(j)}, \psi^{(j)}, h_n^{(j)}, x_n^{(j)}, t_n^{(j)})$, où le reste vérifie

$$\forall T > 0, \quad \lim_{\ell\to\infty} \limsup_{n\to\infty} \|w_n^{(\ell)}\|_{L^\infty([-T,T];L^6(\Omega))} = 0,$$

et où les énergies sont orthogonales, au sens où

$$E_0(v_n) = \sum_{j=1}^{\ell} E_0(p_n^{(j)}) + E_0(w_n^{(\ell)}) + o(1), \quad n \to \infty.$$

Considérons à présent l'équation non linéaire (3). Nous allons commencer par énoncer un théorème comparant une onde de concentration linéaire à la solution de (3) avec même donnée de Cauchy. Dans ce but, donnons quelques définitions nécessaires à l'énoncé du théorème.

Définition 1.2 - *Soit $(\varphi, \psi, h_n, x_n, t_n)$ une donnée concentrante, et soit p_n l'onde de concentration linéaire associée, au sens de la définition 1.1. Alors l'onde de concentration non linéaire associée à p_n est la solution de l'équation suivante:*

$$\left\{ \begin{array}{l} \Box q_n + |q_n|^4 q_n = 0 \quad dans \quad \mathbb{R}_t \times \Omega, \quad q_{n|\mathbb{R}_t \times \partial\Omega} = 0 \\ \qquad (q_n, \partial_t q_n)_{|t=0} = (p_n, \partial_t p_n)_{|t=0} . \end{array} \right. \tag{8}$$

Pour toute suite (h_n, x_n) convergeant vers $(0, x_\infty)$, on définit le domaine remis à l'échelle

$$\Omega_n \stackrel{\text{déf}}{=} \frac{\Omega - x_n}{h_n} . \tag{9}$$

On verra dans la section suivante que Ω_n admet un "domaine limite", quand h_n tend vers zéro, que nous noterons Ω_∞.

Définition 1.3 - *Soient M_n et M deux ouverts de $\mathbb{R}^3$. On dit que M_n converge vers M quand n tend vers l'infini si les deux propriétés suivantes sont vérifiées:*

(i) pour tout compact K de M, il existe $N \in \mathbb{N}$ tel que pour tout $n \geq N, K$ est inclus dans M_n;

(ii) pour tout compact K' de ${}^c\overline{M}$, il existe $N \in \mathbb{N}$ tel que pour tout $n \geq N, K'$ est inclus dans ${}^c\overline{M}_n$.

On montrera dans la section 2.1 le résultat suivant:

Proposition 1.4 - *Soit le domaine Ω_n défini en (9). Alors quand n tend vers l'infini, Ω_n prend l'une des limites suivantes, quitte à extraire une sous–suite: $\emptyset$, $\mathbb{R}^3$ ou un demi–espace H, suivant la position de x_n par rapport à $\partial\Omega$.*

Enfin rappelons que dans [1]-[2], un opérateur de scattering S ainsi que des opérateurs d'onde $W^\pm$ sont étudiés. Dans notre travail, nous aurons besoin d'associer à une onde de concentration linéaire p_n deux ondes $p_n^\pm$, de la manière suivante.

Définition 1.5 - *Soit p_n une onde de concentration linéaire, associée à des données concentrantes $(\varphi, \psi, h_n, x_n, t_n)$. Soit Ω_∞ la limite du domaine Ω_n défini en (9), au sens de la définition 1.3. Si Ω_∞ est un demi–espace, on note R_{Ω_∞} la réflexion par antisymétrie par rapport à $\partial\Omega_\infty$, et P_{Ω_∞} l'opérateur $P_{\Omega_\infty}(f,g) \stackrel{\text{déf}}{=} (\mathcal{P}_{\Omega_\infty} f, \mathbf{1}_{\Omega_\infty} g)$, pour tout $(f,g) \in \dot{H}^1(\mathbb{R}^3) \times L^2(\mathbb{R}^3)$.*

Alors on associe à p_n deux ondes de concentration linéaires $p_n^\pm$, associées aux données concentrantes $(\varphi^\pm, \psi^\pm, h_n, x_n, t_n)$, avec

$$(\varphi^-, \psi^-) \stackrel{\text{déf}}{=} \begin{cases} (\varphi, \psi) & si \quad \lim\limits_{n\to\infty} \dfrac{t_n}{h_n} = +\infty \\ P_{\Omega_\infty} W^- R_{\Omega_\infty}(\varphi, \psi) & si \quad \lim\limits_{n\to\infty} \dfrac{t_n}{h_n} = 0 \\ P_{\Omega_\infty} S^{-1} R_{\Omega_\infty}(\varphi, \psi) & si \quad \lim\limits_{n\to\infty} \dfrac{t_n}{h_n} = -\infty \, , \end{cases}$$

$$(\varphi^+, \psi^+) \stackrel{\text{déf}}{=} \begin{cases} P_{\Omega_\infty} S R_{\Omega_\infty}(\varphi, \psi) & si \quad \lim\limits_{n\to\infty} \dfrac{t_n}{h_n} = +\infty \\ P_{\Omega_\infty} W^+ R_{\Omega_\infty}(\varphi, \psi) & si \quad \lim\limits_{n\to\infty} \dfrac{t_n}{h_n} = 0 \\ (\varphi, \psi) & si \quad \lim\limits_{n\to\infty} \dfrac{t_n}{h_n} = -\infty \, . \end{cases}$$

Théorème 2 - *Soit p_n une onde de concentration linéaire, associée à des données concentrantes $(\varphi, \psi, h_n, x_n, t_n)$, et soit q_n l'onde de concentration non linéaire associée à p_n. Alors q_n est bornée dans $L^5_{loc}(\mathbb{R}, L^{10}(\mathbb{R}^3))$, et vérifie les propriétés suivantes, avec les notations de la définition 1.5: pour tout temps $T > 0$, la limite quand λ tend vers l'infini de*

$$\limsup_{n\to\infty} \left(\sup_{-T \le t \le t_n - \lambda h_n} E_0(p_n^- - q_n, t) + \|p_n^- - q_n\|_{L^5([-T, t_n - \lambda h_n], L^{10}(\Omega))} \right)$$

et de

$$\limsup_{n\to\infty} \left(\sup_{t_n + \lambda h_n \le t \le T} E_0(p_n^+ - q_n, t) + \|p_n^+ - q_n\|_{L^5([t_n + \lambda h_n, T], L^{10}(\Omega))} \right)$$

est nulle.

Remarques. Le résultat que nous démontrerons sera en fait une version plus précise du théorème 2, car nous obtiendrons aussi une description de l'onde de concentration non linéaire q_n autour du temps de concentration t_n, c'est–à–dire pour $t \in [t_n - \lambda h_n, t_n + \lambda h_n]$ avec les notations du théorème 2. Ce résultat, associé au théorème 1, permet d'obtenir une décomposition en profils pour la solution u_n du problème de départ (3) (voir [4]): on utilise pour cela l'orthogonalité des $p_n^{(j)}$ ainsi que des données concentrantes. La décomposition s'obtient facilement pour des normes de Strichartz petites, puis le théorème 2 permet, par un argument de déformation en temps semblable au cas traité dans [1], de déduire la décomposition

$$\forall \ell \in \mathbb{N} \setminus \{0\}, \quad u_n(t,x) = \sum_{j=1}^{\ell} q_n^{(j)}(t,x) + w_n^{(\ell)}(t,x) + r_n^{(\ell)}(t,x),$$

où $r_n^{(\ell)}$ tend vers zéro en norme énergie et en normes de Strichartz. Les $q_n^{(j)}$ sont des ondes de concentration non linéaires, uniformément bornées en énergie et dans $L^5_{loc}(\mathbb{R}; L^{10}(\Omega))$. Nous n'entrerons pas dans les détails ici, mais nous concentrerons sur la démonstration des théorèmes 1 et 2. Nous renvoyons à [4] pour plus de précisions.

D'autre part, comme nous l'avons remarqué plus haut, on s'est restreint au cas de l'extérieur d'un domaine strictement convexe car c'est le seul cas connu pour lequel on dispose d'une estimation de type Strichartz pour les solutions de l'équation des ondes, cruciale dans notre démarche. Néanmoins, à supposer qu'une telle estimation soit vraie dans d'autres cadres, d'autres problèmes apparaissent: notamment l'orthogonalité en temps des ondes de concentration linéaires repose sur un résultat de non–concentration de la norme L^6 de la solution de l'équation linéaire (voir la section 2.3 ci–dessous), pour lequel la forme de Ω choisie ici joue un rôle important.

Le plan de l'étude est le suivant. Dans la section 2 ci–dessous, nous étudions tout d'abord le domaine remis à l'échelle Ω_n défini en (9), puis nous énonçons un résultat de propagation de la (h_n)–oscillation stricte, et enfin nous donnons un résultat fondamental de non concentration de la norme $L^6(\Omega)$ pour des ondes de concentration linéaires. Quelques éléments de démonstrations sont présentés. Dans la section 3, nous démontrons le théorème 1, tandis que le théorème 2 est démontré en section 4.

Nous noterons toutes les constantes universelles par la même lettre C. D'autre part, sauf mention du contraire, toutes les fonctions définies sur un domaine de $\mathbb{R}^3$ seront prolongées par zéro en des fonctions définies dans l'espace $\mathbb{R}^3$ entier.

2. Étude préliminaire

L'objet de cette section est d'énoncer quelques résultats préliminaires auxquels nous ferons appel par la suite. Nous en donnons les idées de démonstration, et renvoyons à [4] pour les détails.

2.1. Passage à la limite dans des domaines remis à l'échelle

Nous avons défini en (9) le domaine remis à l'échelle Ω_n. Nous allons étudier ici l'évolution de ce domaine quand n tend vers l'infini, et notamment démontrer la proposition 1.4; rappelons que la notion de convergence du domaine Ω_n a été donnée en définition 1.3.

La proposition suivante décrit les différentes configurations possibles pour Ω_∞, la limite de Ω_n, et précise la proposition 1.4. Avant d'énoncer le résultat, donnons quelques notations. Rappelons que l'on a noté $x_\infty \in \mathbb{R}^3$ la limite de la suite (x_n). On choisit une fonction Φ telle que

$$\Omega = \{x \in \mathbb{R}^3 \mid \Phi(x) > 0\}.$$

Si $x_\infty \in \partial\Omega$, on note alors $\omega \overset{\text{déf}}{=} \nabla\Phi(x_\infty)$ et $\alpha \overset{\text{déf}}{=} \lim_{n\to\infty} \dfrac{\Phi(x_n)}{h_n}$, limite qui existe dans $\overline{\mathbb{R}}$ quitte à extraire une sous–suite. Enfin on définit les ensembles

$$H^{\alpha,\omega} \overset{\text{déf}}{=} \{y \in \mathbb{R}^3 \mid y \cdot \omega > -\alpha\} \quad \text{si } \alpha \in \mathbb{R}, \tag{10}$$

$$H^{+\infty,\omega} \overset{\text{déf}}{=} \mathbb{R}^3, \quad \text{et} \quad H^{-\infty,\omega} \overset{\text{déf}}{=} \emptyset.$$

Proposition 2.6 - *Le domaine Ω_n défini en (9) admet l'une des limites suivantes quand n tend vers l'infini:*

(i) si $x_\infty \in \Omega$, alors $\Omega_\infty = \mathbb{R}^3$ et pour tous $(\varphi, \psi) \in \dot{H}^1 \times L^2(\mathbb{R}^3)$, pour toute suite $(t_n) \in \mathbb{R}^{\mathbb{N}}$, l'onde de concentration linéaire p_n associée à $(\varphi, \psi, h_n, x_n, t_n)$ vérifie

$$\lim_{n\to\infty} E_0(p_n) = \|\nabla\varphi\|^2_{L^2(\mathbb{R}^3)} + \|\psi\|^2_{L^2(\mathbb{R}^3)} \cdot$$

(ii) si $x_\infty \in {}^c\overline{\Omega}$, alors le domaine limite vérifie $\Omega_\infty = \emptyset$ et toutes les données concentrantes sont équivalentes à zéro.

(iii) si $x_\infty \in \partial\Omega$, alors le domaine limite vérifie $\Omega_\infty = H^{\alpha,\omega}$ et toute donnée concentrante $(\varphi, \psi, h_n, x_n, t_n)$ est équivalente à la donnée concentrante $(\mathcal{P}_{H^{\alpha,\omega}}\varphi, \mathbf{1}_{H^{\alpha,\omega}}\psi, h_n, x_n, t_n)$.

Démonstration. Nous n'allons pas écrire tous les calculs ici: les deux premiers cas se démontrent de façon très simple, nous allons donc seulement donner les arguments permettant d'obtenir le cas (iii).

Par définition, $y \in \mathbb{R}^3$ est un élément de Ω_n si et seulement si $\Phi(x_n + h_n y) > 0$. Commençons par supposer que $\alpha = +\infty$, et soit K un compact de $\mathbb{R}^3$. Alors la formule de Taylor nous permet d'écrire, uniformément en y,

$$\forall y \in K, \quad \frac{\Phi(x_n + h_n y)}{h_n} = \frac{\Phi(x_n)}{h_n} + \nabla\Phi(x_\infty) \cdot y + o(1), \quad n \to \infty.$$

Pour n assez grand, on a donc $\Phi(x_n+h_ny)>0$, donc $y\in\Omega_n$, et donc $\Omega_\infty=\mathbb{R}^3$. Le même type de raisonnement permet de montrer que si $\alpha=-\infty$, alors $\Omega_\infty=\emptyset$.

Reste donc à étudier le cas où $\alpha\in\mathbb{R}$. Si K est un compact de $H^{\alpha,\omega}$, alors par définition de $H^{\alpha,\omega}$ en (10), on a

$$\forall y\in K,\quad \nabla\Phi(x_\infty)\cdot y=-\alpha+m(y),\quad \text{avec}\quad m(y)>0,$$

donc pour n assez grand, uniformément en $y\in K$, on a $\Phi(x_n+h_ny)>0$. Inversement, si K est un compact de ${}^c\overline{H}^{\alpha,\omega}$, alors pour tout $y\in K$, pour n assez grand uniformément en y, on a $\Phi(x_n+h_ny)<0$, et le résultat s'en déduit immédiatement: $\Omega_\infty=H^{\alpha,\omega}$.

Pour conclure la démonstration de la proposition, considérons une donnée concentrante $(\varphi,\psi,h_n,x_n,t_n)$, d'onde de concentration linéaire associée p_n, et soit $p_n^{\alpha,\omega}$ l'onde de concentration linéaire associée à $(\varphi^{\alpha,\omega},\psi^{\alpha,\omega},h_n, x_n,t_n)$, avec $(\varphi^{\alpha,\omega},\psi^{\alpha,\omega})\overset{\text{déf}}{=}(\mathcal{P}_{H^{\alpha,\omega}}\varphi,\mathbf{1}_{H^{\alpha,\omega}}\psi)$.

Par conservation de l'énergie, on peut écrire

$$\begin{aligned}E_0(p_n-p_n^{\alpha,\omega})&=h_n^{-3}\int_{\mathbb{R}^3}\left|\nabla\left(\mathcal{P}_\Omega(\varphi^{\alpha,\omega}(\frac{x-x_n}{h_n}))-\mathcal{P}_\Omega(\varphi(\frac{x-x_n}{h_n}))\right)\right|^2dx\\
&\quad+h_n^{-3}\int_{\mathbb{R}^3}\left|\nabla\left(\mathbf{1}_\Omega(x)\psi^{\alpha,\omega}(\frac{x-x_n}{h_n})-\mathbf{1}_\Omega(x)\psi(\frac{x-x_n}{h_n})\right)\right|^2dx\\
&=\int_{\mathbb{R}^3}\left|\nabla\big(\mathcal{P}_{\Omega_n}(\varphi^{\alpha,\omega}(y))-\mathcal{P}_{\Omega_n}(\varphi(y))\big)\right|^2dy\\
&\quad+\int_{\mathbb{R}^3}\left|\nabla\big(\mathbf{1}_{\Omega_n}(\psi^{\alpha,\omega}(y))-\mathbf{1}_{\Omega_n}(\psi(y))\big)\right|^2dy.\end{aligned}$$

Il suffit donc à présent de démontrer la convergence de $\mathcal{P}_{\Omega_n}f$ vers $\mathcal{P}_{\Omega_\infty}f$ dans $\dot H^1(\mathbb{R}^3)$, ainsi que celle de $\mathbf{1}_{\Omega_n}g$ vers $\mathbf{1}_{\Omega_\infty}g$ dans $L^2(\mathbb{R}^3)$, pour tout couple $(f,g)\in\dot H^1\times L^2(\mathbb{R}^3)$.

Lemme 2.7 - *Soit Ω_n un domaine convergeant vers Ω_∞, dans le sens de la définition 1.3, où Ω_∞ est un demi–espace, $\emptyset$ ou $\mathbb{R}^3$. Alors on a les propriétés suivantes:*

(i) $\dot H_0^1(\Omega_\infty)=\left\{u\in\dot H^1(\mathbb{R}^3)\mid supp\,u\subset\overline{\Omega}_\infty\right\}$;

(ii) $\forall f\in\dot H^1(\mathbb{R}^3),\quad \lim_{n\to\infty}\mathcal{P}_{\Omega_n}f=\mathcal{P}_{\Omega_\infty}f\quad$ *dans* $\dot H^1(\mathbb{R}^3)$;

(iii) $\forall g\in L^2(\mathbb{R}^3),\quad \lim_{n\to\infty}\mathbf{1}_{\Omega_n}g=\mathbf{1}_{\Omega_\infty}g\quad$ *dans* $L^2(\mathbb{R}^3)$.

Démonstration du lemme. Les cas où $\Omega_\infty=\mathbb{R}^3$ ou $\emptyset$ sont évidents. Dans le cas où Ω_∞ est un demi–espace, il est facile de démontrer le point (i): on suppose par exemple que Ω_∞ est défini par $\Omega_\infty = \{x \in \mathbb{R}^3 \mid x_3 > 0\}$, et l'on considère une fonction $\rho \in C_0^\infty(\Omega_\infty)$. Alors pour toute fonction $u \in \dot{H}^1(\mathbb{R}^3)$ supportée dans $\overline{\Omega}_\infty$, la fonction $u^\varepsilon \stackrel{\text{déf}}{=} u * \rho^\varepsilon$, où $\rho^\varepsilon \stackrel{\text{déf}}{=} 1/\varepsilon^3 \rho(\cdot/\varepsilon)$, tend vers u dans $\dot{H}^1(\mathbb{R}^3)$, et est supportée dans Ω_∞, donc (i) est démontré.

Le point (ii) se démontre de la manière suivante: on constate facilement que pour tout domaine M, et pour toute fonction f, la fonction $v \stackrel{\text{déf}}{=} \mathcal{P}_M f$ est définie comme la solution de

$$\begin{cases} \Delta v = \Delta f \quad \text{dans} \quad M \\ \qquad v_{|\partial M} = 0. \end{cases}$$

Alors il suffit de définir $v_n \stackrel{\text{déf}}{=} \mathcal{P}_{\Omega_n} f$, et soit v une de ses limites au sens des distributions; on a pour toute fonction $\phi \in C_0^\infty(\Omega_\infty)$

$$\int_{\mathbb{R}^3} \Delta v_n \overline{\phi}\, dx = \int_{\mathbb{R}^3} \Delta f \overline{\phi}\, dx,$$

puisque pour n assez grand, on a $\phi \in C_0^\infty(\Omega_n)$. Par passage à la limite faible dans la condition aux limites de Dirichlet, on en déduit que $v = \mathcal{P}_{\Omega_\infty} f$. Finalement la convergence forte est due au fait que

$$\begin{aligned} -\int_{\mathbb{R}^3} |\nabla v_n|^2\, dx &= \int_{\mathbb{R}^3} v_n \Delta \overline{v}_n\, dx \\ &= -\int_{\mathbb{R}^3} \nabla v_n \cdot \nabla \overline{f}\, dx, \end{aligned}$$

et donc comme $\displaystyle\int_{\mathbb{R}^3} \nabla v \cdot \nabla \overline{f}\, dx = -\int_{\mathbb{R}^3} |\nabla v|^2\, dx$, le résultat suit.

Le point (iii) s'obtient de manière similaire, nous laissons les détails au lecteur. □

Ce lemme termine la démonstration de la proposition 2.6. □

Le résultat suivant sera utilisé fréquemment dans la suite; sa démonstration est du même type que les calculs ci–dessus conduisant au point (ii) du lemme, nous ne le démontrerons donc pas ici.

Proposition 2.8 - *Soit M_n un ouvert de $\mathbb{R}^3$, de limite M. Soit (f_n^0, f_n^1) une suite bornée de $E(M_n)$, convergeant faiblement (resp. fortement) vers un couple (f^0, f^1) dans $\dot{H}^1(\mathbb{R}^3) \times L^2(\mathbb{R}^3)$, avec $(f^0, f^1) \in \dot{H}^1 \times L^2(M)$. Alors la solution de*

$$\begin{cases} \Box f_n = 0 \quad \text{dans} \quad \mathbb{R}_t \times M_n, \quad f_{n|\mathbb{R}_t \times \partial M_n} = 0 \\ \qquad (f_n, \partial_t f_n)_{|t=0} = (f_n^0, f_n^1) \end{cases}$$

converge faiblement (resp. fortement) vers f dans $L^\infty_{loc}(\mathbb{R}; \dot H^1(\mathbb{R}^3))$, et $\partial_t f_n$ converge faiblement (resp. fortement) vers $\partial_t f$ dans $L^\infty_{loc}(\mathbb{R}; L^2(\mathbb{R}^3))$, où

$$\left\{ \begin{array}{l} \Box f = 0 \quad \text{dans} \quad \mathbb{R}_t \times M, \quad f_{|\mathbb{R}_t \times \partial M} = 0 \\ \qquad\qquad (f, \partial_t f)_{|t=0} = (f^0, f^1). \end{array} \right.$$

D'autre part, si M_n est l'extérieur d'un domaine strictement convexe, alors la convergence a lieu aussi dans $L^5_{loc}(\mathbb{R}; L^{10}(\mathbb{R}^3))$.

2.2. Propagation de la (h_n)–oscillation

Commençons par rappeler la définition d'une fonction strictement (h_n)–oscillante. Dans la suite nous noterons A l'opérateur auto–adjoint non borné suivant:

$$\mathcal{D}(A) = \left\{ (u, \dot u) \in \dot H^1_0 \times L^2(\Omega) \mid \dot u \in H^1_0(\Omega), \Delta u \in L^2(\Omega) \right\},$$

$$A \stackrel{\text{déf}}{=} \begin{pmatrix} 0 & 1 \\ \Delta & 0 \end{pmatrix}.$$

Définition 2.9 - *Soit (h_n) une suite de réels strictement positifs, tendant vers 0, et soit (f_n, g_n) une suite bornée dans $\dot H^1 \times L^2(\mathbb{R}^3)$. La suite (f_n, g_n)est dite (h_n)–oscillante si*

$$\lim_{R \to \infty} \lim_{h \to 0} \left\| \mathbf{1}_{|A| \geq \frac{R}{h_n}} (f_n, g_n) \right\|_{\dot H^1 \times L^2(\mathbb{R}^3)} = 0.$$

La suite (f_n, g_n) est dite strictement (h_n)–oscillante si elle vérifie d'autre part

$$\lim_{\varepsilon \to 0} \lim_{h \to 0} \left\| \mathbf{1}_{|A| \leq \frac{\varepsilon}{h_n}} (f_n, g_n) \right\|_{L^2(\mathbb{R}^3)} = 0.$$

Remarque. Nous dirons qu'une suite (v_n) bornée dans $C^0([0,T], \dot H^1_0(\Omega))$ telle que $\partial_t v_n$ est bornée dans $C^0([0,T], L^2(\Omega))$ est (strictement) (h_n)–oscillante si $(\nabla v_n, \partial_t v_n)$ l'est au sens de la définition 2.9.

Nous ne rappellerons pas ici la notion de composante strictement (h_n)–oscillante d'une suite bornée de $L^2(\mathbb{R}^3)$, et renvoyons à [1], Lemme 3.2 (*iii*), pour une définition. Énonçons maintenant le résultat suivant, de propagation de la (h_n)–oscillation stricte. Nous laissons sa démonstration au lecteur (voir [4]).

Proposition 2.10 - *Soit (φ_n, ψ_n) une suite bornée de $E(\Omega)$, telle que la suite $(\nabla \varphi_n, \psi_n)$ est (strictement) (h_n)–oscillante. Soit v_n la solution de (5) associée; alors $(\nabla v_n, \partial_t v_n)(t)$ est (strictement) (h_n)–oscillante, uniformément en t pour tout $t \in \mathbb{R}$.*

D'autre part, soient v_n et $\widetilde{v}_n$ les solutions de l'équation des ondes linéaire, associées à des données (v_n^0, v_n^1) et $(\widetilde{v}_n^0, \widetilde{v}_n^1)$ respectivement, où $(\widetilde{v}_n^0, \widetilde{v}_n^1)$ est la composante strictement (h_n)–oscillante de (v_n^0, v_n^1). Alors pour toute suite (t_n), $(\nabla\widetilde{v}_n(t_n), \partial_t\widetilde{v}_n(t_n))$ est la composante strictement (h_n)–oscillante de $(\nabla v_n(t_n), \partial_t v_n(t_n))$.

2.3. Un résultat de non concentration

Nous allons dans cette section donner quelques idées de la démonstration du résultat suivant.

Proposition 2.11 - *Soit p_n une onde de concentration linéaire associée à une donnée concentrante $(\varphi, \psi, h_n, x_n, 0)$. Alors pour tout intervalle de temps $I \subset \mathbb{R}$ borné, pour toute suite (ε_n) telle que $\lim_{n\to\infty} \varepsilon_n/h_n = +\infty$, on a*

$$\lim_{n\to\infty} \|p_n\|_{L^\infty(I\setminus[-\varepsilon_n,\varepsilon_n]),L^6(\Omega)} = 0.$$

Démonstration. Nous allons en donner le principe général, mais n'entrerons pas dans les détails des calculs (on renvoie à [4] pour les arguments précis).

Commençons par remarquer que la proposition 2.11 est démontrée si l'on montre que

(i) pour tout temps $T \neq 0$, et pour toute suite $T_n \to T$, on a

$$\lim_{n\to\infty} \|p_n(T_n)\|_{L^6(\Omega)} = 0\,;$$

(ii) pour toute suite $\varepsilon_n \to 0$, avec $\varepsilon_n/h_n \to +\infty$, on a

$$\lim_{n\to\infty} \|p_n(\varepsilon_n)\|_{L^6(\Omega)} = 0\,.$$

Le cas (ii) peut être considéré comme une version dégénérée du cas (i), et se démontre de façon analogue, mais plus simple (par changement d'échelle); nous n'y reviendrons donc pas.

La méthode de démonstration de (i) s'appuie sur le principe de concentration–compacité de P.-L. Lions (voir [11]-[12]): introduisons la densité d'énergie

$$e_n(t,x) \stackrel{\text{déf}}{=} \left(|\partial_t p_n(t,x)|^2 + |\nabla_x p_n(t,x)|^2\right) dx,$$

qui vérifie

$$\partial_t e_n = \operatorname{div}_x(\partial_t p_n \nabla_x p_n).$$

Alors si e_∞ est un point d'adhérence de e_n, alors e_∞ est continue en temps, à valeurs mesure, et par le principe de concentration–compacité, il suffit de démontrer que

$$e_\infty(t)(\{x_0\}) = 0, \quad \forall x_0 \in \overline{\Omega}, \quad \forall t \neq 0. \tag{11}$$

Le calcul de e_∞ se fait par utilisation des mesures semi–classiques de Wigner (voir [7], [8], [13]): soit μ la mesure semi–classique associée à $(\nabla_{t,x} p_n)$. Il est bien connu que

$$\int_{\mathbb{R}_\tau \times \mathbb{R}^3_\xi} \mu(t,x,d\tau,d\xi) \le e_\infty(t)dt,$$

que le support de μ est tel que $(\tau^2 - |\xi|^2)\mu = 0$, et que μ vérifie l'équation de transport

$$(\tau\partial_t - \xi \cdot \nabla_x)\mu = 0.$$

Notre démarche va être d'exhiber une mesure μ_* telle que

$$\mu_* \le \mu, \tag{12}$$

$$\text{et} \quad \int_{\mathbb{R}_\tau \times \mathbb{R}^3_\xi} \mu_*(t,x,d\tau,d\xi) = e_*(t)dt,$$

où e_* est continue en temps, pour $t \neq 0$, et vérifie

$$e_*(t)(\Omega) = \|\nabla \mathcal{P}_{\Omega_\infty}\varphi\|^2_{L^2(\mathbb{R}^3)} + \|\mathbf{1}_{\Omega_\infty}\psi\|^2_{L^2(\mathbb{R}^3)}.$$

Par application de la conservation de l'énergie et de la vitesse finie de propagation, on peut montrer que

$$\forall t \neq 0, \quad e_*(t) = e_\infty(t),$$

et si l'on connaît μ_* explicitement, on peut en déduire (11).

Le calcul de μ_* se fait de manière différente suivant la position de x_∞ par rapport au bord $\partial\Omega$. Rappelons que d'après la proposition 2.6, on peut supposer que $x_\infty \in \overline{\Omega}$. Nous allons expliquer brièvement la méthode d'obtention de μ_*, suivant que $x_\infty \in \Omega$ ou $x_\infty \in \partial\Omega$.

Dans le cas où $x_\infty \in \Omega$, on prolonge p_n par zéro en dehors de Ω, en une suite notée u_n. La mesure μ_* évoquée ci–dessus est la mesure semi–classique associée à $\nabla_{t,x} u_n$, dont on calcule l'expression en utilisant le flot du billard G_t au–dessus de Ω: on peut montrer que

$$\int f\left(G_{-t}(x,\tau,\xi)\right)\mu_*(t,dx,d\tau,d\xi) = \int f(x,\tau,\xi)d\mu_0(x,\tau,\xi)$$

avec

$$d\mu_0(x,\tau,\xi) = \frac{1}{2}\delta(x - x_\infty)\sum_{\pm}\delta(\tau \mp |\xi|)\left|\widehat{\psi}(\xi) \pm i|\xi|\widehat{\varphi}(\xi)\right|^2 \frac{d\xi}{(2\pi)^3} \cdot \tag{13}$$

On montre que

$$e_\infty(t)(\{x_0\}) = \mu_0\left(\{(x,\tau,\xi) \in \Omega \times \mathbb{R} \setminus \{0\} \times \mathbb{R}^3\},\ p_x G_t(x,\tau,\xi) = x_0\right),$$

où p_x est la projection sur x. Alors par (13), on en déduit que

$$e_\infty(t)(\{x_0\}) = \sum_{\pm} \int_{E_\pm(t)} f_\pm(\xi) d\xi,$$

où $f_\pm \in L^1(\mathbb{R}^3)$, et où $E_\pm(t) \stackrel{\text{déf}}{=} \left\{\xi \in \mathbb{R}^3 \mid p_x G_t(x_\infty, \pm|\xi|, \xi) = x_0\right\}$. Il est alors facile de constater, par des arguments géométriques sur Ω, que $E_\pm(t)$ est soit vide, soit une demi–droite, ce qui implique (11).

Dans le cas où $x_\infty \in \partial\Omega$, on montre que

$$\mu_* = \delta\left(x - x_\infty - t\frac{\xi}{\tau}\right)\lambda_0(\tau,\xi), \tag{14}$$

où λ_0 est une mesure, absolument continue par rapport à la mesure de Lebesgue sur le cône $\left\{\tau^2 = |\xi|^2,\ \frac{\xi}{\tau} \cdot N(x_\infty) < 0\right\}$, où $N(x)$ désigne la normale unitaire sortante en $x \in \partial\Omega$. En outre, la mesure λ_0 a pour masse totale

$$TM(\lambda_0) = \|\nabla \mathcal{P}_{\Omega_\infty}\varphi\|^2_{L^2(\mathbb{R}^3)} + \|\mathbf{1}_{\Omega_\infty}\psi\|^2_{L^2(\mathbb{R}^3)}\,. \tag{15}$$

Montrons comment on exhibe cette mesure λ_0: on prolonge p_n par réflexion, dans un voisinage de $x_\infty \in \partial\Omega$. La suite v_n ainsi obtenue vérifie une équation du type

$$\rho\partial_t^2 v_n - \operatorname{div}(A \cdot \nabla v_n) = 0,$$

où ρ et A sont continues, régulières par morceaux, avec $\partial\Omega$ pour interface. En outre, on a $\rho(x) = 1$ et $A(x) = \mathrm{Id}$ dans Ω. On considère alors la mesure semi–classique ν associée à $(\sqrt{\rho}\partial_t v_n, \sqrt{A}\nabla_x v_n)$, dont on montre qu'elle est continue en temps, et vérifie

$$\nu\left(\{x \in \partial\Omega,\ \xi \cdot N(x) \neq 0,\ \tau \neq 0\}\right) = 0.$$

D'autre part, cette mesure vérifie une équation de transport le long du champ Hamiltonien associé au symbole $\frac{1}{2}(\tau^2 - \frac{1}{\rho}A\xi\cdot\xi)$. La difficulté est due au fait que les coefficients de cette équation de transport sont discontinus ($\nabla\rho$ et ∇A). Néanmoins, leur composante normale est continue, ce qui suffit pour définir localement un flot G_t pour cette équation, tel que si $\nu_{|t=0} \geq \nu_*$, alors $\nu(t) \geq G_t(\nu_*)$. On montre alors que $\nu_{|t=0}$ s'écrit $\delta(x - x_\infty)\lambda(\tau,\xi)$, où λ est tel que $\lambda_0(\tau,\xi) \stackrel{\text{déf}}{=} \mathbf{1}_{\frac{\xi}{\tau}\cdot N<0}\lambda(\tau,\xi)$ vérifie (15). Enfin on remarque

que si $\frac{\xi}{\tau} \cdot N < 0$, alors $G_t(x_\infty, \tau, \xi) = (x_\infty + t\dfrac{\xi}{\tau}, \tau, \xi)$ reste au-dessus de Ω, ce qui permet d'en déduire (12), et donc le résultat cherché, avec μ_* définie en (14). Nous renvoyons à [4] pour des détails. ■

3. Le problème linéaire: démonstration du théorème 1

Cette section est dévolue à la démonstration du théorème 1 énoncé en introduction. La démarche, inspirée de [1], est la suivante: dans une première partie, les échelles $h_n^{(j)}$ de concentration sont extraites; dans la seconde partie, nous décomposons une fonction strictement (h_n)-oscillante, au sens de la définition 2.9, en une somme d'ondes de concentration linéaires orthogonales, avec donnée en un temps $t_n^{(k)}$ concentrée en un point $x_n^{(k)}$. Avec les résultats de la section 2, la méthode est alors très semblable au cas de tout l'espace traité dans [1]; nous n'entrerons pas dans les détails ici, et renvoyons à [4] pour les arguments précis.

3.1. Extraction des échelles

D'après le théorème de structure démontré par P. Gérard dans [6], on peut décomposer les données initiales de la manière suivante:

$$\forall \ell \in \mathbb{N} \setminus \{0\}, \quad \nabla \varphi_n(x) = \sum_{j=1}^{\ell} \nabla \widetilde{\varphi}_n^{(j)}(x) + \nabla \Phi_n^{(\ell)}(x),$$

$$\psi_n(x) = \sum_{j=1}^{\ell} \widetilde{\psi}_n^{(j)}(x) + \Psi_n^{(\ell)}(x),$$

où les $\widetilde{\varphi}_n^{(j)}$ et les $\widetilde{\psi}_n^{(j)}$ sont strictement $(h_n^{(j)})$–oscillants, les $h_n^{(j)}$ sont orthogonaux deux-à-deux, où

$$\|\nabla \varphi_n\|_{L^2(\mathbb{R}^3)}^2 = \sum_{j=1}^{\ell} \|\nabla \widetilde{\varphi}_n^{(j)}\|_{L^2(\mathbb{R}^3)}^2 + \|\nabla \Phi_n^{(\ell)}\|_{L^2(\mathbb{R}^3)}^2 + o(1), \quad n \to \infty,$$

$$\|\psi_n\|_{L^2(\mathbb{R}^3)}^2 = \sum_{j=1}^{\ell} \|\widetilde{\psi}_n^{(j)}\|_{L^2(\mathbb{R}^3)}^2 + \|\Psi_n^{(\ell)}\|_{L^2(\mathbb{R}^3)}^2 + o(1), \quad n \to \infty,$$

et où

$$\lim_{\ell \to \infty} \limsup_{n \to \infty} \|\nabla \Phi_n^{(\ell)}\|_{\dot{B}^0_{2,\infty}} = \lim_{\ell \to \infty} \limsup_{n \to \infty} \|\Psi_n^{(\ell)}\|_{\dot{B}^0_{2,\infty}} = 0.$$

Rappelons que pour toute fonction f, on définit

$$\|f\|_{\dot{B}^0_{2,\infty}} \stackrel{\text{déf}}{=} \sup_{k\in\mathbf{Z}} \left(\int_{2^k\leq|\xi|\leq 2^{k+1}} |\widehat{f}(\xi)|^2\, d\xi \right)^{1/2},$$

où $\widehat{f}$ est la transformée de Fourier de f. Mais dans [9], P. Gérard, Y. Meyer et F. Oru ont démontré que pour toute fonction f, on a

$$\|f\|_{L^6(\mathbb{R}^3)} \leq C\|\nabla f\|^{1/3}_{L^2(\mathbb{R}^3)}\|\nabla f\|^{2/3}_{\dot{B}^0_{2,\infty}}.$$

Alors par le résultat de propagation démontré en section 2.2, et par conservation de l'énergie, on peut finalement montrer la proposition suivante.

Proposition 3.12 - *Sous les hypothèses du théorème 1, la fonction v_n peut, quitte à extraire une sous–suite, se décomposer de la manière suivante:*

$$\forall \ell \in \mathbb{N}\setminus\{0\}, \quad v_n(t,x) = \sum_{j=1}^{\ell} v_n^{(j)}(t,x) + \widetilde{w}_n^{(\ell)}(t,x),$$

où $v_n^{(j)}$ est une solution de l'équation des ondes linéaire dans Ω, strictement $(h_n^{(j)})$–oscillante, et où les $h_n^{(j)}$ sont orthogonaux deux–à–deux. On a en outre

$$\forall T>0, \quad \lim_{\ell\to\infty} \limsup_{n\to\infty} \|\widetilde{w}_n^{(\ell)}\|_{L^\infty([-T,T],L^6(\Omega))} = 0,$$

et

$$E_0(v_n) = \sum_{j=1}^{\ell} E_0(v_n^{(j)}) + E_0(\widetilde{w}_n^{(\ell)}) + o(1), \quad n\to\infty.$$

Enfin l'on a

$$\forall T>0, \quad \forall j\neq k, \quad \lim_{n\to\infty} \|v_n^{(j)} v_n^{(k)}\|_{L^\infty([-T,T],L^3(\Omega))} = 0.$$

Démonstration. Seul le dernier résultat de cette proposition reste à être démontré. Il repose sur le lemme suivant.

Lemme 3.13 - *Soient $h_n^{(j)}$ et $h_n^{(k)}$ deux échelles orthogonales, et $(f_n^{(j)})$ et $(f_n^{(k)})$ deux suites de fonctions bornées dans $\dot{H}^1(\mathbb{R}^3)$, respectivement strictement $(h_n^{(j)})$– et $(h_n^{(k)})$–oscillantes au sens où*

$$\lim_{R\to\infty} \limsup_{n\to\infty} \|\mathbf{1}_{|\xi|\geq R/h_n^{(j)}} \widehat{f}_n^{(j)}\|_{L^2(\mathbb{R}^3)} = 0,$$

$$et \quad \lim_{\varepsilon\to 0}\limsup_{n\to\infty}\|\mathbf{1}_{|\xi|\le\varepsilon/h_n^{(j)}}\widehat{f}_n^{(j)}\|_{L^2(\mathbb{R}^3)}=0,$$

et de même en remplaçant j par k. Alors

$$\lim_{n\to\infty}\|f_n^{(j)}f_n^{(k)}\|_{L^3(\Omega)}=0.$$

Ce lemme conduit évidemment directement à la proposition 3.12, puisque les suites $\nabla v_n^{(j)}(t_n^{(j)},\cdot)$ et $\nabla v_n^{(k)}(t_n^{(k)},\cdot)$, pour des suites quelconques $(t_n^{(j)})$ et $(t_n^{(k)})$ de réels, vérifient les hypothèses du lemme 3.13 et la proposition 3.12 est démontrée. ■

Démonstration du lemme 3.13. Supposons par exemple que $\lim_{n\to\infty} h_n^{(k)}/h_n^{(j)} = 0$. Commençons par supposer que les fonctions $f_n^{(j)}$ et $f_n^{(k)}$ ont un spectre tel que respectivement $h_n^{(j)}|\xi|\in\mathcal{C}_0^{(j)}$ et $h_n^{(k)}|\xi|\in\mathcal{C}_0^{(k)}$, où $\mathcal{C}_0^{(j)}$ et $\mathcal{C}_0^{(k)}$ sont des couronnes fixes de $\mathbb{R}$. On a alors

$$\|f_n^{(j)}\|_{L^2(\mathbb{R}^3)}\le Ch_n^{(j)}\|\nabla f_n^{(j)}\|_{L^2(\mathbb{R}^3)}$$

$$\text{et}\quad \|f_n^{(k)}\|_{L^2(\mathbb{R}^3)}\le Ch_n^{(k)}\|\nabla f_n^{(k)}\|_{L^2(\mathbb{R}^3)}$$

et donc

$$\|f_n^{(j)}\|_{L^2(\mathbb{R}^3)}\le Ch_n^{(j)}\quad\text{et}\quad \|f_n^{(k)}\|_{L^2(\mathbb{R}^3)}\le Ch_n^{(k)}.$$

Alors par l'inégalité de Hausdorff–Young, on a

$$\begin{aligned}\|f_n^{(j)}f_n^{(k)}\|_{L^3(\mathbb{R}^3)} &\le C\|\widehat{f}_n^{(j)}*\widehat{f}_n^{(k)}\|_{L^{3/2}(\mathbb{R}^3)}\\ &\le C\|\widehat{f}_n^{(j)}\|_{L^1(\mathbb{R}^3)}\|\widehat{f}_n^{(k)}\|_{L^{3/2}(\mathbb{R}^3)}.\end{aligned}$$

Donc finalement

$$\begin{aligned}\|f_n^{(j)}f_n^{(k)}\|_{L^3(\mathbb{R}^3)} &\le C\left(\frac{1}{h_n^{(j)}}\right)^{3/2}h_n^{(j)}\left(\frac{1}{h_n^{(k)}}\right)^{1/2}h_n^{(k)}\\ &\le C\left(\frac{h_n^{(k)}}{h_n^{(j)}}\right)^{1/2}.\end{aligned}$$

Le lemme est démontré dans le cas d'un spectre localisé. Sinon, les hypothèses de stricte oscillation faites sur $f_n^{(j)}$ et $f_n^{(k)}$ permettent d'approximer ces fonctions dans $\dot{H}^1(\mathbb{R}^3)$ par de telles fonctions localisées, et d'obtenir le résultat. ■

3.2. Extraction des temps et des coeurs de concentration

Chacune des fonctions $v_n^{(j)}$ obtenue dans la proposition 3.12 ci–dessus va être décomposée en une somme d'ondes de concentration linéaires orthogonales. Pour simplifier les notations, nous allons fixer une échelle de concentration h, et considérer une famille de fonctions (v_h), strictement (h)–oscillante.

Proposition 3.14 - *Soit (v_h) une famille de fonctions strictement (h)–oscillante. Quitte à extraire une sous–suite, il existe des ondes de concentration linéaires $p_h^{(k)}$ associées à des données $(\varphi^{(k)}, \psi^{(k)}, h, x_h^{(k)}, t_h^{(k)})$, orthogonales deux–à–deux, telles que*

$$\forall \ell \in \mathbb{N} \setminus \{0\}, \quad v_h(t,x) = \sum_{k=1}^{\ell} p_h^{(k)}(t,x) + \overline{w}_h^{(\ell)}(t,x),$$

avec orthogonalité des énergies:

$$E_0(v_h) = \sum_{k=1}^{\ell} E_0(p_h^{(k)}) + E_0(\overline{w}_h^{(\ell)}) + o(1), \quad h \to 0,$$

et avec

$$\forall T > 0, \quad \lim_{\ell \to \infty} \lim_{h \to 0} \|\overline{w}_h^{(\ell)}\|_{L^\infty([-T,T];L^6(\Omega))} = 0.$$

Démonstration. La méthode de démonstration suit les calculs de [1], qui s'inspirent de la démarche de [14]. On commence par définir, pour tout couple $(x_h^{(j)}, t_h^{(j)})$, l'opérateur $D_h^{(j)}$ suivant:

$$D_h^{(j)} f(y) \stackrel{\text{déf}}{=} \Big(h^{1/2} f(t_h^{(j)}, x_h^{(j)} + hy), h^{3/2} \partial_t f(t_h^{(j)}, x_h^{(j)} + hy) \Big).$$

On définit alors une fonction dite d'exhaustion δ, pour toute famille de fonctions (f_h), de la manière suivante:

$$\delta(f_h) \stackrel{\text{déf}}{=} \sup_{(x_h),(t_h)} \Big\{ \|\nabla\varphi\|^2_{L^2(\mathbb{R}^3)} + \|\psi\|^2_{L^2(\mathbb{R}^3)}, \; D_h f_h \rightharpoonup (\varphi, \psi), \Big\},$$

où la limite ci–dessus s'entend à extraction de sous–suite près.

Lemme 3.15 - *Pour toute famille de fonctions (f_h) bornée en énergie, on a*

$$\forall T > 0, \quad \lim_{h \to 0} \|f_h\|_{L^\infty([-T,T];L^6(\Omega))} \leq C \delta(f_h)^{1/2} E_0(f_h)^{1/6}$$

Démonstration. Il suffit d'utiliser la décomposition de [6] pour $f_h(t_h, \cdot)$, suite bornée dans $\dot{H}^1(\mathbb{R}^3)$, ainsi que l'injection de $\dot{H}^1(\mathbb{R}^3)$ dans $L^6(\mathbb{R}^3)$. Nous laissons les détails au lecteur. ■

La construction des fonctions $p_h^{(k)}$ se fait par récurrence: d'après le lemme 3.15, on peut supposer que $\delta(v_h) > 0$. On choisit alors la donnée concentrante $(\varphi^{(1)}, \psi^{(1)}, x_h^{(1)}, t_h^{(1)})$ tel que

$$\|\nabla\varphi^{(1)}\|_{L^2(\mathbb{R}^3)}^2 + \|\psi^{(1)}\|_{L^2(\mathbb{R}^3)}^2 \geq \frac{1}{2}\delta(v_h)$$

et, quitte à extraire une sous–suite,

$$D_h^{(1)} v_h \rightharpoonup (\varphi^{(1)}, \psi^{(1)}).$$

Alors $p_h^{(1)}$ est l'onde de concentration linéaire associée à la donnée concentrante $(\varphi^{(1)}, \psi^{(1)}, h, x_h^{(1)}, t_h^{(1)})$. Le lemme suivant se démontre par un simple calcul, utilisant la proposition 2.6 et le lemme 2.7.

Lemme 3.16 - *Soit* $w_h^{(1)} \overset{\text{déf}}{=} v_h - v_h^{(1)}$. *Alors*

$$E_0(v_h) = E_0(v_h^{(1)}) + E_0(w_h^{(1)}) + o(1), \quad h \to 0.$$

On poursuit alors le développement par récurrence: on suppose que

$$v_h(t,x) = \sum_{k=1}^{K-1} p_h^{(k)}(t,x) + w_h^{(K-1)}(t,x),$$

avec orthogonalité en énergie, où chacun des $p_h^{(k)}$ est une onde de concentration linéaire avec orthogonalité des temps $t_h^{(k)}$ et des coeurs $x_h^{(k)}$ de concentration. Alors on exhibe par la même technique que ci–dessus une nouvelle onde de concentration linéaire $p_h^{(K)}$, de la manière suivante: on peut, d'après le lemme 3.15, supposer que $\delta(w_h^{(K-1)}) > 0$, et l'on définit la donnée concentrante $(\varphi^{(K)}, \psi^{(K)}, x_h^{(K)}, t_h^{(K)})$ de telle sorte que

$$\|\nabla\varphi^{(K)}\|_{L^2(\mathbb{R}^3)}^2 + \|\psi^{(K)}\|_{L^2(\mathbb{R}^3)}^2 \geq \frac{1}{2}\delta(w_h^{(K-1)}) \tag{16}$$

et, quitte à extraire une sous–suite,

$$D_h^{(K)} w_h^{(K-1)} \rightharpoonup (\varphi^{(K)}, \psi^{(K)}).$$

Alors $p_h^{(K)}$ est l'onde de concentration linéaire associée à la donnée concentrante $(\varphi^{(K)}, \psi^{(K)}, h, x_h^{(K)}, t_h^{(K)})$, et l'on a, comme dans le lemme 3.16

ci–dessus,

$$E_0(v_h) = \sum_{k=1}^{K} E_0(p_h^{(k)}) + E_0(w_h^{(K)}) + o(1), \quad h \to 0,$$

avec $w_h^{(K)} \stackrel{\text{déf}}{=} v_h - \sum_{k=1}^{K} p_h^{(k)}$.

L'orthogonalité des temps $t_h^{(k)}$ de concentration est due à la proposition 2.11. Celle des coeurs est laissée au lecteur: elle se montre par un argument classique de changement d'échelle en espace, en utilisant la proposition 2.8.

Reste alors à démontrer que

$$\forall T > 0, \quad \lim_{K\to\infty} \lim_{h\to 0} \|w_h^{(K)}\|_{L^\infty([-T,T],L^6(\Omega))} = 0.$$

Ce résultat resulte du fait que la série de terme général $\|\nabla\varphi^{(k)}\|^2_{L^2(\mathbb{R}^3)} + \|\psi^{(k)}\|^2_{L^2(\mathbb{R}^3)}$ est convergente, et donc $\|\nabla\varphi^{(k)}\|^2_{L^2(\mathbb{R}^3)} + \|\psi^{(k)}\|^2_{L^2(\mathbb{R}^3)}$ tend vers zéro quand k tend vers l'infini. Mais par définition de $(\varphi^{(K)}, \psi^{(K)})$ en (16), on a

$$\delta(w_h^{(K-1)}) \leq 2\left(\|\nabla\varphi^{(K)}\|^2_{L^2(\mathbb{R}^3)} + \|\psi^{(K)}\|^2_{L^2(\mathbb{R}^3)}\right),$$

donc $\lim_{h\to 0} \delta(w_h^{(K-1)})$ tend vers zéro quand K tend vers l'infini. Le lemme 3.15 donne alors le résultat.

La proposition 3.14 est démontrée. ∎

4. Le problème non linéaire: démonstration du théorème 2

Nous allons procéder en plusieurs étapes, qui nous permettront d'obtenir un résultat plus précis que celui énoncé dans le théorème, avec notamment une étude autour du temps de concentration t_n; comme dans la section précédente, nous allons pour simplifier les notations considérer une famille (p_h) d'ondes de concentration linéaires. D'autre part, nous ne nous intéresserons qu'au seul cas où

$$\lim_{h\to 0} \frac{t_h}{h} = +\infty,$$

et laissons les deux autres cas ($-\infty$ ou 0) au lecteur.

Les lemmes 4.17, 4.18 et 4.19 suivants conduisent directement au théorème.

4.1. Avant le temps de concentration

Lemme 4.17 - *Sous les hypothèses du théorème 2, on a*

$$\lim_{\lambda\to\infty}\lim_{h\to 0}\left(\sup_{-T\leq t\leq t_h-\lambda h} E_0(p_h-q_h,t)+\|p_h-q_h\|_{L^5([-T,t_h-\lambda h],L^{10}(\Omega))}\right)=0.$$

Démonstration. Le résultat est simplement dû à la proposition 2.11 et du théorème de linéarisation de [5]. ■

4.2. Autour du temps de concentration

Avant d'énoncer le résultat que nous nous proposons de démontrer, nous avons besoin d'introduire une onde non linéaire, notée $\bar{q}_h^\lambda$. Tout d'abord, définissons l'onde remise à l'échelle suivante:

$$P_h(s,y)\stackrel{\text{déf}}{=}h^{1/2}p_h(t_h+hs,x_h+hy).$$

La fonction P_h vérifie une équation d'ondes linéaire dans le domaine $\Omega_h \stackrel{\text{déf}}{=} \dfrac{\Omega - x_h}{h}$ · On lui associe la solution de l'équation linéaire dans le domaine limite Ω_∞:

$$\left\{\begin{array}{c}\Box P=0 \quad \text{dans} \quad \mathbb{R}_s\times\Omega_\infty, \quad P_{|\mathbb{R}_s\times\partial\Omega_\infty}=0\\ (P,\partial_s P)_{|s=0}=(\mathcal{P}_{\Omega_\infty}\varphi,\mathbf{1}_{\Omega_\infty}\psi)\,.\end{array}\right.$$

On définit alors Q^λ, la solution de l'équation non linéaire suivante:

$$\left\{\begin{array}{c}\Box Q^\lambda+|Q^\lambda|^4Q^\lambda=0 \quad \text{dans} \quad \mathbb{R}_s\times\Omega_\infty, \quad Q^\lambda_{|\mathbb{R}_s\times\partial\Omega_\infty}=0\\ (Q^\lambda,\partial_s Q^\lambda)_{|s=-\lambda}=(P,\partial_s P)_{|s=-\lambda}\,.\end{array}\right.$$

Enfin $\bar{q}_h^\lambda$ est définie par

$$\bar{q}_h^\lambda(t,x)\stackrel{\text{déf}}{=}\frac{1}{h^{1/2}}Q^\lambda\left(\frac{x-x_h}{h},\frac{t-t_h}{h}\right).$$

Nous allons montrer le résultat suivant.

Lemme 4.18 - *Sous les hypothèses du théorème 2, on a*

$$\lim_{\lambda\to\infty}\lim_{h\to 0}\widetilde{\|}q_h-\bar{q}_h^\lambda\|_{h,\lambda}=0,$$

où l'on a noté

$$\widetilde{\|}f\|_{h,\lambda}\stackrel{\text{déf}}{=}\sup_{t\in[t_h-\lambda h,t_h+\lambda h]}E_0(f,t)+\|f\|_{L^5([t_h-\lambda h,t_h+\lambda h],L^{10}(\mathbb{R}^3))}.$$

Démonstration. Elle va se faire en deux étapes. Commençons par définir la solution du système suivant:

$$\begin{cases} \Box Q_h^\lambda + |Q_h^\lambda|^4 Q_h^\lambda = 0 \quad \text{dans} \quad \mathbb{R}_s \times \Omega_h, \quad Q^\lambda_{h|\mathbb{R}_s\times\partial\Omega_h} = 0 \\ (Q_h^\lambda, \partial_s Q_h^\lambda)_{|s=-\lambda} = (P_h, \partial_s P_h)_{|s=-\lambda} \, . \end{cases}$$

Nous allons commencer par démontrer que

$$\lim_{h\to 0} \widetilde{\|} Q_h^\lambda - Q^\lambda \|_\lambda = 0, \quad \forall \lambda > 0, \tag{17}$$

avec $\widetilde{\|} f\|_\lambda \stackrel{\text{déf}}{=} \sup_{s\in[-\lambda,\lambda]} E_0(f,s) + \|f\|_{L^5([-\lambda,\lambda],L^{10}(\mathbb{R}^3))}$. La convergence forte en norme énergie s'obtient par un argument identique à celui permettant de démontrer la proposition 2.8 ci–dessus (voir [4]). Pour ce qui est de la norme Strichartz, le résultat s'obtient par les calculs suivants: l'inégalité de Hölder fournit

$$\|Q_h^\lambda - Q^\lambda\|^5_{L^5([-\lambda,\lambda],L^{10})} \leq \|Q_h^\lambda - Q^\lambda\|_{L^\infty([-\lambda,\lambda],L^6)} \|Q_h^\lambda - Q^\lambda\|^4_{L^4([-\lambda,\lambda],L^{12})}.$$

Mais Q_h^λ et Q^λ vérifient l'estimation de Strichartz, et comme l'énergie de $Q_h^\lambda - Q^\lambda$ tend vers zéro quand h tend vers zéro, pour tout $\lambda > 0$, on peut écrire

$$\|Q_h^\lambda - Q^\lambda\|^5_{L^5([-\lambda,\lambda],L^{10})} \leq C\varepsilon_\lambda(h) \left(\|Q^\lambda\|^{20}_{L^5([-\lambda,\lambda],L^{10})} + \|Q_h^\lambda\|^{20}_{L^5([-\lambda,\lambda],L^{10})} \right),$$

avec $\lim_{h\to 0} \varepsilon_\lambda(h) = 0$ pour tout $\lambda > 0$. On conclut alors par bootstrap superlinéaire (voir par exemple [1], lemme 2.2) à partir de l'estimation

$$\begin{aligned} \|Q_h^\lambda - Q^\lambda\|^5_{L^5([-\lambda,\lambda],L^{10})} &\leq C\varepsilon_\lambda(h) \|Q^\lambda\|^{20}_{L^5([-\lambda,\lambda],L^{10})} \\ &+ C\varepsilon_\lambda(h) \|Q_h^\lambda - Q^\lambda\|^{20}_{L^5([-\lambda,\lambda],L^{10})}. \end{aligned}$$

Pour terminer la démonstration du lemme, on introduit la fonction $R_h^\lambda \stackrel{\text{déf}}{=} Q_h - Q_h^\lambda$, avec

$$Q_h(s,y) \stackrel{\text{déf}}{=} h^{1/2} q_h(t_h + hs, x_h + hy).$$

Alors R_h^λ vérifie l'équation suivante:

$$\begin{cases} \Box R_h^\lambda + |R_h^\lambda + Q_h^\lambda|^4 (R_h^\lambda + Q_h^\lambda) - |Q_h^\lambda|^4 Q_h^\lambda = 0 \quad \text{dans} \quad \mathbb{R}_s \times \Omega_h, \\ R^\lambda_{h|\mathbb{R}_s\times\partial\Omega_h} = 0 \\ (R_h^\lambda, \partial_s R_h^\lambda)_{|s=-\lambda} = (Q_h, \partial_s Q_h)_{|s=-\lambda} - (P_h, \partial_s P_h)_{|s=-\lambda} \, . \end{cases}$$

D'après le lemme 4.17, on a

$$\lim_{\lambda\to\infty} \lim_{h\to 0} E_0(R_h^\lambda, -\lambda) = 0,$$

donc en notant de manière générique $\varepsilon(\lambda, h)$ une fonction vérifiant

$$\lim_{\lambda\to\infty}\lim_{h\to 0}\varepsilon(\lambda, h) = 0,$$

l'estimation de Strichartz (2) fournit, pour tout $T_0 > -\lambda$,

$$\begin{aligned}\|R_h^\lambda\|_{L^5([-\lambda,T_0],L^{10}(\mathbb{R}^3))} &\leq C\varepsilon(\lambda, h)\\ &+C\varepsilon(\lambda,h)\sum_{j=1}^{5}\|(R_h^\lambda)^j(Q_h^\lambda)^{5-j}\|_{L^1([-\lambda,T_0],L^2(\mathbb{R}^3))}\\ \leq C&\left(\varepsilon(\lambda,h)+\sum_{j=1}^{5}\|R_h^\lambda\|^j_{L^5([-\lambda,T_0],L^{10}(\mathbb{R}^3))}\|Q_h^\lambda\|^{5-j}_{L^5([-\lambda,T_0],L^{10}(\mathbb{R}^3))}\right). \quad (18)\end{aligned}$$

Il est facile de voir, par un changement d'échelle, que la constante C ne dépend ni de h, ni de λ. Mais d'après (17), on peut remplacer dans (18) la fonction Q_h^λ par Q^λ. En outre, la fonction Q^λ converge, quand λ tend vers l'infini, vers Q^- dans $L^5(\mathbb{R}, L^{10}(\mathbb{R}^3))$, avec

$$\begin{cases} \Box Q^- + |Q^-|^4 Q^- = 0 \quad \text{dans} \quad \mathbb{R}_s\times\Omega_\infty, \quad Q^-_{|\mathbb{R}_s\times\partial\Omega_\infty} = 0\\ \lim\limits_{s\to-\infty} E_0(Q^- - P, s) = 0.\end{cases}$$

Cette convergence a lieu pour tout temps, car le domaine limite Ω_∞ est soit l'espace $\mathbb{R}^3$ entier, soit un demi–espace, pour lesquels on sait que les solutions de l'équation des ondes non linéaire critique sont dans $L^5(\mathbb{R}, L^{10}(\Omega_\infty))$ (voir [3], le cas du demi–espace découlant du cas $\mathbb{R}^3$ par réflexion antisymétrique).

On a donc en fait

$$\begin{aligned}\|R_h^\lambda\|_{L^5([-\lambda,T_0],L^{10}(\mathbb{R}^3))} &\leq C\varepsilon(\lambda, h)\\ &+C\sum_{j=1}^{5}\|R_h^\lambda\|^j_{L^5([-\lambda,T_0],L^{10}(\mathbb{R}^3))}\|Q^-\|^{5-j}_{L^5([-\lambda,T_0],L^{10}(\mathbb{R}^3))},\end{aligned}$$

et l'on va conclure par déformation en temps (voir aussi [2]): si le temps T_0 est assez petit, uniformément en h et en λ, alors l'estimation ci–dessus implique directement, par bootstrap superlinéaire, que

$$\lim_{\lambda\to\infty}\lim_{h\to 0}\|R_h^\lambda\|_{L^5([-\lambda,T_0],L^{10}(\mathbb{R}^3))} = 0.$$

Définissons à présent

$$T_{\max} \overset{\text{déf}}{=} \sup\left\{T\in\mathbb{R} \mid \lim_{\lambda\to\infty}\lim_{h\to 0}\|R_h^\lambda\|_{L^5([-\lambda,T],L^{10}(\mathbb{R}^3))} = 0\right\}.$$

On a $T_{\max} \geq T_0$; soit $T_1 < T_{\max}$, alors on peut écrire

$$\|R_h^\lambda\|_{L^5([-\lambda,T_{\max}],L^{10}(\mathbb{R}^3))} \leq C\varepsilon(\lambda,h)$$

$$+C\|R_h^\lambda\|_{L^5([T_1,T_{\max}],L^{10}(\mathbb{R}^3))}\|Q^-\|^4_{L^5([T_1,T_{\max}],L^{10}(\mathbb{R}^3))}.$$

Il suffit alors de choisir T_1 assez proche de $T_{\max}$, uniformément en h et en λ, pour conclure que

$$\lim_{\lambda\to\infty}\lim_{h\to 0}\|R_h^\lambda\|_{L^5([-\lambda,T_{\max}],L^{10}(\mathbb{R}^3))} = 0,$$

et par l'inégalité d'énergie,

$$\lim_{\lambda\to\infty}\lim_{h\to 0}\sup_{-\lambda\leq s\leq T_{\max}} E_0(R_h^\lambda,s) = 0.$$

Le même type de calcul permet de montrer que si $\eta > 0$ est assez petit, alors

$$\lim_{\lambda\to\infty}\lim_{h\to 0}\|R_h^\lambda\|_{L^5([T_{\max},T_{\max}+\eta],L^{10}(\mathbb{R}^3))} = 0,$$

ce qui contredit la maximalité de $T_{\max}$; enfin on montre de la même manière que

$$\lim_{\lambda\to\infty}\lim_{h\to 0}\|R_h^\lambda\|_{L^5([-\lambda,+\infty[,L^{10}(\mathbb{R}^3))} = 0,$$

et avec l'inégalité d'énergie, le lemme est démontré. □

4.3. Après le temps de concentration

Lemme 4.19 - *Sous les hypothèses du théorème 2, on a*

$$\lim_{\lambda\to\infty}\lim_{h\to 0}\left(\sup_{t_h+\lambda h\leq t\leq T} E_0(p_h^+ - q_h,t) + \|p_h^+ - q_h\|_{L^5([t_h+\lambda h,T],L^{10}(\Omega))}\right) = 0.$$

Démonstration. Elle est très simple, au vu des deux lemmes précédents: il suffit d'appliquer les résultats de scattering obtenus dans [1]-[2], qui s'étendent à notre cadre par réflexion antisymétrique si Ω_∞ est un demi–espace. ■

Références

[1] H. Bahouri et P. Gérard, High frequency approximation of solutions to critical nonlinear wave equations, American Journal of Mathematics, 121 (1999), 131–175.

[2] H. Bahouri et P. Gérard, Concentration effects in critical nonlinear wave equations, Geometrical Optics and Related Topics (F. Colombini and N. Lerner eds.), Progress in Nonlinear Differential Equations and Applications, 32 (1997), Birkhäuser, Boston, 17–30.

[3] H. Bahouri et J. Shatah, Global estimate for the critical semilinear wave equation, Annales de l'Institut Henri Poincaré, Analyse non linéaire, 15 (1998), 6, 783–789.

[4] I. Gallagher et P. Gérard, Profile decomposition for the wave equation outside a convex obstacle, Journal de Mathématiques Pures et Appliquées, 80, 1, pages 1–49, 2001.

[5] P. Gérard, Oscillations and concentration effects in semilinear dispersive wave equations, Journal of Functional Analysis, 141 (1996), 60–98.

[6] P. Gérard, Description du défaut de compacité de l'injection de Sobolev, ESAIM Contrôle Optimal et Calcul des Variations, 3 (1998), 213–233, (version électronique: http://www.emath.fr/cocv/).

[7] P. Gérard et E. Leichtnam, Ergodic properties of eigenfunctions for the Dirichlet problem, Duke Mathematical Journal, 71 (1993), 559–607, .

[8] P. Gérard, P. Markowitch, N. Mauser et F. Poupaud: Homogenization limits and Wigner transforms, Communications on Pure and Applied Mathematics, 50 (1997), 323–379, .

[9] P. Gérard, Y. Meyer et F. Oru, Inégalités de Sobolev précisées, Séminaire Équations aux Dérivées Partielles, École Polytechnique, décembre 1996.

[10] J. Ginibre et G. Velo, Generalized Strichartz inequalities for the wave equation, Journal of Functional Analysis, 133 (1995), 1,50–68.

[11] P.-L. Lions, The concentration–compactness principle in the calculus of variations. The locally compact case, I, Annales de l'Institut Henri Poincaré, 1 (1984), 109–145.

[12] P.-L. Lions, The concentration–compactness principle in the calculus of variations. The limit case, II, Revista Matematica Iberoamericana, 1 (1985), 145–201.

[13] P.-L. Lions et T. Paul, Sur les mesures de Wigner, Revista Matematica Iberoamericana, 9 (1993),553–618.

[14] G. Métivier et S. Schochet, Trilinear resonant interactions of semilinear hyperbolic waves, Duke Mathematical Journal, 95 (1998), 2, 241–304.

[15] J. Shatah et M. Struwe, Well–posedness in the energy space for semilinear wave equations with critical growth, International Mathematics Research Notices, 7 (1994), 303–309.

[16] H. Smith et C. Sogge, On the critical semilinear wave equation outside convex obstacles, Journal of the American Mathematical Society, 8 (1995), 4, 879–916.

[17] H. Smith et C. Sogge, Global Strichartz estimates for nontrapping perturbations of the Laplacian, preprint.

Isabelle Gallagher
Département de Mathématiques
Université de Paris-Sud
91405 Orsay Cedex

and

Centre de Mathématiques
École Polytechnique
91128 Palaiseau Cedex, France

E-mail: Isabelle.Gallagher@math.polytechnique.fr

Patrick Gérard
Département de Mathématiques
Université de Paris-Sud
91405 Orsay Cedex, France

E-mail: Patrick.GERARD@math.u-psud.fr

Studies in Mathematics and its Applications, Vol. 31
D. Cioranescu and J.L. Lions (Editors)

Chapter 18

UPWIND DISCRETIZATIONS OF A STEADY GRADE-TWO FLUID MODEL IN TWO DIMENSIONS

V. GIRAULT AND L. R. SCOTT

1. Introduction

A fluid of grade two is a particular non-Newtonian Rivlin-Ericksen fluid (cf. [28]) whose equation of motion is

$$\begin{aligned}\frac{\partial}{\partial t}(\mathbf{u}-\alpha_1\Delta\,\mathbf{u}) - \nu\,\Delta\,\mathbf{u} + \mathbf{curl}(\mathbf{u} - (2\alpha_1+\alpha_2)\Delta\,\mathbf{u}) \times \mathbf{u}\\ - (\alpha_1+\alpha_2)\Delta(\mathbf{u}\cdot\nabla\,\mathbf{u}) + 2(\alpha_1+\alpha_2)\mathbf{u}\cdot\nabla(\Delta\,\mathbf{u}) + \nabla\,\tilde{p} = \mathbf{f}\,.\end{aligned} \tag{1.1}$$

This system of equations is completed by the condition of incompressibility:

$$\operatorname{div}\mathbf{u} = 0\,, \tag{1.2}$$

and suitable initial and boundary conditions. Here $\mathbf{f}$ is an an external force (usually gravity), $\mathbf{u}$ is the velocity, $\tilde{p}$ is the pressure, ν is the viscosity and α_1 and α_2 are material stress moduli, the three parameters being constant and divided by the density. It is considered an appropriate model for the motion of a water solution of polymers.

Dunn and Fosdick prove in [13] that, to be consistent with thermodynamics, the viscosity and normal stress moduli must satisfy

$$\nu \geq 0 \quad , \quad \alpha_1 \geq 0 \quad \text{and } \alpha_1+\alpha_2 = 0\,.$$

The reader can refer to [14] for a thorough discussion on the sign of α_1. With these assumptions, setting $\alpha = \alpha_1$, (1.1) simplifies and leads to the equation of motion

$$\frac{\partial}{\partial t}(\mathbf{u} - \alpha\Delta\,\mathbf{u}) - \nu\Delta\,\mathbf{u} + \mathbf{curl}(\mathbf{u} - \alpha\Delta\,\mathbf{u}) \times \mathbf{u} + \nabla\,p = \mathbf{f}\,,$$

where the modified pressure p is related to $\tilde{p}$ by

$$p = \tilde{p} + \frac{1}{2}\mathbf{u}\cdot\mathbf{u} - \alpha(\mathbf{u}\cdot\Delta\,\mathbf{u} + \frac{1}{4}\mathrm{tr}\,\mathbf{A}_1^2)\,,$$

where $\mathbf{A}_1 = \nabla \mathbf{u} + (\nabla \mathbf{u})^t$ is the symmetric gradient tensor. Interestingly, in [21, 22], Holm, Marsden and Ratiu derive these equations with $\nu = 0$ and $\alpha_1 + \alpha_2 = 0$, as a model of turbulence. They are called averaged-Euler equations, and α_1 is an averaged length scale. It can also be interpreted as a measure of dispersion and in this respect, these equations describe a dispersive fluid model (cf. [18, 21]).

The equations of a grade-two fluid model have been studied by many authors (Videman gives in [31] a very extensive list of references), but the best construction of solutions for the problem, with homogeneous Dirichlet boundary conditions and mildly smooth data, is given by Ouazar in [26] and by Cioranescu and Ouazar in [7, 8]. These authors prove existence of solutions, with H^3 regularity in space, by looking for a velocity $\mathbf{u}$ such that

$$\mathbf{z} = \mathbf{curl}(\mathbf{u} - \alpha \Delta \mathbf{u}), \tag{1.3}$$

has L^2 regularity in space, introducing $\mathbf{z}$ as an auxiliary variable and discretizing the equations of motion (in variational form) by Galerkin's method in the basis of the eigenfunctions of the operator $\mathbf{curl}\,\mathbf{curl}(\mathbf{u} - \alpha \Delta \mathbf{u})$. This excellent choice of basis allows one to recover estimates from the transport equation

$$\alpha \frac{\partial \mathbf{z}}{\partial t} + \nu \mathbf{z} + \alpha(\mathbf{u} \cdot \nabla \mathbf{z} - \mathbf{z} \cdot \nabla \mathbf{u}) = \nu \mathbf{curl}\,\mathbf{u} + \alpha \mathbf{curl}\,\mathbf{f}. \tag{1.4}$$

Whenever $\mathbf{curl}\,\mathbf{f}$ belongs to $L^2(\Omega)^3$, this construction is optimal because, in contrast to fixed-point arguments, it uses all the information conveyed by (1.1)–(1.4). Thus, it allows one to derive global existence of solutions with minimal restrictions on the size of the data, cf. [3] and [9]. A fixed-point argument cannot use all four equations because they are redundant. It is particularly important to preserve (1.1) since it implies that the energy is bounded without restrictions on the data. This point will be crucial for the numerical analysis of schemes discretizing (1.1).

The transport equation (1.4) substantially simplifies in two dimensions since the second nonlinear term $\mathbf{z} \cdot \nabla \mathbf{u}$ vanishes. In this case, $\mathbf{z} = (0, 0, z)$ with $z = \mathrm{curl}(\mathbf{u} - \alpha \Delta \mathbf{u})$, where curl is the operator

$$\mathrm{curl}\mathbf{v} = \frac{\partial v_1}{\partial x_2} - \frac{\partial v_2}{\partial x_1}.$$

Hence $\mathbf{z}$ is necessarily orthogonal to $\mathbf{u} = (u_1, u_2, 0)$. This vanishing term has a very important consequence: all the analysis can be performed without having to derive an a priori estimate for $\mathbf{u}$ in $W^{1,\infty}(\Omega)^2$. The same property will hold in the discrete case, provided the discrete scheme is suitably chosen.

In this article, we propose finite-element schemes for solving numerically the equations of a steady two dimensional grade-two fluid model, with a non-homogeneous tangential boundary condition. Defining $\mathbf{z}$ as above, the equation of motion becomes

$$-\nu\Delta\,\mathbf{u} + \mathbf{z} \times \mathbf{u} + \nabla\, p = \mathbf{f} \quad \text{in } \Omega\,, \tag{1.5}$$

the incompressibility condition is unchanged:

$$\operatorname{div} \mathbf{u} = 0 \quad \text{in } \Omega\,, \tag{1.6}$$

the boundary condition is

$$\mathbf{u} = \mathbf{g} \quad \text{on } \partial\Omega \text{ with } \mathbf{g} \cdot \mathbf{n} = 0\,, \tag{1.7}$$

where $\mathbf{n}$ denotes the unit exterior normal to $\partial\Omega$, and the transport equation becomes

$$\nu\mathbf{z} + \alpha\mathbf{u} \cdot \nabla\mathbf{z} = \nu\operatorname{curl}\mathbf{u} + \alpha\operatorname{curl}\mathbf{f}\,. \tag{1.8}$$

Girault and Scott in [16] prove that (1.5)–(1.8) always has a solution $\mathbf{u}$ in $H^1(\Omega)^2$ and p in $L^2(\Omega)$, on a Lipschitz-continuous domain, without restriction on the size of the data, provided $\operatorname{curl}\mathbf{f}$ belongs to $L^2(\Omega)$, thus extending to rough data a result of Ouazar [26]. This unconditional existence result relies entirely on the fact that $\mathbf{u}$ does not need to be bounded in $W^{1,\infty}(\Omega)^2$. Similarly, our finite-element schemes are chosen so the numerical analysis can be performed without having to derive a uniform $W^{1,\infty}$ estimate for the discrete velocity. As expected, the difficulties arise from the transport equation (1.8). As is observed in [17] and [10], a straightforward argument shows that either the discrete velocity must have exactly zero divergence, or its non-zero divergence must be compensated by an extra stabilizing term in the transport equation or by a compatibility condition between the spaces of discrete pressure and discrete auxiliary variable z. Roughly speaking, let X_h, M_h and Z_h be discrete spaces for the velocity, pressure and variable z and, as usual, let us discretize (1.6) by

$$\forall q_h \in M_h\,,\ (q_h, \operatorname{div}\mathbf{u}_h) = 0\,. \tag{1.9}$$

Clearly, if we want to derive an unconditional a priori estimate from the discrete analogue of (1.8), we must be able to eliminate the nonlinear term. But even in the simplest case, Green's formula gives

$$\int_\Omega (\mathbf{u}_h \cdot \nabla\, z_h) z_h\, d\mathbf{x} = -\frac{1}{2}\int_\Omega (\operatorname{div}\mathbf{u}_h)(z_h)^2\, d\mathbf{x}\,.$$

Hence, we can eliminate this right-hand side either by adding to the left-hand side of (1.8) a stabilizing, consistent term, so that it becomes

$$\nu \mathbf{z} + \alpha \mathbf{u} \cdot \nabla \mathbf{z} + \frac{1}{2}(\operatorname{div} \mathbf{u}) z = \nu \operatorname{curl} \mathbf{u} + \alpha \operatorname{curl} \mathbf{f}\,, \tag{1.10}$$

or by asking that

$$(z_h)^2 \in M_h\,, \tag{1.11}$$

and applying (1.9). Keeping this in mind, we propose to discretize (1.10) or (1.8) by an upwind scheme based on the discontinuous Galerkin method of degree one introduced by Lesaint and Raviart in [23]. This means that in each element of the triangulation, z_h is a polynomial of degree one, without continuity requirement on interelement boundaries. On one hand, if the form (1.10) is used for discretizing the transport equation, then we can approximate the velocity and pressure by the standard $\mathbb{P}_2 - \mathbb{P}_1$ Hood-Taylor scheme, where $\mathbb{P}_k$ denotes the space of polynomials of degree k in two variables (cf. for example [15]). On the other hand, if we discretize the transport equation in the form (1.8), then (1.11) implies that p_h must be a polynomial of degree two, discontinuous across elements. In addition, the fact that the pressure and velocity spaces must satisfy a uniform discrete inf-sup condition implies that each component of $\mathbf{u}_h$ can be a polynomial of degree three plus two bubble functions of degree four, with continuity requirement on interelement boundaries (cf. [15]). Thus, denoting by $c(\mathbf{u}_h; z_h, \theta_h)$ the discrete nonlinear part of (1.10) in variational form (cf. (3.5)), our scheme is: Find $\mathbf{u}_h$ in $X_h + \mathbf{g}_h$, p_h in M_h and z_h in Z_h such that

$$\forall \mathbf{v}_h \in X_h\,,\ \nu(\nabla \mathbf{u}_h, \nabla \mathbf{v}_h) + (\mathbf{z}_h \times \mathbf{u}_h, \mathbf{v}_h) - (p_h, \operatorname{div} \mathbf{v}_h) = (\mathbf{f}, \mathbf{v}_h)\,, \tag{1.12}$$

$$\forall q_h \in M_h\,,\ (q_h, \operatorname{div} \mathbf{u}_h) = 0\,, \tag{1.13}$$

$$\begin{aligned}\forall \theta_h \in Z_h\,,\ \nu\,(z_h, \theta_h) + c(\mathbf{u}_h; z_h, \theta_h) &= \nu\,(\operatorname{curl} \mathbf{u}_h, \theta_h) \\ &\quad + \alpha\,(\operatorname{curl} \mathbf{f}, \theta_h)\,.\end{aligned} \tag{1.14}$$

Here $\mathbf{g}_h$ is a suitable approximation of $\mathbf{g}$ and the functions of X_h vanish on $\partial\Omega$. Without restriction on the size of the data, we establish that this scheme always has a discrete solution in a Lipschitz polygonal domain and that this solution converges strongly to a solution of the exact problem. Furthermore, if the domain is convex and the data small, this solution can be computed by a converging successive approximation algorithm, with arbitrary starting guess. In addition, we prove an error inequality that leads to error estimates when the solution is sufficiently smooth. For both velocity-pressure discretizations, the error is of the order of $h^{3/2}$, a result that remains valid as α tends to zero. With the $\mathbb{P}_2 - \mathbb{P}_1$ Hood-Taylor scheme, this is the best that can be achieved, considering that the discretization of the transport equation loses inevitably a factor $h^{1/2}$. For the $\mathbb{P}_3 - \mathbb{P}_2$ scheme with discontinuous pressure, whose transport equation is simpler, this result is disappointing considering that the interpolation error for the velocity and pressure is of order h^3. These two results complete those of Girault and Scott in [17].

Remark 1.1 Another possibility is that z_h be constant in each element of the triangulation. Then (1.11) implies that p_h must also be a piecewise constant and we can associate with it the incomplete $\mathbb{P}_2$ finite-element of Bernardi and Raugel [15] for the velocity, or even the non-conforminig element of Crouzeix and Raviart [12]. Otherwise, if we use the stabilizing term of (1.10), we can discretize the velocity and pressure with the "mini-element" of Arnold, Brezzi and Fortin [15]. The analysis below extends to these examples and it can be shown that their error is the order of $h^{1/2}$.

Remark 1.2 The results presented here are much more valuable than what Baia and Sequeira derive in [2]. Their analysis is of very limited use because, in order to guarantee the convergence of their algorithm (or even any algorithm), they must start with a first guess that has an error of order $h^{3/2}$. And since they prove no a priori estimate, they cannot construct this first guess, which in fact amounts to solving their problem directly.

The remainder of this paper is divided into three sections. Section 1 briefly recalls the analysis of the exact problem (1.5)–(1.8) and compares it with the formulation proposed by [2]. The finite-element schemes are described in Section 2 and their error is estimated in Section 3.

We end this introduction by recalling some notation and basic functional results. For any non-negative integer m and number $r \geq 1$, recall the classical Sobolev space (cf. Adams [1] or Nečas [25])

$$W^{m,r}(\Omega) = \{v \in L^r(\Omega)\,;\, \partial^k v \in L^r(\Omega)\ \forall |k| \leq m\}\,,$$

equipped with the seminorm

$$|v|_{W^{m,r}(\Omega)} = \Big[\sum_{|k|=m} \int_\Omega |\partial^k v|^r\, d\mathbf{x}\Big]^{1/r},$$

and norm (for which it is a Banach space)

$$\|v\|_{W^{m,r}(\Omega)} = \Big[\sum_{0\leq k\leq m} |v|^r_{W^{k,r}(\Omega)}\Big]^{1/r},$$

with the usual extension when $r = \infty$. The reader can refer to [20] and [24] for extensions of this definition to non-integral values of m. When $r = 2$, this space is the Hilbert space $H^m(\Omega)$. In particular, the scalar product of $L^2(\Omega)$ is denoted by $(\cdot,\cdot)$. The definitions of these spaces are extended straightforwardly to vectors, with the same notation, but with the following modification for the norms in the non-Hilbert case. Let $\mathbf{u} = (u_1, u_2)$; then we set

$$\|\mathbf{u}\|_{L^r(\Omega)} = \left[\int_\Omega \|\mathbf{u}(\mathbf{x})\|^r \, d\mathbf{x}\right]^{1/r} ,$$

where $\|\cdot\|$ denotes the Euclidean vector norm.

For vanishing boundary values, we define

$$H_0^1(\Omega) = \{v \in H^1(\Omega)\,;\, v|_{\partial\Omega} = 0\}\,.$$

We shall often use Sobolev's imbeddings: for any real number $p \geq 1$, there exists a constant S_p such that

$$\forall v \in H_0^1(\Omega)\ ,\ \|v\|_{L^p(\Omega)} \leq S_p |v|_{H^1(\Omega)}\,. \tag{1.15}$$

When $p = 2$, this reduces to Poincaré's inequality and S_2 is Poincaré's constant. For tangential boundary values, we define

$$H_T^1(\Omega) = \{\mathbf{v} \in H^1(\Omega)^2\,;\, \mathbf{v}\cdot\mathbf{n} = 0 \text{ on } \partial\Omega\}\,. \tag{1.16}$$

A straightforward application of Peetre-Tartar's Theorem (cf. [27] and [30] or [15]) shows that the analogue of Sobolev's imbeddings holds in $H_T^1(\Omega)$ for any real number $p \geq 1$:

$$\forall \mathbf{v} \in H_T^1(\Omega)\ ,\ \|\mathbf{v}\|_{L^p(\Omega)} \leq \tilde{S}_p |\mathbf{v}|_{H^1(\Omega)}\,. \tag{1.17}$$

In particular, for $p = 2$, the mapping $\mathbf{v} \mapsto |\mathbf{v}|_{H^1(\Omega)}$ is a norm on $H_T^1(\Omega)$, equivalent to the H^1 norm and $\tilde{S}_2$ is the analogue of Poincaré's constant. We shall also use the standard spaces for Navier-Stokes equations

$$V = \{\mathbf{v} \in H_0^1(\Omega)^2\,;\, \operatorname{div}\mathbf{v} = 0 \text{ in } \Omega\}\,,$$

$$W = \{\mathbf{v} \in H_T^1(\Omega)\,;\, \operatorname{div}\mathbf{v} = 0 \text{ in } \Omega\}\,,$$

$$L_0^2(\Omega) = \{q \in L^2(\Omega)\,;\, \int_\Omega q\, d\mathbf{x} = 0\}\,,$$

and also the space

$$H(\operatorname{curl},\Omega) = \{\mathbf{v} \in L^2(\Omega)^2\,;\, \operatorname{curl}\mathbf{v} \in L^2(\Omega)\}\,.$$

Finally, recall the standard lifting $\mathbf{w_g}$ in W of $\mathbf{g}$: it is the solution of the non-homogeneous Stokes problem:

$$-\Delta\,\mathbf{w_g} + \nabla\,p_{\mathbf{g}} = \mathbf{0} \ \text{ and } \ \operatorname{div}\mathbf{w_g} = 0 \ \text{ in } \Omega \quad , \quad \mathbf{w_g} = \mathbf{g} \text{ on } \partial\Omega\,. \tag{1.18}$$

It satisfies the bound, with a constant T that depends only on the domain (cf. for instance [15])

$$|\mathbf{w_g}|_{H^1(\Omega)} \le T\|\mathbf{g}\|_{H^{1/2}(\partial\Omega)}\,. \tag{1.19}$$

2. The exact problem

Let Ω be a Lipschitz polygon in two dimensions with boundary $\partial\Omega$ (cf. [20]). We consider the steady grade-two fluid model: Find a vector-valued velocity $\mathbf{u}$, a scalar pressure p, and an auxiliary scalar function z, solution of (1.5)–(1.8), namely

$$\begin{aligned} -\nu\Delta\,\mathbf{u} + \mathbf{z}\times\mathbf{u} + \nabla\,p = \mathbf{f} \quad \text{in } \Omega\,,\\ \operatorname{div}\mathbf{u} = 0 \quad \text{in } \Omega\,,\\ \mathbf{u} = \mathbf{g} \quad \text{on } \partial\Omega \text{ with } \mathbf{g}\cdot\mathbf{n} = 0\,,\\ \nu\mathbf{z} + \alpha\mathbf{u}\cdot\nabla\mathbf{z} = \nu\operatorname{curl}\mathbf{u} + \alpha\operatorname{curl}\mathbf{f}\,, \end{aligned}$$

where $\mathbf{z}\times\mathbf{u} = (-zu_2, zu_1)$. For simplicity, we call it *Problem P*. Here $\nu > 0$, and for the sake of generality, we allow α to take negative values. It can be readily checked that $\mathbf{z} = (0,0,z)$ and $\mathbf{u}$ are related by $z = \operatorname{curl}(\mathbf{u} - \alpha\Delta\,\mathbf{u})$, so that (1.5) is indeed the steady-state version of (1.1), with $\alpha_1 + \alpha_2 = 0$. The following theorem is established in [16] and [17].

Theorem 2.3 - *Let Ω be a Lipschitz polygon. For all $\nu > 0$, all real numbers α, all $\mathbf{f} \in H(\operatorname{curl},\Omega)$ and all tangential vectors $\mathbf{g} \in H^{1/2}(\partial\Omega)^2$, Problem P has at least one solution $(\mathbf{u}, p, z)$. All solutions of Problem P satisfy the following estimates:*

$$|\mathbf{u}|_{H^1(\Omega)} \le \frac{S_2}{\nu}\|\mathbf{f}\|_{L^2(\Omega)} + T\|\mathbf{g}\|_{H^{1/2}(\partial\Omega)}\big(1 + \frac{S_4\tilde{S}_4}{\nu}\|z\|_{L^2(\Omega)}\big)\,, \tag{2.1}$$

$$\begin{aligned} \|p\|_{L^2(\Omega)} \le \frac{1}{\beta}\,\big(S_2\|\mathbf{f}\|_{L^2(\Omega)} + \nu T\|\mathbf{g}\|_{H^{1/2}(\partial\Omega)}\\ + S_4\tilde{S}_4|\mathbf{u}|_{H^1(\Omega)}\|z\|_{L^2(\Omega)}\big)\,, \end{aligned} \tag{2.2}$$

$$\|z\|_{L^2(\Omega)} \le \sqrt{2}|\mathbf{u}|_{H^1(\Omega)} + \frac{|\alpha|}{\nu}\|\operatorname{curl}\mathbf{f}\|_{L^2(\Omega)}\,, \tag{2.3}$$

$$\|\alpha\,\mathbf{u}\cdot\nabla z\|_{L^2(\Omega)} \le \nu\sqrt{2}\,|\mathbf{u}|_{H^1(\Omega)} + |\alpha|\,\|\mathrm{curl}\,\mathbf{f}\|_{L^2(\Omega)}\,, \tag{2.4}$$

where $\beta > 0$ is the isomorphism constant of the divergence operator (cf. [15] or [5]), S_p and $\tilde{S}_p$ are defined in (1.15) and (1.17) respectively and T is the constant of (1.19).

Furthermore, for any $t > \frac{1}{2}$, we have

$$\forall\varepsilon > 0\,,\ |\mathbf{u}|_{H^1(\Omega)} \le \frac{S_2}{\nu}\|\mathbf{f}\|_{L^2(\Omega)} + \frac{C}{\varepsilon^t}\|\mathbf{g}\|^{1+t}_{H^{1/2}(\partial\Omega)} + \frac{\varepsilon}{\nu}\|z\|_{L^2(\Omega)}\,, \tag{2.5}$$

$$\begin{aligned}\|z\|_{L^2(\Omega)} \le & 2\frac{|\alpha|}{\nu}\|\mathrm{curl}\,\mathbf{f}\|_{L^2(\Omega)} + 2\frac{\sqrt{2}}{\nu}S_2\|\mathbf{f}\|_{L^2(\Omega)} \\ & + (2\sqrt{2})^{1+t}\frac{C}{\nu^t}\|\mathbf{g}\|^{1+t}_{H^{1/2}(\partial\Omega)}\,,\end{aligned} \tag{2.6}$$

where C depends only on t and Ω.

The above a priori estimates substantially simplify when $\mathbf{g} = \mathbf{0}$; we obtain

$$\|z\|_{L^2(\Omega)} \le \sqrt{2}\frac{S_2}{\nu}\|\mathbf{f}\|_{L^2(\Omega)} + \frac{|\alpha|}{\nu}\|\mathrm{curl}\,\mathbf{f}\|_{L^2(\Omega)}\,, \tag{2.7}$$

$$|\mathbf{u}|_{H^1(\Omega)} \le \frac{S_2}{\nu}\|\mathbf{f}\|_{L^2(\Omega)}\,, \tag{2.8}$$

$$\|p\|_{L^2(\Omega)} \le \frac{1}{\beta}\big(S_2\|\mathbf{f}\|_{L^2(\Omega)} + S_4^2|\mathbf{u}|_{H^1(\Omega)}\|z\|_{L^2(\Omega)}\big)\,. \tag{2.9}$$

It is worthwhile comparing *Problem P* with the formulation proposed by [2] (cf. also [31]) for the problem with homogeneous boundary conditions. This formulation is based on the fact that $\mathbf{f}$ can be written as a divergence:

$$\mathbf{f} = \mathbf{div}\,\mathbf{F}\,.$$

Then, introducing an auxiliary pressure π and tensor σ, the equations are

$$-\Delta\,\mathbf{u} + \nabla\,\pi = \mathbf{div}\,\sigma \quad , \quad \mathrm{div}\,\mathbf{u} = 0\,, \tag{2.10}$$

$$\mathbf{u} = \mathbf{0} \ \text{ on } \partial\Omega\,, \tag{2.11}$$

$$\begin{aligned}\nu\sigma + \alpha\mathbf{u}\cdot\nabla\,\sigma - \alpha\sigma\cdot(\nabla\,\mathbf{u})^t = & \ \mathbf{F} - \alpha\pi(\nabla\,\mathbf{u})^t - \mathbf{u}\otimes\mathbf{u} \\ & + \alpha(\nabla\,\mathbf{u})^t(\nabla\,\mathbf{u} + (\nabla\,\mathbf{u})^t)\,.\end{aligned} \tag{2.12}$$

The equivalence between (2.10)–(2.12) and the original grade-two fluid model is established in [31]. This formulation is not as interesting as ours, even in three dimensions, because it loses the unconditional energy estimate (2.8). Thus, in contrast to *Problem P*, its analysis requires in particular that $\mathbf{u}$ be bounded in $W^{1,\infty}(\Omega)^2$. This has serious consequences: on one hand, existence of solutions cannot be derived without heavy restrictions on the size of the data, and on the other hand, the numerical discretization and programming are unnecessarily difficult, and rigorous proofs for its numerical analysis are non-existent.

3. A discontinuous upwind scheme

Let $h > 0$ be a discretization parameter and let $\mathcal{T}_h$ be a *regular* family of triangulations of $\overline{\Omega}$, consisting of triangles K with maximum mesh size h (cf. [6], [5]): there exists a constant σ_0, independent of h, such that

$$\forall K \in \mathcal{T}_h\,,\ \frac{h_K}{\rho_K} \leq \sigma_0\,, \tag{3.1}$$

where h_K is the diameter and ρ_K is the diameter of the ball inscribed in K.

We first recall how upwinding can be achieved by the discontinuous Galerkin approximation introduced in [23]. Consider the discontinuous finite-element space

$$Z_h = \{\theta_h \in L^2(\Omega)\,;\ \forall K \in \mathcal{T}_h\,,\ \theta_h|_K \in I\!P_1\}\,. \tag{3.2}$$

As interpolant, we shall mostly use an approximation operator (cf. [11], [29], [4]) $R_h \in \mathcal{L}(W^{1,p}(\Omega); Z_h \cap \mathcal{C}^0(\bar{\Omega}))$ for any number $p \geq 1$, such that, for $m = 0, 1$ and $0 \leq l \leq 1$

$$\forall z \in W^{l+1,p}(\Omega)\,,\ |R_h(z) - z|_{W^{m,p}(\Omega)} \leq C\, h^{l+1-m} |z|_{W^{l+1,p}(\Omega)}\,. \tag{3.3}$$

Let $\mathbf{u}_h$ be a discrete velocity in $H^1_T(\Omega)$, and for each triangle K, let

$$\partial K_- = \{\mathbf{x} \in \partial K\,;\ \alpha \mathbf{u}_h \cdot \mathbf{n} < 0\}\,. \tag{3.4}$$

Note that, when running over all triangles K of $\mathcal{T}_h$, ∂K_- only involves interior segments of $\mathcal{T}_h$, because $\mathbf{u}_h \cdot \mathbf{n} = 0$ on $\partial\Omega$. Then we approximate the nonlinear terms $\alpha[(\mathbf{u}\cdot\nabla z,\theta)+\frac{1}{2}((\operatorname{div}\mathbf{u})z,\theta)]$ by

$$\begin{aligned} c(\mathbf{u}_h;z_h,\theta_h) = \frac{\alpha}{2}\int_\Omega (\operatorname{div}\mathbf{u}_h) z_h\theta_h\,d\mathbf{x} \\ + \sum_{K\in\mathcal{T}_h}\Big(\int_K \alpha(\mathbf{u}_h\cdot\nabla z_h)\theta_h\,d\mathbf{x} \\ + \int_{\partial K_-} |\alpha\mathbf{u}_h\cdot\mathbf{n}|(z_h^{\text{int}}-z_h^{\text{ext}})\theta_h^{\text{int}}\,ds\Big)\,, \end{aligned} \tag{3.5}$$

where the superscript int (resp. ext) refers to the trace on the segment ∂K of the function taken inside (resp. outside) K. Note also that when summing over all triangles, ∂K_- is not counted twice because $\mathbf{u}_h\cdot\mathbf{n}$ changes sign across adjacent elements. Rather, in the above sum, the boundary integrations are taken over complete interior segments.

3.1. The Hood-Taylor scheme

Let us first recall the standard Hood-Taylor discretization of the velocity and pressure. The discrete pressure space is

$$M_h = \{q_h\in H^1(\Omega)\cap L^2_0(\Omega)\,;\,\forall K\in\mathcal{T}_h\,,\,q_h|_K\in I\!P_1\}\,, \tag{3.6}$$

and we interpolate the functions of $L^2_0(\Omega)$ by a regularization operator analogous to R_h, $r_h\in\mathcal{L}(L^2_0(\Omega);M_h)$, such that for $0\le l\le 2$,

$$\forall q\in H^l(\Omega)\cap L^2_0(\Omega)\,,\ \|r_h(q)-q\|_{L^2(\Omega)}\le C\,h^l|q|_{H^l(\Omega)}\,. \tag{3.7}$$

The discrete velocity space is

$$\begin{aligned} X_{h,T} &= \{\mathbf{v}\in C^0(\overline{\Omega})^2\,;\,\forall K\in\mathcal{T}_h\,,\,\mathbf{v}|_K\in I\!P_2^2, \mathbf{v}\cdot\mathbf{n}|_{\partial\Omega}=0\}\,,\\ X_h &= X_{h,T}\cap H^1_0(\Omega)^2\,, \end{aligned} \tag{3.8}$$

$$\begin{aligned} W_h &= \{\mathbf{v}\in X_{h,T}\,;\,\forall q\in M_h\,,\,\int_\Omega q\operatorname{div}\mathbf{v}\,d\mathbf{x}=0\}\,,\\ V_h &= W_h\cap H^1_0(\Omega)^2\,. \end{aligned} \tag{3.9}$$

If all triangles K of $\mathcal{T}_h$ have at most one edge on the boundary $\partial\Omega$, the pair of spaces (X_h, M_h) satisfies a uniform discrete inf-sup condition (cf. [15]). But as in [17] and [19], we can obtain better results from a local inf-sup condition that yields an approximation operator satisfying sharp local estimates. Indeed, we can prove that there exists an operator $P_h \in$

$\mathcal{L}(H_0^1(\Omega)^2; X_h) \cap \mathcal{L}(H_T^1(\Omega); X_{h,T})$, such that the support of $P_h(\mathbf{v})$ is contained in a neighborhood of the support of $\mathbf{v}$, and

$$\forall \mathbf{v} \in H_T^1(\Omega)\,,\ \|\mathbf{v} - P_h(\mathbf{v})\|_{L^p(\Omega)} \le C\, h^{2/p} |\mathbf{v}|_{H^1(\Omega)}\,, \tag{3.10}$$

$$\begin{aligned} &\forall \mathbf{v} \in H_T^1(\Omega) \cap W^{s,p}(\Omega)^2\,,\\ &\|\mathbf{v} - P_h(\mathbf{v})\|_{W^{m,p}(\Omega)} \le C\, h^{s-m} \|\mathbf{v}\|_{W^{s,p}(\Omega)}\,, \end{aligned} \tag{3.11}$$

for all numbers $p \ge 2$ and all real numbers s with $1 \le s \le 3$, $m = 0, 1$, and

$$\forall K \in \mathcal{T}_h\,,\ \forall \mathbf{v} \in H_T^1(\Omega)\,,\ \forall q_h \in M_h\,,\ \int_K q_h \operatorname{div}(P_h(\mathbf{v}) - \mathbf{v})\, d\mathbf{x} = 0\,. \tag{3.12}$$

Let $\mathbf{g}_h = P_h(\mathbf{r})$ where $\mathbf{r}$ is any lifting of $\mathbf{g}$ in W. This lifting is only a theoretical convenience because on one hand, $\mathbf{g}_h$ can be constructed directly by interpolating $\mathbf{g}$ appropriately on $\partial\Omega$ and on the other hand, $\mathbf{g}_h$ does not depend on the particular lifting chosen; in addition, $\mathbf{g}_h$ satisfies

$$\begin{aligned} \|\mathbf{g}_h\|_{H^{1/2}(\partial\Omega)} = \|P_h(\mathbf{w_g})\|_{H^{1/2}(\partial\Omega)} &\le \|P_h(\mathbf{w_g})\|_{H^1(\Omega)}\\ &\le C_0\, |\mathbf{w_g}|_{H^1(\Omega)} \le C_0\, T\, \|\mathbf{g}\|_{H^{1/2}(\partial\Omega)}\,, \end{aligned} \tag{3.13}$$

where T is the constant of (1.19) and C_0 and all subsequent constants C_i are independent of h. Then, as written in the introduction, our discrete scheme is: find $\mathbf{u}_h$ in $X_h + \mathbf{g}_h$, p_h in M_h and z_h in Z_h solution of (1.12)–(1.14):

$$\forall \mathbf{v}_h \in X_h\,,\ \nu(\nabla \mathbf{u}_h, \nabla \mathbf{v}_h) + (\mathbf{z}_h \times \mathbf{u}_h, \mathbf{v}_h) - (p_h, \operatorname{div} \mathbf{v}_h) = (\mathbf{f}, \mathbf{v}_h)\,,$$

$$\forall q_h \in M_h\,,\ (q_h, \operatorname{div} \mathbf{u}_h) = 0\,,$$

$$\forall \theta_h \in Z_h\,,\ \nu\,(z_h, \theta_h) + c(\mathbf{u}_h; z_h, \theta_h) = \nu\,(\operatorname{curl} \mathbf{u}_h, \theta_h) + \alpha\,(\operatorname{curl} \mathbf{f}, \theta_h)\,,$$

where c is defined by (3.5).

As in [17], the trace preserving properties of P_h and its sharp local estimates allow one to construct a Leray-Hopf's lifting of $\mathbf{g}_h$ satisfying:

Lemma 3.4 - *For any $\mathbf{g} \in H^{1/2}(\partial\Omega)^2$ such that $\mathbf{g} \cdot \mathbf{n} = 0$ and for any real number $\varepsilon > 0$, there exists a lifting $\mathbf{u}_{h,\mathbf{g}}$ of $\mathbf{g}$ such that*

$$|\mathbf{u}_{h,\mathbf{g}}|_{H^1(\Omega)} \le C\varepsilon^{-1/2-1/t} \|\mathbf{g}\|_{H^{1/2}(\partial\Omega)}\,,\ 2 \le t < \infty\,, \tag{3.14}$$

and if $h_b < \varepsilon$, where h_b denotes the maximum mesh length of triangles in a tubular neighborhood of the boundary, then for all $\mathbf{v} \in H_0^1(\Omega)^2$ and for $t < s < \infty$, (recall that $\|\cdot\|$ denotes the Euclidean vector norm),

$$\big\| \|\mathbf{u}_{h,\mathbf{g}}\|\, \|\mathbf{v}\| \big\|_{L^2(\Omega)} \le C\varepsilon^{1/2-1/s} \|\mathbf{g}\|_{H^{1/2}(\partial\Omega)} |\mathbf{v}|_{H^1(\Omega)}\,, \tag{3.15}$$

where the constants C depend on t or s, but are independent of h, ε and $\mathbf{g}$.

In order to prove existence of solutions of (1.12)–(1.14), let us recall the following identity established by Lesaint and Raviart in [23]:

Lemma 3.5 - *For all* $\mathbf{v}_h$ *in* X_h*, for all* z_h *and* θ_h *in* Z_h*, we have*

$$c(\mathbf{v}_h; z_h, \theta_h) = \sum_{K\in\mathcal{T}_h} \Big(-\int_K \alpha(\mathbf{v}_h \cdot \nabla\, \theta_h) z_h \, d\mathbf{x} + \int_{\partial K_-} |\alpha \mathbf{v}_h \cdot \mathbf{n}| z_h^{\text{ext}} (\theta_h^{\text{ext}} - \theta_h^{\text{int}}) ds\Big) - \frac{\alpha}{2}\int_\Omega (\operatorname{div} \mathbf{u}_h)\theta_h z_h \, d\mathbf{x} . \tag{3.16}$$

Note that when θ_h is in $H^1(\Omega)$, (3.16) reduces to

$$c(\mathbf{v}_h; z_h, \theta_h) = -\int_\Omega \alpha(\mathbf{v}_h \cdot \nabla\, \theta_h) z_h \, d\mathbf{x} - \frac{\alpha}{2}\int_\Omega (\operatorname{div} \mathbf{u}_h)\theta_h z_h \, d\mathbf{x} . \tag{3.17}$$

Note also that, when $\theta_h = z_h \in Z_h$, then

$$c(\mathbf{v}_h; z_h, z_h) = \frac{1}{2}\sum_{K\in\mathcal{T}_h} \int_{\partial K_-} |\alpha \mathbf{v}_h \cdot \mathbf{n}| (z_h^{\text{ext}} - z_h^{\text{int}})^2 ds . \tag{3.18}$$

Therefore, choosing $\theta_h = z_h$ in (1.14) and applying (3.18), we obtain

$$\nu \, \|z_h\|^2_{L^2(\Omega)} + \frac{1}{2}\sum_{K\in\mathcal{T}_h} \int_{\partial K_-} |\alpha \mathbf{u}_h \cdot \mathbf{n}| (z_h^{\text{ext}} - z_h^{\text{int}})^2 ds = \nu \, (\operatorname{curl} \mathbf{u}_h, z_h) + \alpha \, (\operatorname{curl} \mathbf{f}, z_h) . \tag{3.19}$$

Equation (3.19) allows us to prove the following existence theorem.

Theorem 3.6 - *There exists a constant* $C_1 > 0$*, independent of* h*, such that for all* $\nu > 0$ *and* $\alpha \in \mathbb{R}$*, for all* $\mathbf{f}$ *in* $H(curl, \Omega)$ *and all* $\mathbf{g}$ *in* $H^{1/2}(\partial\Omega)^2$ *satisfying* $\mathbf{g} \cdot \mathbf{n} = 0$*, if*

$$h_b < C_1 \nu^{2+t} \|\mathbf{g}\|^{-2-t}_{H^{1/2}(\partial\Omega)} , \quad \textit{for some } t > 0 , \tag{3.20}$$

then the discrete problem (1.12)–(1.14) has at least one solution $(\mathbf{u}_h, p_h, z_h)$ *in* $(X_h + \mathbf{g}_h) \times M_h \times Z_h$ *and every solution satisfies the following a priori estimates:*

$$|\mathbf{u}_h|_{H^1(\Omega)} \le \frac{S_2}{\nu} \|\mathbf{f}\|_{L^2(\Omega)} + K_1(h) C_0 \, T \|\mathbf{g}\|_{H^{1/2}(\partial\Omega)} , \tag{3.21}$$

$$\|p_h\|_{L^2(\Omega)} \le \frac{1}{\beta^*}\big(S_2\|\mathbf{f}\|_{L^2(\Omega)} + \nu C_0 T\|\mathbf{g}\|_{H^{1/2}(\partial\Omega)} + S_4\tilde{S}_4|\mathbf{u}_h|_{H^1(\Omega)}\|z_h\|_{L^2(\Omega)}\big)\,, \tag{3.22}$$

$$\|z_h\|_{L^2(\Omega)} \le \sqrt{2}|\mathbf{u}_h|_{H^1(\Omega)} + \frac{|\alpha|}{\nu}\|\operatorname{curl}\mathbf{f}\|_{L^2(\Omega)}\,, \tag{3.23}$$

$$\frac{1}{2}\sum_{K\in\mathcal{T}_h}\int_{\partial K_-}|\alpha\mathbf{u}_h\cdot\mathbf{n}|\big(z_h^{\text{ext}} - z_h^{\text{int}}\big)^2 ds \le \big(\sqrt{2}\nu|\mathbf{u}_h|_{H^1(\Omega)} + |\alpha|\|\operatorname{curl}\mathbf{f}\|_{L^2(\Omega)}\big)\|z_h\|_{L^2(\Omega)}\,, \tag{3.24}$$

where C_0 is the constant of (3.13), $\beta^\star$ is the constant of the discrete inf-sup condition, and

$$K_1(h) = 1 + \frac{S_4\tilde{S}_4}{\nu}\|z_h\|_{L^2(\Omega)}\,.$$

In addition, we have for any real number $s > \frac{t}{2}$:

$$\|z_h\|_{L^2(\Omega)} \le C_2\nu^{-1-s}\|\mathbf{g}\|^{2+s}_{H^{1/2}(\partial\Omega)} + 2\sqrt{2}\frac{S_2}{\nu}\|\mathbf{f}\|_{L^2(\Omega)} + 2\frac{|\alpha|}{\nu}\|\operatorname{curl}\mathbf{f}\|_{L^2(\Omega)}\,, \tag{3.25}$$

where the constant C_2 depends on s and t, but not on h or ν.

Extracting subsequences (that we still denote by the index h), the uniform a priori estimate (3.25) combined with (3.21), (3.22), (3.24) show that, on one hand, $(\mathbf{u}_h, p_h, z_h)$ converge weakly to functions $(\mathbf{u}, p, z)$ in $H^1_T(\Omega) \times L^2(\Omega) \times L^2(\Omega)$, and on the other hand, the quantity $\sum_{K\in\mathcal{T}_h}\int_{\partial K_-}|\alpha\mathbf{u}_h \cdot \mathbf{n}|(z_h^{\text{ext}} - z_h^{\text{int}})^2 ds$ converges to a non-negative number, say S. Passing to the limit in (1.12), we see that $(\mathbf{u}, p, z)$ satisfies (1.5). Next the strong convergence of $\mathbf{u}_h$ is easily established, as in [17]. Owing to this strong convergence, using $R_h(\theta)$ with smooth θ for test function in (1.14), and applying (3.17), we can readily prove that $(\mathbf{u}, p, z)$ is a solution of *Problem P*. Using again the strong convergence of $\mathbf{u}_h$ and the weak convergence of z_h, and comparing with (1.8), we see that the right-hand side of (3.19) converges to $\nu\|z\|^2_{L^2(\Omega)}$. Here, we need the following consequence of Green's formula established in [16]:

$$\int_\Omega (\mathbf{u}\cdot\nabla z)z\,d\mathbf{x} = 0\,,$$

for z a solution of (1.8). Therefore

$$\lim_{h\to 0} \nu \, \|z_h\|^2_{L^2(\Omega)} + \frac{S}{2} = \nu \, \|z\|^2_{L^2(\Omega)} \, .$$

Then the lower semi-continuity of the norm for the weak topology implies that

$$\lim_{h\to 0} \|z_h\|_{L^2(\Omega)} = \|z\|_{L^2(\Omega)} \, .$$

This yields on one hand the strong convergence of z_h and on the other hand the fact that $S = 0$. The following theorem sums up these convergence results.

Theorem 3.7 - *Suppose that h_b satisfies (3.20), and let $(\mathbf{u}_h, p_h, z_h)$ be a solution of (1.12)–(1.14). Then we can extract a subsequence of h, still denoted by h, such that*

$$\lim_{h\to 0} \mathbf{u}_h = \mathbf{u} \quad \textit{strongly in } W \, ,$$

$$\lim_{h\to 0} p_h = p \quad \textit{strongly in } L^2_0(\Omega) \, ,$$

$$\lim_{h\to 0} z_h = z \quad \textit{strongly in } L^2(\Omega) \, ,$$

$$\lim_{h\to 0} \sum_{K\in\mathcal{T}_h} \int_{\partial K_-} |\alpha \mathbf{u}_h \cdot \mathbf{n}| (z_h^{\text{ext}} - z_h^{\text{int}})^2 ds = 0 \quad \textit{in } I\!R \, ,$$

where $(\mathbf{u}, p, z)$ is a solution of Problem P.

3.2. The $I\!P3 - I\!P2$ scheme with discontinuous pressure

We describe briefly this scheme, since its analysis follows closely that of the Hood-Taylor method. In view of (1.11), we discretize pressures in the space

$$M_h = \{q_h \in L^2_0(\Omega) \, ; \, \forall K \in \mathcal{T}_h \, , \, q_h|_K \in I\!P_2\} \, , \tag{3.26}$$

and we interpolate the functions of $L^2_0(\Omega)$ by a local projection on each triangle K, $r_h \in \mathcal{L}(L^2_0(\Omega); M_h)$, such that for $0 \le l \le 3$,

$$\forall q \in H^l(\Omega) \cap L^2_0(\Omega) \, , \; \|r_h(q) - q\|_{L^2(\Omega)} \le C \, h^l |q|_{H^l(\Omega)} \, . \tag{3.27}$$

Then, in order to satisfy a uniform discrete inf-sup condition, we discretize each component of velocities in each K, in the polynomial space

$$\mathcal{P} = I\!P_3 + b_K \tilde{I\!P}_1 \, , \tag{3.28}$$

where $b_K = \lambda_1\lambda_2\lambda_3$ is a bubble function and

$$\tilde{I\!P}_1 = \{c_1x_1 + c_2x_2\,;\,\forall c_1, c_2 \in I\!R\}\,.$$

Then, we discretize velocities in the space

$$\begin{aligned} X_{h,T} &= \{\mathbf{v} \in C^0(\overline{\Omega})^2\,;\,\forall K \in \mathcal{T}_h\,,\,\mathbf{v}|_K \in \mathcal{P}^2, \mathbf{v}\cdot\mathbf{n}|_{\partial\Omega} = 0\}\,,\\ X_h &= X_{h,T} \cap H_0^1(\Omega)^2\,, \end{aligned} \tag{3.29}$$

with W_h and V_h defined as in (3.9). The fact that the functions of W_h satisfy $\mathbf{v}_h \cdot \mathbf{n} = 0$ implies

$$\forall K \in \mathcal{T}_h\,,\,\forall \mathbf{v}_h \in W_h\,,\,\forall q \in I\!P_2\,,\quad \int_K (\operatorname{div}\mathbf{v}_h)q\,d\mathbf{x} = 0\,. \tag{3.30}$$

Hence

$$\forall K \in \mathcal{T}_h\,,\,\forall \mathbf{v}_h \in W_h\,,\,\forall z_h, \theta_h \in Z_h\,,\quad \int_K (\operatorname{div}\mathbf{v}_h)z_h\theta_h\,d\mathbf{x} = 0\,. \tag{3.31}$$

It is shown in [15] that the pair of spaces (X_h, M_h) satisfies a uniform discrete inf-sup condition, but as in [19], combining the construction of [15] and [17], we can prove that there exists an operator P_h with the same local properties and satisfying (3.12), (3.10) and (3.11) with $1 \le s \le 4$.

By virtue of (3.31), the stabilizing term in c is no longer necessary and (3.5) is replaced by

$$\begin{aligned} c(\mathbf{u}_h; z_h, \theta_h) = \sum_{K\in\mathcal{T}_h} \Big(&\int_K \alpha(\mathbf{u}_h\cdot\nabla\, z_h)\theta_h\,d\mathbf{x}\\ &+ \int_{\partial K_-} |\alpha\mathbf{u}_h\cdot\mathbf{n}|(z_h^{\text{int}} - z_h^{\text{ext}})\theta_h^{\text{int}}ds\Big)\,. \end{aligned} \tag{3.32}$$

With (3.32), the scheme (1.12)–(1.14) is simpler, but the finite-element functions for the velocity and pressure are more complex. All the results of this section remain valid for this scheme; the proof of Theorem 3.7 being easier owing to (3.31).

4. Error estimates

Estimating the error of (1.12)–(1.14) is not straightforward because of the strong nonlinearity of the equations. First, we derive the following bound:

$$\begin{aligned} |\mathbf{u}_h - \mathbf{u}|_{H^1(\Omega)} \le 2\,|P_h(\mathbf{u}) - \mathbf{u}|_{H^1(\Omega)} &+ \frac{S_4}{\nu}\|P_h(\mathbf{u})\|_{L^4(\Omega)}\|z_h - z\|_{L^2(\Omega)}\\ + \frac{S_4}{\nu}\|z\|_{L^2(\Omega)}\|P_h(\mathbf{u}) - \mathbf{u}\|_{L^4(\Omega)} &+ \frac{\sqrt{2}}{\nu}\|r_h(p) - p\|_{L^2(\Omega)}\,. \end{aligned} \tag{4.1}$$

But this is not sufficient. As mentioned in the introduction, we do not need a $W^{1,\infty}$ estimate for $\mathbf{u}_h - \mathbf{u}$ but we do need $W^{1,r}$ estimates with some $2 < r < \infty$. Such estimates stem from the following decoupling error formula, proven in [17], that extends (4.1). It is valid for both schemes, with possibly different constants.

Theorem 4.8 - *Assume that Ω is convex, $\mathbf{g}$ belongs to $H^{1/2+s}(\partial\Omega)^2$ for some $s \in (0, 1/2)$, and in addition to (3.1) and (3.20), $\mathcal{T}_h$ satisfies:*

$$\forall K \in \mathcal{T}_h\,,\ \rho_K \geq \gamma h^6\,, \tag{4.2}$$

with a constant $\gamma > 0$ independent of h. Then there exist constants C_i, independent of h, such that

$$\begin{aligned}|\mathbf{u}_h - \mathbf{u}|_{W^{1,2+1/4}(\Omega)} \leq{}& |P_h(\mathbf{u}) - \mathbf{u}|_{W^{1,2+1/4}(\Omega)}\\ &+ \frac{1}{\rho_{\min}^{1/9}}\Big(C_3|P_h(\mathbf{u}) - \mathbf{u}|_{H^1(\Omega)} + \frac{\sqrt{2}}{\nu}\|r_h(p) - p\|_{L^2(\Omega)}\Big)\\ &+ \|z_h - z\|_{L^2(\Omega)}\big(C_4 + C_5 h^{1/4}\big)\,,\end{aligned} \tag{4.3}$$

where $\rho_{\min}$ is the minimum of ρ_K for all K in $\mathcal{T}_h$.

Remark 4.9 The bound (4.3) is one in a family of estimates established by [17] in $W^{1,r}(\Omega)^2$ for all r in $(2, \infty)$. The assumptions on the boundary data $\mathbf{g}$, the domain and the triangulation depend upon the value of r. The choice $r = 2+1/4$ is convenient here because on one hand, it is sufficient for deriving the error inequalities below and on the other hand, the condition (4.2) allows for a wide range of refinements of the triangulation.

Remark 4.10 Since, $W^{1,2+1/4}(\Omega)$ in continuously imbedded into $L^\infty(\Omega)$, an easy variant of Theorem 4.8 gives with possibly different constants but under the same assumptions:

$$\begin{aligned}\|\mathbf{u}_h - \mathbf{u}\|_{L^\infty(\Omega)} \leq{}& \|P_h(\mathbf{u}) - \mathbf{u}\|_{L^\infty(\Omega)} + \frac{1}{\rho_{\min}^{1/9}}\big(C_3|P_h(\mathbf{u}) - \mathbf{u}|_{H^1(\Omega)}\\ &+ \frac{\sqrt{2}}{\nu}\|r_h(p) - p\|_{L^2(\Omega)}\big) + \|z_h - z\|_{L^2(\Omega)}\big(C_4 + C_5 h^{1/4}\big)\,.\end{aligned} \tag{4.4}$$

Now, in order to prove an error inequality for $z_h - z$, we assume that z belongs to $H^1(\Omega)$. It is shown in [17] that this assumption holds on a convex domain, when $\mathbf{g}$ and $\mathbf{f}$ are sufficiently smooth and small. It is restrictive, but so far, relaxing this assumption appears difficult. The following arguments are written when c is in the form (3.5); they simplify when the stabilizing term is missing. Taking the difference between (1.8) and (1.14), inserting any element ζ_h of Z_h, choosing $\theta_h = z_h - \zeta_h$, and applying (3.16) and (3.18), we obtain

$$
\begin{aligned}
&\nu \|z_h - \zeta_h\|^2_{L^2(\Omega)} \\
&\quad + \sum_{K \in \mathcal{T}_h} \frac{1}{2} \int_{\partial K_-} |\alpha \mathbf{u}_h \cdot \mathbf{n}| ((z_h - \zeta_h)^{\text{ext}} - (z_h - \zeta_h)^{\text{int}})^2 ds \\
&\quad + \sum_{K \in \mathcal{T}_h} \Big(-\alpha \int_K \mathbf{u}_h \cdot \nabla (z_h - \zeta_h)(\zeta_h - z) d\mathbf{x} \\
&\quad + \int_{\partial K_-} |\alpha \mathbf{u}_h \cdot \mathbf{n}| ((z_h - \zeta_h)^{\text{ext}} - (z_h - \zeta_h)^{\text{int}})(\zeta_h - z)^{\text{ext}} ds \Big) \\
&\quad - \frac{\alpha}{2} \int_\Omega \operatorname{div}(\mathbf{u}_h - \mathbf{u})(\zeta_h - z)(z_h - \zeta_h) d\mathbf{x} \\
&\quad + \frac{\alpha}{2} \int_\Omega \operatorname{div}(\mathbf{u}_h - \mathbf{u}) z (z_h - \zeta_h) d\mathbf{x} \\
&\quad + \alpha \int_\Omega (\mathbf{u}_h - \mathbf{u}) \cdot \nabla z (z_h - \zeta_h) d\mathbf{x} \\
&\quad = \nu (z - \zeta_h, z_h - \zeta_h) + \nu (\operatorname{curl}(\mathbf{u}_h - \mathbf{u}), z_h - \zeta_h) .
\end{aligned}
\tag{4.5}
$$

Clearly, the most troublesome terms in the left-hand side of (4.5) are the third and fourth terms. In each element, the third term can be split into:

$$
\begin{aligned}
\int_K \mathbf{u}_h \cdot \nabla (z_h - \zeta_h)(\zeta_h - z) d\mathbf{x} &= \int_K (\mathbf{u}_h - \mathbf{u}) \cdot \nabla (z_h - \zeta_h)(\zeta_h - z) d\mathbf{x} \\
&\quad + \int_K \mathbf{u} \cdot \nabla (z_h - \zeta_h)(\zeta_h - z) d\mathbf{x} .
\end{aligned}
$$

Now, instead of choosing $\zeta_h = R_h(z)$, that clearly cannot give anything better in the second term than an error of the order of h, let us take advantage of the discontinuity of the space Z_h and choose $\zeta_h = \varrho_h(z)$, the L^2 projection of z on $\mathbb{P}_1$ in each triangle K: $\varrho_h(z) \in \mathbb{P}_1$, such that

$$
\forall q \in \mathbb{P}_1 , \int_K (\varrho_h(z) - z) q \, d\mathbf{x} = 0 .
$$

This operator has locally the same accuracy as R_h. Moreover, we have for any constant vector $\mathbf{c}$:

$$
\int_K \mathbf{u} \cdot \nabla (z_h - \varrho_h(z))(\varrho_h(z) - z) d\mathbf{x} = \int_K (\mathbf{u} - \mathbf{c}) \cdot \nabla (z_h - \varrho_h(z))(\varrho_h(z) - z) d\mathbf{x} ,
$$

because $\nabla(z_h - \varrho_h(z))$ is also a constant vector. Thus a local inverse inequality and (3.1) yield:

$$\begin{aligned}
&|\alpha \int_K (\mathbf{u}-\mathbf{c}) \cdot \nabla(z_h - \varrho_h(z))(\varrho_h(z) - z) d\mathbf{x}| \\
&\qquad \leq \frac{c_1}{\rho_K} |\alpha| \|\mathbf{u}-\mathbf{c}\|_{L^\infty(K)} \|\varrho_h(z) - z\|_{L^2(K)} \|z_h - \varrho_h(z)\|_{L^2(K)} \\
&\qquad \leq c_2 \sigma_0 |\alpha| |\mathbf{u}|_{W^{1,\infty}(K)} \|\varrho_h(z) - z\|_{L^2(K)} \|z_h - \varrho_h(z)\|_{L^2(K)} ,
\end{aligned} \tag{4.6}$$

where c_i denote various constants independent of h. Similarly, applying the same local inverse inequality and the approximation property of ϱ_h, we obtain

$$\begin{aligned}
&|\alpha \int_K (\mathbf{u}_h - \mathbf{u}) \cdot \nabla(z_h - \varrho_h(z))(\varrho_h(z) - z) d\mathbf{x}| \\
&\qquad \leq c_3 \sigma_0 |\alpha| |z|_{H^1(K)} \|\mathbf{u}_h - \mathbf{u}\|_{L^\infty(K)} \|z_h - \varrho_h(z)\|_{L^2(K)} .
\end{aligned} \tag{4.7}$$

The fourth term can be split into:

$$\begin{aligned}
&\sum_{K \in \mathcal{T}_h} \int_{\partial K_-} |\alpha \mathbf{u}_h \cdot \mathbf{n}| ((z_h - \varrho_h(z))^{\text{ext}} - (z_h - \varrho_h(z))^{\text{int}})(\varrho_h(z) - z)^{\text{ext}} ds \\
&\qquad \leq \frac{1}{2} \int_{\partial K_-} |\alpha \mathbf{u}_h \cdot \mathbf{n}| ((z_h - \varrho_h(z))^{\text{ext}} - (z_h - \varrho_h(z))^{\text{int}})^2 ds \\
&\qquad + \frac{1}{2} \int_{\partial K_-} |\alpha \mathbf{u}_h \cdot \mathbf{n}| \big((\varrho_h(z) - z)^{\text{ext}}\big)^2 ds .
\end{aligned}$$

The first term cancels with the first boundary term in the left-hand side of (4.5) and by virtue of a scaling argument, the second term is bounded by

$$\begin{aligned}
&\int_{\partial K_-} |\alpha \mathbf{u}_h \cdot \mathbf{n}| \Big((\varrho_h(z) - z)^{\text{ext}}\Big)^2 ds \\
&\qquad \leq h \sigma_0^2 c_4 |\alpha| \, \|\mathbf{u}_h\|_{L^\infty(\omega_K)} |\varrho_h(z) - z|^2_{H^1(\omega_K)} ,
\end{aligned} \tag{4.8}$$

where ω_K denotes the union of triangles adjacent to K. The first term involving the divergence is bounded as follows:

$$\begin{aligned}
&\Big|\frac{\alpha}{2} \int_\Omega \operatorname{div}(\mathbf{u}_h - \mathbf{u})(\varrho_h(z) - z)(z_h - \varrho_h(z)) d\mathbf{x}\Big| \\
&\qquad \leq c_5 \frac{|\alpha|}{2} \|\operatorname{div}(\mathbf{u}_h - \mathbf{u})\|_{L^{2+1/4}(\Omega)} |z|_{H^1(\Omega)} \|z_h - \varrho_h(z)\|_{L^2(\Omega)} ,
\end{aligned} \tag{4.9}$$

with a similar bound for the second term. Finally, the last term is bounded by:

$$\Big|\alpha \int_\Omega (\mathbf{u}_h - \mathbf{u}) \cdot \nabla z (z_h - \varrho_h(z)) d\mathbf{x}\Big| \\ \leq |\alpha|\, |z|_{H^1(\Omega)} \|\mathbf{u}_h - \mathbf{u}\|_{L^\infty(\Omega)} \|z_h - \varrho_h(z)\|_{L^2(\Omega)} . \tag{4.10}$$

Hence collecting (4.6)–(4.10), substituting them into (4.5), and applying Jensen's inequality, we obtain:

$$\begin{aligned} \|z_h - \varrho_h(z)\|^2_{L^2(\Omega)} \leq\ & c_6 \frac{|\alpha|^2}{\nu^2} \big(|z|^2_{H^1(\Omega)} \|\mathbf{u}_h - \mathbf{u}\|^2_{L^\infty(\Omega)} \\ & + |\mathbf{u}|^2_{W^{1,\infty}(\Omega)} \|z - \varrho_h(z)\|^2_{L^2(\Omega)}\big) \\ & + c_7 h \frac{|\alpha|}{\nu} \|\mathbf{u}_h\|_{L^\infty(\Omega)} |z - \varrho_h(z)|^2_{H^1(\Omega)} \\ & + c_8 \frac{|\alpha|^2}{\nu^2} \|z\|^2_{H^1(\Omega)} \|\mathrm{div}(\mathbf{u}_h - \mathbf{u})\|^2_{L^{2+1/4}(\Omega)} \\ & + 2\big(\|z - \varrho_h(z)\|^2_{L^2(\Omega)} + 2|\mathbf{u}_h - \mathbf{u}|^2_{H^1(\Omega)}\big) . \end{aligned} \tag{4.11}$$

By substituting (4.1), (4.3) and (4.4) into (4.11), we derive the following error inequality.

Theorem 4.11 - *Let Ω be a convex polygon and suppose $\mathcal{T}_h$ satisfies (3.1), (4.2) and (3.20). Assume that the data $\mathbf{f}$ and $\mathbf{g}$ are sufficiently smooth and small so that $\mathbf{u} \in W^{1,\infty}(\Omega)^2$ and $z \in H^1(\Omega)$, and α and ν are such that*

$$\frac{|\alpha|}{\nu}(C_4 + C_5 h^{1/4}) \|z\|_{H^1(\Omega)} (\sqrt{c_6} + \sqrt{2c_8}) + 2\frac{S_4}{\nu} \|P_h(\mathbf{u})\|_{L^4(\Omega)} \leq \frac{1}{2}, \tag{4.12}$$

with the constants C_i of (4.3) and (4.4) and c_i of (4.11) (which implies that Problem P has a unique solution), then there exists a constant C, depending on the data but independent of h, such that each solution of (1.12)–(1.14) satisfies the error bound

$$\begin{aligned} \|z_h - z\|_{L^2(\Omega)} \leq\ & C\big(\|P_h(\mathbf{u}) - \mathbf{u}\|_{L^\infty(\Omega)} + |P_h(\mathbf{u}) - \mathbf{u}|_{W^{1,2+1/4}(\Omega)} \\ & + \frac{1}{\rho_{\min}^{1/9}} (|P_h(\mathbf{u}) - \mathbf{u}|_{H^1(\Omega)} + \|r_h(p) - p\|_{L^2(\Omega)}) \\ & + \|P_h(\mathbf{u}) - \mathbf{u}\|_{L^4(\Omega)} + \|\varrho_h(z) - z\|_{L^2(\Omega)} \\ & + h^{1/2} |\varrho_h(z) - z|_{H^1(\Omega)}\big) . \end{aligned} \tag{4.13}$$

Clearly, the last term in (4.13) is dominating; it accounts for the loss of $h^{1/2}$ in the final error term. Thus, under the assumptions of Theorem 4.11, if z

belongs to $H^2(\Omega)$, $\mathbf{u}$ to $H^3(\Omega)^2$ and p to $H^2(\Omega)$, we have

$$|\mathbf{u}_h - \mathbf{u}|_{H^1(\Omega)} + \|p_h - p\|_{L^2(\Omega)} + \|z_h - z\|_{L^2(\Omega)} \leq C\, h^{3/2}\,. \tag{4.14}$$

We finish this section by addressing the computation of $(\mathbf{u}_h, p_h, z_h)$. Consider the following successive-approximations algorithm. Let z_h^0 be an arbitrary element of Z_h (e.g. $z_h^0 = 0$); then for $k \geq 0$ we compute the sequence $\mathbf{u}_h^k \in X_h + \mathbf{g}_h$, $p_h^k \in M_h$ and $z_h^{k+1} \in Z_h$ by solving first the generalized Stokes problem:

$$\forall \mathbf{v}_h \in X_h\,,\; \nu(\nabla\, \mathbf{u}_h^k, \nabla\, \mathbf{v}_h) + (\mathbf{z}_h^k \times \mathbf{u}_h^k, \mathbf{v}_h) - (p_h^k, \operatorname{div} \mathbf{v}_h) = (\mathbf{f}, \mathbf{v}_h)\,,$$

$$\forall q_h \in M_h\,,\; (q_h, \operatorname{div} \mathbf{u}_h^k) = 0\,,$$

and next the linear transport equation:

$$\forall \theta_h \in Z_h\,,\; \nu\,(z_h^{k+1}, \theta_h) + c(\mathbf{u}_h^k; z_h^{k+1}, \theta_h) = \nu\,(\operatorname{curl} \mathbf{u}_h^k, \theta_h) + \alpha\,(\operatorname{curl} \mathbf{f}, \theta_h)\,.$$

As in [17], the uniform a priorí estimate (3.25), combined with (3.21), (3.22), (3.24) allow us to prove that there exists a subsequence of k tending to infinity such that $(\mathbf{u}_h^k, p_h^k, z_h^k)$ converges uniformly without restriction on the data or the domain. Furthermore, under the sufficient conditions of Theorem 4.11, the limit is a solution $(\mathbf{u}_h, p_h, z_h)$ of (1.12)–(1.14).

References

[1] Adams, R. A., Sobolev Spaces, Academic Press, New York, NY, 1975.

[2] Baia, M. and Sequeira, A., A finite element approximation for the steady solution of a second-grade fluid model, J. Comp. and Appl. Math., 111 (1999), 281–295.

[3] Bernard, J. M., Stationary problem of second-grade fluids in three dimensions: existence, uniqueness and regularity, Math. Meth. Appl. Sci., 22 (1999), pp. 655–687.

[4] Bernardi, C. and Girault, V., A local regularization operator for triangular and quadrilateral finite elements, SIAM J. Numer. Anal., 35, 5 (1998), pp. 1893–1916.

[5] Brenner, S. and Scott, L. R., The Mathematical Theory of Finite Element Methods, TAM 15, Springer-Verlag, Berlin, 1994.

[6] Ciarlet, P. G., The Finite Element Method for Elliptic Problems, North-Holland, Amsterdam, 1978.

[7] Cioranescu, D. and Ouazar, E. H., Existence et unicité pour les fluides de second grade, Note CRAS 298, Série I (1984), 285–287.

[8] Cioranescu, D. and Ouazar, E. H., Existence and uniqueness for fluids of second grade, in Nonlinear Partial Differential Equations, Collège de France Seminar, Pitman 109, Boston, MA (1984), 178–197.

[9] Cioranescu, D. and Girault, V., Weak and classical solutions of a family of second grade fluids, Int. J. Non-Linear Mech. 32 (1987), 317–335.

[10] Cioranescu, D., Girault, V., Glowinski, R. and Scott, L. R., Some theoretical and numerical aspects of grade-two fluid models, in Partial Differential Equations - Theory and Numerical Solution, (W. Jaeger, J. Necas, O. John, K. Najzar, J. Stara, eds.), Research Notes in Mathematics 406, Chapman & Hall/CRC, New York, NY, 99–110, 1999.

[11] Clément, P., Approximation by finite element functions using local regularization, RAIRO, Anal. Num. R-2 (1975), 77–84.

[12] M. Crouzeix and P. A. Raviart, Conforming and non-conforming finite element methods for solving the stationary Stokes problem, RAIRO Anal. Numér., 8, (1973), 33–76.

[13] Dunn, J. E. and Fosdick, R. L., Thermodynamics, stability, and boundedness of fluids of complexity two and fluids of second grade, Arch. Rational Mech. Anal. 56, 3 (1974), 191–252.

[14] Dunn, J. E. and Rajagopal, K. R., Fluids of differential type: Critical review and thermodynamic analysis, Int. J. Engng Sci. 33, 5 (1995), 689–729.

[15] Girault, V. and Raviart, P. A., Finite Element Methods for Navier-Stokes Equations. Theory and Algorithms, SCM 5, Springer-Verlag, Berlin, 1986.

[16] Girault, V. and Scott, L. R., Analysis of a two-dimensional grade-two fluid model with a tangential boundary condition, J. Math. Pures Appl. 78, 10 (1999), 981–1011.

[17] Girault, V. and Scott, L. R., Finite-element discretizations of a two-dimensional grade-two fluid model, M2AN 35, 6, (2001), 1007–1053.

[18] Girault, V. and Scott, L. R., Stability of dispersive model equations for fluid flow, to appear in Collected Lectures on the Preservation of Stability Under Discretization, eds. D. Estep and S. Tavener, 2002.

[19] Girault, V. and Scott, L. R., A quasi-local interpolation operator preserving the discrete divergence, in preparation.

[20] Grisvard, P., Elliptic Problems in Nonsmooth Domains, Pitman Monographs and Studies in Mathematics, 24, Pitman, Boston, MA, 1985.

[21] Holm, D. D., Marsden, J. E. and Ratiu, T. S., Euler-Poincaré models of ideal fluids with nonlinear dispersion, Phys. Rev. Lett. 349 (1998), 4173–4177.

[22] Holm, D. D., Marsden, J. E., Ratiu, T. S., The Euler-Poincaré equations and semidirect products with applications to continuum theories, Adv. in Math. 137 (1998), 1–81.

[23] Lesaint, P. and Raviart, P. A., On a finite element method for solving the neutron transport equation, in *Mathematical Aspects of finite Elements in Partial Differential Equations*, 89–122, Academic Press, New York, NY, 1974.

[24] Lions, J. L. and Magenes, E., Problèmes aux Limites non Homogènes et Applications, I, Dunod, Paris, 1968.

[25] Nečas, J., Les Méthodes directes en théorie des équations elliptiques, Masson, Paris, 1967.

[26] Ouazar, E. H., Sur les Fluides de Second Grade, Thèse de 3ème Cycle de l'Université Pierre et Marie Curie, Paris VI, 1981.

[27] Peetre, J., Espaces d'interpolation et théorème de Soboleff, Ann. Inst. Fourier 16 (1966), 279–317.

[28] Rivlin, R. S. and Ericksen, J. L., Stress-deformation relations for isotropic materials, Arch. Rational Mech. Anal. 4 (1955), 323–425.

[29] Scott, L. R. and Zhang, S., Finite element interpolation of non-smooth functions satisfying boundary conditions Math. Comp., 54 (1990), 483–493.

[30] Tartar, L., Topics in Nonlinear Analysis, Publications Mathématiques d'Orsay, Université Paris-Sud, Orsay, 1978.

[31] Videmann, J. H., Mathematical analysis of visco-elastic non-newtonian fluids, Thesis, University of Lisbon, 1997.

Vivette Girault
Laboratoire Jacques-Louis Lions
Université Pierre et Marie Curie
175 rue du Chevaleret
75013 Paris
France

E-mail: girault@ann.jussieu.fr

L. R. Scott
Department of Mathematics
University of Chicago
Chicago, Illinois
USA

E-mail: ridg@cs.uchicago.edu

Studies in Mathematics and its Applications, Vol. 31
D. Cioranescu and J.L. Lions (Editors)

Chapter 19

STABILITY OF THIN LAYER APPROXIMATION OF ELECTROMAGNETIC WAVES SCATTERING BY LINEAR AND NON LINEAR COATINGS

H. HADDAR AND P. JOLY

1. Introduction

The scattering of electromagnetic waves by thin coatings, using effective boundary conditions have been widely studied in the case of harmonic Maxwell equations (see [1], [4], and references therein). We give an extension to the time dependent problem, where obtaining the stability in time is fundamental. We apply a general technique based on an asymptotic expansion of the solution with respect to the thickness of the coating ([17], [1]).

The main motivation for using these effective boundary conditions comes from numerical considerations. Roughly speaking, as the obtained approximate model is posed on the exterior domain (i.e. not including the thin layer), we eliminate the geometrical constraints for meshing. This reduces (especially for layers that are very thin compared with the wavelength of the incident wave) the size of the discrete model, and thus the time of computations. We notice also that this method may constitute a valuable alternative to mesh refinement around the obstacle.

This article is divided into three parts. In the first one, the coating is made of linear material and the boundary is planar. This simplified case enables us to present, in a simple way, the formal construction of the effective boundary conditions. Once stability conditions are derived, one can perform an asymptotic analysis to establish error estimates. In the second part, we treat the case of curved and regular boundaries. We explain how to take into account the geometric characteristics of the boundaries. The expression and the stability of the third order condition constitute the main result of this part. In the last section, our work is generalized to a class of non linear materials of ferromagnetic type. Only the formal derivation and the stability analysis are discussed. The error estimates will be the subject of a forthcoming paper.

2. Thin layer approximation: the linear case

2.1. Description of the model

We consider a linear material characterized by its electric permittivity ε_{f}, and its magnetic permeability μ_{f} that occupies a layer $\Omega_{\mathrm{f}}^{\eta} \subset \mathbf{R}^3$ of a thickness η. This layer separates two media: the exterior domain Ω_v of characteristics (ε_0, μ_0) and the interior domain Ω_0^{η}, namely a perfectly conducting scatterer. We denote by Γ the boundary of Ω_v and Γ^{η} the boundary of Ω_0^{η} (see fig. 1).

We assume that the material is homogeneous and that its characteristics do not depend on η, (the non-homogeneous case where the characteristics depend on the thickness coordinate and vary slowly with respect to the tangential coordinate can be treated with minor modifications).

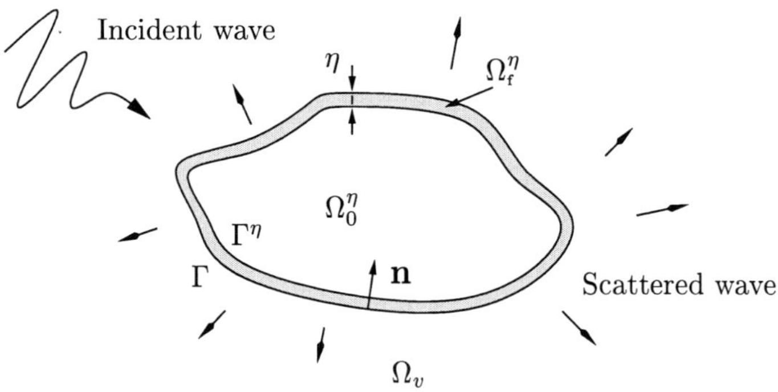

Figure 1: Problem presentation

We denote by $(\mathbf{E}_0, \mathbf{H}_0)$ the electromagnetic field at $t = 0$. We assume that it is compactly supported in Ω_v. Let $(\mathbf{E}_v^{\eta}, \mathbf{H}_v^{\eta})$ (resp. $(\mathbf{E}_{\mathrm{f}}^{\eta}, \mathbf{H}_{\mathrm{f}}^{\eta})$) be the electromagnetic field in Ω_v (resp. in $\Omega_{\mathrm{f}}^{\eta}$). They satisfy:

$$\left\{ \begin{array}{ll} \varepsilon_0 \, \partial_t \mathbf{E}_v^{\eta} - \mathrm{rot}\, \mathbf{H}_v^{\eta} = 0, \quad \mu_0 \, \partial_t \mathbf{H}_v^{\eta} + \mathrm{rot}\, \mathbf{E}_v^{\eta} = 0 & \text{on } \Omega_v \times \mathbf{R}^+, \\ (\mathbf{E}_v^{\eta}, \mathbf{H}_v^{\eta})(t=0) = (\mathbf{E}_0, \mathbf{H}_0) & \text{on } \Omega_v. \end{array} \right. \tag{1}$$

$$\left\{ \mathbf{E}_v^{\eta} \times \mathbf{n} = \mathbf{E}_{\mathrm{f}}^{\eta} \times \mathbf{n}, \quad \mathbf{H}_v^{\eta} \times \mathbf{n} = \mathbf{H}_{\mathrm{f}}^{\eta} \times \mathbf{n} \text{ on } \Gamma \times \mathbf{R}^+, \right. \tag{2}$$

where $\mathbf{n}$ is a unit vector normal to Γ,

$$\left\{ \begin{array}{ll} \varepsilon_{\mathrm{f}} \, \partial_t \mathbf{E}_{\mathrm{f}}^{\eta} - \mathrm{rot}\, \mathbf{H}_{\mathrm{f}}^{\eta} = 0, \quad \mu_{\mathrm{f}} \, \partial_t \mathbf{H}_{\mathrm{f}}^{\eta} + \mathrm{rot}\, \mathbf{E}_{\mathrm{f}}^{\eta} = 0 & \text{on } \Omega_{\mathrm{f}}^{\eta} \times \mathbf{R}^+, \\ (\mathbf{E}_{\mathrm{f}}^{\eta}, \mathbf{H}_{\mathrm{f}}^{\eta})(t=0) = (0, 0) & \text{on } \Omega_{\mathrm{f}}^{\eta}, \end{array} \right. \tag{3}$$

and finally, the "Dirichlet" boundary condition on Γ^η (where $\mathbf{n}$ is a unit vector normal to Γ^η),

$$\mathbf{E}_{\mathrm{f}}^\eta \times \mathbf{n} = 0, \qquad \text{on } \Gamma^\eta \times \mathbb{R}^+. \tag{4}$$

The problem constituted by (1), (2), (3) and (4) is the exact transmission problem. As we consider the situations where $\eta \ll \lambda$ (the incident wavelength), numerical methods applied to this problem will be penalized by the computation of $(\mathbf{E}_{\mathrm{f}}^\eta, \mathbf{H}_{\mathrm{f}}^\eta)$ (that requires fine meshing). The idea to overcome this difficulty, is to replace the equations (2), (3) and (4) by an effective boundary condition posed on the exterior boundary Γ such that: the electromagnetic field, satisfying Maxwell's equations on Ω_v coupled with this effective boundary condition, constitutes a good approximation of $(\mathbf{E}_v^\eta, \mathbf{H}_v^\eta)$. More precisely, our construction leads to the hierarchy of effective boundary conditions indexed by an integer k

$$\widetilde{\mathbf{E}}_v^\eta \times \mathbf{n} = \mathcal{B}_k^\eta(\mathbf{n} \times (\widetilde{\mathbf{H}}_v^\eta \times \mathbf{n})) \quad \text{on} \quad \Gamma \times \mathbb{R}^+, \tag{5}$$

where $(\widetilde{\mathbf{E}}_v^\eta, \widetilde{\mathbf{H}}_v^\eta)$ is the approximate electromagnetic field on Ω_v, satisfying

$$\begin{cases} \varepsilon_0\, \partial_t \widetilde{\mathbf{E}}_v^\eta - \operatorname{rot} \widetilde{\mathbf{H}}_v^\eta = 0, \qquad \mu_0\, \partial_t \widetilde{\mathbf{H}}_v^\eta + \operatorname{rot} \widetilde{\mathbf{E}}_v^\eta = 0 & \text{on } \Omega_v \times \mathbb{R}^+, \\ (\widetilde{\mathbf{E}}_v^\eta, \widetilde{\mathbf{H}}_v^\eta)(t=0) = (\mathbf{E}_0, \mathbf{H}_0) & \text{on } \Omega_v, \end{cases} \tag{6}$$

and where $\mathcal{B}_k^\eta$ is a tangential operator such that (at least in a formal way)

$$\left\| (\mathbf{E}_v^\eta, \mathbf{H}_v^\eta) - (\widetilde{\mathbf{E}}_v^\eta, \widetilde{\mathbf{H}}_v^\eta) \right\|_{L^2} = O(\eta^{k-\frac{1}{2}}). \tag{7}$$

In such a case, the effective boundary condition (5) is said to be of order k. The identity (7) is referred to the **consistency property** of the construction. A fundamental notion linked to this property is the **stability** with respect to the small parameter η. For problem (1), (2), (3), (4) one gets a uniform bound, with respect to $t \geq 0$ and η, of the electromagnetic energy in Ω_v, namely

$$\mathcal{E}_v(\mathbf{E}_v^\eta(t), \mathbf{H}_v^\eta(t)) \leq \mathcal{E}_v(\mathbf{E}_0, \mathbf{H}_0), \tag{8}$$

where, $\mathcal{E}_v(\mathbf{E}, \mathbf{H}) = \frac{\varepsilon_0}{2} \|\mathbf{E}\|^2_{L^2(\Omega_v)} + \frac{\mu_0}{2} \|\mathbf{H}\|^2_{L^2(\Omega_v)}$. Relation (8) is obtained from the energy identity

$$\mathcal{E}_v(\mathbf{E}_v^\eta(t), \mathbf{H}_v^\eta(t)) + \mathcal{E}_{\mathrm{f}}^\eta(\mathbf{E}_{\mathrm{f}}^\eta(t), \mathbf{H}_{\mathrm{f}}^\eta(t)) = \mathcal{E}_v(\mathbf{E}_0, \mathbf{H}_0), \tag{9}$$

where $\mathcal{E}_{\mathrm{f}}^\eta(\mathbf{E}, \mathbf{H}) = \frac{\varepsilon_{\mathrm{f}}}{2} \|\mathbf{E}\|^2_{L^2(\Omega_{\mathrm{f}}^\eta)} + \frac{\mu_0}{2} \|\mathbf{H}\|^2_{L^2(\Omega_{\mathrm{f}}^\eta)}$. We deduce that estimation (7) requires at least a stability identity, for the solution of (6), of the form

$$\mathcal{E}_v(\widetilde{\mathbf{E}}_v^\eta(t), \widetilde{\mathbf{H}}_v^\eta(t)) \leq \mathcal{E}_v(\mathbf{E}_0, \mathbf{H}_0) + O(\eta^{k-\frac{1}{2}}). \tag{10}$$

In fact, identity (10) will constitute the basic requirement of the construction of $\mathcal{B}_k^\eta$.

2.2. Formal construction: the case of a plane boundary

In this section we assume that the unitary normal vector field $\mathbf{n}$ on Γ, outwardly directed to Ω_v, is constant (see figure 2).

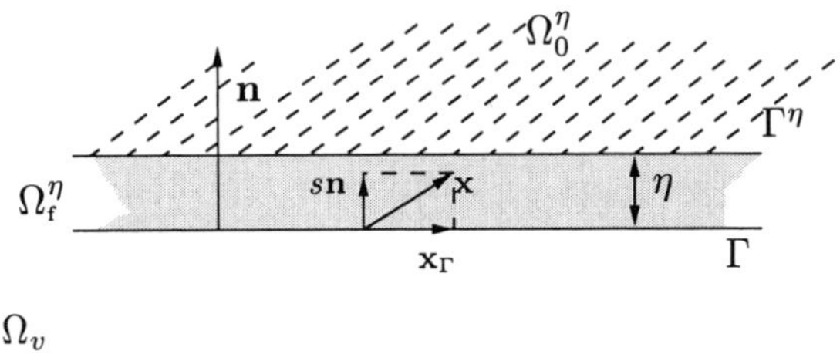

Figure 2: Problem presentation: the case of a plane boundary

2.2.1. Scaling, asymptotic expansion and principle of the effective conditions

We apply first the well known technique of scaling with respect to the thickness coordinate of the thin layer, which removes the dependence of the geometry on the small parameter η. Let us proceed in an intrinsic way, so that the link with the case of curved boundary shall be more clear later. We use the parametric coordinates system of Ω_f^η (fig. 2).

$$\mathbf{x} \in \bar{\Omega}_f^\eta \longmapsto (\mathbf{x}_\Gamma, s) \in \Gamma \times [0, \eta], \text{ such that: } \mathbf{x} = \mathbf{x}_\Gamma + s\,\mathbf{n}. \tag{11}$$

In the case of straight boundary, this map is well defined for every η. The scaling consists on the change of variables:

$$(\mathbf{x}_\Gamma, s) \in \Gamma \times [0, \eta] \longmapsto (\mathbf{x}_\Gamma, \nu) \in \Gamma \times [0, 1] \text{ such that: } s = \eta\,\nu. \tag{12}$$

Some notations. We denote by ∇_Γ the derivative operator with respect to $\mathbf{x}_\Gamma$. For instance, if $(\boldsymbol{\tau}^1, \boldsymbol{\tau}^2) \in \mathbb{R}^3 \times \mathbb{R}^3$ is an orthonormal basis of Γ, and (ξ_1, ξ_2) is the system of coordinates associated to this basis, we have: $\nabla_\Gamma = (\partial_{\xi_1} \cdot)\,\boldsymbol{\tau}^1 + (\partial_{\xi_2} \cdot)\,\boldsymbol{\tau}^2$. It is easy to see then that the expression of the derivative operator ∇ in $\mathbb{R}^3$ becomes in the coordinates system $(\mathbf{x}_\Gamma, \nu)$,

$$\nabla \longmapsto \nabla_\Gamma + \frac{1}{\eta}\,\partial_\nu\,\mathbf{n}$$

Let $\mathbf{v}$ be a tangential vector field defined on Γ. We define the operator div_Γ by:

$$\mathrm{div}_\Gamma\,\mathbf{v} = \nabla_\Gamma \cdot \mathbf{v} = (\partial_{\xi_1}\mathbf{v}) \cdot \boldsymbol{\tau}^1 + (\partial_{\xi_2}\mathbf{v}) \cdot \boldsymbol{\tau}^2.$$

We recall that div_Γ is the adjoint of $-\nabla_\Gamma$ for the L^2 scalar product on Γ. We introduce also:

$$\mathrm{rot}_\Gamma\,\mathbf{v} = \mathrm{div}_\Gamma\,(\mathbf{v}\times\mathbf{n}), \qquad \vec{\mathrm{rot}}_\Gamma\, u = (\nabla_\Gamma\, u)\times\mathbf{n}, \tag{13}$$

where $\mathbf{v}$ (resp. u) is a vectorial (resp. scalar) field defined on Γ. rot_Γ and $\vec{\mathrm{rot}}_\Gamma$ are adjoint operators for the L^2 scalar product on Γ. Furthermore, $\mathrm{div}_\Gamma\,\vec{\mathrm{rot}}_\Gamma = 0$ and $\mathrm{rot}_\Gamma\,\nabla_\Gamma = 0$.

We shall distinguish the normal and the tangential part of a vector field $\mathbf{v}\in\mathbb{R}^3$:

$$\mathbf{v} = \Pi_\parallel\mathbf{v} + \Pi_\perp\mathbf{v}, \text{ where } \Pi_\parallel\mathbf{v} = \mathbf{n}\times(\mathbf{v}\times\mathbf{n}) \text{ and } \Pi_\perp\mathbf{v} = (\mathbf{v}\cdot\mathbf{n})\,\mathbf{n}. \tag{14}$$

Throughout this paper, the variable t will always indicate the time variable but may sometimes be omitted from the list of variables. On the other hand $\mathbf{x}_\Gamma\in\mathbb{R}^3$, will always denote the tangential coordinate belonging to Γ.

Scaled equations. Let $\mathbf{E}$ be a vector field defined on Ω_f^η. We define $\hat{\mathbf{E}}$ on the scaled domain $\Gamma\times]0,1[$ by $\hat{\mathbf{E}}(\mathbf{x}_\Gamma,\nu) = \mathbf{E}(\mathbf{x})$ where $(\mathbf{x}_\Gamma,\nu)$ and $\mathbf{x}$ are linked by (11) and (12). Working in the orthonormal basis $(\boldsymbol{\tau}^1,\boldsymbol{\tau}^2,\mathbf{n})$, one checks that

$$\mathrm{rot}\,\mathbf{E} = \vec{\mathrm{rot}}_\Gamma\,(\hat{\mathbf{E}}\cdot\mathbf{n}) + (\mathrm{rot}_\Gamma\,\hat{\mathbf{E}})\,\mathbf{n} - \tfrac{1}{\eta}\,\partial_\nu(\hat{\mathbf{E}}\times\mathbf{n}). \tag{15}$$

This identity enables us to rewrite equations (2), (3) and (4) in the scaled domain. For simplicity, we shall denote (in abusive way) by $(\mathbf{E}_f^\eta,\mathbf{H}_f^\eta)$ the electromagnetic field in $\Gamma\times]0,1[$,

$$(\mathbf{E}_f^\eta,\mathbf{H}_f^\eta)(\mathbf{x},t) \mapsto (\hat{\mathbf{E}}_f^\eta,\hat{\mathbf{H}}_f^\eta)(\mathbf{x}_\Gamma,\nu,t) \equiv (\mathbf{E}_f^\eta,\mathbf{H}_f^\eta)(\mathbf{x}_\Gamma,\nu,t).$$

So, equations (2), (3) and (4) become respectively,

$$\Pi_\parallel\mathbf{E}_v^\eta(\mathbf{x}_\Gamma,t) = \Pi_\parallel\mathbf{E}_f^\eta(\mathbf{x}_\Gamma,0,t),\;\; \Pi_\parallel\mathbf{H}_v^\eta(\mathbf{x}_\Gamma,t) = \Pi_\parallel\mathbf{H}_f^\eta(\mathbf{x}_\Gamma,0,t) \tag{16}$$

for $(\mathbf{x}_\Gamma,t)\in\Gamma\times\mathbb{R}^+$.

$$\left\{\begin{array}{l} \varepsilon_f\,\partial_t\mathbf{E}_f^\eta - \vec{\mathrm{rot}}_\Gamma\,(\mathbf{H}_f^\eta\cdot\mathbf{n}) - (\mathrm{rot}_\Gamma\,\mathbf{H}_f^\eta)\,\mathbf{n} + \frac{1}{\eta}\,\partial_\nu(\mathbf{H}_f^\eta\times\mathbf{n}) = 0 \\ \mu_f\,\partial_t\mathbf{H}_f^\eta + \vec{\mathrm{rot}}_\Gamma\,(\mathbf{E}_f^\eta\cdot\mathbf{n}) + (\mathrm{rot}_\Gamma\,\mathbf{E}_f^\eta)\,\mathbf{n} - \frac{1}{\eta}\,\partial_\nu(\mathbf{E}_f^\eta\times\mathbf{n}) = 0 \\ \text{on } \Gamma\times\,]0,1[\times\mathbb{R}^+ \\ (\mathbf{E}_f^\eta,\mathbf{H}_f^\eta)(t=0) = (0,0), \text{ on } \Gamma\times\,]0,1[, \end{array}\right. \tag{17}$$

$$\mathbf{E}_f^\eta(\mathbf{x}_\Gamma,\,1)\times\mathbf{n} = 0 \text{ for } (\mathbf{x}_\Gamma,t)\in\Gamma\times\mathbb{R}^+, \tag{18}$$

and the exact transmission problem is constituted now by equations (1), (16), (17) and (18).

Asymptotic expansion. We seek a solution of the form:

$$\left\{\begin{array}{l}(\mathbf{E}_v^\eta, \mathbf{H}_v^\eta) = (\mathbf{E}_v^0, \mathbf{H}_v^0) + \eta\,(\mathbf{E}_v^1, \mathbf{H}_v^1) + \eta^2\,(\mathbf{E}_v^2, \mathbf{H}_v^2) + \cdots \\ (\mathbf{E}_{\mathrm{f}}^\eta, \mathbf{H}_{\mathrm{f}}^\eta) = (\mathbf{E}_{\mathrm{f}}^0, \mathbf{H}_{\mathrm{f}}^0) + \eta\,(\mathbf{E}_{\mathrm{f}}^1, \mathbf{H}_{\mathrm{f}}^1) + \eta^2\,(\mathbf{E}_{\mathrm{f}}^2, \mathbf{H}_{\mathrm{f}}^2) + \cdots\end{array}\right. \tag{19}$$

and proceed to formal identification after the substitution of (19) into (1), (16), (17) and (18). This process allows a recursive determination of the different terms. From (1) and (16) we get for all $k \geq 0$,

$$\left\{\begin{array}{l}\varepsilon_0\,\partial_t \mathbf{E}_v^k - \operatorname{rot}\mathbf{H}_v^k = 0, \qquad \mu_0\,\partial_t \mathbf{H}_v^k + \operatorname{rot}\mathbf{E}_v^k = 0, \\ (\mathbf{E}_v^k, \mathbf{H}_v^k)(t=0) = \left|\begin{array}{ll}(\mathbf{E}_0, \mathbf{H}_0) & \text{if } k = 0, \\ (0,0) & \text{if } k > 0,\end{array}\right. \\ \Pi_{\|}\mathbf{E}_v^k(\mathbf{x}_\Gamma) = \Pi_{\|}\mathbf{E}_{\mathrm{f}}^k(\mathbf{x}_\Gamma, 0) \text{ for } \mathbf{x}_\Gamma \in \Gamma,\end{array}\right. \tag{20}$$

that can be interpreted as Maxwell's equations with non-homogeneous boundary conditions on Γ. They permit to determine $(\mathbf{E}_v^k, \mathbf{H}_v^k)$ for each $k \geq 0$, if we are able to compute the boundary value $\Pi_{\|}\mathbf{E}_{\mathrm{f}}^k(\mathbf{x}_\Gamma, 0)$. This will be possible using (17) and (18). For $k = 0$, we get

$$\left\{\begin{array}{l}\partial_\nu(\mathbf{E}_{\mathrm{f}}^0 \times \mathbf{n}) = 0, \ \ \mathbf{E}_{\mathrm{f}}^0(\mathbf{x}_\Gamma, 1) \times \mathbf{n} = 0, \\ \partial_\nu(\mathbf{H}_{\mathrm{f}}^0 \times \mathbf{n}) = 0, \ \ \mathbf{H}_{\mathrm{f}}^0(\mathbf{x}_\Gamma, 0) \times \mathbf{n} = \mathbf{H}_v^0(\mathbf{x}_\Gamma) \times \mathbf{n},\end{array}\right. \quad \text{for } \mathbf{x}_\Gamma \in \Gamma, \tag{21}$$

while for $k > 0$, we have

$$\left\{\begin{array}{l}\partial_\nu(\mathbf{E}_{\mathrm{f}}^k \times \mathbf{n}) = \mu_{\mathrm{f}}\,\partial_t \mathbf{H}_{\mathrm{f}}^{k-1} + \vec{\operatorname{rot}}_\Gamma\,(\mathbf{E}_{\mathrm{f}}^{k-1} \cdot \mathbf{n}) + (\operatorname{rot}_\Gamma \mathbf{E}_{\mathrm{f}}^{k-1})\,\mathbf{n}, \\ \mathbf{E}_{\mathrm{f}}^k(\mathbf{x}_\Gamma, 1) \times \mathbf{n} = 0, \quad \text{for } \mathbf{x}_\Gamma \in \Gamma.\end{array}\right. \tag{22}$$

and

$$\left\{\begin{array}{l}\partial_\nu(\mathbf{H}_{\mathrm{f}}^k \times \mathbf{n}) = -\varepsilon_{\mathrm{f}}\,\partial_t \mathbf{E}_{\mathrm{f}}^{k-1} + \vec{\operatorname{rot}}_\Gamma\,(\mathbf{H}_{\mathrm{f}}^{k-1} \cdot \mathbf{n}) + (\operatorname{rot}_\Gamma \mathbf{H}_{\mathrm{f}}^{k-1})\,\mathbf{n}, \\ \mathbf{H}_{\mathrm{f}}^k(\mathbf{x}_\Gamma, 0) \times \mathbf{n} = \mathbf{H}_v^k(\mathbf{x}_\Gamma) \times \mathbf{n}, \quad \text{for } \mathbf{x}_\Gamma \in \Gamma.\end{array}\right. \tag{23}$$

Equations (21), (22) and (23) can be solved like the following. For $k = 0$, equation (21) shows that,

$$\left\{\begin{array}{l}\mathbf{E}_{\mathrm{f}}^0(\mathbf{x}_\Gamma, \nu) \times \mathbf{n} = 0, \qquad (\mathbf{x}_\Gamma, \nu) \in \Gamma\times]0, 1[, \\ \mathbf{H}_{\mathrm{f}}^0(\mathbf{x}_\Gamma, \nu) \times \mathbf{n} = \mathbf{H}_v^0(\mathbf{x}_\Gamma) \times \mathbf{n}, \qquad (\mathbf{x}_\Gamma, \nu) \in \Gamma\times]0, 1[.\end{array}\right. \tag{24}$$

For higher order terms, we apply

Lemma 0.1 – *For all $k > 0$, and for all $(\mathbf{x}_\Gamma, \nu) \in \Gamma \times]0, 1[$,*

$$\left\{\begin{array}{l} \partial_t \mathbf{E}_{\mathrm{f}}^k(\mathbf{x}_\Gamma, \nu) \times \mathbf{n} = -\dfrac{1}{\varepsilon_{\mathrm{f}}} \displaystyle\int_\nu^1 \left(\varepsilon_{\mathrm{f}}\, \mu_{\mathrm{f}}\, \partial_{tt} + \vec{\mathrm{rot}}_\Gamma\, \mathrm{rot}_\Gamma\,\right) \Pi_{\|}\, \mathbf{H}_{\mathrm{f}}^{k-1}(\mathbf{x}_\Gamma,\, \xi)\, d\xi, \\ (\mathbf{E}_{\mathrm{f}}^k(\mathbf{x}_\Gamma, \nu) \times \mathbf{n})(t = 0) = 0. \end{array}\right. \tag{25}$$

$$\left\{\begin{array}{l} \partial_t \mathbf{H}_{\mathrm{f}}^k(\mathbf{x}_\Gamma, \nu) \times \mathbf{n} = \partial_t \mathbf{H}_v^k(\mathbf{x}_\Gamma) \times \mathbf{n} \\ -\dfrac{1}{\mu_{\mathrm{f}}} \displaystyle\int_0^\nu \left(\varepsilon_{\mathrm{f}}\, \mu_{\mathrm{f}}\, \partial_{tt} + \vec{\mathrm{rot}}_\Gamma\, \mathrm{rot}_\Gamma\,\right) \Pi_{\|}\, \mathbf{E}_{\mathrm{f}}^{k-1}(\mathbf{x}_\Gamma,\, \xi)\, d\xi, \\ (\mathbf{H}_{\mathrm{f}}^k(\mathbf{x}_\Gamma, \nu) \times \mathbf{n})(t = 0) = 0. \end{array}\right. \tag{26}$$

Proof. Let us prove (25). Applying $\Pi_{\|}$ to the first equation of (22) shows that:

$$\partial_\nu(\mathbf{E}_{\mathrm{f}}^k \times \mathbf{n}) = \mu_{\mathrm{f}}\, \partial_t(\Pi_{\|}\, \mathbf{H}_{\mathrm{f}}^{k-1}) + \vec{\mathrm{rot}}_\Gamma\, (\mathbf{E}_{\mathrm{f}}^{k-1} \cdot \mathbf{n}). \tag{27}$$

We take now the scalar product of the first equation of (23) by $\mathbf{n}$. As $\mathbf{n} \cdot \vec{\mathrm{rot}}_\Gamma\ = 0$ we obtain:

$$\varepsilon_{\mathrm{f}}\, \partial_t\, \left(\mathbf{E}_{\mathrm{f}}^{k-1} \cdot \mathbf{n}\right) = \mathrm{rot}_\Gamma\, \mathbf{H}_{\mathrm{f}}^{k-1}. \tag{28}$$

Combining (27) and (28) yields

$$\partial_\nu(\partial_t \mathbf{E}_{\mathrm{f}}^k \times \mathbf{n}) = \left(\mu_{\mathrm{f}}\, \partial_{tt} + \frac{1}{\varepsilon_{\mathrm{f}}}\, \vec{\mathrm{rot}}_\Gamma\, \mathrm{rot}_\Gamma\,\right) \Pi_{\|}\, \mathbf{H}_{\mathrm{f}}^{k-1}.$$

We get identity (25) by integrating the previous equation between ν and 1 and by using the derivative in time of the boundary condition given in (22). Identity (26) is obtained in a similar way. ■

Remark 2.1. From equation (28) one can easily deduce that the normal components $(\mathbf{E}_{\mathrm{f}}^k \cdot \mathbf{n})$ and $(\mathbf{H}_{\mathrm{f}}^k \cdot \mathbf{n})$, $k \geq 0$, are determined by

$$\left\{\begin{array}{l} \varepsilon_{\mathrm{f}}\, \partial_t\, \left(\mathbf{E}_{\mathrm{f}}^k \cdot \mathbf{n}\right) = \mathrm{rot}_\Gamma\, \Pi_{\|} \mathbf{H}_{\mathrm{f}}^k, \\ \left(\mathbf{E}_{\mathrm{f}}^k \cdot \mathbf{n}\right)(t = 0) = 0. \end{array}\right. \quad \left\{\begin{array}{l} \mu_{\mathrm{f}}\, \partial_t\, \left(\mathbf{H}_{\mathrm{f}}^k \cdot \mathbf{n}\right) = -\mathrm{rot}_\Gamma\, \Pi_{\|} \mathbf{E}_{\mathrm{f}}^k, \\ \left(\mathbf{H}_{\mathrm{f}}^k \cdot \mathbf{n}\right)(t = 0) = 0. \end{array}\right. \tag{29}$$

These equations will be useful later in order to prove the error estimates.

Construction principle of the effective boundary conditions. In order to obtain an approximation of order k of the scattered field $(\mathbf{E}_v^\eta, \mathbf{H}_v^\eta)$ a good candidate would be the truncated expansion

$$(\overline{\mathbf{E}}_v^k, \overline{\mathbf{H}}_v^k) = \sum_{i=0}^{k-1} \eta^i\, (\mathbf{E}_v^i,\, \mathbf{H}_v^i)$$

(cf. theorem 0.2). According to (20), $(\overline{\mathbf{E}}_v^k, \overline{\mathbf{H}}_v^k)$ satisfies Maxwell's equations on Ω_v coupled with the boundary condition on Γ

$$\Pi_{\parallel} \overline{\mathbf{E}}_v^k(\mathbf{x}_\Gamma) = \sum_{i=0}^{k-1} \eta^i \, \Pi_{\parallel} \mathbf{E}_{\mathrm{f}}^i(\mathbf{x}_\Gamma, 0), \quad \mathbf{x}_\Gamma \in \Gamma. \tag{30}$$

On the other hand, we notice that, having $\Pi_{\parallel} \mathbf{E}_{\mathrm{f}}^0$ and $\Pi_{\parallel} \mathbf{H}_{\mathrm{f}}^0$ by (24), equations (25) and (26) are nothing but first order differential equations with respect to ν that permit to compute step by step $\Pi_{\parallel} \mathbf{E}_{\mathrm{f}}^k$ and $\Pi_{\parallel} \mathbf{H}_{\mathrm{f}}^k$ for each order k as functions of the boundary values of $\Pi_{\parallel} \mathbf{H}_v^k$ on Γ. So one can express the right hand side of (30) as a function of the boundary values of $\Pi_{\parallel} \mathbf{H}_v^i$, $0 \le i \le k-1$. Then the question is: can we express it as a function of $(\overline{\mathbf{H}}_v^k \times \mathbf{n})$? The response is no, in general. Meanwhile, we may establish formally the existence of a linear tangential operator $\mathcal{B}_k^\eta$ such that

$$\overline{\mathbf{E}}_v^k \times \mathbf{n} = \mathcal{B}_k^\eta(\Pi_{\parallel} \overline{\mathbf{H}}_v^k) + O(\eta^k) \quad \text{on} \quad \Gamma.$$

To obtain our effective boundary condition of order k, we chose to omit $O(\eta^k)$ in the previous identity. That is why the approximate solution, denoted by $(\widetilde{\mathbf{E}}_v^\eta, \widetilde{\mathbf{H}}_v^\eta)$, will be different from $(\overline{\mathbf{E}}_v^k, \overline{\mathbf{H}}_v^k)$. We shall discuss the validity of this choice via the stability study. We are going to give the details of this construction for $k = 2, 3, 4$.

Remark 2.2. As we assumed that the thin layer characteristics are independent of the the thickness η, the limit problem, when η goes to 0, is obtained by omitting the thin layer and applying the Dirichlet boundary condition (4) on Γ. This coincides with the effective boundary condition of order 1: combine the first equation of (24) with the system (20) written for $k = 0$.

2.2.2. Effective boundary condition of order 2 and 3

According to condition (30) we can derive the effective boundary conditions of order 2 by computing $\mathbf{E}_{\mathrm{f}}^0(\mathbf{x}_\Gamma, 0) \times \mathbf{n}$ and $\mathbf{E}_{\mathrm{f}}^1(\mathbf{x}_\Gamma, 0) \times \mathbf{n}$. $\mathbf{E}_{\mathrm{f}}^0(\mathbf{x}_\Gamma, 0) \times \mathbf{n} = 0$ by (24). From (25), written for $k = 1$, and from the second equation of (24), we deduce that

$$\partial_t \mathbf{E}_{\mathrm{f}}^1(\mathbf{x}_\Gamma, \nu) \times \mathbf{n} = -\frac{1}{\varepsilon_{\mathrm{f}}} (1 - \nu) \left(\varepsilon_{\mathrm{f}} \, \mu_{\mathrm{f}} \, \partial_{tt} + \vec{\mathrm{rot}}_\Gamma \, \mathrm{rot}_\Gamma \right) \Pi_{\parallel} \mathbf{H}_v^0(\mathbf{x}_\Gamma). \tag{31}$$

We have then:

$$\partial_t \overline{\mathbf{E}}_v^2(\mathbf{x}_\Gamma) \times \mathbf{n} = -\frac{\eta}{\varepsilon_{\mathrm{f}}} \left(\varepsilon_{\mathrm{f}} \, \mu_{\mathrm{f}} \, \partial_{tt} + \vec{\mathrm{rot}}_\Gamma \, \mathrm{rot}_\Gamma \right) \Pi_{\parallel} \overline{\mathbf{H}}_v^2(\mathbf{x}_\Gamma) \; + \; O(\eta^2),$$

(where $O(\eta^2) = \frac{\eta^2}{\varepsilon_{\mathrm{f}}} \left(\varepsilon_{\mathrm{f}} \, \mu_{\mathrm{f}} \, \partial_{tt} + \vec{\mathrm{rot}}_\Gamma \, \mathrm{rot}_\Gamma \right) \Pi_{\parallel} \mathbf{H}_v^1(\mathbf{x}_\Gamma)$).

If we denote by $(\widetilde{\mathbf{E}}_v^\eta, \widetilde{\mathbf{H}}_v^\eta)$ the approximate scattered field, the effective boundary condition of order 2 is:

$$\boxed{\partial_t \widetilde{\mathbf{E}}_v^\eta \times \mathbf{n} = -\frac{\eta}{\varepsilon_{\mathrm{f}}} \left(\varepsilon_{\mathrm{f}}\, \mu_{\mathrm{f}}\, \partial_{tt} + \vec{\mathrm{rot}}_\Gamma\, \mathrm{rot}_\Gamma\right) \Pi_{\|} \widetilde{\mathbf{H}}_v^\eta \quad \text{on } \Gamma \times \mathbb{R}^+.} \tag{32}$$

In this particular case of a straight boundary, condition (32) is in fact of order 3 (but will be false with curved boundary or non linear materials (see section or section 0.0.1)). Indeed, in order to compute $(\mathbf{E}_{\mathrm{f}}^2(\mathbf{x}_\Gamma, 0) \times \mathbf{n})$ using equation (25), we need to compute $(\Pi_{\|} \mathbf{H}_{\mathrm{f}}^1(\mathbf{x}_\Gamma, \nu))$. Since $\Pi_{\|} \mathbf{E}_{\mathrm{f}}^0 = 0$, equation (26) shows that (after time integration)

$$\Pi_{\|} \mathbf{H}_{\mathrm{f}}^1(\mathbf{x}_\Gamma, \nu) = \Pi_{\|} \mathbf{H}_v^1(\mathbf{x}_\Gamma), \quad (\mathbf{x}_\Gamma, \nu) \in \Gamma \times]0, 1[. \tag{33}$$

So we have from (25) written for $k = 2$,

$$\partial_t \mathbf{E}_{\mathrm{f}}^2(\mathbf{x}_\Gamma, \nu) \times \mathbf{n} = -\frac{1}{\varepsilon_{\mathrm{f}}} (1 - \nu) \left(\varepsilon_{\mathrm{f}}\, \mu_{\mathrm{f}}\, \partial_{tt} + \vec{\mathrm{rot}}_\Gamma\, \mathrm{rot}_\Gamma\right) \Pi_{\|} \mathbf{H}_v^1(\mathbf{x}_\Gamma). \tag{34}$$

We deduce from (24), (31) and (34) that

$$\partial_t \overline{\mathbf{E}}_v^3(\mathbf{x}_\Gamma) \times \mathbf{n} = -\frac{\eta}{\varepsilon_{\mathrm{f}}} \left(\varepsilon_{\mathrm{f}}\, \mu_{\mathrm{f}}\, \partial_{tt} + \vec{\mathrm{rot}}_\Gamma\, \mathrm{rot}_\Gamma\right) \Pi_{\|} \overline{\mathbf{H}}_v^3(\mathbf{x}_\Gamma) \; + \; O(\eta^3),$$

which gives, when omitting $O(\eta^3)$, the same expression as (32). The fundamental point is that the coupling between this condition and Maxwell's equation leads to a stable problem.

Theorem 0.1 – *Sufficiently regular solutions $(\widetilde{\mathbf{E}}_v^\eta, \widetilde{\mathbf{H}}_v^\eta)$ of {(6), (32)} satisfy the a priori energy estimate:*

$$\frac{d}{dt} \left\{ \mathcal{E}_v(\widetilde{\mathbf{E}}_v^\eta(t), \widetilde{\mathbf{H}}_v^\eta(t)) + \eta\, \mathcal{E}_\Gamma^\eta(t) \right\} = 0, \tag{35}$$

where, if we set, $\varphi^\eta(\mathbf{x}_\Gamma, t) = \Pi_{\|} \widetilde{\mathbf{H}}_v^\eta(\mathbf{x}_\Gamma, t), \; (\mathbf{x}_\Gamma, t) \in \Gamma \times \mathbb{R}^+$,

$$\mathcal{E}_\Gamma^\eta(t) = \frac{\mu_{\mathrm{f}}}{2} \, \|\varphi^\eta(t)\|_{L^2}^2 + \frac{1}{2\,\varepsilon_{\mathrm{f}}} \left\| \int_0^t \mathrm{rot}_\Gamma\, \varphi^\eta(\tau)\, d\tau \right\|_{L^2}^2 .$$

Identity (10) is then well satisfied.

Proof. In the classical energy identity for Maxwell's equation,

$$\frac{d}{dt}\, \mathcal{E}_v^\eta(t) = \int_\Gamma (\widetilde{\mathbf{E}}^\eta \times \mathbf{n}) \cdot \varphi^\eta \, d\mathbf{x}_\Gamma \tag{36}$$

we explicit the right hand side using (32). We integrate in time this condition, take the scalar product with φ^η and integrate over Γ. We use the duality between the operators rot_Γ and $\vec{\mathrm{rot}}_\Gamma$ to deduce that:

$$\int_\Gamma (\widetilde{\mathbf{E}}^\eta \times \mathbf{n}) \cdot \varphi^\eta \, d\mathbf{x}_\Gamma = -\eta \frac{d}{dt} \mathcal{E}_\Gamma^\eta(t). \tag{37}$$

Identity (35) is yielded by (36) and (37). ■

2.2.3. Effective boundary condition of order 4

We need to compute $(\mathbf{E}_\mathrm{f}^3(\mathbf{x}_\Gamma, 0) \times \mathbf{n})$ using equation (25), written for $k = 3$, which requires $(\Pi_\parallel \mathbf{H}_\mathrm{f}^2(\mathbf{x}_\Gamma, \nu))$. Equation (26), implies for $k = 2$

$$\partial_t \mathbf{H}_\mathrm{f}^2(\mathbf{x}_\Gamma, \nu) \times \mathbf{n} = \partial_t \mathbf{H}_v^2(\mathbf{x}_\Gamma) \times \mathbf{n} - \frac{1}{\mu_\mathrm{f}} \int_0^\nu \left(\varepsilon_\mathrm{f}\, \mu_\mathrm{f}\, \partial_{tt} + \vec{\mathrm{rot}}_\Gamma \, \mathrm{rot}_\Gamma\right) \Pi_\parallel \mathbf{E}_\mathrm{f}^1(\mathbf{x}_\Gamma, \xi)\, d\xi. \tag{38}$$

From (31) we have,

$$\partial_t \Pi_\parallel \mathbf{E}_\mathrm{f}^1(\mathbf{x}_\Gamma, \nu) = \frac{1}{\varepsilon_\mathrm{f}} (1 - \nu) \left(\varepsilon_\mathrm{f}\, \mu_\mathrm{f}\, \partial_{tt} - \nabla_\Gamma \, \mathrm{div}_\Gamma \right) (\mathbf{H}_v^0(\mathbf{x}_\Gamma) \times \mathbf{n}). \tag{39}$$

We apply the operator $\left(\varepsilon_\mathrm{f}\, \mu_\mathrm{f}\, \partial_{tt} + \vec{\mathrm{rot}}_\Gamma \, \mathrm{rot}_\Gamma \right)$ to (39) (using $\mathrm{rot}_\Gamma \nabla_\Gamma = 0$) and integrate in time. Hence

$$\begin{aligned} &\left(\varepsilon_\mathrm{f}\, \mu_\mathrm{f}\, \partial_{tt} + \vec{\mathrm{rot}}_\Gamma \, \mathrm{rot}_\Gamma \right) \Pi_\parallel \mathbf{E}_\mathrm{f}^1(\mathbf{x}_\Gamma, \nu) \\ &\quad = (1 - \nu) \mu_\mathrm{f}\, \partial_t \left(\varepsilon_\mathrm{f}\, \mu_\mathrm{f}\, \partial_{tt} - \nabla_\Gamma \, \mathrm{div}_\Gamma + \vec{\mathrm{rot}}_\Gamma \, \mathrm{rot}_\Gamma \right) (\mathbf{H}_v^0(\mathbf{x}_\Gamma) \times \mathbf{n}). \end{aligned} \tag{40}$$

Combining this identity with (38) yields

$$\mathbf{H}_\mathrm{f}^2(\mathbf{x}_\Gamma, \nu) \times \mathbf{n} = \mathbf{H}_v^2(\mathbf{x}_\Gamma) \times \mathbf{n} - \nu \left(1 - \tfrac{\nu}{2}\right) \left(\varepsilon_\mathrm{f}\, \mu_\mathrm{f}\, \partial_{tt} - \vec{\Delta}_\Gamma \right) (\mathbf{H}_v^0(\mathbf{x}_\Gamma) \times \mathbf{n}). \tag{41}$$

where $\vec{\Delta}_\Gamma = \nabla_\Gamma \, \mathrm{div}_\Gamma - \vec{\mathrm{rot}}_\Gamma \, \mathrm{rot}_\Gamma$, is the Laplace-Beltrami operator. From (41) and (25) we finally deduce

$$\begin{aligned} \partial_t \mathbf{E}_\mathrm{f}^3(\mathbf{x}_\Gamma, \nu) \times \mathbf{n} = &-\frac{1}{\varepsilon_\mathrm{f}} (1 - \nu) \left(\varepsilon_\mathrm{f}\, \mu_\mathrm{f}\, \partial_{tt} + \vec{\mathrm{rot}}_\Gamma \, \mathrm{rot}_\Gamma \right) \\ &\left(\Pi_\parallel \mathbf{H}_v^2(\mathbf{x}_\Gamma) - \tfrac{1}{3}\left(1 + \nu - \tfrac{\nu^2}{2}\right) \left(\varepsilon_\mathrm{f}\, \mu_\mathrm{f}\, \partial_{tt} - \vec{\Delta}_\Gamma \right) \Pi_\parallel \mathbf{H}_v^0(\mathbf{x}_\Gamma)\right). \end{aligned} \tag{42}$$

Combining (42) with (24), (31) and (34), we obtain the formal identity:

$$\begin{aligned} &\partial_t \overline{\mathbf{E}}_v^4(\mathbf{x}_\Gamma) \times \mathbf{n} \\ &= -\frac{\eta}{\varepsilon_\mathrm{f}} \left(\varepsilon_\mathrm{f}\, \mu_\mathrm{f}\, \partial_{tt} + \vec{\mathrm{rot}}_\Gamma \, \mathrm{rot}_\Gamma \right) \left(1 - \tfrac{\eta^2}{3} \left(\varepsilon_\mathrm{f}\, \mu_\mathrm{f}\, \partial_{tt} - \vec{\Delta}_\Gamma \right)\right) \Pi_\parallel \overline{\mathbf{H}}_v^4(\mathbf{x}_\Gamma) + O(\eta^4), \end{aligned} \tag{43}$$

which gives the following effective boundary condition of order 4 on $\Gamma \times \mathbb{R}^+$

$$\boxed{\partial_t \widetilde{\mathbf{E}}_v^\eta \times \mathbf{n} = -\frac{\eta}{\varepsilon_f} \left(\varepsilon_f\, \mu_f\, \partial_{tt} + \vec{\mathrm{rot}}_\Gamma\, \mathrm{rot}_\Gamma\,\right) \left(1 - \tfrac{\eta^2}{3} \left(\varepsilon_f\, \mu_f\, \partial_{tt} - \vec{\Delta}_\Gamma\,\right)\right) \Pi_\parallel \widetilde{\mathbf{H}}_v^\eta.} \tag{44}$$

An instability result. Contrary to the case of the second order condition, The condition (44) coupled with Maxwell's equations does not lead to a stable problem. In particular, one cannot obtain a priori estimate like in (10).

Let us study for that the case of the 2D problem. We denote by (x, y, z) the space coordinates system, and assume that all the fields do not depend on the z variable. For sake of simplicity, we make $\varepsilon_0 = \varepsilon_f = 1$ and $\mu_0 = \mu_f = 1$ and assume that $\Omega_v = y < 0$. If we consider $(\widetilde{\mathbf{E}}_v^\eta, \widetilde{\mathbf{H}}_v^\eta)$ satisfying Maxwell's system (6) coupled with the boundary conditions (44), the we observe that the z component of $\widetilde{\mathbf{E}}_v^\eta$ (resp. of $\widetilde{\mathbf{H}}_v^\eta$) is a solution of the initial boundary value problem $\mathcal{P}_d^\eta$ (resp. $\mathcal{P}_n^\eta$).

$$\mathcal{P}_d^\eta \quad \begin{cases} \partial_{tt} u - (\partial_{xx} + \partial_{yy}) u = 0, \quad t > 0,\ y < 0, \\ u + \eta(1 - \frac{\eta^2}{3}(\partial_{tt} - \partial_{xx}))\partial_n u = 0, \quad t > 0,\ y = 0, \\ (u, \partial_t u) = (u_0, u_1), \quad t = 0,\ y < 0. \end{cases} \tag{45}$$

$$\mathcal{P}_n^\eta \quad \begin{cases} \partial_{tt} u - (\partial_{xx} + \partial_{yy}) u = 0, \quad t > 0,\ y < 0, \\ \partial_n u + \eta(\partial_{tt} - \partial_{xx})(1 - \frac{\eta^2}{3}(\partial_{tt} - \partial_{xx})) u = 0, \quad t > 0,\ y = 0, \\ (u, \partial_t u) = (u_0, u_1), \quad t = 0,\ y < 0. \end{cases} \tag{46}$$

These problems correspond to the classical decomposition of the electromagnetic waves into two independent polarizations: T.E. (system $\mathcal{P}_n^\eta$) and T.M. (system $\mathcal{P}_d^\eta$).

Theorem 0.2 – *For fixed η, the problems $\mathcal{P}_d^\eta$ and $\mathcal{P}_n^\eta$ are well posed in the sense of Kreiss [5]. However both of them are strongly unstable in the following sense: there exists $(u_0^\eta(y), u_1^\eta(y))$ a sequence of initial data and $u^\eta(y, t)$, the corresponding solution, such that:*

$$\begin{cases} \|u_0^\eta\|_{H^m(-\infty,0)} + \|u_1^\eta\|_{H^m(-\infty,0)} \le C, \\ \textit{where } C \in \mathbb{R}^+ \textit{ and is independent of } \eta, \\ \lim_{\eta \to 0} \|u^\eta\|_{L^2(]-\infty,0[\times]0,T[)} = +\infty, \quad \forall T > 0. \end{cases}$$

Proof. We seek a solution of $\mathcal{P}_d^\eta$ or $\mathcal{P}_n^\eta$ of the form:

$$u(x, y, t) = Re(\,\hat{u}(y)\, e^{ikx} e^{st}\,),$$

where, $k \in \mathbf{R}$, $s = \alpha + i\omega$, $(\alpha, \omega) \in \mathbf{R}^2$, and $\hat{u} \in L^2(-\infty, 0)$. The wave equation implies that our particular solution is of the form

$$\hat{u}(y) = A\, e^{\left((s^2+k^2)^{\frac{1}{2}} y\right)}, \quad A \in \mathbf{R},$$

with $Re((s^2+k^2)^{\frac{1}{2}}) > 0$. Let us set $\kappa = \eta(s^2+k^2)^{\frac{1}{2}}$. One checks that the effective boundary condition of either $\mathcal{P}_d$ or $\mathcal{P}_n$ leads to the characteristic equation:

$$1 + \kappa\,(1 - \tfrac{1}{3}\kappa^2) = 0. \tag{47}$$

This equation admits one real solution $\kappa_0 > 0$. The two other ones have negative real part, so they are not admissible. Hence acceptable waves satisfy: $s^2 + k^2 = \frac{\kappa_0^2}{\eta^2}$, which is equivalent to

$$\left(\omega = 0 \text{ and } \alpha^2 + k^2 = \frac{\kappa_0^2}{\eta^2}\right) \text{ or } \left(\alpha = 0 \text{ and } k^2 - \omega^2 = \frac{\kappa_0^2}{\eta^2}\right). \tag{48}$$

Let η be fixed. Relation (48) shows that plane wave solutions of $\mathcal{P}_d^\eta$ or $\mathcal{P}_n^\eta$ are such that $Re(s) \leq \frac{\kappa_0}{\eta}$. This means that the boundary value problems $\mathcal{P}_d^\eta$ and $\mathcal{P}_n^\eta$ are well posed in the sense of Kreiss (see [5]).

However, one notices that when η goes to 0, the higher bound of the $Re(s)$ goes to $+\infty$. This means the existence of plane wave solutions that blow up for $t > 0$, when $\eta \to 0$. For instance, let us consider the worst case, corresponding to $k = 0$ (which is nothing but the 1D approximation) and $\alpha = \alpha_0^\eta = \frac{\kappa_0}{\eta}$. We fix $m \in \mathbf{N}$, and set $f^\eta(y) = e^{\alpha_0^\eta y}$, $y < 0$. The sequence of initial data

$$u_0^\eta(y) = \frac{f^\eta(y)}{\alpha_0^\eta\, \|f^\eta(y)\|_{H^m}}, \quad u_1^\eta(y) = \frac{f^\eta(y)}{\|f^\eta\|_{H^m}},$$

satisfy $\{\|u_0\|_{H^m} + \|u_0\|_{H^m}\} \leq 1 + \frac{\eta}{\kappa_0} < C$, when $\eta \to 0$. They correspond to the plane wave solutions:

$$u^\eta(y,t) = \frac{f^\eta(y+t)}{\alpha_0^\eta\, \|f^\eta\|_{H^m}} = u_0^\eta(y) f^\eta(t)$$

We have: $\|u^\eta\|_{L^2(]-\infty,0[\times]0,T[)} = \|u_0^\eta\|_{L^2(]-\infty,0[)}\, \|f^\eta\|_{L^2(0,T)}$.

When $\eta \to 0$: $\quad \|u_0^\eta\|_{L^2} \sim \left(\frac{\eta}{\kappa_0}\right)^{m+1}$ whereas

$$\|f^\eta\|_{L^2(0,T)} \sim \sqrt{\frac{\eta}{2\kappa_0}}\, \exp\left(\frac{\kappa_0\, T}{\eta}\right).$$

Therefore $\|u^\eta\|_{L^2(]-\infty,0[\times]0,T[)} \to +\infty$ when $\eta \to 0$. ■

permits to

Stable condition of order 4

We are going to built, from the unstable fourth order condition (44), another condition that gives the same order of approximation, but leads to stable problems. This is reminiscent of the techniques used for absorbing boundary conditions [3]. Let us consider first the simple case of the 1D approximation: i.e. we omit the tangential derivatives in the effective boundary conditions. We take the problem (45) as example. The boundary condition becomes when we apply Fourier transform with respect to t:

$$\hat{u} + \eta(1 + \tfrac{1}{3}\eta^2\omega^2)\partial_n \hat{u} = 0 \tag{49}$$

where ω is the dual variable of t. Comparing this condition with the stable condition of order 2: $\hat{u} + \eta\partial_n\hat{u} = 0$, we observe that the instability comes from the substitution of $(1 + \frac{1}{3}\eta^2\omega^2)$ for 1. A natural idea is to replace $(1 + \frac{1}{3}\eta^2\omega^2)$ by another function $g(\eta\omega)$ such that: $g(\eta\omega) = 1 + \frac{1}{3}\eta^2\omega^2 + O((\eta\omega)^3)$. As (49) is obtained in a formal way by neglecting the terms of order greater than 4 (with respect to η), we keep unchanged the order of approximation. We chose $g(\eta\omega) = 1/(1 - \frac{1}{3}\eta^2\omega^2)$. Hence, instead of using (49), we suggest to use the condition

$$(1 - \tfrac{1}{3}\eta^2\omega^2)\hat{u} + \eta\partial_n\hat{u} = 0,$$

that corresponds to

$$(1 + \tfrac{1}{3}\eta^2\partial_{tt})u + \eta\partial_n u = 0.$$

This new condition leads to stable initial boundary value problems. This result will be shown later in the general case via energy estimates, but we can already see that the previous construction of unstable plane wave solutions fails: the characteristic equation (47) becomes

$$1 + \tfrac{1}{3}\kappa^2 + \kappa = 0, \quad \kappa = \eta\, s,$$

and has no roots with positive real part (which can easily checked).

For the 3D case, we would like to apply the same ideas to (44). Unfortunately, working directly on the expression of (44) do not lead to the appropriate condition with regards to energy estimates. That is why we begin by writing differently the differential operator in the right hand side of (44). As, $\mathrm{rot}_\Gamma \nabla_\Gamma = 0$, we have

$$\begin{aligned}\left(\varepsilon_f\,\mu_f\,\partial_{tt} + \vec{\mathrm{rot}}_\Gamma\,\mathrm{rot}_\Gamma\right)\left(1 - \tfrac{\eta^2}{3}\left(\varepsilon_f\,\mu_f\,\partial_{tt} - \vec{\Delta}_\Gamma\right)\right) = \\ \varepsilon_f\,\mu_f\,\partial_{tt}\left(1 - \tfrac{\eta^2}{3}\left(\varepsilon_f\,\mu_f\,\partial_{tt} - \vec{\Delta}_\Gamma + \vec{\mathrm{rot}}_\Gamma\,\mathrm{rot}_\Gamma\right)\right) \\ + \vec{\mathrm{rot}}_\Gamma\,\mathrm{rot}_\Gamma\left(1 - \tfrac{\eta^2}{3}\vec{\mathrm{rot}}_\Gamma\,\mathrm{rot}_\Gamma\right)\end{aligned} \tag{50}$$

Let P be one of the two differential operators $\left(\varepsilon_f\,\mu_f\,\partial_{tt} - \vec{\Delta}_\Gamma + \vec{\mathrm{rot}}_\Gamma\,\mathrm{rot}_\Gamma\right)$ and $(\vec{\mathrm{rot}}_\Gamma\,\mathrm{rot}_\Gamma)$. Like previously, we apply to (50) the formal approximation

$$\left(1 - \tfrac{\eta^2}{3}P\right) = \left(1 + \tfrac{\eta^2}{3}P\right)^{-1} + O(\eta^4),$$

and obtain the condition (51), where we need to introduce two new variables ψ^η and ϕ^η on Γ, and where $\varphi^\eta(\mathbf{x}_\Gamma, t) = \Pi_\parallel \widetilde{\mathbf{H}}^\eta(\mathbf{x}_\Gamma, t)$, $(\mathbf{x}_\Gamma, t) \in \Gamma \times \mathbf{R}^+$.

$$\boxed{\begin{aligned} &\partial_t \widetilde{\mathbf{E}}^\eta_v \times \mathbf{n} = -\frac{\eta}{\varepsilon_f}\left(\varepsilon_f\,\mu_f\,\partial_{tt}\psi^\eta + \vec{\mathrm{rot}}_\Gamma\,\mathrm{rot}_\Gamma\,\phi^\eta\right), \\ &\left(1 + \tfrac{\eta^2}{3}\vec{\mathrm{rot}}_\Gamma\,\mathrm{rot}_\Gamma\right)\phi^\eta = \varphi^\eta, \\ &\left(1 + \tfrac{\eta^2}{3}\left(\varepsilon_f\,\mu_f\,\partial_{tt} - \vec{\Delta}_\Gamma + \vec{\mathrm{rot}}_\Gamma\,\mathrm{rot}_\Gamma\right)\right)\psi^\eta = \varphi^\eta, \end{aligned}} \quad \text{on } \Gamma \times \mathbf{R}^+. \tag{51}$$

To the third equation of (51) we associate the initial conditions

$$\psi^\eta(t = 0) = 0, \qquad \partial_t\psi^\eta(t = 0) = 0. \tag{52}$$

Such doing, we have constructed a boundary condition, that when coupled with Maxwell's equations, leads to a stable problem:

Theorem 0.3 – *Sufficiently regular solutions* $(\widetilde{\mathbf{E}}^\eta_v, \widetilde{\mathbf{H}}^\eta_v)$ *of* $\{(6), (51), (52)\}$ *satisfy the a priori energy estimate:*

$$\frac{d}{dt}\left\{\mathcal{E}_v(\widetilde{\mathbf{E}}^\eta_v(t), \widetilde{\mathbf{H}}^\eta_v(t)) + \eta\,\mathcal{E}^\eta_\Gamma(t)\right\} = 0, \ \ \textit{where,}$$

$$\begin{aligned} \mathcal{E}^\eta_\Gamma(t) = \tfrac{\mu_f}{2}\,\|\psi^\eta(t)\|^2_{L^2} + \tfrac{1}{2\,\varepsilon_f}\left\|\textstyle\int_0^t(\mathrm{rot}_\Gamma\,\phi ge)(\tau)\,d\tau\right\|^2_{L^2} + \\ \tfrac{\eta^2}{3}\left(\tfrac{\mu_f}{2}\,\mathcal{E}_1(\psi^\eta(t)) + \tfrac{1}{2\,\varepsilon_f}\,\mathcal{E}_2(\phi^\eta(t))\right), \end{aligned} \tag{53}$$

$$\begin{aligned} \textit{and} \quad \mathcal{E}_1(\psi^\eta(t)) &= \varepsilon_f\,\mu_f\,\|\partial_t\psi^\eta(t)\|^2_{L^2} + \|\mathrm{div}_\Gamma\,\psi^\eta(t)\|^2_{L^2} + 2\,\|\mathrm{rot}_\Gamma\,\psi^\eta(t)\|^2_{L^2}, \\ \mathcal{E}_2(\phi^\eta(t)) &= \left\|\int_0^t\left(\vec{\mathrm{rot}}_\Gamma\,\mathrm{rot}_\Gamma\,\phi^\eta\right)(\tau)\,d\tau\right\|^2_{L^2}. \end{aligned}$$

Proof. We first write the $L^2(\Gamma)$ scalar product of the first equation of (51) by φ^η

$$\begin{aligned} &\textstyle\int_\Gamma(\widetilde{\mathbf{E}}^\eta \times \mathbf{n})\cdot\varphi^\eta\,d\mathbf{x}_\Gamma \\ &= -\eta\left(\mu_f\textstyle\int_\Gamma\varphi^\eta\cdot\partial_t\psi^\eta d\mathbf{x}_\Gamma + \tfrac{1}{\varepsilon_f}\int_\Gamma\varphi^\eta\cdot\int_0^t\vec{\mathrm{rot}}_\Gamma\,\mathrm{rot}_\Gamma\,\phi^\eta\,d\tau\,d\mathbf{x}_\Gamma\right). \end{aligned} \tag{54}$$

When we take $L^2(\Gamma)$ scalar product of the second equation of (51) by $\displaystyle\int_0^t \left(\vec{\mathrm{rot}}_\Gamma\, \mathrm{rot}_\Gamma\, \boldsymbol{\phi}^\eta\right)(\tau)\, d\tau$ and integrate by part, we get

$$\begin{aligned} &\textstyle\int_\Gamma \boldsymbol{\varphi}^\eta \cdot \left(\int_0^t \vec{\mathrm{rot}}_\Gamma\, \mathrm{rot}_\Gamma\, \boldsymbol{\phi}^\eta\, d\tau\right) d\mathbf{x}_\Gamma \\ &= \tfrac{1}{2}\tfrac{d}{dt}\left(\left\|\int_0^t \mathrm{rot}_\Gamma\, \boldsymbol{\phi}^\eta\, d\tau\right\|_{L^2}^2 + \tfrac{\eta^2}{3}\left\|\int_0^t \vec{\mathrm{rot}}_\Gamma\, \mathrm{rot}_\Gamma\, \boldsymbol{\phi}^\eta\, d\tau\right\|_{L^2}^2\right), \end{aligned} \tag{55}$$

and when we take $L^2(\Gamma)$ scalar product of the third equation of (51) by $\partial_t \boldsymbol{\psi}^\eta$ and integrate by part, we get

$$\int_\Gamma \boldsymbol{\varphi}^\eta \cdot \partial_t \boldsymbol{\psi}^\eta\, d\mathbf{x}_\Gamma = \frac{1}{2}\frac{d}{dt}\left(\|\boldsymbol{\psi}^\eta(t)\|_{L^2}^2 + \frac{\eta^2}{3}\mathcal{E}_1(\boldsymbol{\psi}^\eta(t))\right). \tag{56}$$

The energy identity of theorem 0.3 is obtained by using (56) and (55) in (54) and substituting (54) in (36). ■

2.3. Error estimates

Let $k \in \{1, 2, 3, 4\}$. We recall that the effective condition of order k on the boundary Γ is the Dirichlet boundary condition when $k = 1$, the condition (32) when $k = 2, 3$ and the condition (51)-(52) when $k = 4$. We shall determine, using the stability results (theorems 0.1 and 0.3), the order of approximation between the exact solution $(\mathbf{E}_v^\eta, \mathbf{H}_v^\eta)$ and the approximate one $(\widetilde{\mathbf{E}}_v^\eta, \widetilde{\mathbf{H}}_v^\eta)$. Our main result is summarized by theorem 0.4 below.

In fact, rather than working directly on the difference $(\mathbf{E}_v^\eta - \widetilde{\mathbf{E}}_v^\eta, \mathbf{H}_v^\eta - \widetilde{\mathbf{H}}_v^\eta)$, we shall insert the asymptotic expansion (19).

So we consider $(\mathbf{E}_v^k, \mathbf{H}_v^k, \mathbf{E}_f^k, \mathbf{H}_f^k)_{k\geq 0}$ the sequence (of sufficiently regular fields) satisfying (20), (24), (25), (26) and (29).

Remark 2.3. At least, when $(\mathbf{E}_0, \mathbf{H}_0)$ is regular (let say in $\mathcal{D}(\Omega_v)$) the existence of the expansion $(\mathbf{E}_v^k, \mathbf{H}_v^k, \mathbf{E}_f^k, \mathbf{H}_f^k)_{k\geq 0}$ can be easily shown by a recurrence on k, as explained in the construction of the equations (20), (24), (25), (26) and (29). Moreover, one can check that $(\mathbf{E}_f^k, \mathbf{H}_f^k)$ are polynomial functions with respect to ν, of degree lesser than k.

Theorem 0.4 – *Let $k \in \{0, 1, 2, 3\}$. If $(\widetilde{\mathbf{E}}_v^\eta, \widetilde{\mathbf{H}}_v^\eta)$ is a sufficiently regular solution of (6) coupled with the effective condition of order $(k+1)$ and $(\mathbf{E}_v^\eta, \mathbf{H}_v^\eta)$ is a sufficiently regular solution of the exact transmission problem, then, there exists for all $0 < T < +\infty$ a constant $C_k(T)$ independent of η, but depending on $\{(\mathbf{E}_v^0, \mathbf{H}_v^0, \mathbf{E}_f^0, \mathbf{H}_f^0) \cdots (\mathbf{E}_v^k, \mathbf{H}_v^k, \mathbf{E}_f^k, \mathbf{H}_f^k)\}$ and T, such that:*

$$\sup_{0\leq t\leq T}\left\{\left\|\mathbf{E}_v^\eta(t) - \widetilde{\mathbf{E}}_v^\eta(t)\right\|_{L^2(\Omega_v)} + \left\|\mathbf{H}_v^\eta(t) - \widetilde{\mathbf{H}}_v^\eta(t)\right\|_{L^2(\Omega_v)}\right\} \leq C_k(T)\,\eta^{k+\frac{1}{2}}.$$

Proof. The proof of this theorem uses the results of the subsections below: Lemma 0.2 gives estimates on the difference between the exact solution and the truncated series, meanwhile lemmas 0.3, 0.4 and 0.5 give similar estimates but related to the approximate solutions. Admitting (for a while) these results, the present theorem is deduced by

- $k=0$: apply lemma 0.2 to $k=0$.
- $k=1,2,3$: combine the result of lemma 0.2 with the result of lemma 2.k using triangular inequality. ∎

Remark 2.4. Theorem 0.4 gives very little information about the constant $C_k(T)$. One can show for instance that if $\mathbf{E}_0$ and $\mathbf{H}_0$ are in $H_0^{3k+2}(\Omega_v)$ and satisfies $\operatorname{div}(\mathbf{E}_0) = \operatorname{div}(\mathbf{H}_0) = 0$, then we can chose

$$C_k(T) = C(T)\{\|\mathbf{E}_0\|_{H^{3k+2}} + \|\mathbf{H}_0\|_{H^{3k+2}}\},$$

where $C(T)$ is a polynomial function of T. This result (even if it may not be optimal) requires some sharp estimates that are very technical and do not add a substantial interest to the result.

2.3.1. Consistency of the asymptotic expansion (19)

Lemma 0.2 – *Let $0 < T < +\infty$. There exists a constant $C_k(T)$ independent of η such that, $\forall\ 0 < t < T$,*

$$\begin{aligned}&\tfrac{\varepsilon_0}{2}\left\|\mathbf{E}_v^\eta(t) - \textstyle\sum_{i=0}^k \eta^i\,\mathbf{E}_v^i(t)\right\|_{L^2}^2\\ &+\tfrac{\mu_0}{2}\left\|\mathbf{H}_v^\eta(t) - \textstyle\sum_{i=0}^k \eta^i\,\mathbf{H}_v^i(t)\right\|_{L^2}^2 \le C_k(T)\,\eta^{2k+1}.\end{aligned}$$

The constant $C_k(T)$ depends only on T and $(\mathbf{E}_f^k, \mathbf{H}_f^k)$, and can be expressed by (61).

Proof. Let us set:

$$\begin{aligned}\mathbf{e}_v^k &= \mathbf{E}_v^\eta - \textstyle\sum_{i=0}^k \eta^i\,\mathbf{E}_v^i, \quad \mathbf{h}_v^k = \mathbf{H}_v^\eta - \textstyle\sum_{i=0}^k \eta^i\,\mathbf{H}_v^i,\\ \mathbf{e}_f^k &= \mathbf{E}_f^\eta - \textstyle\sum_{i=0}^k \eta^i\,\mathbf{E}_f^i, \quad \mathbf{h}_f^k = \mathbf{H}_f^\eta - \textstyle\sum_{i=0}^k \eta^i\,\mathbf{H}_f^i.\end{aligned}$$

Of course each of $\mathbf{e}_v^k$, $\mathbf{h}_v^k$, $\mathbf{e}_f^k$ and $\mathbf{h}_f^k$ depends on η (which is not indicated explicitly to simplify the notations). Using (1) and (20), one observes that

$$\begin{cases}\varepsilon_0\,\partial_t\mathbf{e}_v^k - \operatorname{rot}\mathbf{h}_v^k = 0, \quad \mu_0\,\partial_t\mathbf{h}_v^k + \operatorname{rot}\mathbf{e}_v^k = 0, \text{ on } \Omega_v,\\ (\mathbf{e}_v^k, \mathbf{h}_v^k)(t=0) = (0,0).\end{cases} \tag{57}$$

while using (17), (21), (22) and (23) leads to

$$\left\{\begin{array}{l}\eta(\varepsilon_{\mathrm{f}}\partial_t \mathbf{e}_{\mathrm{f}}^k - \vec{\mathrm{rot}}_\Gamma\,(\mathbf{h}_{\mathrm{f}}^k\cdot\mathbf{n}) - (\mathrm{rot}_\Gamma\,\mathbf{h}_{\mathrm{f}}^k)\mathbf{n}) + \partial_\nu(\mathbf{h}_{\mathrm{f}}^k\times\mathbf{n}) = \eta^{k+1}F^k,\\ \eta(\mu_{\mathrm{f}}\partial_t \mathbf{h}_{\mathrm{f}}^k + \vec{\mathrm{rot}}_\Gamma\,(\mathbf{e}_{\mathrm{f}}^k\cdot\mathbf{n}) + (\mathrm{rot}_\Gamma\,\mathbf{e}_{\mathrm{f}}^k)\mathbf{n}) - \partial_\nu(\mathbf{e}_{\mathrm{f}}^k\times\mathbf{n}) = \eta^{k+1}G^k,\\ (\mathbf{e}_{\mathrm{f}}^k,\mathbf{h}_{\mathrm{f}}^k)(t=0) = (0,0),\end{array}\right. \tag{58}$$

where we have set

$$\begin{aligned}F^k &= \varepsilon_{\mathrm{f}}\,\partial_t \mathbf{E}_{\mathrm{f}}^k - \vec{\mathrm{rot}}_\Gamma\,(\mathbf{H}_{\mathrm{f}}^k\cdot\mathbf{n}) - (\mathrm{rot}_\Gamma\,\mathbf{H}_{\mathrm{f}}^k)\,\mathbf{n},\\ G^k &= \mu_{\mathrm{f}}\,\partial_t \mathbf{H}_{\mathrm{f}}^k + \vec{\mathrm{rot}}_\Gamma\,(\mathbf{E}_{\mathrm{f}}^k\cdot\mathbf{n}) + (\mathrm{rot}_\Gamma\,\mathbf{E}_{\mathrm{f}}^k)\,\mathbf{n}.\end{aligned}$$

The boundary conditions in (16), (18), (20), (21), (22) and (23) show that for $\mathbf{x}_\Gamma \in \Gamma$

$$\left\{\begin{array}{l}\mathbf{e}_v^k(\mathbf{x}_\Gamma)\times\mathbf{n} = \mathbf{e}_{\mathrm{f}}^k(\mathbf{x}_\Gamma,0)\times\mathbf{n},\ \mathbf{h}_v^k(\mathbf{x}_\Gamma)\times\mathbf{n} = \mathbf{h}_{\mathrm{f}}^k(\mathbf{x}_\Gamma,0)\times\mathbf{n},\\ \mathbf{e}_{\mathrm{f}}^k(\mathbf{x}_\Gamma,1)\times\mathbf{n} = 0, \quad \mathbf{x}_\Gamma\in\Gamma.\end{array}\right. \tag{59}$$

We remark that using (29), one has

$$F^k = \varepsilon_{\mathrm{f}}\,\partial_t \Pi_{\|}\mathbf{E}_{\mathrm{f}}^k - \vec{\mathrm{rot}}_\Gamma\,(\mathbf{H}_{\mathrm{f}}^k\cdot\mathbf{n}) \quad\text{and}\quad G^k = \mu_{\mathrm{f}}\,\partial_t \Pi_{\|}\mathbf{H}_{\mathrm{f}}^k + \vec{\mathrm{rot}}_\Gamma\,(\mathbf{E}_{\mathrm{f}}^k\cdot\mathbf{n}).$$

Let us choose $C_1(T)$ such that

$$\sup_{0<t<T}\ \sup_{0\le\nu\le1}\left(\sqrt{\tfrac{2}{\varepsilon_{\mathrm{f}}}}\,\left\|F^k(\nu,t)\right\|_{L^2(\Gamma)} + \sqrt{\tfrac{2}{\mu_{\mathrm{f}}}}\,\left\|G^k(\nu,t)\right\|_{L^2(\Gamma)}\right) = C_1(T),$$

and consider the classical energy identity of the system {(57), (58), (59)}

$$\begin{aligned}\mathcal{E}(t) = \tfrac{\varepsilon_0}{2}\left\|\mathbf{e}_v^k(t)\right\|_{L^2}^2 + \tfrac{\mu_0}{2}\left\|\mathbf{h}_v^k(t)\right\|_{L^2}^2 + \eta\left\{\tfrac{\varepsilon_{\mathrm{f}}}{2}\left\|\mathbf{e}_{\mathrm{f}}^k(t)\right\|_{L^2}^2 + \tfrac{\mu_{\mathrm{f}}}{2}\left\|\mathbf{h}_{\mathrm{f}}^k(t)\right\|_{L^2}^2\right\}\\ = \eta^{k+1}\int_0^t\left\{\int_\Gamma\int_0^1 F^k\cdot\mathbf{e}_{\mathrm{f}}^k + G^k\cdot\mathbf{h}_{\mathrm{f}}^k\,d\mathbf{x}_\Gamma\,d\nu\right\}\,d\tau.\end{aligned}$$

Applying Cauchy-Schwarz inequality to the right hand side of the previous identity yields

$$\mathcal{E}(t) \le \eta^{k+\frac12}C_1(T)\int_0^t \sqrt{\mathcal{E}(\tau)}\,d\tau. \tag{60}$$

By Gronwall lemma we conclude that $\mathcal{E}(t) \le (\frac12\eta^{k+\frac12}C_1(T)\,t)^2$, which proves the lemma with

$$C_k(T) = (\tfrac12 C_1(T)\,T)^2. \tag{61}$$

■

2.3.2. Consistency of the second and third order conditions

Lemma 0.3 – *Let $(\widetilde{\mathbf{E}}_v^\eta, \widetilde{\mathbf{H}}_v^\eta)$ be a sufficiently regular solution of (6) coupled with the effective condition of order 2. Then, there exists for all $0 < T < +\infty$, a constant $C(T)$ independent of η such that, $\forall\, 0 \le t \le T$,*

$$\tfrac{\varepsilon_0}{2} \left\| \widetilde{\mathbf{E}}_v^\eta(t) - \textstyle\sum_{i=0}^1 \eta^i\, \mathbf{E}_v^i(t) \right\|_{L^2}^2 + \tfrac{\mu_0}{2} \left\| \mathbf{H}_v^\eta(t) - \textstyle\sum_{i=0}^1 \eta^i\, \mathbf{H}_v^i(t) \right\|_{L^2}^2 \le C(T)\, \eta^3.$$

The constant $C(T)$ depends only on T and $\mathbf{H}_v^1$ and can be chosen like in (65).

Proof. We introduce the fields $(\widetilde{\mathbf{e}}_v^1, \widetilde{\mathbf{h}}_v^1)$ (that depend on η)

$$\widetilde{\mathbf{e}}_v^1 = \widetilde{\mathbf{E}}_v^\eta - (\mathbf{E}_v^0 + \eta\, \mathbf{E}_v^1), \quad \widetilde{\mathbf{h}}_v^1 = \widetilde{\mathbf{H}}_v^\eta - (\mathbf{H}_v^0 + \eta\, \mathbf{H}_v^1).$$

and set:
$\widetilde{\varphi}^1(\mathbf{x}_\Gamma, t) = \Pi_\| \widetilde{\mathbf{h}}_v^1(\mathbf{x}_\Gamma, t)$, $\varphi^1(\mathbf{x}_\Gamma, t) = \Pi_\| \mathbf{H}_v^1(\mathbf{x}_\Gamma, t)$ $(\mathbf{x}_\Gamma, t) \in \Gamma \times [0, \infty[$.
Combining (32) with (24) and (31), shows that $(\widetilde{\mathbf{e}}_v^1, \widetilde{\mathbf{h}}_v^1)$ satisfies

$$\left\{ \begin{array}{l} \varepsilon_0\, \partial_t \widetilde{\mathbf{e}}_v^1 - \operatorname{rot} \widetilde{\mathbf{h}}_v^1 = 0, \qquad \mu_0\, \partial_t \widetilde{\mathbf{h}}_v^1 + \operatorname{rot} \widetilde{\mathbf{e}}_v^1 = 0, \quad \text{on } \Omega_v, \\ (\widetilde{\mathbf{e}}_v^1, \widetilde{\mathbf{h}}_v^1)(t=0) = (0,0), \quad \text{on } \Omega_v, \\ \partial_t \widetilde{\mathbf{e}}_v^1 \times \mathbf{n} = -\dfrac{\eta}{\varepsilon_{\mathrm{f}}} \left(\varepsilon_{\mathrm{f}}\, \mu_{\mathrm{f}}\, \partial_{tt} + \vec{\operatorname{rot}}_\Gamma \operatorname{rot}_\Gamma \right) (\widetilde{\varphi}^1 - \eta \varphi^1), \quad \text{on } \Gamma. \end{array} \right. \tag{62}$$

Following the same steps as for the proof of identity (35) of theorem 0.1, leads to

$$\mathcal{E}(t) \equiv$$
$$\tfrac{\varepsilon_0}{2} \left\| \widetilde{\mathbf{e}}_v^1(t) \right\|_{L^2}^2 + \tfrac{\mu_0}{2} \left\| \widetilde{\mathbf{h}}_v^1(t) \right\|_{L^2}^2 + \eta \left\{ \tfrac{\mu_{\mathrm{f}}}{2} \left\| \widetilde{\varphi}^1(t) \right\|_{L^2}^2 + \tfrac{1}{2\,\varepsilon_{\mathrm{f}}} \left\| \textstyle\int_0^t \operatorname{rot}_\Gamma \widetilde{\varphi}^1\, d\tau \right\|_{L^2}^2 \right\}$$
$$= \eta^2 \int_0^t \int_\Gamma \mu_{\mathrm{f}}\, \partial_t \varphi^1 \cdot \widetilde{\varphi}^1 + \tfrac{1}{\varepsilon_{\mathrm{f}}} \left(\textstyle\int_0^\tau \operatorname{rot}_\Gamma \varphi^1\, d\theta \right) \cdot \operatorname{rot}_\Gamma \widetilde{\varphi}^1\, d\mathbf{x}_\Gamma\, d\tau.$$

Integrating by part in time the second term of the right hand side yields

$$\begin{aligned} \mathcal{E}(t) \;=\; & \eta^2 \int_\Gamma \tfrac{1}{\varepsilon_{\mathrm{f}}} \left(\textstyle\int_0^t \operatorname{rot}_\Gamma \varphi^1\, d\tau \right) \left(\textstyle\int_0^t \operatorname{rot}_\Gamma \widetilde{\varphi}^1\, d\tau \right) d\mathbf{x}_\Gamma \\ & + \eta^2 \int_0^t \int_\Gamma \mu_{\mathrm{f}}\, \partial_t \varphi^1 \cdot \widetilde{\varphi}^1 - \tfrac{1}{\varepsilon_{\mathrm{f}}} \cdot \operatorname{rot}_\Gamma \varphi^1 \left(\textstyle\int_0^\tau \operatorname{rot}_\Gamma \widetilde{\varphi}^1\, d\theta \right) d\mathbf{x}_\Gamma\, d\tau. \end{aligned} \tag{63}$$

Let us chose now

$$C_1(T) = \sup_{t \le T} \left(\sqrt{\tfrac{2}{\varepsilon_{\mathrm{f}}}} \left\| \textstyle\int_0^t \operatorname{rot}_\Gamma \varphi^1(\tau)\, d\tau \right\|_{L^2(\Gamma)} \right),$$
$$C_2(T) = \sup_{t \le T} \left(\sqrt{\tfrac{2}{\varepsilon_{\mathrm{f}}}} \left\| \operatorname{rot}_\Gamma \varphi^1(t) \right\|_{L^2(\Gamma)} + \sqrt{2\mu_{\mathrm{f}}} \left\| \partial_t \varphi^1(t) \right\|_{L^2(\Gamma)} \right)$$

Applying Cauchy-Schwarz inequality to (63) shows that

$$\mathcal{E}(t) \leq \eta^{\frac{3}{2}} C_1(T) \sqrt{\mathcal{E}(t)} + \eta^{\frac{3}{2}} C_2(T) \int_0^t \sqrt{\mathcal{E}(\tau)}\, d\tau. \tag{64}$$

Consider now, for $t \geq 0$, the function, $X(t) = |\sqrt{\mathcal{E}(t)} - \frac{1}{2}\eta^{\frac{3}{2}} C_1(T)|^2$, we obtain from (64), that

$$X(t) \leq \eta^3 C_3(T) + \eta^{\frac{3}{2}} C_2(T) \int_0^t \sqrt{(X(\tau))}\, d\tau,$$

where $C_3(T) = \frac{1}{4} C_1(T)^2 + T\, C_1(T) C_2(T)$. Applying Gronwall's lemma to the previous inequality yields

$$X(t) \leq \{\eta^{\frac{3}{2}} (\sqrt{C_3(T)} + \frac{1}{2} C_2(T)\, t)\}^2,$$

which proves the lemma with

$$C(T) = \{\sqrt{C_3(T)} + \frac{1}{2}(T\, C_2(T) + C_1(T))\}^2 \tag{65}$$

■

Lemma 0.4 – *Let $(\widetilde{\mathbf{E}}_v^\eta, \widetilde{\mathbf{H}}_v^\eta)$ be a sufficiently regular solution of (6) coupled with the effective condition of order 3. Then, there exists for all $0 < T < +\infty$, a constant $C(T)$ independent of η such that, $\forall\, 0 \leq t \leq T$,*

$$\frac{\varepsilon_0}{2} \left\| \widetilde{\mathbf{E}}_v^\eta(t) - \textstyle\sum_{i=0}^2 \eta^i\, \mathbf{E}_v^i(t) \right\|_{L^2}^2 + \frac{\mu_0}{2} \left\| \mathbf{H}_v^\eta(t) - \textstyle\sum_{i=0}^2 \eta^i\, \mathbf{H}_v^i(t) \right\|_{L^2}^2 \leq C(T)\, \eta^5.$$

The constant $C(T)$ depends only on T and $\mathbf{H}_v^2$.

Proof. Like in the previous proof, we introduce

$$\widetilde{\mathbf{e}}_v^2 = \widetilde{\mathbf{E}}_v^\eta - (\mathbf{E}_v^0 + \eta\, \mathbf{E}_v^1 + \eta^2\, \mathbf{E}_v^2), \quad \widetilde{\mathbf{h}}_v^2 = \widetilde{\mathbf{H}}_v^\eta - (\mathbf{H}_v^0 + \eta\, \mathbf{H}_v^1 + \eta^2\, \mathbf{H}_v^2).$$

and set: $\widetilde{\varphi}^2(\mathbf{x}_\Gamma, t) = \Pi_\| \widetilde{\mathbf{h}}_v^2(\mathbf{x}_\Gamma, t), \quad \varphi^2(\mathbf{x}_\Gamma, t) = \Pi_\| \mathbf{H}_v^2(\mathbf{x}_\Gamma, t) \quad (\mathbf{x}_\Gamma, t) \in \Gamma \times [0, \infty[$.
Using the relation (34) we see that $(\widetilde{\mathbf{e}}_v^2, \widetilde{\mathbf{h}}_v^2)$ satisfies the following equations, that are similar to (62).

$$\left\{ \begin{array}{l} \varepsilon_0\, \partial_t \widetilde{\mathbf{e}}_v^2 - \mathrm{rot}\, \widetilde{\mathbf{h}}_v^2 = 0, \qquad \mu_0\, \partial_t \widetilde{\mathbf{h}}_v^2 + \mathrm{rot}\, \widetilde{\mathbf{e}}_v^2 = 0, \text{ on } \Omega_v, \\ (\widetilde{\mathbf{e}}_v^2, \widetilde{\mathbf{h}}_v^2)(t=0) = (0,0), \text{ on } \Omega_v, \\ \partial_t \widetilde{\mathbf{e}}_v^2 \times \mathbf{n} = -\dfrac{\eta}{\varepsilon_{\mathrm{f}}} \left(\varepsilon_{\mathrm{f}}\, \mu_{\mathrm{f}}\, \partial_{tt} + \vec{\mathrm{rot}}_\Gamma\, \mathrm{rot}_\Gamma \right) (\widetilde{\varphi}^2 - \eta^2 \varphi^2), \text{ on } \Gamma, \end{array} \right.$$

The proof follows exactly the same steps as the proof of the previous lemma. One has only to substitute η^3 (resp. $\widetilde{\varphi}^2$, φ^2) for η^2 (resp. $\widetilde{\varphi}^1$, φ^1). ■

2.3.3. Consistency of the fourth order condition

Lemma 0.5 – *Let* $(\widetilde{\mathbf{E}}_v^\eta, \widetilde{\mathbf{H}}_v^\eta)$ *be a sufficiently regular solution of (6) coupled with the effective condition of order 4. Then, there exists for all* $0 < T < +\infty$, *a constant* $C(T)$ *independent of* η *such that,* $\forall\, 0 \le t \le T$,

$$\frac{\varepsilon_0}{2} \left\| \widetilde{\mathbf{E}}_v^\eta(t) - \textstyle\sum_{i=0}^{3} \eta^i\, \mathbf{E}_v^i(t) \right\|_{L^2}^2 + \frac{\mu_0}{2} \left\| \mathbf{H}_v^\eta(t) - \textstyle\sum_{i=0}^{3} \eta^i\, \mathbf{H}_v^i(t) \right\|_{L^2}^2 \le C(T)\, \eta^7 .$$

The constant $C(T)$ *depends only on* T *and* $(\mathbf{H}_v^1, \mathbf{H}_v^2, \mathbf{H}_v^3)$ *and can be chosen like in (72).*

Proof. Like previously, we denote

$$\widetilde{\mathbf{e}}_v^3 = \widetilde{\mathbf{E}}_v^\eta - (\mathbf{E}_v^0 + \eta\, \mathbf{E}_v^1 + \eta^2\, \mathbf{E}_v^2 + \eta^3\, \mathbf{E}_v^3),$$

$$\widetilde{\mathbf{h}}_v^3 = \widetilde{\mathbf{H}}_v^\eta - (\mathbf{H}_v^0 + \eta\, \mathbf{H}_v^1 + \eta^2\, \mathbf{H}_v^2 + \eta^3\, \mathbf{H}_v^3).$$

Let us denote also by φ^k the trace of $\Pi_\| \mathbf{H}_v^k$ on Γ, for $k = 0, \cdots, 3$. In order to write the equations satisfied by $(\widetilde{\mathbf{e}}_v^3, \widetilde{\mathbf{h}}_v^3)$ we need to introduce

$$\psi^0 = \varphi^0,\ \ \psi^1 = \varphi^1,\ \ \psi^2 = \varphi^2 - \tfrac{1}{3} \left(\varepsilon_{\mathrm{f}}\, \mu_{\mathrm{f}}\, \partial_{tt} - \vec{\Delta}_\Gamma + \vec{\mathrm{rot}}_\Gamma\, \mathrm{rot}_\Gamma \right) \varphi^0,$$
$$\phi^0 = \varphi^0,\ \ \phi^1 = \varphi^1,\ \ \phi^2 = \varphi^2 - \tfrac{1}{3}\vec{\mathrm{rot}}_\Gamma\, \mathrm{rot}_\Gamma\, \varphi^0,$$

and to set, $\quad \widetilde{\psi}^3 = \psi^\eta - (\psi^0 + \eta\psi^1 + \eta^2\psi^2), \quad \widetilde{\phi}^3 = \phi^\eta - (\phi^0 + \eta\phi^1 + \eta^2\phi^2).$

One notices that ψ^i (resp. ϕ^i), $i = 0, 1, 2$, are independent of η and are nothing but the first three terms of the asymptotic expansion of ψ^η (resp. ϕ^η) introduced by the condition (51). However, $\widetilde{\psi}^3$ and $\widetilde{\phi}^3$ depend on η. We verify, using (42) and the first equation of (51), that

$$\partial_t \widetilde{\mathbf{e}}_v^3 \times \mathbf{n} = -\frac{\eta}{\varepsilon_{\mathrm{f}}} \left(\varepsilon_{\mathrm{f}}\, \mu_{\mathrm{f}}\, \partial_{tt} \widetilde{\psi}^3 + \vec{\mathrm{rot}}_\Gamma\, \mathrm{rot}_\Gamma\, \widetilde{\phi}^3 \right) \quad \text{on } \Gamma \times \mathbf{R}^+. \tag{66}$$

Let $\widetilde{\varphi}^3$ be the trace of $\Pi_\| \widetilde{\mathbf{h}}_v^3$ on Γ. We deduce, from the second equation of (51) and the definition of $\widetilde{\phi}^3$

$$\left(1 + \tfrac{\eta^2}{3} \vec{\mathrm{rot}}_\Gamma\, \mathrm{rot}_\Gamma \right) \widetilde{\phi}^3 = \widetilde{\varphi}^3 + \eta^3\, G_3^\eta \quad \text{on } \Gamma \times \mathbf{R}^+, \tag{67}$$

and from the third equation of (51) and the definition of $\widetilde{\psi}^3$

$$\left(1 + \tfrac{\eta^2}{3} \left(\varepsilon_{\mathrm{f}}\, \mu_{\mathrm{f}}\, \partial_{tt} - \vec{\Delta}_\Gamma + \vec{\mathrm{rot}}_\Gamma\, \mathrm{rot}_\Gamma \right) \right) \widetilde{\psi}^3 = \widetilde{\varphi}^3 + \eta^3\, L_3^\eta \quad \text{on } \Gamma \times \mathbf{R}^+. \tag{68}$$

In (67): $\quad G_3^\eta = \frac{1}{3}\vec{\mathrm{rot}}_\Gamma\, \mathrm{rot}_\Gamma\, (\varphi^1 + \eta\,\varphi^2) - \varphi^3,$

and in (68): $\quad L_3^\eta = \frac{1}{3}\left(\varepsilon_f\,\mu_f\,\partial_{tt} - \vec{\Delta}_\Gamma + \vec{\mathrm{rot}}_\Gamma\,\mathrm{rot}_\Gamma\right)(\varphi^1 + \eta\,\varphi^2) - \varphi^3.$

In conclusion, $(\widetilde{\mathbf{e}}_v^3, \widetilde{\mathbf{h}}_v^3)$ satisfies

$$\left\{\begin{array}{l} \varepsilon_0\,\partial_t\widetilde{\mathbf{e}}_v^3 - \mathrm{rot}\,\widetilde{\mathbf{h}}_v^3 = 0, \qquad \mu_0\,\partial_t\widetilde{\mathbf{h}}_v^3 + \mathrm{rot}\,\widetilde{\mathbf{e}}_v^3 = 0, \ \text{ on } \Omega_v \times \mathbf{R}^+, \\ (\widetilde{\mathbf{e}}_v^3, \widetilde{\mathbf{h}}_v^3)(t=0) = (0,0), \ \text{ on } \Omega_v, \\ \widetilde{\mathbf{e}}_v^3 \times \mathbf{n} \text{ and } \widetilde{\boldsymbol{\varphi}}^3 \text{ verify (66), (67) and (68) on } \Gamma \times \mathbf{R}^+. \end{array}\right. \tag{69}$$

The derivation of an energy estimate for (69) follows the same steps as theorem 0.3. If we set

$$\begin{aligned} &\mathcal{E}(t) \\ &= \tfrac{\varepsilon_0}{2}\left\|\widetilde{\mathbf{e}}_v^3(t)\right\|_{L^2}^2 + \tfrac{\mu_0}{2}\left\|\widetilde{\mathbf{h}}_v^3(t)\right\|_{L^2}^2 + \eta\left\{\tfrac{\mu_f}{2}\left\|\widetilde{\boldsymbol{\psi}}^3(t)\right\|_{L^2}^2 + \tfrac{1}{2\,\varepsilon_f}\left\|\textstyle\int_0^t \mathrm{rot}_\Gamma\,\widetilde{\boldsymbol{\phi}}^3\,d\tau\right\|_{L^2}^2\right\} \\ &\quad + \tfrac{\eta^3\mu_f}{6}\left\{\varepsilon_f\,\mu_f\left\|\partial_t\widetilde{\boldsymbol{\psi}}^3(t)\right\|_{L^2}^2 + \left\|\mathrm{div}_\Gamma\,\widetilde{\boldsymbol{\psi}}^3(t)\right\|_{L^2}^2 + 2\left\|\mathrm{rot}_\Gamma\,\widetilde{\boldsymbol{\psi}}^3(t)\right\|_{L^2}^2\right\} \\ &\qquad\qquad + \tfrac{\eta^3}{6\varepsilon_f}\left\|\textstyle\int_0^t \vec{\mathrm{rot}}_\Gamma\,\mathrm{rot}_\Gamma\,\widetilde{\boldsymbol{\phi}}^3\,d\tau\right\|_{L^2}^2, \end{aligned}$$

then we get $\mathcal{E}(t) = \eta^4 \displaystyle\int_0^t\int_\Gamma \{\mathrm{rot}_\Gamma\, G_3^\eta(\tau)\cdot\textstyle\int_0^\tau \mathrm{rot}_\Gamma\,\widetilde{\boldsymbol{\phi}}^3\,d\kappa + L_3^\eta(\tau)\cdot\partial_t\widetilde{\boldsymbol{\psi}}^3(\tau)\}\,d\mathbf{x}_\Gamma\,d\tau.$
Integrating by parts, with respect to τ, we obtain

$$\begin{aligned} &\mathcal{E}(t) = \eta^4\int_\Gamma L_3^\eta(t)\cdot\widetilde{\boldsymbol{\psi}}^3(t)\ d\mathbf{x}_\Gamma + \\ &\eta^4\ \textstyle\int_0^t\int_\Gamma \{\mathrm{rot}_\Gamma\, G_3^\eta(\tau)\cdot\int_0^\tau \mathrm{rot}_\Gamma\,\widetilde{\boldsymbol{\phi}}^3\,d\kappa - \partial_t L_3^\eta(\tau)\cdot\widetilde{\boldsymbol{\psi}}^3(\tau)\}\,d\mathbf{x}_\Gamma\,d\tau. \end{aligned} \tag{70}$$

If we set

$$\begin{aligned} C_1(T) &= \sup_{\eta\le 1}\ \sup_{t\le T}\left(\sqrt{\tfrac{2}{\mu_f}}\ \|\partial_t L_3^\eta(t)\|_{L^2(\Gamma)}\right), \\ C_2(T) &= \sup_{\eta\le 1}\ \sup_{t\le T}\left(\sqrt{2\varepsilon_f}\ \|\mathrm{rot}_\Gamma\, G_3^\eta(t)\|_{L^2(\Gamma)} + \sqrt{\tfrac{2}{\mu_f}}\ \|\partial_t L_3^\eta(t)\|_{L^2(\Gamma)}\right), \end{aligned}$$

then, applying Cauchy-Schwarz inequality to the right member of (70) shows that, $\forall t > 0$,

$$\mathcal{E}(t) \le \eta^{\frac{7}{2}} C_1(T)\sqrt{\mathcal{E}(t)} + \eta^{\frac{7}{2}} C_2(T)\int_0^t \sqrt{\mathcal{E}(\tau)}\,d\tau. \tag{71}$$

We notice that this identity is similar to (64). The same technique based on Gronwall's lemma enables us to prove the present lemma with

$$\left\{\begin{array}{l} C(T) = \{\sqrt{C_3(T)} + \frac{1}{2}(T\,C_2(T) + C_1(T))\}^2; \\ C_3(T) = \frac{1}{4}C_1(T)^2 + T\,C_1(T)C_2(T). \end{array}\right. \tag{72}$$

■

2.4. Effective construction: regular boundary

The principle of deriving effective boundary conditions in the case of curved boundary is the same as in the case of straight boundary. However, it is technically much more complicated because one needs to take into account the geometrical properties of the boundary.

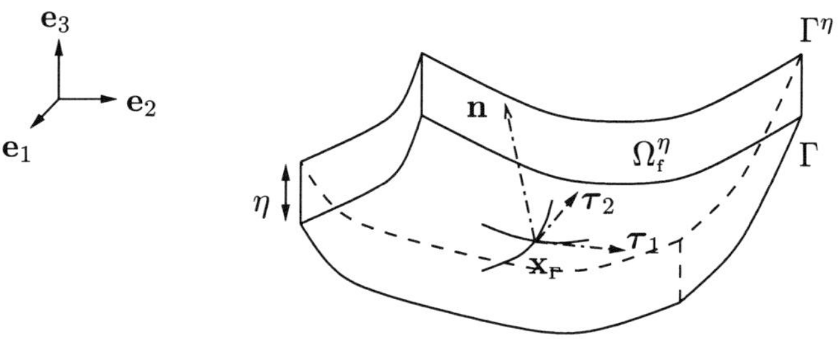

Figure 3: Local description of the thin layer

We assume that Γ is a regular surface (at least of class C^2), and we denote also by $\mathbf{n}$ the unitary normal vector field to Γ outwardly directed to Ω_v.

2.4.1. Some geometrical properties of regular surfaces

let us first introduce some tangential differential operators on on Γ, that we shall use to express the differential operator rot in the parametric coordinates. After giving some recalls of differential geometry (the definitions below are close to that used in [14], see also [15]), we shall prove some specific results in connection with our formal construction of the effective boundary conditions (we have tried to make this section self-contained).

For η small enough (see Remark 2.5. below), the parametric representation (11) of Ω_f^η remains valid. The only difference is that $\mathbf{n}$ depends now on $\mathbf{x}_\Gamma$:

$$\mathbf{x} \in \bar{\Omega}_f^\eta \longmapsto (\mathbf{x}_\Gamma, s) \in \Gamma \times [0, \eta], \text{ such that: } \mathbf{x} = \mathbf{x}_\Gamma + s\,\mathbf{n}(\mathbf{x}_\Gamma), \tag{73}$$

where $\mathbf{x}_\Gamma$ is defined by $|\mathbf{x}_\Gamma - \mathbf{x}| = \min\limits_{\mathbf{y}\in\Gamma} |\mathbf{y} - \mathbf{x}|$. For each $\mathbf{x}_\Gamma \in \Gamma$ we denote by $T_{\mathbf{x}_\Gamma}$ the tangent plane to Γ on $\mathbf{x}_\Gamma$, which is the plane orthogonal to $\mathbf{n}(\mathbf{x}_\Gamma)$. The projection operators $\Pi_\perp$ and $\Pi_\parallel$ are still defined by (14) but they depend $\mathbf{x}_\Gamma$.

Let $\boldsymbol{\xi} = (\xi_1, \xi_2) \longmapsto \mathbf{x}_\Gamma \in \Gamma$ be a parameterization of a neighborhood of $\mathbf{x}_\Gamma$; for $\boldsymbol{\xi} \in \mathcal{O} \subset \mathbb{R}^2$. We introduce the covariant base $(\boldsymbol{\tau}_\alpha)_{\alpha=1,2}$ of $T_{\mathbf{x}_\Gamma}$: $\boldsymbol{\tau}_\alpha = \partial_{\xi_\alpha} \mathbf{x}_\Gamma$. The contravariant (or dual) base $(\boldsymbol{\tau}^\alpha)_{\alpha=1,2}$ is defined by: $\boldsymbol{\tau}^\alpha \cdot \boldsymbol{\tau}_\beta = \delta^\alpha_\beta$ for $\beta = 1, 2$, where δ^α_β is the Kronecker symbol, and $\boldsymbol{\tau}^\alpha \cdot \mathbf{n} = 0$. We shall denote by $(\mathbf{e}_i)_{i=1,2,3}$ $(=(\mathbf{e}^i)_{i=1,2,3}$, by convention) the canonical orthonormal basis of $\mathbb{R}^3$. We use in the sequel the convention of summation for repeated subscripts, where Greek subscripts vary between 1 and 2 and Roman subscripts vary between 1 and 3.

The curvature tensor $\mathcal{C}$: We introduce $\breve{\mathbf{n}}$ the vector field defined on $\bar{\Omega}^\eta_{\mathrm{f}}$ by: $\breve{\mathbf{n}}(\mathbf{x}) = \mathbf{n}(\mathbf{x}_\Gamma)$ where $\mathbf{x} \in \bar{\Omega}^\eta_{\mathrm{f}}$ and $\mathbf{x}_\Gamma$ are linked by (73). We define the curvature tensors $\mathcal{C}_s$, $s \in [0, \eta]$, and $\mathcal{C}$ by

$$\mathcal{C}_s(\mathbf{x}_\Gamma) = (\vec{\nabla}\, \breve{\mathbf{n}})(\mathbf{x}_\Gamma, s), \qquad \mathcal{C} = \mathcal{C}_0. \tag{74}$$

For every $\mathbf{x}_\Gamma \in \Gamma$, the operator $\mathcal{C}(\mathbf{x}_\Gamma)$ is symmetric, and has 0 as an eigenvalue associated to $\mathbf{n}$ and two other eigenvalues (c_1, c_2), called principal curvatures, that are associated to tangential eigenvectors $(\tilde{\boldsymbol{\tau}}_1, \tilde{\boldsymbol{\tau}}_2)$. This operator satisfies also the following useful property, where $H = \frac{1}{2}\, tr\, \mathcal{C}$, is the mean curvature,

$$\mathcal{C} \cdot (\mathbf{v} \times \mathbf{n}) + (\mathcal{C} \cdot \mathbf{v}) \times \mathbf{n} = 2\, H\, \mathbf{v} \times \mathbf{n}, \;\; \forall\, \mathbf{v} \in \mathbb{R}^3, \tag{75}$$

that can be checked out, by using the orthonormal basis $(\tilde{\boldsymbol{\tau}}_1, \tilde{\boldsymbol{\tau}}_2, \mathbf{n})$ of $\mathbb{R}^3$, where $\mathbf{e}_{c_i}$ is a unitary eigenvector of $\mathcal{C}$ associated to c_i $(i = 1, 2)$. Moreover, using the parameterization (ξ_1, ξ_2) and differentiating $\mathbf{n}$ with respect to (ξ_1, ξ_2) we get:

$$\partial_{\xi_\alpha} \mathbf{n} = \mathcal{C}\, \boldsymbol{\tau}_\alpha.$$

Remark 2.5. Assuming Γ regular enough implies in particular the existence of a lower bound, $c > 0$ of $\mathbf{x}_\Gamma \mapsto \min(|c_1(\mathbf{x}_\Gamma)|\,, |c_2(\mathbf{x}_\Gamma)|)$, on Γ. The mapping (73) is then an isomorphism for $\eta < c^{-1}$.

The covariant derivative operator ∇_Γ : Let v be a scalar function defined on Γ. We define $\breve{v}$ on $\bar{\Omega}^\eta_{\mathrm{f}}$ by: $\breve{v}(\mathbf{x}) = v(\mathbf{x}_\Gamma)$, $\mathbf{x}$ and $\mathbf{x}_\Gamma$ related by (73). The operator ∇_Γ is defined by

$$(\nabla_\Gamma\, v)(\mathbf{x}_\Gamma) = (\nabla \breve{v})(\mathbf{x}_\Gamma, 0), \qquad \mathbf{x}_\Gamma \in \Gamma \tag{76}$$

It is a tangential operator that can be expressed in the contravariant basis of $T_{\mathbf{x}_\Gamma}$, by

$$\nabla_\Gamma v = (\partial_{\xi_\alpha} v)\, \boldsymbol{\tau}^\alpha. \tag{77}$$

To prove (77), we use (73) and differentiate $\breve{v}$ with respect to (ξ_1, ξ_2, s). We get

$$\left| \begin{aligned} &(\partial_{\xi_\alpha} \breve{v})(\mathbf{x}_\Gamma, s) = (\nabla \breve{v})(\mathbf{x}_\Gamma, s) \cdot (\boldsymbol{\tau}_\alpha + s\, \mathcal{C} \boldsymbol{\tau}_\alpha), \\ &(\partial_s \breve{v})(\mathbf{x}_\Gamma, s) = (\nabla \breve{v})(\mathbf{x}_\Gamma, s) \cdot \mathbf{n}. \end{aligned} \right.$$

Making $s = 0$ and using $(\partial_s \breve{v}) = 0$ (see the definition of $\breve{v}$) show that $(\nabla_\Gamma v) \cdot \boldsymbol{\tau}_\alpha = \partial_{\xi_\alpha} v$ and $(\nabla_\Gamma v) \cdot \mathbf{n} = 0$. We deduce (77) since $(\boldsymbol{\tau}^1, \boldsymbol{\tau}^2, \mathbf{n})$ is the dual base of $(\boldsymbol{\tau}_1, \boldsymbol{\tau}_2, \mathbf{n})$.

Consider now a tangential vector field $\mathbf{v}$ defined on Γ and a field of symmetric matrices $\mathcal{R}$ defined on Γ such that $\mathcal{R}\mathbf{n} = 0$. We define $(\operatorname{div}_\Gamma \mathbf{v})$, $((\mathcal{R}\nabla_\Gamma) \cdot \mathbf{v})$ and $(\operatorname{div}_\Gamma \mathcal{R})$ by the following expressions, where $\cdot$ denotes the scalar product in $\mathbf{R}^3$ ((79) and (80) are original definitions),

$$\operatorname{div}_\Gamma \mathbf{v} \equiv \big(\nabla_\Gamma (\mathbf{v} \cdot \mathbf{e}^i)\big) \cdot \mathbf{e}_i = (\partial_{\xi_\alpha} \mathbf{v}) \cdot \boldsymbol{\tau}^\alpha, \tag{78}$$

$$(\mathcal{R}\nabla_\Gamma) \cdot \mathbf{v} = \big((\mathcal{R}\nabla_\Gamma)(\mathbf{v} \cdot \mathbf{e}^i)\big) \cdot \mathbf{e}_i \equiv (\partial_{\xi_\alpha} \mathbf{v}) \cdot (\mathcal{R}\boldsymbol{\tau}^\alpha), \tag{79}$$

$$\operatorname{div}_\Gamma \mathcal{R} \equiv (\nabla_\Gamma \cdot (\mathcal{R}\mathbf{e}^i)) \cdot \mathbf{e}_i = (\partial_{\xi_\alpha} \mathcal{R})\, \boldsymbol{\tau}^\alpha. \tag{80}$$

Remark that the definition of $\operatorname{div}_\Gamma \mathbf{v}$ coincides with the definition of $\nabla_\Gamma \cdot \mathbf{v}$ $(\mathcal{R} = \Pi_\parallel)$.

As in the case of straight boundary, the operators $\operatorname{rot}_\Gamma$ and $\overrightarrow{\operatorname{rot}}_\Gamma$ are defined, from $\operatorname{div}_\Gamma$ and ∇_Γ, by relation (13), that we recall here

$$\operatorname{rot}_\Gamma \mathbf{v} = \operatorname{div}_\Gamma (\mathbf{v} \times \mathbf{n}), \qquad \overrightarrow{\operatorname{rot}}_\Gamma u = (\nabla_\Gamma u) \times \mathbf{n},$$

where $\mathbf{v}$ (resp. u) is a vectorial (resp. scalar) field defined on Γ. We still have (we omit the proof): $\operatorname{rot}_\Gamma \nabla_\Gamma = \operatorname{div}_\Gamma \overrightarrow{\operatorname{rot}}_\Gamma = 0$. Also $\operatorname{div}_\Gamma$ (resp. $\operatorname{rot}_\Gamma$) is the adjoint of $-\nabla_\Gamma$ (resp. $\overrightarrow{\operatorname{rot}}_\Gamma$) for the $L^2(\Gamma)$ scalar product.

We prove now the basic result used in the formal derivation of the effective boundary conditions (it differs from the classical representation of the curl operator (see for example [14]) by making explicit the dependence on s).

Lemma 0.6 – *Let* $\mathbf{v}$ *be a vector field defined on* Ω_f^η. *Let* $(\mathbf{x}_\Gamma, s) \in \Gamma \times]0, \eta[$ *be the parametric representation of* Ω_f^η. *Then we have*

$$\operatorname{rot} \mathbf{v} = \mathcal{T}_\Gamma^s\, \mathbf{v} \; - \; \partial_s (\mathbf{v} \times \mathbf{n}),$$

where, $\mathcal{T}_\Gamma^s \mathbf{v} = [(\mathcal{R}_s \nabla_\Gamma) \cdot (\mathbf{v} \times \mathbf{n})]\, \mathbf{n} + [\mathcal{R}_s \nabla_\Gamma (\mathbf{v} \cdot \mathbf{n})] \times \mathbf{n} - (\mathcal{R}_s\, \mathcal{C}\, \mathbf{v}) \times \mathbf{n},$

and $\mathcal{R}_s$ *defined by:* $\quad \mathcal{R}_s\, (\Pi_\parallel + s\, \mathcal{C}) = \Pi_\parallel, \; \mathcal{R}_s\, \mathbf{n} = 0.$

Proof. For a scalar function u defined on $\Omega_{\mathrm{f}}^{\eta}$, one gets (using (73)),

$$\left| \begin{array}{l} \partial_{\xi_\alpha} u = (\nabla u) \cdot (\Pi_{\parallel} + s\,\mathcal{C}) \boldsymbol{\tau}_\alpha, \\ \partial_s u = (\nabla u) \cdot \mathbf{n}. \end{array} \right.$$

As $\mathcal{R}_s(\Pi_{\parallel} + s\,\mathcal{C}) = \Pi_{\parallel}$, $\mathcal{R}_s\,\mathbf{n} = (\Pi_{\parallel} + s\,\mathcal{C})\,\mathbf{n} = 0$, $\mathcal{R}_s$ and $(\Pi_{\parallel} + s\,\mathcal{C})$ are symmetric, we deduce that $(\mathcal{R}_s\boldsymbol{\tau}^1, \mathcal{R}_s\boldsymbol{\tau}^2, \mathbf{n})$ is the dual basis of $((\Pi_{\parallel} + s\,\mathcal{C})\boldsymbol{\tau}_1, (\Pi_{\parallel} + s\,\mathcal{C})\boldsymbol{\tau}_2, \mathbf{n})$. Hence,

$$\nabla\, u = (\partial_{\xi_\alpha} u)\, \mathcal{R}_s \boldsymbol{\tau}^\alpha + (\partial_s u)\, \mathbf{n}. \tag{81}$$

From $\operatorname{rot} \mathbf{v} = \mathbf{e}^i \times \nabla(\mathbf{v} \cdot \mathbf{e}_i)$, we obtain using (81),

$$\operatorname{rot} \mathbf{v} = (\mathcal{R}_s \boldsymbol{\tau}^\alpha) \times (\partial_{\xi_\alpha} \mathbf{v}) - (\partial_s \mathbf{v}) \times \mathbf{n}. \tag{82}$$

Let us set $\mathcal{B}_\Gamma^s\, \mathbf{v} = (\mathcal{R}_s \boldsymbol{\tau}^\alpha) \times (\partial_{\xi_\alpha} \mathbf{v})$ and prove that $\mathcal{B}_\Gamma^s = \mathcal{T}_\Gamma^s$.

$$\begin{aligned} (\mathcal{B}_\Gamma^s\, \mathbf{v}) \times \mathbf{n} = -\left((\partial_{\xi_\alpha} \mathbf{v}) \cdot \mathbf{n}\right)\, \mathcal{R}_s \boldsymbol{\tau}^\alpha \;&=\; -(\partial_{\xi_\alpha}(\mathbf{v} \cdot \mathbf{n}) - \mathcal{C}\boldsymbol{\tau}_\alpha \cdot \mathbf{v})\, \mathcal{R}_s \boldsymbol{\tau}^\alpha \\ &=\; -(\partial_{\xi_\alpha}(\mathbf{v} \cdot \mathbf{n})\, \mathcal{R}_s \boldsymbol{\tau}^\alpha + (\mathcal{C}\boldsymbol{\tau}_\alpha \cdot \mathbf{v})\, \mathcal{R}_s \boldsymbol{\tau}^\alpha. \end{aligned}$$

As the tangential and symmetric tensors $\mathcal{R}_s$ and $\mathcal{C}$ commute, the previous equality yields,

$$\Pi_{\parallel}(\mathcal{B}_\Gamma^s\, \mathbf{v}) = \mathbf{n} \times ((\mathcal{B}_\Gamma^s\, \mathbf{v}) \times \mathbf{n}) = [\mathcal{R}_s\, \nabla_\Gamma\, (\mathbf{v} \cdot \mathbf{n})] \times \mathbf{n} - (\mathcal{R}_s\, \mathcal{C}\, \mathbf{v}). \tag{83}$$

On the other hand, $(\mathcal{B}_\Gamma^s\, \mathbf{v}) \cdot \mathbf{n} = ((\partial_{\xi_\alpha} \mathbf{v}) \times \mathbf{n}) \cdot \mathcal{R}_s \boldsymbol{\tau}^\alpha = (\partial_{\xi_\alpha}(\mathbf{v} \times \mathbf{n})) \cdot \mathcal{R}_s \boldsymbol{\tau}^\alpha + \mathbf{v} \cdot (\mathcal{R}_s \boldsymbol{\tau}^\alpha \times \mathcal{C}\boldsymbol{\tau}_\alpha)$. By applying the lemma 0.7 below, to $\mathcal{A} = \mathcal{C}$ and $\mathcal{B} = \mathcal{R}_s$, we deduce that the last term in the previous equality is 0. Consequently, this equality becomes, using (79),

$$(\mathcal{B}_\Gamma^s\, \mathbf{v}) \cdot \mathbf{n} = (\mathcal{R}_s\, \nabla_\Gamma\,) \cdot (\mathbf{v} \times \mathbf{n}). \tag{84}$$

Identities (83) and (84) show that $\mathcal{B}_\Gamma^s = \mathcal{T}_\Gamma^s$. Hence, lemma 0.6 is proved by (82). ∎

Lemma 0.7 – *Let $\mathcal{A}$ and $\mathcal{B}$ be two symmetric and real matrixes in $\mathcal{L}(\mathbf{R}^3, \mathbf{R}^3)$ such that $\mathcal{A}\,\mathbf{n} = \mathcal{B}\,\mathbf{n} = 0$ and having the same eigenvectors. We have:*

$$\mathcal{A}\boldsymbol{\tau}_\kappa \times \mathcal{B}\boldsymbol{\tau}^\kappa = 0.$$

Proof. Let $(\varepsilon_1, \varepsilon_2)$ be the eigenvectors of $\mathcal{A}$ and $\mathcal{B}$ and (a_1, a_2) (resp. (b_1, b_2)) the corresponding eigenvalues of $\mathcal{A}$ (resp. $\mathcal{B}$). Let us set: $\boldsymbol{\tau}_\kappa = \tau_{\kappa,1}\, \varepsilon_1 + \tau_{\kappa,2}\, \varepsilon_2$ and $\boldsymbol{\tau}^\kappa = \tau^{\kappa,1}\, \varepsilon_1 + \tau^{\kappa,2}\, \varepsilon_2$. Then

$$\begin{aligned} \mathcal{A}\boldsymbol{\tau}_\kappa \times \mathcal{B}\boldsymbol{\tau}^\kappa \;&=\; (a_1\, \tau_{\kappa,1}\, \varepsilon_1 + a_2\, \tau_{\kappa,2}\, \varepsilon_2) \times (b_1\, \tau^{\kappa,1}\, \varepsilon_1 + b_2\, \tau^{\kappa,2}\, \varepsilon_2) \\ &=\; (a_1\, b_2\, \tau_{\kappa,1} \tau^{\kappa,2} - a_2\, b_1\, \tau_{\kappa,2} \tau^{\kappa,1})\, \varepsilon_1 \times \varepsilon_2. \end{aligned}$$

By definition of the dual base $(\boldsymbol{\tau}^1, \boldsymbol{\tau}^2)$, we have, setting $\delta = 1/(\tau_{1,1}\,\tau_{2,2} - \tau_{1,2}\,\tau_{2,1})$,

$$\tau^{1,1} = \delta\,\tau_{2,2}, \quad \tau^{1,2} = -\delta\,\tau_{2,1}, \quad \tau^{2,1} = -\delta\,\tau_{1,2}, \quad \tau^{2,2} = \delta\,\tau_{1,1}.$$

Then, one easily checks that $\tau_{\kappa,1}\,\tau^{\kappa,2} = 0$ and $\tau_{\kappa,2}\,\tau^{\kappa,1} = 0$. ■

We prove finally the following geometrical identity that enables us later to write in symmetric way the effective boundary condition of order three.

Lemma 0.8 – *We have*

$$\mathrm{div}_\Gamma\,(2H\,\Pi_\parallel - \mathcal{C}) = -2\,G\,\mathbf{n},$$

where $G = \det \mathcal{C}$, is the Gaussian curvature.

Proof. Following the definition (80), we have, $\mathrm{div}_\Gamma\,(2H\,\Pi_\parallel - \mathcal{C}) = (\partial_{\xi_\alpha}(2H\,\Pi_\parallel - \mathcal{C}))\boldsymbol{\tau}^\alpha$. So

$$\mathrm{div}_\Gamma\,(2H\,\Pi_\parallel - \mathcal{C})\cdot\mathbf{n} = [\,(\partial_{\xi_\alpha}(2H\,\Pi_\parallel - \mathcal{C}))\,\boldsymbol{\tau}^\alpha\,]\cdot\mathbf{n} = [\,(\partial_{\xi_\alpha}(2H\,\Pi_\parallel - \mathcal{C}))\,\mathbf{n}\,]\cdot\boldsymbol{\tau}$$

$$= [\,(\mathcal{C} - 2H\,\Pi_\parallel)\partial_{\xi_\alpha}\mathbf{n}\,]\cdot\boldsymbol{\tau}^\alpha = [\,(\mathcal{C} - 2H\,\Pi_\parallel)\,\mathcal{C}\,\boldsymbol{\tau}_\alpha\,]\cdot\boldsymbol{\tau}^\alpha = tr(\mathcal{C}^2) - tr(\mathcal{C})^2 = -2$$

To conclude, we are going to show that $\mathrm{div}_\Gamma\,(2H\,\Pi_\parallel - \mathcal{C})\cdot\boldsymbol{\tau}_\beta = 0$, where $\beta \in \{1,2\}$ (which means: $\Pi_\parallel \mathrm{div}_\Gamma\,(2H\,\Pi_\parallel - \mathcal{C}) = 0$).

$$\mathrm{div}_\Gamma\,(2H\,\Pi_\parallel - \mathcal{C})\cdot\boldsymbol{\tau}_\beta = [\,(\partial_{\xi_\alpha}(2H\,\Pi_\parallel - \mathcal{C}))\,\boldsymbol{\tau}^\alpha\,]\cdot\boldsymbol{\tau}_\beta = [\,(\partial_{\xi_\alpha}(2H\,\Pi_\parallel - \mathcal{C}))\,\boldsymbol{\tau}_\beta\,]\cdot$$

$$= \partial_{\xi_\beta}(2H) - [(\partial_{\xi_\alpha}\mathcal{C})\,\boldsymbol{\tau}_\beta\,]\cdot\boldsymbol{\tau}^\alpha$$

We have (using $\partial_{\xi_\beta}\partial_{\xi_\alpha}\mathbf{n} = \partial_{\xi_\alpha}\partial_{\xi_\beta}\mathbf{n}$)

$$\begin{aligned}
[(\partial_{\xi_\alpha}\mathcal{C})\,\boldsymbol{\tau}_\beta\,]\cdot\boldsymbol{\tau}^\alpha &= [\partial_{\xi_\alpha}(\mathcal{C}\,\boldsymbol{\tau}_\beta)]\cdot\boldsymbol{\tau}^\alpha - [\mathcal{C}\partial_{\xi_\alpha}\boldsymbol{\tau}_\beta]\cdot\boldsymbol{\tau}^\alpha \\
&= [\partial_{\xi_\beta}(\mathcal{C}\,\boldsymbol{\tau}_\alpha)]\cdot\boldsymbol{\tau}^\alpha - [\mathcal{C}\partial_{\xi_\alpha}\boldsymbol{\tau}_\beta]\cdot\boldsymbol{\tau}^\alpha \\
&= \partial_{\xi_\beta}(\mathcal{C}\,\boldsymbol{\tau}_\alpha\cdot\boldsymbol{\tau}^\alpha) - \left(\mathcal{C}\,\boldsymbol{\tau}_\alpha\partial_{\xi_\beta}\boldsymbol{\tau}^\alpha + \mathcal{C}\,\boldsymbol{\tau}^\alpha\partial_{\xi_\alpha}\boldsymbol{\tau}_\beta\right),
\end{aligned}$$

whence $\mathrm{div}_\Gamma\,(2H\,\Pi_\parallel - \mathcal{C})\cdot\boldsymbol{\tau}_\beta = -\left(\mathcal{C}\,\boldsymbol{\tau}_\alpha\partial_{\xi_\beta}\boldsymbol{\tau}^\alpha + \mathcal{C}\,\boldsymbol{\tau}^\alpha\partial_{\xi_\alpha}\boldsymbol{\tau}_\beta\right) \equiv -B$.

Let us set: $a^\alpha_\kappa = (\mathcal{C}\,\boldsymbol{\tau}^\alpha)\cdot\boldsymbol{\tau}_\kappa$. Since $\mathcal{C}\,\boldsymbol{\tau}^\alpha = a^\alpha_\kappa\,\boldsymbol{\tau}^\kappa$ and $\mathcal{C}\,\boldsymbol{\tau}_\alpha = a^\kappa_\alpha\,\boldsymbol{\tau}_\kappa$,

$$\begin{aligned}
B &= a^\kappa_\alpha\,\boldsymbol{\tau}_\kappa\,\partial_{\xi_\beta}\boldsymbol{\tau}^\alpha + a^\alpha_\kappa\,\boldsymbol{\tau}^\kappa\,\partial_{\xi_\alpha}\boldsymbol{\tau}_\beta = -a^\kappa_\alpha\,\boldsymbol{\tau}^\alpha\,\partial_{\xi_\beta}\boldsymbol{\tau}_\kappa + a^\alpha_\kappa\,\boldsymbol{\tau}^\kappa\,\partial_{\xi_\alpha}\boldsymbol{\tau}_\beta \\
(= &\quad -\partial_{\xi_\beta}\boldsymbol{\tau}_\alpha + \partial_{\xi_\alpha}\boldsymbol{\tau}_\beta)\;a^\alpha_\kappa\,\boldsymbol{\tau}^\kappa
\end{aligned} \tag{85}$$

But we have $\partial_{\xi_\beta}\boldsymbol{\tau}_\alpha = \partial_{\xi_\beta}(\partial_{\xi_\alpha}\mathbf{x}_\Gamma) = \partial_{\xi_\alpha}\boldsymbol{\tau}_\beta$. Consequently, $B = 0$ and $\mathrm{div}_\Gamma\,(2H\,\Pi_\parallel - \mathcal{C})\cdot\boldsymbol{\tau}_\beta = 0$. ■

2.4.2. Scaling and asymptotic expansion

Scaling. As in the case of the straight boundary, the scaling corresponds to the change of variables (12). According to lemma 0.6, the differential operator rot becomes in the new coordinates systems $(\mathbf{x}_\Gamma, \nu) \in \Gamma \times]0,1[$: $\mathcal{T}_\Gamma^{\eta\nu} + \frac{1}{\eta}\mathbf{n} \times \partial_\nu$. Notice that the operator $\mathcal{T}_\Gamma^{\eta\nu}$ is not singular with respect to η.

Equation (17) becomes

$$\begin{cases} \varepsilon_{\mathrm{f}}\, \partial_t \mathbf{E}_{\mathrm{f}}^\eta - \mathcal{T}_\Gamma^{\eta\nu}\mathbf{H}_{\mathrm{f}}^\eta + \frac{1}{\eta}\,\partial_\nu(\mathbf{H}_{\mathrm{f}}^\eta \times \mathbf{n}) = 0, \\ \mu_{\mathrm{f}}\, \partial_t \mathbf{H}_{\mathrm{f}}^\eta + \mathcal{T}_\Gamma^{\eta\nu}\mathbf{E}_{\mathrm{f}}^\eta - \frac{1}{\eta}\,\partial_\nu(\mathbf{E}_{\mathrm{f}}^\eta \times \mathbf{n}) = 0. \end{cases} \tag{86}$$

Our problem is constituted now by equations (1), (16), (86) and (18).

Asymptotic expansion. We use the ansatz (19). Moreover, we need an analogous expansion of the operator $\mathcal{T}_\Gamma^{\eta\nu}$. We use, for instance, a Taylor expansion of $\mathcal{R}_{\eta\nu}$ with respect to the parameter $(\eta\nu\,\mathcal{C})$:

$$\mathcal{R}_{\eta\nu} = \Pi_{\|} + \textstyle\sum_{i=1}^{\infty}(-\eta\nu\,\mathcal{C})^i,$$

so we have

$$\begin{aligned} \mathcal{T}_\Gamma^{\eta\nu} &= \textstyle\sum_{i=0}^{\infty}(-\eta\nu)^i\,\mathcal{T}_\Gamma^i, \text{ where,} \\ \mathcal{T}_\Gamma^i\mathbf{v} &= [(\mathcal{C}^i\,\nabla_\Gamma\,)\cdot(\mathbf{v}\times\mathbf{n})]\,\mathbf{n} + [\mathcal{C}^i\,\nabla_\Gamma\,(\mathbf{v}\cdot\mathbf{n})]\times\mathbf{n} - (\mathcal{C}^{i+1}\,\mathbf{v})\times\mathbf{n}. \end{aligned} \tag{87}$$

The formal identification lets unchanged equations (20) and (21). However, equations (22) and (23) change to the following ones, where $k \geq 0$,

$$\begin{cases} \partial_\nu(\mathbf{E}_{\mathrm{f}}^{k+1} \times \mathbf{n}) = \mu_{\mathrm{f}}\,\partial_t\mathbf{H}_{\mathrm{f}}^k + \sum_{i=0}^{k}(-\nu)^i\,\mathcal{T}_\Gamma^i\mathbf{E}_{\mathrm{f}}^{k-i}, \\ \mathbf{E}_{\mathrm{f}}^{k+1}(\mathbf{x}_\Gamma, 1)\times\mathbf{n} = 0, \quad \text{for } \mathbf{x}_\Gamma \in \Gamma. \end{cases} \tag{88}$$

$$\begin{cases} \partial_\nu(\mathbf{H}_{\mathrm{f}}^{k+1} \times \mathbf{n}) = -\varepsilon_{\mathrm{f}}\,\partial_t\mathbf{E}_{\mathrm{f}}^k + \sum_{i=0}^{k}(-\nu)^i\,\mathcal{T}_\Gamma^i\mathbf{H}_{\mathrm{f}}^{k-i}, \\ \mathbf{H}_{\mathrm{f}}^{k+1}(\mathbf{x}_\Gamma, 0)\times\mathbf{n} = \mathbf{H}_v^{k+1}(\mathbf{x}_\Gamma)\times\mathbf{n}, \quad \text{for } \mathbf{x}_\Gamma \in \Gamma. \end{cases} \tag{89}$$

The zero order terms are determined by (24). To determine the higher order terms, we use the lemma 0.9. We introduce the notations, where i is an integer,

$$\vec{\mathrm{rot}}_\Gamma^{(i)}\,u = (\mathcal{C}^i\,\nabla_\Gamma\,u)\times\mathbf{n}, \quad \mathrm{rot}_\Gamma^{(i)}\,\mathbf{v} = (\mathcal{C}^i\,\nabla_\Gamma\,)\cdot(\mathbf{v}\times\mathbf{n}).$$

Lemma 0.9 – *For all $k \geq 0$, we have for all $(\mathbf{x}_\Gamma, \nu) \in \Gamma \times]0, 1[$,*

$$\begin{aligned}\partial_t \ \mathbf{E}_{\mathrm{f}}^{k+1} \ (\mathbf{x}_\Gamma, \nu) \times \mathbf{n} = & - \int_\nu^1 \mu_{\mathrm{f}} \, \partial_{tt} \Pi_{\|} \mathbf{H}_{\mathrm{f}}^k (\mathbf{x}_\Gamma, \xi) \, d\xi \\ & - \int_\nu^1 \frac{1}{\varepsilon_{\mathrm{f}}} \sum_{i=0}^k \sum_{j=0}^i (-\xi)^{k-j} \, \vec{\mathrm{rot}}_\Gamma^{(k-i)} \, \mathrm{rot}_\Gamma^{(i-j)} \, \mathbf{H}_{\mathrm{f}}^j (\mathbf{x}_\Gamma, \xi) \, d\xi. \\ & + \int_\nu^1 \sum_{i=0}^k (-\xi)^{k-i} \, (\mathcal{C}^{k-i+1} \, \partial_t \mathbf{E}_{\mathrm{f}}^i (\mathbf{x}_\Gamma, \xi)) \times \mathbf{n} \, d\xi.\end{aligned} \tag{90}$$

$$\begin{aligned}\partial_t \mathbf{H}_{\mathrm{f}}^{k+1} \ (\ \mathbf{x}_\Gamma, \nu) \times \mathbf{n} = & \ \partial_t \mathbf{H}_v^{k+1} (\mathbf{x}_\Gamma) \times \mathbf{n} - \int_0^\nu \varepsilon_{\mathrm{f}} \, \partial_{tt} \Pi_{\|} \mathbf{E}_{\mathrm{f}}^k (\mathbf{x}_\Gamma, \xi) \, d\xi \\ & - \int_0^\nu \frac{1}{\mu_{\mathrm{f}}} \sum_{i=0}^k \sum_{j=0}^i (-\xi)^{k-j} \, \vec{\mathrm{rot}}_\Gamma^{(k-i)} \, \mathrm{rot}_\Gamma^{(i-j)} \, \mathbf{E}_{\mathrm{f}}^j (\mathbf{x}_\Gamma, \xi) \, d\xi \\ & - \int_0^\nu \sum_{i=0}^k (-\xi)^{k-i} \, (\mathcal{C}^{k-i+1} \, \partial_t \mathbf{H}_{\mathrm{f}}^i (\mathbf{x}_\Gamma, \xi)) \times \mathbf{n} \, d\xi\end{aligned} \tag{91}$$

Proof. We are going to prove (90) only. One deduces (91) by analogous arguments. By applying $\Pi_{\|}$ to the first equation of (88) we obtain

$$\begin{aligned}\partial_\nu (\mathbf{E}_{\mathrm{f}}^{k+1} \times \mathbf{n}) = & \ \mu_{\mathrm{f}} \, \partial_t (\Pi_{\|} \mathbf{H}_{\mathrm{f}}^k) + \sum_{i=0}^k (-\nu)^{k-i} \vec{\mathrm{rot}}_\Gamma^{(k-i)} \, (\mathbf{E}_{\mathrm{f}}^i \cdot \mathbf{n}) \\ & - \sum_{i=0}^k (-\nu)^{k-i} \, (\mathcal{C}^{k-i+1} \, \partial_t \mathbf{E}_{\mathrm{f}}^i (\mathbf{x}_\Gamma, \xi)) \times \mathbf{n}.\end{aligned} \tag{92}$$

On the other hand, applying $\Pi_\perp$ to the first equation of (89) shows that for $i \geq 0$,

$$\varepsilon_{\mathrm{f}} \, \partial_t \left(\mathbf{E}_{\mathrm{f}}^i \cdot \mathbf{n} \right) = \sum_{j=0}^i (-\nu)^{i-j} \mathrm{rot}_\Gamma^{(i-j)} \, \mathbf{H}_{\mathrm{f}}^j. \tag{93}$$

Equation (90) is obtained from $\int_\nu^1 \partial_t (92) \, d\xi$, by using the expression of $\partial_t \left(\mathbf{E}_{\mathrm{f}}^i \cdot \mathbf{n} \right)$ in (93), and by using the boundary condition at $\nu = 1$, given in (88). ■

0.0.1 2.4.3. Effective Construction

Of course the principle is the same as the case of the straight boundary.

Condition of order 2. The terms $\Pi_{\|} \mathbf{E}_{\mathrm{f}}^0$ and $\Pi_{\|} \mathbf{H}_{\mathrm{f}}^0$ are given by (24). Making $k = 0$ in (90) and using (24) shows that

$$\partial_t \mathbf{E}_{\mathrm{f}}^1 (\mathbf{x}_\Gamma, \nu) \times \mathbf{n} = - (1 - \nu) \left(\mu_{\mathrm{f}} \, \partial_{tt} + \frac{1}{\varepsilon_{\mathrm{f}}} \vec{\mathrm{rot}}_\Gamma \, \mathrm{rot}_\Gamma \right) \Pi_{\|} \mathbf{H}_v^0 (\mathbf{x}_\Gamma), \tag{94}$$

which is exactly the same expression as (31). Hence, the condition of order 2 in the case of curved boundary is given by (32). Indeed, the stability theorem 0.1 still applies. However, the condition of order 3 has no longer the same expression.

Condition of order 3. Make $k = 0$ in (91). Using (24) we get,

$$\Pi_{\|}\mathbf{H}_{\mathrm{f}}^{1}(\mathbf{x}_{\Gamma},\nu) = \Pi_{\|}\mathbf{H}_{v}^{1}(\mathbf{x}_{\Gamma}) - \nu\,\mathcal{C}\,\mathbf{H}_{v}^{0}(\mathbf{x}_{\Gamma}), \quad (\mathbf{x}_{\Gamma},\nu) \in \Gamma\times]0,1[. \tag{95}$$

Consider (90) when $k = 1$,

$$\begin{aligned}\partial_t\mathbf{E}_{\mathrm{f}}^{2}(\mathbf{x}_{\Gamma},\nu)\times\mathbf{n} \;&=\; -\int_{\nu}^{1}\mu_{\mathrm{f}}\,\partial_{tt}\Pi_{\|}\mathbf{H}_{\mathrm{f}}^{1}(\mathbf{x}_{\Gamma},\xi)\,d\xi + \int_{\nu}^{1}(\mathcal{C}\partial_t\mathbf{E}_{\mathrm{f}}^{1}(\mathbf{x}_{\Gamma},\xi))\times\mathbf{n}\,d\xi\\ &\quad -\int_{\nu}^{1}\tfrac{1}{\varepsilon_f}\,\vec{\mathrm{rot}}_{\Gamma}\,\mathrm{rot}_{\Gamma}\,\mathbf{H}_{\mathrm{f}}^{1}(\mathbf{x}_{\Gamma},\xi)\,d\xi\\ &\quad +\int_{\nu}^{1}\xi\,\tfrac{1}{\varepsilon_f}\,(\,\vec{\mathrm{rot}}_{\Gamma}^{(1)}\,\mathrm{rot}_{\Gamma} + \vec{\mathrm{rot}}_{\Gamma}\,\mathrm{rot}_{\Gamma}^{(1)}\,)\mathbf{H}_{\mathrm{f}}^{0}(\mathbf{x}_{\Gamma},\xi)\,d\xi,\end{aligned} \tag{96}$$

$\Pi_{\|}\mathbf{H}_{\mathrm{f}}^{0}$ is given by (24), $\Pi_{\|}\mathbf{H}_{\mathrm{f}}^{1}$ is given by (95) and we have $\mathcal{C}\partial_t\mathbf{E}_{\mathrm{f}}^{1} = \mathcal{C}\partial_t\Pi_{\|}\mathbf{E}_{\mathrm{f}}^{1}$, where $\partial_t\Pi_{\|}\mathbf{E}_{\mathrm{f}}^{1}$ is given by (39). We obtain after simplifications and using (75),

$$\begin{aligned}&\partial_t\mathbf{E}_{\mathrm{f}}^{2}(\mathbf{x}_{\Gamma},\nu)\times\mathbf{n} =\\ &-\mu_{\mathrm{f}}\,\partial_{tt}\left((1-\nu)\,\Pi_{\|}\mathbf{H}_{v}^{1}(\mathbf{x}_{\Gamma}) - (1-\nu^2)(\mathcal{C}-H)\Pi_{\|}\mathbf{H}_{v}^{0}(\mathbf{x}_{\Gamma})\right)\\ &-\tfrac{1}{\varepsilon_f}\,\vec{\mathrm{rot}}_{\Gamma}\,\mathrm{rot}_{\Gamma}\,\left((1-\nu)\,\Pi_{\|}\mathbf{H}_{v}^{1}(\mathbf{x}_{\Gamma}) - \tfrac{1}{2}(1-\nu^2)\mathcal{C}\Pi_{\|}\mathbf{H}_{v}^{0}(\mathbf{x}_{\Gamma})\right)\\ &+\tfrac{1}{\varepsilon_f}\,\tfrac{1}{2}(1-\nu^2)\,\vec{\mathrm{rot}}_{\Gamma}\,\mathrm{rot}_{\Gamma}^{(1)}\,\mathbf{H}_{v}^{0}(\mathbf{x}_{\Gamma}).\end{aligned} \tag{97}$$

The ultimate simplification is due to lemma 0.8 and is given by the following lemma:

Lemma 0.10 – *Let* $\mathbf{v}$ *be a tangential vector field defined on* Γ. *We have the identity:*

$$\mathrm{rot}_{\Gamma}^{(1)}\,\mathbf{v} + \mathrm{rot}_{\Gamma}\,(\mathcal{C}\,\mathbf{v}) = 2H\,\mathrm{rot}_{\Gamma}\,\mathbf{v}$$

Proof. By definition, $\mathrm{rot}_{\Gamma}^{(1)}\,\mathbf{v} = (\mathcal{C}\,\nabla_{\Gamma})\cdot(\mathbf{v}\times\mathbf{n})$. Hence, using (75),

$$\begin{aligned}\mathrm{rot}_{\Gamma}^{(1)}\,\mathbf{v} + \mathrm{rot}_{\Gamma}\,(\mathcal{C}\,\mathbf{v}) \;&=\; (\mathcal{C}\,\nabla_{\Gamma})\cdot(\mathbf{v}\times\mathbf{n}) + \nabla_{\Gamma}\cdot\left(\,(2H\,\Pi_{\|} - \mathcal{C})(\mathbf{v}\times\mathbf{n})\,\right)\\ &=\; 2H\,\mathrm{div}_{\Gamma}\,(\mathbf{v}\times\mathbf{n}) + (\mathrm{div}_{\Gamma}\,(2H\,\Pi_{\|} - \mathcal{C}))\cdot(\mathbf{v}\times\mathbf{n}),\end{aligned}$$

and according to lemma 0.8, $(\mathrm{div}_{\Gamma}\,(2H\,\Pi_{\|} - \mathcal{C}))\cdot(\mathbf{v}\times\mathbf{n}) = 0$. ■

Applying this lemma to relation (97) yields

$$\begin{aligned}&\partial_t\mathbf{E}_{\mathrm{f}}^{2}(\mathbf{x}_{\Gamma},\nu)\times\mathbf{n}\\ &\quad= -\mu_{\mathrm{f}}\,\partial_{tt}\left((1-\nu)\,\Pi_{\|}\mathbf{H}_{v}^{1}(\mathbf{x}_{\Gamma}) - (1-\nu^2)(\mathcal{C}-H)\Pi_{\|}\mathbf{H}_{v}^{0}(\mathbf{x}_{\Gamma})\right)\\ &\quad\; -\tfrac{1}{\varepsilon_f}\left((1-\nu)\,\vec{\mathrm{rot}}_{\Gamma}\,\mathrm{rot}_{\Gamma}\,\Pi_{\|}\mathbf{H}_{v}^{1}(\mathbf{x}_{\Gamma}) - (1-\nu^2)\,\vec{\mathrm{rot}}_{\Gamma}\,H\mathrm{rot}_{\Gamma}\,\Pi_{\|}\mathbf{H}_{v}^{0}(\mathbf{x}_{\Gamma})\right)\end{aligned} \tag{98}$$

Combining (98) with (94) and (24) leads finally to the following third order condition

$$\boxed{\partial_t \widetilde{\mathbf{E}}_v^\eta \times \mathbf{n} = -\frac{\eta}{\varepsilon_{\mathrm{f}}} \left(\varepsilon_{\mathrm{f}}\, \mu_{\mathrm{f}} \left[1 - \eta(\mathcal{C} - H)\right]\partial_{tt} + \vec{\mathrm{rot}}_\Gamma \left(1 - \eta H\right)\mathrm{rot}_\Gamma\right) \Pi_{\|} \widetilde{\mathbf{H}}_v^\eta,} \tag{99}$$

on $\Gamma \times \mathbb{R}^+$. Notice that this condition differs from the second order one by corrector terms involving the geometry characteristics $\mathcal{C}$ and H, only. In the case of straight boundary, $\mathcal{C} = 0$ and $H = 0$, the two conditions are of course identical. We give now the stability result for this condition.

Theorem 0.5 – *Sufficiently regular solutions* $(\widetilde{\mathbf{E}}_v^\eta, \widetilde{\mathbf{H}}_v^\eta)$ *of* $\{(6), (99)\}$ *satisfy the a priori energy estimate:*

$$\frac{d}{dt}\left\{\mathcal{E}_v(\widetilde{\mathbf{E}}_v^\eta(t), \widetilde{\mathbf{H}}_v^\eta(t)) + \eta\, \mathcal{E}_\Gamma^\eta(t)\right\} = 0,$$

where, if we set, $\varphi^\eta(\mathbf{x}_\Gamma, t) = \Pi_{\|} \widetilde{\mathbf{H}}_v^\eta(\mathbf{x}_\Gamma, t), \ \ \mathbf{x}_\Gamma \in \Gamma,$

$$\mathcal{E}_\Gamma^\eta(t) = \frac{\mu_{\mathrm{f}}}{2} \int_\Gamma [1-\eta(\mathcal{C}-H)]\partial_t \varphi^\eta \cdot \partial_t \varphi^\eta \, d\mathbf{x}_\Gamma + \frac{1}{2\,\varepsilon_{\mathrm{f}}} \int_\Gamma (1-\eta H) \left|\int_0^t \mathrm{rot}_\Gamma\, \varphi^\eta \, d\tau\right|^2 d\mathbf{x}_\Gamma.$$

We have stability as defined by (10) when,

$$\eta \, \inf_\Gamma \left(\max\left(|c_1 + c_2|\,,\ |c_1 - c_2|\right)\right) < 2,$$

and where c_1 *and* c_2 *are the eigenvalues of* $\mathcal{C}$.

Proof. The proof of the a priori estimate is rather straightforward. It follows the same steps as the proof of theorem 0.1 and is based on the symmetry of $[1 - \eta(\mathcal{C} - H)]$.

The stability (defined by (10)) is obtained when the eigenvalues of $1 - \eta(\mathcal{C} - H)$ and the real $(1 - \eta H)$ are positive; i.e. when $\eta \max\left(|c_1 + c_2|\,,\ |c_1 - c_2|\right) < 2$ a.e. on Γ. In this case we have $\mathcal{E}_\Gamma^\eta(t) \geq 0$. ■

3. Thin layer approximation: the non linear case

We generalize the construction of thin layer approximations to the case of non linear materials of ferromagnetic type. We consider here directly the case of a curved boundary.

3.1. Description of the model

We keep the same notation as in section . However the material of the thin layer Ω_f^η is no longer linear and obeys to the following equations:

$$\begin{cases} \varepsilon_f\, \partial_t \mathbf{E}_f^\eta - \operatorname{rot} \mathbf{H}_f^\eta = 0, \qquad \mu_f\, \partial_t(\mathbf{H}_f^\eta + \mathbf{M}_f^\eta) + \operatorname{rot} \mathbf{E}_f^\eta = 0, \ \text{ on } \ \Omega_f^\eta, \\ (\mathbf{E}_f^\eta, \mathbf{H}_f^\eta)(t=0) = (0,0), \end{cases} \tag{100}$$

where $\mathbf{M}_f^\eta$ is called the magnetization field and is linked to $\mathbf{H}_f^\eta$ through the ferromagnetic law $\mathcal{L}^\eta$ as follows:

$$\begin{cases} \partial_t \mathbf{M}_f^\eta = \mathcal{L}^\eta(\mathbf{M}_f^\eta,\, \mathbf{H}_{\text{tot}}^\eta;\, \mathbf{x}), \quad \mathbf{M}_f^\eta(\mathbf{x},0) = \mathbf{M}_0^\eta(\mathbf{x}), \ \ \mathbf{x} \in \Omega_f^\eta, \\ \mathbf{H}_{\text{tot}}^\eta = \mathbf{H}_f^\eta - \nabla_M \Phi^\eta(\mathbf{M}_f^\eta;\, \mathbf{x}), \end{cases} \tag{101}$$

where, for *a.e.* $\mathbf{x} \in \Omega_f^\eta$, $(m,h) \in \mathbb{R}^3 \times \mathbb{R}^3 \longmapsto \mathcal{L}^\eta(m,\, h;\, \mathbf{x}) \in \mathbb{R}^3$ is a C^∞ function and $m \longmapsto \Phi^\eta(m;\, \mathbf{x}) \in \mathbb{R}$ is C^∞ and positive function, $\mathcal{L}^\eta$ satisfying:

$$\begin{aligned} &(i) \quad h \longmapsto \mathcal{L}^\eta(m,h;\,\mathbf{x}) \text{ is linear, } m \in \mathbb{R}^3,\ \mathbf{x} \in \Omega_f^\eta, \\ &(ii) \quad \mathcal{L}^\eta(m,h;\,\mathbf{x}) \cdot m = 0, \ \ m,h \in \mathbb{R}^3,\ \mathbf{x} \in \Omega_f^\eta, \\ &(iii) \quad \mathcal{L}^\eta(m,h;\,\mathbf{x}) \cdot h \geq 0, \ \ m \in \mathbb{R}^3,\ \mathbf{x} \in \Omega_f^\eta. \end{aligned} \tag{102}$$

The diffraction problem is constituted by the equations (1), (2), (100), (101) and (4).

Example. The Landau-Lifshitz law of ferromagnetic materials, without exchange field (see [13] or [16]), is a particular case of this general framework. It corresponds to:

$$\begin{aligned} &\mathcal{L}^\eta(m,h;\,\mathbf{x}) = \gamma\, h \times m + \frac{\alpha^\eta(\mathbf{x})}{|\mathbf{M}_0^\eta(\mathbf{x})|}\, m \times (h \times m), \\ &\Phi^\eta(m;\,\mathbf{x}) = \tfrac{1}{2}\, |\mathbf{H}_s(\mathbf{x}) - m|^2 + \tfrac{1}{2} K_a^\eta(\mathbf{x})\, |m - (p^\eta(\mathbf{x}) \cdot m)\, p^\eta(\mathbf{x})|^2, \end{aligned}$$

where, γ is the absolute value of the gyromagnetic factor, that is a universal constant, K_a^η and α^η are positive scalar functions, $\mathbf{M}_0^\eta$ is the initial magnetization, $\mathbf{H}_s^\eta$ is a given static magnetic field, and finally, p^η is a unit vector called easy axis of the magnetization. One easily checks that this law satisfies the required properties. We have for example:

$$\mathcal{L}^\eta(m,h;\,\mathbf{x}) \cdot h = \frac{\alpha^\eta(\mathbf{x})}{|\mathbf{M}_0^\eta(\mathbf{x})|}\, |h \times m|^2 \ \geq \ 0.$$

Apriori estimates. The stability of the coupling between Maxwell equations and the ferromagnetic law relies on the following fundamental a priori estimates. The first one is a consequence of the nonlinear law itself. According to the property (102) (*ii*), if we take the scalar product of (101) by $\mathbf{M}_f^\eta$, we see that

$$\partial_t |\, \mathbf{M}_f^\eta(\mathbf{x},t)\,|^2 = 0 \implies |\, \mathbf{M}_f^\eta(\mathbf{x},t)\,| = |\, \mathbf{M}_0^\eta(\mathbf{x})\,| \quad \forall t \geq 0 \text{ and } a.e.\ \mathbf{x} \in \Omega_f^\eta. \tag{103}$$

The second one is the equivalent to the classic energy estimate for Maxwell's system:

$$\frac{d}{dt}\left\{ \mathcal{E}_v\left(\mathbf{E}_v^\eta(t), \mathbf{H}_v^\eta(t)\right) + \mathcal{E}_f^\eta\left(\mathbf{E}_f^\eta(t), \mathbf{H}_f^\eta(t)\right) + \mu_f \int_{\Omega_f^\eta} \Phi^\eta(\mathbf{M}_f^\eta(t))\, d\mathbf{x} \right\} \leq 0. \tag{104}$$

Proof. Using the continuity relations (2), we have:

$$\frac{d}{dt}\mathcal{E}_v\left(\mathbf{E}_v^\eta(t), \mathbf{H}_v^\eta(t)\right) + \mathcal{E}_f^\eta\left(\mathbf{E}_f^\eta(t), \mathbf{H}_f^\eta(t)\right) = -\mu_f \int_{\Omega_f^\eta} \partial_t \mathbf{M}_f^\eta \cdot \mathbf{H}_f^\eta\, d\mathbf{x}. \tag{105}$$

On the other hand, property (102) (*iii*) yields, for *a.e.* $\mathbf{x} \in \Omega_f^\eta$,

$$\partial_t \mathbf{M}_f^\eta \cdot \mathbf{H}_f^\eta = \partial_t \mathbf{M}_f^\eta \cdot (\mathbf{H}_{tot}^\eta + \nabla_M \Phi^\eta(\mathbf{M}_f^\eta)) \geq \partial_t \Phi^\eta(\mathbf{M}_f^\eta).$$

Consequently

$$-\mu_f \int_{\Omega_f^\eta} \partial_t \mathbf{M}_f^\eta \cdot \mathbf{H}_f^\eta\, d\mathbf{x} \leq -\mu_f \frac{d}{dt} \int_{\Omega_f^\eta} \Phi^\eta(\mathbf{M}_f^\eta(t))\, d\mathbf{x} \tag{106}$$

Identity (104) follows from (105) and (106). ■

3.2. Construction of the effective boundary conditions

Scaling. As in the linear case, the scaling is done through the transformation: $\mathbf{x} \in \bar{\Omega}_f^\eta \longmapsto (\mathbf{x}_\Gamma, \nu) \in \Gamma \times [0, 1]$, such that $\mathbf{x} = \mathbf{x}_\Gamma + \eta\nu\,\mathbf{n}$. We still keep calling $(\mathbf{E}_f^\eta, \mathbf{H}_f^\eta, \mathbf{M}_f^\eta)$ the unknowns defined on the scaled domain $\Gamma\times]0, 1[$. The system (100) is transformed to:

$$\begin{cases} \varepsilon_f\, \partial_t \mathbf{E}_f^\eta - \mathcal{T}_\Gamma^{\eta\nu} \mathbf{H}_f^\eta + \frac{1}{\eta}\, \partial_\nu (\mathbf{H}_f^\eta \times \mathbf{n}) = 0, \\ \mu_f\, \partial_t (\mathbf{H}_f^\eta + \mathbf{M}_f^\eta) + \mathcal{T}_\Gamma^{\eta\nu} \mathbf{E}_f^\eta - \frac{1}{\eta}\, \partial_\nu (\mathbf{E}_f^\eta \times \mathbf{n}) = 0, \end{cases} \tag{107}$$

where the operator $\mathcal{T}_\Gamma^{\eta\nu}$ is given by the lemma (0.6). To write the ferromagnetic law on the scaled domain, we assume that the characteristics of the thin layer do not depend on the thickness η. That is, there exists, $\mathcal{L}$, Φ, and $\mathbf{M}_0$ functions defined on $\Gamma\times]0,\,1[$ and independent of η, such that: for $\mathbf{x} \in \Omega_{\mathrm{f}}^{\eta}$; $\mathbf{x} = \mathbf{x}_\Gamma + \eta\nu\,\mathbf{n}$,

$$
\begin{aligned}
&\mathcal{L}^{\eta}(m,h;\,\mathbf{x}) = \mathcal{L}(m,h;\,\mathbf{x}_\Gamma,\nu), \quad m,\,h \in \mathbb{R}^3,\\
&\Phi^{\eta}(m;\,\mathbf{x}) = \Phi(m,h;\,\mathbf{x}_\Gamma,\nu), \quad m \in \mathbb{R}^3,\\
&\mathbf{M}_0^{\eta}(\mathbf{x}) = \mathbf{M}_0(\mathbf{x}_\Gamma,\nu)
\end{aligned}
$$

From now on, we will not indicate explicitly the dependence of $\mathcal{L}$ and Φ on the variables $(\mathbf{x}_\Gamma,\nu)$. The system (101) becomes

$$
\left\{
\begin{aligned}
&\partial_t \mathbf{M}_{\mathrm{f}}^{\eta} = \mathcal{L}(\mathbf{M}_{\mathrm{f}}^{\eta},\,\mathbf{H}_{\mathrm{tot}}^{\eta}), \quad \mathbf{M}_{\mathrm{f}}^{\eta}(t=0) = \mathbf{M}_0, \quad \text{on} \quad \Gamma\times]0,\,1[,\\
&\mathbf{H}_{\mathrm{tot}}^{\eta} = \mathbf{H}_{\mathrm{f}}^{\eta} - \nabla_M \Phi(\mathbf{M}_{\mathrm{f}}^{\eta}).
\end{aligned}
\right. \tag{108}
$$

Principle of the construction. We use the ansatz (19) and the analogous one for $\mathbf{M}_{\mathrm{f}}^{\eta}$:

$$
\mathbf{M}_{\mathrm{f}}^{\eta} = \mathbf{M}_{\mathrm{f}}^{0} + \eta\,\mathbf{M}_{\mathrm{f}}^{1} + \eta^2\,\mathbf{M}_{\mathrm{f}}^{2} + \;\cdots \tag{109}
$$

as well as the Taylor expansion (87) of the operator $\mathcal{T}_\Gamma^{\eta\nu}$. To adapt the construction of the linear case to the present one, the idea is to consider first the field $\mathbf{M}_{\mathrm{f}}^{\eta}$ as given source term for (107). This permits to derive effective conditions of order k of the form

$$
\widetilde{\mathbf{E}}_v^{\eta} \times \mathbf{n} = \mathcal{B}_k^{\eta}(\Pi_{\|}\widetilde{\mathbf{H}}_v^{\eta}) + \mathcal{T}_k^{\eta}(\widetilde{\mathbf{M}}_{\mathrm{f}}^{\eta}) \quad \text{on} \quad \Gamma\times\mathbb{R}^+, \tag{110}
$$

where $\mathcal{B}_k^{\eta}$ is the same operator as in the linear case, $\mathcal{T}_k^{\eta}$ is a linear operator and $\widetilde{\mathbf{M}}_{\mathrm{f}}^{\eta}$ is a certain approximation of $\mathbf{M}_{\mathrm{f}}^{\eta}$ defined on $\Gamma\times]0,1[\times\mathbb{R}^+$. Let us recall that

$$
\begin{aligned}
\mathcal{B}_2^{\eta} &= -\frac{\eta}{\varepsilon_{\mathrm{f}}}\;\left(\varepsilon_{\mathrm{f}}\,\mu_{\mathrm{f}}\,\partial_{tt} + \vec{\mathrm{rot}}_\Gamma\,\mathrm{rot}_\Gamma\,\right),\\
\mathcal{B}_3^{\eta} &= -\frac{\eta}{\varepsilon_{\mathrm{f}}}\;\left(\varepsilon_{\mathrm{f}}\,\mu_{\mathrm{f}}\,[1-\eta(\mathcal{C}-H)]\partial_{tt} + \vec{\mathrm{rot}}_\Gamma\,(1-\eta H)\mathrm{rot}_\Gamma\,\right).
\end{aligned} \tag{111}
$$

To determine the operator $\mathcal{T}_k^{\eta}$, we use the same procedure as the construction of $\mathcal{B}_k^{\eta}$, explained in the linear case. So the first step requires the identified equations associated to the asymptotic expansions. The second step of the formal construction is to characterize $\widetilde{\mathbf{M}}_{\mathrm{f}}^{\eta}$ by using the ferromagnetic law. Indeed, the main difficulty consists on proving the stability in time.

Formal identification. Comparing with the linear case, in the identification process, only equation (88) changes to:

$$
\left\{
\begin{aligned}
&\partial_\nu(\mathbf{E}_{\mathrm{f}}^{k+1}\times\mathbf{n}) = \mu_{\mathrm{f}}\,\partial_t(\mathbf{H}_{\mathrm{f}}^{k} + \mathbf{M}_{\mathrm{f}}^{k}) + \textstyle\sum_{i=0}^{k}(-\nu)^i\,\mathcal{T}_\Gamma^{i}\mathbf{E}_{\mathrm{f}}^{k-i}\\
&\mathbf{E}_{\mathrm{f}}^{k+1}(\mathbf{x}_\Gamma,1)\times\mathbf{n} = 0, \quad \text{for} \;\; \mathbf{x}_\Gamma \in \Gamma.
\end{aligned}
\right. \tag{112}
$$

In a similar way, and keeping the notation of the linear case, $\Pi_\| \mathbf{E}_\mathrm{f}^0$ and $\Pi_\| \mathbf{H}_\mathrm{f}^0$ are still given by (24) and

Lemma 0.11 *For all $k \geq 0$, we have for $(\mathbf{x}_\Gamma, \nu) \in \Gamma \times]0,1[$,*

$$\begin{aligned}\partial_t \mathbf{E}_\mathrm{f}^{k+1}(\mathbf{x}_\Gamma, \nu) \times \mathbf{n} = & - \int_\nu^1 \mu_\mathrm{f}\, \partial_{tt} \left(\Pi_\| \mathbf{H}_\mathrm{f}^k + \Pi_\| \mathbf{M}_\mathrm{f}^k\right)(\mathbf{x}_\Gamma, \xi)\, d\xi \\ & - \int_\nu^1 \frac{1}{\varepsilon_\mathrm{f}} \sum_{i=0}^k \sum_{j=0}^i (-\xi)^{k-j}\, \vec{\mathrm{rot}}_\Gamma^{(k-i)}\, \mathrm{rot}_\Gamma^{(i-j)}\, \mathbf{H}_\mathrm{f}^j(\mathbf{x}_\Gamma, \xi)\, d\xi \\ & + \int_\nu^1 \sum_{i=0}^k (-\xi)^{k-i} \left(\mathcal{C}^{k-i+1}\, \partial_t \mathbf{E}_\mathrm{f}^i(\mathbf{x}_\Gamma, \xi)\right) \times \mathbf{n}\, d\xi \end{aligned} \tag{113}$$

$$\begin{aligned}\partial_t \mathbf{H}_\mathrm{f}^{k+1}(\mathbf{x}_\Gamma, \nu) \times \mathbf{n} = & \, \partial_t \mathbf{H}_v^{k+1}(\mathbf{x}_\Gamma) \times \mathbf{n} - \int_0^\nu \varepsilon_\mathrm{f}\, \partial_{tt} \Pi_\| \mathbf{E}_\mathrm{f}^k(\mathbf{x}_\Gamma, \xi)\, d\xi \\ & - \int_0^\nu \frac{1}{\mu_\mathrm{f}} \sum_{i=0}^k \sum_{j=0}^i (-\xi)^{k-j}\, \vec{\mathrm{rot}}_\Gamma^{(k-i)}\, \mathrm{rot}_\Gamma^{(i-j)}\, \mathbf{E}_\mathrm{f}^j(\mathbf{x}_\Gamma, \xi)\, d\xi \\ & - \int_0^\nu \sum_{i=0}^k (-\xi)^{k-i} \left(\mathcal{C}^{k-i+1}\, \partial_t \mathbf{H}_\mathrm{f}^i(\mathbf{x}_\Gamma, \xi)\right) \times \mathbf{n}\, d\xi \\ & - \int_0^\nu \sum_{i=0}^k (-\xi)^{k-i}\, \vec{\mathrm{rot}}_\Gamma^{(k-i)}\, \partial_t (\mathbf{M}_\mathrm{f}^i \cdot \mathbf{n})(\mathbf{x}_\Gamma, \xi)\, d\xi. \end{aligned} \tag{114}$$

3.3. Condition of order 2

We need to find the analogous to (94). Making $k = 0$ in (113) and using (24) yields

$$\begin{aligned} & \partial_t \mathbf{E}_\mathrm{f}^1(\mathbf{x}_\Gamma, \nu) \times \mathbf{n} \\ & \quad = -(1-\nu)\left(\mu_\mathrm{f}\, \partial_{tt} + \frac{1}{\varepsilon_\mathrm{f}} \vec{\mathrm{rot}}_\Gamma\, \mathrm{rot}_\Gamma\right) \Pi_\| \mathbf{H}_v^0(\mathbf{x}_\Gamma) \\ & \qquad - \int_\nu^1 \mu_\mathrm{f}\, \partial_{tt} \Pi_\| \mathbf{M}_\mathrm{f}^0(\mathbf{x}_\Gamma, \xi)\, d\xi \end{aligned} \tag{115}$$

Let $\widetilde{\mathbf{M}}_\mathrm{f}^\eta$ be an approximation of order one, that will be specified later, of the magnetization $\mathbf{M}_\mathrm{f}^\eta$. We can write in a formal way: $\widetilde{\mathbf{M}}^\eta = \mathbf{M}_\mathrm{f}^0 + O(\eta)$. If we combine this with (115) and (24), we can write an effective boundary condition of order 2, like:

$$\partial_t \widetilde{\mathbf{E}}_v^\eta(\mathbf{x}_\Gamma) \times \mathbf{n} = \mathcal{B}_2^\eta(\Pi_\| \widetilde{\mathbf{H}}_v^\eta)(\mathbf{x}_\Gamma) - \eta\, \mu_\mathrm{f} \int_0^1 \partial_{tt} \Pi_\| \widetilde{\mathbf{M}}_\mathrm{f}^\eta(\mathbf{x}_\Gamma, \nu)\, d\nu. \tag{116}$$

Now, the question is how to determine $\widetilde{\mathbf{M}}_\mathrm{f}^\eta$, using the ferromagnetic law $\mathcal{L}$.

Let us give the general principle of the answer (i.e. how to determine $\widetilde{\mathbf{M}}_\mathrm{f}^\eta$, an approximation of order k of $\mathbf{M}_f^\eta$): suppose that $\widetilde{\mathbf{H}}_f^\eta$ is an approximation of order k of $\mathbf{H}_f^\eta$: $\mathbf{H}_f^\eta = \widetilde{\mathbf{H}}_f^\eta + \eta^k \mathbf{H}_0^\eta$, where $\mathbf{H}_0^\eta$ is uniformly bounded with respect to η. Thus

$$\partial_t \mathbf{M}_f^\eta = \mathcal{L}(\mathbf{M}_f^\eta, \widetilde{\mathbf{H}}_f^\eta - \nabla_M \Phi(\mathbf{M}_f^\eta)) + \eta^k \mathcal{L}(\mathbf{M}_f^\eta, \mathbf{H}_0^\eta).$$

As $\mathbf{M}_f^\eta$ has a uniform bound with respect to η (by (103)), we conclude (at least formally) that $\mathcal{L}(\mathbf{M}_f^\eta, \mathbf{H}_0^\eta)$ is also bounded uniformly with respect to η. So $\mathcal{L}(\mathbf{M}_f^\eta, \widetilde{\mathbf{H}}_f^\eta - \nabla_M \Phi(\mathbf{M}_f^\eta))$ constitute an approximation of order k of $\partial_t \mathbf{M}_f^\eta$. A natural way to define $\widetilde{\mathbf{M}}_f^\eta$ is then impose

$$\partial_t \widetilde{\mathbf{M}}_f^\eta = \mathcal{L}(\widetilde{\mathbf{M}}_f^\eta, \widetilde{\mathbf{H}}_f^\eta - \nabla_M \Phi(\widetilde{\mathbf{M}}_f^\eta)), \quad \widetilde{\mathbf{M}}_f^\eta(t=0) = \mathbf{M}_0. \tag{117}$$

Doing so, we see that the definition of $\widetilde{\mathbf{M}}_f^\eta$ amounts the one of $\widetilde{\mathbf{H}}_f^\eta$. This can be done using the asymptotic expansion

$$\mathbf{H}_f^\eta = \sum_{i=0}^{k-1} \eta^i \, \mathbf{H}_f^i + O(\eta^k) \tag{118}$$

To determine $\mathbf{H}_f^k$, we split it into two parts: $\Pi_\| \mathbf{H}_f^k$ that can be computed by (114) and $\Pi_\perp \mathbf{H}_f^k$ that can be computed by applying $\Pi_\perp$ to (112), that yields

$$\partial_t(\mathbf{H}_f^k \cdot \mathbf{n}) = -\partial_t(\mathbf{M}_f^k \cdot \mathbf{n}) - \frac{1}{\mu_f} \sum_{i=0}^{k} (-\nu)^i \, \mathrm{rot}_\Gamma^{(i)} \, \mathbf{E}_f^{k-i}. \tag{119}$$

We are going to see that applying expression (118) is sufficient to derive a stable effective boundary condition of order 2, while the condition of order 3 requires slight modifications.

We come back now to the second order condition, where, according to the previous considerations, we need an approximation of order 1 of $\mathbf{H}_f^\eta$. Combining (24) and (119) applied to $k=0$, we see that

$$\begin{aligned} \mathbf{H}_f^\eta(\mathbf{x}_\Gamma, \nu) &= \Pi_\| \mathbf{H}_v^0(\mathbf{x}_\Gamma) + \Pi_\perp (\mathbf{M}_0 - \mathbf{M}_f^0)(\mathbf{x}_\Gamma, \nu) + O(\eta) \\ &= \Pi_\| \widetilde{\mathbf{H}}_v^\eta(\mathbf{x}_\Gamma) + \Pi_\perp (\mathbf{M}_0 - \widetilde{\mathbf{M}}_f^\eta)(\mathbf{x}_\Gamma, \nu) + O(\eta) \end{aligned} \tag{120}$$

So, setting $\widetilde{\mathbf{H}}_{\mathrm{f}}^{\eta}(\mathbf{x}_\Gamma,\nu) = \Pi_{\|}\widetilde{\mathbf{H}}_v^{\eta}(\mathbf{x}_\Gamma) + \Pi_{\perp}(\mathbf{M}_0 - \widetilde{\mathbf{M}}_{\mathrm{f}}^{\eta})(\mathbf{x}_\Gamma,\nu)$ and applying (116) and (117) leads to following condition

$$\partial_t\widetilde{\mathbf{E}}_v^{\eta}(\mathbf{x}_\Gamma)\times\mathbf{n} = \mathcal{B}_2^{\eta}(\Pi_{\|}\widetilde{\mathbf{H}}_v^{\eta})(\mathbf{x}_\Gamma) - \eta\,\mu_{\mathrm{f}}\int_0^1 \Pi_{\|}\widetilde{\mathbf{M}}_{\mathrm{f}}^{\eta}(\mathbf{x}_\Gamma,\nu)\,d\nu,$$

where $\widetilde{\mathbf{M}}_{\mathrm{f}}^{\eta}$ satisfies for a.e. $(\mathbf{x}_\Gamma,\nu)\in\Gamma\times]0,\,1[$,

$$\begin{cases} \partial_t\widetilde{\mathbf{M}}_{\mathrm{f}}^{\eta} = \mathcal{L}(\widetilde{\mathbf{M}}_{\mathrm{f}}^{\eta},\,\widetilde{\mathbf{H}}_{\mathrm{tot}}^{\eta}), \quad \widetilde{\mathbf{M}}_{\mathrm{f}}^{\eta}(t=0) = \mathbf{M}_0, \\ \widetilde{\mathbf{H}}_{\mathrm{tot}}^{\eta}(\mathbf{x}_\Gamma,\nu) = \Pi_{\|}\widetilde{\mathbf{H}}_v^{\eta}(\mathbf{x}_\Gamma) + \Pi_{\perp}(\mathbf{M}_0 - \widetilde{\mathbf{M}}_{\mathrm{f}}^{\eta})(\mathbf{x}_\Gamma,\nu) - \nabla_{\!M}\Phi(\widetilde{\mathbf{M}}_{\mathrm{f}}^{\eta})(\mathbf{x}_\Gamma,\nu). \end{cases} \tag{121}$$

According to the following theorem, this condition is stable.

Theorem 0.6 – *Regular solutions* $(\widetilde{\mathbf{E}}_v^{\eta},\widetilde{\mathbf{H}}_v^{\eta},\widetilde{\mathbf{M}}_{\mathrm{f}}^{\eta})$ *of* {(6), (121)} *satisfy, for all* $t\geq 0$,

$$|\,\widetilde{\mathbf{M}}_{\mathrm{f}}^{\eta}\,| = |\,\mathbf{M}_0\,| \;\; a.e.\;\; \Gamma\times]0,\,1[\quad and$$

$$\frac{d}{dt}\left\{\mathcal{E}_v(\widetilde{\mathbf{E}}_v^{\eta}(t),\widetilde{\mathbf{H}}_v^{\eta}(t)) + \eta\,\mathcal{E}_\Gamma^{\eta}(t) + \eta\,\mathcal{E}_{\mathrm{f}}^{\eta}(t)\right\} \leq 0.$$

where, if we set, $\varphi^{\eta}(\mathbf{x}_\Gamma,t) = \Pi_{\|}\widetilde{\mathbf{H}}_v^{\eta}(\mathbf{x}_\Gamma,t),\;\; (\mathbf{x}_\Gamma,t)\in\Gamma\times\mathbb{R}^+$,

$$\mathcal{E}_\Gamma^{\eta}(t) = \frac{\mu_{\mathrm{f}}}{2}\,\|\varphi^{\eta}(t)\|_{L^2}^2 + \frac{1}{2\,\varepsilon_{\mathrm{f}}}\left\|\int_0^t(\mathrm{rot}_\Gamma\,\varphi^{\eta})(\tau)\,d\tau\right\|_{L^2}^2,$$

$$\mathcal{E}_{\mathrm{f}}^{\eta}(t) = \frac{\mu_{\mathrm{f}}}{2}\,\|\,\Pi_{\perp}(\mathbf{M}_0 - \widetilde{\mathbf{M}}_{\mathrm{f}}^{\eta})(t)\,\|_{L^2}^2 + \mu_{\mathrm{f}}\int_{\Gamma\times]0,\,1[}\Phi(\widetilde{\mathbf{M}}_{\mathrm{f}}^{\eta}(t))\,d\mathbf{x}_\Gamma\,d\nu.$$

Proof. As (121) has the same structure than (101), property (103) still applies on the scaled domain. From Maxwell's equations and (121), we get (see the proof of theorem 35)

$$\frac{d}{dt}\{\mathcal{E}_v(\widetilde{\mathbf{E}}_v^{\eta}(t),\widetilde{\mathbf{H}}_v^{\eta}(t)) + \eta\,\mathcal{E}_\Gamma^{\eta}(t)\} = -\mu_{\mathrm{f}}\int_\Gamma\int_0^1\partial_t\Pi_{\|}\widetilde{\mathbf{M}}_{\mathrm{f}}^{\eta}(\mathbf{x}_\Gamma,\nu)\cdot\varphi^{\eta}(\mathbf{x}_\Gamma)\,d\nu\,d\mathbf{x}_\Gamma. \tag{122}$$

For *a.e.* $(\mathbf{x}_\Gamma,\nu)\in\Gamma\times]0,\,1[$,

$$\partial_t\Pi_{\|}\widetilde{\mathbf{M}}_{\mathrm{f}}^{\eta}\cdot\varphi^{\eta} = \partial_t\widetilde{\mathbf{M}}_{\mathrm{f}}^{\eta}\cdot\varphi^{\eta} = \partial_t\widetilde{\mathbf{M}}_{\mathrm{f}}^{\eta}\cdot\left(\widetilde{\mathbf{H}}_{\mathrm{tot}}^{\eta} + \nabla_{\!M}\Phi(\widetilde{\mathbf{M}}_{\mathrm{f}}^{\eta}) - \Pi_{\perp}(\mathbf{M}_0 - \widetilde{\mathbf{M}}_{\mathrm{f}}^{\eta})\right)$$

So, using property (102)(ii),

$$\partial_t\Pi_{\|}\widetilde{\mathbf{M}}_{\mathrm{f}}^{\eta}\cdot\varphi^{\eta} \geq \partial_t\{\,\Phi(\widetilde{\mathbf{M}}_{\mathrm{f}}^{\eta}) + \tfrac{1}{2}|\,\Pi_{\perp}(\mathbf{M}_0 - \widetilde{\mathbf{M}}_{\mathrm{f}}^{\eta})\,|^2\,\}$$

Integrating over $\Gamma\times]0,\,1[$, shows

$$-\mu_{\mathrm{f}}\int_\Gamma\int_0^1\partial_t\Pi_{\|}\widetilde{\mathbf{M}}_{\mathrm{f}}^{\eta}(\mathbf{x}_\Gamma,\nu)\cdot\varphi^{\eta}(\mathbf{x}_\Gamma)\,d\nu\,d\mathbf{x}_\Gamma \leq -\frac{d}{dt}\,\mathcal{E}_{\mathrm{f}}^{\eta}(t) \tag{123}$$

The energy estimate of theorem (0.6) is deduced by identifying the right hand side of (122) with the left hand side of (123). ■

3.4. Condition of order 3

Let us begin by the "Maxwell" part (like for the condition of order two). We point out the differences with the linear case. Equation (95) changes to

$$\Pi_{\parallel}\mathbf{H}_{\rm f}^1(\mathbf{x}_\Gamma,\nu) = \Pi_{\parallel}\mathbf{H}_v^1(\mathbf{x}_\Gamma) - \nu\,\mathcal{C}\,\mathbf{H}_v^0(\mathbf{x}_\Gamma) - \int_0^\nu \nabla_\Gamma\left((\mathbf{M}_{\rm f}^0 - \mathbf{M}_0)\cdot\mathbf{n}\right)(\mathbf{x}_\Gamma,\xi)\,d\xi, \tag{124}$$

and relation (98) becomes

$$\begin{aligned}
&\partial_t\mathbf{E}_{\rm f}^2(\mathbf{x}_\Gamma,\nu)\times\mathbf{n} = \\
&-\mu_{\rm f}\,\partial_{tt}\left((1-\nu)\,\Pi_{\parallel}\mathbf{H}_v^1(\mathbf{x}_\Gamma) - (1-\nu^2)(\mathcal{C}-H)\Pi_{\parallel}\mathbf{H}_v^0(\mathbf{x}_\Gamma)\right)\\
&-\tfrac{1}{\varepsilon_f}\left((1-\nu)\,\vec{\mathrm{rot}}_\Gamma\,\mathrm{rot}_\Gamma\,\Pi_{\parallel}\mathbf{H}_v^1(\mathbf{x}_\Gamma) - (1-\nu^2)\,\vec{\mathrm{rot}}_\Gamma\,H\mathrm{rot}_\Gamma\,\Pi_{\parallel}\mathbf{H}_v^0(\mathbf{x}_\Gamma)\right)\\
&-\mu_{\rm f}\,\partial_{tt}\int_\nu^1\left(\Pi_{\parallel}\mathbf{M}_{\rm f}^1(\mathbf{x}_\Gamma,\xi) + \int_\xi^1 (2H-\mathcal{C})\Pi_{\parallel}\mathbf{M}_{\rm f}^0(\mathbf{x}_\Gamma,\tau)\,d\tau\right)d\xi\\
&+\mu_{\rm f}\,\partial_{tt}\int_\nu^1\int_0^\xi \nabla_\Gamma\,(\mathbf{M}_{\rm f}^0\cdot\mathbf{n})(\mathbf{x}_\Gamma,\tau)\,d\tau\,d\xi.
\end{aligned} \tag{125}$$

If we regroup (125), (115) and (24), written for $\nu = 0$, and if we set $\widetilde{\mathbf{M}}_{\rm f}^\eta = \mathbf{M}_{\rm f}^0 + \eta\mathbf{M}_{\rm f}^1 + O(\eta^2)$, an approximation of order 2 of the magnetization $\mathbf{M}_{\rm f}^\eta$ (to be specified later), we can write the condition of order three in the following way.

$$\begin{aligned}
\partial_t\widetilde{\mathbf{E}}_v^\eta(\mathbf{x}_\Gamma)\times\mathbf{n} &= \mathcal{B}_3^\eta(\Pi_{\parallel}\widetilde{\mathbf{H}}_v^\eta)(\mathbf{x}_\Gamma)\\
&-\eta\,\mu_{\rm f}\,\partial_{tt}\int_0^1 [1+\eta\,\nu\,(2H-\mathcal{C})]\Pi_{\parallel}\widetilde{\mathbf{M}}_{\rm f}^\eta(\mathbf{x}_\Gamma,\nu)\,d\nu\\
&+\eta^2\,\mu_{\rm f}\,\partial_{tt}\int_0^1 (1-\nu)\,\nabla_\Gamma\,(\widetilde{\mathbf{M}}_{\rm f}^\eta\cdot\mathbf{n})(\mathbf{x}_\Gamma,\nu)\,d\nu
\end{aligned} \tag{126}$$

We used the fact that $\int_0^1 f(\xi)\left(\int_\xi^1 g(\nu)\,d\nu\right)d\xi = \int_0^1\left(\int_0^\xi f(\nu)\,d\nu\right)g(\xi)\,d\xi$.

To obtain an approximation of order 2 of $\mathbf{M}_{\rm f}^\eta$ we need to compute $\Pi_\perp\mathbf{H}_{\rm f}^1$. Equation (119), written for $k = 1$ and combined with (115), leads to (we use $\mathrm{rot}_\Gamma\,\nabla_\Gamma = 0$)

$$\begin{aligned}
\mathbf{H}_{\rm f}^1(\mathbf{x}_\Gamma,\nu)\cdot\mathbf{n} = -(\mathbf{M}_{\rm f}^1\cdot\mathbf{n})(\mathbf{x}_\Gamma,\nu) + (1-\nu)\,\mathrm{div}_\Gamma\,(\Pi_{\parallel}\mathbf{H}_v^0)(\mathbf{x}_\Gamma) - \\
\int_\nu^1 \mathrm{div}_\Gamma\,(\Pi_{\parallel}(\mathbf{M}_0 - \mathbf{M}_{\rm f}^0))(\mathbf{x}_\Gamma,\xi)\,d\xi.
\end{aligned} \tag{127}$$

We obtain the following approximation of $\mathbf{H}_\mathrm{f}^\eta$ by regrouping (120), (124) and (127):

$$\begin{aligned}&\mathbf{H}_\mathrm{f}^\eta(\mathbf{x}_\Gamma,\nu)\\&=[1-\eta\,\nu\,\mathcal{C}]\,\Pi_\parallel\widetilde{\mathbf{H}}_v^\eta(\mathbf{x}_\Gamma)+\eta\int_0^\nu\nabla_\Gamma\left((\mathbf{M}_0-\widetilde{\mathbf{M}}_\mathrm{f}^\eta)\cdot\mathbf{n}\right)(\mathbf{x}_\Gamma,\xi)\,d\xi,\\&+\Pi_\perp(\mathbf{M}_0-\widetilde{\mathbf{M}}_\mathrm{f}^\eta)(\mathbf{x}_\Gamma,\nu)+\eta\,(1-\nu)\,\{\,\mathrm{div}_\Gamma\,(\Pi_\parallel\widetilde{\mathbf{H}}_v^\eta)(\mathbf{x}_\Gamma)\,\}\,\mathbf{n}\\&-\eta\left(\int_\nu^1\mathrm{div}_\Gamma\,(\Pi_\parallel(\mathbf{M}_0-\widetilde{\mathbf{M}}_\mathrm{f}^\eta))(\mathbf{x}_\Gamma,\xi)\,d\xi\right)\mathbf{n}\;+O(\eta^2).\end{aligned}\tag{128}$$

Contrary to the case of the second order condition, when we use (128) (of course without $O(\eta^2)$) to determine $\widetilde{\mathbf{M}}_\mathrm{f}^\eta$ via (117) and we couple this with (126) in order to built a third order condition, we are unable to prove stability via energy estimate like in theorem 0.6 (this doesn't mean the instability of the obtained condition). However, we are able derive another expression of the third order condition, that differs from the initial one by terms of order η^3, but that enables us to get energy estimates. The next modifications are suggested by difficulties encountered when trying to prove energy estimates.

In (126), we replace the operator $[1+\eta\,\nu\,(2H-\mathcal{C})]$ by $(1+\eta\,\nu\,2H)(1-\eta\,\nu\,\mathcal{C})$. As the two expressions differs by a $O(\eta^2)$ term of order η^2, the new condition differs from the initial one by $O(\eta^3)$ terms, which is consistent with the order of (126). Remark that

$$\eta\,r^\eta\stackrel{\mathrm{def}}{=}\eta\,(1+\eta\,\nu\,2H)\tag{129}$$

is the development of order 3 of the volume element $d\Omega_\mathrm{f}^\eta$. Meanwhile unstead of using (128) we set, using the usual notation $\varphi^\eta(\mathbf{x}_\Gamma,t)=\Pi_\parallel\widetilde{\mathbf{H}}^\eta(\mathbf{x}_\Gamma,t)$,

$$\begin{aligned}\widetilde{\mathbf{H}}_\mathrm{f}^\eta(\mathbf{x}_\Gamma,\nu)\;&=\;[1-\eta\,\nu\,\mathcal{C}]\varphi^\eta(\mathbf{x}_\Gamma)+\frac{\eta}{r^\eta}\int_0^\nu\nabla_\Gamma\,\phi_\mathbf{n}^\eta(\mathbf{x}_\Gamma,\xi)\,d\xi\\&\quad+\left(\phi_\mathbf{n}^\eta(\mathbf{x}_\Gamma,\nu)+\frac{\eta\,(1-\nu)}{r^\eta}\,\mathrm{div}_\Gamma\,\varphi^\eta(\mathbf{x}_\Gamma)\right)\mathbf{n},\end{aligned}\tag{130}$$

where $\phi_\mathbf{n}^\eta$ is a scalar field defined on $\Gamma\times]0,\,1[$ by:

$$\phi_\mathbf{n}^\eta(\mathbf{x}_\Gamma,\nu)=(\mathbf{M}_0-\widetilde{\mathbf{M}}_\mathrm{f}^\eta)\cdot\mathbf{n}-\frac{\eta}{r^\eta}\int_\nu^1\mathrm{div}_\Gamma\,(\Pi_\parallel(\mathbf{M}_0-\widetilde{\mathbf{M}}_\mathrm{f}^\eta))(\mathbf{x}_\Gamma,\xi)\,d\xi.\tag{131}$$

Comparing the right hand side of (128) to the right hand side (130), we see that

$$\mathbf{H}_\mathrm{f}^\eta=\widetilde{\mathbf{H}}_\mathrm{f}^\eta+O(\eta^2),$$

which satisfies the requirement of the construction of the third order condition. In conclusion, if we set

$$\psi^\eta(\mathbf{x}_\Gamma,\nu) = r^\eta(1-\eta\,\nu\mathcal{C})\Pi_{\|}\widetilde{\mathbf{M}}_{\mathrm{f}}^\eta(\mathbf{x}_\Gamma,\nu) - \eta\,(1-\nu)\,\nabla_\Gamma\,(\widetilde{\mathbf{M}}_{\mathrm{f}}^\eta\cdot\mathbf{n})(\mathbf{x}_\Gamma,\nu), \quad (132)$$

the new third order condition can be rewritten as

$$\partial_t\widetilde{\mathbf{E}}_v^\eta(\mathbf{x}_\Gamma)\times\mathbf{n} = \mathcal{B}_3^\eta(\varphi^\eta)(\mathbf{x}_\Gamma) - \eta\,\mu_{\mathrm{f}}\int_0^1 \partial_{tt}\psi^\eta(\mathbf{x}_\Gamma,\nu)\,d\nu,$$

where ψ^η is related to $\widetilde{\mathbf{M}}_{\mathrm{f}}^\eta$ by (132), and $\widetilde{\mathbf{M}}_{\mathrm{f}}^\eta$ satisfies a.e. $\Gamma\times]0,1[$,

$$\partial_t\widetilde{\mathbf{M}}_{\mathrm{f}}^\eta = \mathcal{L}(\widetilde{\mathbf{M}}_{\mathrm{f}}^\eta,\,\widetilde{\mathbf{H}}_{\mathrm{tot}}^\eta),\quad \widetilde{\mathbf{M}}_{\mathrm{f}}^\eta(t=0)=\mathbf{M}_0,\quad \widetilde{\mathbf{H}}_{\mathrm{tot}}^\eta = \widetilde{\mathbf{H}}_{\mathrm{f}}^\eta - \nabla_M\Phi(\mathbf{M}_{\mathrm{f}}^\eta).$$

The field $\widetilde{\mathbf{H}}_{\mathrm{f}}^\eta$ is given by (130).

(133)

The main justification of this condition is the following a priori estimate.

Theorem 0.7 – *For $\eta < \frac{1}{|2H|}$, theorem 0.6 remains valid when we replace the effective boundary condition (121), by condition (133), and the expressions of $\mathcal{E}_\Gamma^\eta$ and $\mathcal{E}_{\mathrm{f}}^\eta$ by*

$$\begin{aligned}\mathcal{E}_\Gamma^\eta(t) &= \tfrac{\mu_{\mathrm{f}}}{2}\textstyle\int_\Gamma[1-\eta(\mathcal{C}-H)]\partial_t\varphi^\eta\cdot\partial_t\varphi^\eta d\mathbf{x}_\Gamma \\ &+\tfrac{1}{2\,\varepsilon_{\mathrm{f}}}\textstyle\int_\Gamma(1-\eta H)\left|\displaystyle\int_0^t(\mathrm{rot}_\Gamma\,\varphi^\eta)(\tau)d\tau\right|^2 d\mathbf{x}_\Gamma\end{aligned} \quad (134)$$

$$\mathcal{E}_{\mathrm{f}}^\eta(t) = \mu_{\mathrm{f}}\int_{\Gamma\times]0,\,1[}\left(\tfrac{1}{2}\,|\phi_{\mathbf{n}}^\eta(t)|^2 + \Phi(\widetilde{\mathbf{M}}_{\mathrm{f}}^\eta(t))\right)\,r^\eta\,d\mathbf{x}_\Gamma\,d\nu.$$

Proof. Like in the case of the second order condition, we have the classical identity

$$\frac{d}{dt}\{\mathcal{E}_v^\eta(t) + \eta\,\mathcal{E}_\Gamma^\eta(t)\} = -\mu_{\mathrm{f}}\int_\Gamma\int_0^1 \partial_t\psi^\eta(\mathbf{x}_\Gamma,\nu)\cdot\varphi^\eta(\mathbf{x}_\Gamma)\,d\nu\,d\mathbf{x}_\Gamma. \quad (135)$$

We are going to compute the right hand side of (135). Using the expression of ψ^η in (132), one has

$$\partial_t\psi^\eta\cdot\varphi^\eta = r^\eta\,\partial_t\widetilde{\mathbf{M}}_{\mathrm{f}}^\eta\cdot[1-\eta\,\nu\,\mathcal{C}]\varphi^\eta - \eta(1-\nu)\nabla_\Gamma\,(\partial_t\widetilde{\mathbf{M}}_{\mathrm{f}}^\eta\cdot\mathbf{n})\cdot\varphi^\eta. \quad (136)$$

$[1-\eta\,\nu\,\mathcal{C}]\varphi^\eta$ can be expressed using (130), where $\widetilde{\mathbf{H}}_{\mathrm{f}}^\eta = \widetilde{\mathbf{H}}_{\mathrm{tot}}^\eta + \nabla_M\Phi(\mathbf{M}_{\mathrm{f}}^\eta)$ as shown by (133). Such doing, we get

$$\begin{aligned}r^\eta\,\partial_t\widetilde{\mathbf{M}}_{\mathrm{f}}^\eta\cdot[1-\eta\,\nu\,\mathcal{C}]\varphi^\eta &= \partial_t\widetilde{\mathbf{M}}_{\mathrm{f}}^\eta\cdot\{r^\eta\,(\widetilde{\mathbf{H}}_{\mathrm{tot}}^\eta \\ -\phi_{\mathbf{n}}^\eta\,\mathbf{n} + \nabla_M\Phi(\mathbf{M}_{\mathrm{f}}^\eta)) &- \eta\left(\textstyle\int_0^\nu \nabla_\Gamma\,\phi_{\mathbf{n}}^\eta\,d\xi\right) - \eta(1-\nu)(\mathrm{div}_\Gamma\,\varphi^\eta)\,\mathbf{n}\}.\end{aligned} \quad (137)$$

For $\eta < \frac{1}{|2H|}$, $r^\eta > 0$. Thus, property (102)(iii) yields $r^\eta\, \partial_t \widetilde{\mathbf{M}}_{\mathrm{f}}^\eta \cdot \widetilde{\mathbf{H}}_{\mathrm{tot}}^\eta \geq 0$. Then using (137) in (136), yields, after integrating over $\Gamma \times [0,1]$ and multiplying by $-\mu_{\mathrm{f}}$,

$$
\begin{aligned}
-\mu_{\mathrm{f}} \int_\Gamma \int_0^1 \partial_t \psi^\eta \cdot \varphi^\eta \, d\nu \, d\mathbf{x}_\Gamma \leq & -\mu_{\mathrm{f}} \frac{d}{dt} \int_\Gamma \int_0^1 \Phi(\widetilde{\mathbf{M}}_{\mathrm{f}}^\eta(t))\, r^\eta \, d\mathbf{x}_\Gamma \, d\nu \\
& + \mu_{\mathrm{f}} \int_\Gamma \int_0^1 \phi_{\mathbf{n}}^\eta \cdot \partial_t (\widetilde{\mathbf{M}}_{\mathrm{f}}^\eta \cdot \mathbf{n})\, r^\eta \, d\nu \, d\mathbf{x}_\Gamma \\
& + \eta\, \mu_{\mathrm{f}} \textstyle\int_\Gamma \int_0^1 \left(\displaystyle\int_0^\nu \nabla_\Gamma \phi_{\mathbf{n}}^\eta \, d\xi \right) \cdot \partial_t \Pi_{\|} \widetilde{\mathbf{M}}_{\mathrm{f}}^\eta \, d\nu \, d\mathbf{x}_\Gamma \\
& + \eta\, \mu_{\mathrm{f}} \int_0^1 (1-\nu) \left\{ \int_\Gamma \nabla_\Gamma (\partial \widetilde{\mathbf{M}}_{\mathrm{f}}^\eta \cdot \mathbf{n}) \cdot \varphi^\eta + (\partial_t \widetilde{\mathbf{M}}_{\mathrm{f}}^\eta \cdot \mathbf{n}) \operatorname{div}_\Gamma (\varphi^\eta) \, d\mathbf{x}_\Gamma \right\} d\nu
\end{aligned}
\tag{138}
$$

The last term in (138) is zero (integrate by parts). Let A be the value of the the second line of (138). Using the expression (131) of $\phi_{\mathbf{n}}^\eta$, one has $A = \mu_{\mathrm{f}} (I_1 + \eta\, I_2)$ where,

$$
\begin{aligned}
I_1 &= \int_\Gamma \int_0^1 \phi_{\mathbf{n}}^\eta \cdot \partial_t (\widetilde{\mathbf{M}}_{\mathrm{f}}^\eta \cdot \mathbf{n})\, r^\eta \, d\nu \, d\mathbf{x}_\Gamma \\
&= -\int_\Gamma \int_0^1 \frac{1}{2} \partial_t \left| \Pi_\perp (\mathbf{M}_0 - \widetilde{\mathbf{M}}_{\mathrm{f}}^\eta) \right|^2 r^\eta \, d\nu \, d\mathbf{x}_\Gamma \\
&\quad + \eta \int_\Gamma \int_0^1 \partial_t ((\mathbf{M}_0 - \widetilde{\mathbf{M}}_{\mathrm{f}}^\eta) \cdot \mathbf{n}) \left(\int_\nu^1 \operatorname{div}_\Gamma (\Pi_{\|} (\mathbf{M}_0 - \widetilde{\mathbf{M}}_{\mathrm{f}}^\eta)) \, d\xi \right) d\nu \, d\mathbf{x}_\Gamma
\end{aligned}
$$

and,

$$
\begin{aligned}
I_2 &= \int_\Gamma \int_0^1 \left(\int_0^\nu \nabla_\Gamma \phi_{\mathbf{n}}^\eta \, d\xi \right) \cdot \partial_t \Pi_{\|} \widetilde{\mathbf{M}}_{\mathrm{f}}^\eta \, d\nu \, d\mathbf{x}_\Gamma \\
&= \int_\Gamma \int_0^1 \phi_{\mathbf{n}}^\eta \cdot \partial_t \left(\int_\nu^1 \operatorname{div}_\Gamma (\Pi_{\|} (\mathbf{M}_0 - \widetilde{\mathbf{M}}_{\mathrm{f}}^\eta)) d\xi \right) d\nu \, d\mathbf{x}_\Gamma \\
&= \int_\Gamma \int_0^1 ((\mathbf{M}_0 - \widetilde{\mathbf{M}}_{\mathrm{f}}^\eta) \cdot \mathbf{n})\, \partial_t \left(\int_\nu^1 \operatorname{div}_\Gamma (\Pi_{\|} (\mathbf{M}_0 - \widetilde{\mathbf{M}}_{\mathrm{f}}^\eta)) \, d\xi \right) d\nu \, d\mathbf{x}_\Gamma \\
&\quad - \eta \int_\Gamma \int_0^1 \frac{1}{r^\eta} \frac{1}{2} \partial_t \left| \int_\nu^1 \operatorname{div}_\Gamma (\Pi_{\|} (\mathbf{M}_0 - \widetilde{\mathbf{M}}_{\mathrm{f}}^\eta)) \, d\xi \right|^2 d\nu \, d\mathbf{x}_\Gamma .
\end{aligned}
$$

This gives

$$
A = -\mu_{\mathrm{f}} \frac{d}{dt} \int_\Gamma \int_0^1 \frac{1}{2} |\phi_{\mathbf{n}}^\eta(t)|^2 \, r^\eta \, d\nu \, d\mathbf{x}_\Gamma
$$

which, combined with (138), shows that

$$
-\mu_{\mathrm{f}} \int_\Gamma \int_0^1 \partial_t \psi^\eta \cdot \varphi^\eta \, d\nu \, d\mathbf{x}_\Gamma \leq -\frac{d}{dt} \mathcal{E}_{\mathrm{f}}(t) \tag{139}
$$

We obtain the desired energy estimate by combining (139) and (135).

References

[1] A. Bendali and K. Lemrabet, The effect of a thin coating on the scattering of a time-harmonic wave for the Helmholtz equation, SIAM J. Appl. Math., 58 (1996), 1664–1693.

[2] F. Collino, Conditions absorbantes d'ordre élevé pour des modèles de propagation d'ondes dans des domaines rectangulaires, I.N.R.I.A, no. 1794 (1992).

[3] B. Engquist and A. Majda, Absorbing boundary conditions for the numerical simulation of waves, Math. Comp., 31 (1977), 629–651.

[4] B. Engquist and J. C. Nedelec, Effective boundary conditions for acoustic and electromagnetic scattering in thin layers,Ecole Polytechnique-CMAP (France),278 (1993).

[5] H. Kreiss, Initial Boundary Value Problems for Hyperbolic Systems, Comm. on Pure and Appl. Math.,13 (1970),277–298.

[6] B. Gustafson, H. Kreiss, and A. Sundstrom, Stability theory of difference approximations for mixed initial boundary value problems, Math. Comp.,26 (1972), 649–686.

[7] A. Haraux, Nonlinear evolution equations-global behavior of solutions, Springer-Verlag, 1981.

[8] J.L. Joly, G. Metivier, and J. Rauch, Global solutions to Maxwell equations in a ferromagnetic medium, Séminaire EDP, Ecole polytechnique (France),1996-1997, no. 11.

[9] K. S. Yee, Numerical solution of initial boundary value problems involving Maxwell's equations, IEEE Trans. Antennas and Propagat., AP-14 (1966), 302–307.

[10] A. Visintin, On Landau-Lifchitz equations for ferromagnetism,Japan J. Appl. Math.,2 (1985), 69–84.

[11] L. N. Trefethen, Group velocity interpretation of the stability theory of Gustafsson, Kreiss and Sundström, J. Comp. Phys., 49 (1983), 199–217.

[12] P. Joly and O. Vacus, Mathematical and numerical studies of 1d non linear ferromagnetic materials, In Numerical Methods in Engineering 96, ECCOMAS, 1996.

[13] P. Joly and O. Vacus, Mathematical and numerical studies of non-linear ferromagnetic materials,M2AN, (1997).

[14] I. Terasse, Résolution mathématique des équations de Maxwell instationnaires par une méthode de potentiels retardés, Ecole polytechnique (France), 1993.

[15] Y. Choquet-Bruhat, Géométrie différentielle et système extérieur, DUNOD, Paris, 1968.

[16] O. Vacus, Modélisation de la propagation d'ondes en milieu ferromagnétique, Ecole Centrale de Paris, 1997.

[17] H. Haddar and P. Joly, An Asymptotic Approach of the Scattering of Electromagnetic Waves by Thin Ferromagnetic Coatings,Mathematical and Numerical Aspects of Wave Propagation,Siam,(1998),June.

[18] H. Haddar and P. Joly,Conditions équivalentes pour des couches minces ferromagnétiques, étude du problème monodimensionnel,I.N.R.I.A.,3431, (1998),May, Thème 4.

[19] H. Haddar and P. Joly, Effective Boundary Conditions For Thin Ferromagnetic Layers; the 1D Model.,(99),Submitted to Siam J. Appl. Math..

[20] H. Haddar and P. Joly, Stability of thin layer approximation of electromagnetic waves scattering by linear and non linear coatings, (2000),preprint.

[21] H. Haddar and P. Joly, Electromagnetic waves in laminar ferromagnetic medium. The Homogenized Problem., Mathematical and Numerical Aspects of Wave Propagation, Siam, (2000) ,July.

Houssem Haddar and Patrick Joly
INRIA, Domaine de Voluceau-Rocquencourt
BP 105
78153 Le Chesnay Cédex
France

E-mail: houssem.haddar@inria.fr, patrick.joly@inria.fr

Studies in Mathematics and its Applications, Vol. 31
D. Cioranescu and J.L. Lions (Editors)

Chapter 20

REMARQUES SUR LA LIMITE $\alpha \to 0$ POUR LES FLUIDES DE GRADE 2

D. IFTIMIE

Résumé. On considère la limite $\alpha \to 0$ dans l'équation des fluides de grade 2. On montre la convergence faible des solutions vers une solution faible de l'équation de Navier-Stokes, en supposant que les données initiales convergent faiblement dans L^2.

Abstract. We consider the limit $\alpha \to 0$ for the equation of the second grade fluids. We prove that weak convergence of the solutions to a weak solution of the Navier-Stokes equation holds by assuming that the initial data weakly converges in L^2.

1. Introduction

Il existe dans la nature des fluides qui n'obéissent pas aux classiques équations de Navier-Stokes. Des modèles plus compliqués ont dû être développés pour les étudier. Ainsi, Rivlin et Ericksen [10] introduisent les fluides de type différentiel. Un cas particulier de ces fluides est constitué par les fluides de grade 2. L'analyse de Dunn et Fosdick [6] montre que l'équation d'un tel fluide est donnée par

$$\begin{aligned} \partial_t(u - \alpha\Delta u) - \nu\Delta u + \sum_j (u - \alpha\Delta u)_j \nabla u_j & \\ + u \cdot \nabla(u - \alpha\Delta u) &= -\nabla p + f, \\ \operatorname{div} u &= 0, \end{aligned} \tag{1}$$

où $\alpha \geq 0$ est une constante matérielle, $\nu > 0$ est la viscosité du fluide, u le champ de vitesses et p la pression. Pour $\alpha = 0$, on obtient les équations classiques de Navier-Stokes

$$\partial_t u - \nu\Delta u + u \cdot \nabla u = -\nabla p + f, \quad \operatorname{div} u = 0, \tag{2}$$

de sorte que l'équation du fluide de grade 2 est une généralisation simple des équations de Navier-Stokes.

Les premiers résultats mathématiques pour les fluides de grade 2 ont été obtenus par Cioranescu et Ouazar [4]. Ces auteurs montrent l'existence et l'unicité des solutions, globale en dimension 2 et locale en dimension 3, pour des données initiales appartenant à H^3. L'existence et l'unicité globale des solutions tridimensionnelles ont été obtenues par Cioranescu et Girault [3] pour des données initiales petites dans H^3. La méthode de démonstration repose sur des estimations d'énergie. Un autre point de vue est adopté par Galdi, Grobbelaar van Dalsen, Sauer [7] et Galdi, Sequeira [8]. Ces travaux utilisent une méthode de point fixe pour obtenir des résultats similaires. Tous ces résultats sont énoncés dans des domaines bornés mais l'extension à $\mathbb{R}^n$ ne semble pas poser de difficulté.

Une question qui se pose naturellement est de savoir si les solutions des équations du fluide de grade 2 convergent vers une solution des équations de Navier-Stokes lorsque $\alpha \to 0$. La réponse n'est pas évidente et ne découle pas des travaux précédents car toutes les estimations précédentes "explosent" lorsque $\alpha \to 0$. Le but de ce travail est de montrer que la convergence vers une solution des équations de Navier-Stokes a bien lieu, et cela sous des hypothèses très générales. La seule hypothèse "artificielle" sera la borne $C\alpha^{-1/2}$ pour la norme H^1 de la donnée initiale.

Avant d'énoncer nos résultats, rappelons un résultat classique d'existence des solutions faibles pour l'équation de Navier-Stokes qui est du à Leray [9], voir aussi [5], [12]. On appelle solution faible des équations de Navier-Stokes sur $[0,T)$ un champ de vecteurs de divergence nulle

$$u \in C_w\big([0,T);L^2\big) \cap L^2_{loc}\big([0,T);H^1\big)$$

qui vérifie l'équation (2) au sens des distributions. Le théorème classique de Leray affirme l'existence d'une telle solution, unique en dimension 2, dès lors que $u_0 \in L^2$, $\operatorname{div} u_0 = 0$ et $f \in L^2_{loc}\big([0,T);H^{-1}\big)$; de plus, on peut supposer que cette solution vérifie l'inégalité d'énergie suivante:

$$\|u(t)\|^2_{L^2} + 2\nu \int_0^t \|\nabla u(\tau)\|^2_{L^2}\, d\tau \le \|u(0)\|^2_{L^2} + 2\int_0^t \langle f(\tau), u(\tau)\rangle\, d\tau, \tag{3}$$

pour tout $t < T$. Une relation similaire a lieu pour le fluide de grade 2. On multiplie (1) par u et l'on intègre. Cela implique, après quelques intégrations par parties, l'estimation H^1 suivante:

$$\|u(t)\|^2_{L^2} + \alpha\|\nabla u(t)\|^2_{L^2} + 2\nu \int_0^t \|\nabla u(\tau)\|^2_{L^2}\, d\tau$$
$$\le \|u_0\|^2_{L^2} + \alpha\|\nabla u_0\|^2_{L^2} + 2\int_0^t \langle f(\tau), u(\tau)\rangle\, d\tau. \tag{4}$$

On voit tout de suite que ces estimations donnent des informations "a priori" pour des normes H^1 en espace seulement, *i.e.* seules les dérivées d'ordre 1 en espace peuvent être "contrôlées". Par conséquent, pour pouvoir passer à la limite dans (1) avec l'information (4) seulement, il faut mettre l'équation sous une forme où les termes non-linéaires soient des produits de dérivées de u d'ordre inférieur où égal à 1 ou des dérivées de tels produits. Cette forme sera la suivante:

$$\partial_t(u-\alpha\triangle u)-\nu\triangle u+u\cdot\nabla u-\alpha\sum_{j,k}\partial_j\partial_k(u_j\partial_k u)+\alpha\sum_{j,k}\partial_j(\partial_k u_j\partial_k u)\\ -\alpha\sum_{j,k}\partial_k(\partial_k u_j\nabla u_j)=-\nabla p+f. \quad (5)$$

Remarquons enfin que, pour montrer que les termes supplémentaires par rapport à l'équation de Navier-Stokes convergent vers 0 au sens des distributions, l'hypothèse $\nu>0$ est importante.

On montre le théorème suivant:

Théorème 1.1 - *Considérons l'équation d'un fluide de grade 2 posée dans $\mathbb{R}^n$, $n\geq 2$. Soient $\nu>0$ et $f\in L^2_{loc}\big([0,T);\dot H^{-1}\big)$ fixés et $u^{\alpha_k}(0)$ une suite de données initiales à divergence nulle correspondant à une suite $\alpha_k\to 0$ tels que*

a) il existe $\widetilde u_0\in L^2$ tel que $u^{\alpha_k}(0)\rightharpoonup\widetilde u_0$ faiblement dans L^2;

b) la suite $\alpha_k^{1/2}u^{\alpha_k}(0)$ est bornée dans H^1;

c) il existe une solution (au sens des distributions) $u^{\alpha_k}\in C_w\big([0,T);H^1\big)$ de (5) avec $\alpha=\alpha_k$, ayant comme donnée initiale $u^{\alpha_k}(0)$ et vérifiant l'inégalité d'énergie (4).

Alors, il existe une solution faible $\widetilde u$ de l'équation de Navier-Stokes sur $[0,T)$ avec donnée initiale $\widetilde u(0)=\widetilde u_0$ et une sous-suite $u^{\alpha_{\varphi(k)}}$ telles que pour tout $\theta<T$ on ait

$$u^{\alpha_{\varphi(k)}}\rightharpoonup\widetilde u \text{ dans } L^\infty(0,\theta;L^2) \text{ faible* et dans } L^2(0,\theta;H^1) \text{ faible.}$$

Remarque 1.2. Le fait que $u^{\alpha_k}\in C_w\big([0,T);H^1\big)$ et que u^{α_k} vérifie l'inégalité d'énergie est automatiquement vérifié pour des solutions obtenues par régularisation et passage à la limite. Ainsi, la seule hypothèse "restrictive" est la borne sur $\|\nabla u^{\alpha_k}(0)\|_{L^2}$. Cette borne ne dit pas que les données initiales restent bornées dans H^1, elle dit seulement que ces données initiales n'explosent pas trop vite en norme H^1 lorsque $\alpha_k\to 0$.

Remarque 1.3. En dimension 2, comme on a unicité des solutions faibles de l'équation de Navier-Stokes, il s'ensuit que la conclusion reste vraie pour toute la suite α_k au lieu d'une sous-suite $\alpha_{\varphi(k)}$.

Remarque 1.4. Pour que la limite $\widetilde{u}$ satisfasse l'inégalité d'énergie (3), il suffit d'ajouter les hypothèses $u^{\alpha_k}(0) \to \widetilde{u}_0$ fortement dans L^2 et $\alpha_k^{1/2} u^{\alpha_k}(0) \to 0$ fortement dans H^1.

Remarque 1.5. La preuve qu'on va donner n'utilise pas de manière essentielle le fait qu'on se place dans l'espace entier au lieu d'un domaine borné. En effet, les estimations "a priori" (4) et l'équation équivalente (5) qui sont les ingrédients essentiels de la démonstration, restent vrais dans des domaines bornés. Ensuite, les techniques du passage à la limite peuvent être remplacées par des techniques similaires adaptées aux domaines bornées sans trop de modifications.

2. Préliminaires

On note par H^s l'espace de Sobolev suivant:

$$H^s = \left\{ g : \mathbb{R}^n \to \mathbb{C}; \quad \|g\|_{H^s}^2 = \int_{\mathbb{R}^n} (1 + |\xi|^2)^s |\widehat{g}(\xi)|^2 \, d\xi < +\infty \right\},$$

où $\widehat{g}$ désigne la transformée de Fourier de g et $|\cdot|$ la norme euclidienne. La version homogène de ces espaces est

$$\dot{H}^s = \left\{ g : \mathbb{R}^n \to \mathbb{C}; \ \widehat{g} \in L^1_{loc}(\mathbb{R}^n) \text{ et } \|g\|_{\dot{H}^s}^2 = \int_{\mathbb{R}^n} |\xi|^{2s} |\widehat{g}(\xi)|^2 \, d\xi < +\infty \right\}.$$

L'espace homogène $\dot{H}^s$ est un espace de Banach pour $s < n/2$. Pour des fonctions à valeurs vectorielles $h : \mathbb{R}^n \to \mathbb{R}^m$, on dira que $h \in H^s$ si et seulement si chaque composante h_i de h appartient à H^s et on notera

$$\|h\|_{H^s}^2 = \sum_{i=1}^{m} \|h_i\|_{H^s}^2.$$

On utilisera la même notation pour les espaces de Sobolev homogènes.

On désigne par $\langle \cdot, \cdot \rangle$ le produit scalaire L^2, le produit de dualité entre H^s et H^{-s} ou encore le produit de dualité entre $\dot{H}^s$ et $\dot{H}^{-s}$. Pour des fonctions à valeurs vectorielles $g, h : \mathbb{R}^n \to \mathbb{R}^m$, on notera

$$\langle g, h \rangle = \sum_{i=1}^{m} \langle g_i, h_i \rangle.$$

Le projecteur de Leray $\mathbb{P}$ désigne la projection orthogonale L^2 sur les champs de vecteurs à divergence nulle.

Le théorème de produit suivant est classique (voir par exemple [1]):

Théorème 2.1 - *Soient s et t des réels tels que $s+t>0$, $s<n/2$, $t<n/2$. Il existe une constante $C>0$ telle que si $u\in H^s$ et $v\in H^t$ alors $uv\in H^{s+t-n/2}$ et*

$$\|uv\|_{H^{s+t-n/2}}\leq C\|u\|_{H^s}\|v\|_{H^t}.$$

Si $|s|<n/2$ et $\epsilon>0$, il existe une constante $C'>0$ telle que si $u\in H^s$ et $v\in H^{-s}$ alors $uv\in H^{-n/2-\epsilon}$ et

$$\|uv\|_{H^{-n/2-\epsilon}}\leq C'\|u\|_{H^s}\|v\|_{H^{-s}}.$$

Enfin, on a le lemme suivant très simple:

Lemme 2.2 - *Soit H un espace de Hilbert et $A\subset H$ un sous ensemble dense. Si u_n est une suite bornée de H telle que $\langle u_n,a\rangle\to\langle v,a\rangle$ pour tout élément a de A, alors u_n converge faiblement vers v.*

Rappelons l'équation du fluide de grade 2:

$$\partial_t v-\nu\Delta u+u\cdot\nabla v+\sum_j v_j\nabla u_j=-\nabla p+f,\quad v=u-\alpha\Delta u. \tag{6}$$

Dans un premier temps, on va donner une autre forme à cette équation où la régularité H^1 en espace pour u suffira pour donner un sens à l'équation. Pour cela, remarquons d'abord que $u\cdot\nabla v$ est un vecteur dont la i-ème composante vaut

$$\begin{aligned}\sum_j u_j\partial_j v_i&=\sum_j u_j\partial_j u_i-\alpha\sum_{j,k}u_j\partial_j\partial_k^2u_i\\&=\sum_j u_j\partial_j u_i-\alpha\sum_{j,k}\partial_j(u_j\partial_k^2u_i)\\&=\sum_j u_j\partial_j u_i-\alpha\sum_{j,k}\partial_j\partial_k(u_j\partial_k u_i)+\alpha\sum_{j,k}\partial_j(\partial_k u_j\partial_k u_i).\end{aligned}\tag{7}$$

De même, la i-ème composante de $\sum_j v_j\nabla u_j$ est égale à

$$\begin{aligned}\sum_j v_j\partial_i u_j&=\sum_j u_j\partial_i u_j-\alpha\sum_{j,k}\partial_k^2u_j\partial_i u_j\\&=\frac12\partial_i|u|^2-\alpha\sum_{j,k}\partial_k(\partial_k u_j\partial_i u_j)+\alpha\sum_{j,k}\partial_k u_j\partial_i\partial_k u_j\\&=\frac12\partial_i(|u|^2+\alpha|\nabla u|^2)-\alpha\sum_{j,k}\partial_k(\partial_k u_j\partial_i u_j).\end{aligned}\tag{8}$$

Les relations (7) et (8) utilisées dans (6) donnent une forme équivalente à (6):

$$\partial_t v - \nu \Delta u + u \cdot \nabla u - \alpha \sum_{j,k} \partial_j \partial_k (u_j \partial_k u) + \alpha \sum_{j,k} \partial_j (\partial_k u_j \partial_k u) - \alpha \sum_{j,k} \partial_k (\partial_k u_j \nabla u_j) = -\nabla p + f, \quad (9)$$

où on a incorporé dans la pression le terme $\frac{1}{2}\nabla(|u|^2 + \alpha |\nabla u|^2)$ qui apparaît dans $\sum_j v_j \nabla u_j$. Remarquons qu'il suffit de supposer que $u \in L^2_{loc}(0, T; H^1)$ pour définir l'équation (9) au sens des distributions. En effet, $\partial_t v$ et Δu sont toujours définis si u est une distribution. Ensuite, un terme du type $Du\, D'u$, où Du et $D'u$ désignent une composante de u ou une dérivée d'ordre 1 d'une telle composante, est dans $L^1_{loc}(0, T; L^1)$ et, par conséquent, définit une distribution. Ses dérivées aussi et on a ainsi épuisé tous les termes de (9). Nous travaillerons désormais sur l'équation (9).

On a mentionné dans l'introduction que toute solution obtenue par un procédé de régularisation vérifie l'inégalité d'énergie (4). En effet, si l'on multiplie formellement (6) par u et qu'on intègre, on trouve, après quelques intégrations par parties, la relation (4) (voir aussi [4]). Rigoureusement, (4) sera vérifiée par la solution approchée (ce sera même une égalité). Le terme de droite passe à la limite sans problème. Pour le terme de gauche, après avoir fait les extractions habituelles, il suffit d'utiliser le fait que si $x_m \to x$ faiblement dans un espace de Hilbert, alors $\|x\| \le \liminf \|x_m\|$.

En ce qui concerne la continuité faible à valeurs dans H^1, on a classiquement que $u \in L^\infty(0, \theta; H^1)$ (par extraction d'une suite convergente dans $L^\infty(0, \theta; H^1)$ faible*). A partir de l'équation, il est facile de voir que $\partial_t u \in L^1(0, \theta; H^{-k})$, pour k assez grand; par conséquent $u \in C(0, \theta; H^{-k})$. Comme H^k est dense dans H^1, le lemme 2.2 implique $u \in C_w([0, \theta]; H^1)$ pour tout $\theta < T$.

3. Preuve du théorème 1.1

La preuve s'inspire fortement de la démonstration de l'existence des solutions faibles de l'équation de Navier-Stokes, et plus précisément de la partie concernant le passage à la limite, tel qu'on peut la trouver dans [2], voir aussi [11]. Ici, la difficulté consiste en l'obtention d'estimations indépendantes de α et de montrer que les termes supplémentaires par rapport à l'équation de Navier-Stokes convergent vers 0 au sens des distributions.

Dans la suite, l'hypothèse $\nu > 0$ joue un rôle important. Ainsi, C désignera une constante indépendante de ν et α, qui peut changer d'une inégalité à l'autre. Pour alléger la rédaction, on notera $\alpha = \alpha_k$ et $u = u^\alpha = u^{\alpha_k}$. Toutes les limites qui suivent ont lieu pour $\alpha = \alpha_k \to 0$.

La première étape consiste en l'obtention d'estimations indépendantes de α.

3.1. Estimations uniformes en α

Soit $\theta < T$ fixé. On va utiliser l'inégalité d'énergie (4). Remarquons que

$$\int_0^t \langle f, u\rangle \leq \int_0^t \|f\|_{\dot{H}^{-1}} \|u\|_{\dot{H}^1} \leq \frac{\nu}{2}\int_0^t \|\nabla u\|_{L^2}^2 + \frac{1}{2\nu}\int_0^t \|f\|_{\dot{H}^{-1}}^2.$$

En utilisant cette relation dans (4) on trouve

$$\|u(t)\|_{L^2}^2 + \alpha\|\nabla u(t)\|_{L^2}^2 + \nu\int_0^t \|\nabla u(\tau)\|_{L^2}^2\, d\tau \leq \|u_0\|_{L^2}^2 + \alpha\|\nabla u_0\|_{L^2}^2 + \frac{1}{\nu}\int_0^t \|f\|_{\dot{H}^{-1}}^2. \quad (10)$$

En tenant compte des hypothèses a), b) et du fait que $f \in L^2_{loc}\big([0,T);\dot{H}^{-1}\big)$ on obtient de l'inégalité ci-dessus que, pour tout $\theta < T$,

$$u \text{ est borné dans } L^\infty(0,\theta;L^2); \quad (11_{\mathrm{a}})$$

$$\nabla u \text{ est borné dans } L^2(0,\theta;L^2); \quad (11_{\mathrm{b}})$$

$$\alpha^{1/2}\nabla u \text{ est borné dans } L^\infty(0,\theta;L^2). \quad (11_{\mathrm{c}})$$

Avant de pouvoir passer à la limite, il nous faut une certaine convergence forte. Une convergence forte peut s'obtenir si l'on dispose de l'équicontinuité en temps qui, à son tour, peut s'obtenir à partir d'estimations sur $\partial_t u$. Revenons à l'équation (9). On va estimer $\partial_t v$ dans un certain espace H^{-k}, k assez grand. Étudions chaque terme de (9), à l'aide des informations (11) et du théorème de produit 2.1:

- $\nu\triangle u$ est borné dans $L^\infty(0,\theta;H^{-2})$ car u l'est dans $L^\infty(0,\theta;L^2)$;
- $\|u\cdot\nabla u\|_{H^{-1-d/2}} \leq C\|u\|_{L^2}\|\nabla u\|_{L^2}$ donc
$$u\cdot\nabla u \quad \text{est borné dans } L^2(0,\theta;H^{-1-d/2});$$
- de même, $u_j\partial_k u$ est borné dans $L^2(0,\theta;H^{-1-d/2})$ donc
$$\alpha\sum_{j,k}\partial_j\partial_k(u_j\partial_k u) \to 0 \quad \text{dans} \quad L^2(0,\theta;H^{-3-d/2});$$

- $\|\partial_k u_j \, \partial_k u\|_{H^{-1-d/2}} \leq C\|\nabla u\|_{L^2}^2$ donc $\alpha^{1/2}\partial_k u_j \, \partial_k u$ est borné dans $L^2(0,\theta;H^{-1-d/2})$, d'où

$$\alpha \sum_{j,k} \partial_j(\partial_k u_j \partial_k u) \to 0 \quad \text{dans} \quad L^2(0,\theta;H^{-2-d/2});$$

- de même,

$$\alpha \sum_{j,k} \partial_k(\partial_k u_j \nabla u_j) \to 0 \quad \text{dans} \quad L^2(0,\theta;H^{-2-d/2}).$$

En ce qui concerne le terme de pression, le plus simple est d'appliquer la projection de Leray $\mathbb{P}$ à (9) pour obtenir

$$\begin{aligned}\partial_t v = \mathbb{P}\Big(\nu\triangle u - u\cdot\nabla u + \alpha\sum_{j,k}\partial_j\partial_k(u_j\partial_k u) - \alpha\sum_{j,k}\partial_j(\partial_k u_j\partial_k u)& \\ + \alpha\sum_{j,k}\partial_k(\partial_k u_j\nabla u_j) + f\Big)&.\end{aligned}$$

Comme $\mathbb{P}$ est une projection orthogonale dans tout espace de Sobolev H^s, il découle de la discussion ci-dessus que

$$\partial_t v \text{ est borné dans } L^2(0,\theta;H^{-3-d/2}). \tag{12}$$

Or, on a la relation suivante:

$$\|\partial_t u\|_{H^s} \leq \|\partial_t v\|_{H^s}, \tag{13}$$

pour tout $s \in \mathbb{R}$. En effet, si l'on note $\breve{v} = \partial_t(1-\triangle)^{s/2}v$ et $\breve{u} = \partial_t(1-\triangle)^{s/2}u$, on a

$$\begin{aligned}\|\partial_t v\|_{H^s}^2 = \|\breve{v}\|_{L^2}^2 &= \|\breve{u} - \alpha\triangle\breve{u}\|_{L^2}^2 \\ &= \langle \breve{u} - \alpha\triangle\breve{u}, \breve{u} - \alpha\triangle\breve{u}\rangle \\ &= \|\breve{u}\|_{L^2}^2 + \alpha^2\|\triangle\breve{u}\|_{L^2}^2 - 2\alpha\langle\breve{u},\triangle\breve{u}\rangle \\ &= \|\breve{u}\|_{L^2}^2 + \alpha^2\|\triangle\breve{u}\|_{L^2}^2 + 2\alpha\langle\nabla\breve{u},\nabla\breve{u}\rangle \\ &= \|\breve{u}\|_{L^2}^2 + \alpha^2\|\triangle\breve{u}\|_{L^2}^2 + 2\alpha\|\nabla\breve{u}\|_{L^2}^2 \geq \|\breve{u}\|_{L^2}^2 = \|\partial_t u\|_{H^s}^2.\end{aligned}$$

On déduit de (12) et (13) que

$$\partial_t u \text{ est borné dans } L^2(0,\theta;H^{-3-d/2}). \tag{14}$$

On dispose maintenant de tous les éléments nécessaires pour passer à la limite.

3.2. Passage à la limite

On a déjà vu que

$$\alpha \sum_{j,k} \partial_j \partial_k (u_j \partial_k u) - \alpha \sum_{j,k} \partial_j (\partial_k u_j \partial_k u) + \alpha \sum_{j,k} \partial_k (\partial_k u_j \nabla u_j) \to 0 \quad \text{dans } L^2(0,\theta; H^{-3-d/2}), \tag{15}$$

donc la convergence a lieu aussi au sens des distributions.

Avec les informations (11_{a}) et (11_{b}) on peut extraire une sous-suite, encore notée u, telle que

$$u \rightharpoonup \widetilde{u} \text{ dans } L^\infty(0,\theta; L^2) \text{ faible*} \tag{16}$$

et

$$u \rightharpoonup \widetilde{u} \text{ dans } L^2(0,\theta; H^1) \text{ faible,} \tag{17}$$

pour tout $\theta < T$. De plus, grâce à (14), on peut aussi supposer que

$$\partial_t u \rightharpoonup \partial_t \widetilde{u} \text{ dans } L^2(0,\theta; H^{-3-d/2}) \text{ faible .} \tag{18}$$

Comme $u \to \widetilde{u}$ au sens des distributions, on aura $\partial_t \Delta u \to \partial_t \Delta \widetilde{u}$ au sens des distributions donc $\alpha \partial_t \Delta u \to 0$ au sens des distributions. Les seuls termes qui restent dans (9) sont exactement les mêmes que ceux de l'équation de Navier-Stokes. Les mêmes arguments s'appliquent donc ici. Il n'est pas nécessaire de les reproduire en détail; on en donnera les grandes lignes seulement.

Soit θ fixé et $t,\ t' \in [0,\theta]$ arbitraires. On part de

$$u(t) - u(t') = \int_t^{t'} \partial_t u,$$

d'où

$$\|u(t) - u(t')\|_{H^{-3-d/2}} \le \int_t^{t'} \|\partial_t u\|_{H^{-3-d/2}} \le |t-t'|^{1/2} \|\partial_t u\|_{L^2(0,\theta;H^{-3-d/2})}.$$

En se rappelant (14), il s'ensuit que les $u = u^\alpha$ sont équicontinus (par rapport à α) dans $C([0,\theta]; H^{-3-d/2})$. Par le théorème d'Ascoli, pour tout compact K on peut extraire une sous-suite, encore notée u, telle que $u|_K$ converge fortement dans $C([0,\theta]; H^{-4-d/2}(\overset{\circ}{K}))$. En prenant une suite croissante de réels $\theta_m \to T$, de compacts $K_m = \overline{B(0,m)}$, des sous-suites successives et en extrayant une suite diagonale, on voit qu'on peut supposer que

$u|_U$ converge fortement dans $C([0,\theta]; H^{-4-d/2}(U))$ pour tout $\theta < T$ et U ouvert relativement compact. Une inégalité d'interpolation simple ainsi que la relation (11_{a}) montrent maintenant que

$$u|_U \to \widetilde{u}|_U \text{ fortement dans } C([0,\theta]; H^s(U)) \tag{19}$$

pour tout $s < 0$, $\theta < T$, et U ouvert relativement compact. Soit $\varphi \in C_0^\infty\big((0,T) \times \mathbb{R}^n\big)$. On a, après une intégration par parties,

$$\int_0^T \int_{\mathbb{R}^n} u \cdot \nabla u \, \varphi = -\sum_{i,j} \int_0^T \int_{\mathbb{R}^n} u_i u_j \partial_i \varphi_j.$$

Comme le domaine d'intégration ci-dessus est un compact de $(0,T) \times \mathbb{R}^n$, on peut utiliser les relations (17) et (19) pour déduire que

$$\int_0^T \int_{\mathbb{R}^n} u \cdot \nabla u \, \varphi \to \int_0^T \int_{\mathbb{R}^n} \widetilde{u} \cdot \nabla \widetilde{u} \, \varphi,$$

ce qui revient à dire que

$$u \cdot \nabla u \to \widetilde{u} \cdot \nabla \widetilde{u}$$

au sens des distributions. Enfin, on a clairement que $\partial_t u \to \partial_t \widetilde{u}$ et $\Delta u \to \Delta \widetilde{u}$ au sens des distributions. Comme la limite d'une suite de gradients est un gradient, on obtient, après passage à la limite dans (9),

$$\partial_t \widetilde{u} - \nu \Delta \widetilde{u} + \widetilde{u} \cdot \nabla \widetilde{u} = -\nabla p + f,$$

donc $\widetilde{u}$ est une solution de l'équation de Navier-Stokes qui vérifie

$$\widetilde{u} \in L^\infty_{loc}\big([0,T); L^2\big) \cap L^2_{loc}\big([0,T); H^1\big), \quad \partial_t \widetilde{u} \in L^2_{loc}\big([0,T); H^{-3-d/2}\big),$$

où on a utilisé (16), (17) et (18). La continuité faible résulte de la continuité forte $\widetilde{u} \in C\big([0,T); H^{-3-d/2}\big)$ (conséquence de $\partial_t \widetilde{u} \in L^2_{loc}\big([0,T); H^{-3-d/2}\big)$), de l'appartenance de $\widetilde{u}$ à $L^\infty_{loc}\big([0,T); L^2\big)$ et du lemme 2.2:

$$\widetilde{u} \in C_w\big([0,T); L^2\big).$$

La donnée initiale de $\widetilde{u}$ vérifie, grâce à la relation (19),

$$u(0)|_U \to \widetilde{u}(0)|_U,$$

au sens des distributions pour tout U ouvert relativement compact. Comme on a aussi que

$$u(0) \to \widetilde{u}_0,$$

au sens des distributions, il s'ensuit que $\widetilde{u}_0|_U = \widetilde{u}(0)|_U$ pour tout U ouvert relativement compact. Par conséquent, $\widetilde{u}_0 = \widetilde{u}(0)$. Pour terminer la preuve, il reste à montrer la Remarque 1.4. Supposons que $u(0) \to \widetilde{u}_0$ fortement dans L^2 et $\alpha^{1/2}u(0)$ tend vers 0 dans H^1. Sous ces hypothèses, en utilisant aussi (17), la partie droite de (4) tend vers

$$||\widetilde{u}_0||_{L^2}^2 + 2\int_0^t \langle f(\tau), \widetilde{u}(\tau)\rangle \, d\tau.$$

Quant à la partie de gauche, remarquons d'abord que par (11_a), (19) et par le lemme 2.2 on a que $u(t) \rightharpoonup \widetilde{u}(t)$ faiblement dans L^2 pour tout $t \in [0,T)$. Rappelons maintenant que si $x_m \to x$ faiblement dans un espace de Hilbert, alors $||x|| \leq \liminf ||x_m||$. Il ne reste plus qu'à utiliser (17) pour minorer la limite supérieure du terme de gauche de (4) par

$$||\widetilde{u}(t)||_{L^2}^2 + 2\nu \int_0^t ||\nabla \widetilde{u}(\tau)||_{L^2}^2 \, d\tau,$$

ce qui conclut la preuve de l'inégalité d'énergie (3) pour $\widetilde{u}$. Le théorème 1.1 est complètement démontré. ■

Remarque 3.1. D'autres régularités en espace peuvent être considérées pour le terme de force f. Si l'on supposait par exemple que $f \in L^2_{loc}([0,T); L^2)$, il suffirait de majorer $\int_0^t \langle f, u\rangle \leq \int_0^t ||u||_{L^2}^2 + \int_0^t ||f||_{L^2}^2$, de modifier ensuite l'inégalité (10) et d'appliquer le lemme de Gronwall pour obtenir les mêmes estimations a priori (11).

Références

[1] J.-Y. Chemin, Fluides parfaits incompressibles. Astérisque, 230, 1995.

[2] J.-Y. Chemin, Méthodes mathématiques en mécanique des fluides, I. 1997. Cours de DEA et Preprint Laboratoire d'Analyse Numérique A97004.

[3] D. Cioranescu et V. Girault, Weak and classical solutions of a family of second grade fluids. Internat. J. Non-Linear Mech., 32, 2 (1997), 317–335.

[4] D. Cioranescu et E. H. Ouazar, Existence and uniqueness for fluids of second grade, in Nonlinear partial differential equations and their applications. Collège de France seminar, Vol. VI (Paris, 1982/1983), 178–197. Boston, MA, Pitman, 1984.

[5] P. Constantin et C. Foiaş, Navier-Stokes equations. Chicago, University of Chicago Press, 1988.

[6] J.E. Dunn et R.L. Fosdick, Thermodynamics, stability, and boundedness of fluids of complexity 2 and fluids of second grade. Arch. Rational Mech. Anal., 56 (1974), 191–252.
[7] G. P. Galdi, M. Grobbelaar-van Dalsen, et N. Sauer, Existence and uniqueness of classical solutions of the equations of motion for second-grade fluids. Arch. Rational Mech. Anal., 124, 3 (1993), 221–237.
[8] G. P. Galdi et A. Sequeira, Further existence results for classical solutions of the equations of a second-grade fluid. Arch. Rational Mech. Anal., 128, 4 (1994), 297–312.
[9] J. Leray, Sur le mouvement d'un liquide visqueux emplissant l'espace. Acta Math., 63 (1934), 193–248.
[10] R.S. Rivlin et J.L. Ericksen, Stress-deformation relations for isotropic materials. J. Rational Mech. Anal., 4 (1955), 323–425.
[11] M. E. Taylor, Partial differential equations. III, New York, Springer-Verlag, 1997.
[12] R. Temam, Navier-Stokes equations, Amsterdam, North-Holland, 1984.

Dragos Iftimie
IRMAR
Université de Rennes 1
Campus de Beaulieu
35042 Rennes Cedex
France
E-mail: iftimie@maths.univ-rennes1.fr

Studies in Mathematics and its Applications, Vol. 31
D. Cioranescu and J.L. Lions (Editors)

Chapter 21

REMARKS ON THE KOMPANEETS EQUATION, A SIMPLIFIED MODEL OF THE FOKKER-PLANCK EQUATION

O. KAVIAN

1. Introduction

In this talk we present a few results regarding the Kompaneets equation described below. These results are part of a joint work with C.D. Levermore, which will appear in more details elsewhere.

The equation we study concerns a physical model in which radiation interacts with the electrons and ions of a fully ionized plasma, primarily through scattering with electrons and emission-absorption. In each of these processes energy is exchanged between the photon and particle fields. In the low density-high temperature regime, scattering is the dominant mechanism for the energy exchange. This process was quantified by A.S. Kompaneets [6] for the weakly relativistic limit.

In this model, for a spatially homogeneous plasma held at a constant temperature θ, the radiation distribution f at time t is only a function of $x := h\nu/\theta$, which is the photon energy normalized by the temperature (h is Planck's constant and ν the photon frequency). This radiation distribution f is defined so that the total photon number density is given by:

$$N[f] := \int_0^\infty f(t,x)x^2 dx. \tag{1.1}$$

For a scattering process this quantity must be conserved. The Kompaneets equation governing f is:

$$\partial_t f = x^{-2}\partial_x[\alpha(x)(\partial_x f + f + f^2)] \qquad \text{for } 0 < x < \infty \tag{1.2}$$

where $\alpha(x) > 0$ on $(0, \infty)$. Classically $\alpha(x) = x^4$ and that is the case we will treat here. Physically f must satisfy the boundary conditions :

$$\lim_{x\to 0} \alpha(x)(\partial_x f + f + f^2)(x) = \lim_{x\to\infty} \alpha(x)(\partial_x f + f + f^2)(x) = 0, \tag{1.3}$$

which express the conservation of photon number density. These nonlinear boundary conditions can be satisfied by seeking a solution in an appropriate functional space.

We point out here the fact that equation (1.2), written formally in an expanded form, reads:

$$\partial_t f = x^2 \partial_{xx} f + (x^2 \partial_x f + 4x \partial_x f) + 4xf + x^2 \partial_x (f^2) + 4x^2 f^2. \tag{1.4}$$

As far as the order of the principal part of the differential operator and the degree and the type of nonlinearity are considered, equation (1.4) reminds of the Burger's equation

$$\partial_t u = \partial_{xx} u + \partial_x (u^2).$$

However it is clear that the singular term $x^2 \partial_{xx} f$, as $x \to 0^+$, makes equation (1.4) more delicate to handle. But we attract the reader's attention to the fact that this is not the only difficulty with equation (1.4). Indeed, the terms $x^2 \partial_x f$ and xf add some degree of difficulty to the appropriate study of that equation, especially if one wishes to prove a local or global existence theorem in a Banach space of the type $L^p(d\mu)$ for some $p \in [1, +\infty]$ and some positive measure $d\mu$. In an attempt to convince the reader, we may give the following example of a *linear* equation, which contains terms of the above mentioned nature (however slightly different because we want to show an explicite example). For $x \in \mathbb{R}$ consider the linear equation

$$\begin{cases} \partial_t u(t,x) = \partial_{xx} u - x \partial_x u + \dfrac{1}{4} x^2 u \\ u(0,x) = u_0(x). \end{cases} \tag{1.5}$$

Now the principal part of the differential operator is $\partial_{xx} u$, which is not singular, but if one tries to prove a global existence result for a given $u_0 \in L^p(\mathbb{R})$ for $1 \le p \le +\infty$, one sees that the usual techniques do not apply. For example when $u_0(x) := \exp(-x^2/4)$, then one may show that the unique solution of (1.5) in $L^p(\mathbb{R})$ is given by

$$u(t,x) := (2t+1)^{-1/2} e^{t/2} \exp\left(\frac{(2t-1)x^2}{4(2t+1)} \right).$$

Clearly this solution blows up in $L^p(\mathbb{R})$, and this for *any* p, at time $T_{\max} = 1/2$.

This is to emphasize that, in some cases, it is not a good idea to consider such terms as $x \partial_x u$ or $x^2 u/4$ as perturbations of $\partial_{xx} u$, the principal part of the differential operator. It is also clear that if one considers the solution of (1.5) in the weighted Lebesgue space $L^p(d\mu)$, where $d\mu := \exp(-x^2/2)$, with the linear operator $u \mapsto \partial_{xx}u - x\partial_x u$, such difficulties are easily avoided.

Regarding the Kompaneets equation (1.2)–(1.3), we are going to construct a solution in the Hilbert space

$$L_2 := \left\{ u \,;\, \int_0^\infty |u(x)|^2 x e^x \, dx < \infty \right\} \tag{1.6}$$

endowed with its natural norm and scalar product. To do so we consider the unbounded linear operator $(B, D(B))$ defined by

$$\begin{aligned} Bu &:= x\partial_x u + (x-1)u = e^{-x} x^2 \partial_x (x^{-1} e^x u), \\ D(B) &:= \{ u \in L_2 \,;\, Bu \in L_2 \}. \end{aligned} \tag{1.7}$$

We will show that B is closed and that $D(B) = D(B^*)$. Then, as one can be convinced after some simple calculations using the fact that

$$B^* u = -x\partial_x u - 3u, \qquad \text{and} \qquad B^* B u = -x^2 \partial_{xx} u - (x^2 + 3x)\partial_x u - (4x-3)u,$$

we write equation (1.2) in the following form:

$$\begin{cases} \partial_t f + B^* B f = -B^*(f + x f^2) & \text{on } (0,T) \times (0,\infty) \\ f(t,\cdot) \in D(B^* B) & \text{for } t \in (0,T) \\ f(0,\cdot) = f_{\text{Init}}(\cdot). \end{cases} \tag{1.8}$$

In this setting the boundary conditions (1.3) are derived from the fact that we seek solutions satisfying $f(t) \in D(B^* B)$, when f_{Init} is an appropriate given initial data. Our main result is the following.

Theorem 1.1. – *For any given $f_{\text{Init}} \in D(B^* B))$, there exists $T > 0$ such that equation (1.8) has a unique solution $f \in C([0,T], D(B^* B))$. Denoting by $T_{\max} := T_{\max}(f_{\text{Init}})$ the supremum of all such T's, then either $T_{\max} = +\infty$ or*

$$T_{\max} < \infty \qquad \text{and} \qquad \lim_{t \to T_{\max}} \|x u(t)\|_\infty = +\infty.$$

Also a comparison principle holds for solutions of (1.8), i.e. if two intial data $f_{\text{Init}}, \widetilde{f}_{\text{Init}} \in D(B^* B)$ satisfy $f_{\text{Init}} \leq \widetilde{f}_{\text{Init}}$ then the corresponding solutions satisfy

$$f(t) \leq \widetilde{f}(t) \qquad \text{for} \qquad t < \min(T_{\max}(f_{\text{Init}}), T_{\max}(\widetilde{f}_{\text{Init}})).$$

We prove also that unlike Burger's equation, the Kompaneets equation has solutions which blow up in finite time:

Proposition1.2. – *Assume that $f_{\text{Init}} \in D(B^*B))$, is such that $f_{\text{Init}} \geq 0$ and*

$$\int_0^\infty f_{\text{Init}}(x)x^4 \mathrm{e}^{-x} dx > 2,$$

then the corresponding solution blows up in a finite time $T_{\max} < \infty$.

One may prove global existence of solutions for a large class of initial data. First recall that the equilibria for (1.2) are obtained by setting

$$\partial_x f + f + f^2 = 0,$$

yielding the one parameter family of positive solutions $f = \varphi_\mu$

$$\varphi_\mu(x) := \frac{1}{\mathrm{e}^{x+\mu} - 1} \qquad \text{for } \mu \geq 0,$$

which are the classical Bose–Einstein equilibria distributions. Now it is clear that by the comparison principle stated in theorem 1.1, if $0 \leq f_{\text{Init}} \leq \varphi_\mu$ for some $\mu > 0$ then $T_{\max} = +\infty$. However we point out the fact that our functional space setting is not optimal, as the Bose–Einstein equilibrium φ_0 does not belong to L_2.

Before going into the details of the proof of these results, we would like to mention that the main ingredient in the proof of Theorem 1.1 is the following *Nash type* lemma, which, we believe, is interesting in its own right.

Lemma 1.3. – *Let H be a Hilbert space, $(A, D(A))$ a densely defined, closed linear operator acting on H and a Banach space X such that $D(A) \cap X$ is dense in H. Assume that A^*A generates the continuous semigroup $S(t) := \exp(-tA^*A)$ which satisfies*

$$\forall u_0 \in X, \qquad \|S(t)u_0\|_X \leq c_0 \|u_0\|_X. \tag{1.9}$$

Then the following properties are equivalent:

(i) *There exist $c_1 > 0$ and $\beta > 0$ such that for all $u \in D(A) \cap X$ one has*

$$\|u\|_H^{2+\beta} \leq c_1 \|Au\|_H^2 \|u\|_X^\beta. \tag{1.10}$$

(ii) *There exist $c_2 > 0$ and $\beta > 0$ such that for all $u_0 \in X$ one has*

$$\forall t > 0, \qquad \|S(t)u_0\|_H \leq c_2\, t^{-1/\beta} \|u_0\|_X. \tag{1.11}$$

This result is essentially well known, although we give here a slightly different and general expression for it. Indeed since the works of J. Nash [7], M. Fukushima [4], N. Varopoulos [10] and E.B Fabes & W. Strook [3], it is known that there is an equivalence between the regularizing effect (or more precisely the ultracontractivity) of a heat semigroup on one hand, and Nash or Sobolev inequalities on the other hand.

More precisely (see for instance E.B. Davies [2]), assume that $L = L^*$ is a positive selfadjoint operator acting on $L^2(d\mu)$ and such that the semigroup $S(t) := \mathrm{e}^{-tL}$ is positivity preserving and maps continuously $L^1(d\mu)$ into itself, then denoting $\|\cdot\|_p$ the norm in $L^p(d\mu)$, the following properties are equivalent:

1). there exist two constants $c_1 > 0$ and $\beta > 0$ such that for all $\varphi \in D(L) \cap L^1(d\mu)$ one has

$$\|\varphi\|_2^{(2\beta+4)/\beta} \leq c_1 \|\varphi\|_1^{4/\beta} (L\varphi|\varphi) ;$$

2). there exist two constants $c_2 > 0$ and $\beta > 0$ such that for all $\varphi \in L^1(d\mu)$ and all $t > 0$ one has

$$\|S(t)\varphi\|_\infty \leq c_2 t^{-\beta/2} \|\varphi\|_1 .$$

Moreover if in either of above properties one has $\beta > 2$, then they are equivalent to the following Sobolev embedding:

3). there exist two constants $c_3 > 0$ and $\beta > 2$ such that for all $\varphi \in D(L)$ one has

$$\|\varphi\|_{2\beta/(\beta-2)} \leq c_3 (L\varphi|\varphi).$$

(The number β may be thought of as a geometrical dimension for the underlying measure space).

The remainder of this note is organized as follows. In Section 2 we establish some of the properties of the operator B defined in (1.7) and we prove lemma 1.3.

In Section 3 we prove local existence in time and uniqueness in an appropriate space using semigroups and fixed point theorem.

In Section 4 we study the qualitative properties of the solutions and we discuss global existence for some solutions, as well as demonstrating finite time blow-up for some initial data.

2. Preliminary results

In order to solve the Kompaneets equation (1.2), we introduce the Hilbert space

$$L_2 := \left\{ u \; ; \; \int_0^\infty |u(x)|^2 x \mathrm{e}^x \, dx < \infty \right\} \tag{2.1}$$

with its natural inner product $(u|v) := \int_0^\infty u(x)v(x)x\mathrm{e}^x \, dx$ and associated norm. On this space we consider the differential operator $(B, D(B))$ acting on L_2 and defined by

$$Bu := \mathrm{e}^{-x} x^2 \partial_x \left(\frac{\mathrm{e}^x}{x} u \right) = x \partial_x u + (x-1)u \tag{2.2}$$

with domain

$$D(B) := \left\{ u \in L_2 \; ; \; \int_0^\infty x^5 \mathrm{e}^{-x} \left| \partial_x \left(\frac{\mathrm{e}^x u(x)}{x} \right) \right|^2 dx < \infty \right\}. \tag{2.3}$$

In order to determine the adjoint of $(B, D(B))$ we introduce an operator $(B_1, D(B_1))$ (which actually is the formal adjoint of B) by setting

$$\begin{aligned} B_1 u &:= -(x\partial_x u + 3u) = -x^{-2} \partial_x (x^3 u), \\ D(B_1) &:= \{ u \in L_2 \; ; \; B_1 u \in L_2 \}. \end{aligned} \tag{2.4}$$

As usual for a closed operator $(B, D(B))$ we define its graph norm by

$$\|u\|_{D(B)} := \left(\|u\|^2 + \|Bu\|^2 \right)^{1/2} \qquad \text{for} \qquad u \in D(B).$$

Lemma 2.1. – *$(B, D(B))$ and $(B_1, D(B_1))$ are densely defined closed operators in L_2 and $C_c^\infty(0,\infty)$ is dense in $D(B)$ and in $D(B_1)$ endowed with their respective graph norms. Moreover for any $u \in D(B) \cup D(B_1)$ one has the boundary limits*

$$\lim_{x \to 0} x u(x) = 0 = \lim_{x \to \infty} \mathrm{e}^{x/2} x^{3/2} u(x). \tag{2.5}$$

In particular there exists a constant $C > 0$ such that for $u \in D(B)$ (resp. $u \in D(B_1)$) one has $\|(x + x^{3/2}\mathrm{e}^{x/2})u\|_\infty \le C\|u\|_{D(B)}$ (resp. $\|(x + x^{3/2}\mathrm{e}^{x/2})u\|_\infty \le C\|u\|_{D(B_1)}$).

Proof. The density of $C_c^\infty(0,\infty)$ in $D(B)$, or $D(B_1)$, equipped with the graph norm is straightforward. Let u be in $D(B)$; for any $0 < x_1 < x_2 < \infty$ we have the basic estimate :

$$\begin{aligned} \left| \frac{\mathrm{e}^{x_2} u(x_2)}{x_2} - \frac{\mathrm{e}^{x_1} u(x_1)}{x_1} \right| &= \left| \int_{x_1}^{x_2} \partial_x \left(\frac{\mathrm{e}^x u(x)}{x} \right) dx \right| = \left| \int_{x_1}^{x_2} Bu(x) x^{-3} \, x \mathrm{e}^x \, dx \right| \\ &\le \left(\int_{x_1}^{x_2} \frac{\mathrm{e}^x}{x^5} dx \right)^{1/2} \cdot \left(\int_{x_1}^{x_2} |Bu|^2 x \mathrm{e}^x \, dx \right)^{1/2}. \end{aligned} \tag{2.6}$$

Using the rough estimate obtained from

$$\left(\int_{x_1}^{x_2} \frac{\mathrm{e}^x}{x^5} dx\right)^{1/2} \leq \mathrm{e}^{x_2/2}\left(\int_{x_1}^{\infty} \frac{dx}{x^5}\right)^{1/2} = \frac{\mathrm{e}^{x_2/2}}{2x_1^2}, \tag{2.7}$$

we begin with the small x limit. From the above rough estimate and (2.6) we get

$$x_1|u(x_1)| = x_1^2 \frac{\mathrm{e}^{x_1}|u(x_1)|}{x_1} \leq x_1^2 \left|\frac{\mathrm{e}^{x_2}u(x_2)}{x_2}\right| + \frac{1}{2}\mathrm{e}^{x_2/2}\Big(\int_{x_1}^{x_2} |Bu|^2 x \mathrm{e}^x dx\Big).$$

Letting $x_1 \to 0$ in the right hand side, followed by $x_2 \to 0$, yields the result

$$\lim_{x \to 0} xu(x) = 0.$$

The large x limit is obtained in two steps. First, (2.7) and (2.6) give

$$\begin{aligned}\frac{\mathrm{e}^{x_2/2}|u(x_2)|}{x_2} &= \mathrm{e}^{-x_2/2}\frac{\mathrm{e}^{x_2}u(x_2)}{x_2} \\ &\leq \mathrm{e}^{-x_2/2}\left|\frac{\mathrm{e}^{x_1}u(x_1)}{x_1}\right| + \frac{1}{2x_1^2}\Big(\int_{x_1}^{x_2} |Bu(x)| x \mathrm{e}^x dx\Big)^{1/2}.\end{aligned}$$

Letting $x_2 \to \infty$, followed by $x_1 \to \infty$, we obtain the intermediate result

$$\lim_{x \to \infty} \frac{\mathrm{e}^{x/2}|u(x)|}{x} = 0.$$

Returning to the basic estimate (2.6), set $x_1 := x_2/2$ and multiply by $x_2^{5/2}\mathrm{e}^{-x_2/2}$ to get

$$\begin{aligned}x_2^{3/2}\mathrm{e}^{x_2/2}|u(x_2)| \leq & 2x_2^{3/2}\left|u\Big(\frac{x_2}{2}\Big)\right| \\ & + \left(x_2^5 \mathrm{e}^{-x_2} \int_{\frac{x_2}{2}}^{x_2} \frac{\mathrm{e}^x}{x^5} dx\right)^{1/2} \cdot \left(\int_{\frac{x_2}{2}}^{x_2} |Bu|^2 x \mathrm{e}^x dx\right)^{1/2}\end{aligned}$$

Letting $x_2 \to \infty$, the first term on the right hand side vanishes by the above intermediate result, while the second term vanishes because $Bu \in L_2$ and

$$\limsup_{x_2 \to \infty} x_2^5 \mathrm{e}^{-x_2} \int_{\frac{x_2}{2}}^{x_2} \frac{\mathrm{e}^x}{x^5} dx < \infty.$$

This establishes the result for $u \in D(B)$; if $u \in D(B_1)$ the proof is much the same and based on the basic estimate (2.6). ■

Lemma 2.2. – *The adjoint B^* of B in L_2 is given by*

$$B^*u = \frac{-1}{x^2}\partial_x(x^3 u) = -(x\partial_x u + 3u) \tag{2.8}$$

and its domain is

$$D(B^*) := \left\{ u \in L_2 : \int_0^\infty \frac{e^x}{x^3} \left|\partial_x(x^3 u)\right|^2 dx < \infty \right\}. \tag{2.9}$$

and $(B^*, D(B^*)) = (B_1, D(B_1))$. In particular any $u \in D(B^*)$ satisfies the boundary conditions (2.5).

Proof. From lemma 2.1, we have that for any $u \in D(B)$ and $v \in D(B_1)$:

$$\begin{aligned}(v|Bu) - (B_1 v|u) &= \int_0^\infty \left[x^3 v \partial_x\left(\frac{e^x u}{x}\right) + \frac{e^x u(x)}{x}\partial_x(xv^3)\right] dx \\ &= \left[x^2 e^x v(x) u(x)\right]_0^\infty = 0.\end{aligned}$$

Therefore we may conclude that $v \in D(B^*)$ and $u \in D(B_1^*)$, thus the inclusions $(B_1, D(B_1)) \subset (B^*, D(B^*)$ and $(B, D(B)) \subset (B_1, D(B_1^*))$ in the sense of operators.

Now let $v \in D(B^*)$; then for any $\varphi \in C_c^\infty(0,\infty)$ we have $(v|B\varphi) = (B^* v|\varphi)$ by the definition of B^*. In particular this implies $v \in H^1_{\rm loc}(0,\infty)$ (i.e. $\partial_x v$ is locally square integrable for the Lebesgue measure on $(0,\infty)$). Repeating the calculation above, one sees that $(v|B\varphi) = (B_1 v|\varphi)$; from which we infer $B_1 v \in L_2$, i.e. $v \in D(B_1)$. ■

Lemma 2.3. – *We have $D(B) = D(B^*)$ and for any $u \in D(B)$ one has*

$$\|B^*u\|^2 = \|Bu\|^2 + \|x^{1/2}u\|^2, \qquad \|B^*u\|^2 \le 2\|Bu\|^2 + 5\|u\|^2. \tag{2.10}$$

Also for $u \in D(BB^*)$ we have $BB^*u = B^*Bu + xu$.

Proof. For $u \in C_c^\infty(0,\infty)$ we have $BB^*u = B^*Bu + xu$, which yields the first identity for $u \in C_c^\infty(0,\infty)$. On the other hand note that

$$(B^* + B)u = -4u + xu,$$

and

$$(B^* - I)(B - I)u + xu = B^*Bu - (B^* + B)u + u + xu = B^*Bu + 5u$$

which implies that

$$\|x^{1/2}u\|^2 \le \|(B - I)u\|^2 + \|x^{1/2}u\|^2 = \|Bu\|^2 + 5\|u\|^2.$$

This, combined with the first identity, gives the claimed inequality. The operators B^* and B being closed, by a density argument one sees that $D(B) = D(B^*)$ and that relations (2.10) are valid for all $u \in D(B)$. ■

We have also a Poincaré inequality for B^*, namely we have:

Lemma 2.4. – *For any $u \in D(B^*)$ one has*

$$3\|u\|^2 \leq \|B^*u\|^2.$$

Proof. For $u \in C_c^\infty(0,\infty)$, the result follows from the identity $BB^*u = 3u + B_2^*B_2u$, where $B_2u := x\mathrm{e}^{-x}\partial_x(\mathrm{e}^{-x}u)$ and $B_2^*u = -x^{-1}\partial_x(x^2u)$ (recall that the adjoints are taken with respect to scalar product in L_2). ■

From the above we can conclude that the selfadjoint operators B^*B and BB^* generate contractive analytic semigroups on L_2; they are denoted by e^{-tB^*B} and e^{-tBB^*} respectively. Some important properties of these analytic semigroups are gathered in the following lemma.

Lemma 2.5. – *(i) For any $u \in D(B^*) = D(B)$ one has $\mathrm{e}^{-tBB^*}u \in D(B^*)$ and*

$$B^*\mathrm{e}^{-tBB^*}u = \mathrm{e}^{-tB^*B}B^*u.$$

(ii) For any $u \in L_2$ and $t > 0$ one has $\mathrm{e}^{-tBB^*}u \in D(B^*)$ and $\mathrm{e}^{-tB^*B}u \in D(B)$, with the estimates

$$\|B^*\mathrm{e}^{-tBB^*}u\| \leq ct^{-1/2}\|u\|, \qquad \text{and} \qquad \|B\mathrm{e}^{-tB^*B}u\| \leq ct^{-1/2}\|u\|.$$

(iii) If $u \in L_2$, and $u \geq 0$ then $\mathrm{e}^{-tBB^*}u \geq 0$ and $\mathrm{e}^{-tB^*B}u \geq 0$.

We omit the proof, which is standard (see for instance M. Reed & B. Simon [9]).

Using the operators B and B^*, equation (1.2) (with $\alpha(x) := x^4$) can be written in the form

$$\begin{cases} \partial_t f = -B^*(Bf + f + xf^2) \\ f(0,x) = f_{\mathrm{Init}}(x) \geq 0, \end{cases} \tag{2.11}$$

and in its mild formulation

$$f(t) = \mathrm{e}^{-tB^*B}f_{\mathrm{Init}} - \int_0^t B^*\mathrm{e}^{-t-s)BB^*}(f + xf^2)(s)ds, \tag{2.12}$$

where we are using lemma 2.5 (i) to help write the integral term, but more is needed to control the nonlinear term. Actually we will need to prove a regularizing property of the semigroup e^{-tBB^*}, namely that it is ultracontractive in an appropriate sense. As it is known, a useful property of the classical heat semigroup in $\mathbb{R}^N$ is that any initial data in $L^1(\mathbb{R}^N)$ is mapped into $L^\infty(\mathbb{R}^N)$ for any time $t > 0$. Using this property one can develop existence and uniqueness theories for semilinear perturbations of the heat equation. This so-called ultracontractivity property may be proved by knowing the heat kernel and using Young's inequality for convolutions. However (see E.B. Davies [2] for a general discussion) for general semigroups one may use the equivalence of properties 1), 2) and 3) given at the end of section § 1. Here we establish a Nash inequality related to B^* in order to prove the ultracontactivity of the semigroup e^{-tBB^*}.

We begin with the proof of the general lemma 1.3.
Proof of Lemma 1.3. Assume that property (i) of lemma 1.3 is satisfied. Let $u_0 \in X \cap D(A)$ be given and set $u(t) := \mathrm{e}^{-tA^*A}u_0$ be the solution of

$$\partial_t u = -A^*Au, \qquad u(0) = u_0.$$

Then if $\varphi(t) := \|u(t)\|_H^2$, one has $\varphi'(t) = -2\|Au(t)\|^2$. Using (1.10) and the fact that $\|u(t)\|_X \leq c_0\|u_0\|_X$, we have:

$$-\varphi'(t) \geq 2c_1^{-1}\|u(t)\|_X^{-\beta}\|u(t)\|_H^{2+\beta} \geq 2c_1^{-1}c_0^{-\beta}\|u_0\|_X^{-\beta}\,\varphi(t)^{(2+\beta)/2}.$$

This can be written in the form

$$2\beta^{-1}(\varphi(t)^{-\beta/2})' \geq 2c_1^{-1}c_0^{-\beta}\|u_0\|_X^{-\beta},$$

from which we infer that

$$\varphi(t)^{-\beta/2} \geq \beta c_1^{-1}c_0^{-\beta}\|u_0\|_X^{-\beta}t + \varphi(0)^{-\beta/2} \geq \beta c_1^{-1}c_0^{-\beta}\|u_0\|_X^{-\beta}t.$$

Clearly this yields (1.11), that is $\|u(t)\|_H \leq c_2 t^{-1/\beta}\|u_0\|_X$ with $c_2 := (c_1/\beta)^{1/\beta}c_0$.

Conversely, assume that property (ii) of lemma 1.3 is satisfied. Then for $u_0 \in X \cap D(A)$ we have, with the above notation for $u(t)$

$$\begin{aligned} c_2^2 t^{-2/\beta}\|u_0\|_X^2 \geq (u(t)|u(t))_H &= \|u_0\|_H^2 + \int_0^t \frac{d}{d\sigma}\|u(\sigma)\|_H^2 d\sigma \\ &\geq \|u_0\|_H^2 - 2\int_0^t \|AS(\sigma)\|_H^2 d\sigma \\ &\geq \|u_0\|_H^2 - 2t\|Au_0\|_H^2 \end{aligned}$$

where we have used the fact that $u(t)$ satisfies $\partial_t u = -A^*Au$ and therefore $\|Au(t)\|_H \leq \|Au_0\|_H$ (it is enough to take the scalar product with $\partial_t u$). Finally we obtain

$$\|u_0\|_H^2 \leq 2t\|Au_0\|_H^2 + c_2^2 t^{-2/\beta}\|u_0\|_X^2.$$

Upon minimizing the right hand side for $t > 0$, i.e. setting

$$t := \left(\frac{c_2^2 \|u_0\|_X^2}{\beta \|Au_0\|_H^2} \right)^{\beta/(\beta+2)},$$

one obtains inequality (1.10) of the lemma. ■

In order to get the regularizing property of the semigroup e^{-tBB^*}, we apply lemma 1.3 to $A := B^*$ by proving inequality (1.10) when

$$X := \left\{ u \,;\ \|u\|_X := \int_0^\infty |u(x)| e^{x/2} dx < \infty \right\}. \tag{2.13}$$

We begin by seeking a pointwise estimate for $u \in D(B^*)$.

Lemma 2.6. – *For any $u \in D(B^*) = D(B)$ we have*

$$|u(x)| \leq \sqrt{2}\, \frac{e^{-x/2}}{x} \|B^* u\|^{1/2} \, \|u\|^{1/2}.$$

Proof. For $u \in C_c^\infty(0, \infty)$ we have

$$\begin{aligned}(x^3 u(x))^2 &= -2 \int_x^\infty y^3 u(y) \partial_y (y^3 u(y)) dy = 2 \int_x^\infty y^5 u(y) B^* u(y) dy \\ &\leq 2 \int_x^\infty y^4 e^{-y} \cdot |u(y) B^* u(y)| y e^y dy.\end{aligned}$$

Now for $4 \leq x \leq y$ one has $y^4 e^{-y} \leq x^4 e^{-x}$. So for $x \geq 4$ we get

$$|u(x)|^2 \leq 2x^{-2} e^{-x} \|u\| \|B^* u\|,$$

which is the desired result for $x \geq 4$. When $x \leq 4$ we argue similarly beginning with the identity

$$(x^3 u(x))^2 = 2 \int_0^x y^3 u(y) \partial_y (y^3 u(y)) dy.$$

■

Next we prove that the semigroup e^{-tBB^*} acts on X.

Lemma 2.7. – *The semigroup e^{-tBB^*} induces a contractive, positivity preserving semigroup on X; more precisely we have*

$$\|e^{-tBB^*} u_0\|_X \leq e^{-7t/4} \|u_0\|_X.$$

Proof. It is sufficient to prove the lemma for $u_0 \in X$ such that $u_0 \geq 0$. Now for $u_0 \in C_c^\infty(0,\infty)$, and $u_0 \geq 0$ set $u(t) := \mathrm{e}^{-tBB^*}u_0$, that is $\partial_t u = -BB^*u$ and $u(0) = u_0$. One has $u(t) \geq 0$ and $u(t) \in D(BB^*)$ while

$$\begin{aligned}\frac{d}{dt}\int_0^\infty u(t)\mathrm{e}^{x/2}dx &= -\int_0^\infty BB^*u\,\mathrm{e}^{x/2}dx\\ &= \int_0^\infty x^2\mathrm{e}^{-x/2}\partial_x\Big(\frac{\mathrm{e}^x}{x^3}\partial_x(x^3u)\Big)dx.\end{aligned}$$

But integrating by parts twice yields

$$\begin{aligned}\frac{d}{dt}\int u(t,x)\mathrm{e}^{x/2}dx &= \int_0^\infty u(t,x)x^3\partial_x\Big(\frac{1}{x^2}\mathrm{e}^{x/2}(2-\frac{x}{2})\Big)dx\\ &= -\int_0^\infty u(t,x)\left[\frac{x}{2}+\frac{(x-4)^2}{4}\right]\mathrm{e}^{x/2}dx\\ &\leq \frac{-7}{4}\int_0^\infty u(t,x)\mathrm{e}^{x/2}dx,\end{aligned}$$

and the lemma follows for $u_0 \geq 0$. ■

Finally in the following lemma we state the desired regularizing property of the semigroup e^{-tBB^*}.

Lemma 2.8. – *For any $u \in X \cap D(B^*)$ one has*

$$\|u\|^6 \leq 4\|B^*u\|^2\|u\|_X^4.$$

and for all $t > 0$ and $u_0 \in X$ one has

$$\|\mathrm{e}^{-tBB^*}u_0\| \leq t^{-1/4}\,\|u_0\|_X. \tag{2.14}$$

Proof. Using the uniform estimate given in lemma 2.6 we have

$$\begin{aligned}\|u\|^2 = \int_0^\infty |u(x)|^2 x\mathrm{e}^x dx &= \int_0^\infty x|u(x)|\,|u(x)|\mathrm{e}^x dx\\ &\leq \sqrt{2}\|B^*u\|^{1/2}\|u\|^{1/2}\int_0^\infty |u(x)|\mathrm{e}^{x/2}dx,\end{aligned}$$

and hence $\|u\|^{3/2} \leq \sqrt{2}\|B^*u\|^{1/2}\|u\|_X$. Next, using lemma 2.7, as a corollary of lemma 1.3 we obtain (2.14). ■

Combining this result and lemma 2.5 (ii) we have the following regularization result:

Corollary 2.9 . – *For $u_0 \in X$ and all $t > 0$ one has*

$$\|B^*\mathrm{e}^{-tBB^*}u_0\| \leq ct^{-3/4}\|u_0\|_X.$$

3. Existence of local solutions

In order to solve the mild formulation of the Kompaneets equation given by (2.12) we use the classical approach of the fixed point method, when the time t belongs to a sufficiently small interval $[0,T]$. Given $T>0$, we denote by $C([0,T],L_2)$ the space of continuous functions from $[0,T]$ into L_2, equipped with the norm

$$\|u\|_{C([0,T])} := \sup_{0\leq t\leq T} \|u(t)\|.$$

For $u \in L_2$ we define the mapping:

$$(\Phi(u))(t) := \int_0^t B^* \mathrm{e}^{-(t-s)BB^*}(u(s) + xu^2(s))ds. \tag{3.1}$$

Lemma 3.1. – *For any $T>0$, Φ is a locally Lipschitz map from $C([0,T],L_2)$ into itself and*

$$\|\Phi(u) - \Phi(v)\| \leq \ell(T, \|u+v\|_{C_T}) \cdot \|u-v\|_{C_T}$$

with

$$\ell(T, \|u+v\|) \leq c(T^{1/2} + T^{1/4}\|u+v\|_{C([0,T])}).$$

Proof. Using lemma 2.5 (ii) we bound the linear term of the integrand by:

$$\|B^* \mathrm{e}^{-(t-s)BB^*} u(s)\| \leq c\,(t-s)^{-1/2}\|u(s)\|$$

and the nonlinear term by using corollary 2.9

$$\|B^* \mathrm{e}^{-(t-s)BB^*} xu^2(s)\| \leq c\,(t-s)^{-3/4}\|xu^2(s)\|_X \leq c\,(t-s)^{-3/4}\,\|u(s)\|^2.$$

From this one infers that $t \mapsto (\Phi(u))(t)$ is Hölder continuous of exponent $1/4$. Now for $u, v \in C([0,T],L_2)$, using the above estimates we may write:

$$\begin{aligned}\|\Phi(u)(t) - \Phi(v)(t)\| \leq & cT^{1/2}\,\|u-v\|_{C([0,T])} \\ & + cT^{1/4}\|u+v\|_{C([0,T])}\|u-v\|_{C([0,T])}.\end{aligned}$$

(Here we have estimated roughly all the constants, which are not relevant in this study). ■

We can now state the existence and uniqueness result for the mild formulation (1.8), the proof of which is classical.

Theorem 3.2. – *For any $f_{\rm Init} \in L_2$ there exists $T > 0$ such that a unique solution f of the mild formulation (2.12) exists and $f \in C([0,T], L_2)$.*

Concerning existence of strong solutions when $f_{\rm Init} \in D(B^*)$, we proceed as follows. First we set :

$$\Psi(u) := B^*(u + xu^2), \tag{1.13}$$

then we recall that (2.12) can be written as

$$f(t) = \mathrm{e}^{-tB^*B} f_{\rm Init} - \int_0^t \mathrm{e}^{-(t-s)B^*B} \Psi(f(s))ds. \tag{3.2}$$

Using standard Hilbert space techniques, which we omit (cf. M. Reed and B. Simon [9]), we have the following characterization of the domain of $(BB^*)^{1/2}$:

Lemma. *Define $A := (BB^*)^{1/2}$. Then $D(B) = D(A)$, the inverse A^{-1} exists and*

$$\|B^* A^{-1} u\| = \|u\|.$$

With a slight modification of the proof given in A. Pazy [8] (theorem 3.1, section 6.3) one may state the following

Theorem 3.3. – *Let Ψ and A be defined as above. Then for any $f_{\rm Init} \in D(B)$, there exists $T > 0$ such that equation (2.12) has a unique solution $f \in C([0,T], L_2) \cap C^1((0,T), L_2)$. Moreover $f \in C([0,T], D(B))$ and in particular f is a strong solution to*

$$\begin{cases} \partial_t f = -B^*(Bf + f + xf^2) & \text{on } (0,T) \times (0,\infty) \\ f(0,x) = f_{\rm Init}(x). & \text{on } (0,\infty) \end{cases}$$

Moreover if $T_{\max}$ is the maximal time of existence, then either $T_{\max} = +\infty$ or else one has $T_{\max} < +\infty$ and $\lim_{t \uparrow T_{\max}} \|Bf(t)\| = +\infty$.

Proof. We begin by checking that $\Psi : D(A) \to L_2$ is locally Lipschitz. Using the pointwise estimate of lemma 2.6 and the fact that $B^*(xu^2) = 2xuB^*u + u$, we have easily

$$\|\Psi(u) - \Psi(v)\| \le C\,(1 + \|B^*u\| + \|B^*v\|)\,\|B^*u - B^*v\|$$

and also

$$\|\Psi(B^{-1}u) - \Psi(B^{-1}v)\| \le C\,(1 + \|u\| + \|v\|)\,\|u - v\| \tag{3.3}$$

for some constant $C > 0$ (we use also the fact that $\|u\| \leq \|B^*u\|$ by lemma 2.4). Now for $u \in C([0,T], L_2)$ we define:

$$(F(u))(t) := A\mathrm{e}^{-tBB^*} f_{\mathrm{Init}} - \int_0^t A\mathrm{e}^{-(t-s)B^*B}\Psi(A^{-1}u(s))ds.$$

It is straightforward to show that

$$\|F(u)(t) - F(v)(t)\| \leq C \cdot (1 + \|u\|_{C([0,T])} + \|v\|_{C([0,T])}) \\ \times \|u - v\|_{C([0,T])} \int_0^t (t-s)^{-1/2} ds$$

and therefore

$$\|F(u) - F(v)\|_{C_T} \leq C\, \left(1 + \|u\|_{C([0,T])} + \|v\|_{C([0,T])}\right) T^{1/2}\, \|u - v\|_{C([0,T])}.$$

From this one classically infers that, for T small enough, there exists a unique $u \in C([0,T], L_2)$ such that

$$u(t) = A\mathrm{e}^{-tB^*B} f_{\mathrm{Init}} - \int_0^t A\mathrm{e}^{-(t-s)A^*A}\Psi(A^{-1}u(s))ds. \tag{3.4}$$

Now one may check that as $t \mapsto u(t)$ is Hölder continuous on $[0,T] \to L_2$, and Ψ is locally Lipschitz continuous, $t \mapsto \Psi(A^{-1}u(t))$ is also Hölder continuous.
But knowing that e^{-tB^*B} is an analytic semigroup the function

$$f(t) := \mathrm{e}^{-tB^*B} f_{\mathrm{Init}} - \int_0^t \mathrm{e}^{-(t-s)B^*B}\Psi(A^{-1}u(s))ds \tag{3.5}$$

is a strong solution to the initial value problem

$$\begin{cases} \partial_t f + B^*Bf = \Psi(A^{-1}u(t)) \\ f(0) = f_{\mathrm{Init}}. \end{cases}$$

(This means that $f \in C([0,T], L_2) \cap C^1((0,T), L_2)$). One sees also that because of the uniqueness of the solution to (3.4), f defined by (3.5) satisfies $f(t) = A^{-1}u(t)$ and therefore

$$f(t) = \mathrm{e}^{-tB^*B} f_{\mathrm{Init}} - \int_0^t \mathrm{e}^{-(t-s)B^*B}\Psi(f(s))ds$$

and the proof of the theorem is done. ■

4. Qualitative properties, global existence or blow up

In order to prove a comparison result, we observe first that B being a first order differential operator, if $\varphi \in D(B)$, then $\varphi^+ := \max(\varphi, 0) \in D(B)$ and $B(\varphi^+) = B\varphi 1_{[\varphi>0]}$. Recall also that if $\varphi \in D(B)$, then by lemma 2.1 we have $\|x\varphi\|_\infty \leq C\|\varphi\|_{D(B)}$, for some constant $C > 0$.

Proposition 4.1. – *Assume that $f_{1,\text{Init}} \leq f_{2,\text{Init}}$ are two initial data in $D(B)$ and denote by f_1 and f_2 the corresponding solutions in $C^1((0,T), L_2)$ $\cap C([0,T], D(B))$ for some $T > 0$. Then one has $f_1 \leq f_2$ on $[0,T] \times (0,\infty)$. In particular if $f_{\text{Init}} \geq 0$, then $f \geq 0$.*

Proof. The argument is a classical one: set $g := f_1 - f_2$ and $q(t,x) := x(f_1 + f_2)$; recall that due to the above remarks $q \in L^\infty((0,T) \times (0,\infty))$. As g satisfies the equation

$$\partial_t g + B^* Bg = -B^*(g + qg),$$

upon multplying this by g^+ in the sense of the scalar product of L_2 we get

$$\begin{aligned}
\frac{d}{dt}\int_0^\infty |g^+(t,x)|^2 x e^x dx + \int_0^\infty |B(g^+)(t,x|^2 x e^x dx = \\
-\int_0^\infty g^+ B(g^+)(t,x) x e^x dx - \int_0^\infty q(t,x) g^+ B(g^+)(t,x) x e^x dx \\
\leq \varepsilon \int_0^\infty |B(g^+)(t,x|^2 x e^x dx \\
+ (C(\varepsilon) + \|q(t,\cdot)\|_\infty) \int_0^\infty |g^+(t,x)|^2 x e^x dx
\end{aligned}$$

where we have used Young's inequality $\alpha\beta \leq \varepsilon\alpha^2 + C_\varepsilon \beta^2$. Choosing $\varepsilon = 1/2$ for instance, Gronwall lemma and the fact that $g^+(0,x) \equiv 0$ imlpy $g^+(t,x) \equiv 0$, that is $f_1 \leq f_2$. ■

The same observation leads to the following:

Lemma 4.2. – *If f is a solution such that $T_{\max} < \infty$, then*

$$\limsup_{t \uparrow T_{\max}} \|xf(t,\cdot)\|_\infty = +\infty.$$

Proof. Indeed, assuming that $T_{\max} < \infty$ and $\|xf\|_\infty \leq C$ for some constant $C > 0$ and all $t < T_{\max}$, multiplying equation (2.11) by f, using the approach developed above we get

$$\frac{d}{dt}\int_0^\infty |f(t,x)|^2 x e^x dx + \frac{1}{2}\int_0^\infty |Bf(t,x)|^2 x e^x dx \leq C\int_0^\infty |f(t,x)|^2 x e^x dx,$$

which first implies, via the Gronwall lemma, an upper bound for $\|f(t)\| \leq M := M(T_{\max}) < \infty$ for $t < T_{\max}$. Now using the fact that $B^*(xf^2) = -6xf^2 + 2x^2f^2 - 2xfBf$, we obtain, via the estimates of lemma 2.1, for some constant $C > 0$ depending on M, $\|\Psi(f(s))\| \leq C(1 + \|Bf(s)\|)$, and

$$\|Bf(t)\| \leq \|B\mathrm{e}^{-tB^*B} f_{\mathrm{Init}}\| + C \int_0^t (t-s)^{-1/2} \|Bf(s)\| ds,$$

where we have also used the regularizing property of the semigroup e^{-tB^*B} given by lemma 2.5 (ii). It is clear that the generalized Gronwall lemma yields a bound on $\|Bf(t)\|$ for $t < T_{\max}$. This contradicts the assumption $T_{\max} < \infty$. ■

We point out that the comparison result together with lemma 4.2 gives also a criteria for global existence. Indeed if for instance for some $\mu > 0$ we have $f_{\mathrm{Init}} \leq \varphi_\mu(x) := (\mathrm{e}^{x+\mu} - 1)^{-1}$, then the solution is global, because this implies that $f(t,x) \leq \varphi_\mu(x)$ and therefore for all $t > 0$ we have $\|xf(t,\cdot)\| \leq \|x\varphi_\mu\| < \infty$.

Regarding the blow up results, we give here a simple criteria for blow up of solutions, and we refer to [5] for other results. It is clear that the criteria below applies not only to the solutions constructed above, but also to any solution for which the quantities which are manipulated in the following lemma, make sense.

Lemma 4.3. – *Denote $\varphi(x) := \mathrm{e}^{-x}$. Assume that f is a nonnegative solution of equation (2.11), such that $t \mapsto \int_0^\infty f(t,x)\varphi(x)x^2 dx$ is of classe C^1 on an interval $[0,T]$ and such that*

$$\int_0^\infty (1 + f(t,x)) f(t,x) \varphi(x) x^4 dx < \infty$$

for all $t \in [0,T]$. Then if

$$a_0 := a_0(f_{\mathrm{Init}}) := \int_0^\infty f_{\mathrm{Init}}(x)\varphi(x)x^2 dx > 2$$

there exists $T_*(a_0) < \infty$ such that $T < T_*(a_0)$.

Proof. Let $h(t) := \int_0^\infty f(t,x)\varphi(x)x^2 dx$. One checks that

$$\begin{aligned} h'(t) &= -\int_0^\infty x^4(\partial_x f + f + f^2)\partial_x\varphi(x) dx \\ &= -\int_0^\infty f(t,x)\partial_x(x^4\varphi(x)) dx + \int_0^\infty (f + f^2) x^4 \varphi(x) dx \end{aligned}$$

where we have performed another integration by parts on the first term involving $\partial_x f$. As $\partial_x(x^4\varphi(x)) = (4x^3 - x^4)\varphi(x)$, we get finally

$$h'(t) = \int_0^\infty f(t,x)(2x^2 - 4x)x^2\varphi(x)dx + \int_0^\infty f(t,x)^2 x^4\varphi(x)dx.$$

Upon using the fact that $2x^2 - 4x \geq -2$ in the first term and Jensen's inequality

$$\int_0^\infty f(t,x)^2 x^4 \varphi(x)dx \geq \left(\int_0^\infty f(t,x)x^2\varphi(x)dx\right)^2,$$

in the second term, we end up with the differential inequality $h'(t) \geq -2h(t) + h(t)^2$, or equivalently

$$\left(\mathrm{e}^{2t}h(t)\right)' \geq \mathrm{e}^{-2t}\left(\mathrm{e}^{2t}h(t)\right)^2.$$

This implies that

$$0 < \frac{\mathrm{e}^{-2t}}{h(t)} \leq \frac{1}{h(0)} - \frac{1}{2} + \frac{\mathrm{e}^{-2t}}{2},$$

and therefore $T < T_*(a_0) := -\log((h(0)-2)/h(0))/2$.

One may check also that the test function $\varphi(x) := \mathrm{e}^{-x}$ may be replaced by a function of the type $\varphi(x) := \mathrm{e}^{-\lambda x}$ for some $\lambda > 0$. In this case one may see that if the initial data satisfies:

$$\int_0^\infty f_{\mathrm{Init}}(x)\mathrm{e}^{-\lambda x}\, x^2 dx > \frac{4}{\lambda(\lambda+1)},$$

then the corresponding solution blows up in a finite time. ■

References

[1]. R.E. Caflisch and C.D. Levermore, Equilibrium for radiation in a homgeneous plasma Phys. Fluids, 29 (1986), 748–752

[2]. E.B. Davies, Heat Kernels and Spectral Theory Cambridge Tracts in Mathematics 92, Cambridge University Press, Cambridge, UK, 1989

[3]. E.B. Fabes and D.W. Strook, A new proof of Moser's parabolic Harnack inequality via the old ideas of Nash, Arch. Rat. Mech. Analysis, 96 (1986), 327–338

[4]. M. Fukushima, On an L^p estimate of resolvents of Markov processes, Research Inst. Math. Science, Kyoto Univ., 13 (1977), 277–284

[5]. O. Kavian and C.D. Levermore, On the Kompaneets equation, a simplified model of the Fokker–Planck equationIn preparation

[6]. A.S. Kompaneets, The establishment of thermal equilibrium between quanta and electrons, Soviet Physics JETP, 4 (1957), 730–737 (translated from J. Exptl. Theoret. Phys. (USSR), 31 (1956), 876–885)

[7]. J. Nash, Continuity of solutions of parabolic and elliptic equations, Amer. J. Math., 80 (1958), 931–954

[8]. A. Pazy, Semigroups of linear operators and applications to partial differential equations, Applied Math. Science 44, Springer, New-York, 1983

[9]. M. Reed and B. Simon, Methods of Modern Mathematical Physics, (volume IV, Analysis of Operators)Academic Press, New York 1978

[10]. N.Th. Varopoulos, Hardy-Littlewood theory for semigroups, J. Func. Analysis, 63 (1985), 240–260

Otared Kavian
Laboratoire de Mathématiques Appliquées (UMR 7641)
Université de Versailles
45, avenue des Etats Unis
78035 Versailles cedex
France
E-mail:

Studies in Mathematics and its Applications, Vol. 31
D. Cioranescu and J.L. Lions (Editors)

Chapter 22

SINGULAR PERTURBATIONS WITHOUT LIMIT IN THE ENERGY SPACE. CONVERGENCE AND COMPUTATION OF THE ASSOCIATED LAYERS

D. LEGUILLON, E. SANCHEZ-PALENCIA and C. DE SOUZA

1. Introduction

We consider a class of elliptic singular perturbation problems depending on a small parameter ε. The energy space V for $\varepsilon > 0$ is strictly contained in the energy space V_a for the limit problem ($\varepsilon = 0$). Obviously,the dual spaces are such that V'_a is strictly contained in V'. Classical singular perturbation theory is concerned with the case when the loading f is contained in V'_a, so that the problem for $\varepsilon > 0$, as well as the limit problem $\varepsilon = 0$ make sense in the variational formulation. We consider here the case when $f \in V'$ but $f \notin V'_a$, where the variational formulation makes sense for $\varepsilon > 0$ but does not for $\varepsilon = 0$. The energy of the solution tends to infinity as ε tends to 0.

Two examples in dimension one, borrowed from [9] are considered in sections 3 and 5, using the formal asymptotic method of " matched asymptotic expansions" [14]. The solutions exhibit an important layer phenomenon, and the energy of the solution concentrates in such a layer as $\varepsilon = 0$.

Sections 4 and 6 are devoted to a rigorous justification of the above formal results. To this end, the exact problem for u^ε in the variable x is written in terms of a new variable $y = x/\varepsilon$, the " inner variable" for the description of the layer. The perturbation problem changes drastically, becoming a sequence of problems in domains depending on ε which tend to an unbounded limit domain. Moreover, the different terms of the expression of the energy are changed in different ways, so that a new concept of energy appears in the variable y, and this energy remains bounded (after an appropriate scaling of the unknown u^ε, in the example of sections 5 and 6). Then the convergence is proved using an elementary estimate of the new energy.

The problems examined here may be considered as a first attempt to understand asymptotic properties of thin elastic shells, where the structure of the limit space V_a depends on the shape and the fixation conditions of the shell. In some cases, the space V_a is very large, going even out of the space of distribution and accordingly V_a' is very small, so that loadings f do not belong to V_a' (see [6,11,12], and [7] for a model problem). The corresponding difficulties of the numerical computation are mentioned in [4,13].

The special structure of the solutions u^ε with small ε needs a mesh refinement in the region of the layers, as explained in [9] for the one-dimensional examples of sections 3 and 5. A first attempt to elaborate strategies of numerical computation for the problems of shells is done in section 8 for a problem analogous to that of section 3 in dimension 2. The loading f is a δ' distribution localized along a curve $\mathcal{C}$. One may think to use refined meshes in the vicinity of $\mathcal{C}$, in particular anisotropic meshes involving flatened triangles in the direction normal to the layer as used in other problems involving layers [1,2,3]. Instead of this, we used an iterative adaptive mesh procedure. Starting from a conventional F.E. mesh, new refined meshes are sucessively derived using numerical estimates of the computed solution on the previous mesh [5,8]. Such a procedure has been successfully used for computing fluid mechanics problems involving shock waves, boundary layers and wakes.

Traditionally, automatic mesh generator produce "isotropic" meshes where triangles are as close as possible to equilateral ones. In the present case it will lead to meshes containing a drastic large number of elements. Thus, we select a mesh generator (BL2D [8]) able to constructanisotropic meshes, i.e. triangles having a large aspect ratio.

2. Singular perturbations

Let V be a real Hilbert space, $a(u, v)$ and $b(u, v)$ two continuous and symmetric bilinear forms on V. In addition, the form b is coercive, so that it may be taken as scalar product in V. Let the form a satisfy

$$\begin{aligned} &a(v,v) \geq 0, \\ &a(v,v) = 0 \Longrightarrow v = 0. \end{aligned} \tag{2.1}$$

In other words, $a^{1/2}$ is a norm on V.

Let V_a be the completion of V with this norm. Obviously, V, V_a and their duals V', V_a' satisfy

$$V \subset V_a \ , \ V_a' \subset V' \tag{2.2}$$

with dense and continuous embedings.

We consider the following family of problems with parameter $\varepsilon \in (0, 1]$:

Problem P_ε : Let $f \in V'$, find $u^\varepsilon \in V$ satisfying

$$a(u^\varepsilon, v) + \varepsilon^2 b(u^\varepsilon, v) = < f, v >, \quad \forall v \in V. \tag{2.3}$$

Obviously, u^ε exists and is unique. Its energy is defined by

$$E(u^\varepsilon) = \frac{1}{2}\left[a(u^\varepsilon, u^\varepsilon) + \varepsilon^2 b(u^\varepsilon, u^\varepsilon)\right]. \tag{2.4}$$

In usual examples, this energy is an integral on a certain domain, and the energy in a part of this domain makes sense.

Problem P_0 : let $f \in V_a'$. Find $u \in V_a$ satisfying

$$a(u, v) = < f, v >, \quad \forall v \in V_a. \tag{2.5}$$

with the energy

$$E(u) = \frac{1}{2}a(u, u). \tag{2.6}$$

Theorem 1.1 - *Let $f \in V_a'$ be fixed. The solutions u^ε and u of P_ε and P_0 are then well defined and satisfy*

$$u^\varepsilon \longrightarrow u \quad \text{in } V_a \text{ strongly} \tag{2.7}$$

$$E(u^\varepsilon) \longrightarrow E(u) \tag{2.8}$$

We shall not give here the proof of this classical result which may be found, for instance in [13] sect. VI.1.4. In particular we note that (2.8) is a corollary of (2.7), taking $v = u^\varepsilon$ or u in (2.3) and (2.5) respectively. Moreover, we have (see [6] or[7])

Theorem 1.2 - *Let $f \in V'$ and u^ε be the solution of P_ε, then*
a) $E(u^\varepsilon)$ is bounded iff $f \in V_a'$,
b) If $f \notin V_a'$, then $E(u^\varepsilon) \longrightarrow \infty$.

Remark 1.3. When $f \notin V_a'$, the limit problem does not make sense as a variational problem, but it may happen in elliptic problems that it does in the Lions and Magenes sense [10].

3. First example in dimension one

Let us consider $V = H_0^2(0, 1)$ and the forms

$$a(u, v) = \int_0^1 u'(x)v'(x)dx, \tag{3.1}$$

$$b(u, v) = \int_0^1 u''(x)v''(x)dx. \tag{3.2}$$

The completion of V is $V_a = H_0^1(0,1)$. We shall identify $H = L^2(0,1)$ to its dual, so that

$$V' = H^{-2}(0,1), \qquad V_a' = H^{-1}(0,1). \tag{3.3}$$

Let us take $f = \delta'_{1/2}$, where $\delta_{1/2}$ denotes the Dirac distribution at $x = 1/2$. We note that

$$\delta'_{1/2} \in V', \quad \delta'_{1/2} \notin V_a'. \tag{3.4}$$

and we are in the situation of Theorem 1.2.b. The problem P_ε is:

$$\left(-\frac{d^2}{dx^2} + \varepsilon^2 \frac{d^4}{dx^4}\right) u^\varepsilon = \delta'_{1/2} \tag{3.5}$$

$$u^\varepsilon(0) = u^\varepsilon(1) = \frac{du^\varepsilon}{dx}(0) = \frac{du^\varepsilon}{dx}(1) = 0 \tag{3.6}$$

The limit problem P_0 does not make sense as a variational problem because of (3.4). Nevertheless, as $\delta'_{1/2}$ is smooth near the boundary, it belongs to the space Ξ^{-2} of [10] and P_0 is a "Lions–Magenes problem" (see also Remark 1.3) :

$$-\frac{d^2}{dx^2} u = \delta'_{1/2} \tag{3.7}$$

$$u(0) = u(1) = 0 \tag{3.8}$$

whose solution is

$$u(x) = \begin{cases} x & \text{for } 0 < x < 1/2 \\ x - 1 & \text{for } 1/2 < x < 1 \end{cases} \tag{3.9}$$

which presents a " Heaviside step" at $x = 1/2$. The formalasymptotic expansion of u^ε is easily obtained by the method of the matched asymptotic expansions [14]. It appears that u^ε exhibits an internal layer in the vicinity of $x = 1/2$ which is described by an inner asymptotic expansion in terms of the "inner variable"

$$y = \frac{x - 1/2}{\varepsilon}, \tag{3.10}$$

the leading term of this asymptotic expansion is

$$v^0(y) = \begin{cases} -\frac{1}{2}e^{y} + \frac{1}{2} & \text{for } y < 0 \\ \frac{1}{2}e^{-y} - \frac{1}{2} & \text{for } y > 0 \end{cases} \tag{3.11}$$

which is the " smoothed" step.

There are also " small boundary layers" in the vicinity of $x = 0$ and $x = 1$ described in terms of x/ε and $(1-x)/\varepsilon$ respectively, accounting for the lost boundary conditions on the derivative (see (3.6) and (3.8)). Out of these three layers, the convergence of u^ε to u is uniform. It is then seen that the energies of u^ε in the layer (3.11), in the small layers and out of the layers are of orders $O(\varepsilon^{-1})$, $O(\varepsilon)$ and $O(1)$ respectively. We then see that the total energy tends to infinity, according to Theorem 1.2.b ; moreover it " concentrates asymptotically" in the layer (3.11).

4. Convergence to the layer

The layer (an internal layer) (3.11) is a well defined function of the variable y. We show that, writting down the problemP_ε in the variable y, it is possible to prove that the solutions converge to (3.11); this convergence holds in the topology of some energy space of functions of y, which does not coincide (at least concerning limits as $\varepsilon \to 0$) with the energy spaces for the variable x.

Let us write explicitly the variational formulation of P_ε in the case of Section 3 :

$$\begin{aligned} &\text{Find } u^\varepsilon \in H_0^2(0,1) \text{ such that}, \forall v \in H_0^2(0,1) \\ &\int_0^1 (u^{\varepsilon\prime} v' + \varepsilon^2 u^{\varepsilon\prime\prime} v'')dx = -\frac{dv}{dx}(1/2) \end{aligned} \tag{4.1}$$

Then we define

$$U^\varepsilon(y) = u^\varepsilon(\varepsilon y + 1/2). \tag{4.2}$$

The problem for U^ε is obtained from (4.1) using (4.2) and an analogous formula for the test function. This gives, after multiplying by ε

$$\begin{aligned} &\text{Find} u^\varepsilon \in V_\varepsilon \text{ such that}, \forall W \in V_\varepsilon \\ &\int_{-1/2\varepsilon}^{1/2\varepsilon} [(\partial_y U^\varepsilon)(\partial_y W) + (\partial_y^2 U^\varepsilon)(\partial_y^2 W)]\, dy = -\partial_y W(0) \end{aligned} \tag{4.3}$$

where $V_\varepsilon = H_0^2(-1/(2\varepsilon), 1/(2\varepsilon))$.

Here, it is useful to consider the functions of this space continued with value zero for $\mid y \mid > 1/(2\varepsilon)$. In order to define an appropriate " limit space" we note that, as ε tends to zero, the " boundary conditions" are sent to infinity, which is not very easy to handle. We also note that the right hand side of (4.3) only contains the trace of the first derivative, so that it is not modified by adding a constant to the test functions. Consequently, we are passing to the limit" up to additive constants". To this end, for each V_ε we also consider the space $\tilde{V}_\varepsilon$ of the functions of $H_0^2(-1/2\varepsilon, 1/2\varepsilon)$ defined up to an additive constant. The spaces are ordered by embeding as ε decreases. Then we define the " limit space" $\mathcal{V}$ as the completion for the norm

$$\| W \|_v^2 = \int_{-\infty}^{+\infty} \left[(\partial_y^2 W)^2 + (\partial_y W)^2 \right] dy, \tag{4.4}$$

of the space

$$\bigcup_\varepsilon \tilde{V}_\varepsilon . \tag{4.5}$$

The " limit problem" writes: Find $U \in \mathcal{V}$ such that $\forall W \in \mathcal{V}$

$$\int_{-\infty}^{+\infty} \left[(\partial_y U)(\partial_y W) + (\partial_y^2 U)(\partial_y^2 W) \right] dy = -\partial_y W(0) \,. \tag{4.6}$$

We note that it is a classical variational problem in the Hilbert space $\mathcal{V}$ as the right hand side is obviously a continuous functional on it. Then the solution U is well defined (we shall see later that it is the layer (3.11) up to an additive constant). Then we have

Theorem 4.1 - *Let $\tilde{U}^\varepsilon \in \tilde{V}_\varepsilon$ be the solution of (4.3) defined up to an additive constant, and U the solution of (4.6). Then :*

$$\tilde{U}^\varepsilon \longrightarrow U \quad \text{in } \mathcal{V} \text{ strongly} \tag{4.7}$$

Proof. Take $W = U^\varepsilon$ in (4.3). Then, considering for each U^ε the corresponding equivalence class $\tilde{U}^\varepsilon$ defined up to an additive constant, we have

$$\| \tilde{U}^\varepsilon \|_{\mathcal{V}} \le C \tag{4.8}$$

and after extracting a subsequence :

$$\tilde{U}^\varepsilon \longrightarrow U^* \quad \text{in } \mathcal{V} \text{ weakly} \tag{4.9}$$

which implies that the first and second derivatives converge in $L^2(\mathbb{N})$ weakly.

Let us fix W belonging to a certain V_ε (and then to the V_ε with smaller ε) in (4.3). We may write $\tilde{U}^\varepsilon$ and $\tilde{W}$ instead of U^ε and W. Then, passing to the limit (4.9) we obtain (4.6) with U^* instead of U.As the considered W are in a space dense in $\mathcal{V}$, we see that $U^* = U$, i.e. the subsequence (and then the whole sequence) tends to U. It only remains to prove that the convergence in (4.9) is strong. Let us denote by B the bilinear form in the left hand side of either (4.3) or (4.6). We have :

$$\begin{aligned} \| \tilde{U}^\varepsilon - U \|_{\mathcal{V}}^2 &= B(\tilde{U}^\varepsilon - U, \tilde{U}^\varepsilon - U) \\ &= B(\tilde{U}^\varepsilon, \tilde{U}^\varepsilon) - 2B(U, \tilde{U}^\varepsilon) + B(U,U) \\ &= -\partial_y \tilde{U}^\varepsilon(0) + 2\partial_y \tilde{U}^\varepsilon(0) - \partial_y U(0) \end{aligned} \tag{4.10}$$

where we used (4.3) with $W = U^\varepsilon$ and (4.6) with $W = U^\varepsilon$ and $W = U$. But the right hand side of (4.10) tends to zero by virtue of (4.9) as it involves a continuous functional on $\mathcal{V}$. ■

It is not hard to check that (3.11) (up to an aditive constant) is the solution of (4.6). Indeed, the equation associated with (4.6) is

$$\left(-\partial_y^2 + \partial_y^4\right) U = \delta'(y) \tag{4.11}$$

and we note that (3.11) is a solution of (4.11) for $y \neq 0$. Moreover, at $y = 0,$, $v^0(y)$ has a discontinuity of the second derivative, which implies a δ' term for the fourth derivative, so that $v^0(y)$ solves (4.11).

Finally, we must check that v^0 in (3.11) (up to an additive constant) isan element of $\mathcal{V}$. Indeed, the completion process passing from (4.5) to $\mathcal{V}$ allows functions tending to two different constants at $+\infty$ and $-\infty$, whereas (4.5) only contains functions vanishing (i.e. equal to a single constant) for large $\mid y \mid$. Let U tend to the different constants at $+\infty$ and $-\infty$. It may be approximated by functions U_L tending to the same constant using a matching on a large interval of length $L \to \infty$; the first and second derivatives of U_L are of order $O(L^{-1})$ and $O(L^{-2})$, respectively. Then

$$\| U - U_L \|_{\mathcal{V}}^2 = O\left[L(L^{-2} + L^{-4})\right] \to 0. \tag{4.12}$$

As a result, the boundary layer (3.11) is the limit of the functions $\tilde{U}^\varepsilon$ in the topology of $\mathcal{V}$, i.e. in the energy of the " inner problem" for the variable y (which is not the energy for the variable x).

5. Second example in dimension one

We are now considering a second example with (3.1)–(3.3), when the right hand side is given by

$$f(x) = x^{-p-2} + (1-x)^{-p-2}. \tag{5.1}$$

with some $p \in (0, 1/2)$. We note that in this case f is a second derivative of a function of L^2, so that it belongs to $V' = H^{-2}$. Nevertheless, it is singular at the boundary of the domain, and then it does not belong to the Ξ^{-2} space and the Lions-Magenens theory does not apply. Moreover, we shall see that the limit of u^ε does not exist in a usualsense. We must perform a re-scaling to prove the convergence to the corresponding boundary layer. The equation is always (3.5) with the right hand side (5.1).

The (formal) asymptotic expansion of the solution u^ε takes the form :

$$u^\varepsilon_{out} = \varepsilon^{-p}\gamma + ... \tag{5.2}$$

$$u^\varepsilon_{in} = \varepsilon^{-p}v^0(y) + ..., \qquad y = x/\varepsilon \tag{5.3}$$

where "out" and "in" denote the outer and inner matched expansions. Of course there is an analogous inner expansion in the vicinity of $x = 1$. Note that the first one is such that its leading term is constant with respect to x; γ is a constant coming from the study of the boundary layer(5.3). As for the inner expansion, $v^0(y)$ is the unique solution of to

$$\left(-\frac{d^2}{dy^2} + \frac{d^4}{dy^4}\right) v^0(y) = y^{-p-2}, \quad y \in (0, +\infty) \tag{5.4}$$

$$v^0(0) = \frac{dv^0}{dy}(0) = 0 \tag{5.5}$$

$$v^0 \text{ is bounded on } (0, \infty). \tag{5.6}$$

It appears that (5.4)–(5.6) has a unique solution which tends to a certain constant γ as $y \to +\infty$. This is the constant which appears in the outer expansion (5.2).

It is apparent that the boundary layer is somehow "autonomous", as (5.4)–(5.6) is a well-posed problem with a given right hand side. The outer expansion (5.2) is in some sort a "sequel" of the boundary layers: in fact its leading term is nothing but the horizontal asymptotic of the function $v^0(y)$. It is noticeable that an accurate finite element computation of u^ε needs a very small mesh step h in the layers, whereas a coarse mesh out of the layers works very well. Moreover, if the mesh step in the layers is not sufficiently fine, the computation of the layer is obviously poor, but, in addition, the region out of the layers (which depends on them, as we just pointed out) is also inaccurately computed.

Concerning the energy, a simple computation of (5.3) shows that the energy in the layers is of order $O(\varepsilon^{-2p-1})$, whereas out of the layers,as the leading term in (5.2) has a vanishing energy, it is of order $o(\varepsilon^{-2p})$. We observe again that the total energy tends to infinity and it concentrates in the layers.

6. Convergence in the second example

First we give a variational formulation of the problem (5.4)–(5.6).

Let $\mathcal{V}$ be the completion of the space of function of $H_0^2(0,\infty)$ which vanish for sufficiently large y, with the norm

$$\| w \|_v^2 = \int_0^\infty \left[(\partial_y w)^2 + (\partial_y^2 w)^2 \right] dy \ . \tag{6.1}$$

It is easily checked as at the end of Section 4, that $\mathcal{V}$ contains functions tending to a constant different from zero at infinity. Then, the variational formulation of (5.4)–(5.6) is :

$$\begin{aligned} &\text{Find } v^0 \in \mathcal{V} \text{ such that } \quad \forall w \in \mathcal{V} \\ &(v^0, w)_{\mathcal{V}} = \int_0^{+\infty} y^{-p-2} w(y)\, dy \end{aligned} \tag{6.2}$$

We must check the following result:

Lemma 6.1 - *The right hand side of (6.2) is a continuous functional on* $\mathcal{V}$.

Proof. As the function y^{-p-2} is localy of class H^{-2}, its behaviour at infinity must only be checked. We may consider

$$\int_0^{+\infty} \varphi(y) w(y)\, dy \tag{6.3}$$

where φ is a smooth function, equal to y^{-p-2} for sufficiently large y. We may choose φ such that

$$\int_0^{+\infty} \varphi(y) dy = 0. \tag{6.4}$$

Let us construct

$$\Phi(y) = \int_0^y \varphi(\lambda) d\lambda. \tag{6.5}$$

which is smooth and satisfies

$$\Phi(0) = 0 \tag{6.6}$$

$$\Phi(y) = -\frac{y^{-p-2}}{p+1} \qquad \text{for large } y. \tag{6.7}$$

Let us take $w \in H_0^2(0,\infty)$ vanishing for large y. We have

$$\left|\int_0^\infty \varphi(y)w(y)dy\right| = \left|\int_0^\infty \Phi'(y)w(y)dy\right| = \left|\int_0^\infty \Phi(y)w'(y)dy\right| \\ \leq \| \Phi \|_{L^2} \| w' \|_{L^2} \leq C \| w \|_{\mathcal{V}} \tag{6.8}$$

which proves the lemma as w is any function in a dense set of $\mathcal{V}$. ■

In order to prove the convergence to the layer at $x = 0$, we write the problem P_ε ((3.1)–(3.3)) with the right hand side (5.1) after the change :

$$u^\varepsilon(x) = \varepsilon^{-p} U^\varepsilon(y), \quad y = x/\varepsilon \tag{6.9}$$

namely: Find $U^\varepsilon \in H_0^2(0,1/\varepsilon)$ such that for all $W \in H_0^2(1,1/\varepsilon)$

$$\int_0^{1/\varepsilon} [(\partial_y U^\varepsilon)(\partial_y W) + (\partial_y^2 U^\varepsilon)(\partial_y^2 W)] \, dy \\ = \int_0^{1/\varepsilon} \left[y^{-p-2} + \left(\frac{1}{\varepsilon} - y\right)^{-p-2} \right] W dy \tag{6.10}$$

The function U^ε may be continued with value zero for $y > 1/\varepsilon$, so that it is element of $\mathcal{V}$. The proof of the convergence is then analogous to that of Section 4, and even simpler, as we do not deal with equivalence classes.

Theorem 6.1 - *Let U^ε and v^0 be the solutions to (6.10) and (6.2) respectively. Then,*

$$U^\varepsilon \to v^0 \text{ in } \mathcal{V} \text{ weakly} \tag{6.11}$$

Remark 6.2. In Theorem 6.1 the convergence is only weak, and the reasoning (4.10) for the strong convergence does not work because of the presence of the term f near $y = 1/\varepsilon$. Nevertheless, by the linearity of the problem, we may decompose it into two problems for the two terms of f in (5.1). For each one (the solution of which is somehow analogous to that of Section 5), we may prove strong convergence in $\mathcal{V}$ to the corresponding layer. Oppositely, when we consider simultaneously the two terms in (5.1), strong convergence does not hold, as the limit is the layer in the neighbourhood of $x = 1$, whereas U^ε bears the energy of both layers.

7. An example in dimension two

In this section we consider a problem analogous to that of Section 3 but in $\Omega = (0,1) \times (0,1)$ of the (x_1, x_2) plane. The equation is

$$(\varepsilon^2 \Delta^2 - \Delta) u^\varepsilon = f. \tag{7.1}$$

The chosen boundary conditions are not the Dirichlet ones, but special ones allowing on one hand anti-symmetry continuation for $x \in (1,2)$ and on the other hand, such that the solutions for f independent of y is it self independent of y. The sake of such a choice is obviously to compare two-dimensional solutions with the previously obtained one-dimensional ones. This conditions are

$$u = 0 \quad \text{for } x = 0, \tag{7.2}$$

$$\varepsilon(\partial u/\partial n) = 0 \quad \text{for } x = 0 \tag{7.3}$$

$$u = 0 \quad \text{for } x = 1 \tag{7.4}$$

$$\varepsilon^2 \Delta u = 0 \quad \text{for } x = 1 \tag{7.5}$$

$$\partial u/\partial u = 0 \quad \text{for } y = 0,\ \ y = 1 \tag{7.6}$$

$$\varepsilon^2(\partial \Delta u/\partial n) = 0 \quad \text{for } y = 0,\ y = 1 \tag{7.7}$$

Clearly conditions with the factor ε only are concerned with $\varepsilon > 0$. The spaces V and V_a are :

$$V = \{v \in H^2(\Omega),\ u \text{ satisfies } (7.2),\ (7.3),\ (7.4),\ (7.6)\} \tag{7.8}$$

$$V_a = \{v \in H^1(\Omega),\ u \text{ satisfies } (7.2),\ (7.4)\} \tag{7.9}$$

Let us define the bilinear form

$$a^\varepsilon(u, v) = \varepsilon^2 \int_\Omega \Delta u . \Delta v \, dx + \int_\Omega \nabla u . \nabla v \, dx. \tag{7.10}$$

The space V_a is classical. As for V, we have

Lemma 7.1 - *There exist two constants c_1 and c_2 such that for any u, v in V :*

$$| \, a^\varepsilon(u, v) \, | \leq c_1 \left(\varepsilon^2 \parallel u \parallel_V \parallel v \parallel_V + \parallel u \parallel_{V_a} \parallel v \parallel_{V_a}\right) \tag{7.11}$$

$$a^\varepsilon(v, v) \geq c_2 \left(\varepsilon^2 \parallel v \parallel_V^2 + \parallel v \parallel_{V_a}^2\right) \tag{7.12}$$

Proof. Inequality (7.11) is obvious. To prove (7.12), we note that if $v \in V$, then

$$\Delta v = \varphi \in L^2(\Omega) \tag{7.13}$$

and v satisfies (7.2), (7.4), (7.6). Each one of these conditions fulfils the Shapiro-Lopatinskii condition for (7.13) (i.e. it is a " good boundary condition for the elliptic problem (7.13)") so that the classical inequalities of regularity theory for elliptic problems hold true and we have

$$\varepsilon^2 \parallel v \parallel_{H^2}^2 \leq \varepsilon^2 c_3 \left(\parallel \Delta v \parallel_{L^2}^2 + \parallel v \parallel_{L^2}^2\right). \tag{7.14}$$

Moreover, for $v \in V$ the Poincaré inequality gives :

$$\| v \|_{L^2}^2 \leq c_4 \| \nabla v \|_{L^2}^2 \tag{7.15}$$

and (7.12) follows from (7.14) and (7.15). It should be noticed that regularity theory holds true in the above problem, even if the boundary of Ω is not smooth. Indeed, the corners have amplitude $\pi/2$, and a reasoning of continuation by symmetry or antisymmetry shows that the problem is not singular, and behaves as on a straight boundary. ■

As for the " loading" f, we shall take a $\delta'(\mathcal{C})$ distribution on a curve $\mathcal{C}$, so that

$$< f, v >= - \int_{\mathcal{C}} (\partial v / \partial n) ds, \tag{7.16}$$

where s and n denote the arc and the normal to $\mathcal{C}$. Obviously, by the trace theorem,

$$\delta'(\mathcal{C}) \in V', \quad \delta'(\mathcal{C}) \notin V_a' \tag{7.17}$$

8. Numerical approximation

The two-idimensional problem (7.1)–(7.7) with the right handside member (7.16), (7.17) is still a "Lions-Magenes" problem (see Section 3 and remark 1.3). As in the one dimensional case [9], it can be reasonably expected that an almost classical finite element procedure will solve the problem, provided some precautions.

First of all, the element must be able to take into account the special right handside member (7.16). Obviously, Lagrange elements are not well-fitted to this formulation. Zienkiewicz and Taylor [15] propose an Hermite triangular element with corner nodes (9DOF) which "performs excellently" in problems of plate and shell bending. It does not ensure the C^1 continuity, normal slope is discontinuous through the triangles edges, however it includes derivatives as degrees of freedom at nodes and allows us to compute expression (7.16).

As emphasized in the one-dimensional case [9], the second precaution is to select an appropriate mesh, the problem being of course more complicated in two dimensions. It will be the aim of the following in the present section.

Computations are carried out on the square $[0, 1] \times [0, 1]$. As a test, the line $\mathcal{C}$ in (7.16) is the right edge $x_1 = 1$ of the square. Thus, it can be easily seen that the solution is independent of x_2 and coincides with the one-dimensional solution derived from (3.5), (3.6) (see Section 7). It must be pointed out that conditions (7.4) and (7.5) play the role of anti-symmetry conditions, thus the solution extended by anti-symmetry to the domain $[0,2]\times[0,1]$ is exactly that exhibited in Section 3 after having being stretched. The computation of (7.16) does not afford special difficulties since $\mathcal{C}$ is made of the successive triangle edges forming the right side of the square. Moreover, there is no lack of normal slope continuity through $\mathcal{C}$ in the present anti-symmetrical case. Thus, $\partial v/\partial n$ in (7.16) is numerically defined without ambiguity.

A first computation has been carried out on a coarse mesh with 155 triangles (figure 4-(0)), for $\varepsilon = 0.01$. As anticipated, results are far from being satisfactory (as observed for instance from the isovalues of Figure 5-(0), corresponding to the second numerical example). Obviously, it is the coarseness which is responsible for the inaccuracy. A strongly and uniformly refined mesh with 25600 triangles, as illustrated in Figure 1, would be able to provide us with better results, but the computation cost would increase drastically.

Figure 1: A uniformly refined isotropic mesh (25600 triangles)

The right approach is clearly to refine the mesh only in the vicinity of the layer which location is well known in the present case. It can be done (Figure 2) using a structured mesh with 684 triangles.

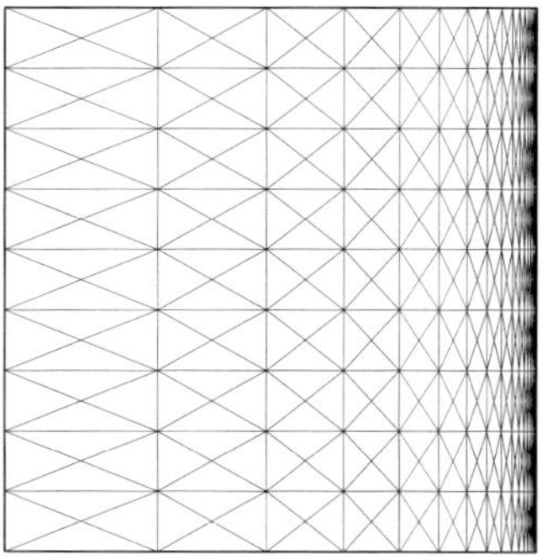

Figure 2: A refined structured anisotropic mesh (684 triangles)

A similar result can be achieved by using an automatic mesh generator (BL2D [8]) with a prescribed map of triangle sizes. It is illustrated in Figure 3 with a mesh made of 3031 triangles. There is a strong refinement along the right edge which covers the singular layer. A smaller refinement has also been required along the left edge to take into account the weaker boundary layer in this area (as already mentioned in Section 3). It is to be pointed out that triangles in the right part of Figure 3 are by far smaller than those of Figure 1, although the total number of triangles is itself smaller.

Figure 3: An automatically (BL2D) refined isotropic mesh (3031 triangles)

Independently of the procedure used to generate the meshes, there is a fundamental difference between Figures 2 and 3. For most of mesh generators, as the one used to produce Figure 3, more a triangle is equilateral and more it is optimal, this a consequence of an interpolation error estimate. Thus triangles have roughly the same diameter in both x_1 and x_2 directions.

Using a different procedure, Figure 2 has been obtained by subdividing the edges in an appropriate way. A refined mesh is prescribed on the right side by small divisions of the horizontal axis, since it is known that it is the location of the layer. On the other hand, the knowledge of the test solution (independent of x_2) brings us to define coarse divisions on the vertical axis. It is clear, in this very particular case, that the high aspect ratio of the elements located within the layer is not at the origin of additional errors (as a matter of fact, a single vertical element would suffice to get a reliable solution). Such a mesh is called *anisotropic*, there is a privileged direction. On the contrary, a mesh favouring equilateral triangles is baptised *isotropic*.

This remark and the observation of meshes obtained by [5, Chap.12], concerning problems of fluid mechanics with shock waves, give raise to a new question: could it be possible to generate automatically anisotropic meshes taking into account such layers, in the general case, i.e. even if their location is unknown ?

The procedure proposed in [5, Chap.12] and [8], is an iterative one. A first calculation is performed on a coarse isotropic mesh. A metric is extracted at each node from the computed solution. This metric consists in a 2×2 symmetric positive definite matrix. Its orthogonal eigenvectors form a local basis and the inverse of the eigenvalues define shrink coefficients. In such a metric, if the eigenvalues differ significantly, shrinked equilateral triangles become thin in the direction of the eigenvector associated with the larger eigenvalue. Triangles seem to arrange in a parrallel direction to the other eigenvector. Based on these metrics, an anisotropic new mesh is generated and then an (expected) improved solution is computed and so on. It must be pointed out that such a mesh generation is a complicated procedure, in particular "interpolated" metrics must be defined between nodes (see the guide [8] for a complete explanation).

The main difficulty to use the above iterative procedure is then to define at each node a metric extracted from a computed solution. As suggested in [5, Chap.12], the hessian matrix of the solution (i.e., the matrix of the second order derivatives) will be used. Thus, in a first step the second order derivatives of the solution $\partial^2 u/\partial x_1^2$, $\partial^2 u/\partial x_2^2$ and $\partial^2 u/\partial x_1 \partial x_2$ are computed at each node (an average between neighbouring elements). Then, eigenvectors v_1 and v_2 and eigenvalues λ_1 and λ_2 are derived. Since the hessian is not necessarily a positive matrix, in a third step, a new matrix is built having the eigenvectors v_1 and v_2 and the eigenvalues $|\lambda_1|$ and $|\lambda_2|$. This matrix calculated at each node provides the procedure with appropriate metrics. From a mechanical point of view, it means that the meaningful parameter to refine a mesh is the stress gradient (i.e., the second order derivatives of the displacements).

The above procedure has been checked in the present case of singular layers. The line $\mathcal{C}$ in (7.16) is chosen within the domain of computation, it is defined by $x_2 = -3x_1 + 2$. The computation of (7.16) is carried out on the successive segments at the intersection of $\mathcal{C}$ and the triangles using a three-point quadrature formula. Of course, we consider at the beginning of the process that the location of the layer around $\mathcal{C}$ is not known. Thus, no special mesh refinement can be initially prescribed.

Figure 4 shows a "film" of the 4 iterations required to generate a satisfying mesh for $\varepsilon = 0.01$ (the initial isotropic step called iteration 0 is included in the account). Meshes contain a reasonably low number of elements, respectively 155, 197, 314 and 685 triangles.

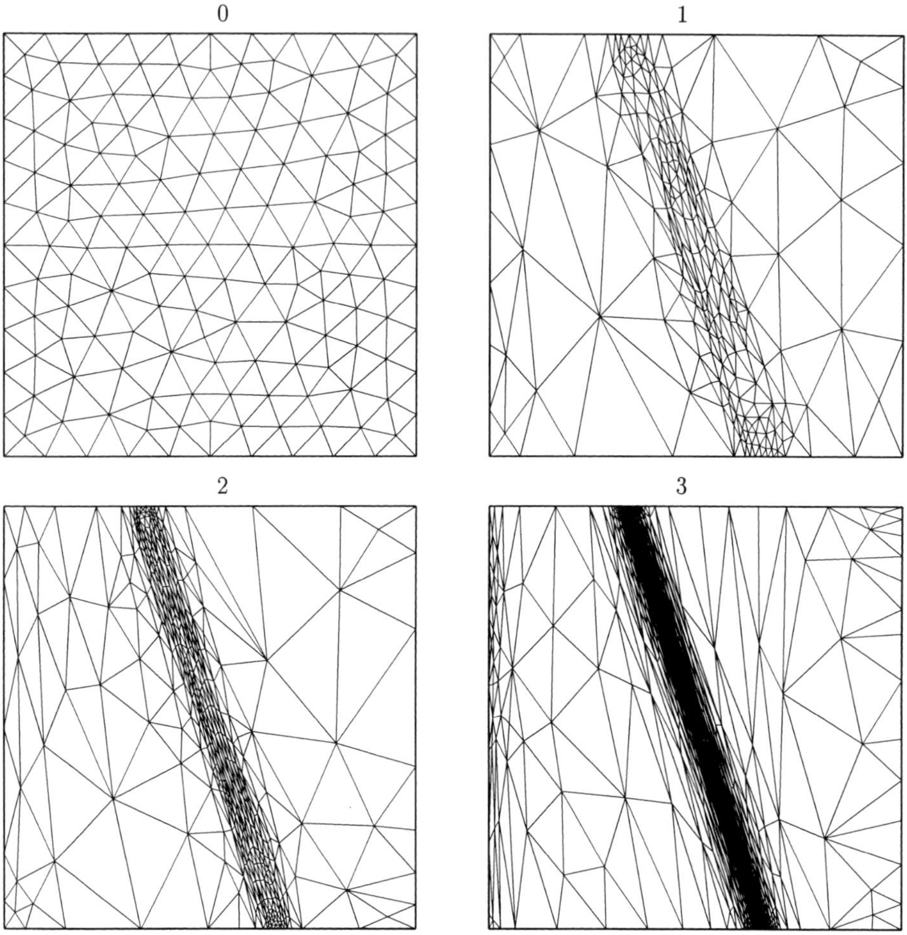

Figure 4: The four succesive meshes generated by the iterative process (BL2D), the first one is isotropic the next ones become more and more anisotropic ((0) 155, (1) 197, (2) 314 and *(3) 685* triangles*)*

Figure 5 gives an insight into the numerical solution improvement which can be expected from such an iterative remeshing procedure.

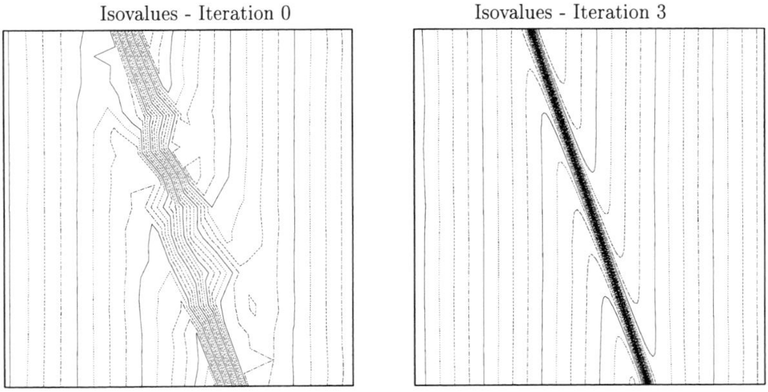

Figure 5: The isovalues of the computed solution corresponding to the first (coarse isotropic mesh 4 (0)) and last ("converged" anisotropic mesh 4 (3)) iterations

The last Figure 6 exhibits the final 11th iteration in the situation of a thiner layer arising from the case $\varepsilon = 0.001$ (528 triangles). Despite the high aspect ratio of the triangles, never any warning, pointing out the matrix ill-conditioning, has been displayed during computations (performed in double precision).

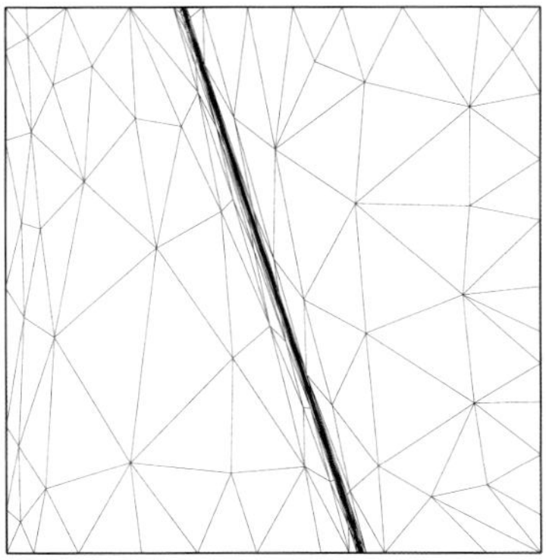

Figure 6: The 11th anisotropic iteration (BL2D) in case of a very thin layer (528 triangles)

Two difficulties remain in this procedure, first there is no criterion to stop automatically the iterations, such a criterion would probably imply the concept of optimal mesh which seems to be out of reach (at the time being). Iterations have been stopped at a guess. Secondly, a parameter is missing to define the triangles. In addition to the metrics, the procedure requires an estimation of a reference element size to perform the shrinks. An arbitrary parameter selected by the user has been added in oder to get the above results. This parameter differs in the two cases $\varepsilon = 0.01$ and $\varepsilon = 0.001$.

Acknowledgements. The authors are indebted to P. L. George and P. Laug from I.N.R.I.A. (Rocquencourt, France) for fruitful discussions and explanations on the automatic mesh refinement procedures. P. Laug has initiated and designed BL2D and P. L. George is the author of successful books on mesh generation procedures.

References

[1] Apel T. and Lube G., Anisotropic mesh refinement in stabilizedGalerkin methods, Numer. Math., 74 (1996), 261–282.

[2] Apel T. and Lube G., Anisotropic mesh refinement for a singularly perturbed reaction-diffusion model problem, Appl. Numer. Math., 26 (1998), 415–433.

[3] Apel T. and Nicaise S., Elliptic problems in domains with edges: anisotropic regularity and anisotropic finite element meshes, in Partial differential equations and functional analysis, in memory of P. Grisvard, J. Céa, D. Chesnais, G. Geymonat and J.L. Lions eds., Birkhauser, Boston, 18–34 (1996).

[4] Chapelle D. and Bathe K. J., Fundamental considerations for thefinite element analysis of shell structures, Comp. and Structures, 66 (1998), 19–36.

[5] George P. L. and Bourouchaki H., Triangulation de Delaunay et maillage. Application aux éléments finis, Hermes, Paris 1997.

[6] Gerard P. and Sanchez-Palencia E., Sensitivity phenomena for certain thin elastic shells with edges, Math. Meth. Appl. Sci. (to appear).

[7] Karamian P., Sanchez-Hubert J. and Sanchez-Palencia E., A model problem for boundary layers on thin elastic shells, Math. Modelling and Num. Anal. (to appear).

[8] Laug P. and Bourouchaki H., The BL2D mesh generator: beginner'sguide, user's and programmer's manual, Rapport technique n°0194, I.N.R.I.A., Rocquencourt, France (1996)

[9] Leguillon D., Sanchez-Hubert J. and Sanchez-Palencia E., Model problem of singular perturbation without limit in the space of finite energy and its computation, C. R. Acad. Sci. Paris, série IIb, 327 (1999), 485-492.

[10] Lions J.L. and Magenes E., Problèmes aux limites nonhomogènes et applications, vol. 1, Dunod, Paris 1968.

[11] Lions J.L. and Sanchez-Palencia E., Problèmes sensitifs et coques élastiques minces, in Partial differential equations and functional analysis, in memory of P. Grisvard, J. Céa, D. Chesnais, G. Geymonat and J.L. Lions eds., Birkhauser, Boston, 207-220 (1996).
[12] Sanchez-Hubert J. and Sanchez-Palencia E., Coques élastiques minces. Propriétés asymptotiques, Masson, Paris 1997.
[13] Sanchez-Hubert J. and Sanchez-Palencia E., Pathological phenomena in computation of thin elastic shells, Trans. Canad. Soc. Mech. Engng., 2(4B) (1999), 435-446.
[14] Van Dyke, Perturbation methods in fluid mechanics, Academic Press, New-York 1964.
[15] Zienkiewicz O. C. and Taylor R. L., The finite element method. Fourth edition , Vol.2, Mac Graw Hill, London 1991.

Dominique Leguillon, Evariste Sanchez-Palencia and Carlos de Souza
Laboratoire de Modélisation en Mécanique UMR 7607
CNRS–Université Pierre et Marie Curie
4 place Jussieu
75252 Paris Cedex 05
France
E-mail: dol@ccr.jussieu.fr, sanchez@lmm.jussieu.fr,
souza@lmm.jussieu.fr

Studies in Mathematics and its Applications, Vol. 31
D. Cioranescu and J.L. Lions (Editors)

Chapter 23

OPTIMAL DESIGN OF GRADIENT FIELDS WITH APPLICATIONS TO ELECTROSTATICS

ROBERT LIPTON AND ANI P. VELO

1. Introduction

Consider a two dimensional design domain Ω, containing two isotropic dielectric materials. The dielectric permittivity is specified by $\varepsilon(x)$ and is piece-wise constant taking the values α and β where $\beta > \alpha > 0$. For a prescribed charge density f the associated electric potential φ satisfies the Poisson equation given by

$$-\text{div } (\varepsilon(x)\nabla\varphi) = f, \tag{1}$$

and $\varphi = 0$ on the boundary of Ω. In order to include the broadest class of charge densities we suppose that f lies in $W^{-1,2}(\Omega)$ and that φ is a $W_0^{1,2}(\Omega)$ solution of the Poisson equation. The associated electric field in the domain is $-\nabla\varphi$. We introduce a "target" electric field $\hat{\mathbf{E}}$. For a given charge density, our objective is to design a two phase dielectric that supports an electric field $-\nabla\varphi$ that is as close as possible to $\hat{\mathbf{E}}$. Here $\hat{\mathbf{E}} = -\nabla\hat{\varphi}$, where $\hat{\varphi}$ is a potential in $W^{1,2}(\Omega)$. Placing a constraint on the amount of the better dielectric β, the design problem is to minimize the difference

$$\int_\Omega |\nabla\varphi - \nabla\hat{\varphi}|^2 \, dx, \tag{2}$$

over all configurations of the two dielectrics.

In general, material layout problems of this type fail to have an optimal design given by a configuration of the two materials. Instead one must study the behavior of minimizing sequences of configurations. The purpose of the analysis given here is to provide the methodology for the recovery of optimal configurations when they exist and to identify minimizing sequences of configurations for (2) otherwise. We introduce a tractable method for the numerical computation of minimizing sequences of configurations. These minimizing sequences are associated with materials with graded dielectric properties that may exhibit a fine scale structure composed of layers of the two dielectrics. Moreover, for a dense class of target fields we are able to characterize all fine scale structure that can appear in minimizing sequences of configurations, see Theorem 1 of this Section. We illustrate our approach through numerical examples provided in Section 6. The examples illustrate how the electric field can be controlled using functionally graded materials.

2. Background and main theoretical results

The nonexistence of an optimal configuration for the design problem coincides with the appearance of minimizing sequences containing regions of finite measure where the dielectric permittivity becomes highly oscillatory. As one follows these minimizing sequences the dielectric permittivity oscillates between the values α and β on progressively finer scales. To describe this mathematically we denote the subset of the design domain Ω containing the β dielectric by ω. The characteristic function of this set is written as χ where $\chi = 1$ for x in ω and $\chi = 0$ otherwise. The piece-wise constant dielectric permittivity is given by

$$\varepsilon(x) = \varepsilon(\chi) \triangleq \beta\chi + \alpha(1 - \chi). \tag{3}$$

Oscillation of a sequence of designs $\{\omega^\nu\}_{\nu=1}^\infty$ is described by the weak $L^\infty(\Omega)$ **star** convergence of the associated sequence of characteristic functions $\{\chi^\nu\}_{\nu=1}^\infty$ to a density θ in $L^\infty(\Omega)$ where $0 \le \theta \le 1$. The issue of nonexistence of optimal configurations for problems of material layout has been the object of much interest. The classic example is illustrated in the problem of minimizing the energy dissipation associated with configurations of two materials. In the context of two phase dielectric materials the energy dissipation for a configuration is given by

$$\int_\Omega \varepsilon(x)\nabla\varphi \cdot \nabla\varphi \, dx.$$

The problem of nonexistence was resolved in an elegant fashion by extending the design space to include all effective dielectric permittivities that could be obtained through oscillation, see [10], [11] and [6]. The crucial connection between minimizing sequences of configurations and optimal designs in the extended design space is established through a continuity property of the energy dissipation given in [3]. This continuity property is an example of the theory of compensated compactness developed in [7] and [12]. Here continuity is given in the context of *weakly* convergent sequences in $L^2(\Omega)^2$. Indeed, consider sequences $\{\varepsilon^\nu(x)\nabla\varphi^\nu\}_{\nu=1}^\infty$ and $\{\nabla\varphi^\nu\}_{\nu=1}^\infty$, such that $-\text{div}\,(\varepsilon(x)^\nu\nabla\varphi^\nu) = f$. If these sequences weakly converge to the limits

$\varepsilon^\infty \nabla \varphi^\infty$ and $\nabla \varphi^\infty$ where ε^∞ is an effective tensor in the extended space of designs then one has the continuity expressed by

$$\lim_{\nu \to \infty} \int_\Omega \varepsilon^\nu(x) \nabla \varphi^\nu \cdot \nabla \varphi^\nu \, dx = \int_\Omega \varepsilon^\infty(x) \nabla \varphi^\infty \cdot \nabla \varphi^\infty \, dx.$$

For the design problem treated here we can attempt to resolve the nonexistence problem by extending the design space to include effective properties. However unlike the energy dissipation and other continuous functionals treated earlier, the objective functional given by (2) is not continuous with respect to weak convergence. Thus additional theoretical work is required to provide the connection between an extended space of designs and minimizing sequences of configurations. In this presentation we outline a methodology for the identification of minimizing sequences of configurations. The method is based on a careful extension of the design space and by replacing (2) with a suitable "relaxed" functional that is associated with the extended design space.

It is evident that any attempt to identify minimizing sequences of configurations must account for the possibility of oscillations in the sequence of gradients associated with minimizing sequences of designs. As above an oscillatory sequence of gradients is characterized by the weak convergence of the sequence in $L^2(\Omega)^2$. To fix ideas, let $\{\nabla \varphi_\nu\}_{\nu=1}^\infty$ denote a weakly converging sequence of gradients associated with a minimizing sequence of designs. The weak limit of the sequence is denoted by $\nabla \tilde{\varphi}$, and one has

$$\begin{aligned} &\lim_{\nu \to \infty} \int_\Omega |\nabla \varphi_\nu - \nabla \hat{\varphi}|^2 \, dx \\ &\quad = \lim_{\nu \to \infty} \int_\Omega |\nabla \varphi_\nu - \nabla \tilde{\varphi}|^2 \, dx + \int_\Omega |\nabla \tilde{\varphi} - \nabla \hat{\varphi}|^2 \, dx. \end{aligned} \tag{4}$$

The oscillatory behavior of minimizing sequences is naturally linked to the dependence of the limit

$$\lim_{\nu \to \infty} \int_\Omega |\nabla \varphi_\nu - \nabla \tilde{\varphi}|^2 \, dx, \tag{5}$$

on the weak limits $\nabla \tilde{\varphi}$, θ, together with other moments of measures associated with weak limits of geometric quantities, (e.g., the H measure introduced by L. Tartar [14]). Our methodology for identifying minimizing sequences is based upon on writing (5) as an explicit function of the relevant weak limits. Although at this time we are unable to produce a formula for every type of oscillation we note that an explicit closed form expression is available when the oscillations consist of layers of the two materials. The formula follows directly from the corrector theory of homogenization given in [1] and [9]. To be precise we introduce the characteristic function $\chi(x, t)$. Here $\chi(x, t)$ is piece-wise constant in the first variable and unit periodic in the scalar t variable. As is done in homogenization theory [1], we define a locally layered material by $\chi(x) = \chi(x, x \cdot n)$. Here $n = n(x)$ represents the normal to the layers given by $n = (\cos(\gamma(x)), \sin(\gamma(x)))$. An oscillatory sequence is given by $\chi^\nu(x) = \chi(x, \nu x \cdot n)$, $\nu = 1, 2 \ldots, \infty$. For this case we have the closed form expression given by

$$\begin{aligned} &\lim_{\nu\to\infty} \textstyle\int_\Omega |\nabla\varphi_\nu - \nabla\tilde{\varphi}|^2 \, dx \\ &= \textstyle\int_\Omega R(\gamma(x))H(\theta(x))R^T(\gamma(x))\nabla\tilde{\varphi}(x) \cdot \nabla\tilde{\varphi}(x) \, dx. \end{aligned} \tag{6}$$

Here $R(\gamma)$ is the orthogonal matrix associated with a rotation of γ radians and the matrix $H(\theta)$ is a function of the density θ given by

$$H(\theta) = \begin{pmatrix} (\frac{1}{\alpha} - \frac{1}{\beta})^2\theta(1-\theta)h_\theta^2 & 0 \\ 0 & 0 \end{pmatrix}, \tag{7}$$

where $h_\theta = (\frac{1-\theta}{\alpha} + \frac{\theta}{\beta})^{-1}$ is the harmonic mean of the two dielectric permittivities. Here, the sequence of gradients $\{\nabla\varphi_\nu\}_{\nu=1}^\infty$ is related to the sequence of configurations through the equilibrium condition

$$-\text{div } (\varepsilon(\chi^\nu)\nabla\varphi_\nu) = f. \tag{8}$$

The "homogenized " equilibrium equation satisfied by the weak limit $\tilde{\varphi}$ is given by

$$-\text{div } (\varepsilon(\theta(x), \gamma(x))\nabla\tilde{\varphi}) = f, \tag{9}$$

where

$$\varepsilon(\theta(x), \gamma(x)) = R(\gamma(x)) \; \Lambda(\theta(x)) \; R^T(\gamma(x)), \tag{10}$$

and the diagonal tensor $\Lambda(\theta)$ is given by

$$\Lambda(\theta) = \begin{pmatrix} h_\theta & 0 \\ 0 & m_\theta \end{pmatrix},$$

with $m_\theta = \alpha \; (1 - \theta) + \beta \; \theta$. Here the tensor $\varepsilon(\theta, \gamma)$ is the G-limit associated with the sequence of dielectric tensors $\{\varepsilon(\chi^\nu)\}_{n=1}^\infty$, see [9]. (Since the dielectric tensors are symmetric, the G-convergence [18] and the H-convergence [9] of any sequence of dielectric tensors is the same.)

The methodology presented here uses the explicit formula given by (6). Our approach is to replace χ and $\varepsilon(\chi)$ with the new design variables θ,

γ, and $\varepsilon(\theta,\gamma)$ given by (10). In addition we introduce the new objective functional given by

$$RF(\theta,\gamma,\varepsilon(\theta,\gamma),\nabla\hat{\varphi}) = \int_\Omega |\nabla\varphi - \nabla\hat{\varphi}|^2\,dx + \int_\Omega R(\gamma(x))H(\theta(x))R^T(\gamma(x))\nabla\varphi\cdot\nabla\varphi\,dx, \tag{11}$$

where the state variable is the $W_0^{1,2}$ solution of

$$-\text{div}\,(\varepsilon(\theta(x),\gamma(x))\nabla\varphi) = f. \tag{12}$$

In order to state our results we formulate the original design problem in a precise way. We introduce the constant Θ, such that $0 < \Theta < 1$. The space of admissible configurations and associated dielectric permittivities is denoted by ad_Θ, and

$$ad_\Theta = \{\chi : \int_\Omega \chi\,dx \leq \Theta\ meas(\Omega)\}. \tag{13}$$

The objective functional is denoted by $F(\chi,\varepsilon(\chi),\nabla\hat{\varphi})$ and is given by

$$F(\chi,\varepsilon(\chi),\nabla\hat{\varphi}) = \int_\Omega |\nabla\varphi - \nabla\hat{\varphi}|^2\,dx, \tag{14}$$

where the state variable φ is a solution of (1). The original design problem is formulated as

$$P = \inf_{\chi\in ad_\Theta} F(\chi,\varepsilon(\chi),\nabla\hat{\varphi}). \tag{15}$$

The admissible space of designs for the new design problem is given by

$$D_\Theta = \{\ (\theta,\gamma,\varepsilon(\theta,\gamma))\,|\,\theta\in L^\infty(\Omega;[0,1]);\ \gamma\in L^\infty(\Omega;[0,2\pi]) : \int_\Omega \theta\,dx \leq \Theta\ meas\Omega\ ;\ \varepsilon(\theta(x),\gamma(x)) = R(\gamma(x))\ \Lambda(\theta(x))\ R^T(\gamma(x))\ \}, \tag{16}$$

and the new design problem is formulated as

$$RP = \inf_{(\theta,\gamma,\varepsilon(\theta,\gamma))\in D_\Theta} RF(\theta,\gamma,\varepsilon(\theta,\gamma),\nabla\hat{\varphi}). \tag{17}$$

We point out that the extended space of designs D_Θ contains the original space of designs ad_Θ. Indeed, choosing $\theta = \chi$ we have $\varepsilon(\theta,\gamma) = \varepsilon(\chi)$, $H(\theta) = 0$ and

$$F(\chi,\varepsilon(\chi),\nabla\hat{\varphi}) = RF(\theta,\gamma,\varepsilon(\theta,\gamma),\nabla\hat{\varphi}). \tag{18}$$

The first result that we describe is given in Theorem 8. It states that for every $\hat{\varphi}$ in $W^{1,2}(\Omega)$ and for every f in $W^{-1,2}(\Omega)$ that

$$P = RP. \tag{19}$$

In deriving (19) we identify a class of minimizing sequences for P that is tractable for numerical computation. Our approach is based upon a discrete approximation of the design space D_Θ. We consider any partition T_κ of Ω consisting of a finite number of pair-wise disjoint subdomains $\Omega_i \subset \Omega$, $i = 1, \ldots, N(\kappa)$ such that:

$$\Omega = \bigcup_{i=1}^{N(\kappa)} \Omega_i \text{ and } \max_{i=1,\ldots N(\kappa)} (\text{diam}(\Omega_{\text{i}})) \leq \kappa. \tag{20}$$

We fix the partition T_κ and the discrete approximation D_Θ^κ is given by the piece-wise constant functions $\theta^\kappa(x)$, $\gamma^\kappa(x)$ taking constant values in each subdomain. We denote the restriction of $\theta^\kappa(x)$ and $\gamma^\kappa(x)$ to Ω_i by θ_i^κ and γ_i^κ respectively. Here $0 \leq \theta_i^\kappa \leq 1$, $0 \leq \gamma_i^\kappa \leq 2\pi$, and

$$\Sigma_{i=1}^{N(\kappa)} (\theta_i^\kappa \text{ meas } (\Omega_i)) = \Theta \text{ meas } (\Omega). \tag{21}$$

The piece-wise constant dielectric permittivity tensor is given by

$$\varepsilon(\gamma^\kappa(x), \theta^\kappa(x)) = R(\gamma^\kappa(x))\Lambda(\theta^\kappa(x))R^T(\gamma^\kappa(x)), \tag{22}$$

and the associated state variable φ^κ solves the Poisson equation

$$-\text{div } (\varepsilon(\gamma^\kappa(x), \theta^\kappa(x))\nabla\varphi^\kappa) = f. \tag{23}$$

It is clear that D_Θ^κ is contained in the larger set D_Θ and the design problem posed on this smaller set of designs is written

$$RP^\kappa = \inf_{(\theta^\kappa,\gamma^\kappa,\varepsilon(\theta^\kappa,\gamma^\kappa))\in D_\Theta^\kappa} RF(\theta^\kappa, \gamma^\kappa, \varepsilon(\theta^\kappa, \gamma^\kappa), \nabla\hat{\varphi}). \tag{24}$$

It is shown that a minimizing vector of design variables $(\overline{\theta}^\kappa, \overline{\gamma}^\kappa)$ exists for this problem, see Theorem 4. Most importantly, it follows from (6) that there exists a recovery sequence of configurations χ^ν in ad_Θ for which

$$\lim_{\nu\to\infty} F(\chi^\nu, \varepsilon(\chi^\nu), \nabla\hat{\varphi}) = RP^\kappa, \tag{25}$$

see Theorem 5. For any given partition T_κ we consider its refinements, i.e., the nested family of partitions $\{T_\epsilon\}_{\epsilon>0}$ that includes T_κ. We show that the optimal design vectors associated with the refinements represent a minimizing sequence of designs for the problem RP, see Theorem 7. Moreover, using (25) we are able to recover the explicit form of minimizing sequences for the original problem P, see Theorem 5 and equation (57). We point out that the choice of the initial partition T_κ is arbitrary so this method generates minimizing sequences of designs for any initial choice of partition.

It is instructive to write $RF(\theta,\gamma,\varepsilon(\theta,\gamma),\nabla\hat{\varphi})$ in a form where $\varepsilon(\theta,\gamma)$ appears explicitly. Manipulation gives

$$R(\gamma)H(\theta)R^T(\gamma) = \frac{(m_\theta I - \varepsilon(\theta,\gamma))^2}{(1-\theta)\beta(\beta-\alpha)} + \frac{(m_\theta I - \varepsilon(\theta,\gamma))}{\beta}. \tag{26}$$

It is clear from (11) and (26) that if $\varepsilon(\theta,\gamma)\nabla\varphi = m_\theta\nabla\varphi$ then

$$\int_\Omega R(\gamma)H(\theta)R^T(\gamma)\nabla\varphi\cdot\nabla\varphi\,dx = 0.$$

It now follows from (6) that the gradients $\nabla\varphi^\nu$ associated with a recovery sequence of configurations $\{\varepsilon(\chi_\omega^\nu)\}_{\nu=0}^\infty$ G-converging to $\varepsilon(\theta,\gamma)$ converge strongly to $\nabla\varphi$. Conversely if the gradients $\nabla\varphi^\nu$ associated with a recovery sequence of configurations converge strongly to $\nabla\varphi$ then the term

$$\int_\Omega R(\gamma)H(\theta)R^T(\gamma)\nabla\varphi\cdot\nabla\varphi\,dx$$

vanishes and $\varepsilon(\theta,\gamma)\nabla\varphi = m_\theta\nabla\varphi$ follows from (26). In this context we mention that the earlier work of [5] focuses on the energy dissipation to show that the condition $\varepsilon^e\nabla\varphi = \varepsilon^*\nabla\varphi$ is necessary and sufficient for the strong convergence of gradients associated with sequences $\{\varepsilon_\nu^e\}_{\nu=0}^\infty$ G-converging to ε^e and weak L^∞ **star** converging to ε^*.

For a dense set of target fields we show that our method accounts for all oscillations appearing in minimizing sequences of configurations. We consider target fields of the form

$$-\nabla\hat{\varphi},\ \hat{\varphi}\in W_0^{1,2}(\Omega) \tag{27}$$

and the relaxed version of the original problem is given by

Theorem 1. – *There exists a dense G_δ subset K of $W_0^{1,2}(\Omega)$ such that for $\hat{\varphi}$ in K,*

(1) There exists a minimizer of RP in D_Θ,

(2) $P = RP$,

(3) Any cluster point of any minimizing sequence in ad_Θ of P is a minimizer of RP and any minimizer of RP in D_Θ is a limit of a minimizing sequence for P.

(4) *Let $\bar{\varphi}$ be the potential associated with the minimizer $(\bar{\theta}, \bar{\gamma}, \varepsilon(\bar{\theta}, \bar{\gamma}))$ of RP, then*

$$RF(\bar{\theta}, \bar{\gamma}, \varepsilon(\bar{\theta}, \bar{\gamma}), \nabla\hat{\varphi}) = \int_\Omega |\nabla\bar{\varphi} - \nabla\hat{\varphi}|^2 \, dx \quad \text{and} \quad \varepsilon(\bar{\theta}, \bar{\gamma})\nabla\bar{\varphi} = m_{\bar{\theta}}\nabla\bar{\varphi}.$$

Here the convergence of sequences of designs are with respect to the G-convergence [18].

The class of targets appearing in Theorem 1 is motivated by the following theorem of L. Tartar [2], [13], which is an improvement of a result of M. Edelstein [4].

Theorem 2. – *Let S be a non-empty strongly closed subset of a Hilbert space H. Then there exists a dense G_δ subset K of H such that for any $x \in \mathrm{K}$, the minimizing sequences $\{c_n\}_{n=1}^{\infty} \in \mathrm{S}$ of the function $c \to \|x - \mathrm{c}\|$ are Cauchy sequences. In particular the subset of points of H with a unique projection on S contains a dense G_δ subset, as it contains K.*

With Theorem 2 in mind we can take advantage of the geometry of the set of effective tensors for two dimensional problems and establish Theorem 1. This topic is taken up in Section 5 where Theorem 1 is proved.

The recent work of P. Pedregal [15], [16] approaches similar design problems from a different perspective. In that work the equilibrium equation for the potential, together with the resource constraint is incorporated into the cost functional and a new type of envelope for the augmented functional is introduced. The envelope is shown to be weakly lower semicontinuous in $W_0^{1,2}(\Omega) \times L^\infty(\Omega)$ see [15], [16] and can be thought of as a constrained quasiconvexification of the original augmented functional. The constrained quasiconvex envelope can be expressed in terms of a class $\mathcal{A}$ of gradient Young measures, see [15]. To proceed further, the envelope needs to be given in terms of explicit formulas. This requires knowledge of the set $\mathcal{A}$. However at this stage the characterization of $\mathcal{A}$ is not known. In principle the methods of this paper can be used to deduce the part of $\mathcal{A}$ containing Young measures associated with simple one scale laminates, see [15]. Theorem 1 shows that the knowledge of gradient Young measures associated with layered microstructures (a.k.a. laminates) is sufficient for the computation of the constrained quasi convex hull when the target fields are in the class $K \subset W_0^{1,2}$.

We point out that the discrete problem given by (24) is of interest on its own right. From a practical perspective there is a prohibitive manufacturing cost incurred when attempting to make a graded material with possibly different anisotropic dielectric properties at every point. Instead there is a smallest scale κ over which the dielectric properties change. The scale is set by the manufacturing cost. Practically speaking one partitions the design domain into subdomains of diameter κ and inside these subdomains one optimizes the dielectric properties. This approach to the design of graded materials is naturally incorporated in the formulation of the discrete problem given here and is discussed in the context of the numerical examples given in Section 6.

3. The discrete problem

In this Section we analyze the design problem on the discretized space of designs. The existence of an optimal design is established in this space. Next we apply the corrector theory to exhibit a recovery sequence of configurations of the two dielectrics.

We consider any partition T_κ of Ω consisting of a finite number of pairwise disjoint subdomains $\Omega_i \subset \Omega$, $i = 1, \ldots, N(\kappa)$ such that:

$$\Omega = \bigcup_{i=1}^{N(\kappa)} \Omega_i \quad \text{and} \quad \max_{i=1,\ldots N(\kappa)} (diam(\Omega_i)) \leq \kappa.$$

We fix the partition T_κ and the discrete approximation D_Θ^κ is described by equations (20–23) given in the introduction. The design problem over the discrete space is given by (24). Existence of the optimal design is established using the direct method of the calculus of variations. We start by introducing the type of convergence relevant to the discrete problem. A design $(\theta^\kappa, \gamma^\kappa, \varepsilon(\theta^\kappa, \gamma^\kappa))$ in D_Θ^κ can be identified with the vector $(\theta_i^\kappa, \gamma_i^\kappa)$ for $i = 1, \ldots, N(\kappa)$ in $R^{2N(\kappa)}$. Thus D_Θ^κ is identified with a compact subset of $R^{2N(\kappa)}$ and convergence of designs in D_Θ^κ is given by sequential convergence in $R^{2N(\kappa)}$. Existence of an optimal design will follow once we show that the functional $RF(\theta^\kappa, \gamma^\kappa, \varepsilon(\theta^\kappa, \gamma^\kappa), \nabla\hat{\varphi})$ is continuous with respect to sequential convergence in $R^{2N(\kappa)}$.

Theorem 3. – *Given a sequence of designs $\{(\theta^{\kappa,n}, \gamma^{\kappa,n})\}_{n=1}^{\infty}$ and a design $(\bar{\theta}^\kappa, \bar{\gamma}^\kappa)$ such that*

$$\lim_{n\to\infty} (\theta^{\kappa,n}, \gamma^{\kappa,n}) = (\bar{\theta}^\kappa, \bar{\gamma}^\kappa) \tag{28}$$

as elements of $R^{2N(\kappa)}$, then

$$\lim_{n\to\infty} RF(\theta^{\kappa,n}, \gamma^{\kappa,n}, \varepsilon(\theta^{\kappa,n}, \gamma^{\kappa,n}), \nabla\hat{\varphi}) = RF(\bar{\theta}^\kappa, \underline{a}r\gamma^\kappa, \varepsilon(\bar{\theta}^\kappa, \bar{\gamma}^\kappa), \nabla\hat{\varphi}). \tag{29}$$

Proof. The state variable associated with the limit design $(\bar{\theta}^\kappa, \bar{\gamma}^\kappa)$ is denoted by $\bar{\varphi}^\kappa$ and is the $W_0^{1,2}(\Omega)$ solution of

$$-\text{div}\left(\varepsilon(\bar{\theta}^\kappa, \bar{\gamma}^\kappa)\nabla\bar{\varphi}^\kappa\right) = f. \tag{30}$$

The convergence of the sequence of designs given by (28) implies that the associated conductivities $\{\varepsilon(\theta^{\kappa,n},\gamma^{\kappa,n})\}_{n=1}^{\infty}$ converge to $\varepsilon(\bar{\theta}^{\kappa},\bar{\gamma}^{\kappa})$ almost everywhere. From the theory of G-convergence [18] we also know that the sequence G-converges to $\varepsilon(\bar{\theta}^{\kappa},\bar{\gamma}^{\kappa})$. The definition of G-convergence implies that the state variables φ_n^{κ} associated with the sequence $\{\varepsilon(\theta^{\kappa,n},\gamma^{\kappa,n})\}_{n=1}^{\infty}$ converge weakly in $W_0^{1,2}(\Omega)$ to $\bar{\varphi}^{\kappa}$. In order to establish the continuity given by (29) we first show that the sequence φ_n^{κ} converges strongly to $\bar{\varphi}^{\kappa}$ in $W_0^{1,2}(\Omega)$. To do this we recall the formula (10) for $\varepsilon(\theta,\gamma)$ to easily see that

$$0<\alpha\leq\varepsilon(\theta^{\kappa,n},\gamma^{\kappa,n})\leq\beta$$

We apply this estimate to obtain,

$$\int_{\Omega}\alpha|\nabla\varphi_n^{\kappa}-\nabla\bar{\varphi}^{\kappa}|^2dx \tag{31}$$

$$\leq\int_{\Omega}\varepsilon(\theta^{\kappa,n},\gamma^{\kappa,n})(\nabla\varphi_n^{\kappa}-\nabla\bar{\varphi}^{\kappa})\cdot(\nabla\varphi_n^{\kappa}-\nabla\bar{\varphi}^{\kappa})dx \tag{32}$$

$$=\int_{\Omega}\varepsilon(\theta^{\kappa,n},\gamma^{\kappa,n})\nabla\varphi_n^{\kappa}\cdot\nabla\varphi_n^{\kappa}dx-2\int_{\Omega}\varepsilon(\theta^{\kappa,n},\gamma^{\kappa,n})\nabla\varphi_n^{\kappa}\cdot\nabla\bar{\varphi}^{\kappa}dx$$

$$+\int_{\Omega}\varepsilon(\theta^{\kappa,n},\gamma^{\kappa,n})\nabla\bar{\varphi}^{\kappa}\cdot\nabla\bar{\varphi}^{\kappa}dx. \tag{33}$$

Passing to the limit as $n\to\infty$ in (32,33) , we apply the well known properties of G-convergence together with the almost everywhere convergence of $\{\varepsilon(\theta^{\kappa,n},\gamma^{\kappa,n})\}_{n=1}^{\infty}$ and the Lebesgue convergence theorem to find that

$$\lim_{n\to\infty}\parallel\nabla\varphi_n^{\kappa}-\nabla\bar{\varphi}^{\kappa}\parallel_{L^2}^2=0, \tag{34}$$

and strong convergence of φ_n^{κ} to $\bar{\varphi}^{\kappa}$ in $W_0^{1,2}$ follows.

From (7) we easily obtain the following estimate for the sequence

$$\{R(\gamma^{\kappa,n}(x))H(\theta^{\kappa,n}(x))R^T(\gamma^{\kappa,n}(x))\}_{n=1}^{\infty}$$

given by

$$R(\gamma^{\kappa,n}(x))H(\theta^{\kappa,n}(x))R^T(\gamma^{\kappa,n}(x))\eta\cdot\eta\leq\frac{\beta^2}{4}(\frac{1}{\alpha}-\frac{1}{\beta})\ |\eta|^2, \tag{35}$$

for every vector η in R^2. Moreover, from the convergence given by (28) the sequence converges almost everywhere to

$$R(\bar{\gamma}^{\kappa}(x))H(\bar{\theta}^{\kappa}(x))R^T(\bar{\gamma}^{\kappa}(x)).$$

Finally since $\nabla\varphi_n^\kappa \to \nabla\bar{\varphi}^\kappa$ strongly in L^2 we apply (35) together with the Lebesgue convergence theorem to conclude that:

$$\lim_{n\to\infty} RF(\theta^{\kappa,n}, \gamma^{\kappa,n}, \varepsilon(theta^{\kappa,n}, \gamma^{\kappa,n}), \nabla\hat{\varphi}) =$$

$$= \lim_{n\to\infty}\left[\int_\Omega |\nabla\varphi_n^\kappa - \nabla\hat{\varphi}|^2\, dx + \int_\Omega R(\gamma^{\kappa,n})H(\theta^{\kappa,n})R^T(\gamma^{\kappa,n})\ \nabla\varphi_n^\kappa\cdot\nabla\varphi_n^\kappa\, dx\right]$$

$$\begin{aligned} &= \int_\Omega |\nabla\bar{\varphi}^\kappa - \nabla\hat{\varphi}|^2\, dx + \int_\Omega R(\bar{\gamma}^\kappa)H(\bar{\theta}^\kappa)R^T(\bar{\gamma}^\kappa)\ \nabla\bar{\varphi}^\kappa\cdot\nabla\bar{\varphi}^\kappa\, dx \\ &= RF(\bar{\theta}^\kappa, \bar{\gamma}^\kappa, \varepsilon(\bar{\theta}^\kappa, \bar{\gamma}^\kappa), \nabla\hat{\varphi}). \end{aligned} \tag{36}$$

which proves the theorem.

Collecting our results we have shown that

Theorem 4. – *There exists an optimal design $(\bar{\theta}^\kappa, \bar{\gamma}^\kappa, \varepsilon(\bar{\theta}^\kappa, \bar{\gamma}^\kappa))$ in D_Θ^κ for the discrete problem, i.e.,*

$$\begin{aligned} RP^\kappa &= RF(\bar{\theta}^\kappa, \bar{\gamma}^\kappa, \varepsilon(\bar{\theta}^\kappa, \bar{\gamma}^\kappa), \nabla\hat{\varphi}) \\ &= \min_{(\theta^\kappa, \gamma^\kappa, \varepsilon(\theta^\kappa \gamma^\kappa)) \in D_\Theta^\kappa} RF(\theta^\kappa, \gamma^\kappa, \varepsilon(\theta^\kappa \gamma^\kappa), \nabla\hat{\varphi}). \end{aligned} \tag{37}$$

Now we show how to construct a sequence of configurations described by with the sequence of characteristic functions $\{\chi^n\}_{n=1}^\infty$ for which

$$\lim_{n\to\infty} F(\chi^n, \varepsilon(\chi^n), \nabla\hat{\varphi}) = RP^\kappa. \tag{38}$$

In view of Theorem 4 it is sufficient to consider any design given by $(\theta^\kappa, \gamma^\kappa)$ in D_θ^κ and show how to construct a sequence $\{\chi^{\kappa,n}\}_{n=1}^\infty$ for which

$$\lim_{n\to\infty} F(\chi^{\kappa,n}, \varepsilon(\chi^{\kappa,n}), \nabla\hat{\varphi}) = RF(\theta^\kappa, \gamma^\kappa, \varepsilon(\theta^\kappa, \gamma^\kappa), \nabla\hat{\varphi}). \tag{39}$$

We start by observing that for $\theta = 0$ or $\theta = 1$ that $\varepsilon(\theta, \gamma) = \alpha I$ or βI respectively, where I is the 2×2 identity. Thus for a design specified by $(\theta^\kappa, \gamma^\kappa)$ we proceed to construct the sequence $\{\chi^{\kappa,n}\}_{n=1}^\infty$ in the following way. In the subdomains Ω_i for which $\theta_i^\kappa = 0$ we set $\chi^{\kappa,n} = 0$, $n = 1, 2\ldots\infty$ and in the subdomains Ω_i for which $\theta_i^\kappa = 1$ we set $\chi^{\kappa,n} = 1$, $n = 1, 2\ldots\infty$. Next we consider the subdomains Ω_i where $0 < \theta_i^\kappa < 1$. In these subdomains we have $0 \le \gamma_i^k \le 2\pi$ and we set $\chi^{\kappa,n} = \mu(nx{\cdot}\mathbf{n}(\gamma_i^\kappa))$, where $\mu(t)$ is a periodic function on the real line of period unity taking the values 1 for $0 \le t \le \theta_i^\kappa$

and 0 for $\theta_i^\kappa < t < 1$ and $\mathbf{n}(\gamma_i^\kappa) = (\cos\gamma_i^k, \sin\gamma_i^k)$. We summarize our construction in the following equation,

$$\chi^{\kappa,n} = \begin{cases} 0, & \text{in } \Omega_i \text{ for which } \theta_i^\kappa = 0, \\ 1, & \text{in } \Omega_i \text{ for which } \theta_i^\kappa = 1, \\ \mu(nx \cdot \mathbf{n}(\gamma_i^\kappa)), & \text{in } \Omega_i \text{ for which } 0 < \theta_i^\kappa < 1. \end{cases} \tag{40}$$

The associated dielectric permittivity $\varepsilon(\chi^{\kappa,n})$ corresponds to pure α dielectric in the subdomains Ω_i where $\theta^\kappa = 0$, pure β dielectric in the subdomains Ω_i where $\theta^\kappa = 1$, and layers of α and β dielectric with layer normal in the direction $(\cos\gamma_i^k, \sin\gamma_i^k)$ in the subdomains where $0 < \theta_i^\kappa < 1$. The associated state variables $\varphi^{\kappa,n}$ are the $W_0^{1,2}(\Omega)$ solutions to the equilibrium equation

$$-\text{div}\,(\varepsilon(\chi^{\kappa,n})\nabla\varphi^{\kappa,n}) = f. \tag{41}$$

The sequence $\varepsilon(\chi^{\kappa,n})$ G-converges to $\varepsilon(\theta^\kappa, \gamma^\kappa)$ [9], hence the sequence $\varphi^{\kappa,n}$ converges weakly in $W_0^{1,2}(\Omega)$ to the state variable φ^κ associated with the design $(\theta^\kappa, \gamma^\kappa)$. With this construction in mind we state the following Theorem that guarantees the existence of recovery sequences of configurations.

Theorem 5. – *Given a design $(\theta^\kappa, \gamma^\kappa)$ in D_θ^κ the sequence of configurations $\{\chi^{\kappa,n}\}_{n=1}^\infty$ given by (40) is a recovery sequence, i.e.,*

$$\lim_{n\to\infty} F(\chi^{\kappa,n}, \varepsilon(\chi^{\kappa,n}), \nabla\hat{\varphi}) = RF(\theta^\kappa, \gamma^\kappa, \varepsilon(\ theta^\kappa, \gamma^\kappa), \nabla\hat{\varphi}). \tag{42}$$

Proof. We have

$$\lim_{n\to\infty} F(\chi^{\kappa,n}, \varepsilon(\chi^{\kappa,n})) = \int_\Omega |\nabla\varphi^\kappa - \nabla\hat{\varphi}|^2\,dx + \lim_{n\to\infty}\int_\Omega |\nabla\varphi^{\kappa,n} - \nabla\varphi^\kappa|^2\,dx \tag{43}$$

Applying the corrector theory of homogenization given in [9] we can write

$$\lim_{n\to\infty}\int_\Omega |\nabla\varphi^{\kappa,n} - \nabla\varphi^\kappa|^2\,dx = \int_\Omega |(P^{\kappa,n} - I)\nabla\varphi^\kappa + z^{\kappa,n}|^2\,dx. \tag{44}$$

$$\lim_{n\to\infty}\int_\Omega |\nabla\varphi^{\kappa,n} - \nabla\varphi^\kappa|^2\,dx = \int_\Omega |(P^{\kappa,n} - I)\nabla\varphi^\kappa + z^{\kappa,n}|^2\,dx, \tag{45}$$

where $P^{\kappa,n}$ is the corrector matrix associated with $\chi^{\kappa,n}$ and on each subdomain Ω_i it is given by:

$$P^{\kappa,n} = R(\gamma_i^\kappa)\begin{pmatrix} \frac{h_{\theta_i^\kappa}}{[\ \alpha(1-\chi^{\kappa,n})+\beta\chi^{\kappa,n}\ ]} & 0 \\ 0 & 1 \end{pmatrix} R^T(\gamma_i^\kappa), \text{ and } P^{\kappa,n} \rightharpoonup I \text{ in } L^2$$

as $n \to \infty$. Since $P^{\kappa,n} \in L^{\infty}(\Omega)^{2\times 2}$ it follows from the corrector theorem of F. Murat and L. Tartar [9] that $z^{\kappa,n} \to 0$ strongly in L^2. As a consequence

$$\lim_{n\to\infty} \int_{\Omega} |(P^{\kappa,n} - I)\nabla\varphi^{\kappa} + z^{\kappa,n}|^2 \, dx =$$

$$\lim_{n\to\infty} \sum_{i=1}^{N(\kappa)} \left[\int_{\Omega_i} R(\gamma_i^{\kappa}) \begin{pmatrix} (\frac{h_{\theta_i^{\kappa}}}{[\alpha(1-\chi^{\kappa,n})+\beta\chi^{\kappa,n}]} - 1)^2 & 0 \\ 0 & 0 \end{pmatrix} R^T(\gamma_i^{\kappa}) \nabla\varphi^{\kappa} \cdot \nabla\varphi^{\kappa} dx \right]$$

$$= \int_{\Omega} R(\gamma^{\kappa}) H(\theta^{\kappa}) R^T(\gamma^{\kappa}) \nabla\varphi^{\kappa} \cdot \nabla\varphi^{\kappa} dx, \tag{46}$$

and the Theorem follows.

4. Minimizing sequences of configurations

In this Section we identify minimizing sequences of design vectors for the problem RP. These sequences are associated with the refinements of a given partition T_{κ}. Next we employ Theorem 5 to deduce that $RP = P$ and identify a special class of minimizing sequences of configurations for P.

We recall that a nested family of partitions $\{T_{\kappa}\}_{\kappa \leq \epsilon}$ of Ω is a family that satisfies:

$$\kappa_1 < \kappa_2 \;\Rightarrow\; \forall \Omega_i^1 \in T_{\kappa_1} \;\; \exists \Omega_j^2 \in T_{\kappa_2} : \; \Omega_i^1 \subset \Omega_j^2. \tag{47}$$

For any given partition $T_{\tilde{\kappa}}$ the sequence of refinements of this partition is denoted by $\{T_{\kappa}\}_{\kappa \leq \tilde{\kappa}}$ and is a nested family of partitions as described by (47). We show that the space of discrete designs D_{Θ}^{κ} associated with the refinements of $T_{\tilde{\kappa}}$ is dense in D_{Θ}.

Theorem 6. – *The system of designs $\{D_{\Theta}^{\kappa}\}_{\kappa\to 0}$ is dense in D_{Θ}. Indeed, for every $(\theta, \gamma, \varepsilon(\theta,\gamma)) \in D_{\Theta}$, there exists a sequence $(\theta^{\kappa}, \gamma^{\kappa}, \varepsilon(\theta^{\kappa}, \gamma^{\kappa})) \in D_{\Theta}^{\kappa}$ for which*

$$\theta^{\kappa} \to \theta, \; \gamma^{\kappa} \to \gamma \textit{ a.e. in } \Omega \textit{ and } \varepsilon(\theta^{\kappa},\gamma^{\kappa}) \textit{ G-converges to } \varepsilon(\theta,\gamma) \textit{ as } \kappa \to 0, \tag{48}$$

furthermore:

$$\lim_{\kappa\to 0} RF(\theta^{\kappa}, \gamma^{\kappa}, \varepsilon(\theta^{\kappa},\gamma^{\kappa}), \nabla\hat{\varphi}) = RF(\theta, \gamma, \varepsilon(\theta,\gamma), \nabla\hat{\varphi}) \,. \tag{49}$$

Proof. For a given design $(\theta, \gamma, \varepsilon(\theta, \gamma)) \in D_\Theta$ we choose any partition $T_{\tilde{\kappa}}$ of Ω and consider its refinements $\{T_\kappa\}_{\kappa \leq \tilde{\kappa}}$. For any refinement T_κ , we construct $(\theta^\kappa, \gamma^\kappa, \varepsilon(\theta^\kappa, \gamma^\kappa)) \in D_\Theta^\kappa$ as described below:

$$\theta_i^\kappa = \frac{1}{meas\ \Omega_i} \int_{\Omega_i} \theta(x)\ dx$$

$$\gamma_i^\kappa = \frac{1}{meas\ \Omega_i} \int_{\Omega_i} \gamma(x)\ dx$$

$$\varepsilon(\theta^\kappa, \gamma^\kappa) = R(\gamma_i^\kappa)\Lambda(\theta_i^\kappa)R^T(\gamma_i^\kappa) \quad \text{on } \Omega_i.$$

We consider the intersection of Lebesgue points for the functions $\theta(x)$ and $\gamma(x)$. On this set we have:

$$\theta^\kappa(x) \to \theta(x), \quad \gamma^\kappa(x) \to \gamma(x) \qquad as\ \kappa \to 0.$$

This delivers the convergence

$$\varepsilon(\theta^\kappa(x), \gamma^\kappa(x)) \to \varepsilon(\theta(x), \gamma(x)) = R(\gamma)\Lambda(\theta)R^T(\gamma)\ a.e.\ in\ \Omega\ as\ \kappa \to 0$$

and

$$R(\gamma^\kappa)H(\theta^\kappa)R^T(\gamma^\kappa) \to R(\gamma)H(\theta)R^T(\gamma)\ a.e.\ in\ \Omega\ as\ \kappa \to 0.$$

From the properties of G-convergence [18] we deduce as in Theorem 3 that $\varepsilon(\theta^\kappa(x), \gamma^\kappa(x))$ G-converges to $\varepsilon(\theta(x), \gamma(x))$ and this establishes (48). This implies that the sequence of state variables φ^κ, satisfying $\varphi^\kappa \in W_0^{1,2}(\Omega)$ and

$$-\text{div}\ (\varepsilon(\theta^\kappa, \gamma^\kappa)\nabla\varphi^\kappa) = f, \tag{50}$$

converges weakly in $W_0^{1,2}(\Omega)$ to the $W_0^{1,2}(\Omega)$ solution φ of

$$-\text{div}\ (\varepsilon(\theta, \gamma)\nabla\varphi) = f. \tag{51}$$

Following the same arguments given in the proof of Theorem 3, we see that the sequence $\{\varphi^\kappa\}_{\kappa>0}$ converges strongly in $W_0^{1,2}(\Omega)$ to φ. Moreover, the same estimate as given in (34) holds for the sequence

$$\{R(\gamma^\kappa)H(\theta^\kappa)R^T(\gamma^\kappa)\}_{\kappa>0}$$

and we can proceed along the same lines as in the proof of Theorem 3 to show that

$$\lim_{\kappa \to 0} RF(\theta^\kappa, \gamma^\kappa, \varepsilon(\theta^\kappa, \gamma^\kappa), \nabla\hat{\varphi}) = RF(\theta, \gamma, \varepsilon(\theta, \gamma), \nabla\hat{\varphi}). \tag{52}$$

We now identify minimizing sequences of designs for the RP problem. We consider any nested family of partitions denoted by $\{T_\kappa\}_{\kappa>0}$. For each value of κ we consider the optimal design for the discrete problem RP^κ denoted by $(\bar{\theta}^\kappa, \bar{\gamma}^\kappa, \varepsilon(\bar{\theta}^\kappa, \bar{\gamma}^\kappa))$.

Theorem 7. *– The sequence $\{(\overline{\theta}^\kappa, \overline{\gamma}^\kappa, \varepsilon(\bar{\theta}^\kappa, \bar{\gamma}^\kappa))\}_{\kappa>0}$, is a minimizing sequence for the RP problem and satisfies the monotonicity condition:*

$$\text{for } \kappa < \kappa',\ RP^\kappa = RF(\overline{\theta}^\kappa, \overline{\gamma}^\kappa, \varepsilon(\bar{\theta}^\kappa, \bar{\gamma}^\kappa), \nabla\hat{\varphi})$$
$$\leq RP^{\kappa'} = RF(\overline{\theta}^{\kappa'}, \overline{\gamma}^{\kappa'}, \varepsilon(\bar{\theta}^{\kappa'}, \bar{\gamma}^{\kappa'}), \nabla\hat{\varphi}),$$

and

$$\lim_{\kappa \to 0} RF(\overline{\theta}^\kappa, \overline{\gamma}^\kappa, \varepsilon(\bar{\theta}^\kappa, \bar{\gamma}^\kappa), \nabla\hat{\varphi}) = RP.$$

Proof. The monotonicity follows immediately from the fact that $\kappa < \kappa'$ implies that $D_\Theta^{\kappa'} \subset D_\Theta^\kappa$. We note that the monotonicity property implies the existence of the limit

$$\lim_{\kappa \to 0} RF(\overline{\theta}^\kappa, \overline{\gamma}^\kappa, \varepsilon(\bar{\theta}^\kappa, \bar{\gamma}^\kappa), \nabla\hat{\varphi}).$$

Since $D_\Theta^\kappa \subset D_\Theta$ we have:

$$RP \leq RF(\bar{\theta}^\kappa, \bar{\gamma}^\kappa, \varepsilon(\bar{\theta}^\kappa, \bar{\gamma}^\kappa), \nabla\hat{\varphi}), \tag{53}$$

for every $\kappa > 0$. On the other hand, for a nested family of partitions $\{T_\kappa\}_{\kappa>0}$ and for any given $(\theta, \gamma, \varepsilon(\theta, \gamma))$ in D_Θ, it follows from Theorem 6 that there exists a sequence $\{(\theta^\kappa, \gamma^\kappa, \varepsilon(\theta^\kappa, \gamma^\kappa))\}_{\kappa>0}$ for which:

$$RF(\overline{\theta}^\kappa, \overline{\gamma}^\kappa, \varepsilon(\bar{\theta}^\kappa, \bar{\gamma}^\kappa), \nabla\hat{\varphi}) \leq RF(\theta^\kappa, \gamma^\kappa, \varepsilon(\theta^\kappa, \gamma^\kappa), \nabla\hat{\varphi}), \tag{54}$$

and

$$\begin{aligned} \lim_{\kappa \to 0+} RF(\overline{\theta}^\kappa, \overline{\gamma}^\kappa, \varepsilon(\bar{\theta}^\kappa, \bar{\gamma}^\kappa), \nabla\hat{\varphi}) &\leq \lim_{\kappa \to 0+} RF(\theta^\kappa, \gamma^\kappa, \varepsilon(\theta^\kappa, \gamma^\kappa), \nabla\hat{\varphi}) \\ &= RF(\theta, \gamma, \varepsilon(\theta, \gamma), \nabla\hat{\varphi}). \end{aligned} \tag{55}$$

It is now evident that: It is now evident that

$$\lim_{\kappa \to 0+} RF(\overline{\theta}^\kappa, \overline{\gamma}^\kappa, \varepsilon(\bar{\theta}^\kappa, \bar{\gamma}^\kappa), \nabla\hat{\varphi})$$
$$\leq \inf_{(\theta,\gamma,\varepsilon(\theta,\gamma)) \in D_\Theta} RF(\theta, \gamma, \varepsilon(\theta, \gamma), \nabla\hat{\varphi}) = RP \tag{56}$$

and the theorem follows from (53) and (56). ■

With Theorems 5 and 7 in hand it is possible to identify a sequence of configurations specified by χ^j for which

$$RP = \lim_{j\to\infty} F(\chi^j, \varepsilon(\chi^j), \nabla\hat{\varphi}). \tag{57}$$

Indeed we consider a minimizing sequence for RP as given by Theorem 7. To each element $(\overline{\theta}^\kappa, \overline{\gamma}^\kappa, \varepsilon(\bar{\theta}^\kappa, \bar{\gamma}^\kappa))$ of the sequence we can apply Theorem 5 to find a recovery sequence of configurations $\{\chi^{\kappa,n}\}_{n=1}^\infty$. In this way we see that

$$RP = \lim_{\kappa\to\infty} \lim_{n\to\infty} F(\chi^{\kappa,n}, \varepsilon(\chi^{\kappa,n}), \nabla\hat{\varphi}), \tag{58}$$

and it follows that we can extract a sequence of configurations $\{\chi^{\kappa_j,n_j}\}_{j=1}^\infty$ for which

$$RP = \lim_{j\to\infty} F(\chi^{\kappa_j,n_j}, \varepsilon(\chi^{\kappa_j,n_j}), \nabla\hat{\varphi}). \tag{59}$$

We now establish the following result. We now establish the following result.

Theorem 8. – *One has*

$$P = RP$$

i.e.,

$$\inf_{(\chi,\varepsilon(\chi))\in ad_\Theta} F(\chi, \varepsilon(\chi), \nabla\hat{\varphi}) = \inf_{(\theta,\gamma,\varepsilon(\theta,\gamma))\in D_\Theta} RF(\theta, \gamma, \varepsilon(\theta,\gamma), \nabla\hat{\varphi}). \tag{60}$$

Proof. Since $ad_\Theta \subset D_\Theta$ and from (18) it follows that $P \geq RP$. Moreover from Theorem 7 and (59), there exist $\{(\overline{\theta}^\kappa, \overline{\gamma}^\kappa, \varepsilon(\bar{\theta}^\kappa, \bar{\gamma}^\kappa)\}_{\kappa>0} \in \{D_\Theta^\kappa\}_{\kappa>0}$, such that

$$RP = \lim_{\kappa\to 0} RF(\overline{\theta}^\kappa, \overline{\gamma}^\kappa, \varepsilon(\bar{\theta}^\kappa, \bar{\gamma}^\kappa), \nabla\hat{\varphi}) = \lim_{j\to\infty} F(\chi^{\kappa_j,n_j}, \varepsilon(\chi^{\kappa_j,n_j}), \nabla\hat{\varphi}). \tag{61}$$

On the other hand

$$F(\chi^{\kappa_j,n_j}, \varepsilon(\chi^{\kappa_j,n_j}), \nabla\hat{\varphi}) \geq P, \text{ for all } j. \tag{62}$$

Thus (61) and (62) imply that $RP \geq P$, and we conclude that $RP = P$.

The results presented in this Section provide the way for the identification of a class of minimizing sequences of configurations of the two conductors. Theorem 7 shows how to generate a minimizing sequence of generalized designs coming from discrete problems. Theorem 5 and (39) provide the methodology for constructing an optimizing sequence of configurations based upon the information given in the solution of the generalized design problem. These results give rigorous rules of thumb for the design of two phase conductors. The numerical implementation is given in Section 6.

5. A complete characterization of minimizing sequences

In this Section we provide the proof of Theorem 1. We consider a dense class of target fields for which we can account for all oscillations in minimizing sequences of designs. $L^2(\Omega)$ closure of gradients of solutions of (1) with coefficients $\varepsilon(x)$ given by (3). For $\chi \in ad_\theta$ we introduce the set of gradients given by

$$\mathcal{S}_\Theta = \begin{cases} \nabla u \mid u \text{ is a } W_0^{1,2}(\Omega) \text{ solution of } -\operatorname{div}(\varepsilon(\chi)\nabla u) = f, \\ \chi \in ad_\theta. \end{cases} \tag{63}$$

The strong $L^2(\Omega)$ closure of the set $\mathcal{S}_\Theta$ is denoted by the set $\bar{\mathcal{S}}_\Theta$. It is evident that

$$\begin{aligned} P = \inf_{\chi \in ad_\Theta} F(\chi, \varepsilon(\chi), \nabla\hat{\varphi}) &= \inf_{\nabla u \in \mathcal{S}_\Theta} \int_\Omega |\nabla u - \nabla\hat{\varphi}|^2 \, dx \\ &= \inf_{\nabla u \in \bar{\mathcal{S}}_\Theta} \int_\Omega |\nabla u - \nabla\hat{\varphi}|^2 \, dx. \end{aligned} \tag{64}$$

In light of (64) and the definition of $\bar{\mathcal{S}}_\Theta$ we apply Theorem 2 of the introduction to conclude the existence of a G_δ subset K of $W_0^{1,2}(\Omega)$ such that

Theorem 9. *–Given a target field $\hat{\varphi} \in$ K and a minimizing sequence*

$$\{(\chi^n, \varepsilon(\chi^n))\}_{n=1}^\infty \in ad_\Theta$$

for P then the associated sequence of state variables $\{\varphi^n\}_{n=1}^\infty$ solving the equilibrium equation

$$-\operatorname{div}(\varepsilon(\chi^n)\nabla\varphi^n) = f \tag{65}$$

is Cauchy in the $W_0^{1,2}(\Omega)$ norm given by $\|u\|^2 = \int_\Omega |\nabla u|^2 \, dx$.

From the completeness of $W_0^{1,2}(\Omega)$ there exists a potential $\bar{\varphi} \in W_0^{1,2}(\Omega)$ such that $\lim_{n\to\infty} \varphi^n = \bar{\varphi}$ strongly in $W_0^{1,2}(\Omega)$. Passing to subsequences if necessary, the sequence $\{\chi^n\}_{n=1}^\infty$ weak $L^\infty(\Omega)$ **star** converges to a density $\bar{\theta}$ and the compactness property of G-convergence implies that the sequence $\{\varepsilon(\chi^n)\}_{n=1}^\infty$ G-converges to an effective tensor ε^e where

$$-\operatorname{div}(\varepsilon^e \nabla\bar{\varphi}) = f. \tag{66}$$

¿From the results given in [6] and [11] we have that the set of effective tensors associated with the density $\bar{\theta}(x)$ are all the symmetric 2×2 matrices with eigenvalues λ_1, λ_2 lying in the set $\mathrm{K}_{\bar{\theta}}$ for almost all x in Ω. The set $\mathrm{K}_{\bar{\theta}}$ is given by the inequalities

$$\sum_{k=1}^{2} \frac{1}{\lambda_j - \alpha} \leq \frac{1}{h_{\bar\theta} - \alpha} + \frac{1}{m_{\bar\theta} - \alpha}$$
$$\sum_{k=1}^{2} \frac{1}{\beta - \lambda_j} \leq \frac{1}{\beta - h_{\bar\theta}} + \frac{1}{\beta - m_{\bar\theta}}. \tag{67}$$

On the other hand the work of J. Dvorak, J. Haslinger, and M. Miettinen [5] shows that the strong convergence of the sequence $\{\varphi^n\}_{n=1}^{\infty}$ delivers the local relation

$$\varepsilon^e \nabla\bar\varphi = m_{\bar\theta} \nabla\bar\varphi, \text{ a.e.} \tag{68}$$

This implies that $m_{\bar\theta}$ is an eigenvalue of ε^e. The constraints on the eigenvalues of ε^e given by (67) together with (68) allows us to uniquely identify ε^e as the effective tensor given by

$$\varepsilon^e = R(\bar\gamma)\Lambda(\bar\theta)R^T(\bar\gamma), \tag{69}$$

where the angle $\bar\gamma$ is chosen according to the requirement given by (68). For this choice of angle we also have the local relation

$$R(\bar\gamma)H(\bar\theta)R^T(\bar\gamma)\nabla\bar\varphi = 0, \text{ a.e.,} \tag{70}$$

and we conclude that

$$P = \int_\Omega |\nabla\bar\varphi - \nabla\hat\varphi|^2\, dx = RF(\bar\theta, \bar\gamma, \varepsilon(\bar\theta, \bar\gamma), \nabla\hat\varphi). \tag{71}$$

In view of Theorem 8 we deduce that the design $(\bar\theta, \bar\gamma, \varepsilon(\bar\theta, \bar\gamma))$ is the optimal design for the problem RP. This establishes parts (1), (2), and (4) of Theorem 1.

To proceed we recall the notion of a cluster point (θ, ε^e) for a sequence of designs associated with a sequence configurations $\{\chi^n\}_{n=1}^{\infty}$. The definition of a cluster point (θ, ε^e) implies the existence of a subsequence $\{\chi^{n_j}, \varepsilon(\chi^{n_j})\}_{j=1}^{\infty}$ such that $\{\chi^{n_j}\}_{j=1}^{\infty}$ weak $L^\infty(\Omega)$ **star** converges to θ and $\{\varepsilon(\chi^{n_j})\}_{j=1}^{\infty}$ G-converges to ε^e. Arguments identical to those given above show that any cluster point of any minimizing sequence for the problem P is a minimizing design for the problem RP. This establishes the first part of (3) of Theorem 1. The second part of (3) of Theorem 1 follows immediately from the construction of a recovery sequence of configurations based upon Theorems 5 and 7, see equations (59).

Part (3) of Theorem 1 together with (68) and (70) point out what kinds of oscillations can occur in minimizing sequences of configurations. In the subregion of Ω where the minimizing sequence oscillates, i.e., the region where $0 < \bar{\theta} < 1$, we see that the oscillations are in the form of layers of the two conductors. The layers are asymptotically parallel to the optimal gradient $\nabla\bar{\varphi}$. This configuration allows for the best effective conductivity properties to be aligned with the direction of the gradient. This is consistent with physical intuition.

6. Numerical solution and a practical approach to design of graded materials

We provide an outline of the method used for the numerical solution of the discrete design problem. For convenience the objective functional is denoted by $E(\theta,\gamma)$ and

$$\begin{aligned} E(\theta,\gamma) &= RF(\theta,\gamma,\varepsilon(\theta,\gamma),\nabla\hat{\varphi}) \\ &= \int_\Omega |\nabla\varphi - \nabla\hat{\varphi}|^2\,dx \qquad (72) \\ &+ \int_\Omega R(\gamma(x))H(\theta(x))R^T(\gamma(x))\nabla\varphi\cdot\nabla\varphi\,dx, \qquad (73) \end{aligned}$$

where the state variable φ solves the equilibrium equation (12). For a given partition, the number of subdomains is $N(\kappa)$ and the design variable (θ,γ) is a vector of length $2N(\kappa)$. The components of (θ,γ) are the constant values (θ_i,γ_i) taken in each subdomain Ω_i. The components of the design vector are subject to the box constraints:

$$\begin{aligned} &0 \le \theta_i \le 1,\ i = 1\ldots,N(\kappa), \\ &0 \le \gamma_i \le 2\pi,\ i = 1\ldots,N(\kappa). \end{aligned} \qquad (74)$$

We include the resource constraint $\int_\Omega \theta\,dx \le \Theta\,meas(\Omega)$ by adding a penalty term

$$\ell \times \left(\int_\Omega \theta\,dx - \Theta\,meas(\Omega)\right),$$

for $\ell \ge 0$. The discrete design problem is written

$$\min_{(\theta,\gamma)} E(\theta,\gamma) + \ell \times \left(\int_\Omega \theta\,dx - \Theta\,meas(\Omega)\right), \qquad (75)$$

where (θ,γ) are subject to the constraints given by (74). The numerical procedure is a straight forward application of the steepest decent method, see [17]. Gradients of the objective $E(\theta,\gamma)$ are computed and increments of the design variables $(\delta\theta, \delta\gamma)$ are chosen to insure $E(\theta,\gamma) \geq E(\theta+\delta\theta, \gamma+\delta\gamma)$. The advantage of this procedure is that it is monotone and convergence is assured.

We provide numerical examples that illustrate how electrostatic fields can be controlled using functionally graded materials. For all examples the design domain is chosen to be the square centered at the origin given by $\Omega = (-1,1) \times (-1,1)$ and we choose the target field to be zero, i.e., $\nabla\hat{\varphi} = (0,0)$. The discrete design is associated with a partition of Ω into $20,000$ subdomains of diameter on the order of 10^{-2}.

For the first two examples the charge distribution is taken to be uniform in Ω and given by $f = 1$. We choose $\alpha = 1$ and $\beta = 2$ and constrain the amount of good dielectric to be 40% of the design domain. The density distribution, $\theta(x)$, of the better dielectric material in the optimized discrete design is given in Figure 1a. Here the darkest regions consist of pure β dielectric, the white regions are occupied by pure α dielectric and the regions of graded conductivity properties are given by the intermediate shades. The layer normals in the graded parts of the design are given by the arrows in Figure 1a. The contours are the level lines of the electric potential. Note that the layer normals are tangential to the level lines, hence perpendicular to the electric field. We emphasize that Figure 1a gives the necessary geometric information for manufacturing graded materials. Indeed, given $\theta(x), \gamma(x)$ we can we apply (39) to construct a sequence of graded materials. Because of the continuity expressed by Theorem 5 we are guaranteed that we can construct a two phase configuration thats nearly optimal.

For the second example we consider a subdomain D of the design domain Ω. Here we take $D = \Omega \backslash \{(-1/2,1/2) \times (-1/2,1/2)\}$. We consider the problem

$$P = \inf_{\chi \in ad_\Theta} \int_D |\nabla\varphi|^2 \, dx. \tag{76}$$

The theory presented in this paper easily generalizes to this case and the relaxed problem is

$$\begin{aligned} RP \quad &= \inf_{(\theta,\gamma,\varepsilon(\theta,\gamma))\in D_\Theta} \left\{ \int_D |\nabla\varphi|^2 \, dx \right. \\ &\left. + \int_D R(\gamma(x))H(\theta(x))R^T(\gamma(x))\nabla\varphi \cdot \nabla\varphi \, dx \right\}, \end{aligned} \tag{77}$$

and $P = RP$.

Here the goal is to screen as much electric field away from the domain D as possible. The good dielectric is constrained to occupy 40% of Ω. The density distribution of the good dielectric in the optimal design is given in Figure 1b. We point out that we allow the two dielectrics to be placed

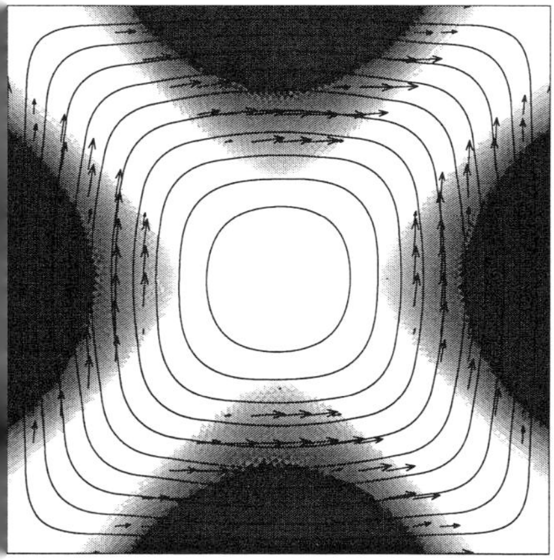

Figure 1: a.

Figure 1: b.

anywhere in Ω, however the algorithm automatically uses the good dielectric only in D. This is consistent with intuition.

For the next example we take the charge distribution to be 1 everywhere outside of D and zero inside D. As before we take $\alpha = 1$ and $\beta = 2$.

The good dielectric is constrained to occupy 15% of the design domain. The density distribution for the optimal design is given in Figure 2a. In Figure 2b we plot the level lines of the potential and the electric field associated with the design. Last we consider the same layout as in Figure 2a but with $\alpha = 1$ and $\beta = 1000$ and we plot the electric field for this case in Figure 3. For this layout and choice of β we see that the electric field has been screened away from D.

Acknowledgments. This work is supported by AFOSR Grant F49620-99-1-0009 and NSF Grants DMS-9700638 and DMS-9403866.

References

[1] Bensoussan A., Lions J. L., and Papanicolaou G., Asymptotic analysis for periodic structures, Studies in Mathematics and its Applications, 5 North Holland, Amsterdam 1978.

[2] Bidaut M., Théorèms d'existence et d'existence en général d'un contrôle optimal pour des systemes régis par des équations aux dérivées partielles non linéaires, PhD thesis, Université Paris VI, June 1973.

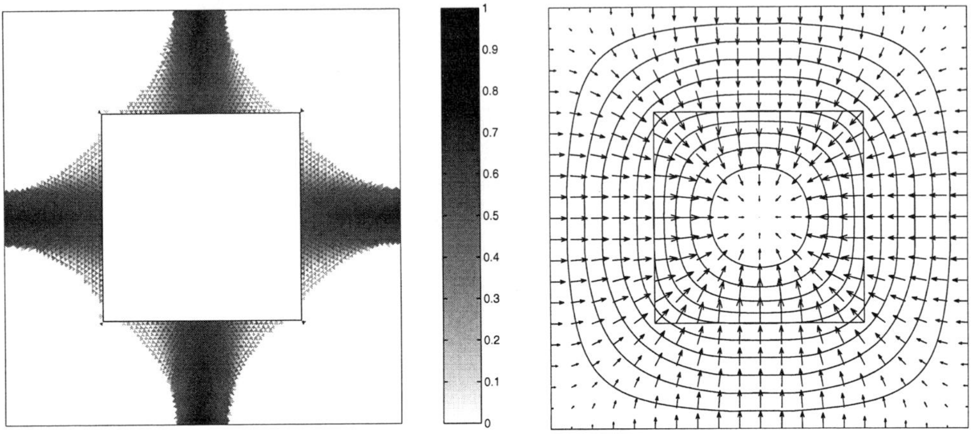

Figure 2: a.

Figure 2: b.

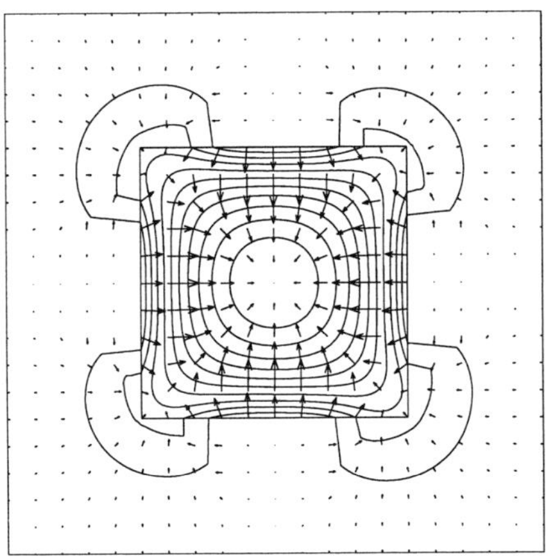

Figure 3: Electric field for $\beta = 1000$.

[3] De Giorgi E. and Spagnolo S. Sulla convergenza degli integrali dell'energia per operatori ellittici del secondo ordine, Boll. U.M.I. 8 (1973), 391–411.

[4] Edelstein M, On nearest points of sets in uniformly convex Banach spaces, J. London Math. Soc., 43 (1968), 375–377.

[5] Dvorak J., Haslinger J., and Miettinen M., On the Problem of Optimal Material Distribution, Preprint, (1996).

[6] Lurie K. A. and Cherkaev A. V., Exact estimates of conductivity of composites formed by two isotropically conducting media taken in prescribed proportion, Proc. Roy. Soc. Edinburgh, 99 A (1984), 71–87.

[7] Murat F., Compacité par compensation, Ann. Sc. Norm. Sup. Pisa, 5 (1978), 489–507.

[8] Murat F. and Tartar L., On the control of coefficients in partial differential equations. In Topics in the mathematical modeling of composite materials, Cherkaev A. and Kohn R. eds., Birkhauser, Boston, (1997), 1–8.

[9] Murat F. and Tartar L., H-convergence. In Topics in the mathematical modeling of composite materials, Cherkaev A. and Kohn R. eds, Birkhauser, Boston, (1997), 21–43.

[10] Murat F. and Tartar L., Calcul des variations et homogénéisation. In Les Méthodes de l'homogénéisation: Théorie et applications en physique. Collection de la Direction des Etudes et Recherches d'Electricité de France, 57, Eyrolles, Paris, (1985), 319-369.

[11] Tartar L., Estimations fines de coefficients homogénéisés. In Ennio De Giorgi Colloquium, Kree P. ed., Research Notes in Mathematics 125, Pitman, London 1985, 168–187.

[12] Tartar L., Compensated compactness and applications to partial differential equations, In Nonlinear Analysis and Mechanics, Heriot-Watt Symposium, Volume IV, Knops R.J., ed., Research Notes in Mathematics, 39. Pitman, Boston 1979, 136–212.

[13] Tartar L., Remarks on optimal design problems. In Calculus of Variations, Homogenization and Continuum Mech., Buttazzo G., Bouchitte G., and Suquet P. eds., World Scientific, Singapore, (1994), 279–296.

[14] Tartar L., H-measures, a new approach for studying homogenization, oscillations and concentration effects in partial differential equations, Proc. Roy. Soc. Edinb., 115A (1990), 193–230.

[15] Pedregal P., Optimal design and constrained quasiconvexity, Preprint, (1998).

[16] Pedregal P., Optimization, Relaxation and Young Measures, Bull. Amer. Math. Soc., 36 (1999), 27–58.

[17] Pironneau O., Optimal shape design for elliptic systems. Springer Verlag, New York 1984.

[18] Spagnolo S., Convergence in energy for elliptic operators. In Proceedings of the third symposium on numerical solutions of partial differential equations, (College Park,1975), Hubbard B. ed. , Academic Press, New York, 496–498.

Robert Lipton
Department of Mathematics
Louisiana State University
Baton Rouge, LA 70803-4918
USA

E-mail: lipton@math.lsu.edu

Ani P. Velo
Department of Mathematical Sciences
United States Military Academy
West Point, NY 10996
USA

E-mail: aa0349@exmail.usma.army.mil

Studies in Mathematics and its Applications, Vol. 31
D. Cioranescu and J.L. Lions (Editors)

Chapter 24

A BLACKBOX REDUCED-BASIS OUTPUT BOUND METHOD FOR NONCOERCIVE LINEAR PROBLEMS

Y. MADAY, A. T. PATERA AND D. V. ROVAS

1. Introduction

Reduced-basis methods [2, 5, 9, 19, 20, 24] are very attractive techniques for the efficient prediction of the parametric dependence of ***outputs*** — functionals of solutions — of partial differential equations. In particular, reduced-basis techniques enjoy a "state-space" optimality property that ensures rapid convergence even in higher dimensional parameter spaces; very good accuracy can thus be obtained with relatively few modes (basis functions), and hence at relatively low cost. In many cases — in which the operator depends affinely on the parameters — the efficiency can be further enhanced with "blackbox" superposition formulations [12], in which the operation count for the "on-line" stage of the procedure depends only on the dimension of the reduced-basis space and the complexity of the parametric representation. In the limit in which one is interested in rapid evaluations at many points in parameter space (e.g., as in design, optimization, and real-time control), reduced-basis approaches are clearly the most efficient choice.

Although there are several ***a priori*** error estimates for reduced-basis methods [5, 7, 25] that support the empirically observed exponential convergence rates, there are no sharp and rigorous ***a posteriori*** bounds (though see [20] for early efforts in this direction). It is thus difficult, if not impossible, to determine how many empirical basis functions suffice for any given problem, output, and prescribed accuracy; as a result, the application of reduced-basis methods has been rather limited. To remedy the situation, we propose in [12, 17] a blackbox reduced-basis output bound procedure that provides rigorous error bounds for the particular quantities of interest. These methods (as well as an alternative approach to reduced-basis a posteriori error estimation described in [14]) are related to earlier output bound procedures [13, 16, 18, 21], which in turn are based upon classical finite element ***a posteriori*** error estimation theory [1, 4, 6, 10].

In [12, 17] we address either coercive elliptic partial differential equations or very mildly noncoercive equations (in particular, the symmetric positive-definite eigenvalue problem). In the current paper we consider *noncoercive* problems; see also [11] for related work within the iterative context. In Section 2 we present the problem formulation; in Section 3 we describe the reduced-basis output approximation and associated error estimator; in Section 4 we prove the optimality of the output approximation as well as the optimality and asymptotic bounding property of the error estimator; in Section 5 we describe the blackbox computational procedure, and present the operation count and storage requirements for each step of the process; and in Section 6 we give numerical results for a particular instantiation, the Helmholtz equation (a future paper will discuss the incompressible Stokes problem).

2. Problem description

2.1. Nomenclature

For a general Hilbert space Z, we denote the associated inner product and induced norm by $(\cdot,\cdot)_Z$ and $\|\cdot\|_Z$, respectively; we identify the corresponding dual space as Z', with norm $\|\cdot\|_{Z'}$ given by

$$\|f\|_{Z'} = \sup_{v\in Z} \frac{f(v)}{\|v\|_Z}.$$

From the Riesz representation theorem we know that for every $f \in Z'$ there exists a $\rho_f^Z \in Z$ such that

$$(\rho_f^Z, v)_Z = f(v), \ \forall v \in Z.$$

It is then readily deduced that

$$\rho_f^Z = \arg\sup_{v\in Z} \frac{f(v)}{\|v\|_Z},$$

and

$$\|f\|_{Z'} = \|\rho_f^Z\|_Z,$$

which we will use repeatedly in what follows.

2.2. Problem statement

Given a Hilbert space Y of dimension $\mathcal{N}$ (possibly infinite), a linear functional $\ell \in Y'$, and a parameter μ in a set $\mathcal{D} \subset I\!R^P$, we look for $u(\mu) \in Y$ such that

$$a(u(\mu), v; \mu) = \ell(v), \ \forall v \in Y, \tag{1}$$

where $a(\cdot,\cdot;\mu)$ is a bilinear form the assumptions on which are detailed below. We further prescribe an *output functional* $\ell^O \in Y'$, in terms of which we can evaluate our output of interest $s(\mu)$ as

$$s(\mu) = \ell^O(u(\mu)). \tag{2}$$

We also introduce a dual, or adjoint, problem associated with ℓ^O: find $\psi(\mu) \in Y$ such that

$$a(v,\psi(\mu);\mu) = -\ell^O(v), \ \forall v \in Y.$$

The relevance of this dual problem will become clear in what follows. It is relatively simple to permit μ-dependence in ℓ and ℓ^O as well, however for clarity of exposition we do not consider this here.

We shall assume (though this is essential for only some of our arguments) that a is symmetric,

$$a(w,v;\mu) = a(v,w;\mu), \ \forall w,v \in Y^2, \ \forall \mu \in \mathcal{D}.$$

We further assume that $a(w,v;\mu)$ is uniformly continuous,

$$|a(w,v;\mu)| \le \gamma \|w\|_Y \|v\|_Y, \ \forall w,v \in Y^2, \ \forall \mu \in \mathcal{D},$$

and that $a(w,v;\mu)$ satisfies a uniform inf-sup condition,

$$0 < \beta_0 \le \beta(\mu) = \inf_{w\in Y} \sup_{v\in Y} \frac{a(w,v;\mu)}{\|w\|_Y \|v\|_Y} = \inf_{w\in Y} \frac{\|a(w,\cdot;\mu)\|_{Y'}}{\|w\|_Y}, \ \forall \mu \in \mathcal{D}; \tag{3}$$

it is classical that these latter two conditions are required for the well-posedness of our primal and dual problems. Finally, we shall make the assumption that our bilinear form a is affine in the parameter μ in the sense that, for some finite integer Q,

$$a(w,v;\mu) = \sum_{q=1}^{Q} \alpha_q(\mu) a_q(w,v), \ \forall w,v \in Y^2, \tag{4}$$

where the a_q are bilinear forms. The assumption (4) permits our blackbox approach; non-blackbox variants of the methods described here — in which (4) is relaxed — can also be developed, and will be discussed in a future paper.

2.3. Inf-sup supremizers and infimizers

We can rephrase (3) as: for every $w \in Y$, there exists an element $T_\mu w$ in Y, such that

$$\beta(\mu)\|w\|_Y \|T_\mu w\|_Y \le a(w,T_\mu w;\mu), \tag{5}$$

where $T_\mu w$ is the *supremizer* associated with $\|a(w,\cdot;\mu)\|_{Y'}$. It follows from Section 2.1 that $T_\mu w = \rho^Y_{a(w,\cdot;\mu)}$, that is,

$$(T_\mu w, v)_Y = a(w,v;\mu),\ \forall v \in Y.$$

It is thus clear that T_μ: $Y \to Y$ is a linear operator; we can also readily show that T_μ is symmetric (since a is symmetric); furthermore, T_μ is bounded, since

$$\|T_\mu w\|_Y^2 = a(w, T_\mu w; \mu) \le \gamma \|w\|_Y \|T_\mu w\|_Y,$$

and hence

$$\frac{\|T_\mu w\|_Y}{\|w\|_Y} \le \gamma. \tag{6}$$

Finally, we can now express our inf-sup parameter in terms of T_μ as

$$\beta(\mu) = \inf_{w\in Y} \frac{\|T_\mu w\|_Y}{\|w\|_Y} = \frac{\|T_\mu \chi(\mu)\|_Y}{\|\chi(\mu)\|_Y}, \tag{7}$$

where

$$\chi(\mu) = \arg \inf_{w\in Y} \frac{\|T_\mu w\|_Y}{\|w\|_Y}$$

is our *infimizer*; we thus also have

$$\beta(\mu) = \frac{a(\chi(\mu), T_\mu \chi(\mu))}{\|\chi(\mu)\|_Y \|T_\mu \chi(\mu)\|_Y}.$$

It will be useful in the subsequent analysis to recognize that $\beta(\mu)$ and $\chi(\mu)$ are related to the minimum eigenvalue and associated eigenfunction of a symmetric positive-definite eigenproblem. In particular we look for $(\theta, \lambda) \in (Y \times \mathbb{R})$ solution of

$$\mathcal{A}(\theta(\mu), v; \mu) = \lambda(\mu)(\theta(\mu), v)_Y,\ \forall v \in Y, \text{ and } \|\theta(\mu)\|_Y = 1, \tag{8}$$

where

$$\mathcal{A}(w, v; \mu) = (T_\mu w, T_\mu v)_Y,\ \forall w, v \in Y^2; \tag{9}$$

we denote the resulting eigenpairs as $(\theta_i(\mu), \lambda_i(\mu))$, $i = 1, \ldots$, with $\lambda_{\min} \equiv \lambda_1 \le \lambda_2 \le \cdots$. It follows immediately from Rayleigh quotient arguments that

$$\lambda_{\min}(\mu) = \min_{w\in Y} \frac{\mathcal{A}(w,w;\mu)}{(w,w)_Y} = \min_{w\in Y} \frac{\|T_\mu w\|_Y^2}{\|w\|_Y^2} = \beta^2(\mu),$$

and thus $\beta(\mu) = \sqrt{\lambda_{\min}(\mu)}$ and $\theta_{\min}(\mu) \equiv \theta_1(\mu) = \chi(\mu)$.

To further understand the relationship between the infimizers and supremizers, we consider a second symmetric positive-definite eigenproblem: find $(\Upsilon \times \omega) \in (Y \times I\!R)$ such that

$$2\gamma(\Upsilon(\mu), v)_Y = \omega(\mu)\mathcal{B}(\Upsilon(\mu), v; \mu), \tag{10}$$

where

$$\mathcal{B}(w, v; \mu) = 2\gamma(w, v)_Y - a(w, v; \mu);$$

note that $\mathcal{B}$ is symmetric and coercive. By the usual arguments (and appropriate normalization), $2\gamma(\Upsilon_i, \Upsilon_j)_Y = \omega_i \mathcal{B}(\Upsilon_i, \Upsilon_j) = \omega_i \delta_{ij}$, with $0 < \omega_1 \leq \omega_2 \leq \cdots$; here δ_{ij} is the Kronecker delta symbol, and (Υ_i, ω_i) refers to the ith eigenpair associated with (10). We can then write

$$(T_\mu w, v)_Y = 2\gamma(w, v)_Y - \mathcal{B}(w, v; \mu),$$

expand

$$w = \sum_{i=1}^{\mathcal{N}} c_i \Upsilon_i,$$

and exploit orthogonality to deduce that

$$T_\mu w = \sum_{i=1}^{\mathcal{N}} d_i \Upsilon_i$$

for $d_i = 2\gamma c_i(\omega_i - 1)/\omega_i$. Thus

$$\frac{\|T_\mu w\|_Y^2}{\|w\|_Y^2} = \frac{4\gamma^2 \displaystyle\sum_{i=1}^{\mathcal{N}} \left| \frac{\omega_i(\mu) - 1}{\omega_i(\mu)} \right|^2 \omega_i(\mu) c_i^2}{\displaystyle\sum_{i=1}^{\mathcal{N}} \omega_i(\mu) c_i^2},$$

again by orthogonality. We conclude that

$$\beta(\mu) = 2\gamma \left| \frac{\omega_{i^*(\mu)}(\mu) - 1}{\omega_{i^*(\mu)}(\mu)} \right|,$$

and $\chi(\mu) = c_{i^*(\mu)} \Upsilon_{i^*(\mu)}$, where

$$i^*(\mu) = \arg \min_{i \in \{1, \ldots, \mathcal{N}\}} \left| \frac{\omega_i(\mu) - 1}{\omega_i(\mu)} \right|. \tag{11}$$

We thus observe that

$$T_\mu \chi = 2\gamma \left(\frac{\omega_{i^*(\mu)}(\mu) - 1}{\omega_{i^*(\mu)}(\mu)} \right) \chi,$$

which states that $T_\mu \chi$ and χ are *collinear*; in general, $T_\mu w$ and w will be linearly dependent only if w is proportional to a *single* eigenfunction Υ_i.

3. Reduced-basis output bound formulation

3.1. Approximation space

We select M points in our parameter set $\mathcal{D}$, $\mu^m \in \mathcal{D}, m = 1, \ldots, M$, the collection of which we denote

$$\mathcal{S}^M = \{\mu^1, \ldots, \mu^M\}.$$

We then introduce three "Lagrangian" spaces [25],

$$W_u^M = \text{span}\{u(\mu^m), m = 1, \ldots, M\}, \tag{12}$$

$$W_\psi^M = \text{span}\{\psi(\mu^m), m = 1, \ldots, M\}, \tag{13}$$

and

$$W_\chi^M = \text{span}\{\chi(\mu^m), m = 1, \ldots, M\}, \tag{14}$$

associated with our primal solutions, dual solutions, and infimizers, respectively.

We shall consider two approximation spaces W^N. In the first, we set $N = 2M$ and choose $W^N = W_0^N$, where

$$\begin{aligned} W_0^N &= W_u^M + W_\psi^M \\ &= \text{span}\{u(\mu^1), \psi(\mu^1), \ldots, u(\mu^M), \psi(\mu^M)\} \\ &\equiv \text{span}\{\zeta_1, \ldots, \zeta_N\}. \end{aligned} \tag{15}$$

In the second case we set $N = 3M$ and choose $W^N = W_1^N$, where

$$\begin{aligned} W_1^N &= W_u^M + W_\psi^M + W_\chi^M \\ &= \text{span}\{u(\mu^1), \psi(\mu^1), \chi(\mu^1), \ldots, u(\mu^M), \psi(\mu^M), \chi(\mu^M)\} \\ &\equiv \text{span}\{\zeta_1, \ldots, \zeta_N\}. \end{aligned} \tag{16}$$

(Obviously the ζ_N — the reduced-basis functions — take different meanings in the two cases.) The role of each of the components of the W^N shall become clear later in our development.

Remark. Compliance. In the case in which $\ell^O = \ell$, then $\psi(\mu) = -u(\mu)$; we thus redefine $W_0^N = W_u^M$ (with $N = M$) and $W_1^N = W_u^M + W_\chi^M$ (with $N = 2M$). This will of course result in computational savings. Note that if ℓ^O is close to ℓ then W_0^N of (15) and W_1^N of (16) can lead to ill-conditioned systems; we shall not address this in the current paper (or the more general issue of ill-conditioning arising in the reduced-basis context). ■

3.2. Output Prediction

We shall see shortly that the W^N will play the role of the infimizing space. We also need supremizing spaces. To that end, we introduce $V^N \subseteq Y$, with $(\cdot,\cdot)_{V^N} = (\cdot,\cdot)_Y$ and hence $\|\cdot\|_{V^N} = \|\cdot\|_Y$. We shall consider two spaces in particular: $V^N = W^N$ ($= W_0^N$ or W_1^N), and hence of dimension N; and $V^N = Y$, and hence of dimension $\mathcal{N}$.

We next define, for all $w^N \in W^N$ and for all $\varphi^N \in W^N$, the primal and dual residuals, $R^{pr}(\cdot; w^N; \mu) \in Y'$ and $R^{du}(\cdot; \varphi^N; \mu) \in Y'$, respectively:

$$\begin{aligned} R^{pr}(v; w^N; \mu) &\equiv \ell(v) - a(w^N, v; \mu), \ \forall v \in Y, \\ R^{du}(v; \varphi^N; \mu) &\equiv -\ell^O(v) - a(v, \varphi^N; \mu), \ \forall v \in Y. \end{aligned}$$

It follows from our primal and dual problem statements that

$$\begin{aligned} R^{pr}(v; w^N; \mu) &= a(u - w^N, v; \mu), \\ R^{du}(v; \varphi^N; \mu) &= a(v, \psi - \varphi^N; \mu), \end{aligned} \tag{17}$$

which is the standard residual-error relation evoked in most *a posteriori* frameworks.

We then look for $u^N(\mu) \in W^N, \psi^N(\mu) \in W^N$, such that

$$\begin{aligned} u^N(\mu) &= \arg \inf_{w^N \in W^N} \|R^{pr}(\cdot; w^N; \mu)\|_{(V^N)'} \\ &= \arg \inf_{w^N \in W^N} \sup_{v \in V^N} \frac{R^{pr}(v; w^N; \mu)}{\|v\|_Y}, \end{aligned} \tag{18}$$

and

$$\begin{aligned} \psi^N(\mu) &= \arg \inf_{\varphi^N \in W^N} \|R^{du}(\cdot; \varphi^N; \mu)\|_{(V^N)'} \\ &= \arg \inf_{\varphi^N \in W^N} \sup_{v \in V^N} \frac{R^{du}(v; \varphi^N; \mu)}{\|v\|_Y}, \end{aligned} \tag{19}$$

which is a minimum-residual (or least-squares) projection; see also [5] for discussion of various projections within the reduced-basis context.

Our output approximation is then given by:

$$s^N(\mu) = \ell^O(u^N) - R^{pr}(\psi^N; u^N; \mu); \tag{20}$$

the additional adjoint terms will improve the accuracy [8, 15, 23].

It will be convenient to express our minimum-residual approximation in terms of affine supremizing operators $P_\mu^N : W^N \to V^N, D_\mu^N : W^N \to V^N$, defined by

$$\begin{aligned} P_\mu^N w^N &= \rho^{V^N}_{R^{pr}(\cdot;w^N;\mu)}, \\ D_\mu^N w^N &= \rho^{V^N}_{R^{du}(\cdot;w^N;\mu)}, \end{aligned}$$

that is

$$(P_\mu^N w^N, v)_Y = R^{pr}(v; w^N; \mu), \ \forall v \in V^N, \tag{21}$$

$$(D_\mu^N w^N, v)_Y = R^{du}(v; w^N; \mu), \ \forall v \in V^N, \tag{22}$$

for any $w^N \in W^N$. In particular, it follows from Section 2.1 that we can now write

$$u^N(\mu) = \arg \inf_{w^N \in W^N} \|P_\mu^N w^N\|_Y, \quad \psi^N(\mu) = \arg \inf_{\varphi^N \in W^N} \|D_\mu^N \varphi^N\|_Y; \tag{23}$$

the μ dependence is through P_μ^N and D_μ^N.

3.3. Error bound prediction

We first define $\beta^N(\mu) \in I\!R$ as

$$\beta^N(\mu) = \inf_{w^N \in W^N} \sup_{v \in V^N} \frac{a(w^N, v; \mu)}{\|w^N\|_Y \|v\|_Y} = \inf_{w^N \in W^N} \frac{\|a(w^N; \cdot; \mu)\|_{(V^N)'}}{\|w^N\|_Y}. \tag{24}$$

We can rephrase (24) as: for any $w^N \in W^N$, there exists an element $T_\mu^N w^N$ in V^N, such that

$$\beta^N(\mu) \|w^N\|_Y \|T_\mu^N w^N\|_Y \le a(w^N, T_\mu^N w^N; \mu), \tag{25}$$

where $T_\mu^N w^N$ is the supremizer associated with $\|a(w^N, \cdot; \mu)\|_{(V^N)'}$. It follows from Section 2.1 that $T_\mu^N : W^N \to V^N$ is given by $\rho^{V^N}_{a(w^N, \cdot; \mu)}$, or more explicitly,

$$(T_\mu^N w^N, v)_Y = a(w^N, v; \mu), \ \forall v \in V^N,$$

for any $w^N \in W^N$. We can now express $\beta^N(\mu)$ as

$$\beta^N(\mu) = \inf_{w^N \in W^N} \frac{\|T_\mu^N w^N\|_Y}{\|w^N\|_Y} = \frac{\|T_\mu^N \chi^N(\mu)\|_Y}{\|\chi^N(\mu)\|_Y},$$

where

$$\chi^N(\mu) = \arg \inf_{w^N \in W^N} \frac{\|T_\mu^N w^N\|_Y}{\|w^N\|_Y}$$

is our infimizer over W^N; we thus also have

$$\beta^N(\mu) = \frac{a(\chi^N(\mu), T_\mu^N \chi^N(\mu))}{\|\chi^N(\mu)\ |_Y \|T_\mu^N \chi^N(\mu)\|_Y}.$$

Then, given $u^N(\mu), \psi^N(\mu)$, and a real constant σ, $0 < \sigma < 1$, we compute

$$\Delta^N(\mu) = \frac{1}{\sigma \beta^N(\mu)} \|R^{pr}(\cdot; u^N; \mu)\|_{Y'} \|R^{du}(\cdot; \psi^N; \mu)\|_{Y'}, \tag{26}$$

which will serve as our *a posteriori* error bound for $|(s - s^N)(\mu)|$.

Remark. Output Bounds. We can of course translate our error bound $\Delta^N(\mu)$ into (symmetric) upper and lower bounds for s, $s_+^N(\mu) = s^N(\mu) + \Delta^N(\mu)$, $s_-^N(\mu) = s^N(\mu) - \Delta^N(\mu)$. In our earlier work [17, 23] we focus on these output bounds, $s_\pm^N$, rather that on the error bound (or bound gap), Δ_N, since for coercive problems the output bounds are in fact nonsymmetric — due to a shift which also effectively reduces the bound gap by a factor of two relative to the noncoercive case. ■

4. Error analysis

In Section 4.1 we analyze the accuracy of our reduced-basis output prediction s^N, and in Section 4.2 we consider the properties of our error estimator Δ^N. In both Section 4.1 and 4.2 we make certain hypotheses about $\beta^N(\mu)$ that we then prove — at least for the case $V^N = Y$, $W^N = W_1^N$ — in Section 4.3. Note that, for convenience of exposition, we shall not always explicitly indicate the μ dependence of all quantities.

4.1. A priori theory

We first prove that our discrete approximation is well-defined, as summarized in

Lemma 1 - *If $\beta^N(\mu) \geq \tilde{\beta}_0 > 0$, $\forall \mu \in \mathcal{D}$, then the discrete problems (18) and (19) are well-posed. If furthermore $V^N = W^N$, the minimum-residual statement is equivalent to standard Galerkin approximation: $u^N(\mu) = u^{N,Gal}(\mu)$, where $u^{N,Gal}(\mu) \in W^N$ satisfies*

$$R^{pr}(v; u^{N,Gal}(\mu); \mu) = 0, \ \forall v \in W^N. \tag{27}$$

An analogous result applies for the dual.

Proof. We consider the primal problem (18); analysis of the dual problem (19) is similar. To begin, we recall that $\rho_\ell^{V^N} \in V^N$ satisfies

$$(\rho_\ell^{V^N}, v)_Y = \ell(v), \ \forall v \in V^N.$$

It thus follows that, for any $w^N \in W^N$,

$$P_\mu^N w^N = \rho_\ell^{V^N} - T_\mu^N w^N;$$

from our minimum-residual statement (23) we then know that $u^N \in W^N$ satisfies

$$(T_\mu^N u^N, T_\mu^N v)_Y = \ell(T_\mu^N v), \ \forall v \in W^N. \tag{28}$$

We now choose $v = u^N$ in (28) and note that, since $T_\mu^N u^N$ is the supremizer over V^N associated with u^N,

$$(T_\mu^N u^N, T_\mu^N u^N)_Y = a(u^N, T_\mu^N u^N; \mu) \geq \beta^N \|u^N\|_Y \|T_\mu^N u^N\|_Y,$$

and thus

$$\|u^N\|_Y \leq \frac{1}{\beta^N}\|\ell\|_{Y'} \leq \frac{1}{\tilde{\beta}_0}\|\ell\|_{Y'}.$$

We have thus proven stability; uniqueness follows in the usual way by considering two candidate solutions.

Finally, we consider $V^N = W^N$: from standard arguments we know that, for $\beta^N(\mu) \geq \tilde{\beta}_0 > 0$, the Galerkin approximation (27) admits a unique solution $u^{N,Gal}$. But since $\|P_\mu^N u^{N,Gal}\|_Y = 0$, $u^{N,Gal}$ must be the (unique) residual minimizer, and hence $u^N = u^{N,Gal}$. ■

We can then prove that u_N, ψ_N are optimal. Indeed, we have

Lemma 2 - *If* $\beta^N(\mu) \geq \tilde{\beta}_0 > 0$, $\forall \mu \in \mathcal{D}$, *then*

$$\|u(\mu) - u^N(\mu)\|_Y \leq \left(1 + \frac{2\gamma}{\tilde{\beta}_0}\right) \min_{w^N \in W^N} \|u(\mu) - w^N\|_Y,$$

with an analogous result for the dual.

Proof. Since for any $w^N \in W^N$, $w^N - u^N$ is an element of W^N, we have from (25) that

$$\begin{aligned} &\beta^N \|w^N - u^N\|_Y \|T_\mu^N (w^N - u^N)\|_Y \\ &\qquad \leq \ a(w^N - u^N, T_\mu^N (w^N - u^N); \mu) \end{aligned}$$

$$
\begin{aligned}
&= a(w^N - u + u - u^N, T_\mu^N(w^N - u^N); \mu) \\
&\leq |a(w^N - u, T_\mu^N(w^N - u^N); \mu)| + |a(u - u^N, T_\mu^N(w^N - u^N); \mu| \\
&= |R^{pr}(T_\mu^N(w^N - u^N); w^N; \mu)| + |R^{pr}(T_\mu^N(w^N - u^N); u^N; \mu)| \\
&\leq (\|P_\mu^N w^N\|_Y + \|P_\mu^N u^N\|_Y)\ \|T_\mu^N(w^N - u^N)\|_Y \\
&\leq 2\|P_\mu^N w^N\|_Y \|T_\mu^N(w^N - u^N)\|_Y, \qquad (29)
\end{aligned}
$$

where the last three steps follow from (17), (21), and (23), respectively. We now take $v = P_\mu^N w^N \in V^N$ in (21) and apply (17) and continuity to obtain

$$\|P_\mu^N w^N\|_Y \leq \gamma \|u - w^N\|_Y, \qquad (30)$$

which then yields, with (29),

$$\|w^N - u^N\|_Y \leq \frac{2\gamma}{\beta^N} \|u - w^N\|_Y,\ \forall w^N \in W^N. \qquad (31)$$

The desired result then follows by expressing $\|u - u^N\|_Y$ as $\|u - w^N + w^N - u^N\|_Y$ and applying the triangle inequality, (31), and our hypothesis on $\beta^N(\mu)$. ■

Finally, we can prove that our output prediction is optimal in the following result:

Theorem 1 - *If $\beta^N(\mu) \geq \tilde{\beta}_0 > 0,\ \forall \mu \in \mathcal{D}$,*

$$|(s - s^N)(\mu)| \leq \gamma \|u - u^N\|_Y \|\psi - \psi^N\|_Y;$$

if furthermore $V^N \supseteq W^N$, which is always satisfied for our choices,

$$|(s - s^N)(\mu)| \leq \gamma \left(1 + \frac{4\gamma}{\tilde{\beta}_0}\right) \inf_{w^N \in W^N} \|u - w^N\|_Y \inf_{\varphi^N \in W^N} \|\psi - \varphi^N\|_Y.$$

Proof. We have that

$$
\begin{aligned}
|(s - s^N)(\mu)| &= |\ell^O(u) - \ell^O(u^N) + \ell(\psi^N) - a(u^N, \psi^N; \mu)| \\
&= |-a(u - u^N, \psi; \mu) + a(u - u^N, \psi^N; \mu)| \\
&= |a(u - u^N, \psi - \psi^N; \mu)| \qquad (32) \\
&\leq \gamma \|u - u^N\|_Y \|\psi - \psi^N\|_Y,
\end{aligned}
$$

which proves the first result.

We also know that, for all $\varphi^N \in W^N, w^N \in W^N$,

$$
\begin{aligned}
&|a(u - u^N, \psi - \psi^N; \mu)| \\
&\quad = |a(u - u^N, \psi - \varphi^N + \varphi^N - \psi^N; \mu)| \\
&\quad \leq |a(u - u^N, \psi - \varphi^N; \mu)| + |a(u - u^N, \varphi^N - \psi^N; \mu)| \\
&\quad \leq \gamma \|u - u^N\|_Y \|\psi - \varphi^N\|_Y + |R^{pr}(\varphi^N - \psi^N; u^N; \mu)| \\
&\quad \leq \gamma \left(1 + \frac{2\gamma}{\tilde{\beta}_0}\right) \|u - w^N\|_Y \|\psi - \varphi^N\|_Y \\
&\qquad + \|\varphi^N - \psi^N\|_Y \sup_{v \in V^N} \frac{R^{pr}(v; u^N; \mu)}{\|v\|_Y}, \qquad (33)
\end{aligned}
$$

where we have evoked continuity, (17), Lemma 2, and $W^N \subseteq V^N$. But from (21), (23), (30), and the dual version of (31)

$$
\begin{aligned}
\sup_{v \in V^N} \frac{R^{pr}(v; u^N; \mu)}{\|v\|_Y} \|\varphi^N - \psi^N\|_Y &\leq \|P_\mu^N u^N\|_Y \frac{2\gamma}{\tilde{\beta}_0} \|\psi - \varphi^N\|_Y \\
&\leq \gamma \|u - w^N\|_Y \frac{2\gamma}{\tilde{\beta}_0} \|\psi - \varphi^N\|_Y,
\end{aligned}
$$

which with (32) and (33) proves the second result. ■

Theorem 1 indicates in what sense reduced-basis methods yield optimal interpolations in parameter space. We could of course predict $s(\mu)$ at some new value of μ as some interpolant or fit to the $s(\mu^m), m = 1, \ldots, M$; however, it is not clear how to choose, or whether one has chosen, the best combination of the $s(\mu^m)$, in particular in higher dimensional parameter spaces. In contrast, Theorem 1 states that, by predicting $s(\mu)$ via a state space (W^N), and by ensuring stability ($\tilde{\beta}_0 > 0$ independent of N), we obtain in some sense a best approximation — the correct weights for each of the reduced-basis components. With sufficient smoothness, this best approximation will converge very rapidly with increasing N [7, 25] (see also Section 4.3). Note the importance of W_ψ^M in W^N in ensuring that $\inf_{\varphi^N \in W^N} \|\psi - \varphi^N\|_Y$ is small — had we included only W_u^M in W^N, this would not be the case, since reduced-basis spaces have no general approximation properties (that is, for arbitrary functions in Y).

Of course, Theorem 1 only tells us that we are doing as well as possible; it does not tell us *how* well we are doing — our *a posteriori* indicators are required for that purpose.

4.2. A posteriori theory

We can directly show that, under certain hypotheses, our error estimators are indeed *error bounds.*

Theorem 2 - *If $\beta^N \to \beta$ as $N \to \infty$, then there exists an $N^*(\mu)$ such that, $\forall N \geq N^*(\mu)$,*

$$|(s - s^N)(\mu)| \leq \Delta^N(\mu),$$

for $\Delta^N(\mu)$ as given in (26).

Proof. We first note, evoking symmetry and our inf-sup condition (5), that

$$\begin{aligned}
\beta(\mu)\|\psi - \psi^N\|_Y \|T_\mu(\psi - \psi^N)\|_Y &\leq a(T_\mu(\psi - \psi^N), \psi - \psi^N; \mu) \\
&= R^{du}(T_\mu(\psi - \psi^N), \psi^N; \mu) \\
&\leq \|R^{du}(\cdot; \psi^N; \mu)\|_{Y'} \|T_\mu(\psi - \psi^N)\|_Y,
\end{aligned}$$

or

$$\|\psi - \psi^N\|_Y \leq \frac{1}{\beta(\mu)} \|R^{du}(\cdot; \psi^N; \mu)\|_{Y'}.$$

We then write, from (32) of Theorem 1,

$$\begin{aligned}
|(s - s^N)(\mu)| &= |a(u - u^N, \psi - \psi^N; \mu)| \\
&= |R^{pr}(\psi - \psi^N; u^N; \mu)| \\
&\leq \|R^{pr}(\cdot; u^N; \mu)\|_{Y'} \|\psi - \psi^N\|_Y \\
&\leq \frac{1}{\beta(\mu)} \|R^{pr}(\cdot; u^N; \mu)\|_{Y'} \|R^{du}(\cdot; \psi^N; \mu)\|_{Y'}.
\end{aligned}$$

The result then directly follows: for $\sigma < 1$, we have from our hypothesis on β^N that $\sigma\beta^N(\mu) \leq \beta(\mu)$ for N sufficiently large, say $N \geq N^*(\mu)$, and thus

$$\begin{aligned}
|(s - s^N)(\mu)| &\leq \frac{1}{\beta(\mu)} \|R^{pr}(\cdot; u^N; \mu)\|_{Y'} \|R^{du}(\cdot; \psi^N; \mu)\|_{Y'} \\
&\leq \frac{1}{\sigma\beta^N(\mu)} \|R^{pr}(\cdot; u^N; \mu)\|_{Y'} \|R^{du}(\cdot; \psi^N; \mu)\|_{Y'} \\
&= \Delta^N(\mu),
\end{aligned}$$

for $N \geq N^*(\mu)$. ∎

It is not only important to determine that $\Delta^N(\mu)$ is a bound for the error, but also that it is a *good* bound. As a measure of bound quality, we introduce the usual *a posteriori* effectivity,

$$\eta^N(\mu) = \frac{\Delta^N(\mu)}{|s(\mu) - s^N(\mu)|}. \tag{34}$$

Under the hypothesis of Theorem 2 we know that $\eta^N(\mu) \geq 1$ as $N \to \infty$, providing us with the desired bounds; to ensure that the bound is tight, we would also like to verify that $\eta^N(\mu) \leq$ Const (independent of N) as $N \to \infty$. We have no proof for this result, but it is certainly plausible given the demonstration of Theorem 2, and is in fact confirmed by the numerical experiments of Section 6.2.

4.3. The discrete inf-sup parameter

It should be clear that good behavior of the discrete inf-sup parameter is the essential hypothesis of Theorems 1 and 2. If β^N vanishes, or becomes very small relative to β, Theorems 1 and 2 indicate we risk that $|(s-s^N)(\mu)|$ and Δ^N will both become very large: accuracy of our predictions thus requires $\beta^N \geq \tilde{\beta}_0 > 0$. However, too much stability is not desirable, either. If β^N is large compared to β as $N \to \infty$, Theorem 2 indicates we risk that $\Delta^N(\mu)$ will not bound $|(s - s^N)(\mu)|$: certainty in our predictions thus requires β^N close to β. It is clear that the best behavior is $\beta^N \to \beta$ from above as $N \to \infty$.

We now discuss several possible choices for V^N, W^N, and the extent to which each — either provably or intuitively — meets our desiderata.

4.3.1. The choice $V^N = Y, W^N = W_1^N$. Method 1

It is simple in this case to prove stability:

Lemma 3 - *For $V^N = Y$ (and any space $W^N \subset Y$),*

$$\beta^N(\mu) \geq \beta(\mu) \geq \beta_0 > 0,$$

for all $\mu \in \mathcal{D}$.

Proof. We have

$$\begin{aligned} \beta^N(\mu) &= \inf_{w^N \in W^N} \frac{\|T_\mu^N w^N\|_Y}{\|w^N\|_Y} = \inf_{w^N \in W^N} \frac{\|T_\mu w^N\|_Y}{\|w^N\|_Y} \\ &\geq \inf_{w \in Y} \frac{\|T_\mu w\|_Y}{\|w\|_Y} = \beta(\mu) \geq \beta_0 > 0, \end{aligned}$$

as desired. ∎

Thus, for $V^N = Y$, the hypothesis of Theorem 1 is satisfied with $\tilde{\beta}_0 = \beta_0$; we are guaranteed stability.

To ensure accuracy of the inf-sup parameter — and hence asymptotic error *bounds* from Theorem 2 — we shall first need

Lemma 4 - *If $\mathcal{S}^M$ is chosen such that*

$$\sup_{\mu \in \mathcal{D}} \inf_{m \in \{1,\ldots,M\}} \|\mu - \mu^m\| \to 0$$

as $M \to \infty$, and if $\chi(\mu)$ is sufficiently smooth in the sense that

$$\| \sup_{\mu \in \mathcal{D}} \|\nabla_\mu \chi\| \, \|_Y < \infty,$$

then

$$\inf_{w^N \in W_1^N} \|\chi(\mu) - w^N\|_Y \to 0, \; \forall \mu \in \mathcal{D}, \tag{35}$$

as M (and hence N)$\to \infty$. Note $\| \cdot \|$ refers to the usual Euclidean norm.

Proof. Recalling that $\chi(\mu)$, the infimizer, is defined in (7), we next introduce $\tilde{\chi}^N(\mu) \in W_1^N$ as

$$\tilde{\chi}^N(\mu) = \chi(\mu^{m^*(\mu)}), \; m^*(\mu) = \arg \min_{m \in \{1,\dots,M\}} |\mu - \mu^m|.$$

Thus

$$\begin{aligned}
\|\chi(\mu) - \tilde{\chi}^N(\mu)\|_Y &\le \Big(\inf_{m \in \{1,\dots,M\}} \|\mu - \mu^m\|\Big) \, \| \sup_{\mu \in \mathcal{D}} \|\nabla_\mu \chi\| \, \|_Y \\
&\le \Big(\sup_{\mu \in \mathcal{D}} \inf_{m \in \{1,\dots,M\}} \|\mu - \mu^m\|\Big) \| \sup_{\mu \in \mathcal{D}} \|\nabla_\mu \chi\| \, \|_Y, \; \forall \mu \in \mathcal{D},
\end{aligned}$$

and therefore for all $\mu \in \mathcal{D}$,

$$\begin{aligned}
\inf_{w^N \in W^N} \|\chi(\mu) - w^N\|_Y &\le \|\chi(\mu) - \tilde{\chi}^N(\mu)\|_Y \\
&\le \Big(\sup_{\mu \in \mathcal{D}} \inf_{m \in \{1,\dots,M\}} \|\mu - \mu^m\|\Big) \, \| \sup_{\mu \in \mathcal{D}} \|\nabla_\mu \chi\| \, \|_Y,
\end{aligned}$$

which tends to zero as M (and hence N) tends to infinity from our hypotheses on $\mathcal{S}^M$ and the smoothness of $\chi(\mu)$. ■

Clearly, with sufficient smoothness, we can develop higher order interpolants [25], suggesting correspondingly higher rates of convergence. For our purposes here, (35) suffices; the method itself will choose a best approximation, typically much closer to χ than our simple candidate above. The essential point is the inclusion of W_χ^M in W_1^N, which provides the necessary approximation properties within our reduced-basis space.

We can now prove that, for $V^N = Y$, $W^N = W_1^N$, $\beta^N(\mu)$ is an accurate approximation to $\beta(\mu)$.

Theorem 3 - *For $V^N = Y$, $W^N = W_1^N$,*

$$\beta^N(\mu^m) = \beta(\mu^m), \; m = 1, \dots, M. \tag{36}$$

Furthermore, under the hypotheses of Lemma 4, there exists a C independent of N and an $N^{**}(\mu)$ such that

$$|\beta(\mu) - \beta^N(\mu)| \le C \frac{\gamma^2}{2\beta(\mu)} \inf_{w^N \in W^N} \|\chi(\mu) - w^N\|_Y^2, \; \forall N \ge N^{**}, \tag{37}$$

and thus from Lemma 4,

$$\beta^N(\mu) \to \beta(\mu) \text{ as } N \to \infty, \ \forall \mu \in \mathcal{D}. \tag{38}$$

Proof. To prove (36), we note that, since $\chi(\mu^m) \in W^N$,

$$\beta^N(\mu^m) = \inf_{w^N \in W^N} \frac{\|T_{\mu^m} w^N\|_Y}{\|w^N\|_Y} \leq \frac{\|T_{\mu^m}\chi(\mu^m)\|_Y}{\|\chi(\mu^m)\|_Y} = \beta(\mu^m);$$

but $\beta^N(\mu^m) \geq \beta(\mu^m)$ from Lemma 3, and thus $\beta^N(\mu^m) = \beta(\mu^m)$.

To prove (37), we introduce the discrete eigenproblem analogous to (8): find $(\theta^N, \lambda^N) \in (W_1^N \times I\!R)$ such that

$$\mathcal{A}(\theta^N(\mu), v; \mu) = \lambda^N(\mu)(\theta^N(\mu), v)_Y, \ \forall v \in W_1^N, \quad \|\theta^N(\mu)\|_Y = 1;$$

by arguments identical to those of Section 2.3 it is simple to show that $\beta^N(\mu) = \sqrt{\lambda^N_{\min}(\mu)}$. We can now apply the standard theory for Galerkin approximation of symmetric positive-definite eigenproblems. To wit, from Theorem 9 of [3] and (35) of our Lemma 4, there exists an $N^{**}(\mu)$ such that, $\forall N \geq N^{**}(\mu)$,

$$|\lambda_{\min}(\mu) - \lambda^N_{\min}(\mu)| \leq C\mathcal{A}(\theta_{\min} - w^N, \theta_{\min} - w^N; \mu), \ \forall w^N \in W^N,$$

for C independent of N. (One can in fact show that C may be taken as $(1 + 2\beta^3)^2$.)

But from (6) and (9) of Section 2.3, we know that

$$\mathcal{A}(\theta_{\min} - w^N, \theta_{\min} - w^N; \mu) = \|T_\mu(\theta_{\min} - w^N)\|_Y^2 \leq \gamma^2 \|\theta_{\min} - w^N\|_Y^2.$$

The result (37) then follows by recalling that $\theta_{\min} = \chi$, $(\beta^N)^2 = \lambda^N_{\min}$, $(\beta)^2 = \lambda_{\min}$, and noting that

$$|(\beta^N)^2 - \beta^2| = |(\beta^N - \beta)|\,|(\beta^N + \beta)| \geq |(\beta^N - \beta)| 2\beta$$

since $\beta^N \geq \beta$ from Lemma 3. ■

The hypothesis of Theorem 2 is thus verified for the case $V^N = Y$, $W^N = W_1^N$. The quadratic convergence of β^N is very important: it suggests an accurate prediction for β — and hence bounds — even if W^N is rather marginal.

4.3.2. The Choice $V^N = Y$, $W^N = W_0^N$ — Method 2

In this case the $\chi(\mu^m), m = 1, \ldots, M$, are no longer members of W^N. We see that Lemma 3 still obtains, and thus the method is stable — in fact, always *at least as stable* as $W^N = W_1^N$. Furthermore, since W_0^N still contains W_u^M and W_ψ^M, we expect $\|u - u^N\|_Y$ and $\|\psi - \psi^N\|_Y$ to be small, and hence from Theorem 1 $|(s - s^N)(\mu)|$ should also be small. There is no difficulty at the level of *stability* or *accuracy of our output.*

However, Lemma 4 can no longer be proven. Thus not only is (36) of Theorem 3 obviously not applicable, but — and even more importantly — (38) no longer obtains: we can not expect $\beta^N(\mu)$ to tend to $\beta(\mu)$ as $N \to \infty$. In short, the scheme may be too stable, β^N may be too large, and hence for any fixed $\sigma < 1$ we may not obtain bounds even as $N \to \infty$. In short, in contrast to the choice $W^N = W_1^N$, the choice $W^N = W_0^N$ no longer ensures that $\beta^N(\mu)$ is sufficiently accurate.

In practice, however, $\beta^N(\mu)$ may be sufficiently close to $\beta(\mu)$ that $\sigma\beta^N(\mu) \leq \beta(\mu)$ for some suitably small σ. To understand why, we observe that, in terms of our eigenpairs (Υ_i, ω_i) of Section 2.3,

$$u(\mu^m) = \sum_{i=1}^{\mathcal{N}} \frac{\ell(\Upsilon_i)}{\omega_i(\mu^m) - 1} \Upsilon_i. \tag{39}$$

For "generic" ℓ, $u(\mu^m)$ will thus contain a significant component of $\Upsilon_{i^*(\mu)}$ and hence $\chi(\mu^m)$. It is possible to construct ℓ such that $\ell(\Upsilon_{i^*}) = 0$, and hence we cannot in general *count on* $\chi(\mu^m)$ being predominantly present in W_0^N; however, in practice, ℓ will typically be broadband, and thus $W = W_0^N$ may sometimes be sufficient. Obviously, for greater certainty that our error bound is, indeed, a bound, $W = W_1^N$ is unambiguously preferred over $W = W_0^N$.

4.3.3. The Choice $V^N = W_1^N$, $W^N = W_1^N$ — Method 3

We know from Lemma 1 that this case corresponds to Galerkin approximation, but with W_χ^M present in our spaces. We first note that not only does Lemma 3 not apply, but unfortunately we can prove that $\beta^N(\mu^m) \leq \beta(\mu^m)$, $m = 1, \ldots, M$:

$$\begin{aligned} \beta(\mu^m) &= \sup_{v \in Y} \frac{a(\chi(\mu^m), v; \mu^m)}{\|\chi(\mu^m)\|_Y \|v\|_Y} \geq \sup_{v \in W_1^N} \frac{a(\chi(\mu^m), v; \mu^m)}{\|\chi(\mu^m)\|_Y \|v\|_Y} \\ &\geq \inf_{w \in W_1^N} \sup_{v \in W_1^N} \frac{a(w, v; \mu^m)}{\|w\|_Y \|v\|_Y} = \beta^N(\mu^m), \end{aligned} \tag{40}$$

since $\chi(\mu^m) \in W_1^N \subset Y$. Stability and accuracy of the output could thus be an issue, though not necessarily so if $\beta^N(\mu)$ is close to $\beta(\mu)$. As regards the accuracy of $\beta^N(\mu)$, Lemma 4 still applies, however (36), (37) (and hence (38)) of Theorem 3 can no longer be readily proven.

Nevertheless, in practice, β^N may be quite close to β. To understand why, we recall from Section 2.3 that $\chi(\mu^m)$ is not only our infimizer, but also proportional to $T_{\mu^m}\chi(\mu^m)$. It follows that *if* $\chi(\mu^m)$ is the most dangerous mode in the sense that

$$\sup_{v\in W_1^N} \frac{a(\chi(\mu^m), v; \mu^m)}{\|\chi(\mu^m)\|_Y \|v\|_Y} \leq \sup_{v\in W_1^N} \frac{a(w, v; \mu^m)}{\|w\|_Y \|v\|_Y}, \ \forall w \in W_1^N, \tag{41}$$

then

$$\beta^N(\mu^m) = \sup_{v\in W_1^N} \frac{a(\chi(\mu^m), v; \mu^m)}{\|\chi(\mu^m)\|_Y \|v\|_Y} \geq \frac{a(\chi(\mu^m), T_{\mu^m}\chi(\mu^m); \mu^m)}{\|\chi(\mu^m)\|_Y \|T_{\mu^m}\chi(\mu^m)\|_Y} = \beta(\mu^m),$$

since both $\chi(\mu^m)$ and $T_{\mu^m}(\chi(\mu^m))$ are in W_1^N; note that (41) is a conjecture, since the supremizing space here is W_1^N, *not* Y as in Section 2.3. Under our assumption (41), we thus conclude from (40) that

$$\beta^N(\mu^m) = \beta(\mu^m). \tag{42}$$

By similar arguments we might expect $\beta^N(\mu)$ to be quite accurate even for general $\mu \in \mathcal{D}$, as both $\chi(\mu)$ and $T_\mu\chi$ are well represented in the basis. (From this discussion we infer that for *nonsymmetric* problems a Petrov-Galerkin formulation is desirable.)

The above arguments are clearly speculative. In order to more rigorously guide our choice of V^N, we can prove an illustrative relationship between the Galerkin $V^N = W_1^N$, $W^N = W_1^N$ (superscript "*Gal*") and minimum residual $V^N = Y$, $W^N = W_1^N$ (superscript "*MR*") approximations:

Theorem 4 - *For all* $\mu \in \mathcal{D}$,

$$\Delta^{N,MR}(\mu) \leq \Delta^{N,Gal}(\mu), \tag{43}$$

where $\Delta^{N,MR}$ *and* $\Delta^{N,Gal}$ *refer to (26) for the minimum-residual and Galerkin cases, respectively.*

Proof. We first note that

$$\beta^{N,Gal} = \inf_{w\in W_1^N} \sup_{v\in W_1^N} \frac{a(w, v; \mu)}{\|w\|_Y \|v\|_Y} \leq \inf_{w\in W_1^N} \sup_{v\in Y} \frac{a(w, v; \mu)}{\|w\|_Y \|v\|_Y} = \beta^{N,MR}, , \tag{44}$$

for all $\mu \in \mathcal{D}$. We then note from (23) that

$$\begin{aligned}
\Delta^{N,MR} &\equiv \frac{1}{\sigma\beta^{N,MR}} \|R^{pr}(\cdot, u^{N,MR}; \mu)\|_{Y'} \|R^{du}(\cdot, \psi^{N,MR}; \mu)\|_{Y'} \\
&= \frac{1}{\sigma\beta^{N,MR}} \|P_\mu^N u^{N,MR}\|_Y \|D_\mu^N \psi^{N,MR}\|_Y \\
&\leq \frac{1}{\sigma\beta^{N,MR}} \|P_\mu^N w^N\|_Y \|D_\mu^N \varphi^N\|_Y, \ \forall w^N \in W_1^N, \ \forall \varphi^N \in W_1^N,
\end{aligned}$$

where $P_\mu^N\colon W_1^N \to Y$ and $D_\mu^N\colon W_1^N \to Y$ are here defined for $V^N = Y$. Thus

$$\begin{aligned}
\Delta^{N,MR} &\leq \frac{1}{\sigma\beta^{N,MR}}\|P_\mu^N u^{N,Gal}\|_Y\|D_\mu^N\psi^{N,Gal}\|_Y \\
&= \frac{1}{\sigma\beta^{N,MR}}\|R^{pr}(\cdot,u^{N,Gal};\mu)\|_{Y'}\|R^{du}(\cdot,\psi^{N,Gal};\mu)\|_{Y'} \\
&\leq \frac{1}{\sigma\beta^{N,Gal}}\|R^{pr}(\cdot,u^{N,Gal};\mu)\|_{Y'}\|R^{du}(\cdot,\psi^{N,Gal};\mu)\|_{Y'} = \Delta^{N,Gal},
\end{aligned}$$

where the last step follows from (44). ■

We thus see that, in general, $V^N = Y$ will provide sharper error estimates: minimum residual is in fact equivalent to minimum error bound. Conversely, we might expect the Galerkin approximation to be more conservative, providing bounds even when the minimum-residual approach may not (e.g. for N very small).

The Galerkin method with $W^N = W_1^N$ thus has some redeeming features. However, there is the possibility of a loss of accuracy in both s^N and Δ^N, reflected in (43) and (44) of Theorem 4.

4.3.4. The Choice $V^N = W_0^N$, $W^N = W_0^N$ — Method 4

This case corresponds to Galerkin approximation, but now with W_χ^M absent. Here (40) no longer applies: $\beta^N(\mu)$ may be greater or less than $\beta(\mu)$. Furthermore, accuracy of $\beta^N(\mu)$ now relies on two fortuitous events — the "selective amplification" of (39) and the "most dangerous mode" of (41). Again, in practice, the method may perform well, but it is now more likely that either β^N will be too small and hence $(s - s^N)(\mu)$ and Δ^N too large, or β^N will be too large and hence $\eta^N < 1$ (no bounds).

We are able to prove a result analogous to Theorem 4, but now comparing $V^N = Y$, $W^N = W_0^N$ to $V^N = W_0^N$, $W^N = W_0^N$: the β^N (respectively Δ^N) for the former will be larger (respectively smaller) than the corresponding quantities for the latter. We thus expect that $V^N = W_0^N$, $W^N = W_0^N$ will yield rather poor accuracy.

5. Computational procedure

For clarity we shall present the detail for the most "rigorous" and general scheme, $V^N = Y$, $W^N = W_1^N$ (Method 1 of Section 4.3.1). As we proceed, we indicate simplifications for the other schemes, and at the conclusion we give a comparison of computational complexity.

5.1. An algebraic representation

5.1.1. Preliminaries

We assume here that Y is finite dimensional, with associated basis $\xi_i, i = 1, \dots, \mathcal{N}$. We also recall that W^N can be expressed as span$\{\zeta_i, i = 1, \dots, N\}$; we implicitly assume independence of the reduced-basis functions. A member $w \in Y$ is expressed as $\underline{w}^t\xi$, $\underline{w} \in I\!R^{\mathcal{N}}$; a member $w \in W^N$ is expressed as $\underline{w}^t\zeta$, $\underline{w} \in I\!R^N$. Here t denotes the transpose.

We next introduce the matrices $\underline{A}^{Y,Y} \in I\!R^{\mathcal{N}\times\mathcal{N}}$, $\underline{A}^{W^N,W^N} \in I\!R^{N\times N}$, $\underline{A}^{Y,W^N} \in I\!R^{\mathcal{N}\times N}$, $\underline{B}^{Y,Y} \in I\!R^{\mathcal{N}\times\mathcal{N}}$, $\underline{B}^{W^N,W^N} \in I\!R^{N\times N}$, given by

$$
\begin{aligned}
A^{Y,Y}_{i,j}(\mu) &= a(\xi_j, \xi_i; \mu), 1 \le i,j \le \mathcal{N},\\
A^{W^N,W^N}_{i,j}(\mu) &= a(\zeta_j, \zeta_i; \mu), 1 \le i,j \le N,\\
A^{Y,W^N}_{i,j}(\mu) &= a(\zeta_j, \xi_i; \mu), 1 \le i \le \mathcal{N}, 1 \le j \le N,\\
B^{Y,Y}_{i,j}(\mu) &= (\xi_j, \xi_i)_Y, 1 \le i,j \le \mathcal{N},\\
B^{W^N,W^N}_{i,j}(\mu) &= (\zeta_j, \zeta_i)_Y, 1 \le i,j \le N.
\end{aligned}
$$

From these matrices we can derive four further matrices, $\underline{Z}^{Y,Y} \in I\!R^{\mathcal{N}\times\mathcal{N}}$, $\underline{Z}^{W^N,Y} \in I\!R^{N\times\mathcal{N}}$, $\underline{S}^{Y,Y} \in I\!R^{\mathcal{N}\times\mathcal{N}}$, $\underline{S}^{W^N,W^N} \in I\!R^{N\times N}$, as

$$
\begin{aligned}
\underline{Z}^{Y,Y}(\mu) &= \underline{A}^{Y,Y}(\mu)(\underline{B}^{Y,Y})^{-1}\\
\underline{Z}^{W^N,Y}(\mu) &= (\underline{A}^{Y,W^N}(\mu))^t(\underline{B}^{Y,Y})^{-1}\\
\underline{S}^{Y,Y}(\mu) &= (\underline{A}^{Y,Y}(\mu))^t(\underline{B}^{Y,Y})^{-1}\underline{A}^{Y,Y}(\mu)\\
\underline{S}^{W^N,W^N}(\mu) &= (\underline{A}^{Y,W^N}(\mu))^t(\underline{B}^{Y,Y})^{-1}\underline{A}^{Y,W^N}(\mu),
\end{aligned}
$$

where ^{t} denotes matrix transpose.

These matrices are representations — in terms of our bases — of the operators introduced earlier. For example, if $w \in Y$ and $v \in Y$ are expressed as $\underline{w}^t\xi$ and $\underline{v}^t\xi$, then $\underline{w}^t\underline{B}^{Y,Y}\,\underline{v}$ is $(w,v)_Y$; $((\underline{Z})^{Y,Y}(\mu))^t$ is our representation of T_μ; and for $w \in Y$ expressed as $\underline{w}^t\xi$, $\underline{w}^t\ \underline{S}^{Y,Y}(\mu)\underline{w}$ is $(T_\mu w, T_\mu w)_Y$. It follows that

$$
(\beta(\mu))^2 = \min_{\underline{w}\in I\!R^{\mathcal{N}}} \frac{\underline{w}^t\underline{S}^{Y,Y}(\mu)\underline{w}}{\underline{w}^t\underline{B}^{Y,Y}\underline{w}}. \tag{45}
$$

Similarly, for $w \in W^N$ expressed as $\underline{w}^t\zeta$, $\underline{w}^t\ \underline{S}^{W^N,W^N}\underline{w}$ is $(T^N_\mu w, T^N_\mu w)$, and

$$
(\beta^N(\mu))^2 = \min_{\underline{w}\in I\!R^{N}} \frac{\underline{w}^t\underline{S}^{W^N,W^N}(\mu)\underline{w}}{\underline{w}^t\underline{B}^{W^N,W^N}\underline{w}}. \tag{46}
$$

Note $\underline{S}^{W^N,W^N}(\mu)$ represents the normal equations associated with the least-squares approach, and β^N is the "generalized" smallest singular value of $\underline{A}^{Y,W^N}$; see [5] for an earlier discussion of singular values and stability in the reduced-basis context.

Finally, we shall need the vectors $\underline{L}^{pr,Y} \in I\!R^{\mathcal{N}}, \underline{L}^{pr,W^N} \in I\!R^N, \underline{L}^{du,Y} \in I\!R^{\mathcal{N}}, \underline{L}^{du,W^N} \in I\!R^N$, defined by

$$\begin{aligned} L_i^{pr,Y} &= \ell(\xi_i), 1 \leq i \leq \mathcal{N}, \\ L_i^{pr,W^N} &= \ell(\zeta_i), 1 \leq i \leq N, \\ L_i^{du,Y} &= \ell^O(\xi_i), 1 \leq i \leq \mathcal{N}, \\ L_i^{du,W^N} &= \ell^O(\zeta_i), 1 \leq i \leq N, \end{aligned}$$

which are the obvious representations of our primal and dual linear functionals.

5.1.2. Reduced basis

We first find, for $m = 1, \ldots, M$, $\underline{u}^m \in I\!R^{\mathcal{N}}, \underline{\psi}^m \in I\!R^{\mathcal{N}}$, solution of

$$\underline{A}^{Y,Y}(\mu^m)\underline{u}^m = \underline{L}^{pr,Y} \quad (\text{or } \underline{S}^{Y,Y}(\mu^m)\underline{u}^m = \underline{Z}^{Y,Y}\underline{L}^{pr,Y}), \tag{47}$$
$$\underline{A}^{Y,Y}(\mu^m)\underline{\psi}^m = -\underline{L}^{du,Y} \quad (\text{or } \underline{S}^{Y,Y}(\mu^m)\underline{\psi}^m = -\underline{Z}^{Y,Y}\underline{L}^{du,Y}). \tag{48}$$

We further obtain $(\underline{\chi}^m, \kappa^m_{\min})$ as the first eigenpair $(\underline{\theta}, \kappa) \in (I\!R^N \times I\!R)$ of the symmetric positive-definite problem

$$\underline{S}^{Y,Y}(\mu^m)\underline{\theta} = \kappa \underline{B}^{Y,Y}\underline{\theta}, \quad \underline{\theta}^t \underline{B}^{Y,Y}\underline{\theta} = 1. \tag{49}$$

(Note that in some cases it may be preferable to find $\underline{\chi}^m$ by considering the $(\cdot,\cdot)_Y - \mathcal{B}(\cdot,\cdot;\mu)$ eigenproblem of Section 2.3; an inverse iteration with shift (of unity, initially) may work well.)

It can then readily be shown that $u(\mu^m), \psi(\mu^m), \chi(\mu^m)$, $m = 1, \ldots, M$, of (12), (13), and (14) are given by

$$u(\mu^m) = \sum_{j=1}^{\mathcal{N}} u_j^m \xi_j \equiv (\underline{u}^m)^t \underline{\xi},$$

$$\psi(\mu^m) = \sum_{j=1}^{\mathcal{N}} \psi_j^m \xi_j \equiv (\underline{\psi}^m)^t \underline{\xi},$$

$$\chi(\mu^m) = \sum_{j=1}^{\mathcal{N}} \chi_j^m \xi_j \equiv (\underline{\chi}^m)^t \underline{\xi},$$

where this last result can be readily motivated from (45); furthermore, $\kappa^m_{\min} = (\beta(\mu^m))^2$, though we shall not have direct need of this result in the construction of W^N. Note that for $W^N = W^N_0$ (Methods 2 and 4) we may omit (49); this constitutes significant "off-line" savings (see Section 5.2.2 below).

5.1.3. Output prediction

We first find, for given $\mu \in \mathcal{D}$, $\underline{u}^N(\mu) \in \mathbb{R}^N, \underline{\psi}^N(\mu) \in \mathbb{R}^N$, solution of

$$\begin{aligned} \underline{S}^{W^N,W^N}(\mu)\underline{u}^N(\mu) &= \underline{Z}^{W^N,Y}\underline{L}^{pr,Y} \\ \underline{S}^{W^N,W^N}(\mu)\underline{\psi}^N(\mu) &= -\underline{Z}^{W^N,Y}\underline{L}^{du,Y}; \end{aligned} \tag{50}$$

it is readily shown that $u^N(\mu)$ and $\psi^N(\mu)$ of (18) and (19) of Section 3 are given by

$$\begin{aligned} u^N(\mu) &= \sum_{j=1}^{N} u_j^N(\mu)\zeta_j \\ \psi^N(\mu) &= \sum_{j=1}^{N} \psi_j^N(\mu)\zeta_j. \end{aligned}$$

We can then evaluate s^N of (20) as

$$s^N(\mu) = (\underline{u}^N(\mu))^t \underline{L}^{du,W^N} - (\underline{\psi}^N(\mu))^t(\underline{L}^{pr,W^N} - \underline{A}^{W^N,W^N}(\mu)\underline{u}^N(\mu)). \tag{51}$$

Note that for the Galerkin formulations, $V^N = W^N$, we may replace (50) with $\underline{A}^{W^N,W^N}\underline{u}^N = \underline{L}^{pr,W^N}$; but since we will need $\underline{S}^{W^N,W^N}$ in the error prediction step below, this is not a significant simplification.

5.1.4. Error bound prediction

We first calculate $\beta^N(\mu) = \sqrt{\kappa^N_{\min}(\mu)}$, where $\kappa^N_{\min}(\mu)$ is the eigenvalue associated with the first eigenpair $(\underline{\theta}^N(\mu), \kappa^N(\mu)) \in (\mathbb{R}^N \times \mathbb{R})$ of

$$\underline{S}^{W^N,W^N}(\mu)\underline{\theta}^N(\mu) = \kappa^N(\mu)\underline{B}^{W^N,W^N}\underline{\theta}^N(\mu), \quad (\underline{\theta}^N(\mu))^t\underline{B}^{W^N,W^N}\underline{\theta}^N(\mu) = 1,$$

as motivated by (46) above. Note in this "integrated formulation" that the *same* reduced-basis matrix, $\underline{S}^{W^N,W^N}$, serves to determine u^N, ψ^N, and β^N.

We next compute $\underline{\tau}^{pr}(\mu) \in \mathbb{R}^{\mathcal{N}}, \underline{\tau}^{du}(\mu) \in \mathbb{R}^{\mathcal{N}}$, solution of

$$\underline{B}^{Y,Y}\underline{\tau}^{pr}(\mu) = \underline{L}^{pr,Y} - \underline{A}^{Y,W^N}(\mu)\underline{u}^N(\mu) \tag{52}$$

$$\underline{B}^{Y,Y}\underline{\tau}^{du}(\mu) = -\underline{L}^{du,Y} - \underline{A}^{Y,W^N}(\mu)\underline{\psi}^N(\mu). \tag{53}$$

It can be readily shown that $P_\mu^N u^N(\mu), D_\mu^N \psi^N(\mu)$ as defined by (21) and (22) of Section 3 (for $V^N = Y$) are given by

$$\begin{aligned} P_\mu^N u^N(\mu) &= \sum_{j=1}^{\mathcal{N}} \tau_j^{pr}(\mu)\xi_j \equiv (\underline{\tau}^{pr}(\mu))^t \underline{\xi} \\ D_\mu^N \psi^N(\mu) &= \sum_{j=1}^{\mathcal{N}} \tau_j^{du}(\mu)\xi_j \equiv (\underline{\tau}^{du}(\mu))^t \underline{\xi}, \end{aligned}$$

respectively. Note that these calculations (52), (53) are required even for the Galerkin approach: we must compute the Y' norm of the residual to estimate the error.

Lastly, it then follows that $\Delta^N(\mu)$ of (26) can be evaluated as

$$\Delta^N(\mu) = \frac{1}{\sigma\beta^N}((\underline{\tau}^{pr}(\mu))^t \underline{B}^{Y,Y} \underline{\tau}^{pr}(\mu))^{1/2}((\underline{\tau}^{du}(\mu))^t \underline{B}^{Y,Y} \underline{\tau}^{du}(\mu))^{1/2}, \quad (54)$$

which completes the procedure.

5.2. Blackbox approach

5.2.1. Preliminaries

The blackbox approach comprises two separate stages [12, 22]. The first stage, or *off-line* stage, or pre-processing stage, invokes various solutions and evaluations associated with the space Y; it will typically be expensive, since for many problems $\mathcal{N}$ is very large. The quantities produced by this off-line stage are passed to the second stage, or *on-line* stage, which serves to predict $s^N(\mu)$ and $\Delta^N(\mu)$ for any new value of μ; this stage invokes only low-dimensional entities — that is, the computational complexity for each new value of μ is *independent* of $\mathcal{N}$, and depends only on N, the dimension of the reduced-basis space, and Q (defined by (4)), a measure of the "complexity" of the parametric dependence. Our "affine" assumption of (4) plays a critical role in permitting this two-stage decomposition and associated linear superposition.

It follows that for N and Q small, as is often the case, the on-line reduced-basis predictions for $s^N(\mu)$ will be much less costly than direct appeal to $u(\mu)$ in (1) and subsequently $s(\mu)$ in (2). (In fact, very often $\underline{A}^{Y,Y}$ has structure and sparsity not shared by $\underline{S}^{W^N,W^N}$, which will typically be dense; nevertheless, for $N \ll \mathcal{N}$, as is usually the case, the on-line reduced-basis prediction is still much faster than direct appeal to $u(\mu)$ and $s(\mu)$.) Reduced-basis methods are thus very advantageous in situations such as control, in which real-time response is required, and in situations such as design or optimization, in which many parameter values are visited (such that the off-line effort can be amortized). Critical to this argument is the ability to assess the error in the reduced-basis predictions — through $\Delta^N(\mu)$ — so that the minimal number of modes N may be (*safely*) retained.

To describe the blackbox procedure, and demonstrate the $\mathcal{N}$-independence of the on-line stage, we shall need a few additional definitions. First, we recognize that the $\zeta_i, i = 1, \ldots, N$, can be represented in terms of the ξ_j, which we express as

$$\zeta_i = \underline{z}_i^t \underline{\xi},$$

where $\underline{z}_i \in I\!R^{\mathcal{N}}, i = 1, \ldots, N$. Second, we need to introduce the matrices $\underline{A}_q^{Y,Y} \in I\!R^{\mathcal{N} \times \mathcal{N}}, q = 1, \ldots, Q$, given by

$$(\underline{A}_q^{Y,Y})_{i,j} = a_q(\xi_j, \xi_i), 1 \leq i, j \leq \mathcal{N},$$

where the $a_q(\cdot, \cdot)$ are defined (implicitly) in (4).

We shall summarize the computational effort in the *off-line* stage in terms of $\underline{A}^{Y,Y}$-solves, $\underline{S}^{Y,Y}$-eigensolves, $\underline{B}^{Y,Y}$-solves, $\underline{A}^{Y,Y}$-actions (matrix vector products), $\underline{B}^{Y,Y}$-actions, and Y-inner products (inner products between two $\mathcal{N}$-vectors). Note for simplicity we assume that $\underline{A}_q^{Y,Y}$-actions are roughly equivalent to $\underline{A}^{Y,Y}$-actions. In many problems of interest, in particular in which there is underlying sparsity in $\underline{A}^{Y,Y}$, $\underline{S}$-eigensolves will be the most expensive, then $\underline{A}$-solves, then $\underline{B}^{Y,Y}$-solves (less costly because the equations are symmetric positive-definite), and then the "actions" (often only $O(\mathcal{N})$) and Y-inner products.

We shall summarize the *on-line* computational effort directly in terms of N and Q (albeit somewhat imprecisely, sometimes considering a multiplication and an addition as a single operation). Note that in the on-line stage we are not compelled to exploit all N basis functions computed in the off-line stage, and thus N in the on-line stage may be replaced with $N_{used}(\mu)$, with the error bound $\Delta^N(\mu)$ guiding the choice of minimal N_{used} [26]; this can significantly reduce the cost of the on-line predictions.

As regards storage, we shall report, in the off-line stage, both Temporary Storage (required just during the off-line stage) and Permanent Storage (quantities passed by the off-line stage to the on-line stage). All quantities stored in the on-line stage originate in the off-line stage. (We note that the on-line code is completely segregated from the original solution method on Y — the link is only the relatively few stored quantities — and furthermore comprises only very elementary (and very few) operations that can be simply implemented and shared.) The simplifications to the procedure in the case of "compliance" should be clear.

5.2.2. Off-line stage

1. Compute reduced-basis vectors: $\underline{u}^m \in \mathbb{R}^{\mathcal{N}}, \underline{\psi}^m \in \mathbb{R}^{\mathcal{N}}, \underline{\chi}^m \in \mathbb{R}^{\mathcal{N}}, m = 1, \ldots, M$, from (47), (48), and (49). Recall that the $\underline{\chi}^m$ are not needed for $W^N = W_0^N$.

Complexity: $2M\ \underline{A}^{Y,Y}$-solves, $M\ \underline{S}^{Y,Y}$-eigensolves.

2. Compute $\underline{B}^{W^N,W^N} \in \mathbb{R}^{N \times N}$ as

$$B_{i,j}^{W^N,W^N} = (\zeta_j, \zeta_i)_Y,$$

or

$$B_{i,j}^{W^N,W^N} = \underline{z}_i^t \underline{B}^{Y,Y} \underline{z}_j.$$

Complexity: $N \underline{B}^{Y,Y}$-actions, N^2 Y-inner products.
Temporary Storage: $N\mathcal{N}$.
Permanent Storage: N^2.

3. Compute $\underline{\mathcal{U}}_{qi} \in \mathbb{R}^{\mathcal{N}}, 1 \le q \le Q, 1 \le i \le N$, where

$$(\mathcal{U}_{qi})_k = a_q(\zeta_i, \xi_k), 1 \le k \le \mathcal{N},$$

or

$$\underline{\mathcal{U}}_{qi} = \underline{A}_q^{Y,Y} \underline{z}_i.$$

Complexity: $NQ\ \underline{A}^{Y,Y}$-actions.
Temporary Storage: $NQ\mathcal{N}$.

4. Compute $\underline{\mathcal{V}}_{qi} \in \mathbb{R}^{\mathcal{N}}, 1 \le q \le Q, 1 \le i \le N$, and $\underline{\mathcal{V}}_0^{pr} \in \mathbb{R}^{\mathcal{N}}, \underline{\mathcal{V}}_0^{du} \in \mathbb{R}^{\mathcal{N}}$, solutions of

$$\underline{B}^{Y,Y} \underline{\mathcal{V}}_{qi} = \underline{\mathcal{U}}_{qi},\ \underline{B}^{Y,Y} \underline{\mathcal{V}}_0^{pr} = \underline{L}^{pr,Y},\ \underline{B}^{Y,Y} \underline{\mathcal{V}}_0^{du} = \underline{L}^{du,Y}.$$

Complexity: $(NQ + 2)\ \underline{B}^{Y,Y}$-solves.
Temporary Storage: $(NQ + 2)\mathcal{N}$.

5. Compute $\Gamma_{qq'ii'} (= \Gamma_{q'qi'i}), 1 \le q, q' \le Q, 1 \le i, i' \le N$, as

$$\Gamma_{qq'ii'} = \underline{\mathcal{U}}_{qi}^t \underline{\mathcal{V}}_{q'i'}.$$

Complexity: N^2Q^2 Y-inner products.
Permanent Storage: N^2Q^2.

6. Compute $\Lambda_{qi}^{pr}, \Lambda_{qi}^{du}, 1 \leq q \leq Q, 1 \leq i \leq N$, as

$$\Lambda_{qi}^{pr} = \underline{\mathcal{U}}_{qi}^t \underline{\mathcal{V}}_0^{pr}, \Lambda_{qi}^{du} = \underline{\mathcal{U}}_{qi}^t \underline{\mathcal{V}}_0^{du}.$$

Complexity: $2NQ$ Y-inner products.
Permanent Storage: $2NQ$.

7. Compute $c_0^{pr} \in \mathbb{R}$, $c_0^{du} \in \mathbb{R}$, as

$$c_0^{pr} = (\underline{L}^{pr,Y})^t \underline{\mathcal{V}}_0^{pr}, \; c_0^{du} = (\underline{L}^{du,Y})^t \underline{\mathcal{V}}_0^{du}.$$

Complexity: 2 Y-inner products.
Permanent Storage: 2.

8. Compute $\underline{L}^{pr,W^N} \in \mathbb{R}^N, \underline{L}^{du,W^N} \in \mathbb{R}^N$, as

$$L_i^{pr,W^N} = (\underline{L}^{pr,Y})^t \underline{z}_i, \; L_i^{du,W^N} = (\underline{L}^{du,Y})^t \underline{z}_i, \; i = 1, \ldots, N.$$

Complexity: $2N$ Y-inner products.
Permanent Storage: $2N$.

9. Compute $\Xi_{qii'}, 1 \leq q \leq Q, 1 \leq i \leq N$, as

$$\begin{aligned} \Xi_{qii'} &= a_q(\zeta_i, \zeta_{i'}) \\ &= \underline{z}_{i'}^t \underline{\mathcal{U}}_{qi}. \end{aligned}$$

Complexity: N^2Q Y-inner products.
Permanent Storage: N^2Q.

5.2.3. On-line stage

Given a new value of the parameter $\mu \in \mathcal{D}$:

1. Form $\underline{S}^{W^N,W^N}(\mu) \in \mathbb{R}^{N \times N}$ as

$$S_{i,i'}^{W^N,W^N} = \sum_{q=1}^{Q} \sum_{q'=1}^{Q} \alpha_q(\mu) \alpha_{q'}(\mu) \Gamma_{qq'ii'}, \; 1 \leq i, i' \leq N.$$

Complexity: N^2Q^2.

2. Form the necessary "right-hand" sides:

$$\begin{aligned} \sum_{j=1}^{\mathcal{N}} Z_{i,j}^{W^N,Y} L_j^{pr,Y} &= \sum_{q=1}^{Q} \alpha_q(\mu) \Lambda_{qi}^{pr}, \; 1 \leq i \leq N, \\ \sum_{j=1}^{\mathcal{N}} Z_{i,j}^{W^N,Y} L_j^{du,Y} &= \sum_{q=1}^{Q} \alpha_q(\mu) \Lambda_{qi}^{du}, \; 1 \leq i \leq N. \end{aligned}$$

Complexity: $2NQ$.

3. Find $\underline{u}^N(\mu) \in I\!R^N, \underline{\psi}^N(\mu) \in I\!R^N, \beta^N(\mu) \in I\!R$ solution of (47), (48), and (50).

Complexity: $O(N^3)$.

4. Compute $s^N(\mu)$ of (51) as

$$s^N(\mu) = (\underline{L}^{du,W^N})^t \underline{u}^N(\mu) - (\underline{L}^{pr,W^N})^t \underline{\psi}^N(\mu) + \sum_{q=1}^{Q}\sum_{i=1}^{N}\sum_{i'=1}^{N} \alpha_q(\mu) u_i^N(\mu) \psi_{i'}^N(\mu) \Xi_{qii'} .$$

Complexity: $2N + N^2 Q$.

5. Compute $\Delta^N(\mu)$ of (54) as

$$\begin{aligned}\Delta^N(\mu) &= \frac{1}{\sigma\beta^N(\mu)}\Big[c_0^{pr} - 2\sum_{q=1}^{Q}\sum_{i=1}^{N} \alpha_q(\mu) u_i^N(\mu) \Lambda_{qi}^{pr} \\ &\quad + \sum_{q=1}^{Q}\sum_{q'=1}^{Q}\sum_{i=1}^{N}\sum_{i'=1}^{N} \alpha_q(\mu)\alpha_{q'}(\mu) u_i^N(\mu) u_{i'}^N(\mu) \Gamma_{qq'ii'}\Big]^{1/2} \times \\ &\quad \Big[c_0^{du} + 2\sum_{q=1}^{Q}\sum_{i=1}^{N} \alpha_q(\mu) \psi_i^N(\mu) \Lambda_{qi}^{du} \\ &\quad + \sum_{q=1}^{Q}\sum_{q'=1}^{Q}\sum_{i=1}^{N}\sum_{i'=1}^{N} \alpha_q(\mu)\alpha_{q'}(\mu) \psi_i^N(\mu) \psi_{i'}^N(\mu) \Gamma_{qq'ii'}\Big]^{1/2} .\end{aligned}$$

Complexity: $2(N^2Q^2 + NQ + 1)$.

We now briefly discuss the computational complexity of the different schemes. The first comparison is between minimum residual ($V^N = Y$) and Galerkin ($V^N = W^N$) approaches. The important point to note is that the quantity — $\Gamma_{qq'ii'}$ — required by Method 1 ($V^N = Y$, $W^N = W_1^N$) to form the projection matrix $\underline{S}^{W^N,W^N}(\mu)$ is the same quantity required by *all* the methods to compute the error bound $\Delta^N(\mu)$; in both capacities, $\Gamma_{qq'ii'}$ represents the calculation of the necessary Y' norm. In this sense (see Theorem 4) the arguably better scheme $V^N = Y$, and somewhat riskier scheme $V^N = W^N$, have *similar complexity*, and we contend that $V^N = Y$ is thus preferred. (We note, however, that faster blackbox schemes and related non-blackbox variants *require* a Galerkin approach, and we thus retain Galerkin projection as an important alternative.) The second comparison is between W_0^N and W_1^N. For the on-line component, *the difference is not large* — $N = 3M$ *vs.* $N = 2M$; however, for the off-line component, the calculations of $\chi(\mu^m)$ can indeed be onerous, and its omission thus welcome. However, there is a corresponding rather significant loss of security, since the accuracy of β^N is *no longer controlled* (See Sections 4.3.2 and 4.3.4).

6. The Helmholtz problem

6.1. Formulation

We take here $Y = H_0^1(\Omega)$, where Ω is a suitably smooth domain in $\mathbb{R}^d$, $d = 1, 2$, or 3, and $H_0^1(\Omega)$ is the space of functions which are square integrable over Ω, which have square integrable (distributional) first derivatives over Ω, and which vanish on the boundary of Ω, $\partial\Omega$. For our inner product we take

$$(w, v)_Y = \int_\Omega \nabla w \cdot \nabla v + wv \ ,$$

which induces the norm

$$\|w\|_Y = (\int_\Omega \nabla w \cdot \nabla w + ww)^{1/2}.$$

It is important to remark that we may substitute for $(\cdot,\cdot)_Y$ any inner product which induces a norm equivalent to the $H^1(\Omega)$-norm — for example, a good preconditioner. The latter will of course greatly reduce the off-line computational cost, as $\underline{B}^{Y,Y}$-solves will now be much less expensive.

For our bilinear form we take

$$a(w, v; \mu) = \int_\Omega \nabla w \cdot \nabla v - g(x; \mu) wv \ ,$$

where we assume that $g(x; \mu)$ satisfies

$$|g(x; \mu)| \leq g_{\max}, \ \ \forall x \in \Omega, \ \forall \mu \in \mathcal{D},$$

and furthermore can be expressed as

$$g(x; \mu) = \sum_{q=1}^{\overline{Q}} F_q(\mu) G_q(x), \tag{55}$$

where $G_q \in L^\infty(\Omega), q = 1, \ldots, \overline{Q}$. The difficult case, on which we focus here, is of course when $g(x;\mu)$ is positive, as in the reduced-wave (Helmholtz) equation.

The decomposition (55) is, in fact, reasonably general. We shall consider the situation in which $P = 2$ with parameter $\mu = (k_1, k_2)$, $\overline{Q} = 2$, $F_1(\mu) = k_1^2$, $F_2(\mu) = k_2^2$, and

$$G_1(x) = \begin{cases} 1 & x \in \Omega_1 \\ 0 & x \in \Omega_2 \end{cases},$$

$$G_2(x) = \begin{cases} 0 & x \in \Omega_1 \\ 1 & x \in \Omega_2 \end{cases},$$

where $\overline{\Omega} = \overline{\Omega}_1 \cup \overline{\Omega}_2$; this represents variable "frequencies" in two subdomains. It can be shown that the regularity of χ required in Lemma 4 follows from the smoothness of the $F_q(\mu)$ and the interpretation of χ as $\theta_{\min}$ of (8).

It is simple to see that our requirement (4) is readily satisfied for $Q = 3$: we choose $\alpha_q(\mu) = F_q(\mu), q = 1,\ 2$, and $\alpha_3(\mu) = 1$, with $a_q(w,v) = -\int_\Omega G_q(x) wv$, $q = 1,\ 2$, and $a_3(w,v) = \int_\Omega \nabla w \cdot \nabla v$. It is similarly easy to show that a is symmetric, and also uniformly continuous with (say) $\gamma = 1 + g_{\max}$. The inf-sup condition will be satisfied so long as we exclude from $\mathcal{D}$ neighborhoods of points μ for which there exists a w such that $a(w, v; \mu) = 0,\ \forall v \in Y$. In general, if the inf-sup condition (3) is thus satisfied, and ℓ and ℓ^O are any bounded linear functionals, then our theoretical results of Section 4 will obtain.

We make two points of a more practical nature. First, in practice, we will of course not know where resonances occur, and thus we will typically posit a parameter domain which does indeed contain several points at which the inf-sup condition does not hold. However, unless driven to such a point by a design or optimization process, it is unlikely that a particular μ will coincide exactly with an eigenvalue, and thus for some sufficiently small β_0 our hypotheses will be "in practice" satisfied. (Obviously the physical model may also be made more elaborate, for example by including damping that will regularize the resonances.) Second, in practice, we choose not $Y = H_0^1(\Omega)$, but rather $Y = X_{\mathcal{N}}$, a suitably fine (say finite element) approximation of finite (albeit very large) dimension $\mathcal{N}$. As we are more and more conservative in defining this "truth" approximation, that is, as we increase $\mathcal{N}$, the off-line computational effort will of course increase; however, thanks to the blackbox formulation, the on-line computational effort is *independent* of the dimension $\mathcal{N}$.

6.2. Numerical Results

We take here $d = 1$ and $\Omega =]0, 1[$ (though obviously the computational savings provided by the reduced-basis approach will only be realized for more complicated multidimensional ($d > 1$) problems). Our truth space $X_{\mathcal{N}}$ is a linear finite element approximation with 200 elements. We consider the two-parameter Helmholtz equation defined in Section 6.1, with $\Omega_1 =]0, 0.5[$ and $\Omega_2 =]0.5, 1.0[$. For simplicity, we present a "compliance" case in which

$$\ell(v) = \ell^O(v) = \int_{0.45}^{0.55} v,$$

corresponding to an imposed (oscillatory) distributed force for the input and an associated average displacement amplitude measurement for the output.

In the below we shall consider the four methods associated with the four choices of spaces of Sections 4.3.1, 4.3.2, 4.3.3, and 4.3.4: Method 1 refers to the choice of Section 4.3.1, $V^N = Y, W^N = W_1^N$, and will be denoted by the line pattern (—) in all plots; Method 2 refers to the choice of Section 4.3.2, $V^N = Y, W^N = W_0^N$, and will be denoted by the line pattern (-·-); Method 3 refers to the choice of Section 4.3.3, $V^N = W_1^N, W^N = W_1^N$, and will be denoted by the line pattern (- -); and Method 4 refers to the choice of Section 4.3.4, $V^N = W_0^N, W^N = W_0^N$, and will be denoted by the line pattern (···). In general, subscripts, unless otherwise indicated, refer to the Method index above. Throughout this section we take $\sigma = (1.1)^{-1}$: it follows from Theorem 2 that a sufficient (though not necessary) condition for bounds is that β^N be within 10% of β. For most of the results of this section, we choose an effectively one–dimensional parameter space $\mathcal{D}$ which is the subset of $\mathcal{D}' =]11, 11[\times]1, 20[$ in which neighborhoods of the two resonance points $\mu \equiv (k_1, k_2) = (11, 7.5)$ and $\mu \equiv (k_1, k_2) = (11, 14.4)$ have been excised such that $\beta_0 = 0.02$. (Of course, in practice, we would not know the location of these resonance points, and we would thus consider $\mathcal{D} = \mathcal{D}'$ — which would only satisfy our inf-sup stability condition, "in practice," as discussed in the previous section.)

To begin, we fix $M = 3$, and hence $N = 2M = 6$ since we are in "compliance," with $\mu^1 = (11, 2)$, $\mu^2 = (11, 8)$ and thus $\mu^3 = (11, 14)$, and thus $\mathcal{S}^M = \{\mu^1, \mu^2, \mu^3\}$; we shall denote this the "$M = 3$" case. We first investigate the behavior of the discrete inf-sup parameter, the accuracy of which is critical for both the accuracy and bounding properties of our output prediction. In Figures 1 and 2 we plot the discrete inf-sup parameter β_i^N, $i = 1, \ldots, 4$, and the ratio β_i^N/β, $i = 1, \ldots, 4$, respectively, as a function of k_2 for fixed k_1 (see Section 6.1); recall that the index i refers to the method under consideration. We first confirm those aspects of the behavior that we have previously demonstrated. First, β_1^N and β_2^N are never less than β^N, as shown in Lemma 3 and Section 4.3.2, respectively; and $\beta_2^N \geq \beta_1^N$, as must be the case since the inf space is smaller. The choice $V^N = Y$ ensures stability. Second, we see that $\beta_1^N \geq \beta_3^N$ and $\beta_2^N \geq \beta_4^N$, as demonstrated in

(44) and Section 4.3.4 respectively; the methods with smaller supremizing spaces are perforce less stable. Third, we see (by closer inspection of the numerical values) that β_3^N is never greater than β at the sample points, consistent with (40); in fact, we observe that equality obtains at the sample points, (42), and hence at least in this particular case the conjecture (41) appears valid. Fourth, we notice that β_4^N can be either below or above β, and is clearly the least "controlled" of the four approximations. (Indeed, for other parameter values we observe near zero values of β_4^N at points quite far away from the true resonances of the system.)

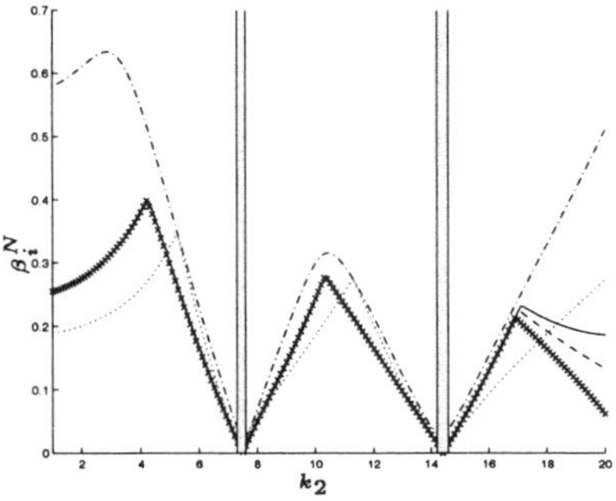

Figure 1: The discrete inf-sup parameter for Methods 1, 2, 3, and 4 as a function of k_2 (see text for legend). The symbol $\times$ denotes the exact value of β.

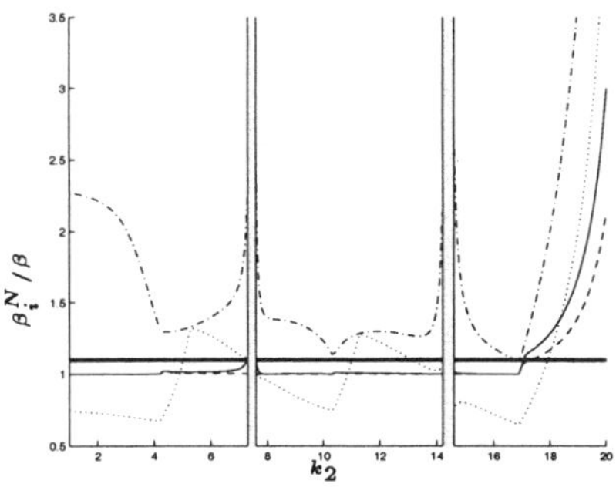

Figure 2: The ratio of the discrete inf-sup parameter to the exact inf–sup parameter for Methods 1, 2, 3, and 4, as a function of k_2 (see text for legend). The thick line denotes the "sufficient" limit: if $\beta^N < 1.1\beta$, bounds are guaranteed.

It is clear from Figures 1 and 2 that β_i^N is indeed a very accurate predictor of β over most of $\mathcal{D}$ for Methods 1 and 3; we anticipated this result in Theorem 3 and the discussion of Section 4.3.3. We now study the convergence of β_i^N to β as N increases. For this test we consider a sample $\mathcal{S}^M = \{\mu^m, m = 1, \ldots, M\}$, with the μ^m randomly drawn from $\mathcal{D}$; the particular parameter points selected are given in the second column of Table 1. (Note that for a given M, indicated in the first column of Table 1, $\mathcal{S}^M$ consists of all $\mu^m, m = 1, \ldots, M$.) We present in Table 1 the values of $\beta_i^N - \beta$ for Methods i =1, 2, 3, and 4 for $k_2 = 11$ (and hence $\mu = (11, 11)$). We note that, indeed, β_i^N converges very rapidly for $i = 1$ and $i = 3$ — the two methods in which we include the infimizers in V^N — whereas for $i = 2$ and $i = 4$ we do not obtain convergence — not surprising given the discussion

Table 1: The error $\beta_i^N - \beta$ for Methods i =1, 2, 3, and 4, for $k_2 = 11$, as a function of M.

		$\beta_i^N - \beta$			
M	μ^M	$i=1$	$i=2$	$i=3$	$i=4$
1	(11, 4.7351)	$1.81e-01$	$2.92e+00$	$-1.88e-01$	$1.08e+00$
2	(11,19.0928)	$1.70e-01$	$4.28e-01$	$-2.03e-01$	$4.14e-01$
3	(11,11.4848)	$3.52e-04$	$2.76e-01$	$-1.24e-04$	$-9.85e-02$
4	(11,13.6038)	$9.25e-06$	$5.76e-02$	$3.99e-06$	$5.31e-02$
5	(11, 1.4975)	$6.57e-09$	$4.91e-02$	$2.43e-09$	$4.13e-02$
6	(11, 2.6998)	$1.90e-11$	$4.81e-02$	$4.49e-08$	$3.83e-02$

of Section 4.3 . Note also that whereas the convergence of Method 1 is (and must be) monotonic, this is not necessarily the case for Method 3.

We conclude that Method 2 and in particular Method 4 are not very reliable: we can certainly not guarantee asymptotic bounds for any given fixed $\sigma < 1$; for this reason we do not recommend these techniques, and we focus primarily on Methods 1 and 3 in the remainder of this section. However, in practice, all four methods may perform reasonably well for some smaller σ, in particular since the accuracy of the inf-sup parameter is only a sufficient and not a necessary condition for bounds. Indeed, for our $M = 3$ case of Figures 1 and 2, Methods 1, 2, and 4 produce bounds for all k_2 less than approximately 18 and Method 3 in fact produces bounds for all k_2 in $\mathcal{D}$; consistent with Theorem 2, bounds are always obtained for all methods so long as $\sigma\beta^N \leq \beta$. The breakdown of bounds for Method 1 (which in fact directly correlates with $\sigma\beta_1^N > \beta$) is due to the poor infimizer approximation properties of W_1^N for larger k_2; if we include an additional sample point, $\mu^4 = (11, 20)$, we recover bounds for all $\mathcal{D}$. (In fact, even for lower k_2 the infimizer approximation is not overly good; but thanks to the quadratic convergence proven in Theorem 3 the inf-sup parameter remains quite accurate.)

The fact that Method 3 produces bounds over the entire range is consistent with the "less stable" arguments of Section 4.3.3. However, by these same arguments, in particular Theorem 4, we expect that the bound gap — the controllable error in the output prediction — will be larger for Method 3 than for Method 1. To demonstrate this empirically, we plot in Figure 3 $\Delta_i^N/|s|, i = 1$ and $i = 3$, as a function of k_2, for the $M = 3$ case of Figures 1 and 2. We observe that, indeed, the bound gap is significantly smaller for Method 1 than for Method 3. Note also that the normalized bound gap is quite large for the k_2 at which we no longer obtain bounds for Method 1; no doubt these predictions would be rejected as overly inaccurate and

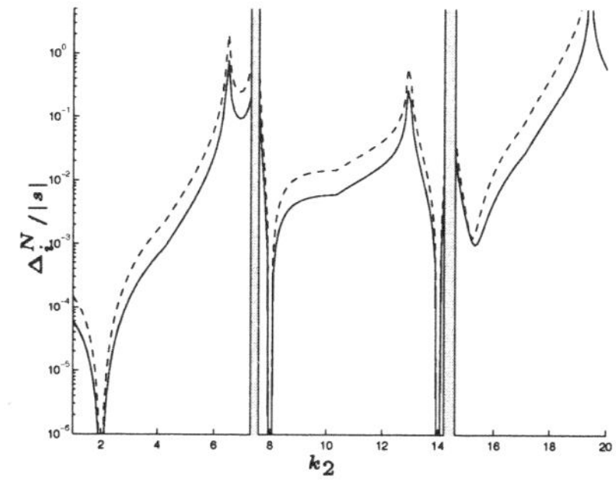

Figure 3: The normalized bound gap $\Delta_i^N/|s|$ for Methods i =1 and i =3 as a function of k_2.

requiring further expansion of the reduced–basis space (thus also recovering the inf-sup parameter accuracy and hence bounds).

As regards the convergence of the bound gap, we present in Table 2 convergence results for $\Delta_i^N, i = 1$ and $i = 3$, for $k_2 = 11$ (and hence $\mu = (11, 11)$), as a function of M (analogous to Table 1 for the inf-sup parameter). (Note for Methods 2 and 4 the convergence is slower, with the bound gap typically an order of magnitude *larger* than for Methods 1 and 3; this suggests that the inclusion of the infimizers can, in fact, reduce the approximation error — as might be anticipated from (39).) We observe that the differences between Methods 1 and 3 become smaller as M increases; however it is precisely for smaller M that reduced-basis methods are most interesting. We conclude — given that the two methods are of comparable cost - that Method 1 is perhaps preferred, in particular because we can also better guarantee the behavior of the inf–sup parameter. Note that the difference in the true error for Methods 1 and 3 is much smaller than the difference in the error bound for the two methods; this is expected, since the inf-sup parameters do not differ appreciably. It follows that the effectivity (defined in (34)) of Method 1 is lower (and hence better) than the effectivity of Method 3; this is not surprising, since for Method 1 the approximation is *designed* to minimize the bound gap.

We close by considering a second set of numerical results included to demonstrate the rapid convergence of the reduced–basis prediction as N increases even in higher dimensional parameter spaces: we now consider $\mathcal{D} =]1, 20[\times]1, 20[$ (without excision of resonances, and hence satisfying our inf-sup condition only "in practice"). In particular we repeat, the convergence

Table 2: The bound gap for Methods i =1 and $i=3$, for $k_2 = 11$, as a function of M.

M	μ^M	Δ_i^N $i=1$	Δ_i^N $i=3$
1	(11, 4.7351)	$2.23e-04$	$1.73e-01$
2	(11, 19.0928)	$2.13e-04$	$6.21e-01$
3	(11, 11.4848)	$5.37e-06$	$2.68e-05$
4	(11, 13.6038)	$4.01e-08$	$5.80e-08$
5	(11, 1.4975)	$6.43e-11$	$6.50e-11$
6	(11, 2.6998)	$1.65e-14$	$1.62e-14$

Table 3: The bound gap and effectivity at $\mu = (11, 17)$, as a function of M, for Methods i =1 and $i=3$, for the two–dimensional parameter space $\mathcal{D} =]1,20[\times]1,20[$.

M	μ^M	Δ_1^N	η_1^N	Δ_3^N	η_3^N
1	(7.5388, 14.2564)	$1.30e-04$	24.20	$6.09e-04$	12.18
2	(2.9571, 7.1526)	$1.22e-04$	3.51	$3.05e-04$	11.88
3	(4.2387, 9.0533)	$5.82e-05$	1.08	$8.77e-05$	1.49
4	(17.7486, 15.9503)	$1.10e-05$	2.56	$1.24e-05$	3.42
5	(9.7456, 14.0523)	$2.99e-08$	3.47	$3.07e-08$	3.71
6	(3.8279, 16.3388)	$2.81e-08$	3.71	$2.88e-08$	3.97
7	(11.2113, 8.0970)	$5.40e-12$	3.78	$5.44e-12$	3.85

scenario of Table 2, but now choose our random sample from the two–dimensional space $\mathcal{D} =]1,20[\times]1,20[$; we present, in Table 3, the bound gap and effectivity (defined in (34)) for Methods 1 and 3 for a particular "new" parameter point $\mu = (11, 17)$. We observe, first, that we obtain bounds in all cases ($\eta^N \geq 1$) — indicative of an accurate inf-sup parameter prediction; second, that the error (true and estimated) tends to zero very rapidly with increasing M, even in this two–dimensional parameter space; and third, that Method 1 again provides smaller bound gaps (and lower effectivities) than Method 3, consistent with Theorem 4 — though the difference is only significant for very small M. Note that u and the output s are order 10^{-3}, so the relative errors are roughly 1000 times larger than the absolute errors in the table. Results similar to those reported in Table 3 are also obtained if we consider the error over a random ensemble of test points μ rather than a single test point.

Acknowledgements. We would like to thank Dr. Luc Machiels of

Lawrence Livermore National Laboratories and Prof. Jaume Peraire of M.I.T. for numerous helpful discussions, and for suggestions leading to simpler and more illustrative proofs. This work was supported by the Singapore-MIT Alliance, by AFOSR Grant F49620-97-1-0052, and by NASA Grant NAG1-1978.

References

[1] M. Ainsworth and J. T. Oden. A posteriori error estimation in finite element analysis. *Comp. Meth. Appl. Mech. Engrg.*, 142:1–88, 1997.

[2] B. O. Almroth, P. Stern, and F. A. Brogan. Automatic choice of global shape functions in structural analysis. *AIAA Journal*, 16:525–528, May 1978.

[3] I. Babuska and J. Osborn. Eigenvalue problems. In *Handbook of numerical analysis*, volume II, pages 641–787. Elsevier, 1991.

[4] R. E. Bank and A. Weiser. Some a posteriori error estimators for elliptic partial differential equations. *Math. Comput.*, 44(170):283–301, 1985.

[5] A. Barret and G. Reddien. On the reduced basis method. *Z. Angew. Math. Mech.*, 75:543–549, 1995.

[6] R. Becker and R. Rannacher. A feedback approach to error control in finite element method: Basic analysis and examples. *East - West J. Numer. Math.*, 4:237–264, 1996.

[7] J. P. Fink and W. C. Rheinboldt. On the error behaviour of the reduced basis technique for nonlinear finite element approximations. *Z. Angew. Math. Mech.*, 63:21–28, 1983.

[8] M. B. Giles and N. A. Pierce. Superconvergent lift estimates through adjoint error analysis. Technical report, Oxford University Computing Laboratory, 1998.

[9] M. D. Gunzburger. *Finite element methods for viscous incompressible flows*. Academic Press, 1989.

[10] P. Ladeveze and D. Leguillon. Error estimation procedures in the finite element method and applications. *SIAM J. Numer. Anal.*, 20:485–509, 1983.

[11] L. Machiels. Output bounds for iterative solutions of linear partial differential equations. Preprint.

[12] L. Machiels, Y. Maday, I. B. Oliveira, A. T. Patera, and D. V. Rovas. Output bounds for reduced-basis approximations of symmetric positive definite eigenvalue problems. *C. R. Acad. Sci. Paris, Série I*, to appear.

[13] L. Machiels, Y. Maday, and A. T. Patera. A "flux-free" nodal Neumann

subproblem approach to output bounds for partial differential equations. *C. R. Acad. Sci. Paris, Série I*, 330(3):249–254, Feb 1 2000.

[14] L. Machiels, Y. Maday, and A. T. Patera. Output bounds for reduced-order approximations of elliptic partial differential equations. *Comp. Meth. Appl. Mech. Engrg.*, to appear.

[15] L. Machiels, A. T. Patera, J. Peraire, and Y. Maday. A general framework for finite element a posteriori error control: Application to linear and nonlinear convection-dominated problems. In *ICFD Conference on numerical methods for fluid dynamics*, Oxford, England, 1998.

[16] L. Machiels, J. Peraire, and A. T. Patera. A posteriori finite element output bounds for the incompressible navier-stokes equations; application to a natural convection problem. *Journal of Computational Physics*, submitted.

[17] Y. Maday, L. Machiels, A. T. Patera, and D. V. Rovas. Blackbox reduced-basis output bound methods for shape optimization. In *Proceedings* 12th *International Domain Decomposition Conference*, Japan, 2000. To appear.

[18] Y. Maday, A. T. Patera, and J. Peraire. A general formulation for a posteriori bounds for output functionals of partial differential equations; application to the eigenvalue problem. *C. R. Acad. Sci. Paris, Série I*, 328:823–828, 1999.

[19] D. A. Nagy. Modal representation of geometrically nonlinear behaviour by the finite element method. *Computers and Structures*, 10:683–688, 1979.

[20] A. K. Noor and J. M. Peters. Reduced basis technique for nonlinear analysis of structures. *AIAA Journal*, 18(4):455–462, April 1980.

[21] M. Paraschivoiu and A. T. Patera. A hierarchical duality approach to bounds for the outputs of partial differential equations. *Comp. Meth. Appl. Mech. Engrg.*, 158(3-4):389–407, June 1998.

[22] M. Paraschivoiu, J. Peraire, Y. Maday, and A. T. Patera. Fast bounds for outputs of partial differential equations. In J. Borgaard, J. Burns, E. Cliff, and S. Schreck, editors, *Computational methods for optimal design and control*, pages 323–360. Birkhäuser, 1998.

[23] A. T. Patera and E. M. Rønquist. A general output bound result: application to discretization and iteration error estimation and control. *Math. Models Methods Appl. Sci.*, 2000. To appear.

[24] J. S. Peterson. The reduced basis method for incompressible viscous flow calculations. *SIAM J. Sci. Stat. Comput.*, 10(4):777–786, July 1989.

[25] T. A. Porsching. Estimation of the error in the reduced basis method solution of nonlinear equations. *Mathematics of Computation*, 45(172):487–496, October 1985.

[26] R. von Kaenel. Reduced basis methods and output bounds for partial differential equations. Dipl. thesis, MIT, EPFL, 2000.

Yvon Maday
Laboratoire Jacques-Louis Lions
Université Pierre et Marie Curie
Boîte courrier 187
75252 Paris Cedex 05
France
E-mail: maday@ann.jussieu.fr

Anthony T. Patera
Massachusetts Institute of Technology
Department of Mechanical Engineering
Room: 3-266
Cambridge, MA 02139-4307
U.S.A
E-mail: patera@mit.edu

Dimitrios V. Rovas
Massachusetts Institute of Technology
Department of Mechanical Engineering
Room: 3-264
Cambridge, MA 02139-4307
U.S.A
E-mail: rovas@mit.edu

Studies in Mathematics and its Applications, Vol. 31
D. Cioranescu and J.L. Lions (Editors)

Chapter 25

SIMULATION OF FLOW IN A GLASS TANK

V. NEFEDOV AND R. M. M. MATTHEIJ

1. Introduction

The manufacturing of glass is a complicated and expensive process. The glass is produced in a so-called glass oven or tank. The raw materials such as soda and sand are dumped on one side of the oven, which is heated from above by gas burners. Ovens are constructed in such a way that the glass stays inside for some time, and after about 20 hours flows via feeders to the production lines on the other side. There are, therefore, several processes involved, such as flow and heat transfer and various chemical reactions. In order to study how different oven configurations affect the production, numerical simulation is required, as full scale experimental studies are too expensive to carry out.

Since the complexity of the problem is quite appalling, existing codes for tackling this problem are in for improvement and further sophistication, like adaptivity in gridding, to make the simulations more feasible (with respect to memory and computing time). In [6] a local refinement procedure was combined with a glass oven model. The method proposed was based on staggered grid finite volumes [7]. In this paper we study a different way to simulate the glas flow, based on locally uniform grids. The next section describes the mathematical model, namely the equations we are dealing with, the boundary conditions and the specific physical parameters. In the third section we discuss the discretisation procedure. In section 4 the solution method is described, including some implementation details. Section 5 deals with a local refinement technique called local defect correction (LDC). Its application to the glass flow is discussed in the two last sections, viz. a stirred flow in section 6 and bubbling in section 7.

2. Mathematical model

The oven in fact consists of an oven proper (melting tank) and a second part, connected by a small channel, from which the glass is going to a feeder channel for further processing, leading to the actual product eventually, see Fig.1.

The mathematical model consists of equations, describing physical processes occuring in the glass flow, complemented by suitable boundary conditions. In order to restrict this still general model to our particular case of interest we perform a dimension analysis based on glass glass flow properties.

2.1. Main equations

Since melted glass can be considered a viscous Newtonian fluid, we can model it by the (incompressible) Navier-Stokes equations

$$\begin{cases} (\rho(\mathbf{x})\mathbf{u}, \nabla\mathbf{u}) = \mathbf{F} - \nabla p + \nabla \cdot (\mu(\mathbf{x})\nabla\mathbf{u}), \\ \nabla \cdot (\rho(\mathbf{x})\mathbf{u}) = 0. \end{cases} \tag{1}$$

Here μ, ρ are the viscosity and the density of the glass respectively, and $\mathbf{F} = (0, 0, -\rho g)^T$ is a gravitational force. Since the glass flow exibits steady state behaviour for large time scales, these equations are written in a time independent form.

The motion of the glass is caused by temperature differences; thus the set of equations is incomplete without the energy equation. For a steady low-velocity flow with negligible dissipation we can derive the energy equation in the following form

$$\nabla \cdot (C_p \rho(\mathbf{x})\mathbf{u}T) = \nabla \cdot (k(\mathbf{x})\nabla T), \tag{2}$$

where k and C_p are the conductivity and the heat capacity, respectively.

2.2. Boundary conditions

A typical glass oven configuration is depicted in Figure 1. We can define the boundary values for this geometry as follows: the velocity has Dirichlet values on the inflow, outflow and, assuming no-slip, on the walls too. The top layer is modelled as a symmetry plane, that is the normal component is zero; for the rest we prescribe homogeneous Neumann boundary conditions.

For the temperature the situation is different, since the heat loss from the walls and heat influx from the top layer are not known a priori, but can be expressed via a Robin-type boundary condition. In particular we can write

$$-(\mathbf{u}, \mathbf{n})|_{\Gamma_{inflow}} = u_0,$$
$$-(\mathbf{u}, \mathbf{n})|_{\Gamma_{outflow}} = u_1,$$

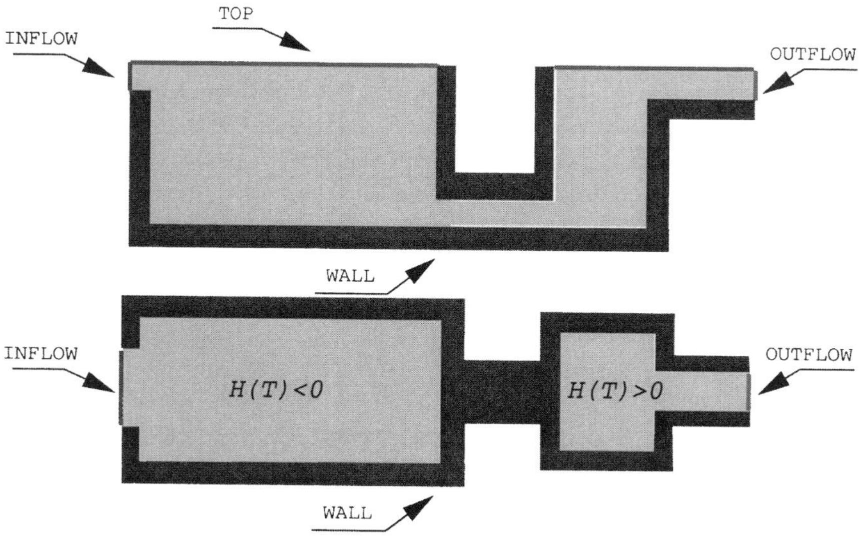

Figure 1: Sketch of the glass tank (horizontal and vertical cross sections)

$$
\begin{gathered}
(\mathbf{u},\tau)|_{\Gamma_{inflow}} = (\mathbf{u},\tau)|_{\Gamma_{outflow}} = 0,\\
\mathbf{u}|_{\Gamma_{w,g}} = 0,\\
(\mathbf{u},\mathbf{n})|_{\Gamma_{top}} = 0, \quad \frac{\partial(\mathbf{u},\tau)}{\partial \mathbf{n}}|_{\Gamma_{top}} = 0,\\
T|_{\Gamma_{inflow}} = T_0, \quad \frac{\partial T}{\partial \mathbf{n}}|_{\Gamma_{outflow}} = 0,\\
\frac{\partial T}{\partial \mathbf{n}}|_{\Gamma_{top}} = H(T), \quad \frac{\partial T}{\partial \mathbf{n}}|_{\Gamma_{wall}} = Q(T).
\end{gathered}
$$

2.3. Physical parameters

The above described equations, together with the boundary conditions, are still too general, since they might as well represent any viscous flow in a tank. In order to resctrict the model to the particular case of the glass flow we are interested in, we need to specify the parameters of the equation, in particualr the viscosity, the density, the heat conductivity and the heat capacity.

The most important of these coefficients are the viscosity and the density. The viscosity does significantly affect the flow pattern; it decays exponentially as the temperature grows (Vogel Fucher Tamman law)

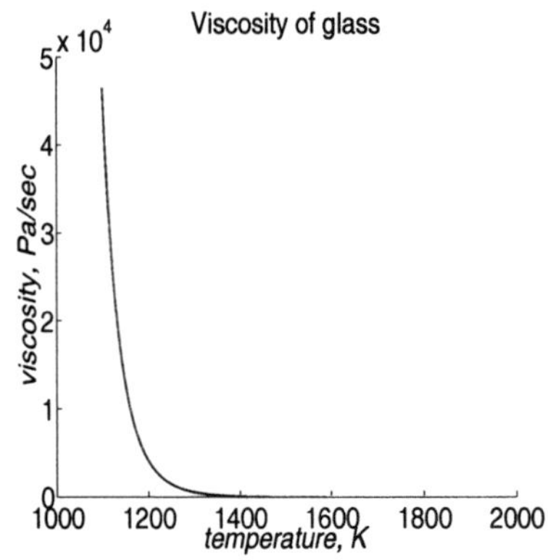

Figure 2: Viscosity of tv-glass

$$\mu_{glass}(\mathbf{x}) = \mu_{glass}(T(\mathbf{x}) = a_\mu e^{b_\mu/(T-c_\mu)}. \tag{3}$$

The coefficients a_μ, b_μ, c_μ in (3) are specific for the glass type (tv-glass, window-glass etc).

The most significant factor in a glass flow computation is the density. It may be modelled as a linear function of the temperature, more precisely

$$\rho_{glass}(T) = a_\rho(1 - b_\rho(T - c_\rho)), \tag{4}$$

and changes by only about 10%; nevertheless it drives the flow via the convective term and the gravitation.

2.4. Dimensional analysis

In order to examine the behaviour of the momentum and the energy equations, in particular to determine whether diffusion or convection prevails, we make the equations dimensionless. Let us rewrite them as new variables

$$\tilde{\mathbf{x}} := \frac{1}{X}\mathbf{x}, \quad \tilde{\mathbf{u}} := \frac{1}{U}\mathbf{u}, \quad \tilde{T} := \frac{T - T_{min}}{\Delta T}.$$

Here X, U are characteristic length and velocity respectively and $\Delta T = T_{max} - T_{min}$ is a maximum possible temperature difference. The gradient and the divergence operators in the old and the new variables are related as

$$\tilde{\nabla} := \frac{1}{X}\nabla, \quad \tilde{\nabla}\cdot := \frac{1}{X}\cdot\nabla.$$

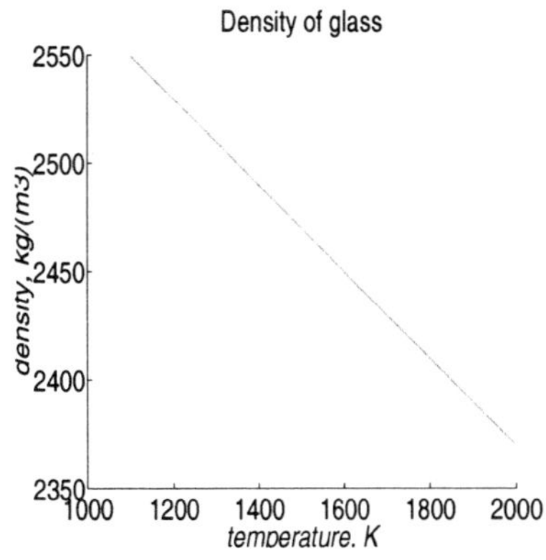

Figure 3: Density of tv-glass

A tilde over the gradient and the divergence thus indicates that it is taken with respect to the new variables. After subsitution the Navier-Stokes and the energy equations will look like

$$\begin{cases} \dfrac{U^2}{X}(\rho\tilde{\mathbf{u}}, \tilde{\nabla}\tilde{\mathbf{u}}) = F - \dfrac{1}{X}\tilde{\nabla}p + \dfrac{U}{X^2}\tilde{\nabla}\cdot(\mu\tilde{\nabla}\tilde{\mathbf{u}}), \\ \tilde{\nabla}\cdot(\rho\tilde{\mathbf{u}}) = 0, \end{cases} \tag{5}$$

$$\frac{\Delta T U}{X}C_p(\rho\tilde{\mathbf{u}}, \tilde{\nabla}\tilde{T}) = \frac{\Delta T}{X^2}\tilde{\nabla}\cdot(k\tilde{\nabla}\tilde{T}). \tag{6}$$

Since the density ρ changes by only about 10% we assume ρ to be constant. After dividing both parts of the first equation of (5) by $\rho U^2/X$ we obtain an equation in dimensionless variables

$$(\tilde{\mathbf{u}}, \tilde{\nabla}\tilde{\mathbf{u}}) = \tilde{F} - \tilde{\nabla}\tilde{p} + \tilde{\nabla}\cdot(\frac{1}{Re}\tilde{\nabla}\tilde{\mathbf{u}}).$$

Here

$$\tilde{F} = \frac{X}{\rho U^2}F, \quad \tilde{p} = \frac{1}{\rho U^2}p.$$

By analogy we divide both parts of (6) by $\Delta T U C_p/X$ to obtain

$$(\tilde{\mathbf{u}}, \tilde{\nabla}\tilde{T}) = \tilde{\nabla}\cdot(\frac{1}{Re\cdot Pr}\tilde{\nabla}\tilde{T}).$$

Here the *Reynolds* (Re) and *Prandtl* (Pe) numbers are defined by

$$Re := \frac{\rho U X}{\mu}, \quad Pr := \frac{\mu C_p}{k}$$

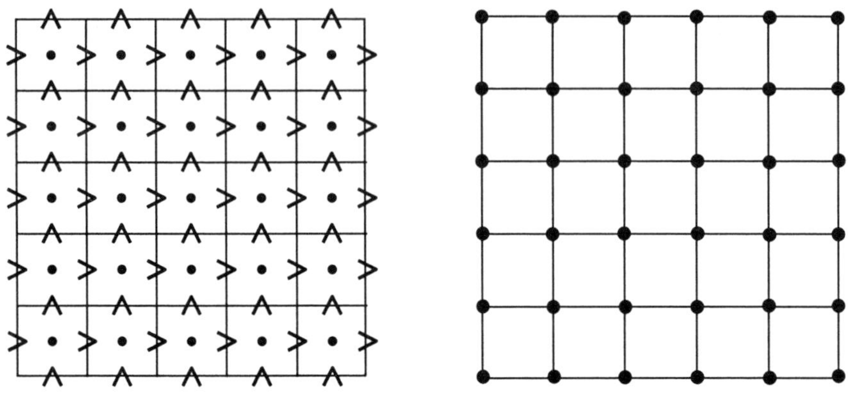

Figure 4: Staggered (left) and collocated (right) grids

The Reynolds number expresses the ratio between convection and diffusion, while the product of Reynolds and Prandtl numbers indicates correspondence between radiative and convective heat transfer. In the case of the glass tank (ommiting physical units)

$$X = 20, \;\; U = 0.01, \;\; \rho \sim 2490, \;\; 25 \leq \mu \leq 250, \;\; 1.2 \leq k \leq 64, \;\; C_p = 1000.$$

That is

$$0.2 \leq Re \leq 20, \quad 7812.5 \leq RePr \leq 416667.$$

3. Discretisation

The next step towards the solution is a discretisation of the continuous problem. We have opted for the finite volume method, because it ensures conservation of a number of important properties, such as momentum and mass. After the discretisation procedure has been fixed, we need to choose the grid on which we will discretise the equations.

3.1. Collocated grid versus staggered grid

The natural choice of the grid for a finite volume method is a *staggered grid* [7]. It is called staggered because the velocity components are staggered with respect to the pressure, which is placed in the center of the cells. The staggered grid ensures that the resulting discrete system will not be singular. In some sense it plays the same role as the LBB or Inf-Sup condition in the finite element method [8]. Despite the fact that using a staggered grid we always end up with a non-singular system, the programming process using such a grid is not a triviality, especially in 3D.

We will use another type of grid, namely a collocated grid, that is the grid where all variables are computed at the same locations. It is far more convenient for programming, but requires a special discretisation procedure to ensure non-singularity of the system.

3.2. Discretisation of the momentum equations

Let us consider the momentum equation corresponding to the horizontal component of the velocity

$$-\nabla \cdot (\mu(T)\nabla u) + (\rho \mathbf{u}, \nabla u) + \frac{\partial p}{\partial x} = 0. \tag{7}$$

For the sake of simplicity we consider here the two dimensional case. Momentum and body force are discretised according to a standard finite volume procedure [3]. Let us restrict our attention to the pressure gradient. Integrating the first component over a control volume, with node C say,

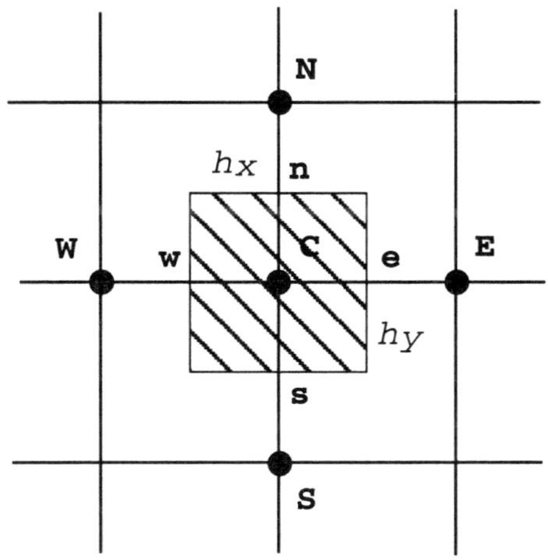

Figure 5: Pressure stencil

$$\int\limits_{V_c} \frac{\partial p}{\partial x} dV \doteq h_y(p_e - p_w) \doteq h_y(p_E - p_C). \tag{8}$$

We use downstream values of the pressure in (8) instead of midvalues. It leads to first order approximation instead of a possibly second order but, as it will be explained later, this will guarantee the non-singularity of the system.

3.3. Discretisation of the continuity equation

Since we are using a collocated grid, the continuity equation is integrated over the same control volumes as the momentum equations.

$$\int_{V_c} \left(\frac{\partial(\rho u)}{\partial x} + \frac{\partial(\rho v)}{\partial y} \right) dV \doteq h_y[(\rho u)_e - (\rho u)_w] + h_x[(\rho v)_n - (\rho v)_s] \doteq$$

$$h_y[(\rho u)_C - (\rho u)_W] + h_x[(\rho v)_C - (\rho v)_S]. \tag{9}$$

Here the upstream values of the velocity are used for the same reason as in the pressure gradient approximation.

4. PISO method

A solution method for the Navier-Stokes equations coupled with the energy equation was suggested by Issa [4]. It is an operator splitting approach, which is closely related to the well-known family of SIMPLE/SIMPLER algorithms by Patankar [7].

Let us assume now that after discretisation the system has the following form

$$\begin{pmatrix} \mathbf{M}(\mathbf{u},\mathbf{T}) & \mathbf{B}^{\mathbf{p}} & \mathbf{0} \\ \mathbf{B}^{\mathbf{u}}\rho(\mathbf{T}) & \mathbf{0} & \mathbf{0} \\ \mathbf{0} & \mathbf{0} & \mathbf{E}(\mathbf{u},\mathbf{T}) \end{pmatrix} \begin{pmatrix} \mathbf{u} \\ \mathbf{p} \\ \mathbf{T} \end{pmatrix} = \begin{pmatrix} \mathbf{F} \\ \mathbf{0} \\ \mathbf{0} \end{pmatrix}, \tag{10}$$

Here $\mathbf{M}(\mathbf{u},\mathbf{T})$ is a discrete momentum operator, $\mathbf{E}(\mathbf{u},\mathbf{T})$ — discrete energy operator, both depending non-linearly on the velocity and the temperature. $\mathbf{B}^{\mathbf{p}}$ and $\mathbf{B}^{\mathbf{u}}$ are the discrete gradient and the discrete divergence, respectively.

4.1. Description of the algorithm

First we append to the momentum equations a time dependent term. Iterants $\mathbf{u}^n$, $\mathbf{T}^n$ and $\mathbf{p}^n$ are assumed to be known. After solving the prediction step

$$\frac{\rho(\mathbf{T}^n)\mathbf{u}^* - \rho(\mathbf{T}^n)\mathbf{u}^n}{\Delta t} = \mathbf{F} - \mathbf{M}(\mathbf{u}^n,\mathbf{T}^n)\mathbf{u}^* - \mathbf{B}^{\mathbf{p}}\mathbf{p}^n \tag{11}$$

we find the velocity field $\mathbf{u}^*$ which is our first guess. Since it might not satisfy the continuity equation we correct it in the following way

$$\frac{\rho(\mathbf{T}^n)\mathbf{u}^{**} - \rho(\mathbf{T}^n)\mathbf{u}^n}{\Delta t} = \mathbf{F} - \mathbf{M}(\mathbf{u}^*,\mathbf{T}^n)\mathbf{u}^* - \mathbf{B}^{\mathbf{p}}\mathbf{p}^*, \tag{12}$$

where $\rho\mathbf{u}^{**}$ should satisfy the continuity equation. The new value of the pressure $\mathbf{p}^*$ is not known yet, but we can get around this problem by taking the divergence of (12), and applying the continuity equation to the new velocity field $\mathbf{u}^{**}$. The equation for the new pressure will look like

$$-\mathbf{B}^{\mathbf{u}}\mathbf{B}^{\mathbf{p}}\mathbf{p}^* = -\mathbf{B}^{\mathbf{u}}\left[\mathbf{F} - \mathbf{M}(\mathbf{u}^*, \mathbf{T}^n)\mathbf{u}^* + \frac{\rho(\mathbf{T}^n)\mathbf{u}^n}{\Delta t}\right]. \tag{13}$$

The matrix $\mathbf{B}^{\mathbf{u}}\mathbf{B}^{\mathbf{p}}$ is a Laplace-type operator with Neumann boundary conditions. We shall examine it further in the next subsection. After the new pressure has been found we can apply (12) to correct the velocity field. At the next step we compute the new temperature

$$\frac{C_p\rho(\mathbf{T}^n)\mathbf{T}^{n+1} - C_p\rho(\mathbf{T}^n)\mathbf{T}^n}{\Delta t} = -\mathbf{E}(\mathbf{u}^{**}, \mathbf{T}^n)\mathbf{T}^{n+1}. \tag{14}$$

In his paper Issa showed that two correction steps give the optimal new value of the velocity for the time-dependent computations. Although we are dealing with steady-state computations, numerical tests show that two correction steps are also more efficient in the steady-state case. The second correction step will then look like

$$\frac{\rho(\mathbf{T}^{n+1})\mathbf{u}^{n+1} - \rho(\mathbf{T}^n)\mathbf{u}^n}{\Delta t} = \mathbf{F} - \mathbf{M}(\mathbf{u}^{**}, \mathbf{T}^{n+1})\mathbf{u}^{**} - \mathbf{B}^{\mathbf{p}}\mathbf{p}^{n+1},$$

and the corresponding pressure system is

$$-\mathbf{B}^{\mathbf{u}}\mathbf{B}^{\mathbf{p}}\mathbf{p}^{n+1} = -\mathbf{B}^{\mathbf{u}}\left[\mathbf{F} - \mathbf{M}(\mathbf{u}^{**}, \mathbf{T}^n)\mathbf{u}^{**} + \frac{\rho(\mathbf{T}^n)\mathbf{u}^n}{\Delta t}\right].$$

The PISO method can be now formulated as follows:

0. Let $\mathbf{u}^n$, $\mathbf{T}^n$, $\mathbf{p}^n$ be given
1. If the residuals

$$\mathbf{r}^{\mathbf{u}} := \mathbf{F} - \mathbf{M}(\mathbf{u}^n, \mathbf{T}^n)\mathbf{u}^n - \mathbf{B}^{\mathbf{p}}\mathbf{p}^n,$$
$$\mathbf{r}^{\mathbf{T}} := -\mathbf{E}(\mathbf{u}^n, \mathbf{T}^n)\mathbf{T}^n$$

are not small enough proceed to step 2.

2. Compute $\mathbf{u}^*$ from

$$\frac{\rho(\mathbf{T}^n)\mathbf{u}^* - \rho(\mathbf{T}^n)\mathbf{u}^n}{\Delta t} = \mathbf{F} - \mathbf{M}(\mathbf{u}^n, \mathbf{T}^n)\mathbf{u}^* - \mathbf{B}^{\mathbf{p}}\mathbf{p}^n$$

3. Solve

$$-\mathbf{B}^{\mathbf{u}}\mathbf{B}^{\mathbf{p}}\mathbf{p}^* = -\mathbf{B}^{\mathbf{u}}\left[\mathbf{F} - \mathbf{M}(\mathbf{u}^*, \mathbf{T}^n)\mathbf{u}^* + \frac{\rho(\mathbf{T}^n)\mathbf{u}^n}{\Delta t}\right]$$

4. `Compute` $\mathbf{u}^{**}$ `from`

$$\frac{\rho(\mathbf{T}^n)\mathbf{u}^{**} - \rho(\mathbf{T}^n)\mathbf{u}^n}{\Delta t} = \mathbf{F} - \mathbf{M}(\mathbf{u}^*, \mathbf{T}^n)\mathbf{u}^* - \mathbf{B}^{\mathbf{p}}\mathbf{p}^*$$

5. `Compute` $\mathbf{T}^{n+1}$ `from`

$$\frac{C_p\rho(\mathbf{T}^n)\mathbf{T}^{n+1} - C_p\rho(\mathbf{T}^n)\mathbf{T}^n}{\Delta t} = -\mathbf{E}(\mathbf{u}^{**}, \mathbf{T}^n)\mathbf{T}^{n+1}$$

6. `Solve`

$$-\mathbf{B}^{\mathbf{u}}\mathbf{B}^{\mathbf{p}}\mathbf{p}^{n+1} = -\mathbf{B}^{\mathbf{u}}\left[\mathbf{F} - \mathbf{M}(\mathbf{u}^{**}, \mathbf{T}^n)\mathbf{u}^{**} + \frac{\rho(\mathbf{T}^n)\mathbf{u}^n}{\Delta t}\right]$$

7. `Compute` $\mathbf{u}^{n+1}$ `from`

$$\frac{\rho(\mathbf{T}^{n+1})\mathbf{u}^{n+1} - \rho(\mathbf{T}^n)\mathbf{u}^n}{\Delta t} = \mathbf{F} - \mathbf{M}(\mathbf{u}^{**}, \mathbf{T}^{n+1})\mathbf{u}^{**} - \mathbf{B}^{\mathbf{p}}\mathbf{p}^{n+1}$$

8. `Increase n; return to step 1`

4.2. The system for the pressure

In the previous subsection we derived the system for the pressure, the matrix of this system being a product of two matrices $\mathbf{B}^{\mathbf{u}}$ - discrete divergence and $\mathbf{B}^{\mathbf{p}}$ - discrete gradient. Let us determine how this matrix acts on a pressure vector in the two-dimensional case. Define

$$\mathbf{B}^{\mathbf{p}} := \begin{pmatrix} \mathbf{B}^{\mathbf{p}}_x \\ \mathbf{B}^{\mathbf{p}}_y \end{pmatrix},$$

where the block matrices $\mathbf{B}^{\mathbf{p}}_x$, $\mathbf{B}^{\mathbf{p}}_y$ are the approximations of the components of the gradient operator. Bearing this in mind we have

$$\begin{gathered}(-\mathbf{B}^{\mathbf{u}}\mathbf{B}^{\mathbf{p}}p)_C = h_y[(\mathbf{B}^{\mathbf{p}}_x\mathbf{p})_W - (\mathbf{B}^{\mathbf{p}}_x\mathbf{p})_C] + h_x[(\mathbf{B}^{\mathbf{p}}_y\mathbf{p})_S - (\mathbf{B}^{\mathbf{p}}_y\mathbf{p})_C] = \\ \frac{h_y}{h_x}[(p_C - p_W) + (p_C - p_E)] + \frac{h_x}{h_y}[(p_C - p_S) + (p_C - p_N)] = \\ 2\left(\frac{h_y}{h_x} + \frac{h_x}{h_y}\right)p_C - \frac{h_y}{h_x}p_W - \frac{h_y}{h_x}p_E - \frac{h_x}{h_y}p_S - \frac{h_x}{h_y}p_N\end{gathered}$$

This is a standard 5-point finite volume (or which is the same in this case, finite element) discretisation of the Laplace operator. It can be shown that the discrete pressure system differs from the 5-point approximation of the Laplace operator only at the boundary nodes. Connectivity graphs of the pressure matrix and the discrete Laplacian are shown in Figure 6.

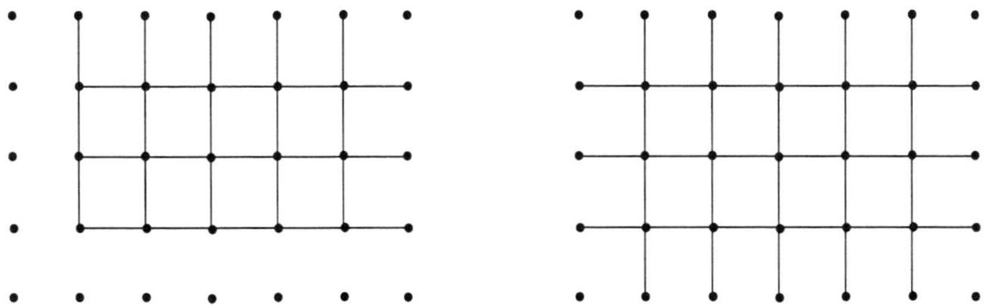

Figure 6: Connectivity graph of the pressure matrix (left) and discrete Laplacian (right)

4.3. Compatibility condition for the pressure system

By construction, the pressure matrix is symmetric, positive-semidefinite, since the diagonal entries are positive, the off-diagonal entries are negative and the row sums are equal to zero. Furthermore this matrix is singular, and the kernel is formed by the constant vectors. Thus before solving the system we need to check whether it is compatible. The compatibility condition can be derived by summing up all equations of the pressure system

$$\sum_C(-\mathbf{B^u B^p})\mathbf{p})_C = \sum_C(-\mathbf{B^u}[\mathbf{F} - \mathbf{M}(\mathbf{u},\mathbf{T})\mathbf{u} + \frac{\rho(\mathbf{T})\mathbf{u}}{\Delta t}])_C. \qquad (15)$$

Since the matrix is symmetric and all row sums are zero, the left-hand side of the equality is zero. At the boundary point C we have

$$\left(\mathbf{F} - \mathbf{M}(\mathbf{u},\mathbf{T})\mathbf{u} + \frac{\rho(\mathbf{T})\mathbf{u}}{\Delta t}\right)_C = \left(\frac{\rho(\mathbf{T})\mathbf{u}}{\Delta t}\right)_C,$$

whence

$$\sum_C(\frac{\rho(\mathbf{T})\mathbf{u}}{\Delta t})_C = \frac{1}{\Delta t}[\sum_{\Gamma_w} h_y(\rho(T)u)_C - \sum_{\Gamma_e} h_y(\rho(T)u)_C + \sum_{\Gamma_s} h_x(\rho(T)v)_C - \sum_{\Gamma_n} h_x(\rho(T)v)_C].$$

This expression should be equal to zero. After multiplying both sides byΔt we obtain the compatibility condition

$$\sum_{\Gamma_w} h_y(\rho(T)u)_C \quad - \quad \sum_{\Gamma_e} h_y(\rho(T)u)_C + \sum_{\Gamma_s} h_x(\rho(T)v)_C$$

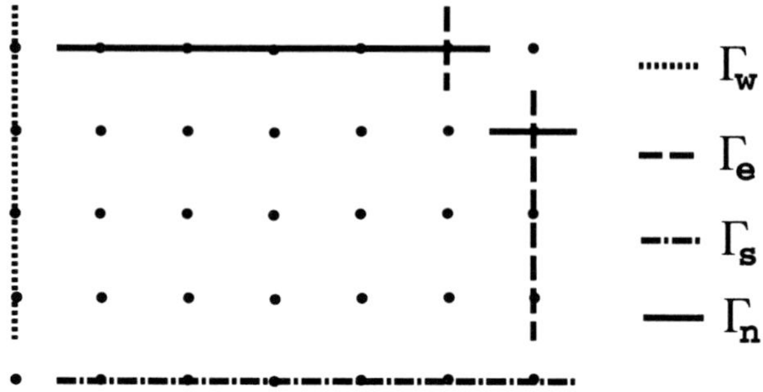

Figure 7: Boundary points involved in the compatibility condition

$$- \sum_{\Gamma_n} h_x(\rho(T)v)_C = 0. \tag{16}$$

Summation is done over four sets of the boundary points, as is shown in Figure 7. The compatibility condition can be interpeted as an approximation of

$$\oint_{\partial\Omega} (\rho(T)\mathbf{u}, \mathbf{n})d\Gamma = 0,$$

which in turn means that no mass is generated inside the domain. A consequence of (16) is the following. Let us assume that the exact solution is known and satisfies the continuity equation. We use it to obtain the boundary values; but since they also should satisfy (16) it might happen that we cannot use the values of the exact solution at the boundary points. To put it another way, even though the continuous solution satisfies the continuity equation, a test problem constructed from it might not have a solution. The same is true for any solution method for the Stokes/Navier-Stokes equations which makes use of the pressure matrix, though the compatibility condition will perhaps look less "exotic" than for the method considered. One of the reasons why (16) (or its variants) is sometimes ignored, is the correction of the pressure system. The pressure system is singular and since the pressure is determined up to a constant, we can fix the pressure at one point, which means that the solution is fully determined despite the fact that the original system might be incompatible. The result of that will be the new velocity field that does not satisfy the continuity equation. The last remark about the compatibility condition is that it is not known a priori, but follows from the discrete momentum and the continuity equations.

5. Local defect correction

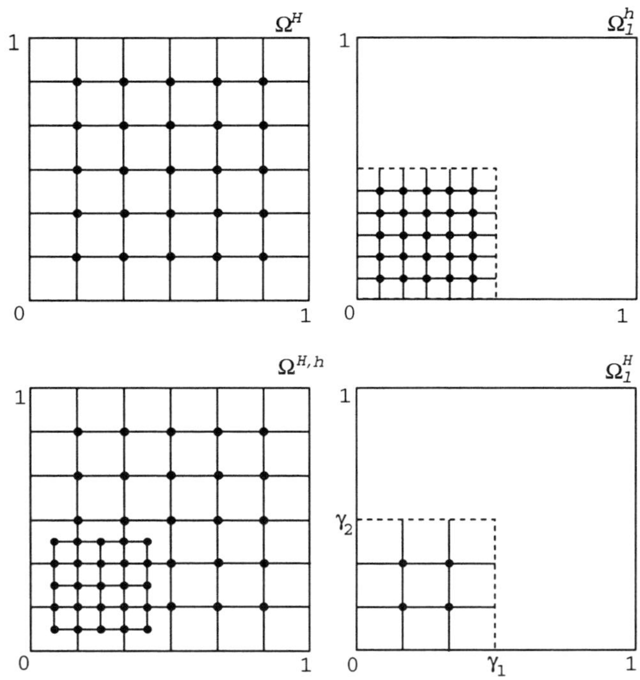

Figure 8: The uniform grids Ω^H, Ω_l^h and Ω_l^H, and the composite grids $\Omega^{H,h}$ for $H = 1/6$, $\sigma = 2$ and $\gamma_1 = \gamma_2 = 1/2$.

Due to the local nature of some processes in the oven, we need a higher resolution in areas where such processes take place. The method we use for local refinement is called *Local Defect Correction* (LDC) [2]. LDC is an iterative procedure which is able to accurately combine solutions on finer local and coarser global grids. In order to describe the method we consider a model boundary value problem

$$Lu = f \text{ in } \Omega = (0,1) \times (0,1), \tag{17}$$

$$u = \varphi \text{ on } \partial\Omega. \tag{18}$$

Here L is an arbitrary elliptic operator, f is a given function on Ω, φ a given function on the boundary $\partial\Omega$.

The model composite grid $\Omega^{H,h}$ (see Fig.8)is composed of a global coarse grid and a local fine grid. The *global coarse grid* Ω^H is a uniform grid with grid size H

$$\Omega^H = \{(x_i, y_j)|\ x_i = iH,\ y_j = jH,\ 0 < i < 1/H,\ 0 < j < 1/H\}.$$

We suppose that $1/H \in \mathbf{N}$. The *local fine grid* Ω^h_{loc} is a uniform grid with grid size $h < H$, covering the region of local refinement $\Omega_{loc} = (0, \gamma_1) \times (0, \gamma_2) \subset \Omega$

$$\Omega^h_{loc} = \{(x_i, y_j)|\ x_i = ih,\ y_j = jh,\ 0 < i < 1/h,\ 0 < j < 1/h\}.$$

We assume $\gamma_1/H \in \mathbf{N}$, $\gamma_2/H \in \mathbf{N}$ and $H/h \in \mathbf{N}$. The interface Γ is defined as the part of the boundary $\partial\Omega_{loc}$ that lies inside Ω,

$$\Gamma = \partial\Omega_{loc} \cap \Omega.$$

We assume also that the high activity region of the boundary value problem lies inside the subregion Ω_{loc}. The *composite grid* $\Omega^{H,h}$ is defined by

$$\Omega^{H,h} = \Omega^H \cup \Omega^h_{loc}.$$

The refinement factor σ is defined as the ratio of the coarse grid size H and the fine grid size h,

$$\sigma = H/h.$$

Besides the uniform coarse grid Ω^H and the uniform local grid Ω^h_{loc}, there is one more important grid, namely the *local coarse grid* Ω^H_{loc} which is defined by

$$\Omega^H_{loc} := \Omega^H \cap \Omega^h_{loc}$$

In the LDC method one starts by discretising the boundary value problem on the global coarse grid Ω^H. This yields the basic discretisation

$$L^H u_0^H = f^H \text{ on } \Omega^H. \tag{19}$$

The grid function u_0^H is an approximation of the solution of the boundary value problem (17), (18). This grid function is used for discretising the boundary value problem on the local fine grid Ω^H_{loc}. Boundary conditions for this problem are obtained from u_0^H by means of interpolation. The discrete problem on the local fine grid is denoted by

$$L^h_{loc} u^h_{loc,0} = f^h_{loc}(u_0^H) \text{ on } \Omega^h_{loc}. \tag{20}$$

The dependence of f^h_{loc} on the aproximation u_0^H is denoted explicitly in (20). The approximations u_0^H and $u^h_{loc,0}$ are used to define a *composite grid approximation* $u_0^{H,h}$

$$u_0^{H,h}(x) = \begin{cases} u^h_{loc,0}(x) & x \in \Omega^h_{loc} \\ u_0^H(x) & x \in \Omega^{H,h} \backslash \Omega^h_{loc} \end{cases}.$$

In the local defect correction method the fine grid approximation $u_{loc,0}^h$ is used to correct the basic discretisation (19) in the following way. The global coarse grid approximation u_0^H and the local fine approximation $u_{loc,0}^h$ are combined to define the global coarse grid function w^H,

$$w^H(x) = \begin{cases} u_{loc,0}^h(x) & x \in \Omega_{loc}^H \\ u_0^H(x) & x \in \Omega^H \backslash \Omega_{loc}^h \end{cases} .$$

Substituting this function into the basic discretisation yields a *residual grid function* or *defect*,

$$d^H = L^H w^H - f^H.$$

The values of this defect at grid points inside Ω_{loc} are used to update the right hand side f^H of the basic discretisation,

$$\bar{f}^H(x) := \begin{cases} f^H(x) + d^H(x) & x \in \Omega_{loc}^H \\ f^H(x) & x \in \Omega^H \backslash \Omega_{loc}^h \end{cases} .$$

The updated coarse grid problem reads

$$L^H u_1^H = \bar{f}^H(u_{loc,0}^h, u_0^H) \text{ on } \Omega^H. \tag{21}$$

Equation (21) yields an approximation u_1^H of the solution of the boundary value problem (17), (18) on the global coarse grid. Like the approximation u_0^H, the approximation u_1^H is used to define artificial Dirichlet boundary values on the interface. The related discrete problem on the local fine grid reads

$$L_{loc}^h u_{loc,1}^h = f_{loc}^h(u_1^H) \text{ on } \Omega_{loc}^h.$$

The approximations u_1^H and $u_{l,1}^h$ are used to define a composite grid approximation $u_1^{H,h}$ of the solution of the boundary value problem

$$u_1^{H,h}(x) = \begin{cases} u_{loc,1}^h(x) & x \in \Omega_{loc}^h \\ u_1^H(x) & x \in \Omega^{H,h} \backslash \Omega_{loc}^h \end{cases} .$$

The LDC method can be summarised as the following scheme:

1. Solve

$$L^H u_n^H = f_n^H \text{ on } \Omega^H, \; f_0^H = f^H.$$

2. Solve

$$L_{loc}^h u_{loc,n}^h = f_{loc,n}^h(u_n^H) \text{ on } \Omega_{loc}^h.$$

(Boundary conditions for this problem are obtained from u_n^H by means of interpolation.)

3. Construct the composite approximation $u_n^{H,h}(x)$

$$u_n^{H,h}(x) = \begin{cases} u_{loc,n}^h(x) & x \in \Omega_{loc}^h, \\ u_n^H(x) & x \in \Omega^{H,h} \backslash \Omega_{loc}^h \end{cases} .$$

4. Construct the global coarse grid function w^H

$$w^H(x) = \begin{cases} u^h_{loc,n}(x) & x \in \Omega^H_{loc}, \\ u^H_n(x) & x \in \Omega^H \backslash \Omega^h_{loc} \end{cases} .$$

5. Compute the defect d^H

$$d^H = L^H w^H - f^H.$$

6. Update the right-hand part of the global problem

$$f^H_{n+1}(x) = \begin{cases} f^H_n(x) + d^H(x) & x \in \Omega^H_{loc}, \\ f^H_n(x) & x \in \Omega^H \backslash \Omega^h_{loc} \end{cases} .$$

7. Check convergence; return to step 1, if not accurate enough.

6. Stirred flow

One of the situations where we can use the advantages of LDC is the modelling of a stirrer. Usually the stirrer is located in a so-called *doghouse*, a small tank from which glass flows to production lines. We model the stirrer as a small rectangular solid piece, with velocity vectors prescribed in such a way that they move the flow around. The problem with computations involving a stirrer is that the size of the stirrer is normally much smaller than the characteristic length. The standard way to resolve the problem is to refine globally in the vicinity of the stirrer. Results of this approach are shown in Figure 9 (since original computations are carried out in 3D, we plot the results for a fixed vertical coordinate) This approach gives the

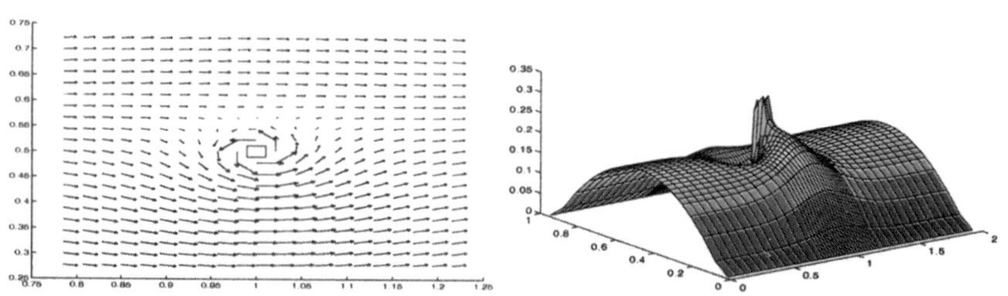

Figure 9: Solution obtained by means of non-uniform global refinement - magnified velocity field (left), absolute value of velocity (right) (horizontal crossection)

best possible accuracy, but has some strong disadvantages. First, by refining globally we obtain a considerable amount of grid points, in which much the information is irrelevant. In 3D this leads to a significant and superfluous increase in memory and computational time. Secondly, the condition number of the discrete system depends on the mesh size ratio.

We can avoid this problems by refining locally, and use LDC. The result is depicted in Figure 10. One of the main advantages of LDC is that we don't

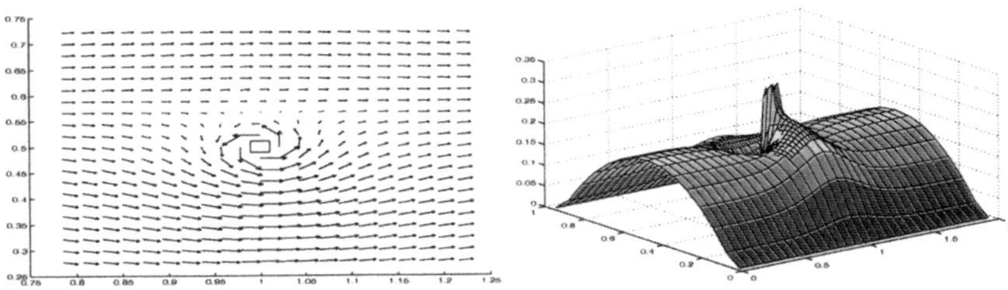

Figure 10: Solution obtained with LDC - magnified velocity field (left), absolute value of velocity (right) (horizontal cross section)

need a special data structure for the composite grid, since it is constructed of regular global and local grids. Furthermore, since the systems used in LDC are of smaller size and better conditioned due to smaller mesh size ratios, the solution process of linear systems is considerably more efficient. Though LDC is an iterative procedure, it converges very fast. We use the

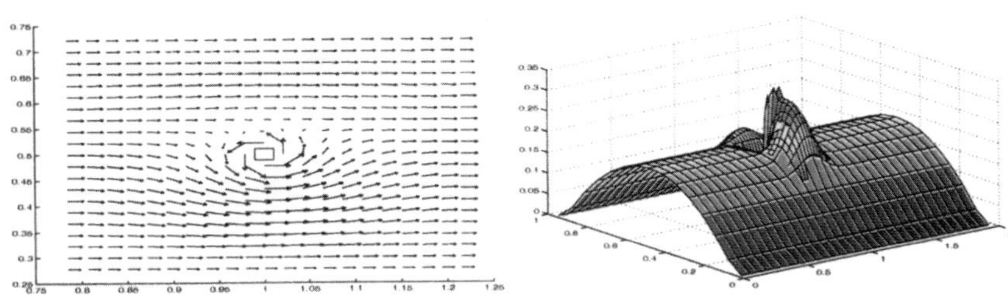

Figure 11: Solution after zero iteration - magnified velocity field (left), absolute value of velocity (right) (horizontal cross section)

results from the global refinement $\hat{\mathbf{u}}$ in order to estimate convergence and accuracy of LDC. $\mathbf{u}^{H,h}$ denotes the solution on the composite grid (see Table 1).

#LDC iter.	$\frac{\|\hat{\mathbf{u}}-\mathbf{u}^{H,h}\|_\infty}{\|\hat{\mathbf{u}}\|_\infty}$
0	$7.90 \cdot 10^{-2}$
1	$3.79 \cdot 10^{-2}$
2	$2.15 \cdot 10^{-2}$
3	$1.26 \cdot 10^{-2}$

Table 1: LDC iteration results

Another specific feature of LDC is the use of the defect. Via the defect local pertubations affect the solution globally. If we would stop the procedure after the initial step, that is without updating the global problem, the results would not be accurate enough, see Figure 11

7. Bubbling

Another situation in the glass oven where we can successfully use LDC, is a bubbling process. The air bubbles are injected into the glass tank from the bottom, and while moving to the top, they attract other bubbles, thus removing the air from the glass.

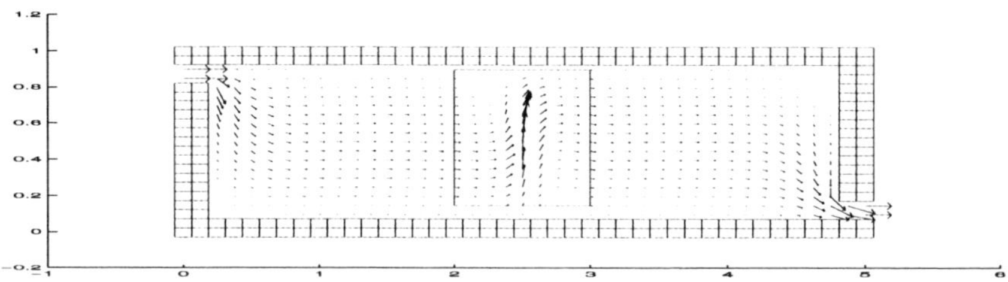

Figure 12: Global velocity field (without LDC) (vertical cross section)

Suppose we have with an initial volume per bubble V_0. Since the hydrostatic pressure decreases going from the bottom to the top, the bubble diameter changes to

$$V_b = V_0 \frac{1 + \rho g H / p_0}{1 + \rho g (H - z)/p_0},$$

where H is a total glass height, z the glass height and p_0 the athmospheric pressure. Suppose the total bubbling volume flow is Q. Then the distance between the bubbles is $d = vV_0/Q$, where v is the bubble rising velocity,

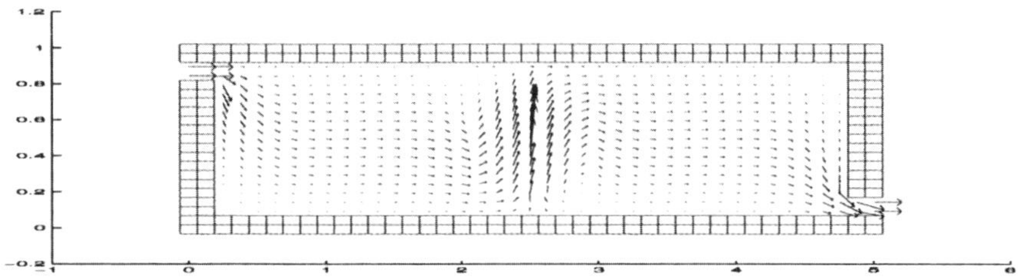

Figure 13: Global velocity field (with LDC) (vertical cross section)

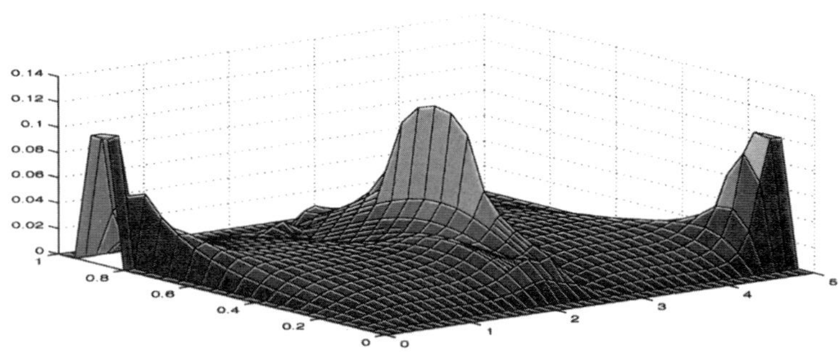

Figure 14: Absolute value of the velocity (vertical cross section)

i.e. the sum of the undisturbed vertical glass velocity and the relative glass-bubble velocity. From one bubble, a force acting on the glass is

$$F_b = -V_b \rho g.$$

Since per vertical meter, there are on the average 1/d bubbles pushing on the glass, the average force per meter height is:

$$F_a = F_b/d = -\frac{V_b \rho g}{v V_0} Q.$$

After that we correct the right-hand side of the vertical momentum equation as follows:

$$F_C = h_z F_a - h_x h_y h_z \rho(T_c) g$$

The computations on a coarse grid appear to be not accurate enough, see Fig.12. In order to improve the solution we use LDC, see Fig.13. The absolute value of the composite solution and the velocity field in the region of the local refinement are depicted in Figure 14 and 15 respectively.

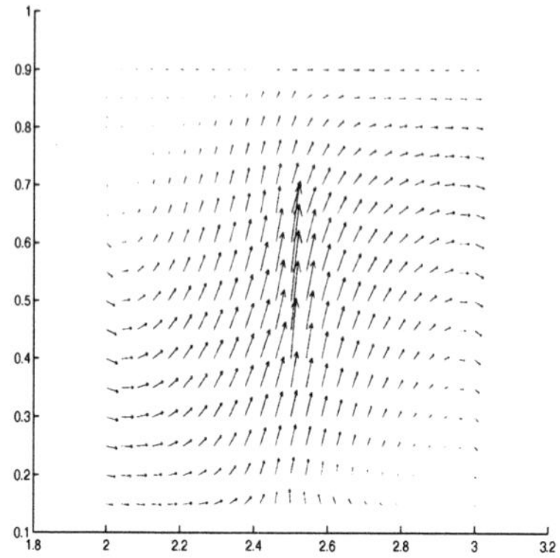

Figure 15: Local velocity field (vertical cross section)

References

[1] G. K. Batchelor, *An introduction to fluid dynamics*, Cambridge University Press, (1970).
[2] P. J. J. Ferket and A. A. Reusken, Further analysis of the local defect correction method, Computing, 56 (1996), 117–139.
[3] J. H. Ferziger, M. Peric, Computational methods for fluid dynamics, Springer (1996).
[4] R. I. Issa, Solution of the Implicitly Discretised Fluid Flow Equations by Operator-Splitting, Journal of Comput. Physics, 62 (1985), 40–65.
[5] P. K. Kundu, Fluid Mechanics, Academic Press, San Diego (1990).
[6] S. Nefedov, Simulation of glass flow in an oven, in Progress in Industrial Mathematics at ECMI98, B. G. Teubner, (1999), 106–113.
[7] S. V. Patankar, Numerical heat transfer and fluid flow, Hemisphere Publishing Corporation (1980).
[8] A. Quarteroni and A. Valli, Numerical approximation of partial differential equations, Springer (1994).

V.Nefedov and R.M.M.Mattheij
Scientific Computing group
Department of Mathematics and Computing Science
Eindhoven University of Technology, PO BOX 513
5600MB Eindhoven
The Netherlands
E-mail: nefedov@win.tue.nl; E-mail: mattheij@win.tue.nl

Studies in Mathematics and its Applications, Vol. 31
D. Cioranescu and J.L. Lions (Editors)

Chapter 26

CONTROL LOCALIZED ON THIN STRUCTURES FOR SEMILINEAR PARABOLIC EQUATIONS

P. A. NGUYEN AND J. -P. RAYMOND

1. Introduction

Let us consider the heat equation with Stefan-Boltzmann boundary conditions

$$\frac{\partial y}{\partial t} - \Delta y = 0 \text{ in } Q, \qquad \frac{\partial y}{\partial n} + |y|^3 y = u\delta_\gamma \text{ on } \Sigma, \qquad y(0) = y_0 \text{ in } \Omega,$$

where Ω is a bounded open subset in R^N ($N \geq 2$) with a regular boundary Γ, $Q = \Omega\times]0,T[$, $\Sigma = \Gamma\times]0,T[$, γ is a regular manifold in Γ of dimension $D \leq N-2$, δ_γ denotes the Dirac measure supported by γ, and the control variable u belongs to a closed convex subset K_U of $L^q(0,T;L^\sigma(\gamma))$. For example if $N = 3$ and $D = 1$, γ is a heating wire. This example is a particular case of equations studied in this paper. We address the two following control problems:

• Find a control $u \in K_U$ such that (y_u, u) minimize the functional

$$I(y,u) = C_Q \int_Q |\nabla y - V_d|^\kappa dxdt + C_\Omega \int_\Omega |y(T) - y_d|^\theta dx$$

$$+ C_\gamma \int_0^T (\int_\gamma |u|^\sigma d\zeta)^{\frac{q}{\sigma}} dt.$$

(y_u is the solution of the above equation associated with u, and C_Q, C_Ω, C_γ are nonnegative constants.) What are the necessary optimality conditions for the solutions (y_u, u) ?

• Let us suppose that $D = 1$, that is γ is a curve, and that u is a control only depending on the time variable. Let $y_{u,\gamma}$ be the solution of above equation associated with (u, γ). What is the derivative of the functional

$$J(\gamma, u) = \int_\Omega |y_{u,\gamma}(T) - y_d|^\theta dx$$

with respect to a field of deformations of γ ? This question is connected with the characterization of the optimal shape of the curve γ in a set of admissible curves.

More generally, we are concerned with control problems for parabolic equations of the form:

$$\frac{\partial y}{\partial t} + Ay + \Phi(\cdot, y) = 0 \text{ in } Q, \quad \frac{\partial y}{\partial n_A} + \Psi(\cdot, y) = u\delta_\gamma \text{ on } \Sigma,$$
$$y(0) = y_0 \text{ in } \Omega, \tag{1.1}$$

where A is a second order elliptic operator, Φ is a Carathéodory function from $Q \times R$ into R, Ψ is a Carathéodory function from $\Sigma \times R$ into R. We suppose that $\gamma \subset \Gamma$, which corresponds to the most difficult problem, because the boundary condition is not necessarily linear. But the case where $\gamma \subset \overline{\Omega}$ may also be considered. We study the control problem

$$(P_1) \quad \inf\{I(y, u) \mid (y, u) \in L^\kappa(0, T; W^{1,\kappa}(\Omega)) \times K_U, \ (y, u) \text{ satisfies (1.1)}\}.$$

There is an important literature on pointwise control problems, they correspond to $D = 0$ ([3], [4], [9], [12], [13], [20], [22]). Some problems in three dimensional domain ($N = 3$), with a control localized on a curve ($D = 1$), have recently been studied for the linear heat equation in [17] (see also [16], [19]). In our knowledge the case of nonlinear equations is not yet studied in the literature.

For simplicity, first consider the following typical situation

$$\Phi(\cdot, y) = |y|^{\mu-1}y, \quad \Psi(\cdot, y) = |y|^{\nu-1}y.$$

Existence of solutions for equation (1.1). Before studying the control problem (P_1), we have to clarify the conditions on $(\sigma, q, N, D, \mu, \nu)$ for which equation (1.1) admits solutions.

Parabolic equations with measure as data have been studied in the case of homogeneous Dirichlet boundary conditions in [7], [6]. In particular Baras and Pierre [6] have proved that the equation

$$\frac{\partial y}{\partial t} + Ay + |y|^{\mu-1}y = \lambda \text{ in } Q, \qquad y = 0 \text{ on } \Sigma, \qquad y(0) = y_0 \text{ in } \Omega, \tag{1.2}$$

with $\lambda \in \mathcal{M}_b(Q)$, and $y_0 \in L^1(\Omega)$, admits a solution if, and only if, the measure λ does not charge the sets of $C_{2,1,\mu'}$-capacity zero. For example if

λ is of the form $\lambda = u\delta_\gamma$, with $u \in L^1(0,T;L^\sigma(\gamma))$, and $\gamma \subset \Omega$ is a manifold of dimension $D \leq N-2$, then equation (1.2) admits a solution if, and only if, $\mu < \frac{N-D}{N-D-2}$. In the case of equation (1.1), when $\Phi(\cdot,y) = |y|^{\mu-1}y$, and $\Psi(\cdot,y) = |y|^{\nu-1}y$, we prove the existence of solutions when $\mu < \frac{N-D}{N-D-2}$ and $\nu < \frac{N-D-1}{N-D-2}$ (see Remark 3.1). From the analysis of equations with measures as data, we know that the solution y to (1.1) belongs to $L^\delta(0,T;W^{1,d}(\Omega))$ for every $\frac{1}{\delta} + \frac{N}{2d} > \frac{N}{2} + \frac{1}{2}$. It can also be proved that $y(T)$ belongs to $L^1(\Omega)$ ([24]). These regularity results are not sufficient to attack the control problem (P_1).

With more restrictive conditions on μ, ν and y_0 (assumptions (A5) and (A6) stated below), we can improve these results (Theorem 3.1), and we are able to prove the existence of solutions for (P_1) for some $\theta > 1$ and $\kappa > 1$ (Theorem 4.1).

Optimality conditions for (P_1). The obtention of optimality conditions is more delicate. Indeed we have to prove that the adjoint state associated with a solution of (P_1) has a trace on $\gamma\times]0,T[$, which belongs to $L^{q'}(0,T;L^{\sigma'}(\gamma))$. The adjoint state p can be written in the form $p = p_\kappa + p_\theta$, where p_κ corresponds to the contribution of the term $\int_Q |\nabla y - V_d|^\kappa dxdt$ in the adjoint equation, and p_θ corresponds to the contribution of the term $\int_\Omega |y(T) - y_d|^\theta dx$. The trace of p_κ on $\gamma\times]0,T[$ belongs to $L^{q'}(0,T;L^{\sigma'}(\gamma))$ only if κ satisfies some conditions depending on σ, q, D, μ, ν, and ρ. These conditions are stated in assumptions (A10)-(A12). The corresponding regularity results for p_κ are stated in Theorems 4.2 and 4.3. The conditions on θ for which the trace of p_θ belongs to $L^{q'}(0,T;L^{\sigma'}(\gamma))$ are stated in assumption (A13). Regularity results for p_θ are stated in Theorem 4.4. Some examples satisfying the technical assumptions (A11)-(A13) are given at the end of Section 4.3.

Finally the derivative of J with respect to deformation of γ is obtained in Theorem 5.2.

2. Preliminary estimates for some linear equations

2.1. Notations and assumptions

We denote by T_γ (respectively T_Γ, $T_{\gamma\times]0,T[}$, T_Σ) the trace mapping on γ (respectively on Γ, $\gamma\times]0,T[$, Σ). Throughout the paper, we denote by C, C_i, K, K_i for $i \in N$, various positive constants depending on known quantities. The same letter may be used for different constants.

(A1) - The elliptic operator A is defined by $Ay = -\sum_{i,j=1}^N D_i(a_{ij}(x)D_j y) + a_0(x)y$. The coefficient a_0 is positive and belongs to $C(\overline{\Omega})$, the coefficients

a_{ij} belong to $C^{1,\nu}(\overline{\Omega})$ with $0 < \nu \leq 1$, $a_{ij} = a_{ji}$, and they satisfy

$$\sum_{i,j=1}^{N} a_{ij}(x)\xi_i\xi_j \geq m_0|\xi|^2 \quad \text{for every } \xi \in R^N \text{ and every } x \in \overline{\Omega}, \text{ with } m_0 > 0.$$

(A2) - Γ is of class $C^{\bar{k}}$, with $\bar{k} = \max\big(2, [\frac{N-D}{\sigma'}]+1\big)$, and γ is a submanifold of Γ of class $C^{\bar{k}}$, of dimension $D \leq N-2$.
(A3) - Φ is a Carathéodory function from $Q \times R$ into R. For almost all $(x,t) \in Q$, $\Phi(x,t,.)$ is of class C^1, and the following estimates hold:

$$|\Phi(x,t,y)| \leq C(1+|y|^{\mu}), \quad 0 \leq \Phi'_y(x,t,y) \leq C(1+|y|^{\mu-1}).$$

For simplicity in the writing of equations we also suppose that $\Phi(x,t,0) = 0$. Still for simplicity, we shall write $\Phi(y)$ in place of $\Phi(x,t,y)$.
(A4) - Ψ is a Carathéodory function from $\Sigma \times R$ into R. For almost all $(s,t) \in \Sigma$, $\Psi(s,t,\cdot)$ is of class C^1, and we have

$$|\Psi(s,t,y)| \leq C(1+|y|^{\nu}), \quad 0 \leq \Psi'_y(s,t,y) \leq C(1+|y|^{\nu-1}).$$

As for Φ, we suppose that $\Psi(s,t,0) = 0$, and we write $\Psi(y)$ in place of $\Psi(s,t,y)$.
(A5) - We make the following assumption on μ and ν:

$$1 < \mu < \inf\{\frac{N+2}{N-\frac{D}{\sigma'}-\frac{2}{q'}}, \frac{N}{N-\frac{D}{\sigma'}-2}, \frac{N-D+2}{N-D-\frac{2}{q'}}, \frac{N-D}{N-D-2}\}, \tag{2.1}$$

and

$$1 < \nu < \inf\{\frac{N+1}{N-\frac{D}{\sigma'}-\frac{2}{q'}}, \frac{N-1}{N-\frac{D}{\sigma'}-2}, \frac{N-D+1}{N-D-\frac{2}{q'}}, \frac{N-D-1}{N-D-2}\}. \tag{2.2}$$

(A6) - The initial condition y_0 belongs to $L^{\rho}(\Omega)$ with $\rho \geq \max\left(\frac{q}{2}, \frac{N}{N-\frac{D}{\sigma'}-\frac{2}{q'}}\right)$, K_U is a closed convex subset in $L^q(0,T;L^{\sigma}(\gamma))$, and $2 \leq \sigma \leq q < \infty$.

2.2. Estimates for linear equations

First recall some results for analytic semigroups. We denote by $\tilde{A}$ the operator defined by

$$\mathcal{D}(\tilde{A}) = \left\{ y \in C^2(\overline{\Omega}) \mid \frac{\partial y}{\partial n_A} = 0 \text{ on } \Gamma \right\}, \quad \tilde{A}y = Ay.$$

For $1 \le \ell < \infty$, we denote by A_ℓ the closure of $\tilde{A}$ in $L^\ell(\Omega)$. The operator $-A_\ell$ is the generator of a strongly continuous analytic semigroup $S_\ell(t)_{t\ge 0}$ in $L^\ell(\Omega)$ [2]. For $1 < \ell < \infty$ the domain of A_ℓ is

$$\mathcal{D}(A_\ell) = \Big\{ y \in W^{2,\ell}(\Omega) \mid \frac{\partial y}{\partial n_A} = 0 \text{ on } \Gamma \Big\}.$$

For $\ell = 1$, $\mathcal{D}(A_1)$ is the set of functions y in $L^1(\Omega)$ such that there exists $z \in L^1(\Omega)$ satisfying $\int_\Omega z(x)v(x)dx = \int_\Omega y(x)Av(x)dx$ for all $v \in D(\tilde{A})$. For any $1 \le \ell < \infty$, 0 belongs to the resolvent of $-A_\ell$ and there exists $\delta > 0$ such $\operatorname{Re}\sigma(A_\ell) \ge \delta$ (it is a consequence of (A1) and of the fact that $\sigma(A_\ell)$ is independent of ℓ). Therefore, for $\alpha > 0$, there exists a constant $K = K(\ell, \alpha)$ such that

$$\|A_\ell^\alpha S_\ell(t)\varphi\|_{L^\ell(\Omega)} \le K t^{-\alpha} \|\varphi\|_{L^\ell(\Omega)},$$

Henry, [23], A_ℓ^α is the α-power of A_ℓ). Thanks to this result the following lemma can be established.

Lemma 2.1 [2], [26] – *For every $1 \le \ell \le \lambda \le \infty$ with $\ell < \infty$, there exists a constant $K_1 = K_1(\lambda, \ell)$ such that*

$$\|S_\ell(t)\varphi\|_{L^\lambda(\Omega)} \le K_1 t^{-\frac{N}{2}(\frac{1}{\ell}-\frac{1}{\lambda})} \|\varphi\|_{L^\ell(\Omega)} \tag{2.3}$$

for every $\varphi \in L^\ell(\Omega)$ and every $t > 0$. For every $1 \le \ell \le \lambda \le \infty$ with $\ell < \infty$, and every $\alpha > 0$, there exists a constant $K_2 = K_2(\lambda, \ell, \alpha)$ such that

$$\|A_\ell^\alpha S_\ell(t)\varphi\|_{L^\lambda(\Omega)} \le K_2 t^{-\frac{N}{2}(\frac{1}{\ell}-\frac{1}{\lambda})-\alpha} \|\varphi\|_{L^\ell(\Omega)} \tag{2.4}$$

for every $\varphi \in L^\ell(\Omega)$ and every $t > 0$.

Lemma 2.2 *(*[1], Theorem 7.58*) –If assumption (A2) is satisfied, then T_γ is a continuous linear operator from $W^{m,p}(\Omega)$ into $L^q(\gamma)$ for all (m,p,q) such that $0 \le m \le \bar{k}$, $0 < N - mp < D$, $p \le q < \frac{Dp}{N-mp}$.*

Throughout the sequel, Θ denotes a diffeomorphism of class $C^{\bar{k}}$, from $\gamma \times (]-1,1[)^{N-D-1}$ onto an open set $\mathcal{O}$ of Γ, such that $J\Theta$ (the Jacobian matrix of Θ) and $H\Theta$ (the Hessian matrix of Θ) are uniformly bounded on $\gamma \times \left(]-1,1[\right)^{N-D-1}$, and such that $\Theta(\zeta, 0) = \zeta$ for every $\zeta \in \gamma$. For shortening we write $]-1,1[^{N-D-1}$ for $\left(]-1,1[\right)^{N-D-1}$. For each $\varsigma \in\,]-1,1[^{N-D-1}$, we denote by γ_ς the manifold $\Theta(\gamma, \varsigma)$.

Lemma 2.3 – *There exists a constant C such that*

$$\|w(\Theta(\cdot,\varsigma))\|_{L^{\sigma'}(\gamma)} \leq C\|w\|_{W^{k,\sigma'}(\Omega)}, \tag{2.5}$$

$\forall w \in W^{k,\sigma'}(\Omega)$, *and all* $\varsigma \in]-1,1[^{N-D-1}$, *where* $\frac{N-D}{\sigma'} < k < [\frac{N-D}{\sigma'}] + 1$.

Proof. The proof may be performed with a system of local charts, in order to use Theorem 7.58 in [1]. ∎

Let J_Θ be the Jacobian of Θ, that is the determinant of $J\Theta$. We introduce an approximation of $u\delta_\gamma$, denoted by $u_n f_n$, and defined as follows. Set

$$\mathcal{O}_n = \Theta\Big(\gamma\times]-\frac{1}{n},\frac{1}{n}[^{N-D-1}\Big), \quad f_n = \Big(\frac{n}{2}\Big)^{N-D-1}\frac{1}{|J_\Theta \circ \Theta^{-1}|}\chi_{\mathcal{O}_n},$$

where $\chi_{\mathcal{O}_n}$ is the characteristic function of $\mathcal{O}_n$. We define u_n on $\mathcal{O}_n$ by $u_n(\Theta(\zeta,\varsigma),t) = u(\Theta(\zeta,0),t) = u(\zeta,t)$, $\forall \zeta \in \gamma$, $\forall \varsigma \in]-\frac{1}{n},\frac{1}{n}[^{N-D-1}$, and $\forall t \in [0,T]$.

Lemma 2.4 – *Let u_n and f_n be defined as above. There exists a constant C such that*

$$\Big|\int_\Gamma u_n(s,t)f_n(s)w(s)\,ds\Big| \leq C\Big(\int_\gamma |u(\zeta,t)|^\sigma d\zeta\Big)^{\frac{1}{\sigma}}\|w\|_{W^{k,\sigma'}(\Omega)},$$

for almost every $t \in]0,T[$, for all $w \in W^{k,\sigma'}(\Omega)$, with $\frac{N-D}{\sigma'} < k < [\frac{N-D}{\sigma'}]+1$.

Proof. We conclude with Lemma 2.3, since from the definition of u_n and f_n, one has

$$\begin{aligned}
&\Big|\int_\Gamma u_n(s,t)f_n(s)w(s)\,ds\Big| \\
&= \Big(\frac{n}{2}\Big)^{N-D-1}\Big|\int_{\Theta(\gamma\times]-\frac{1}{n},\frac{1}{n}[^{N-D-1})}\frac{u_n(s,t)}{|J_\Theta(\Theta^{-1}(s))|}w(s)ds\Big| \\
&= \Big(\frac{n}{2}\Big)^{N-D-1}\Big|\int_{\gamma\times]-\frac{1}{n},\frac{1}{n}[^{N-D-1}} u_n(\Theta(\zeta,\varsigma),t)w(\Theta(\zeta,\varsigma))d\zeta d\varsigma\Big| \\
&\leq \Big(\frac{n}{2}\Big)^{N-D-1}\Big(\int_{\gamma\times]-\frac{1}{n},\frac{1}{n}[^{N-D-1}} |u(\zeta,t)|^\sigma d\zeta d\varsigma\Big)^{\frac{1}{\sigma}} \times \\
&\qquad \times\Big(\int_{\gamma\times]-\frac{1}{n},\frac{1}{n}[^{N-D-1}} |w(\Theta(\zeta,\varsigma))|^{\sigma'} d\zeta d\varsigma\Big)^{\frac{1}{\sigma'}} \\
&\leq \Big(\frac{n}{2}\Big)^{N-D-1}\Big(\frac{2}{n}\Big)^{N-D-1}\Big(\int_\gamma |u(\zeta,t)|^\sigma d\zeta\Big)^{\frac{1}{\sigma}} \times \\
&\qquad \times \sup_{\varsigma\in]-1,1[^{N-D-1}} \|w(\Theta(\cdot,\varsigma))\|_{L^{\sigma'}(\gamma)}.
\end{aligned}$$

∎

Lemma 2.5 – *Let u_n and f_n be defined as above. The sequence $(u_n f_n)_n$ converges to $u\delta_\gamma$ in the weak-star topology of $L^q(0,T;\mathcal{M}_\lfloor(-))$.*

Proof. We prove that $|\int_0^T \int_\Gamma u_n f_n \varphi ds dt - \int_0^T \int_\gamma u(\zeta,t)\varphi(\zeta,t)d\zeta dt| \longrightarrow 0$ when $n \to \infty$, for all φ belonging to $L^{q'}(0,T;C(\Gamma))$. We have

$$\begin{aligned}
&\Big|\int_0^T \int_\Gamma u_n(s,t) f_n(s)\varphi(s,t)\, ds dt - \int_0^T \int_\gamma u(\zeta,t)\varphi(\zeta,t)d\zeta dt\Big| \\
&= \Big|\left(\frac{n}{2}\right)^{N-D-1} \int_0^T \int_{\Theta(\gamma\times]-\frac{1}{n},\frac{1}{n}[^{N-D-1})} \frac{u_n(s,t)}{|J_\Theta(\Theta^{-1}(s))|}\varphi(s,t)ds dt \\
&\qquad - \int_0^T \int_\gamma u(\zeta,t)\varphi(\zeta,t)d\zeta dt\Big| \\
&= \Big|\left(\frac{n}{2}\right)^{N-D-1} \int_0^T \int_{\gamma\times]-\frac{1}{n},\frac{1}{n}[^{N-D-1}} u(\zeta,t)\varphi(\Theta(\zeta,\varsigma),t)d\zeta d\varsigma dt \\
&\qquad - \int_0^T \int_\gamma u(\zeta,t)\varphi(\Theta(\zeta,0),t)d\zeta dt\Big| \\
&\leq \int_0^T \int_\gamma |u(\zeta,t)| \sup_{\varsigma\in]-\frac{1}{n},\frac{1}{n}[^{N-D-1}} |\varphi(\Theta(\zeta,\varsigma),t) - \varphi(\Theta(\zeta,0),t)|d\zeta dt \longrightarrow 0,
\end{aligned}$$

as $n \to \infty$. The proof is complete. ■

Proposition 2.1 – *Let ϕ be in $\mathcal{D}(\Omega)$, and w be the solution of the Cauchy problem:*

$$\frac{\partial w}{\partial t} + Aw = 0 \ in\ Q, \qquad \frac{\partial w}{\partial n_A} = 0 \ on\ \Sigma, \qquad w(0) = A^\alpha_{d'}\phi \ in\ \Omega, \quad (2.6)$$

where $0 \leq \alpha \leq 1$. The mapping that associates w with ϕ is continuous from $L^{d'}(\Omega)$ into $L^i(0,T;W^{k,j}(\Omega))$, for all (α,j,i,k,d) satisfying:

$$2 > k > 0, \quad i \geq 1, \quad j \geq d', \quad \alpha + \frac{k}{2} + \frac{N}{2d'} < \frac{1}{i} + \frac{N}{2j}. \qquad (2.7)$$

Consequently, the mapping that associates $T_{\gamma\times]0,T[}(w)$ with ϕ is continuous from $L^{d'}(\Omega)$ into $L^i(0,T;L^j(\gamma))$ for all (α,j,i,d) satisfying:

$$i \geq 1, \quad j \geq d', \quad \alpha + \frac{N}{2d'} < \frac{1}{i} + \frac{D}{2j}. \qquad (2.8)$$

Proof. Let (α, k, j, i, d) satisfy (2.7). Since the operator $(-A_{d'})$ is the infinitesimal generator of an analytic semigroup $(S_{d'}(t))_{t\geq 0}$ in $L^{d'}(\Omega)$, we have:

$$w(t) = S_{d'}(t)A^{\alpha}_{d'}\phi = A^{\alpha}_{d'}S_{d'}(t)\phi.$$

Observe that if β obeys $k < 2\beta$, then $D(A^{\beta}_{j}) \hookrightarrow W^{k,j}(\Omega)$. Moreover, if $d' \leq j < \infty$, with (2.3) and (2.4) in Lemma 2.1, we have:

$$\begin{aligned}\|w(t)\|_{W^{k,j}(\Omega)} = \|A^{\alpha}_{d'}S_{d'}(t)\phi\|_{W^{k,j}(\Omega)} &\leq C\|A^{\alpha+\beta}_{d'}S_{d'}(t)\phi\|_{L^{j}(\Omega)} \\ &\leq Ct^{-\alpha-\beta-\frac{N}{2}(\frac{1}{d'}-\frac{1}{j})}\|\phi\|_{L^{d'}(\Omega)}.\end{aligned}$$

Due to (2.7), we can choose $\beta > k/2$ to have $\alpha+\beta+\frac{N}{2}(\frac{1}{d'}-\frac{1}{j}) < 1$. Since the mapping $t \mapsto t^{-\alpha-\beta-\frac{N}{2}(\frac{1}{d'}-\frac{1}{j})}$ belongs to $L^{i}(0,T)$ for all i satisfying $\alpha+\beta+\frac{N}{2}(\frac{1}{d'}-\frac{1}{j}) < \frac{1}{i} \leq 1$, the mapping $\phi \mapsto w$ is continuous from $L^{d'}(\Omega)$ into $L^{i}(0,T;W^{k,j}(\Omega))$.

Let (α, j, i, d) satisfying (2.8). There exists k obeying (2.7), such that $kj > N - D$. Thus $T_{\gamma}(W^{k,j}(\Omega)) \hookrightarrow L^{j}(\gamma)$, and the proof is complete. ■

Proposition 2.2 – *Let f be in $\mathcal{D}(Q)$, and z be the solution to the equation:*

$$\frac{\partial z}{\partial t} + Az = f \text{ in } Q, \qquad \frac{\partial z}{\partial n_A} = 0 \text{ on } \Sigma, \qquad z(0) = 0 \text{ in } \Omega. \tag{2.9}$$

The mapping that associates z with f is continuous from $L^{\tilde{\eta}}(0,T;L^{\eta}(\Omega))$ into $L^{\delta}(0,T;W^{\beta,d}(\Omega))$ for all $(\beta,\tilde{\eta},\eta,\delta,d)$ satisfying:

$$0 \leq \beta < 2,\ 1 \leq \tilde{\eta} \leq \delta,\ 1 < \eta \leq d,\ \frac{\beta}{2} + \frac{1}{\tilde{\eta}} + \frac{N}{2\eta} < \frac{1}{\delta} + \frac{N}{2d} + 1. \tag{2.10}$$

Proof. Due to (2.10), there exists α such that

$$\frac{\beta}{2} < \alpha < 1,\ \frac{\beta}{2} < \alpha < \frac{1}{\delta} - \frac{1}{\tilde{\eta}} + 1 - \frac{N}{2\eta} + \frac{N}{2d}. \tag{2.11}$$

Thus $D(A^{\alpha}_{d}) \hookrightarrow W^{\beta,d}(\Omega)$. Let w be the solution of the Cauchy problem (2.6), then:

$$\begin{aligned}\int_{\Omega} A^{\alpha}_{d}z(x,t)\phi(x)\,dx &= \int_{\Omega} A^{\alpha}_{d'}\phi(x)z(x,t)\,dx \\ &= \int_0^t \frac{d}{d\tau}\Big(\int_{\Omega} w(x,t-\tau)z(x,\tau)dx\Big)d\tau \\ &= \int_0^t\int_{\Omega} \big(Aw(x,t-\tau)z(x,\tau) - w(x,t-\tau)Az(x,\tau)\big)dxd\tau \\ &\qquad + \int_0^t\int_{\Omega} w(x,t-\tau)f(x,\tau)dxd\tau.\end{aligned}$$

Hence

$$\int_\Omega A_d^\alpha z(x,t)\phi(x)\,dx = \int_0^t \int_\Omega w(x,t-\tau)f(x,\tau)dxd\tau$$
$$= \int_0^t \int_\Omega A_{d'}^\alpha S_{d'}(t-\tau)\phi(x)f(x,\tau)dxd\tau.$$

Using (2.4) in Lemma 2.1 with $1 \le d' \le \eta' \le \infty, d' < \infty$, we have:

$$\|z(t)\|_{W^{\beta,d}(\Omega)} \le C \sup\left\{|\int_\Omega A_d^\alpha z(t)\phi(x)\,dx|, \|\phi\|_{L^{d'}(\Omega)} = 1\right\}$$
$$= C \sup\left\{|\int_0^t \int_\Omega A_{d'}^\alpha S_{d'}(t-\tau)\phi\, f(\tau)dxd\tau, \|\phi\|_{L^{d'}(\Omega)} = 1\right\}$$
$$\le K \int_0^t (t-\tau)^{-\alpha-\frac{N}{2}(\frac{1}{\eta}-\frac{1}{d})}\|f(\tau)\|_{L^\eta(\Omega)}d\tau.$$

The mapping $\tau \mapsto \|f(\tau)\|_{L^\eta(\Omega)}$ belongs to $L^{\tilde{\eta}}(0,T)$. We set $\frac{1}{i} = 1 + \frac{1}{\delta} - \frac{1}{\tilde{\eta}}$. Due to (2.10) and (2.11), we have $i \ge 1$ and $\alpha + \frac{N}{2}(\frac{1}{\eta} - \frac{1}{d}) < \frac{1}{i}$. Thus the mapping $t \mapsto t^{-\alpha-\frac{N}{2}(\frac{1}{\eta}-\frac{1}{d})}$ belongs to $L^i(0,T)$. Therefore, using estimates on convolution, we have:

$$\Big(\int_0^T \Big(\int_0^t (t-\tau)^{-\alpha-\frac{N}{2}(\frac{1}{\eta}-\frac{1}{d})}\|f(\tau)\|_{L^\eta(\Omega)}d\tau\Big)^\delta dt\Big)^{\frac{1}{\delta}} \le C\|f\|_{L^{\tilde{\eta}}(0,T;L^\eta(\Omega))},$$

and the proof is complete. ∎

2.3. Linear equations with measure

We first consider the equation

$$\frac{\partial y}{\partial t} + Ay = 0 \text{ in } Q, \qquad \frac{\partial y}{\partial n_A} = u\delta_\gamma \text{ on } \Sigma, \qquad y(0) = 0 \text{ in } \Omega, \tag{2.12}$$

where u belongs to $L^q(0,T;L^\sigma(\gamma))$.

Proposition 2.3 – *Equation (2.12) admits a unique solution y_u in the space $L^1(0,T;W^{1,1}(\Omega))$. The mapping that associates y_u with u is continuous from $L^q(0,T;L^\sigma(\gamma))$ into $L^{\tilde{r}}(0,T;L^r(\Omega))$ for every $(\tilde{r},r)$ satisfying:*

$$\begin{aligned} &q \le \tilde{r} \le \infty, \quad \sigma \le r < \tfrac{N}{N-\frac{D}{\sigma'}-2}, \quad \tfrac{N-D}{2} + \tfrac{D}{2\sigma} + \tfrac{1}{q} < \tfrac{N}{2r} + \tfrac{1}{\tilde{r}} + 1, \\ &\qquad\qquad \textit{if } \sigma < \tfrac{N-D}{N-D-2}, \\ &q \le \tilde{r} \le \infty, \quad 1 \le r < \tfrac{N-D}{N-D-2}, \quad \tfrac{N-D}{2} + \tfrac{1}{q} < \tfrac{N-D}{2r} + \tfrac{1}{\tilde{r}} + 1, \\ &\qquad\qquad \textit{if } \sigma \ge \tfrac{N-D}{N-D-2}. \end{aligned} \tag{2.13}$$

The mapping $u \mapsto y_u$ from $L^q(0,T;L^\sigma(\gamma))$ into $L^{\delta_1}(0,T;W^{1,d_1}(\Omega))$ is also continuous for every (δ_1, d_1) satisfying

$$\begin{aligned} &q \le \delta_1, \quad \sigma \le d_1 < \tfrac{N}{N-\frac{D}{\sigma'}-1}, \quad \tfrac{N-D}{2} + \tfrac{D}{2\sigma} + \tfrac{1}{q} < \tfrac{1}{\delta_1} + \tfrac{N}{2d_1} + \tfrac{1}{2}, \\ &\qquad \textit{if } \sigma < \tfrac{N-D}{N-D-1}, \\ &q \le \delta_1, \quad 1 < d_1 < \tfrac{N-D}{N-D-1}, \quad \tfrac{N-D}{2} + \tfrac{1}{q} < \tfrac{1}{\delta_1} + \tfrac{N-D}{2d_1} + \tfrac{1}{2}, \\ &\qquad \textit{if } \sigma \ge \tfrac{N-D}{N-D-1}. \end{aligned} \tag{2.14}$$

The mapping $u \mapsto T_\Sigma(y_u)$ from $L^q(0,T;L^\sigma(\gamma))$ into $L^{\tilde{s}}(0,T;L^s(\Gamma))$ is continuous for every $(\tilde{s}, s)$ satisfying:

$$\begin{aligned} &q \le \tilde{s} \le \infty, \quad \sigma \le s < \tfrac{N-1}{N-\frac{D}{\sigma'}-2}, \quad \tfrac{N-D}{2} + \tfrac{D}{2\sigma} + \tfrac{1}{q} < \tfrac{1}{\tilde{s}} + \tfrac{N-1}{2s} + 1, \\ &\qquad \textit{if } \sigma < \tfrac{N-D-1}{N-D-2}, \\ &q \le \tilde{s} \le \infty, \quad s < \tfrac{N-D-1}{N-D-2}, \quad \tfrac{N-D}{2} + \tfrac{1}{q} < \tfrac{1}{\tilde{s}} + \tfrac{N-D-1}{2s} + 1, \\ &\qquad \textit{if } \sigma \ge \tfrac{N-D-1}{N-D-2}. \end{aligned} \tag{2.15}$$

Proof. 1 - Let us prove that the mapping $u \mapsto y_u$ is continuous from $L^q(0,T;L^\sigma(\gamma))$ into $L^{\delta_1}(0,T;W^{1,d_1}(\Omega))$. For this we proceed by duality. Let w be the solution of the Cauchy problem (2.6) with $\alpha = \frac{1}{2}$, $d = d_1$. By a straightforward calculation, we can prove that

$$\int_\Omega A_{d_1}^{\frac{1}{2}} y(x,t)\phi(x)\,dx = \int_0^t \int_\gamma u(\zeta,\tau)w(\zeta,t-\tau)d\zeta d\tau.$$

Therefore we have

$$\begin{aligned} \|y(t)\|_{W^{1,d_1}(\Omega)} &\le C\|A_{d_1}^{\frac{1}{2}} y(t)\|_{L^{d_1}(\Omega)} \\ &\le C \sup\left\{ \Big| \int_0^t \int_\gamma u(\tau)w(\zeta,t-\tau)d\zeta d\tau \Big|, \|\phi\|_{L^{d_1'}(\Omega)} = 1 \right\} \\ &\le C \sup\left\{ \Big| \int_0^t \|u(\tau)\|_{L^\sigma(\gamma)} \|w(t-\tau)\|_{L^{\sigma'}(\gamma)} d\tau \Big|, \|\phi\|_{L^{d_1'}(\Omega)} = 1 \right\}. \end{aligned}$$

Observe that if $\frac{N-D}{\sigma'} < k < 2\beta$, then $T_\gamma(D(A_{\sigma'}^\beta)) \hookrightarrow T_\gamma(W^{k,\sigma'}(\Omega)) \hookrightarrow L^{\sigma'}(\gamma)$. By using estimate (2.4) in Lemma 2.1, with $1 \le d_1' \le \sigma' \le \infty$, $d_1' < \infty$, and $\alpha = \frac{1}{2}$, we obtain:

$$\|y(t)\|_{W^{1,d_1}(\Omega)} \le C \int_0^t (t-\tau)^{-\frac{N}{2}(\frac{1}{\sigma}-\frac{1}{d_1})-\frac{1}{2}-\beta} \|u(\tau)\|_{L^\sigma(\gamma)} d\tau.$$

If $\sigma < \frac{N-D}{N-D-1}$, due to (2.14), we can choose $\beta > \frac{N-D}{2\sigma'}$ to have $\frac{N-D}{2} + \frac{D}{2\sigma} - \frac{N}{2d_1} + \frac{1}{2} < \frac{N}{2}(\frac{1}{\sigma} - \frac{1}{d_1}) + \frac{1}{2} + \beta < 1 + \frac{1}{\delta_1} - \frac{1}{q}$. From estimates for convolution, we deduce

$$\|y\|_{L^{\delta_1}(0,T;W^{1,d_1}(\Omega))} \leq C\|u\|_{L^q(0,T;L^\sigma(\gamma))}.$$

The case when $\sigma \geq \frac{N-D}{N-D-1}$ may be deduced from the previous one.
2 - Let us prove that the mapping that associates y_u, the solution to equation (2.12), with u is continuous from $L^q(0,T;L^\sigma(\gamma))$ into $L^{\tilde{r}}(0,T;L^r(\Omega))$. Let w be the solution of the Cauchy problem (2.6) with $\alpha = 0$, $d = r$. If $\sigma < \frac{N-D}{N-D-2}$, due to (2.13), we can choose $\beta > \frac{N-D}{2\sigma'}$ to have $\frac{N-D}{2} + \frac{D}{2\sigma} - \frac{N}{2r} < \frac{N}{2}(\frac{1}{\sigma} - \frac{1}{r}) + \beta < 1 + \frac{1}{\tilde{r}} - \frac{1}{q}$. We can conclude with the same arguments as in Step 1. The continuity of the mapping $u \mapsto T_\Sigma(y_u)$ may be obtained in the same manner. ■

Corollary 2.1 – *Let y be the solution to equation (2.12). The mapping that associates y with u, is continuous from $L^q(0,T;L^\sigma(\gamma))$ into $L^r(Q)$ for every r satisfying*

$$1 \leq r < \inf\Big\{\frac{N+2}{(N - \frac{2}{q'} - \frac{D}{\sigma'})}, \frac{N}{(N-2-\frac{D}{\sigma'})}, \frac{N-D+2}{(N-D-\frac{2}{q'})}, \frac{N-D}{N-D-2}\Big\}. \quad (2.16)$$

The mapping that associates $y|_\Sigma$ with u, is continuous from $L^q(0,T;L^\sigma(\gamma))$ into $L^s(\Sigma)$ for every s satisfying

$$1 \leq s < \inf\Big\{\frac{N+1}{(N - \frac{2}{q'} - \frac{D}{\sigma'})}, \frac{N-1}{(N-2-\frac{D}{\sigma'})}, \frac{N-D+1}{(N-D-\frac{2}{q'})}, \frac{N-D-1}{N-D-2}\Big\}. \quad (2.17)$$

Proof. To prove the first result, we distinguish four cases.
1 - If $q < \frac{N}{(N-2-\frac{D}{\sigma'})}$ and if $\sigma < \frac{N-D+2}{(N-D-\frac{2}{q'})}$, then $q < \frac{N-D}{(N-D-2)}$ and $\sigma < \frac{N-D}{(N-D-2)}$. Due to Proposition 2.3, y belongs to $L^r(Q)$ for all r satisfying $q \leq r$, $\sigma \leq r$, $\frac{N-D}{2} + \frac{D}{2\sigma} + \frac{1}{q} < \frac{N}{2r} + \frac{1}{r} + 1$,, i.e., $\sup\{q, \sigma\} \leq r < \frac{N+2}{(N-\frac{2}{q'}-\frac{D}{\sigma'})}$.
2 - If $q \geq \frac{N}{(N-2-\frac{D}{\sigma'})}$ and if $\sigma < \inf\{\frac{N-D+2}{(N-D-\frac{2}{q'})}, \frac{N-D}{(N-D-2)}\}$, then u belongs to $L^r(0,T;L^\sigma(\gamma))$ for all $r < \frac{N}{(N-2-\frac{D}{\sigma'})}$. Due to Proposition 2.3, y belongs to $L^r(Q)$ for all r satisfying $\sigma \leq r$, $\frac{N-D}{2} + \frac{D}{2\sigma} < \frac{N}{2r} + 1$, that is equivalent to $\sigma \leq r < \frac{N}{(N-2-\frac{D}{\sigma'})}$.
3 - If $q < \inf\{\frac{N}{(N-2-\frac{D}{\sigma'})}, \frac{N-D}{(N-D-2)}\}$ and if $\sigma \geq \frac{N-D+2}{(N-D-\frac{2}{q'})}$, then u belongs to $L^q(0,T;L^r(\gamma))$ for all $r < \frac{N-D+2}{(N-D-\frac{2}{q'})}$. Due to Proposition 2.3, y belongs to $L^r(Q)$ for all r satisfying $q \leq r$, $\frac{N-D}{2} + \frac{D}{2r} + \frac{1}{q} < \frac{1}{r} + \frac{N}{2r} + 1$, that is equivalent to $q \leq r < \frac{N-D+2}{(N-D-\frac{2}{q'})}$.

4 - If $q \geq \frac{N-D}{(N-D-2)}$ and if $\sigma \geq \frac{N-D}{(N-D-2)}$, then u belongs to $L^r(0,T;L^r(\gamma))$ for all $r < \frac{N-D}{(N-D-2)}$. Due to Proposition 2.3, y belongs to $L^r(Q)$ for all r satisfying $\frac{N-D}{2}+\frac{D}{2r}+\frac{1}{r} < \frac{1}{r}+\frac{N}{2r}+1$, that is equivalent to $1 \leq r < \frac{N-D}{(N-D-2)}$. The second result for the trace of y on Σ can be obtained in the same manner. ∎

Proposition 2.4 – *Let $u_n f_n$ be the approximation of $u\delta_\gamma$ defined in Section 2.2. Let y_n be the solution to the equation:*

$$\frac{\partial y}{\partial t} + Ay = 0 \ in \ Q, \qquad \frac{\partial y}{\partial n_A} = u_n f_n \ on \ \Sigma, \qquad y(0) = 0 \ in \ \Omega. \tag{2.18}$$

There exist constants $C(\tilde{r}, r)$, $C(\tilde{s}, s)$, $C(\delta_1, d_1)$, independent of n, such that

$$\|y_n\|_{L^{\tilde{r}}(0,T;L^r(\Omega))} \leq C(\tilde{r}, r)\|u\|_{L^q(0,T;L^\sigma(\gamma))}$$

for every $(\tilde{r}, r)$ satisfying (2.13),

$$\|y_n|_\Sigma\|_{L^{\tilde{s}}(0,T;L^s(\Gamma))} \leq C(\tilde{s}, s)\|u\|_{L^q(0,T;L^\sigma(\gamma))}$$

for every $(\tilde{s}, s)$ satisfying (2.15),

$$\|y_n\|_{L^{\delta_1}(0,T;W^{1,d_1}(\Omega))} \leq C(\delta_1, d_1)\|u\|_{L^q(0,T;L^\sigma(\gamma))}$$

for every (δ_1, d_1) satisfying (2.14).

Proof. Let w be the solution of (2.6). Let us prove the estimate in the space $L^{\delta_1}(0,T;W^{1,d_1}(\Omega))$. With Lemma 2.4 and the identity:

$$\int_\Omega A_{d_1}^\alpha y_n(x,t)\phi(x)\,dx = \int_0^t \int_\Gamma u_n(s,\tau)f_n(s)w(s,t-\tau)ds d\tau \quad \text{for } \alpha = \frac{1}{2},$$

we can write:

$$\begin{aligned} &\|y_n(t)\|_{W^{1,d_1}(\Omega)} \\ &\leq C \sup\left\{\int_0^t \|u(\tau)\|_{L^\sigma(\gamma)}\|w(t-\tau)\|_{W^{k,\sigma'}(\Omega)}d\tau \mid \|\phi\|_{L^{d_1'}(\Omega)} = 1\right\}, \end{aligned}$$

where $k > \frac{N-D}{\sigma'}$. The rest of the proof is similar to that of the previous proposition. ∎

Proposition 2.5 – *Let y be the solution to the equation*

$$\frac{\partial y}{\partial t} + Ay = 0 \ in \ Q, \qquad \frac{\partial y}{\partial n_A} = 0 \ on \ \Sigma, \qquad y(0) = y_0 \ in \ \Omega. \tag{2.19}$$

The mapping $y_0 \mapsto y$ is continuous from $L^\rho(\Omega)$ into $L^{\tilde r}(0,T;L^r(\Omega)) \cap L^{\delta_4}(0,T;W^{1,d_4}(\Omega))$, for every $(\tilde r, r)$ satisfying (2.13), and for every (δ_4, d_4) satisfying

$$1 < \delta_4 < 2, \quad \rho \le d_4 < \frac{N\rho}{N-\rho}, \quad 1 + \frac{N}{2\rho} < \frac{1}{\delta_4} + \frac{N}{2d_4} + \frac{1}{2}. \tag{2.20}$$

Consequently, y belongs to $L^r(Q)$ for all r satisfying (2.16). Moreover, the mapping $y_0 \mapsto y|_\Sigma$ is continuous from $L^\rho(\Omega)$ into $L^{\tilde s}(0,T;L^s(\Gamma))$ for all $(\tilde s, s)$ satisfying (2.15), such that $\tilde s < 2\rho$. Consequently, $y|_\Sigma$ belongs to $L^s(\Sigma)$ for all s satisfying (2.17).

Proof. 1 - To prove that $y \in L^{\delta_4}(0,T;W^{1,d_4}(\Omega))$ for every (δ_4, d_4) satisfying (2.20), it is sufficient to use Proposition 2.1 with $\alpha = 0$ and $k = 1$.
2 - Let $(\tilde r, r)$ be a pair satisfying (2.13). If $\sigma < \frac{N-D}{N-D-2}$ and if $r \ge \rho$, then $1 + \frac{N}{2\rho} \le 1 + \frac{N}{2} - \frac{D}{2\sigma'} - \frac{1}{q'} < \frac{N}{2r} + \frac{1}{\tilde r} + 1$. From Proposition 2.1 with $k = 0$, it follows that y belongs to $L^{\tilde r}(0,T;L^r(\Omega))$. If $r < \rho$, then $y \in L^\infty(0,T;L^\rho(\Omega)) \hookrightarrow L^{\tilde r}(0,T;L^r(\Omega))$.
3 - Let $(\tilde s, s)$ be a pair satisfying (2.15), such that $\tilde s < 2\rho$. If $\sigma < \frac{N-D-1}{N-D-2}$ and if $s \ge \rho$, then we have $1 + \frac{N}{2\rho} \le 1 + \frac{N}{2} - \frac{D}{2\sigma'} - \frac{1}{q'} < \frac{N-1}{2s} + \frac{1}{\tilde s} + 1$. From Proposition 2.1, it follows that $y|_\Sigma$ belongs to $L^{\tilde s}(0,T;L^s(\Gamma))$. If $s < \rho$, we have $1 + \frac{N}{2\rho} < \frac{N-1}{2\rho} + \frac{1}{\tilde s} + 1$ (because $\tilde s < 2\rho$). Thus $y|_\Sigma$ belongs to $L^{\tilde s}(0,T;L^\rho(\Gamma)) \hookrightarrow L^{\tilde s}(0,T;L^s(\Gamma))$.
4 - Since $1 + \frac{N}{2\rho} < \frac{N-1}{2\rho} + \frac{1}{\rho} + 1$, the trace of y on Σ belongs to $L^\rho(\Sigma)$. If s satisfies (2.17), then $s \le \rho$, and the trace of y on Σ belongs to $L^s(\Sigma)$. ■

2.4. Linearized equation

Let $\bar y$ be the solution to equation (1.1). Then $\bar y$ is also a solution to the linear equation

$$\frac{\partial y}{\partial t} + Ay + ay = 0 \text{ in } Q, \quad \frac{\partial y}{\partial n_A} + by = u\delta_\gamma \text{ on } \Sigma, \quad y(0) = y_0 \text{ in } \Omega, \tag{2.21}$$

where $a = \int_0^1 \Phi'(\chi \bar y) d\chi$ and $b = \int_0^1 \Psi'(\chi \bar y) d\chi$. (We have supposed that $\Phi(0) = 0$ and $\Psi(0) = 0$, see (A3) and (A4).) Observe that, in the case where $u \ge 0$ and $y_0 \ge 0$, with a comparison principle, we can prove that

$$0 \le \bar y \le \hat y \ \text{ a. e. in } Q, \quad \text{and} \quad 0 \le \bar y|_\Sigma \le \hat y|_\Sigma \ \text{ a. e. in } \Sigma,$$

where $\hat y$ is the solution of (2.21) corresponding to $a \equiv 0$ and $b \equiv 0$. Hence it is natural to look for a solution $\bar y$ to (1.1), belonging to $L^{\tilde r}(0,T;L^r(\Omega))$ for every $(\tilde r, r)$ satisfying (2.13), and such that $\bar y|_\Sigma$ belongs to $L^{\tilde s}(0,T;L^s(\Gamma))$

for every $(\tilde{s}, s)$ satisfying (2.15) and $\tilde{s} < 2\rho$. In this case, we can verify that the coefficient a belongs to $L^{\tilde{k}}(0,T;L^k(\Omega))$ for every $(\tilde{k}, k)$ satisfying

$$\begin{aligned} &\tfrac{q}{\mu-1} \le \tilde{k},\ \tfrac{\sigma}{\mu-1} \le k,\ \tfrac{N-D}{2} + \tfrac{D}{2\sigma} + \tfrac{1}{q} < \tfrac{N}{2(\mu-1)k} + \tfrac{1}{(\mu-1)\tilde{k}} + 1 \\ &\qquad \text{if} \quad \sigma < \tfrac{N-D}{N-D-2}, \\ &\tfrac{q}{\mu-1} \le \tilde{k},\ 1 \le k,\ \tfrac{N-D}{2} + \tfrac{1}{q} < \tfrac{N-D}{2(\mu-1)k} + \tfrac{1}{(\mu-1)\tilde{k}} + 1 \\ &\qquad \text{if} \quad \sigma \ge \tfrac{N-D}{N-D-2}, \end{aligned} \tag{2.22}$$

and b belongs to $L^{\tilde{\ell}}(0,T;L^\ell(\Gamma))$ for every $(\tilde{\ell}, \ell)$ satisfying

$$\begin{aligned} &\tfrac{q}{\nu-1} \le \tilde{\ell} < \tfrac{2\rho}{\nu-1},\ \tfrac{\sigma}{\nu-1} \le \ell,\ \tfrac{N-D}{2} + \tfrac{D}{2\sigma} + \tfrac{1}{q} < \tfrac{N-1}{2(\nu-1)\ell} + \tfrac{1}{(\nu-1)\tilde{\ell}} + 1 \\ &\qquad \text{if}\ \sigma < \tfrac{N-D-1}{N-D-2}, \\ &\tfrac{q}{\nu-1} \le \tilde{\ell} < \tfrac{2\rho}{\nu-1},\ 1 \le \ell\ \tfrac{N-D}{2} + \tfrac{1}{q} < \tfrac{N-D-1}{2(\nu-1)\ell} + \tfrac{1}{(\nu-1)\tilde{\ell}} + 1 \\ &\qquad \text{if}\ \sigma \ge \tfrac{N-D-1}{N-D-2}. \end{aligned} \tag{2.23}$$

Proposition 2.6 – *Assume that a is a nonnegative function belonging to $L^{\tilde{k}}(0,T;L^k(\Omega))$ for every $(\tilde{k},k)$ satisfying (2.22), and b is a nonnegative function belonging to $L^{\tilde{\ell}}(0,T;L^\ell(\Gamma))$ for every $(\tilde{\ell},\ell)$ satisfying (2.23). Equation (2.21) admits a unique solution y in $L^1(0,T;W^{1,1}(\Omega))$. This solution belongs to $L^{\tilde{r}}(0,T;L^r(\Omega))$ for every $(\tilde{r},r)$ satisfying (2.13), and there exists a constant $C(\tilde{r},r)$, not depending on the functions a and b, such that*

$$\|y\|_{L^{\tilde{r}}(0,T;L^r(\Omega))} \le C(\tilde{r},r)(\|u\|_{L^q(0,T;L^\sigma(\gamma))} + \|y_0\|_{L^p(\Omega)}). \tag{2.24}$$

In particular for $\tilde{r} = \infty$ and for every $1 < r < \inf\{\frac{N-D}{N-D-\frac{2}{q'}}, \frac{N}{N-\frac{D}{\sigma'}-\frac{2}{q'}}\}$, y belongs to $C([0,T];L^r_w(\Omega))$ and there exists a constant $C(r)$, not depending on the functions a and b, such that

$$\|y\|_{L^\infty(0,T;L^r(\Omega))} \le C(r)(\|u\|_{L^q(0,T;L^\sigma(\gamma))} + \|y_0\|_{L^p(\Omega)}). \tag{2.25}$$

($C([0,T];L^r_w(\Omega))$ denotes the space of continuous functions from $[0,T]$ into $L^r(\Omega)$ endowed with its weak topology.)

Moreover, the trace $y|_\Sigma$ belongs to $L^{\tilde{s}}(0,T;L^s(\Gamma))$ for every $(\tilde{s},s)$ satisfying (2.15) and $\tilde{s} < 2\rho$. There exists a constant $C(\tilde{s},s)$, not depending on the functions a and b, such that

$$\|y|_\Sigma\|_{L^{\tilde{s}}(0,T;L^s(\Gamma))} \le C(\tilde{s},s)(\|u\|_{L^q(0,T;L^\sigma(\gamma))} + \|y_0\|_{L^p(\Omega)}). \tag{2.26}$$

Proof. 1 - Let $\hat{y}$ be the solution to (2.21) corresponding to a, $b \equiv 0$. For $\hat{y}$, estimates (2.24), (2.25) and (2.26) follow from Propositions 2.3 and 2.5.

2 - Since a and b are nonnegative functions, with a comparison principle we can prove that $\|y\|_{L^{\tilde r}(0,T;L^r(\Omega))} \leq 2\|\hat y\|_{L^{\tilde r}(0,T;L^r(\Omega))}$, $\|y\|_{L^\infty(0,T;L^r(\Omega))} \leq 2\|\hat y\|_{L^\infty(0,T;L^r(\Omega))}$, and $\|y|_\Sigma\|_{L^{\tilde s}(0,T;L^s(\Gamma))} \leq 2\|\hat y|_\Sigma\|_{L^{\tilde s}(0,T;L^s(\Gamma))}$.
3 - In particular, when $\tilde r = \infty$, we can proceed as in the proof of Proposition 2.1 in [13], to conclude that y belongs to $C([0,T];L^r_w(\Omega))$. ∎

Now, we want to prove that the solution of equation (2.21) belongs to the space $L^{\delta_1}(0,T;W^{1,d_1}(\Omega)) + L^{\delta_2}(0,T;W^{1,d_2}(\Omega)) + L^{\delta_3}(0,T;W^{1,d_3}(\Omega)) + L^{\delta_4}(0,T;W^{1,d_4}(\Omega))$, where (δ_1,d_1) obeys (2.14), (δ_4,d_4) obeys (2.20), (δ_2,d_2) and (δ_3,d_3) satisfy:

$$\begin{aligned} &\tfrac{q}{\mu} \leq \delta_2, \quad \tfrac{\sigma}{\mu} \leq d_2, \quad \mu\left(\tfrac{N-D}{2} + \tfrac{D}{2\sigma} + \tfrac{1}{q} - 1\right) < \tfrac{1}{\delta_2} + \tfrac{N}{2d_2} + \tfrac{1}{2}, \\ &\qquad \text{if} \quad \sigma < \tfrac{N-D}{N-D-2}, \\ &\tfrac{q}{\mu} \leq \delta_2, \quad 1 \leq d_2, \quad \tfrac{\mu}{q} + \tfrac{\mu N(N-D-2)}{2(N-D)} < \tfrac{N}{2d_2} + \tfrac{1}{\delta_2} + \tfrac{1}{2} \\ &\qquad \text{if} \quad \sigma \geq \tfrac{N-D}{N-D-2}, \end{aligned} \tag{2.27}$$

$$\begin{aligned} &\tfrac{q}{\nu} \leq \delta_3 < \tfrac{2\rho}{\nu}, \quad \tfrac{\sigma}{\nu} \leq d_3, \quad \nu\left(\tfrac{N-D}{2} + \tfrac{D}{2\sigma} + \tfrac{1}{q} - 1\right) < \tfrac{1}{\delta_3} + \tfrac{N}{2d_3} \\ &\qquad \text{if} \quad \sigma < \tfrac{N-D-1}{N-D-2}, \\ &\tfrac{q}{\nu} \leq \delta_3, \quad 1 \leq d_3, \quad \tfrac{\nu}{q} + \tfrac{\nu(N-1)(N-D-2)}{2(N-D-1)} < \tfrac{N}{2d_3} + \tfrac{1}{\delta_3} \\ &\qquad \text{if} \quad \sigma \geq \tfrac{N-D-1}{N-D-2}. \end{aligned} \tag{2.28}$$

Proposition 2.7 – *Assume that a and b satisfy the assumptions of Proposition 2.6. Let y be the solution of equation (2.21). Denote by ξ_1 the solution of equation (2.12), by ξ_4 the solution of equation (2.19). Let ξ_2 and ξ_3 be the solutions to the equations*

$$\frac{\partial \xi_2}{\partial t} + A\xi_2 = -ay \text{ in } Q, \qquad \frac{\partial \xi_2}{\partial n_A} = 0 \text{ on } \Sigma, \qquad \xi_2(0) = 0 \text{ in } \Omega,$$

$$\frac{\partial \xi_3}{\partial t} + A\xi_3 = 0 \text{ in } Q, \qquad \frac{\partial \xi_3}{\partial n_A} = -by \text{ on } \Sigma, \qquad \xi_3(0) = 0 \text{ in } \Omega .$$

Then $y = \xi_1 + \xi_2 + \xi_3 + \xi_4$, and the following estimates hold for some $(\tilde k, k)$ satisfying (2.22), and some $(\tilde\ell, \ell)$ satisfying (2.23):

$$\|\xi_2\|_{L^{\delta_2}(0,T;W^{1,d_2}(\Omega))} \leq C\|a\|_{L^{\tilde k}(0,T;L^k(\Omega))}(\|u\|_{L^q(0,T;L^\sigma(\gamma))} + \|y_0\|_{L^\rho(\Omega)}),$$

$$\|\xi_3\|_{L^{\delta_3}(0,T;W^{1,d_3}(\Omega))} \leq C(\|u\|_{L^q(0,T;L^\sigma(\gamma))} + \|y_0\|_{L^\rho(\Omega)}).$$

Proof. Due to Proposition 2.6, y belongs to $L^{\tilde r}(0,T;L^r(\Omega))$ for every $(\tilde r, r)$ satisfying (2.13), and $y|_\Sigma$ belongs to $L^{\tilde s}(0,T;L^s(\Gamma))$ for every $(\tilde s, s)$ satisfying (2.15) and $\tilde s < 2\rho$.

Let (δ_2, d_2) be a pair satisfying (2.27). We take $\tilde{k} = \delta_2\mu'$, $k = d_2\mu'$, $\tilde{r} = \delta_2\mu$, $r = d_2\mu$. Then $(\tilde{k}, k)$ satisfies (2.22), and $(\tilde{r}, r)$ satisfies (2.13). Moreover, we have

$$\frac{1}{\tilde{k}} + \frac{1}{\tilde{r}} = \frac{1}{\delta_2}, \quad \frac{1}{k} + \frac{1}{r} = \frac{1}{d_2}, \text{ and } \frac{1}{\tilde{k}} + \frac{1}{\tilde{r}} + \frac{N}{2}(\frac{1}{k} + \frac{1}{r}) < \frac{1}{\delta_2} + \frac{N}{2d_2} + \frac{1}{2}.$$

Due to Proposition 2.2, it follows that

$$\|\xi_2\|_{L^{\delta_2}(0,T;W^{1,d_2}(\Omega))} \le C\|a\|_{L^{\tilde{k}}(0,T;L^k(\Omega))}\|y\|_{L^{\tilde{r}}(0,T;L^r(\Omega))}.$$

Now, let (δ_3, d_3) be a pair satisfying (2.28). We take $\tilde{\ell} = \delta_3\nu'$, $k = d_3\nu'$, $\tilde{r} = \delta_3\nu$, $r = d_3\nu$. Then $(\tilde{\ell}, \ell)$ satisfies (2.23), and $(\tilde{s}, s)$ satisfies (2.15) and $\tilde{s} < 2\rho$. Moreover, we have

$$\frac{1}{\tilde{\ell}} + \frac{1}{\tilde{s}} = \frac{1}{\delta_3}, \quad \frac{1}{\ell} + \frac{1}{s} = \frac{1}{d_3}, \text{ and } \frac{1}{\tilde{\ell}} + \frac{1}{\tilde{s}} + \frac{N-1}{2}(\frac{1}{\ell} + \frac{1}{s}) < \frac{1}{\delta_3} + \frac{N}{2d_3}.$$

Using the same arguments as in the proof of Proposition 3.2 in [27] (see also Lemma 4.3), we obtain

$$\|\xi_3\|_{L^{\delta_3}(0,T;W^{1,d_3}(\Omega))} \le C\|b\|_{L^{\tilde{\ell}}(0,T;L^{\ell}(\Gamma))}\|y|_\Sigma\|_{L^{\tilde{s}}(0,T;L^s(\Gamma))},$$

and the proof is complete. ■

3. State equation

Definition 3.1 *We say that $y \in L^1(0,T;W^{1,1}(\Omega))$ is a weak solution of equation (1.1) if $\Phi(\cdot, y(\cdot))$ belongs to $L^1(Q)$, $\Psi(\cdot, y(\cdot))$ belongs to $L^1(\Sigma)$, and*

$$-\int_Q y\frac{\partial\phi}{\partial t}\, dxdt + \int_Q \sum_{i,j=1}^N a_{ij}(x)D_j y D_i\phi\, dxdt + \int_Q a_0 y\phi\, dxdt$$

$$+\int_Q \Phi(y)\phi\, dxdt + \int_\Sigma \Psi(y)\phi\, dsdt = \int_0^T\int_\gamma u\phi\, d\zeta\, dt + \int_\Omega \phi(0)y_0\, dx$$

for every $\phi \in C^1(\overline{Q})$ such that $\phi(T) = 0$ on $\overline{\Omega}$.

3.1. Compactness results

To study equation (1.1), we need the following lemma:

Lemma 3.1 – *Let λ be a measure belonging to $\mathcal{M}_b(\overline{Q} \setminus (\overline{\Omega} \times \{T\}))$. Let y_λ be the solution to the equation:*

$$\frac{\partial y}{\partial t} + Ay = \lambda_Q \text{ in } Q, \qquad \frac{\partial y}{\partial n_A} = \lambda_\Sigma \text{ on } \Sigma, \qquad y(0) = \lambda_{\overline{\Omega}} \text{ in } \overline{\Omega}, \tag{3.1}$$

where λ_Q (respectively λ_Σ, $\lambda_{\overline{\Omega}}$) denotes the restriction of λ to Q (respectively Σ, $\overline{\Omega} \times \{0\}$). The mapping $\Lambda : \lambda \mapsto (y_\lambda, y_\lambda|_\Sigma)$ is a compact linear operator from $\mathcal{M}_b(\overline{Q} \setminus (\overline{\Omega} \times \{T\}))$ into $L^1(Q) \times L^1(\Sigma)$.

Proof. Let m and n be such that $m > \frac{N}{2} + 1$ and $n > N + 1$. The operator Λ is continuous from $\mathcal{M}_b(\overline{Q} \setminus (\overline{\Omega} \times \{T\}))$ into $L^{m'}(Q) \times L^{n'}(\Sigma)$ (see [24]). To prove the compactness we proceed by transposition. Let z be the solution to

$$-\frac{\partial z}{\partial t} + Az = \phi \text{ in } Q, \qquad \frac{\partial z}{\partial n_A} = \psi \text{ on } \Sigma, \qquad z(T) = 0 \text{ in } \Omega.$$

With a Green formula (see Theorem 4.1 in [24]), we have

$$\int_{\overline{Q} \setminus (\overline{\Omega} \times \{T\})} z d\lambda = \int_Q y\phi \, dxdt + \int_\Sigma y\psi \, dsdt$$

for all $\lambda \in \mathcal{M}_b(\overline{Q} \setminus (\overline{\Omega} \times \{T\}))$, all $\phi \in L^m(Q)$, all $\psi \in L^n(\Sigma)$.

Set $L(\phi, \psi) = z$. Using the same arguments as in [25], proof of Proposition 3.2, we prove that L is continuous from $L^m(Q) \times L^n(\Sigma)$ into $C^{0,\alpha}(\overline{Q}) \cap C_0(\overline{Q} \setminus (\overline{\Omega} \times \{T\}))$, for some $\alpha > 0$. So, L is a compact operator from $L^m(Q) \times L^n(\Sigma)$ into $C_0(\overline{Q} \setminus (\overline{\Omega} \times \{T\}))$. Since $\Lambda = L^*$, the proof is complete. ∎

3.2. Nonlinear equation

Lemma 3.2 – *If the function a belongs to $L^{\tilde{k}}(0, T; L^k(\Omega))$ for every $(\tilde{k}, k)$ satisfying (2.22), and if y belongs to $L^r(Q)$ for every r satisfying (2.16), then ay belongs to $L^\alpha(Q)$ for every α such that*

$$1 \leq \alpha < \inf\left\{\frac{N+2}{\mu(N - \frac{D}{\sigma'} - \frac{2}{q'})}, \frac{N}{\mu(N - \frac{D}{\sigma'} - 2)}, \frac{N-D+2}{\mu(N - D - \frac{2}{q'})}, \frac{N-D}{\mu(N-D-2)}\right\}. \tag{3.2}$$

Proof. As in the proof of Corollary 2.1, we can prove that a belongs to $L^{\frac{r}{\mu-1}}(Q)$ for every r satisfying (2.16). Thus ay belongs to $L^\alpha(Q)$ for all α such that $\frac{1}{\alpha} = \frac{\mu-1}{r} + \frac{1}{r} = \frac{\mu}{r}$, for some r obeying (2.16). The proof is complete. ∎

Lemma 3.3 – *If the function b belongs to $L^{\tilde{\ell}}(0,T;L^{\ell}(\Gamma))$ for every $(\tilde{\ell},\ell)$ satisfying (2.23), and if $y|_\Sigma$ belongs to $L^s(\Sigma)$ for every s satisfying (2.17), then $by|_\Sigma$ belongs to $L^\beta(\Sigma)$ for every β such that*

$$1 \le \beta < \inf\left\{\frac{N+1}{\nu(N-\frac{D}{\sigma'}-\frac{2}{q'})}, \frac{N-1}{\nu(N-\frac{D}{\sigma'}-2)}, \frac{N-D+1}{\nu(N-D-\frac{2}{q'})}, \frac{N-D-1}{\nu(N-D-2)}\right\}. \tag{3.3}$$

Proof. Using the second part of Corollary 2.1, the proof is similar to the previous one. ∎

Proposition 3.8 – *Assume that a and b satisfy the assumptions of Proposition 2.6. Let $u_n f_n$ be the approximation of $u\delta_\gamma$ defined in Section 2.2. Let y_n be the solution of the equation*

$$\frac{\partial y}{\partial t} + Ay + ay = 0 \text{ in } Q, \quad \frac{\partial y}{\partial n_A} + by = u_n f_n \text{ on } \Sigma, \quad y(0) = y_0 \text{ in } \Omega. \tag{3.4}$$

Then the following estimates hold:

$$\|y_n\|_{L^{\tilde{r}}(0,T;L^r(\Omega))} \le C(\|u\|_{L^q(0,T;L^\sigma(\gamma))} + \|y_0\|_{L^p(\Omega)}),$$

for all $(\tilde{r},r)$ satisfying (2.13);

$$\|y_n|_\Sigma\|_{L^{\tilde{s}}(0,T;L^s(\Gamma))} \le C(\|u\|_{L^q(0,T;L^\sigma(\gamma))} + \|y_0\|_{L^p(\Omega)}),$$

for all $(\tilde{s},s)$ satisfying (2.15) and $\tilde{s} < 2\rho$.

In the above estimates, the constants C do not depend on a and b.

Proof. This result can be obtained as in the proof of Proposition 2.6. ∎

Theorem 3.1 – *The state equation (1.1) admits a unique solution, it belongs to $L^{\tilde{r}}(0,T;L^r(\Omega))$ for every $(\tilde{r},r)$ satisfying (2.13) and we have*

$$\|y\|_{L^{\tilde{r}}(0,T;L^r(\Omega))} \le C(\|u\|_{L^q(0,T;L^\sigma(\gamma))} + \|y_0\|_{L^p(\Omega)}).$$

The solution of (1.1) also belongs to $C([0,T];L^r_w(\Omega))$ for every $1 \le r < \inf\{\frac{N-D}{N-D-\frac{2}{q'}}, \frac{N}{N-\frac{D}{\sigma'}-\frac{2}{q'}}\}$, and we have

$$\|y\|_{L^\infty(0,T;L^r(\Omega))} \le C(\|u\|_{L^q(0,T;L^\sigma(\gamma))} + \|y_0\|_{L^p(\Omega)}).$$

The solution y of (1.1) also belongs to $L^{\delta_1}(0,T;W^{1,d_1}(\Omega)) + L^{\delta_2}(0,T;W^{1,d_2}(\Omega)) + L^{\delta_3}(0,T;W^{1,d_3}(\Omega)) + L^{\delta_4}(0,T;W^{1,d_4}(\Omega))$, for all (δ_1,d_1) obeying (2.14), all (δ_2,d_2) obeying (2.27), all (δ_3,d_3) obeying (2.28), and all (δ_4,d_4) obeying (2.20).

Moreover, we have

$$\|y - \xi_4\|_{L^{\delta_1}(0,T;W^{1,d_1}(\Omega)) + L^{\delta_2}(0,T;W^{1,d_2}(\Omega)) + L^{\delta_3}(0,T;W^{1,d_3}(\Omega))}$$

$$+\|\xi_4\|_{L^{\delta_4}(0,T;W^{1,d_4}(\Omega))} \leq C(\|u\|_{L^q(0,T;L^\sigma(\gamma))} + \|u\|^\mu_{L^q(0,T;L^\sigma(\gamma))}$$

$$+\|u\|^\nu_{L^q(0,T;L^\sigma(\gamma))} + \|y_0\|^\mu_{L^\rho(\Omega)} + \|y_0\|^\nu_{L^\rho(\Omega)} + \|y_0\|_{L^\rho(\Omega)}),$$

where ξ_4 *is the solution of*

$$\frac{\partial \xi_4}{\partial t} + A\xi_4 = 0 \ in\ Q, \qquad \frac{\partial \xi_4}{\partial n_A} = 0 \ on\ \Sigma, \qquad \xi_4(0) = y_0 \ in\ \Omega.$$

Proof. We approximate $u\delta_\gamma$ by the sequence of functions $(u_n f_n)_n$ defined in Section 2.2. We denote by y_n the solution of (1.1) corresponding to $u_n f_n$.
1 - Estimate in $L^{\tilde{r}}(0,T;L^r(\Omega))$.

By setting $a_n = \int_0^1 \Phi'(\chi y_n)d\chi$ and $b_n = \int_0^1 \Psi'(\chi y_n)d\chi$, from Proposition 3.8, it follows that $(y_n)_n$ is bounded in $L^{\tilde{r}}(0,T;L^r(\Omega))$ for all $(\tilde{r},r)$ satisfying (2.13), $(y_n|_\Sigma)_n$ is bounded in $L^{\tilde{s}}(0,T;L^s(\Gamma))$ for all $(\tilde{s},s)$ satisfying (2.15) and $\tilde{s} < 2\rho$. Therefore, with assumption (A4), the sequence $(a_n)_n$ is bounded in $L^{\tilde{k}}(0,T;L^k(\Omega))$ for all $(\tilde{k},k)$ satisfying (2.22), and the sequence $(b_n)_n$ is bounded in $L^{\tilde{\ell}}(0,T;L^\ell(\Gamma))$ for all $(\tilde{\ell},\ell)$ satisfying (2.23). Due to Lemmas 3.2 and 3.3, the sequence $(a_n y_n)_n$ is bounded in $L^\alpha(Q)$ for all α satisfying (3.2), and the sequence $(b_n y_n|_\Sigma)_n$ is bounded in $L^\beta(\Sigma)$ for all β satisfying (3.3). From Lemma 3.1, it follows that the sequence $(y_n)_n$ is relatively compact in $L^1(Q)$, and the sequence $(y_n|_\Sigma)_n$ is relatively compact in $L^1(\Sigma)$.
2 - Estimate in $\sum_{j=1}^4 L^{\delta_j}(0,T;W^{1,d_j}(\Omega))$. Let ξ_{1n}, ξ_{2n}, ξ_{3n} be the solutions of the following equations

$$\frac{\partial \xi_{1n}}{\partial t} + A\xi_{1n} = 0 \text{ in } Q, \qquad \frac{\partial \xi_{1n}}{\partial n_A} = u_n f_n \text{ on } \Sigma, \qquad \xi_{1n}(0) = 0 \text{ in } \Omega,$$

$$\frac{\partial \xi_{2n}}{\partial t} + A\xi_{2n} = -a_n y_n \text{ in } Q, \qquad \frac{\partial \xi_{2n}}{\partial n_A} = 0 \text{ on } \Sigma, \qquad \xi_{2n}(0) = 0 \text{ in } \Omega,$$

$$\frac{\partial \xi_{3n}}{\partial t} + A\xi_{3n} = 0 \text{ in } Q, \qquad \frac{\partial \xi_{3n}}{\partial n_A} = -b_n y_n \text{ on } \Sigma, \qquad \xi_{3n}(0) = 0 \text{ in } \Omega.$$

Then we have $y_n = \xi_{1n} + \xi_{2n} + \xi_{3n} + \xi_4$. From Propositions 2.4,we have

$$\|\xi_{1n}\|_{L^{\delta_1}(0,T;W^{1,d_1}(\Omega))} \leq C(\delta_1, d_1)\|u\|_{L^q(0,T;L^\sigma(\gamma))},$$

for all (δ_1, d_1) satisfying (2.14).

Let (δ_2, d_2) satisfy (2.27). As in Proposition 2.7, we can find $(\tilde{k}, k)$ satisfying (2.22) such that

$$\|\xi_{2n}\|_{L^{\delta_2}(0,T;W^{1,d_2}(\Omega))} \le C\|a_n\|_{L^{\tilde{k}}(0,T;L^k(\Omega))}(\|u\|_{L^q(0,T;L^\sigma(\gamma))} + \|y_0\|_{L^\rho(\Omega)}).$$

From assumption (A4), and due to Proposition 3.8, we can write

$$\|\xi_{2n}\|_{L^{\delta_2}(0,T;W^{1,d_2}(\Omega))} \le C(\|u\|_{L^q(0,T;L^\sigma(\gamma))} + \|y_0\|_{L^\rho(\Omega)} + \|u\|^\mu_{L^q(0,T;L^\sigma(\gamma))} + \|y_0\|^\mu_{L^\rho(\Omega)}),$$

for all (δ_2, d_2) satisfying (2.27), and all n. With the same arguments, we obtain

$$\|\xi_{3n}\|_{L^{\delta_3}(0,T;W^{1,d_3}(\Omega))} \le C(\|u\|_{L^q(0,T;L^\sigma(\gamma))} + \|y_0\|_{L^\rho(\Omega)} + \|u\|^\nu_{L^q(0,T;L^\sigma(\gamma))} + \|y_0\|^\nu_{L^\rho(\Omega)}),$$

for all (δ_3, d_3) satisfying (2.28), and all n. Due to Proposition 2.5, we also have

$$\|\xi_4\|_{L^{\delta_4}(0,T;W^{1,d_4}(\Omega))} \le C\|y_0\|_{L^\rho(\Omega)}, \qquad \text{for all } (\delta_4, d_4) \text{ satisfying (2.20).}$$

3 - Passage to the limit. There exists a subsequence, still indexed by n, such that $(y_n)_n$ converges to some y in $L^1(Q)$ and almost everywhere in Q, $(y_n|_\Sigma)_n$ converges to some z in $L^1(\Sigma)$ and almost everywhere in Σ. It is clear that $z = y|_\Sigma$. Due to Lemma 4.1 in [13], we can also prove that $(y_n)_n$ converges to y in $L^r(Q)$ for all r satisfying (2.16), and $(y_n|_\Sigma)_n$ converges to $y|_\Sigma$ in $L^s(\Sigma)$ for all s satisfying (2.17). Thus $(a_n y_n)_n$ converges to $\Phi(y)$ in $L^\alpha(Q)$ for all α satisfying (3.2), and $(b_n y_n)_n$ converges to $\Psi(y)$ in $L^\beta(\Sigma)$ for all β satisfying (3.3). Now by passing to the limit in the variational formulation satisfied by y_n, we can show that y is a solution of (1.1).
We also obtain estimates of y in $L^{\delta_1}(0,T;W^{1,d_1}(\Omega)) + L^{\delta_2}(0,T;W^{1,d_2}(\Omega)) + L^{\delta_3}(0,T;W^{1,d_3}(\Omega)) + L^{\delta_4}(0,T;W^{1,d_4}(\Omega))$, for all (δ_1, d_1) obeying (2.14), all (δ_2, d_2) obeying (2.27), all (δ_3, d_3) obeying (2.28), and all (δ_4, d_4) obeying (2.20), by passing to the limit in Step 2.

Remark 3.1. Due to Theorem 3.1, if u belongs to $L^\infty(0,T;L^\infty(\gamma))$, if $\mu < \frac{N-D}{N-D-2}$ and $\nu < \frac{N-D-1}{N-D-2}$, then equation (1.1) admits a unique weak solution. Now consider the case when u belongs to $L^1(0,T;L^1(\gamma))$, $\mu < \frac{N-D}{N-D-2}$ and $\nu < \frac{N-D-1}{N-D-2}$. Set $\tilde{u}_n := \max(\min(n,u), -n)$, and denote by y_n the solution of (1.1) corresponding to $\tilde{u}_n$. As in [8], using monotonicity arguments, we can prove that $(y_n)_n$ converges to some function y in $L^1(Q)$, $(y_n|_\Sigma)_n$ converges to $y|_\Sigma$ in $L^1(\Sigma)$, $y \in L^\mu(Q)$, $y \in L^\nu(\Sigma)$, and y is the weak solution of (1.1) corresponding to u. Therefore, we have proved the existence of a unique solution even if μ and ν do not satisfy (2.1) and (2.2), but only $\mu < \frac{N-D}{N-D-2}$ and $\nu < \frac{N-D-1}{N-D-2}$. However the regularity results stated in Theorem 3.1 are not true in this case.

4. Control problem ($\mathbf{P}_1$)

In all the sequel, we suppose that the control set K_U is a closed convex subset of $L^q(0,T;L^\sigma(\gamma))$.

4.1. Existence of solutions to problem ($\mathbf{P}_1$)

Theorem 4.1 – *Assume that hypotheses (A1) to (A6) are fulfilled. Suppose that y_d belongs to $L^\theta(\Omega)$, and V_d belongs to $L^\kappa(0,T;(L^\kappa(\Omega))^N)$. Suppose that there exist (δ_1,d_1) satisfying (2.14), (δ_2,d_2) satisfying (2.27), (δ_3,d_3) satisfying (2.28) and (δ_4,d_4) satisfying (2.20), such that $1<\kappa<\delta_i$ and $\kappa<d_i$ for $i=1,...,4$. Suppose in addition that $1<\theta<\rho$, and that either K_U is bounded in $L^q(0,T;L^\sigma(\gamma))$, or $C_\gamma>0$. Then the control problem (P_1) admits solutions.*

Proof. Let $(u_n)_n$ be a minimizing sequence in K_U. Then $(u_n)_n$ is bounded in $L^q(0,T;L^\sigma(\gamma))$. We can suppose that $(u_n)_n$ converges to some u for the weak-star topology of $L^q(0,T;L^\sigma(\gamma))$. Since K_U is convex and closed in $L^q(0,T;L^\sigma(\gamma))$, then $u\in K_U$. Let y_n be the solution of (1.1) corresponding to u_n. Due to Theorem 3.1, the sequence $(y_n)_n$ is bounded in $L^\kappa(0,T;W^{1,\kappa}(\Omega))$, and the sequence $(y_n(T))_n$ is bounded in $L^\theta(\Omega)$. Let y_u be the solution of (1.1) corresponding to u. Using the same arguments as in the proof of Theorem 3.1, we can prove that $(y_n)_n$ converges to y_u for the weak topology of $L^\kappa(0,T;W^{1,\kappa}(\Omega))$, and $(y_n(T))_n$ converges to $y_u(T)$ for the weak topology of $L^\theta(\Omega)$. By classical arguments, we can prove that u is a solution of (P_1). ■

4.2. Regularity results for the adjoint equation

First, to study the adjoint equation for the control problem (P_1), we have to suppose that there exist $(\tilde{k},k)$ obeying (2.22), and $(\tilde{\ell},\ell)$ obeying (2.23), such that

$$\frac{1}{\tilde{k}}+\frac{N}{2k}<1,\quad \frac{1}{\tilde{\ell}}+\frac{N-1}{2\ell}<\frac{1}{2}. \tag{4.1}$$

Setting $\tilde{k}=\frac{q}{\mu-1}$ and $\tilde{\ell}=\frac{q}{\nu-1}$, the above condition will be satisfied if assumption (A7), stated as below, holds.

(A7) - In addition to conditions (2.1) and (2.2), we suppose that μ and ν satisfy

$$\begin{aligned}
&\mu\left(\tfrac{N-D}{2}+\tfrac{D}{2\sigma}+\tfrac{1}{q}-1\right)<\tfrac{N-D}{2}+\tfrac{D}{2\sigma}+\tfrac{1}{q} &&\text{if}\quad \sigma<\tfrac{N-D}{N-D-2},\\
&\mu\left(\tfrac{N-D}{2}+\tfrac{D}{2\bar{\sigma}}+\tfrac{1}{q}-1\right)<\tfrac{N-D}{2}+\tfrac{D}{2\bar{\sigma}}+\tfrac{1}{q} &&\text{if}\quad \sigma\geq\bar{\sigma}=\tfrac{N-D}{N-D-2},
\end{aligned} \tag{4.2}$$

and

$$\begin{aligned} &\nu(\tfrac{N-D}{2}+\tfrac{D}{2\sigma}+\tfrac{1}{q}-1)<\tfrac{N-D}{2}+\tfrac{D}{2\sigma}+\tfrac{1}{q}-\tfrac{1}{2} \\ &\qquad \text{if} \quad \sigma<\tfrac{N-D-1}{N-D-2}\,, \\ &\nu(\tfrac{N-D}{2}+\tfrac{D}{2\sigma_1}+\tfrac{1}{q}-1)<\tfrac{N-D}{2}+\tfrac{D}{2\sigma_1}+\tfrac{1}{q}-\tfrac{1}{2} \\ &\qquad \text{if} \quad \sigma\geq\sigma_1=\tfrac{N-D-1}{N-D-2}\,. \end{aligned} \tag{4.3}$$

Next, to simplify the analysis of Problem (P_1), we suppose that μ, ν, q, σ and ρ satisfy the following additionnal conditions.
(A8) - If $\frac{N-D}{N-D-1} \leq \sigma < \frac{N-D}{N-D-2}$, then

$$\mu(\frac{N-D}{2\sigma'}-\frac{1}{q'})<\frac{1}{q}+\frac{1}{2}. \tag{4.4}$$

If $\frac{N-D}{N-D-1} \leq \sigma < \frac{N-D-1}{N-D-2}$, then

$$\nu(\frac{N-D}{2\sigma'}-\frac{1}{q'})<\frac{1}{q}. \tag{4.5}$$

(A9) - We suppose that $\rho > \max\,(N-D, \frac{q\nu}{2}, \frac{N}{N-\frac{D}{\sigma'}-\frac{2}{q'}})$. The function V_d belongs to $L^q(0,T;(L^{\frac{N-D}{N-D-1}}(\Omega))^N)+(L^2(0,T;(L^\rho(\Omega))^N)$. The function y_d belongs to $L^{\hat{r}}(\Omega)$, where $\hat{r}=\inf\{\frac{N-D}{(N-D-\frac{2}{q'})}, \frac{N}{(N-\frac{D}{\sigma'}-\frac{2}{q'})}\}$.

Moreover, to study the control problem (P_1), we look for solution to equation (1.1) belonging to $L^\kappa(0,T;\,W^{1,\kappa}(\Omega))$. Therefore we must have $\kappa\leq\delta_i$, $\kappa\leq d_i$ for $1\leq i\leq 4$, where (δ_i,d_i) satisfy the conditions stated in Section 2. Observe that if (δ_1,d_1) satisfies (2.14), then $2\leq q\leq\delta_1$, and $d_1<\frac{N-D}{N-D-1}\leq 2$ (because $D\leq N-2$). This simple observation leads to simplifications stated in Proposition 4.1.

Proposition 4.1 – *Assume that (A7) and (A8) hold. If $q\geq 2$, if $\delta_1=q$, and (δ_1,d_1) obeys (2.14), then there exist (δ_2,d_2) obeying (2.27), and (δ_3,d_3) obeying (2.28), such that*

$$\delta_2=\delta_3=q,\quad d_2\geq d_1,\quad d_3\geq d_1.$$

Proof. We only prove the existence of (δ_2,d_2), the rest of the proof can be performed in a similar manner.
If $\sigma<\frac{N-D}{N-D-1}$, then $d_1<\frac{N}{N-\frac{D}{\sigma'}-1}$. We set $\delta_2=\delta_1=q$, $d_2=d_1$. Then $\delta_2\geq q>\frac{q}{\mu}$, and $d_2=d_1\geq\sigma>\frac{\sigma}{\mu}$. Moreover, with assumption (4.2), we have

$$\mu(\frac{N-D}{2}+\frac{D}{2\sigma}+\frac{1}{q}-1)<\frac{N-D}{2}+\frac{D}{2\sigma}+\frac{1}{q}<\frac{N}{2d_2}+\frac{1}{q}+\frac{1}{2}.$$

Thus the pair (δ_2, d_2) obeys (2.27). If $\frac{N-D}{N-D-1} \le \sigma < \frac{N-D}{N-D-2}$, then we have $d_1 < \frac{N-D}{N-D-1}$. We choose $\delta_2 = \delta_1 = q$. Due to assumptions (4.2) and (4.4), there exists $d_2 \ge d_1$ such that

$$\mu(\frac{N-D}{2} + \frac{D}{2\sigma} + \frac{1}{q} - 1) < \frac{N}{2d_2} + \frac{1}{q} + \frac{1}{2} \le \frac{N}{2d_1} + \frac{1}{q} + \frac{1}{2},$$

and

$$\mu(\frac{N-D}{2} + \frac{D}{2\sigma} + \frac{1}{q} - 1) < \frac{N}{2d_2} + \frac{1}{q} + \frac{1}{2} \le \frac{N\mu}{2\sigma} + \frac{1}{q} + \frac{1}{2}.$$

Then (δ_2, d_2) satisfies (2.27). If $\sigma \ge \frac{N-D}{N-D-2}$, then $d_1 < \frac{N-D}{N-D-1}$. We choose $\delta_2 = \delta_1 = q$ and $d_2 = d_1$. With assumption (4.2), we have

$$\begin{aligned}\mu(\frac{N-D}{2} + \frac{D}{2\sigma} + \frac{1}{q} - 1) &\le \mu(\frac{N-D}{2} + \frac{D(N-D-2)}{2(N-D)} + \frac{1}{q} - 1)\\ &< \frac{N-D}{2} + \frac{D(N-D-2)}{2(N-D)} + \frac{1}{q} < \frac{N}{2d_2} + \frac{1}{q} + \frac{1}{2}.\end{aligned}$$

Thus (2.27) is satisfied by (δ_2, d_2). ∎

Due to Proposition 4.1, with assumption (A9), we notice that $\frac{N-D}{N-D-1} \le 2 \le \sigma$, and that the solution to equation (1.1) belongs to $L^q(0,T;W^{1,d}(\Omega)) + L^{\delta_4}(0,T;W^{1,\rho}(\Omega))$, for all $d < \frac{N-D}{N-D-1}$ and all $\delta_4 < 2$. The assumption (A10) below is needed in the proof of Theorems 4.2, 4.3, 4.4.

(**A10**) - The exponents κ, μ, ν satisfy

$$\kappa < 1 + \frac{1}{N-D-1}, \quad \kappa < q, \quad \nu < q, \quad \mu < q, \tag{4.6}$$

$$1 < \nu < 1 + \frac{N-1}{\sigma'(N - \frac{D}{\sigma'} - 2)} \quad \text{if} \quad \sigma < \frac{N-D-1}{N-D-2}, \tag{4.7}$$

$$1 < \mu < 1 + \frac{N}{\sigma'(N - \frac{D}{\sigma'} - 2)} \quad \text{if} \quad \sigma < \frac{N-D}{N-D-2}. \tag{4.8}$$

Let u be a solution of (P_1), y_u be the solution of equation (1.1) corresponding to u, and set

$$a = \Phi'(y_u), \quad b = \Psi'(y_u), \quad \vec{h} = \kappa C_Q |\nabla y_u - V_d|^{\kappa-2}(\nabla y_u - V_d),$$
$$p_T = \theta C_\Omega |y_u(T) - y_d|^{\theta-2}(y_u(T) - y_d).$$

The adjoint equation for (P_1) associated with (y_u, u) is

$$\begin{aligned}&-\frac{\partial p}{\partial t} + Ap + ap = -\operatorname{div} \vec{h} \text{ in } Q, \quad \frac{\partial p}{\partial n_A} + bp = \vec{h}\cdot\vec{n} \text{ on } \Sigma,\\ &p(T) = p_T \text{ in } \Omega,\end{aligned} \tag{4.9}$$

where $\vec{n}$ is the unit normal to Γ outward Ω. When $\vec{h} \cdot \vec{n}$ is not defined, equation (4.9) is a formal writing for the variational equation

$$\int_Q (p\frac{\partial y}{\partial t} + \sum_{i,j=1}^{N} a_{ij} D_j p D_i y + a_0 py + apy)\, dxdt + \int_\Sigma bpy\, dsdt$$

$$= \int_Q \vec{h} \cdot \nabla y dxdt + \int_\Omega y(T) p_T dx,$$

for all $y \in C^1(\overline{Q})$ such that $y(0) = 0$.

Lemma 4.1 – *One has*

$$|\nabla y_u - V_d|^{\kappa-2}(\nabla y_u - V_d) \in L^{\frac{q}{\kappa-1}}(0,T;(L^{\frac{d}{\kappa-1}}(\Omega))^N) + L^{\frac{\delta_4}{\kappa-1}}(0,T;(L^{\frac{\rho}{\kappa-1}}(\Omega))^N)$$

for all $d < \frac{N-D}{N-D-1}$*, and all* $\delta_4 < 2$.

Proof. We know that $(\nabla y_u - V_d)$ belongs to $L^q(0,T;(L^d(\Omega))^N) + L^{\delta_4}(0,T;(L^\rho(\Omega))^N)$, for all $d < \frac{N-D}{N-D-1}$, and all $\delta_4 < 2$. To prove the Lemma, we verify that $|\nabla y_u - V_d|^{\kappa-1}$ belongs to $L^{\frac{q}{\kappa-1}}(0,T;L^{\frac{d}{\kappa-1}}(\Omega)) + L^{\frac{\delta_4}{\kappa-1}}(0,T;L^{\frac{\rho}{\kappa-1}}(\Omega))$. Set $\nabla y_u - V_d = \vec{g}_1 + \vec{g}_2$, where $\vec{g}_1$ belongs to $L^q(0,T;(L^d(\Omega))^N)$, and $\vec{g}_2$ belongs to $L^{\delta_4}(0,T;(L^\rho(\Omega))^N)$. Since $|\nabla y_u - V_d|^{\kappa-1} \le C(|\vec{g}_1|^{\kappa-1} + |\vec{g}_2|^{\kappa-1})$, we can prove that the mapping $f \mapsto \int_Q |\nabla y_u - V_d|^{\kappa-1} f\, dxdt$ belongs to $((L^{\frac{q}{\kappa-1}}(0,T;L^{\frac{d}{\kappa-1}}(\Omega)))' \cap (L^{\frac{\delta_4}{\kappa-1}}(0,T;L^{\frac{\rho}{\kappa-1}}(\Omega)))')'$.

But (see [30], Page 69) $((L^{\frac{q}{\kappa-1}}(0,T;L^{\frac{d}{\kappa-1}}(\Omega)))' \cap (L^{\frac{\delta_4}{\kappa-1}}(0,T;L^{\frac{\rho}{\kappa-1}}(\Omega)))')' \equiv (L^{\frac{q}{\kappa-1}}(0,T;L^{\frac{d}{\kappa-1}}(\Omega)) + L^{\frac{\delta_4}{\kappa-1}}(0,T;L^{\frac{\rho}{\kappa-1}}(\Omega)))$, the function $y_u \mapsto |\nabla y_u - V_d|^{\kappa-1}$ belongs to $L^{\frac{q}{\kappa-1}}(0,T;L^{\frac{d}{\kappa-1}}(\Omega)) + L^{\frac{\delta_4}{\kappa-1}}(0,T;L^{\frac{\rho}{\kappa-1}}(\Omega))$, and the proof is complete. ■

The equation (4.9) is linear, we can set $p = p_\kappa + p_\theta$, where p_κ is the solution to

$$-\frac{\partial p}{\partial t} + Ap + ap = -\operatorname{div} \vec{h} \text{ in } Q, \qquad \frac{\partial p}{\partial n_A} + bp = \vec{h}\cdot\vec{n} \text{ on } \Sigma, \qquad p(T) = 0 \text{ in } \Omega,$$

and p_θ is the solution to

$$-\frac{\partial p}{\partial t} + Ap + ap = 0 \text{ in } Q, \qquad \frac{\partial p}{\partial n_A} + bp = 0 \text{ on } \Sigma, \qquad p(T) = p_T \text{ in } \Omega.$$

Due to Lemma 4.1, $\vec{h} = \kappa|\nabla y_u - V_d|^{\kappa-2}(\nabla y_u - V_d) = \vec{h}^1 + \vec{h}^2$, where $\vec{h}^1$ belongs to $L^{\frac{q}{\kappa-1}}(0,T;(L^{\frac{d}{\kappa-1}}(\Omega))^N)$ for all $d < \frac{N-D}{N-D-1}$, and $\vec{h}^2$ belongs to

$L^{\frac{\delta_4}{\kappa-1}}(0,T;(L^{\frac{\rho}{\kappa-1}}(\Omega))^N)$ for all $\delta_4 < 2$. We can write $p_\kappa = p_\kappa^1 + p_\kappa^2$, where p_κ^1 is the solution corresponding to $\vec{h}^1$, and p_κ^2 is the solution corresponding to $\vec{h}^2$.

To establish optimality conditions for (P_1), we have to prove that the trace of p on $\gamma\times]0,T[$ belongs to $L^{q'}(0,T;L^{\sigma'}(\gamma))$. In assumptions (A10) and (A11), we state sufficient conditions on κ so that the trace of p_κ^1 on $\gamma\times]0,T[$ exists and belongs to $L^{q'}(0,T;L^{\sigma'}(\gamma))$ (Theorem 4.2). A similar result for p_κ^2 is established in Theorem 4.3 when assumptions (A10) and (A12) are fulfilled. The trace of p_θ on $\gamma\times]0,T[$ is studied in Theorem 4.4 under assumptions (A10) and (A13).

(A11) - Assumptions needed in the proof of Theorem 4.2.
(A11a) - Assumptions needed to estimate the term bp.

$$\frac{\kappa-1}{q} + \frac{(\kappa-1)(N-D-1)}{2(N-D)} < \frac{1}{q'} - \frac{\nu-1}{q} + \frac{1}{2}\,. \tag{4.10}$$

If $\sigma < \frac{N-D-1}{N-D-2}$, then

$$\frac{\kappa-1}{q} + \frac{N(\kappa-1)(N-D-1)}{2(N-D)} < \frac{1}{q'} - \frac{\nu-1}{q} + \frac{N-1}{2\sigma'} - \frac{(\nu-1)(N-\frac{D}{\sigma'}-2)}{2} + \frac{1}{2}\,, \tag{4.11}$$

$$\frac{N(\kappa-1)(N-D-1)}{2(N-D)} < \frac{N-1}{2\sigma'} - \frac{(\nu-1)(N-\frac{D}{\sigma'}-2)}{2} + \frac{1}{2}\,. \tag{4.12}$$

If $\sigma \geq \frac{N-D-1}{N-D-2}$, then

$$\frac{\kappa-1}{q} + \frac{N(\kappa-1)(N-D-1)}{2(N-D)} < \frac{1}{q'} - \frac{\nu-1}{q} + \frac{N-1}{2(N-D-1)} - \frac{(\nu-1)(N-1)(N-D-2)}{2(N-D-1)} + \frac{1}{2}\,, \tag{4.13}$$

$$\frac{N(\kappa-1)(N-D-1)}{2(N-D)} < \frac{N-1}{2(N-D-1)} - \frac{(\nu-1)(N-1)(N-D-2)}{2(N-D-1)} + \frac{1}{2}\,. \tag{4.14}$$

(A11b) - Assumptions needed to estimate the term ap.

$$\frac{\kappa-1}{q} < \frac{1}{q'} - \frac{\mu-1}{q} + \frac{1}{2}\,. \tag{4.15}$$

If $\sigma < \frac{N-D}{N-D-2}$, then

$$\frac{\kappa-1}{q} + \frac{N(\kappa-1)(N-D-1)}{2(N-D)} < \frac{1}{q'} - \frac{\mu-1}{q} + \frac{N}{2\sigma'} - \frac{(\mu-1)(N-\frac{D}{\sigma'}-2)}{2} + \frac{1}{2}, \tag{4.16}$$

$$\frac{N(\kappa-1)(N-D-1)}{2(N-D)} < \frac{N}{2\sigma'} - \frac{(\mu-1)(N-\frac{D}{\sigma'}-2)}{2} + \frac{1}{2}. \tag{4.17}$$

If $\sigma \geq \frac{N-D}{N-D-2}$, then

$$\frac{\kappa-1}{q} + \frac{N(\kappa-1)(N-D-1)}{2(N-D)} < \frac{1}{q'} - \frac{\mu-1}{q} + \frac{N}{N-D} - \frac{(\mu-1)N(N-D-2)}{2(N-D)} + \frac{1}{2}, \tag{4.18}$$

$$\frac{N(\kappa-1)(N-D-1)}{2(N-D)} < \frac{N}{N-D} - \frac{(\mu-1)N(N-D-2)}{2(N-D)} + \frac{1}{2}. \tag{4.19}$$

Lemma 4.2 – *Let $\vec{h}$ be in $L^{\tilde{\eta}}(0,T;(L^{\eta}(\Omega))^N)$, where $\tilde{\eta} > 1$ and $\eta > 1$. We consider the equation:*

$$-\frac{\partial p}{\partial t} + Ap = -\ div\, \vec{h}\ in\ Q, \quad \frac{\partial p}{\partial n_A} = \vec{h}\cdot\vec{n}\ on\ \Sigma, \quad p(T) = 0\ in\ \Omega. \tag{4.20}$$

The solution p to equation (4.20) belongs to $L^1(0,T;W^{1,1}(\Omega)) \cap L^{\tilde{r}}(0,T;L^r(\Omega))$ for all $(\tilde{r}, r)$ satisfying:

$$\tilde{\eta} \leq \tilde{r}, \quad \eta \leq r, \quad \frac{N}{2\eta} + \frac{1}{\tilde{\eta}} < \frac{N}{2r} + \frac{1}{\tilde{r}} + \frac{1}{2}. \tag{4.21}$$

The trace of p on Σ belongs to $L^{\tilde{s}}(0,T;L^s(\Gamma))$ for all $(\tilde{s}, s)$ satisfying:

$$\tilde{\eta} \leq \tilde{s}, \quad \eta \leq s, \quad \frac{N}{2\eta} + \frac{1}{\tilde{\eta}} < \frac{N-1}{2s} + \frac{1}{\tilde{s}} + \frac{1}{2}. \tag{4.22}$$

If moreover $\eta > N-D$, the trace of p on $\gamma\times]0,T[$ belongs to $L^{\tilde{\tau}}(0,T;L^{\tau}(\gamma))$ for all $(\tilde{\tau}, \tau)$ satisfying:

$$\tilde{\eta} \leq \tilde{\tau}, \quad \eta \leq \tau, \quad \frac{N}{2\eta} + \frac{1}{\tilde{\eta}} < \frac{D}{2\tau} + \frac{1}{\tilde{\tau}} + \frac{1}{2}. \tag{4.23}$$

Proof. 1 - We know that p belongs to $L^1(0,T;W^{1,1}(\Omega))$ (see [31]).
2 - We want to prove that p belongs to $L^{\tilde{r}}(0,T;L^r(\Omega))$. Let w be the solution to the equation

$$-\frac{\partial w}{\partial t}+Aw=0 \text{ in } Q, \qquad \frac{\partial w}{\partial n_A}=0 \text{ on } \Sigma, \qquad w(T)=\phi \text{ in } \Omega. \quad (4.24)$$

Then

$$\int_\Omega p(x,t)\phi(x)\,dx=\int_t^T\int_\Omega \nabla w(x,T+t-\tau)\cdot\vec{h}(x,\tau)dxd\tau.$$

Using (2.4) in Lemma 2.1 with $1\le r'\le\eta'\le\infty, r'<\infty$, and $\alpha=\frac{1}{2}$, we have:

$$\|p(t)\|_{L^r(\Omega)}\le C\,\sup\{|\int_\Omega p(x,t)\phi(x)\,dx|, \|\phi\|_{L^{r'}(\Omega)}=1\}$$

$$=C\,\sup\{|\int_t^T\int_\Omega \nabla w(x,T+t-\tau)\cdot\vec{h}(\tau)dxd\tau|, \|\phi\|_{L^{r'}(\Omega)}=1\}$$

$$\le C\int_t^T (T+t-\tau)^{-\frac{1}{2}-\frac{N}{2}(\frac{1}{\eta}-\frac{1}{r})}\|\vec{h}(\tau)\|_{(L^\eta(\Omega))^N}d\tau.$$

The mapping $\tau\mapsto\|\vec{h}(\tau)\|_{(L^\eta(\Omega))^N}$ belongs to $L^{\tilde{\eta}}(0,T)$. We set $\frac{1}{i}=1+\frac{1}{\tilde{r}}-\frac{1}{\tilde{\eta}}$. Due to (4.21), $i\ge 1$ and $\frac{1}{2}+\frac{N}{2}(\frac{1}{\eta}-\frac{1}{r})<\frac{1}{i}$. Thus the mapping $t\mapsto t^{-\frac{1}{2}-\frac{N}{2}(\frac{1}{\eta}-\frac{1}{r})}$ belongs to $L^i(0,T)$. Therefore, with estimates on convolution, we have

$$\Big(\int_0^T\Big(\int_t^T (T+t-\tau)^{-\frac{1}{2}-\frac{N}{2}(\frac{1}{\eta}-\frac{1}{r})}\|\vec{h}(\tau)\|_{(L^\eta(\Omega))^N}d\tau\Big)^{\tilde{r}}dt\Big)^{\frac{1}{\tilde{r}}}$$

$$\le C\|\vec{h}\|_{L^{\tilde{\eta}}(0,T;(L^\eta(\Omega))^N)},$$

and p belongs to $L^{\tilde{r}}(0,T;L^r(\Omega))$.
3 - We want to prove that the trace of p on Σ belongs to $L^{\tilde{s}}(0,T;L^s(\Gamma))$. Due to (4.22), there exists β such that

$$\frac{1}{2s}<\beta<1, \quad \beta\le\frac{1}{\tilde{s}}-\frac{1}{\tilde{\eta}}+\frac{1}{2}-\frac{N}{2\eta}+\frac{N}{2s}. \quad (4.25)$$

Thus $T_\Gamma(D(A_s^\beta))\hookrightarrow L^s(\Gamma)$. Let $\tilde{w}$ be the solution of the equation

$$-\frac{\partial\tilde{w}}{\partial t}+A\tilde{w}=0 \text{ in } Q, \quad \frac{\partial\tilde{w}}{\partial n_A}=0 \text{ on } \Sigma, \quad \tilde{w}(T)=A_{s'}^\beta\phi \text{ in } \Omega. \quad (4.26)$$

Then

$$\int_\Omega A_s^\beta p(x,t)\phi(x)\,dx = \int_t^T \int_\Omega \nabla\tilde{w}(x,T+t-\tau)\cdot\vec{h}(x,\tau)dxd\tau.$$

Using (2.4) in Lemma 2.1 with $1 \leq s' \leq \eta' \leq \infty, s' < \infty$, and $\alpha = \frac{1}{2}$, we have

$$\|p(t)|_\Gamma\|_{L^s(\Gamma)} \leq C \sup\left\{|\int_\Omega A_s^\beta p(x,t)\phi(x)\,dx|, \|\phi\|_{L^{s'}(\Omega)} = 1\right\}$$

$$= C \sup\left\{|\int_t^T \int_\Omega \nabla\tilde{w}(x,T+t-\tau)\cdot\vec{h}(x,\tau)dxd\tau|, \|\phi\|_{L^{s'}(\Omega)} = 1\right\}$$

$$\leq C\int_t^T (T+t-\tau)^{-\frac{1}{2}-\beta-\frac{N}{2}(\frac{1}{\eta}-\frac{1}{s})}\|\vec{h}(\tau)\|_{(L^\eta(\Omega))^N}d\tau.$$

The mapping $\tau \mapsto \|\vec{h}(\tau)\|_{(L^\eta(\Omega))^N}$ belongs to $L^{\tilde{\eta}}(0,T)$. We set $\frac{1}{i} = 1+\frac{1}{\tilde{s}}-\frac{1}{\tilde{\eta}}$. Due to (4.22) and (4.25), $i \geq 1$ and $\frac{1}{2}+\beta+\frac{N}{2}(\frac{1}{\eta}-\frac{1}{s}) < \frac{1}{i}$. Thus the mapping $t \mapsto t^{-\frac{1}{2}-\beta-\frac{N}{2}(\frac{1}{\eta}-\frac{1}{s})}$ belongs to $L^i(0,T)$. Therefore, using estimates on convolution, we have

$$\Big(\int_0^T \Big(\int_t^T (T+t-\tau)^{-\frac{1}{2}-\beta-\frac{N}{2}(\frac{1}{\eta}-\frac{1}{s})}\|\vec{h}(\tau)\|_{(L^\eta(\Omega))^N}d\tau\Big)^{\tilde{s}}dt\Big)^{\frac{1}{\tilde{s}}}$$

$$\leq C\|\vec{h}\|_{L^{\tilde{\eta}}(0,T;(L^\eta(\Omega))^N)},$$

and the proof is complete.
4 - We can proceed in the same manner to prove that the trace of p on $\gamma\times]0,T[$ belongs to $L^{\tilde{\tau}}(0,T;L^\tau(\gamma))$. ■

Lemma 4.3 – *Let g be in $L^{\tilde{\beta}}(0,T;L^\beta(\Gamma))$, where $\tilde{\beta} \geq 1$ and $\beta \geq 1$. We consider the equation:*

$$-\frac{\partial p}{\partial t} + Ap = 0 \ in\ Q, \qquad \frac{\partial p}{\partial n_A} = g \ on\ \Sigma, \qquad p(T) = 0 \ in\ \Omega. \qquad (4.27)$$

The solution p of the equation (4.27) belongs to $L^{\tilde{r}}(0,T;L^r(\Omega))$ for all $(\tilde{r},r)$ satisfying:

$$\tilde{\beta} \leq \tilde{r}, \quad \beta \leq r, \quad \frac{N-1}{2\beta} + \frac{1}{\tilde{\beta}} < \frac{N}{2r} + \frac{1}{\tilde{r}} + \frac{1}{2}. \qquad (4.28)$$

The solution p of the equation (4.27) belongs to $L^\delta(0,T;W^{1,d}(\Omega))$ for all (δ,d) satisfying:

$$\tilde{\beta} \le \delta, \quad \beta \le d, \quad \frac{N-1}{2\beta} + \frac{1}{\tilde{\beta}} < \frac{N}{2d} + \frac{1}{\delta}. \tag{4.29}$$

The trace of p on Σ belongs to $L^{\tilde{s}}(0,T;L^s(\Gamma))$ for all $(\tilde{s},s)$ satisfying:

$$\tilde{\beta} \le \tilde{s}, \quad \beta \le s, \quad \frac{N-1}{2\beta} + \frac{1}{\tilde{\beta}} < \frac{N-1}{2s} + \frac{1}{\tilde{s}} + \frac{1}{2}. \tag{4.30}$$

If moreover $\beta > N-D-1$, the trace of p on $\gamma\times]0,T[$ belongs to $L^{\tilde{\tau}}(0,T;L^\tau(\gamma))$ for all $(\tilde{\tau},\tau)$ satisfying:

$$\tilde{\beta} \le \tilde{\tau}, \quad \beta \le \tau, \quad \frac{N-1}{2\beta} + \frac{1}{\tilde{\beta}} < \frac{D}{2\tau} + \frac{1}{\tilde{\tau}} + \frac{1}{2}. \tag{4.31}$$

Proof. We use the same arguments as in the proof of Proposition 2.3. ■

Lemma 4.4 – *Let $\vec{h}$ be in $L^{\tilde{\eta}}(0,T;(L^\eta(\Omega))^N)$, where $\tilde{\eta} > 1$ and $\eta > 1$. Let b be a nonnegative function belonging to $L^{\tilde{\ell}}(0,T;L^\ell(\Gamma))$ for some $(\tilde{\ell},\ell)$ satisfying (2.23) and (4.1). Consider the equation*

$$-\frac{\partial p}{\partial t} + Ap = -\ \mathrm{div}\,\vec{h} \text{ in } Q, \quad \frac{\partial p}{\partial n_A} + bp = \vec{h}\cdot\vec{n} \text{ on } \Sigma, \quad p(T) = 0 \text{ in } \Omega, \tag{4.32}$$

The solution p to equation (4.32) belongs to $L^1(0,T;W^{1,1}(\Omega)) \cap L^{\tilde{r}}(0,T;L^r(\Omega))$ for every $(\tilde{r},r)$ satisfying (4.21). Moreover, the trace of p on Σ belongs to $L^{\tilde{s}}(0,T;L^s(\Gamma))$ for every $(\tilde{s},s)$ satisfying

$$\tilde{\eta} \le \tilde{s}, \quad \tilde{\ell}' \le \tilde{s}, \quad \eta \le s, \quad \ell' \le s, \quad \frac{N}{2\eta} + \frac{1}{\tilde{\eta}} < \frac{N-1}{2s} + \frac{1}{\tilde{s}} + \frac{1}{2}. \tag{4.33}$$

Remark. In Lemma 4.4, we implicitly assume that the set of pairs $(\tilde{s},s)$ satisfying (4.33) is non empty. This assumption must be checked every time that we use Lemma 4.4.

Proof. 1 - Let $(\tilde{s},s)$ be a pair obeying (4.33). By a fixed point argument, we prove that the equation

$$\begin{aligned} &-\frac{\partial p}{\partial t} + Ap = -\ \mathrm{div}\,\vec{h} \text{ in } \Omega\times]T-\bar{t},T[, \\ &\frac{\partial p}{\partial n_A} + bp = \vec{h}\cdot\vec{n} \text{ on } \Gamma\times]T-\bar{t},T[, \quad p(T) = 0 \text{ in } \Omega \end{aligned} \tag{4.34}$$

admits a solution for $\bar{t}$ small enough. Set

$$V(T-\bar{t},T) := \{p \in L^1(T-\bar{t},T;W^{1,1}(\Omega)) \mid p|_{\Gamma\times]T-\bar{t},T[} \in L^{\tilde{s}}(T-\bar{t},T;L^s(\Gamma))\}.$$

It is clear that $V(T-\bar{t},T)$ is a Banach space for the norm $\|p\|_{V(T-\bar{t},T)} := \|p\|_{L^1(T-\bar{t},T;W^{1,1}(\Omega))} + \|p|_{\Gamma\times]T-\bar{t},T[}\|_{L^{\bar{s}}(T-\bar{t},T;L^s(\Gamma))}$. Let ξ belong to $V(T-\bar{t},T)$, and p_ξ be the solution of the equation:

$$-\frac{\partial p}{\partial t} + Ap = -\operatorname{div}\vec{h} \text{ in } \Omega\times]T-\bar{t},T[,$$
$$\frac{\partial p}{\partial n_A} = \vec{h}\cdot\vec{n} - b\xi \text{ on } \Gamma\times]T-\bar{t},T[, \quad p(T)=0 \text{ in } \Omega. \tag{4.35}$$

We have $p_\xi = p_1 + p_2$, where p_1 is the solution to

$$-\frac{\partial p}{\partial t}+Ap = -\operatorname{div}\vec{h} \text{ in } \Omega\times]T-\bar{t},T[, \qquad \frac{\partial p}{\partial n_A} = \vec{h}\cdot\vec{n} \text{ on } \Gamma\times]T-\bar{t},T[,$$
$$p(T)=0 \text{ in } \Omega,$$

and p_2 is the solution to

$$-\frac{\partial p}{\partial t} + Ap = 0 \text{ in } \Omega\times]T-\bar{t},T[, \qquad \frac{\partial p}{\partial n_A} = -b\xi \text{ on } \Gamma\times]T-\bar{t},T[,$$
$$p(T)=0 \text{ in } \Omega.$$

Using Lemma 4.2, from condition (4.33), it follows that p_1 belongs to $L^1(T-\bar{t},T;W^{1,1}(\Omega))$, and $p_1|_{\Gamma\times]T-\bar{t},T[}$ belongs to $L^{\bar{s}}(T-\bar{t},T;L^s(\Gamma))$. Observe that $b\xi|_{\Gamma\times]T-\bar{t},T[}$ belongs to $L^{\frac{\bar{\ell}\bar{s}}{\bar{\ell}+\bar{s}}}(T-\bar{t},T;L^{\frac{\ell s}{\ell+s}}(\Gamma))$, where $\frac{\bar{\ell}\bar{s}}{\bar{\ell}+\bar{s}} \geq 1$, $\frac{\ell s}{\ell+s} \geq 1$. Using Lemma 4.3, p_2 belongs to $L^{\frac{\bar{\ell}\bar{s}}{\bar{\ell}+\bar{s}}}(T-\bar{t},T;\ W^{1,\frac{\ell s}{\ell+s}}(\Omega))$. With condition (4.1), and still from Lemma 4.3, $p_2|_{\Gamma\times]T-\bar{t},T[}$ belongs to $L^{\bar{s}}(T-\bar{t},T;L^s(\Gamma))$. Thus p belongs to $V(T-\bar{t},T)$.

Let ξ_1 and ξ_2 belong to $V(T-\bar{t},T)$, let p_{ξ_1} and p_{ξ_2} be the solutions of equation (4.35) corresponding to ξ_1 and ξ_2. Still from condition (4.1), and using Lemma 4.3, we have:

$$\|p_{\xi_1}-p_{\xi_2}\|_{L^1(T-\bar{t},T;W^{1,1}(\Omega))} + \|(p_{\xi_1}-p_{\xi_2})|_{\Gamma\times]T-\bar{t},T[}\|_{L^{\bar{s}}(T-\bar{t},T;L^s(\Gamma))}$$
$$\leq C\|b\|_{L^{\bar{\ell}}(T-\bar{t},T;L^\ell(\Gamma))}\|(\xi_1-\xi_2)|_{\Gamma\times]T-\bar{t},T[}\|_{L^{\bar{s}}(T-\bar{t},T;L^s(\Gamma))},$$

where C can be chosen depending on T, but independent of $\bar{t}$. The mapping $t \mapsto C^{\bar{\ell}}\int_{T-t}^{T}\|b(s,\tau)\|_{L^\ell(\Gamma)}^{\bar{\ell}}d\tau$ is absolutely continuous, then there exists $\bar{t}>0$ such that $C(\int_{\max\{t-\bar{t},0\}}^{t}\|b(\cdot,\tau)\|_{L^\ell(\Gamma)}^{\bar{\ell}}d\tau)^{\frac{1}{\bar{\ell}}} \leq C = \frac{1}{2}$ for all $t\in[0,T]$. Thus the mapping $\xi \mapsto p_\xi$ is a contraction in the Banach space $V(T-\bar{t},T)$, and admits a unique fixed point. Therefore equation (4.32) admits a local solution p.

2 - Existence of a global solution. We prove that a solution exists in the space $V(T-2\bar{t},T)$, by repeating the above process. Let (ξ_1,ξ_2) belong to $\{\pi \in V(T-2\bar{t},T) \mid \pi = p \text{ on }]T-\bar{t},T[\}$. We still denote by p_{ξ_1} and p_{ξ_2} the solutions to equation (4.35) on $(T-2\bar{t},T)$ corresponding to ξ_1 and ξ_2. As in Step 1, we can prove that p_{ξ_1} and p_{ξ_2} belong to $V(T-2\bar{t},T)$. Moreover, we have

$$\begin{aligned}
&\|p_{\xi_1}-p_{\xi_2}\|_{L^1(T-2\bar{t},T;W^{1,1}(\Omega))} + \|(p_{\xi_1}-p_{\xi_2})|_{\Gamma\times]T-2\bar{t},T[}\|_{L^{\tilde{s}}(T-2\bar{t},T;L^s(\Gamma))} \\
&\qquad \le C\|b\|_{L^{\tilde{\ell}}(T-2\bar{t},T-\bar{t};L^\ell(\Gamma))}\big(\|\xi_1-\xi_2\|_{L^1(T-2\bar{t},T;W^{1,1}(\Omega))} \\
&\qquad\qquad +\|(\xi_1-\xi_2)|_{\Gamma\times]T-2\bar{t},T[}\|_{L^{\tilde{s}}(T-2\bar{t},T;L^s(\Gamma))}\big) \\
&\le \frac{1}{2}\big(\|\xi_1-\xi_2\|_{L^1(T-2\bar{t},T;W^{1,1}(\Omega))} + \|(\xi_1-\xi_2)|_{\Gamma\times]T-2\bar{t},T[}\|_{L^{\tilde{s}}(T-2\bar{t},T;L^s(\Gamma))}\big).
\end{aligned}$$

Thus the mapping $\xi \mapsto p_\xi$ admits a unique fixed point in the metric space $\{\pi \in V(T-2\bar{t},T) \mid \pi = p \text{ on }]T-\bar{t},T[\}$ (the distance is defined by the norm of $V(T-2\bar{t},T)$). After a finite number of iterations, we prove that equation (4.32) admits a global solution in $V(0,T)$.

3 - Uniqueness. If we consider the equation

$$-\frac{\partial p}{\partial t} + Ap = 0 \text{ in } Q, \qquad \frac{\partial p}{\partial n_A} + bp = 0 \text{ on } \Sigma, \qquad p(T) = 0 \text{ in } \Omega,$$

we can apply the above fixed point method to prove that $p \equiv 0$ is the unique solution to this equation in the space $\{p \in L^1(0,T;W^{1,1}(\Omega)) \mid p|_\Sigma \in L^{\tilde{\ell}'}(0,T;L^{\ell'}(\Gamma))\}$.

4 - We see that p is the solution of the equation

$$-\frac{\partial p}{\partial t} + Ap = -\operatorname{div}\vec{h} \text{ in } Q, \qquad \frac{\partial p}{\partial n_A} = \vec{h}\cdot\vec{n} - bp \text{ on } \Sigma, \qquad p(T) = 0 \text{ in } \Omega.$$

Let $(\tilde{r},r)$ be a pair obeying (4.21). Then there exists $(\tilde{s},s)$ satisfying (4.33) and such that

$$\frac{\tilde{\ell}\tilde{s}}{\tilde{\ell}+\tilde{s}} \le \tilde{r}, \quad \frac{\ell s}{\ell+s} \le r, \quad \frac{1}{\tilde{\ell}} + \frac{N-1}{2\ell} + \frac{1}{\tilde{s}} + \frac{N-1}{2s} < \frac{1}{\tilde{r}} + \frac{N}{2r} + \frac{1}{2}.$$

Therefore $bp|_\Sigma \in L^{\frac{\tilde{\ell}\tilde{s}}{\tilde{\ell}+\tilde{s}}}(0,T;L^{\frac{\ell s}{\ell+s}}(\Gamma))$, and $p \in L^{\tilde{r}}(0,T;L^r(\Omega))$. ∎

Lemma 4.5 – *Let f be in $L^{\tilde{\alpha}}(0,T;L^\alpha(\Omega))$, where $\tilde{\alpha} \ge 1$ and $\alpha \ge 1$. Let b be a nonnegative function belonging to $L^{\tilde{\ell}}(0,T;L^\ell(\Gamma))$ for some $(\tilde{\ell},\ell)$ satisfying (2.23) and (4.1). Consider the equation*

$$-\frac{\partial p}{\partial t} + Ap = f \text{ in } Q, \qquad \frac{\partial p}{\partial n_A} + bp = 0 \text{ on } \Sigma, \qquad p(T) = 0 \text{ in } \Omega. \quad (4.36)$$

The solution p to equation (4.36) belongs to $L^{\tilde{r}}(0,T;L^r(\Omega))$ *for every* $(\tilde{r},r)$ *satisfying*

$$\tilde{\alpha} \le \tilde{r}, \quad \alpha \le r, \quad \frac{N}{2\alpha} + \frac{1}{\tilde{\alpha}} < \frac{N}{2r} + \frac{1}{\tilde{r}} + 1.$$

The solution p to equation (4.36) belongs to $L^{\delta}(0,T;W^{1,d}(\Omega))$ *for every* (δ,d) *satisfying*

$$\tilde{\alpha} \le \delta, \quad \alpha \le d, \quad \frac{N}{2\alpha} + \frac{1}{\tilde{\alpha}} < \frac{N}{2d} + \frac{1}{\delta} + \frac{1}{2}.$$

If moreover $\alpha > \frac{N-D}{2}$, *the trace of p on* $\gamma\times]0,T[$ *belongs to* $L^{\tilde{\tau}}(0,T;L^{\tau}(\gamma))$ *for all* $(\tilde{\tau},\tau)$ *satisfying:*

$$\tilde{\alpha} \le \tilde{\tau}, \quad \alpha \le \tau, \quad \frac{N}{2\alpha} + \frac{1}{\tilde{\alpha}} < \frac{D}{2\tau} + \frac{1}{\tilde{\tau}} + 1.$$

Proof. The proof can be established as in Lemma 4.2. ■

Lemma 4.6 *Let* $\vec{h}$ *be in* $L^{\tilde{\eta}}(0,T;(L^{\eta}(\Omega))^N)$, *where* $\tilde{\eta} > 1$ *and* $\eta > 1$. *Let b be a nonnegative function belonging to* $L^{\tilde{\ell}}(0,T;L^{\ell}(\Omega))$ *for some* $(\tilde{\ell},\ell)$ *satisfying (2.23) and (4.1). Let* $a \in$ *be a nonnegative function belonging to* $L^{\tilde{k}}(0,T;L^k(\Omega))$ *for some* $(\tilde{k},k)$ *satisfying (2.22) and (4.1). Consider the equation*

$$\begin{aligned} -\frac{\partial p}{\partial t} + Ap + ap &= -\ div\ \vec{h}\ in\ Q, \\ \frac{\partial p}{\partial n_A} + bp &= \vec{h}\cdot\vec{n}\ on\ \Sigma,\ \ p(T) = 0\ in\ \Omega. \end{aligned} \tag{4.37}$$

Its solution $p \in L^1(0,T;W^{1,1}(\Omega))\cap L^{\tilde{r}}(0,T;L^r(\Omega))$ *for every* $(\tilde{r},r)$ *satisfying*

$$\tilde{\eta} \le \tilde{r}, \quad \tilde{k}' \le \tilde{r}, \quad \eta \le r, \quad k' \le r, \quad \frac{N}{2\eta} + \frac{1}{\tilde{\eta}} < \frac{N}{2r} + \frac{1}{\tilde{r}} + \frac{1}{2}. \tag{4.38}$$

The trace of p on Σ *belongs to* $L^{\tilde{s}}(0,T;L^s(\Gamma))$, $\forall(\tilde{s},s)$ *satisfying (4.33).*

Remark. In Lemma 4.6, we implicitly assume that the set of pairs $(\tilde{s},s)$ satisfying (4.33) and the set of pairs $(\tilde{r},r)$ satisfying (4.38) are non empty. This assumption must be checked every time that we use Lemma 4.6.

Proof. 1 - Let $(\tilde{r},r)$ be a pair obeying (4.38). By a fixed point argument, we prove that the following equation admits a solution for $\bar{t}$ small enough:

$$\begin{aligned} -\frac{\partial p}{\partial t} + Ap + ap &= -\ \text{div}\ \vec{h} \quad \text{in}\ \Omega\times]T-\bar{t},T[, \\ \frac{\partial p}{\partial n_A} + bp &= \vec{h}\cdot\vec{n}\ \text{on}\ \Gamma\times]T-\bar{t},T[, \quad p(T) = 0\ \text{in}\ \Omega \end{aligned} \tag{4.39}$$

Let ξ belong to $L^1(T-\bar{t},T;W^{1,1}(\Omega))\cap L^{\tilde{r}}(T-\bar{t},T;L^r(\Omega))$, and p_ξ be the solution of the equation

$$-\frac{\partial p}{\partial t}+Ap=-\operatorname{div}\vec{h}-a\xi \quad \text{in } \Omega\times]T-\bar{t},T[,$$
$$\frac{\partial p}{\partial n_A}+bp=\vec{h}\cdot\vec{n} \text{ on } \Gamma\times]T-\bar{t},T[, \quad p(T)=0 \text{ in } \Omega. \tag{4.40}$$

We have $p_\xi=p_1+p_2$, where p_1 is the solution to

$$-\frac{\partial p}{\partial t}+Ap=-\operatorname{div}\vec{h} \quad \text{in } \Omega\times]T-\bar{t},T[,$$
$$\frac{\partial p}{\partial n_A}+bp=\vec{h}\cdot\vec{n} \text{ on } \Gamma\times]T-\bar{t},T[, \quad p(T)=0 \text{ in } \Omega,$$

and p_2 is the solution to

$$-\frac{\partial p}{\partial t}+Ap=-a\xi \text{ in } \Omega\times]T-\bar{t},T[,$$
$$\frac{\partial p}{\partial n_A}+bp=0 \text{ on } \Gamma\times]T-\bar{t},T[, \quad p(T)=0 \text{ in } \Omega.$$

Using Lemma 4.4, p_1 belongs to $L^1(T-\bar{t},T;W^{1,1}(\Omega))\cap L^{\tilde{r}}(T-\bar{t},T;L^r(\Omega))$. We know that $a\xi\in L^{\frac{\tilde{k}\tilde{r}}{\tilde{k}+\tilde{r}}}(T-\bar{t},T;\ L^{\frac{kr}{k+r}}(\Omega))$, where $\frac{\tilde{k}\tilde{r}}{\tilde{k}+\tilde{r}}\geq 1$, $\frac{kr}{k+r}\geq 1$. From Lemma 4.5, and under condition (4.1), it follows that p_2 belongs to $L^{\frac{\tilde{k}\tilde{r}}{\tilde{k}+\tilde{r}}}(T-\bar{t},T;W^{1,\frac{kr}{k+r}}(\Omega))\cap L^{\tilde{r}}(T-\bar{t},T;L^r(\Omega))$. Then

$$p_\xi\in L^1(T-\bar{t},T;W^{1,1}(\Omega))\cap L^{\tilde{r}}(T-\bar{t},T;L^r(\Omega)).$$

Let ξ_1 and ξ_2 belong to $L^1(T-\bar{t},T;W^{1,1}(\Omega))\cap L^{\tilde{r}}(T-\bar{t},T;L^r(\Omega))$, and let p_{ξ_1} and p_{ξ_2} be the solutions of equation (4.40) corresponding to ξ_1 and ξ_2. Still from Lemma 4.5 and under condition (4.1), we have:

$$\begin{aligned}&\|p_{\xi_1}-p_{\xi_2}\|_{L^1(0,\bar{t};W^{1,1}(\Omega))\cap L^{\tilde{r}}(T-\bar{t},T;L^r(\Omega))}\\&\qquad\leq C\|a\|_{L^{\tilde{k}}(T-\bar{t},T;L^k(\Omega))}\|\xi_1-\xi_2\|_{L^1(T-\bar{t},T;W^{1,1}(\Omega))\cap L^{\tilde{r}}(T-\bar{t},T;L^r(\Omega))},\end{aligned}$$

where C depends on T, but independent of $\bar{t}$.

The mapping $t\mapsto C^{\tilde{k}}\int_{T-t}^{T}\|a(x,\tau)\|_{L^k(\Omega)}^{\tilde{k}}d\tau$ is absolutely continuous, then $\exists\bar{t}>0$ such that $C(\int_{\max\{t-\bar{t},0\}}^{t}\|a(\cdot,\tau)\|_{L^k(\Omega)}^{\tilde{k}}d\tau)^{\frac{1}{\tilde{k}}}\leq C=\frac{1}{2}$ for all $t\in[0,T]$. Thus $\xi\mapsto p_\xi$ is a contraction in the Banach space $L^1(T-\bar{t},T;W^{1,1}(\Omega))\cap L^{\tilde{r}}(T-\bar{t},T;L^r(\Omega))$, and it admits a unique fixed point, so equation (4.37) admits a local solution p.

As in Step 2 in the proof of Lemma 4.4, we can iterate the process on $]T-2\bar{t},T[$ to obtain a fixed point in $\{\pi \in L^1(T-2\bar{t},T;W^{1,1}(\Omega)) \cap L^{\tilde{r}}(T-2\bar{t},T;L^r(\Omega)) \mid \pi = p \text{ on }]T-\bar{t},T[\}$. After a finite number of iterations, we prove that (4.37) admits a global solution in $L^1(0,T;W^{1,1}(\Omega)) \cap L^{\tilde{r}}(0,T;L^r(\Omega))$.
2 - Uniqueness. If we consider the equation

$$-\frac{\partial p}{\partial t} + Ap + ap = 0 \text{ in } Q, \qquad \frac{\partial p}{\partial n_A} + bp = 0 \text{ on } \Sigma, \qquad p(T) = 0 \text{ in } \Omega,$$

we can apply the above fixed point method to prove that $p \equiv 0$ is the unique solution to this equation in $L^1(0,T;W^{1,1}(\Omega)) \cap L^{\tilde{k}'}(0,T;L^{k'}(\Omega))$.
3 - We see that p is the solution of the equation

$$-\frac{\partial p}{\partial t} + Ap = -\operatorname{div}\vec{h} - ap \text{ in } Q, \cdot \frac{\partial p}{\partial n_A} + bp = \vec{h}\cdot\vec{n} \text{ on } \Sigma, \ p(T) = 0 \text{ in } \Omega.$$

Let $(\tilde{s}, s)$ be a pair obeying (4.33). Under condition (4.1), there exists $(\tilde{r}, r)$ satisfying (4.38) and such that

$$\frac{\tilde{k}\tilde{r}}{\tilde{k}+\tilde{r}} \le \tilde{s}, \qquad \frac{kr}{k+r} \le s, \qquad \frac{1}{\tilde{k}} + \frac{N}{2k} + \frac{1}{\tilde{r}} + \frac{N}{2r} < \frac{1}{\tilde{s}} + \frac{N-1}{2s} + 1.$$

Therefore $ap \in L^{\frac{\tilde{k}\tilde{r}}{\tilde{k}+\tilde{r}}}(0,T;L^{\frac{kr}{k+r}}(\Omega))$ and $p|_\Sigma \in L^{\tilde{s}}(0,T;L^s(\Gamma))$. ■

Theorem 4.2 – *Assume that a and b satisfy the assumptions of Proposition 2.6. Let p be the solution to equation (4.37), where $\vec{h}$ belongs to the space $L^{\frac{q}{\kappa-1}}(0,T;(L^{\frac{d}{\kappa-1}}(\Omega))^N)$ for all $d < \frac{N-D}{N-D-1}$. Under assumptions (A1)-(A11), the trace of p on $\gamma\times]0,T[$ belongs to $L^{q'}(0,T;L^{\sigma'}(\gamma))$.*

Proof. 1 - We first prove that there exists $(\tilde{\ell}, \ell)$ obeying (2.23) and (4.1), such that the set of pairs $(\tilde{s}, s)$ satisfying

$$\tilde{s} \ge \frac{q}{\kappa-1}, \ \tilde{s} \ge \tilde{\ell}', \ s \ge \frac{d}{\kappa-1}, \ s \ge \ell', \ (\kappa-1)\left(\frac{1}{q}+\frac{N}{2d}\right) < \frac{1}{\tilde{s}} + \frac{N-1}{2s} + \frac{1}{2}, \tag{4.41}$$

$$\frac{1}{\tilde{s}} + \frac{1}{\tilde{\ell}} \le \frac{1}{q'}, \quad \frac{1}{s} + \frac{1}{\ell} \le \frac{1}{\sigma'}, \quad \frac{1}{s} + \frac{1}{\ell} < \frac{1}{N-D-1}, \tag{4.42}$$

is non empty for all $d < \frac{N-D}{N-D-1}$, big enough. We distinguish two cases.

First case: $\sigma < \frac{N-D-1}{N-D-2}$, choose $\tilde{\ell} = \frac{q}{\nu-1}$. Due to (4.3), (4.7), (4.11) and (4.12), we can choose ℓ such that $\sigma' < \ell < \frac{(N-1)}{(\nu-1)(N-\frac{D}{\sigma'}-2)}$,

$$\frac{\nu-1}{q} + \frac{N-1}{2\ell} < \frac{1}{2}, \tag{4.43}$$

$$\frac{\kappa-1}{q}+\frac{N(\kappa-1)(N-D-1)}{2(N-D)}<\frac{1}{q'}+\frac{N-1}{2\sigma'}-\frac{\nu-1}{q}-\frac{N-1}{2\ell}+\frac{1}{2}, \quad (4.44)$$

and

$$\frac{N(\kappa-1)(N-D-1)}{2(N-D)}<\frac{N-1}{2\sigma'}-\frac{N-1}{2\ell}+\frac{1}{2}\,. \quad (4.45)$$

Observe that $\frac{1}{q'}-\frac{1}{\ell}>0$ because $\nu<q$. Now we set $\frac{1}{\tilde{s}}=\inf\{\frac{\kappa-1}{q},\frac{1}{q'}-\frac{1}{\ell}\}$, and $\frac{1}{s}=\inf\{\frac{(\kappa-1)(N-D-1)}{N-D},\frac{1}{\sigma'}-\frac{1}{\ell}\}$. Let us prove that the pair $(\tilde{s},s)$ obeys

$$\frac{\kappa-1}{q}+\frac{N(\kappa-1)(N-D-1)}{2(N-D)}<\frac{1}{\tilde{s}}+\frac{N-1}{2s}+\frac{1}{2}. \quad (4.46)$$

If $\frac{1}{\tilde{s}}=\frac{1}{q'}-\frac{1}{\ell}$ and $\frac{1}{s}=\frac{(\kappa-1)(N-D-1)}{N-D}$, condition (4.46) follows from assumption (4.10). If $\frac{1}{\tilde{s}}=\frac{1}{q'}-\frac{1}{\ell}$ and $\frac{1}{s}=\frac{1}{\sigma'}-\frac{1}{\ell}$, condition (4.46) follows from (4.44). If $\frac{1}{\tilde{s}}=\frac{\kappa-1}{q}$ and $\frac{1}{s}=\frac{(\kappa-1)(N-D-1)}{N-D}$, condition (4.46) follows from assumption (4.6). If $\frac{1}{\tilde{s}}=\frac{\kappa-1}{q}$ and $\frac{1}{s}=\frac{1}{\sigma'}-\frac{1}{\ell}$, condition (4.46) follows from (4.45).
Due to (4.46), we can choose $d<\frac{N-D}{N-D-1}$, big enough, so that (4.41) and (4.42) be satisfied.

Second case: $\sigma\geq\frac{N-D-1}{N-D-2}$, choose $\tilde{\ell}=\frac{q}{\nu-1}$. Since $\nu<\frac{N-D-1}{N-D-2}$, due to (4.3), (4.13) and (4.14), we can choose ℓ satisfying (4.43), such that $N-D-1<\ell<\frac{N-D-1}{(\nu-1)(N-D-2)}$,

$$\frac{\kappa-1}{q}+\frac{N(\kappa-1)(N-D-1)}{2(N-D)}<\frac{1}{q'}+\frac{N-1}{2(N-D-1)}-\frac{\nu-1}{q}-\frac{N-1}{2\ell}+\frac{1}{2}\,, \quad (4.47)$$

and

$$\frac{N(\kappa-1)(N-D-1)}{2(N-D)}<\frac{N-1}{2(N-D-1)}-\frac{N-1}{2\ell}+\frac{1}{2}\,. \quad (4.48)$$

Observe that $\frac{1}{q'}-\frac{1}{\ell}>0$ because $\nu<q$, and set $\frac{1}{\tilde{s}}=\inf\{\frac{\kappa-1}{q},\frac{1}{q'}-\frac{1}{\ell}\}$. We distinguish two subcases.
When $\frac{(\kappa-1)(N-D-1)}{N-D}<\frac{1}{N-D-1}-\frac{1}{\ell}$, we set $\frac{1}{s}=\frac{(\kappa-1)(N-D-1)}{N-D}$. Let us prove that the pair $(\tilde{s},s)$ obeys (4.46). If $\frac{1}{\tilde{s}}=\frac{1}{q'}-\frac{1}{\ell}$ condition (4.46) follows from assumption (4.10). If $\frac{1}{\tilde{s}}=\frac{\kappa-1}{q}$, condition (4.46) follows from assumption (4.6).
When $\frac{(\kappa-1)(N-D-1)}{N-D}\geq\frac{1}{N-D-1}-\frac{1}{\ell}$, due to (4.47) and (4.48) (depending on the value of $\frac{1}{\tilde{s}}$), we can choose s such that $0<\frac{1}{s}<\frac{1}{N-D-1}-\frac{1}{\ell}$ and such that $(\tilde{s},s)$ satisfy (4.46).
Due to (4.46), we can choose $d<\frac{N-D}{N-D-1}$, big enough, so that (4.41) and (4.42) be satisfied.
2 - In the same way, using assumptions (4.2), (A10) and (A11b), we can

prove that there exists $(\tilde{k}, k)$ obeying (2.22) and (4.1), such that the set of pairs $(\tilde{r}, r)$ satisfying

$$\tilde{r} \geq \frac{q}{\kappa-1}, \ \tilde{r} \geq \tilde{k}', \ r \geq \frac{d}{\kappa-1}, \quad r \geq k', \ (\kappa-1)(\frac{1}{q}+\frac{N}{2d}) < \frac{1}{\tilde{r}}+\frac{N}{2r}+\frac{1}{2}, \quad (4.49)$$

$$\frac{1}{\tilde{r}}+\frac{1}{\tilde{k}} \leq \frac{1}{q'}, \quad \frac{1}{r}+\frac{1}{k} \leq \frac{1}{\sigma'}, \quad \frac{1}{r}+\frac{1}{k} < \frac{2}{N-D},$$

is non empty for all $d < \frac{N-D}{N-D-1}$, big enough.
3 - From Lemma 4.6, it follows that the trace of the solution p to equation (4.37) belongs to $L^{\tilde{s}}(0,T;L^s(\Gamma))$, where $(\tilde{s}, s)$ is the pair defined in Step 1. We set $\frac{1}{\tilde{\beta}} = \frac{1}{\tilde{s}} + \frac{1}{\ell}, \frac{1}{\beta} = \frac{1}{s} + \frac{1}{\ell}$, then $bp \in L^{\tilde{\beta}}(0,T;L^{\beta}(\Gamma))$. Let π be the solution to the equation

$$-\frac{\partial \pi}{\partial t} + A\pi = 0 \ \text{ in } \ Q, \ \frac{\partial \pi}{\partial n_A} = -bp \ \text{ on } \ \Sigma, \ \pi(T) = 0 \ \text{ in } \ \Omega \qquad (4.50)$$

If $\sigma < \frac{N-D-1}{N-D-2}$ (resp. $\sigma \geq \frac{N-D-1}{N-D-2}$), then $\tilde{\beta} \geq q'$, and $\beta \geq \sigma' > N-D-1$ (resp. $\beta > N-D-1 \geq \sigma'$). Using Lemma 4.3, we deduce that the trace of π on $\gamma\times]0,T[$ exists, and belongs to $L^{q'}(0,T;L^{\sigma'}(\gamma))$.
4 - From Lemma 4.6, it follows that the solution p to equation (4.37) belongs to $L^{\tilde{r}}(0,T;L^r(\Omega))$, where $(\tilde{r}, r)$ is defined in Step 2. Let $\hat{\pi}$ be the solution to the equation

$$-\frac{\partial \hat{\pi}}{\partial t} + A\hat{\pi} = -ap \ \text{ in } Q, \ \frac{\partial \hat{\pi}}{\partial n_A} = 0 \ \text{ on } \Sigma, \ \hat{\pi}(T) = 0 \ \text{ in } \Omega, \qquad (4.51)$$

we can prove that the trace of $\hat{\pi}$ on $\gamma\times]0,T[$ belongs to $L^{q'}(0,T;L^{\sigma'}(\gamma))$.
5 - Let $\tilde{\pi}$ be the solution to the equation

$$-\frac{\partial \tilde{\pi}}{\partial t} + A\tilde{\pi} = -\operatorname{div}\vec{h} \text{ in } Q, \ \frac{\partial \tilde{\pi}}{\partial n_A} = \vec{h}\cdot\vec{n} \text{ on } \Sigma, \ \tilde{\pi}(T) = 0 \text{ in } \Omega. \quad (4.52)$$

Since $\kappa < 1 + \frac{1}{N-D-1}$, for $d < \frac{N-D}{N-D-1}$ big enough, we have $\frac{d}{\kappa-1} > N-D$. From Lemma 4.2, setting $\tilde{\tau} = \tilde{\eta} = \frac{q}{\kappa-1}$ in (4.23), it follows that $\tilde{\pi}|_{\gamma\times]0,T[}$ belongs to $L^{\frac{q}{\kappa-1}}(0,T;L^{\frac{d}{\kappa-1}}(\gamma))$. From (4.6), we have $\frac{q}{\kappa-1} > q'$. Since $\frac{d}{\kappa-1} > N-D \geq 2 \geq \sigma'$, the trace of $\tilde{\pi}$ on $\gamma\times]0,T[$ belongs to $L^{q'}(0,T;L^{\sigma'}(\gamma))$. Notice that $p = \pi + \hat{\pi} + \tilde{\pi}$. The proof is complete. ■

(A12) - Assumptions needed in the proof of Theorem 4.3.
(A12a) - Assumptions needed to estimate the term bp.

$$\frac{\kappa-1}{2} + \frac{\kappa-1}{2\rho} < \frac{1}{q'} - \frac{\nu-1}{q} + \frac{1}{2} \,. \qquad (4.53)$$

If $\sigma < \frac{N-D-1}{N-D-2}$, then

$$\frac{\kappa-1}{2}+\frac{N(\kappa-1)}{2\rho}<\frac{1}{q'}-\frac{\nu-1}{q}+\frac{N-1}{2\sigma'}-\frac{(\nu-1)(N-\frac{D}{\sigma'}-2)}{2}+\frac{1}{2}, \quad (4.54)$$

$$\frac{N(\kappa-1)}{2\rho}<\frac{N-1}{2\sigma'}-\frac{(\nu-1)(N-\frac{D}{\sigma'}-2)}{2}+\frac{1}{2}. \quad (4.55)$$

If $\sigma \geq \frac{N-D-1}{N-D-2}$, then

$$\frac{\kappa-1}{2}+\frac{N(\kappa-1)}{2\rho}<\frac{1}{q'}-\frac{\nu-1}{q}+\frac{N-1}{2(N-D-1)} -\frac{(\nu-1)(N-1)(N-D-2)}{2(N-D-1)}+\frac{1}{2}, \quad (4.56)$$

$$\frac{N(\kappa-1)}{2\rho}<\frac{N-1}{2(N-D-1)}-\frac{(\nu-1)(N-1)(N-D-2)}{2(N-D-1)}+\frac{1}{2}, \quad (4.57)$$

(A12b) - Assumptions needed to estimate the term ap.

$$\frac{\kappa-1}{2}<\frac{1}{q'}-\frac{\mu-1}{q}+\frac{1}{2}. \quad (4.58)$$

If $\sigma < \frac{N-D}{N-D-2}$, then

$$\frac{\kappa-1}{2}+\frac{N(\kappa-1)}{2\rho}<\frac{1}{q'}-\frac{\mu-1}{q}+\frac{N}{2\sigma'}-\frac{(\mu-1)(N-\frac{D}{\sigma'}-2)}{2}+\frac{1}{2}, \quad (4.59)$$

$$\frac{N(\kappa-1)}{2\rho}<\frac{N}{2\sigma'}-\frac{(\mu-1)(N-\frac{D}{\sigma'}-2)}{2}+\frac{1}{2}. \quad (4.60)$$

If $\sigma \geq \frac{N-D}{N-D-2}$, then

$$\frac{\kappa-1}{2}+\frac{N(\kappa-1)}{2\rho}<\frac{1}{q'}-\frac{\mu-1}{q}+\frac{N}{N-D}-\frac{(\mu-1)N(N-D-2)}{2(N-D)}+\frac{1}{2}, \quad (4.61)$$

$$\frac{N(\kappa-1)}{2\rho}<\frac{N}{N-D}-\frac{(\mu-1)N(N-D-2)}{2(N-D)}+\frac{1}{2}. \quad (4.62)$$

Theorem 4.3 *—Assume that a and b satisfy the assumptions of Proposition 2.6. Let p be the solution to the equation (4.37), where $\vec{h}$ belongs to $L^{\frac{\delta_4}{\kappa-1}}(0,T;L^{\frac{\rho}{\kappa-1}}(\Omega))$ for every $1<\delta_4<2$. Under assumptions (A1)-(A10), (A12), the trace of p on $\gamma\times]0,T[$ belongs to $L^{q'}(0,T;L^{\sigma'}(\gamma))$.*

Proof. 1 - We first prove that there exists $(\tilde{\ell},\ell)$ obeying (2.23) and (4.1), such that the set of pairs $(\tilde{s},s)$ satisfying

$$\tilde{s}\geq\frac{\delta_4}{\kappa-1},\quad \tilde{s}\geq\tilde{\ell}',\quad s\geq\frac{\rho}{\kappa-1},\quad s\geq\ell',$$
$$(\kappa-1)(\frac{1}{\delta_4}+\frac{N}{2\rho})<\frac{1}{\tilde{s}}+\frac{N-1}{2s}+\frac{1}{2}, \quad (4.63)$$

$$\frac{1}{\tilde{s}}+\frac{1}{\tilde{\ell}}\leq\frac{1}{q'},\quad \frac{1}{s}+\frac{1}{\ell}\leq\frac{1}{\sigma'},\quad \frac{1}{s}+\frac{1}{\ell}<\frac{1}{N-D-1},\tag{4.64}$$

is non empty for all $\delta_4<2$, big enough. We distinguish two cases.

• First case: $\sigma<\frac{N-D-1}{N-D-2}$. We choose $\tilde{\ell}=\frac{q}{\nu-1}$. Due to (4.3), (4.7), (4.54) and (4.55), we can choose ℓ satisfying (4.43),

$$\sigma'<\ell<\frac{(N-1)}{(\nu-1)(N-\frac{D}{\sigma'}-2)},$$

$$\frac{\kappa-1}{2}+\frac{N(\kappa-1)}{2\rho}<\frac{1}{q'}+\frac{N-1}{2\sigma'}-\frac{\nu-1}{q}-\frac{N-1}{2\ell}+\frac{1}{2},\tag{4.65}$$

and

$$\frac{N(\kappa-1)}{2\rho}<\frac{N-1}{2\sigma'}-\frac{N-1}{2\ell}+\frac{1}{2}\,.\tag{4.66}$$

Observe that $\frac{1}{q'}-\frac{1}{\tilde{\ell}}>0$ because $\nu<q$. Now we set $\frac{1}{\tilde{s}}=\inf\{\frac{\kappa-1}{2},\frac{1}{q'}-\frac{1}{\tilde{\ell}}\}$, and $\frac{1}{s}=\inf\{\frac{\kappa-1}{\rho},\frac{1}{\sigma'}-\frac{1}{\ell}\}$. Let us prove that the pair $(\tilde{s},s)$ obeys

$$\frac{\kappa-1}{2}+\frac{N(\kappa-1)}{2\rho}<\frac{1}{\tilde{s}}+\frac{N-1}{2s}+\frac{1}{2}.\tag{4.67}$$

If $\frac{1}{\tilde{s}}=\frac{1}{q'}-\frac{1}{\tilde{\ell}}$ and $\frac{1}{s}=\frac{\kappa-1}{\rho}$, condition (4.67) follows from assumption (4.53). If $\frac{1}{\tilde{s}}=\frac{1}{q'}-\frac{1}{\tilde{\ell}}$ and $\frac{1}{s}=\frac{1}{\sigma'}-\frac{1}{\ell}$, condition (4.67) follows from (4.65). If $\frac{1}{\tilde{s}}=\frac{\kappa-1}{2}$ and $\frac{1}{s}=\frac{\kappa-1}{\rho}$, condition (4.67) follows from the fact that $(\kappa-1)<1<N-D\leq\rho$. If $\frac{1}{\tilde{s}}=\frac{\kappa-1}{2}$ and $\frac{1}{s}=\frac{1}{\sigma'}-\frac{1}{\ell}$, condition (4.67) follows from (4.66). Due to (4.67), we can choose $1<\delta_4<2$, close enough to 2, so that (4.63) and (4.64) be satisfied.

• Second case: $\sigma\geq\frac{N-D-1}{N-D-2}$. We choose $\tilde{\ell}=\frac{q}{\nu-1}$. Since $\nu<\frac{N-D-1}{N-D-2}$, due to (4.3), (4.56) and (4.57), we can choose ℓ satisfying (4.43),

$$N-D-1<\ell<\frac{N-D-1}{(\nu-1)(N-D-2)},$$

$$\frac{\kappa-1}{2}+\frac{N(\kappa-1)}{2\rho}<\frac{1}{q'}+\frac{N-1}{2(N-D-1)}-\frac{\nu-1}{q}-\frac{N-1}{2\ell}+\frac{1}{2}\,,\tag{4.68}$$

and

$$\frac{N(\kappa-1)}{2\rho}<\frac{N-1}{2(N-D-1)}-\frac{N-1}{2\ell}+\frac{1}{2}\,.\tag{4.69}$$

Observe that $\frac{1}{q'}-\frac{1}{\tilde{\ell}}>0$ because $\nu<q$. We set $\frac{1}{\tilde{s}}=\inf\{\frac{\kappa-1}{2},\frac{1}{q'}-\frac{1}{\tilde{\ell}}\}$, and consider two subcases.

When $\frac{\kappa-1}{\rho} < \frac{1}{N-D-1} - \frac{1}{\ell}$, we set $\frac{1}{s} = \frac{\kappa-1}{\rho}$. Let us prove that $(\tilde{s}, s)$ obeys (4.67). If $\frac{1}{\tilde{s}} = \frac{1}{q'} - \frac{1}{\ell}$, condition (4.67) follows from assumption (4.53). If $\frac{1}{\tilde{s}} = \frac{\kappa-1}{2}$, condition (4.67) follows from the fact that $(\kappa-1) < 1 < N-D \leq \rho$. When $\frac{\kappa-1}{\rho} \geq \frac{1}{N-D-1} - \frac{1}{\ell}$, due to (4.68) and (4.69) (depending on the value of $\frac{1}{\tilde{s}}$), we can choose s such that $0 < \frac{1}{s} < \frac{1}{N-D-1} - \frac{1}{\ell}$, and such that $(\tilde{s}, s)$ satisfy (4.67).

Due to (4.67), it follows that we can choose $1 < \delta_4 < 2$, close enough to 2, so that (4.63) and (4.64) are satisfied.

2 - In the same way, using assumptions (4.2), (A10) and (A12b), we can prove that there exists $(\tilde{k}, k)$ obeying (2.22) and (4.1), such that the set of pairs $(\tilde{r}, r)$ satisfying

$$\tilde{r} \geq \frac{\delta_4}{\kappa - 1}, \quad \tilde{r} \geq \tilde{k}', \quad r \geq \frac{\rho}{\kappa - 1}, \quad r \geq k',$$
$$(\kappa - 1)(\frac{1}{\delta_4} + \frac{N}{2\rho}) < \frac{1}{\tilde{r}} + \frac{N}{2r} + \frac{1}{2}, \tag{4.70}$$
$$\frac{1}{\tilde{r}} + \frac{1}{\tilde{k}} \leq \frac{1}{q'}, \quad \frac{1}{r} + \frac{1}{k} \leq \frac{1}{\sigma'}, \quad \frac{1}{r} + \frac{1}{k} < \frac{2}{N-D},$$

is non empty for all $\delta_4 < 2$, big enough.

3 - From Lemma 4.6, it follows that the trace of the solution p to equation (4.37) belongs to $L^{\tilde{s}}(0, T; L^s(\Gamma))$ for the pair $(\tilde{s}, s)$ chosen in Step 1. We set $\frac{1}{\tilde{\beta}} = \frac{1}{\tilde{s}} + \frac{1}{\ell}, \frac{1}{\beta} = \frac{1}{s} + \frac{1}{\ell}$. Then $bp \in L^{\tilde{\beta}}(0, T; L^{\beta}(\Gamma))$. Let π be the solution to equation (4.50). If $\sigma < \frac{N-D-1}{N-D-2}$ (resp. $\sigma \geq \frac{N-D-1}{N-D-2}$), then $\tilde{\beta} \geq q'$, and $\beta \geq \sigma' > N - D - 1$ (resp. $\beta > N - D - 1 \geq \sigma'$). Using Lemma 4.3, we deduce that the trace of π on $\gamma\times]0, T[$ exists, and belongs to $L^{q'}(0, T; L^{\sigma'}(\gamma))$.

4 - From Lemma 4.6, it follows that the solution p to equation (4.37) belongs to $L^{\tilde{r}}(0, T; L^r(\Omega))$, where $(\tilde{r}, r)$ is chosen in Step 2. Let $\hat{\pi}$ be the solution to equation (4.51), we can prove that the trace of $\hat{\pi}$ on $\gamma\times]0, T[$ belongs to $L^{q'}(0, T; L^{\sigma'}(\gamma))$.

5 - Let $\tilde{\pi}$ be the solution to equation (4.52). Since $\frac{\rho}{\kappa-1} > N - D$, from Lemma 4.2, it follows that $\tilde{\pi}|_{\gamma\times]0,T[}$ belongs to $L^{\frac{\delta_4}{\kappa-1}}(0, T; L^{\frac{\rho}{\kappa-1}}(\gamma))$. We have $\sigma' \leq 2 \leq \rho < \frac{\rho}{\kappa-1}$. Moreover, we can choose $\delta_4 < 2$ big enough, to obtain $q' \leq 2 \leq \frac{\delta_4}{\kappa-1}$. Thus $\tilde{\pi}|_{\gamma\times]0,T[}$ belongs to $L^{q'}(0, T; L^{\sigma'}(\gamma))$. Notice that $p = \pi + \hat{\pi} + \tilde{\pi}$. The proof is complete. ■

In the sequel, we set $\tilde{\sigma} = \min\, (\sigma, \frac{N-D}{N-D-\frac{2}{q'}})$.

(A13) - Assumptions needed in the proof of Theorem 4.4.

$$\theta < \inf\{1 + \frac{N}{(N-D)} \frac{\frac{2}{q'}}{(N - \frac{D}{\tilde{\sigma}'} - \frac{2}{q'})}, 1 + \frac{\frac{D}{\sigma'} + \frac{2}{q'}}{(N - \frac{D}{\tilde{\sigma}'} - \frac{2}{q'})}\}. \tag{4.71}$$

(A13a) - Assumptions needed to estimate the term bp.

$$\frac{(\theta-1)(N-\frac{D}{\tilde{\sigma}'}-\frac{2}{q'})}{2N}+\frac{\nu-1}{q}<1. \tag{4.72}$$

If $\sigma < \frac{N-D-1}{N-D-2}$, then

$$\frac{(\theta-1)(N-\frac{D}{\tilde{\sigma}'}-\frac{2}{q'})}{2}<1-\frac{\nu-1}{q}-\frac{(\nu-1)(N-\frac{D}{\sigma'}-2)}{2}+\frac{N-1}{2(N-D-1)}, \tag{4.73}$$

If $\sigma \geq \frac{N-D-1}{N-D-2}$, then

$$\frac{(\theta-1)(N-\frac{D}{\tilde{\sigma}'}-\frac{2}{q'})}{2}<1-\frac{\nu-1}{q}-\frac{(\nu-1)(N-1)(N-D-2)}{2(N-D-1)} + \frac{N-1}{2(N-D-1)}, \tag{4.74}$$

(A13b) - Assumptions needed to estimate the term ap.
If $\sigma < \frac{N-D}{N-D-2}$, then

$$\frac{(\theta-1)(N-\frac{D}{\tilde{\sigma}'}-\frac{2}{q'})}{2}<1-\frac{\mu-1}{q}-\frac{(\mu-1)(N-\frac{D}{\sigma'}-2)}{2}+\frac{N}{(N-D)}. \tag{4.75}$$

If $\sigma \geq \frac{N-D}{N-D-2}$, then

$$\frac{(\theta-1)N(N-D-\frac{2}{q'})}{2(N-D)}<1-\frac{\mu-1}{q}-\frac{(\mu-1)N(N-D-2)}{2(N-D)}+\frac{N}{(N-D)}. \tag{4.76}$$

Lemma 4.7 – *Let p be the solution to the equation*

$$-\frac{\partial p}{\partial t}+Ap=0 \ in \ Q, \qquad \frac{\partial p}{\partial n_A}=0 \ on \ \Sigma, \qquad p(T)=p_T \ in \ \Omega, \tag{4.77}$$

where p_T belongs to $L^\epsilon(\Omega)$ for some $\frac{N-D}{2}<\epsilon<\frac{N}{(N-\frac{D}{\tilde{\sigma}'}-\frac{2}{q'})(\theta-1)}$.

Then $p|_{\gamma\times]0,T[}$ belongs to $L^{\tilde{\tau}}(0,T;L^\tau(\gamma))$ for all $(\tilde{\tau},\tau)$ satisfying

$$\tilde{\tau}\geq 1, \quad \tau\geq\epsilon, \quad \frac{N}{2\epsilon}<\frac{1}{\tilde{\tau}}+\frac{D}{2\tau}.$$

Proof. This is a straightforward consequence of Proposition 2.1. ■

Lemma 4.8 – *Let p_T be in $L^\epsilon(\Omega)$ for some $\epsilon<\frac{N}{(N-\frac{D}{\tilde{\sigma}'}-\frac{2}{q'})(\theta-1)}$. Let b be a nonnegative function belonging to $L^{\tilde{\ell}}(0,T;L^\ell(\Omega))$ for some $(\tilde{\ell},\ell)$ satisfying (2.23) and (4.1). Let a be a nonnegative function belonging to*

$L^{\tilde{k}}(0,T;L^k(\Omega))$ for some $(\tilde{k},k)$ satisfying (2.22) and (4.1). Consider the equation

$$-\frac{\partial p}{\partial t}+Ap+ap=0 \ \ in \ \ Q,$$
$$\frac{\partial p}{\partial n_A}+bp=0 \ on \ \Sigma, \qquad p(T)=p_T \ in \ \Omega. \tag{4.78}$$

The solution p to equation (4.78) belongs to $L^1(0,T;W^{1,1}(\Omega))\cap L^{\tilde{r}}(0,T;L^r(\Omega))$ for every $(\tilde{r},r)$ satisfying

$$\tilde{k}'\le\tilde{r}, \quad \epsilon\le r, \quad k'\le r, \quad \frac{N}{2\epsilon}<\frac{N}{2r}+\frac{1}{\tilde{r}}. \tag{4.79}$$

The trace of p on Σ belongs to $L^{\tilde{s}}(0,T;L^s(\Gamma))$ for every $(\tilde{s},s)$ satisfying

$$\tilde{\ell}'\le\tilde{s}, \quad \epsilon\le s, \quad k'\le s, \quad \frac{N}{2\epsilon}<\frac{N-1}{2s}+\frac{1}{\tilde{s}}. \tag{4.80}$$

Proof. We can proceed as in the proof of Proposition 2.6. ■

Theorem 4.4 – *Assume that a and b satisfy the assumptions of Proposition 2.6. Let p be the solution to equation (4.78), where p_T belongs to $L^\epsilon(\Omega)$ for all $\epsilon<\frac{N}{(N-\frac{D}{\tilde{\sigma}'}-\frac{2}{q'})(\theta-1)}$. Under assumptions (A1)-(A10), (A13), the trace of p on $\gamma\times]0,T[$ belongs to $L^{q'}(0,T;L^{\sigma'}(\gamma))$.*

Proof. 1 - Let p be the solution corresponding to $a\equiv 0$ and $b\equiv 0$. Due to condition (4.71), there exists $\epsilon<\frac{N}{(N-\frac{D}{\tilde{\sigma}'}-\frac{2}{q'})(\theta-1)}$, such that $\frac{N}{2\epsilon}<\frac{1}{q'}+\frac{D}{2\epsilon}$ and $\frac{N}{2\epsilon}<\frac{1}{q'}+\frac{D}{2\sigma'}$. If $\sigma'\le\epsilon$, due to $\frac{N}{2\epsilon}<\frac{1}{q'}+\frac{D}{2\epsilon}$, the trace of p on $\gamma\times]0,T[$ belongs to $L^{q'}(0,T;L^\epsilon(\gamma))\hookrightarrow L^{q'}(0,T;L^{\sigma'}(\gamma))$. If $\sigma'>\epsilon$, due to $\frac{N}{2\epsilon}<\frac{1}{q'}+\frac{D}{2\sigma'}$, the trace of p on $\gamma\times]0,T[$ belongs to $L^{q'}(0,T;L^{\sigma'}(\gamma))$.
2 - Now suppose that a and b are not necessarily equal to zero.
2a - We first prove that there exists $(\tilde{\ell},\ell)$ satisfying (2.23) and (4.1), such that the set of pairs $(\tilde{s},s)$ obeying

$$\tilde{s}\ge\tilde{\ell}', \quad s\ge\epsilon, \quad s\ge\ell', \quad \frac{1}{\tilde{s}}+\frac{1}{\tilde{\ell}}\le 1,$$
$$\frac{1}{s}+\frac{1}{\ell}<\frac{1}{N-D-1}, \frac{N}{2\epsilon}<\frac{1}{\tilde{s}}+\frac{N-1}{2s}, \tag{4.81}$$

is non empty for all $\epsilon<\frac{N}{(N-\frac{D}{\tilde{\sigma}'}-\frac{2}{q'})(\theta-1)}$, big enough. We distinguish two cases.

First case: $\sigma<\frac{N-D-1}{N-D-2}$. We choose $\tilde{\ell}=\frac{q}{\nu-1}$. From (4.7), we have $\nu<$

$1+\frac{N-1}{\sigma'(N-\frac{D}{\sigma'}-2)} < 1+\frac{N-1}{(N-D-1)(N-\frac{D}{\sigma'}-2)}$. Therefore, from (4.3) and (4.73), we can choose ℓ satisfying (4.43), such that $N-D-1<\ell<\frac{(N-1)}{(\nu-1)(N-\frac{D}{\sigma'}-2)}$, and

$$\frac{(\theta-1)(N-\frac{D}{\tilde{\sigma}'}-\frac{2}{q'})}{2} < 1-\frac{\nu-1}{q}-\frac{N-1}{2\ell}+\frac{N-1}{2(N-D-1)}. \tag{4.82}$$

Now we set $\frac{1}{\tilde{s}}=1-\frac{1}{\tilde{\ell}}$, and $\frac{1}{s}=\inf\{\frac{(\theta-1)(N-\frac{D}{\tilde{\sigma}'}-\frac{2}{q'})}{N}, \frac{1}{N-D-1}-\frac{1}{\ell}\}$. From (4.72), it follows that $\frac{1}{\tilde{s}} > \frac{(\theta-1)(N-\frac{D}{\tilde{\sigma}'}-\frac{2}{q'})}{2N} > \frac{1}{2\epsilon}$. Let us prove that the pair $(\tilde{s}, s)$ obeys

$$\frac{(\theta-1)(N-\frac{D}{\tilde{\sigma}'}-\frac{2}{q'})}{2} < \frac{1}{\tilde{s}}+\frac{N-1}{2s}. \tag{4.83}$$

If $\frac{1}{s}=\frac{1}{N-D-1}-\frac{1}{\ell}$, condition (4.83) follows from (4.82). If $\frac{1}{s}=\frac{(\theta-1)(N-\frac{D}{\tilde{\sigma}'}-\frac{2}{q'})}{N}$, condition (4.83) follows from assumption (4.72). Then, due to (4.83), we can choose $\epsilon < \frac{N}{(N-\frac{D}{\tilde{\sigma}'}-\frac{2}{q'})(\theta-1)}$, big enough, so that (4.81) is satisfied.

Second case: $\sigma \geq \frac{N-D-1}{N-D-2}$. We choose $\tilde{\ell}=\frac{q}{\nu-1}$. Since $\nu < \frac{N-D-1}{N-D-2}$, due to (4.3) and (4.74), we can choose ℓ satisfying (4.43), such that $N-D-1 < \ell < \frac{N-D-1}{(\nu-1)(N-D-2)}$, and ℓ satisfies (4.82).

We set $\frac{1}{\tilde{s}} = 1-\frac{1}{\tilde{\ell}}$. Due to (4.72), we have $\frac{1}{\tilde{s}} > \frac{(\theta-1)(N-\frac{D}{\tilde{\sigma}'}-\frac{2}{q'})}{2N}$. If $\frac{1}{N-D-1}-\frac{1}{\ell} \leq \frac{(N-\frac{D}{\tilde{\sigma}'}-\frac{2}{q'})(\theta-1)}{N}$, thanks to (4.82), we can choose s such that $0<\frac{1}{s}<\frac{1}{N-D-1}-\frac{1}{\ell}$ and s satisfies (4.83). If $\frac{1}{N-D-1}-\frac{1}{\ell} > \frac{(\theta-1)(N-\frac{D}{\tilde{\sigma}'}-\frac{2}{q'})}{N}$, we set $\frac{1}{s} = \frac{(\theta-1)(N-\frac{D}{\tilde{\sigma}'}-\frac{2}{q'})}{N}$. Condition (4.83) follows from (4.72). Due to (4.83), we can choose $\epsilon < \frac{N}{(N-\frac{D}{\tilde{\sigma}'}-\frac{2}{q'})(\theta-1)}$, big enough, so that (4.81) is satisfied.

2b - In the same way, using assumptions (4.2), (A10) and (A13b), we can prove that there exists $(\tilde{k}, k)$ satisfies (2.22) and (4.1), such that the set of pairs $(\tilde{r}, r)$ satisfying

$$\tilde{r} \geq \tilde{k}', \quad r \geq \epsilon, \quad r \geq k', \quad \frac{1}{\tilde{r}}+\frac{1}{\tilde{k}} \leq 1, \quad \frac{1}{r}+\frac{1}{k} < \frac{2}{N-D}, \quad \frac{N}{2\epsilon} < \frac{1}{\tilde{r}}+\frac{N}{2r}, \tag{4.84}$$

is non empty for all $\epsilon < \frac{N}{(N-\frac{D}{\tilde{\sigma}'}-\frac{2}{q'})(\theta-1)}$, big enough.

2c - From Lemma 4.8, it follows that the trace of the solution p to equation (4.78) belongs to $L^{\tilde{s}}(0,T;L^s(\Gamma))$, where $(\tilde{s}, s)$ is chosen in Step 1. We set $\frac{1}{\tilde{\beta}}=\frac{1}{\tilde{s}}+\frac{1}{\tilde{\ell}}, \frac{1}{\beta}=\frac{1}{s}+\frac{1}{\ell}$. Then $bp \in L^{\tilde{\beta}}(0,T;L^{\beta}(\Gamma))$ with $\beta > N-D-1$. Let

π be the solution to equation (4.50). Using Lemma 4.3, we deduce that the trace of π on $\gamma\times]0,T[$ exists in $L^1(0,T;L^1(\gamma))$.
2d - From Lemma 4.8, it follows that the solution p to equation (4.37) belongs to $L^{\tilde{r}}(0,T;L^r(\Omega))$, where $(\tilde{r},r)$ is chosen in Step 2. Let $\hat{\pi}$ be the solution to equation (4.51), we can prove that the trace of $\hat{\pi}$ on $\gamma\times]0,T[$ belongs to $L^1(0,T;L^1(\gamma))$.
3 - From Steps 1, 2b and 2c, it follows that the trace $p|_{\gamma\times]0,T[}$ of the solution to equation (4.78) belongs to $L^1(0,T;L^1(\gamma))$. By a comparison principle and using Step 1, we can prove that it belongs to $L^{q'}(0,T;L^{\sigma'}(\gamma))$. □

Examples. Let us give examples in the three dimensional case, for which (A5), (A7)-(A13) are satisfied.

• Suppose that $N=3$, $D=1$, $q=\sigma=2$. We have $\frac{N-D}{N-D-1}=2\leq\sigma<\frac{N-D}{N-D-2}=\infty$ and $2\leq\sigma<\frac{N-D-1}{N-D-2}=\infty$. Conditions (4.2) and $\mu<q$ correspond to $\mu<2$. Condition (4.3) is satisfied if $\nu<\frac{5}{3}$. Conditions (4.11) and (4.16) on κ are satisfied if $\kappa<\frac{9}{5}$. Condition (4.71) on θ is satisfied if $\theta<2$. If $\rho>2$, (A9) is satisfied. Thus, if $\mu<2$, $\nu<\frac{5}{3}$, $\kappa<\frac{9}{5}$ and $\rho>2$, all the conditions are satisfied.

• Suppose that $N=3$, $D=1$, and that K_U is bounded in $L^\infty(\gamma\times]0,T[)$. We do not set $q=\sigma=\infty$ since the regularity results in Section 2 are stated in the case when q and σ are finite, but we can take q and σ as big as we want. We have $\sigma<\frac{N-D}{N-D-2}=\infty$ and $\sigma<\frac{N-D-1}{N-D-2}=\infty$. Conditions (2.1), (4.2), (4.4), (4.6) and (4.8) correspond to $\mu<\infty$. Conditions (2.2), (4.3), (4.5), (4.6) and (4.7) are satisfied if $\nu<\infty$. Conditions (4.6) and (4.12) on κ are satisfied if $\kappa<2$. Condition (A13) on θ is satisfied if $\theta<\infty$. If $\rho=\infty$, (A9) is satisfied. Thus, if $\mu<\infty$, $\nu<\infty$, $\kappa<2$, $\theta<\infty$ and $\rho=\infty$, all the conditions are satisfied.

• Suppose that $N=3$, $D=0$, $q=\sigma=2$. We have $\frac{N-D-1}{N-D-2}=2\leq\sigma<\frac{N-D}{N-D-2}=3$. Conditions (4.2) and $\mu<q$ correspond to $\mu<2$. Condition (4.3) is satisfied if $\nu<\frac{3}{2}$. Condition (4.6) on κ is satisfied if $\kappa<\frac{3}{2}$. Condition (4.71) on θ is satisfied if $\theta<\frac{3}{2}$. If $\rho>3$, (A9) is satisfied. Thus, if $\mu<2$, $\nu<\frac{3}{2}$, $\kappa<\frac{3}{2}$, $\theta<2$ and $\rho>3$, all the conditions are satisfied.

• Suppose that $N=3$, $D=0$, and that K_U is bounded in $L^\infty(\gamma\times]0,T[)$. Therefore we can take σ and q as big as we want. We have $\sigma\geq\frac{N-D}{N-D-2}=3$ and $\sigma\geq\frac{N-D-1}{N-D-2}=2$. Conditions (2.1) and (4.2) are satisfied if $\mu<3$. Conditions (2.2) and (4.3) are satisfied if $\nu<2$. Conditions (4.6), (4.14) and (4.19) on κ correspond to $\kappa<\frac{3}{2}$. Conditions (4.71), (4.74) and (4.76) on θ correspond to $\theta<3$. If $\rho=\infty$, (A9) is satisfied. Thus, if $\mu<3$, $\nu<2$, $\kappa<\frac{3}{2}$, $\theta<3$ and $\rho=\infty$, all the conditions are satisfied.

4.1. Optimality conditions for (P_1)

Lemma 4.9 – *Let $(a_n)_n$ be a sequence of nonnegative functions converging to a in $L^{\tilde{k}}(0,T;L^k(\Omega))$ for every $(\tilde{k},k)$ satisfying (2.22). Let $(b_n)_n$ be a sequence of nonnegative functions converging to b in $L^{\tilde{\ell}}(0,T;L^{\ell}(\Gamma))$ for every $(\tilde{\ell},\ell)$ satisfying (2.23). Let z_n be the solution of*

$$\frac{\partial z}{\partial t} + Az + a_n z = 0 \text{ in } Q, \quad \frac{\partial z}{\partial n_A} + b_n z = u\delta_\gamma \text{ on } \Sigma, \quad z(0) = 0 \text{ in } \Omega,$$

and let z be the solution of

$$\frac{\partial z}{\partial t} + Az + az = 0 \text{ in } Q, \quad \frac{\partial z}{\partial n_A} + bz = u\delta_\gamma \text{ on } \Sigma, \quad z(0) = 0 \text{ in } \Omega. \tag{4.85}$$

Then $(z_n)_n$ converges to z in $L^{\tilde{r}}(0,T;L^r(\Omega))$ for all $(\tilde{r},r)$ satisfying (2.13), and in $L^q(0,T;W^{1,d}(\Omega))$ for all $d < \frac{N-D}{N-D-1}$. Moreover $(z_n(T))_n$ converges to $z(T)$ for the weak topology of $L^r(\Omega)$ for all $1 \leq r < \frac{N}{(N-\frac{D}{\tilde{\sigma}'}-\frac{2}{q'})}$ (where $\tilde{\sigma} = \min\,(\sigma, \frac{N-D}{(N-D-\frac{2}{q'})})$).

Proof. Due to Proposition 2.6, we have

$$\|z_n\|_{L^{\tilde{r}}(0,T;L^r(\Omega))} + \|z_n|_\Sigma\|_{L^{\tilde{s}}(0,T;L^s(\Gamma))} \leq C\|u\|_{L^q(0,T;L^\sigma(\gamma))}$$

for all $(\tilde{r},r)$ satisfying (2.13), and all $(\tilde{s},s)$ satisfying (2.15) such that $\tilde{s} < 2\rho$, where C does not depend on a_n and b_n. We see that $w = z_n - z$ is solution of the equation

$$\begin{aligned}
&\frac{\partial w}{\partial t} + Aw + aw = (a - a_n)z_n \text{ in } Q,\\
&\frac{\partial w}{\partial n_A} + bw = (b - b_n)z_n \text{ on } \Sigma, \quad w(0) = 0 \text{ in } \Omega.
\end{aligned} \tag{4.86}$$

Due to Propositions 2.6, 2.7, and 4.1, there exist $(\tilde{k},k)$ satisfying (2.22) and $(\tilde{\ell},\ell)$ satisfying (2.23), such that:

$$\begin{aligned}
&\|z_n - z\|_{L^{\tilde{r}}(0,T;L^r(\Omega))} + \|z_n - z\|_{L^q(0,T;W^{1,d}(\Omega))}\\
&\leq C(\|a_n - a\|_{L^{\tilde{k}}(0,T;L^k(\Omega))}\|z_n\|_{L^{\tilde{r}}(0,T;L^r(\Omega))}\\
&+\|b_n - b\|_{L^{\tilde{\ell}}(0,T;L^{\ell}(\Gamma)}\|z_n|_\Sigma\|_{L^{\tilde{s}}(0,T;L^s(\Gamma))})\\
&\leq C(\|a_n - a\|_{L^{\tilde{k}}(0,T;L^k(\Omega))} + \|b_n - b\|_{L^{\tilde{\ell}}(0,T;L^{\ell}(\Gamma))})\|u\|_{L^q(0,T;L^\sigma(\gamma))}.
\end{aligned}$$

Thus the convergence results for $(z_n)_n$ are established.

Since $(z_n(T))_n$ is bounded in $L^r(\Omega)$ for all $1 < r < \frac{N}{(N-\frac{D}{\tilde{\sigma}'}-\frac{2}{q'})}$, we can suppose that $(z_n(T))_n$ converges to some $\tilde{z}$ for the weak topology of $L^r(\Omega)$ for all $1 < r < \frac{N}{(N-\frac{D}{\tilde{\sigma}'}-\frac{2}{q'})}$. By passing to the limit in the Green formula

$$\int_Q (-z_n \frac{\partial \phi}{\partial t} + \sum_{i,j=1}^{N} a_{ij} D_j z_n D_i \phi + a_0 z_n \phi + a_n z_n \phi) dx dt + \int_\Sigma b_n z_n \phi ds dt$$
$$= \int_0^T \int_\gamma u \phi d\xi dt - \int_\Omega \phi(T) z_n(T) dx,$$

we obtain $\int_\Omega \phi(T)\tilde{z}dx = \int_\Omega \phi(T)z(T)dx$ for every $\phi \in C^1(\overline{Q})$, where z is the solution of (4.85). Thus $\tilde{z} = z(T)$. ■

Theorem 4.5 – *Let z be the solution to the equation (4.85). Let $\vec{h} = \vec{h}^1 + \vec{h}^2$, where $\vec{h}^1$ belongs to $L^{\frac{q}{\kappa-1}}(0,T;(L^{\frac{d}{\kappa-1}}(\Omega))^N)$ for all $1 < d < \frac{N-D}{N-D-1}$, and $\vec{h}^2$ belongs to $L^{\frac{\delta_4}{\kappa-1}}(0,T;(L^{\frac{\rho}{\kappa-1}}(\Omega))^N)$ for all $1 < \delta_4 < 2$. Let p be the solution to equation (4.9) corresponding to $\vec{h}$. If assumptions (A7)-(A13) are satisfied, then we have the Green formula*

$$\int_0^T \int_\gamma p(s,t)u(s,t)\; ds dt = \int_Q \nabla z \cdot \vec{h}\; dx dt + \int_\Omega p_T(x) z(x,T)\; dx. \tag{4.87}$$

Proof. Observe that $z(T)$ belongs to $L^r(\Omega)$ for all $1 \le r < \frac{N}{N-\frac{D}{\tilde{\sigma}'}-\frac{2}{q'}}$. With (4.71), we can find ϵ satisfying $\epsilon < \frac{N}{(N-\frac{D}{\tilde{\sigma}'}-\frac{2}{q'})(\theta-1)}$ such that $z(T)$ belongs to $L^\infty(0,T;L^{\epsilon'}(\Omega))$. Thus $p_T z(T)$ belongs to $L^1(\Omega)$.

Let $(\vec{h}^1_m)_m$ (respectively $(\vec{h}^2_m)_m$) be a sequence of regular functions converging to $\vec{h}^1$ (respectively $\vec{h}^2$) in $L^{\frac{q}{\kappa-1}}(0,T;L^{\frac{d}{\kappa-1}}(\Omega))$ for all $1 < d < \frac{N-D}{N-D-1}$ (respectively $L^{\frac{\delta_4}{\kappa-1}}(0,T;L^{\frac{\rho}{\kappa-1}}(\Omega))$ for all $1 < \delta_4 < 2$). Let $(p^m_T)_m$ be a sequence of regular functions converging to p_T in the space $L^\epsilon(\Omega)$ for all $\epsilon < \frac{N}{(N-\frac{D}{\tilde{\sigma}'}-\frac{2}{q'})(\theta-1)}$. Let $\vec{h}_m = \vec{h}^1_m + \vec{h}^2_m$, and let p_m be the solution of the equation

$$-\frac{\partial p}{\partial t} + Ap + ap = -\text{ div } \vec{h}_m \text{ in } Q,$$
$$\frac{\partial p}{\partial n_A} + bp = \vec{h}_m \cdot \vec{n} \text{ on } \Sigma, \qquad p(T) = p^m_T \text{ in } \Omega. \tag{4.88}$$

Using Theorem 4.2 in [24] for equations (4.85) and (4.88), we obtain:

$$\int_0^T \int_\gamma p_m(s,t)u(s,t) ds dt = \int_Q \nabla z \cdot \vec{h}_m dx dt + \int_\Omega p^m_T(x) z(x,T) dx. \tag{4.89}$$

Due to estimates established in the proofs of Theorems 4.2, 4.3 and 4.4, we can prove that $(p_m|_{\gamma\times]0,T[})_m$ converges to $p|_{\gamma\times]0,T[}$ for the weak-star topology of $L^{q'}(0,T;L^{\sigma'}(\gamma))$. Therefore the Green formula (4.87) may be obtained by passing to the limit in (4.89). ■

Theorem 4.6 – *If u is a solution of (P_1), then*

$$\int_0^T\int_\gamma p(v-u)d\zeta dt + qC_\gamma\int_0^T\Big(\int_\gamma |u|^\sigma d\zeta\Big)^{\frac{q}{\sigma}-1}\Big(\int_\gamma |u|^{\sigma-2}u(v-u)\,d\zeta\Big)\,dt \geq 0$$

for every $v \in K_U$, where p is the solution of

$$-\frac{\partial p}{\partial t} + Ap + ap = -\kappa C_Q\ div\ (|\nabla y_u - V_d|^{\kappa-2}(\nabla y_u - V_d))\ in\ Q,$$
$$\frac{\partial p}{\partial n_A} + bp = \kappa C_Q(|\nabla y_u - V_d|^{\kappa-2}(\nabla y_u - V_d))\cdot\vec{n}\ on\ \Sigma,$$
$$p(T) = \theta C_\Omega|y_u(T) - y_d|^{\theta-2}(y_u(T) - y_d)\ in\ \Omega,$$

with $a=\Phi'(y_u)$, $b=\Psi'(y_u)$, and y_u is the solution of (1.1) corresponding to u.

Proof. 1 - Let v be in K_U, and $\lambda > 0$. Denote by y_λ the solution of (1.1) corresponding to $u + \lambda(v-u)$. Let us set $w_\lambda = y_\lambda - y_u$, then w_λ is the solution of the equation:

$$\frac{\partial w}{\partial t}+Aw+a_\lambda w = 0 \text{ in } Q,\quad \frac{\partial w}{\partial n_A}+b_\lambda w = \lambda(v-u)\delta_\gamma \text{ on } \Sigma,\quad w(0) = 0 \text{ in } \Omega,$$

where $a_\lambda = \int_0^1 \Phi'(y_u + \vartheta(y_\lambda - y_u))d\vartheta$, and $b_\lambda = \int_0^1 \Psi'(y_u + \vartheta(y_\lambda - y_u))d\vartheta$. Applying Proposition 2.6 to the above equation, we obtain

$$\|w_\lambda\|_{L^{\tilde r}(0,T;L^r(\Omega))} + \|w_\lambda|_\Sigma\|_{L^{\tilde s}(0,T;L^s(\Gamma))} \leq C\lambda\|v-u\|_{L^q(0,T;L^\sigma(\gamma))}$$

for every $(\tilde r, r)$ satisfying (2.13), and every $(\tilde s, s)$ satisfying (2.15) and $\tilde s < 2\rho$. Therefore, when λ tends to zero, the sequence $(y_\lambda)_\lambda$ tends to y_u in $L^{\tilde r}(0,T;L^r(\Omega))$ for all $(\tilde r, r)$ satisfying (2.13), and the sequence $(y_\lambda|_\Sigma)_\lambda$ tends to $y_u|_\Sigma$ in $L^{\tilde s}(0,T;L^s(\Gamma))$ for all $(\tilde s, s)$ satisfying (2.15) and $\tilde s < 2\rho$. Thus $(a_\lambda)_\lambda$ tends to a in $L^{\tilde k}(0,T;L^k(\Omega))$ for every $(\tilde k, k)$ satisfying (2.22). And $(b_\lambda)_\lambda$ tends to b in $L^{\tilde\ell}(0,T;L^\ell(\Gamma))$ for every $(\tilde\ell, \ell)$ satisfying (2.23). For all $1 \leq r < \frac{N}{(N-\frac{D}{\bar\sigma'}-\frac{2}{q'})}$, we have the estimate:

$$\|w_\lambda(T)\|_{L^r(\Omega)} \leq \|w_\lambda\|_{L^\infty(0,T;L^r(\Omega))} \leq C\lambda\|v-u\|_{L^q(0,T;L^\sigma(\gamma))}.$$

Thus $(y_\lambda(T))_\lambda$ tends to $y_u(T)$ in $L^r(\Omega)$ for all $1 \leq r < \frac{N}{(N-\frac{D}{\bar\sigma'}-\frac{2}{q'})}$.

Due to the previous convergence results for $(a_\lambda)_\lambda$, $(b_\lambda)_\lambda$, and $(y_\lambda)_\lambda$, using Lemma 4.9, we can prove that $(y_\lambda)_\lambda$ tends to y_u in $L^q(0,T;W^{1,d}(\Omega)) + L^{\delta_4}(0,T;W^{1,\rho}(\Omega))$ for all $d < \frac{N-D}{N-D-1}$ and all $\delta_4 < 2$.
2 - Now we set $z_\lambda = (y_\lambda - y_u)/\lambda$, and we denote by z the solution of

$$\frac{\partial z}{\partial t} + Az + az = 0 \text{ in } Q, \qquad \frac{\partial z}{\partial n_A} + bz = (v-u)\delta_\gamma \text{ on } \Sigma, \qquad z(0) = 0 \text{ in } \Omega.$$

Due to Lemma 4.9, we can prove that the sequence $(z_\lambda)_\lambda$ tends to z in $L^q(0,T;W^{1,d}(\Omega))$ for every $d < \frac{N-D}{N-D-1}$, and $(z_\lambda(T))_\lambda$ tends to $z(T)$ for the weak topology of $L^r(\Omega)$, for all $1 \le r < \frac{N}{(N-\frac{D}{\sigma'}-\frac{2}{q'})}$.
3 - We want to calculate the gradient of the cost functional. From the convexity of the mapping $y \mapsto \int_\Omega |y-y_d|^\theta dx + \int_Q |\nabla y - V_d|^\kappa dxdt$, we have

$$\begin{aligned}
&\theta C_\Omega \int_\Omega |y_u(T) - y_d|^{\theta-2}(y_u(T) - y_d) z_\lambda(T) dx \\
&\qquad + \kappa C_Q \int_Q |\nabla y_u - V_d|^{\kappa-2}(\nabla y_u - V_d)\nabla z_\lambda dx \\
&\le \frac{1}{\lambda}(I(u + \lambda(v-u)) - I(u)) \\
&\le \theta C_\Omega \int_\Omega |y_\lambda(T) - y_d|^{\theta-2}(y_\lambda(T) - y_d) z_\lambda(T) dx \\
&\qquad + \kappa C_Q \int_Q |\nabla y_\lambda - V_d|^{\kappa-2}(\nabla y_\lambda - V_d)\nabla z_\lambda dx.
\end{aligned}$$

Let us set $F(u) = I(y_u, u)$. Due to the previous convergence results on $(y_\lambda)_\lambda$, and $(z_\lambda)_\lambda$, by passing to the limit in the above inequalities, we obtain

$$\begin{aligned}
F'(u)(v-u) = &\int_\Omega \theta |y_u(T) - y_d|^{\theta-2}(y_u(T) - y_d) z(T) dx \\
&+ \kappa \int_Q |\nabla y_u - V_d|^{\kappa-2}(\nabla y_u - V_d)\nabla z dx \\
&+ qC_\gamma \int_0^T \Big(\int_\gamma |u|^\sigma d\zeta\Big)^{\frac{q}{\sigma}-1}\Big(\int_\gamma |u|^{\sigma-2}u(v-u)d\zeta\Big)\, dt.
\end{aligned}$$

Finally we conclude with the Green formula (4.87). ∎

5. Derivative of the cost functional with respect to deformations of γ

In this section we consider the case when γ is a curve, that is $D = 1$. Since we suppose that $N \ge D + 2$, we only consider the case where $N \ge 3$.

We calculate the derivative of the cost functional with respect to fields of deformation of the curve γ. For simplicity we suppose that the control u is only depending on the time variable. Let I be a bounded interval in R, and let $(g(\vartheta))_{\vartheta \in I}$ be a regular parametrization of γ. More precisely we suppose that g is a mapping of class $C^{\bar{k}}$ from I to R^N such that $\gamma = (g(\vartheta))_{\vartheta \in I}$, and such that $|g'(\vartheta)|$, the Euclidean norm of $g'(\vartheta)$, is different from zero for all $\vartheta \in I$. Let $(G(\zeta))_{\zeta \in \gamma}$ be a regular field of deformation of γ such that G is a mapping of class $C^{\bar{k}}$ from γ into R^N, and $G(\zeta)$ is a tangent vector to Γ at ζ. We suppose in addition that there exists a family $(g_\lambda)_{\lambda \in]0,1[}$ of mappings from I into Γ, such that $\gamma_\lambda = (g_\lambda(\vartheta))_{\vartheta \in I}$ is a curve of class $C^{\bar{k}}$, and

$$g_\lambda(\vartheta) = g(\vartheta) + \lambda G(g(\vartheta)) + \lambda \varepsilon(\lambda, \vartheta)$$

and

$$g'_\lambda(\vartheta) = g'(\vartheta) + \lambda G'(g(\vartheta))g'(\vartheta) + \lambda \varepsilon_1(\lambda, \vartheta),$$

where $\lim_{\lambda \to 0} \|\varepsilon(\lambda, \cdot)\|_{L^\infty(I;R^N)} = 0$, and $\lim_{\lambda \to 0} \|\varepsilon_1(\lambda, \cdot)\|_{L^\infty(I;R^N)} = 0$. Let y_{u,γ_λ} (resp. y_{u,γ_λ}) be the solution of (1.1) corresponding to u and γ_λ (resp. γ_λ). Recall that u only depends on t. Set

$$F(\gamma) = J(y_{u,\gamma}, u) = \int_\Omega |y_{u,\gamma}(T) - y_d|^\theta dx \ .$$

We want to calculate

$$\lim_{\lambda \to 0} \frac{F(\gamma_\lambda) - F(\gamma)}{\lambda} \ .$$

For this calculation, we suppose that the assumptions (A1), (A3) are satisfied. The initial condition y_0 belongs to $L^\rho(\Omega)$ for $\rho \geq \frac{N}{N-3}$. Some additional assumptions stated below are needed.

(A8') - $\Psi \equiv 0$.

(A9') - The control variable u belongs to $L^\infty(0,T)$. For the regularity of the adjoint state, we make the following assumptions on θ and μ:

$$\theta < 1 + \frac{N-1}{(N-3)N} \ , \quad \mu < 1 + \frac{N-1}{(N-3)N} \ .$$

(Notice that we can use the regularity results in Section 2.2 with $D = 1$, q and σ are as big as we want.)

5.1. Regularity results of the adjoint equation

We consider the following terminal boundary value problem

$$-\frac{\partial p}{\partial t} + Ap + ap = 0 \ \text{ in } \ Q, \quad \frac{\partial p}{\partial n_A} = 0 \ \text{ on } \ \Sigma, \quad p(T) = p_T \ \text{ in } \ \Omega, \ (5.1)$$

where a is a nonnegative function belonging to $L^{\tilde{k}}(0,T;L^k(\Omega))$ for every $(\tilde{k},k)$ satisfying

$$\tilde{k} < \infty, \quad k < \frac{N-1}{(N-3)(\mu-1)}, \tag{5.2}$$

and where p_T belongs to $L^\epsilon(\Omega)$ for all $1 \leq \epsilon < \frac{N-1}{(N-3)(\theta-1)}$. If $\bar{y}$ is the weak solution of (1.1) corresponding to u and γ, if we set $a = \Phi'(\bar{y})$, $p_T = \theta|\bar{y}(T)-y_d|^{\theta-2}(\bar{y}(T)-y_d)$, then equation (5.1) corresponds to the adjoint equation for J associated with $(\gamma,\bar{y})$.

Theorem 5.1 – *Let p be the weak solution to equation (5.1), where p_T belongs to $L^\epsilon(\Omega)$. For all $1 \leq \epsilon < \frac{N-1}{(N-3)(\theta-1)}$, the mapping that associates p with p_T is continuous from $L^\epsilon(\Omega)$ into $L^{\tilde{m}}(0,T;L^m(\Omega))$ for all $(\tilde{m},m)$ satisfying*

$$\tilde{m} > 1, \quad m \geq \epsilon, \quad \frac{N}{2\epsilon} < \frac{1}{\tilde{m}} + \frac{N}{2m}. \tag{5.3}$$

We have the estimate

$$\|p\|_{L^{\tilde{m}}(0,T;L^m(\Omega))} \leq C(\tilde{m},m,\epsilon)\|p_T\|_{L^\epsilon(\Omega)},$$

where $C(\tilde{m},m,\epsilon)$ does not depend on a. Moreover the solution p to equation (5.1) belongs to $L^1(0,T;C^{1,\alpha}(\overline{\Omega}))$ for some $0<\alpha<1$, and

$$\|p\|_{L^1(0,T;C^{1,\alpha}(\overline{\Omega}))} \leq C(1+\|a\|_{L^{\tilde{k}}(0,T;L^k(\Omega))})\|p_T\|_{L^\epsilon(\Omega)},$$

for some $\tilde{k} < \infty$, $k < \frac{N-1}{(N-3)(\mu-1)}$, and $\epsilon < \frac{N-1}{(N-3)(\theta-1)}$.

Proof. 1 - When $a \equiv 0$, the estimate of p in $L^{\tilde{m}}(0,T;L^m(\Omega))$ can be obtained in a classical manner. When $a \not\equiv 0$, this result can be obtained with a comparison principle.
2 - Let π be the solution to

$$-\frac{\partial \pi}{\partial t} + A\pi = -ap \text{ in } Q, \quad \frac{\partial \pi}{\partial n_A} = 0 \text{ on } \Sigma, \quad \pi(T) = 0 \text{ in } \Omega. \tag{5.4}$$

Due to (A9'), we can find a pair (ξ_1, j_1), $1<\xi_1<2$, $j_1>N$, satisfying

$$\sup\left\{\frac{N(N-3)(\theta-1)}{2(N-1)}, \frac{N(N-3)(\mu-1)}{2(N-1)}\right\} < 1 + \frac{N}{2j_1} - \frac{\xi_1}{2} < \frac{1}{2}. \tag{5.5}$$

Next, we can choose $\epsilon < \frac{N-1}{(N-3)(\theta-1)}$ and $k < \frac{N-1}{(N-3)(\mu-1)}$, such that

$$0 < \frac{N}{2\epsilon} < 1 + \frac{N}{2j_1} - \frac{\xi_1}{2},$$

$$\sup\{\frac{N(N-3)(\mu-1)}{2(N-1)}, \frac{N}{2j_1}-\frac{N}{2\epsilon}\} < \frac{N}{2k} < 1+\frac{N}{2j_1}-\frac{\xi_1}{2}.$$

Observe that $\frac{N}{2\epsilon} < 1+\frac{N}{2j_1}-\frac{\xi_1}{2} < \frac{1}{2}$, and $\frac{N}{2k} < 1+\frac{N}{2j_1}-\frac{\xi_1}{2} < \frac{1}{2}$. Thus $\epsilon > N$ and $k > N$. Since $1+\frac{N}{2j_1}-\frac{\xi_1}{2} < \frac{1}{2} < 1-\frac{N}{2\epsilon} < 1+\frac{N}{2}-\frac{N}{2\epsilon}$, and $1+\frac{N}{2j_1}-\frac{\xi_1}{2} < 2+\frac{N}{2j_1}-\frac{\xi_1}{2}-\frac{N}{2\epsilon}$, we also have $\frac{N}{2k} < 1+\frac{N}{2}-\frac{N}{2\epsilon}$ and $\frac{N}{2k} < 2+\frac{N}{2j_1}-\frac{\xi_1}{2}-\frac{N}{2\epsilon}$. Thus there exists $\tilde{m} > 1$ such that

$$\sup\{0, \frac{N}{2\epsilon}-\frac{N}{2}+\frac{N}{2k}, \frac{N}{2\epsilon}-1-\frac{N}{2j_1}+\frac{\xi_1}{2}+\frac{N}{2k}\} < \frac{1}{\tilde{m}} < 1.$$

Finally, we can find $m \geq \epsilon$ such that

$$\sup\{0, \frac{N}{2\epsilon}-\frac{1}{\tilde{m}}, \frac{N}{2j_1}-\frac{N}{2k}\} < \frac{N}{2m} < \inf\{\frac{N}{2\epsilon}, \frac{N}{2}-\frac{N}{2k}, 1+\frac{N}{2j_1}-\frac{\xi_1}{2}-\frac{N}{2k}\}.$$

Then $(\tilde{m}, m)$ satisfy (5.3). Due to Step 1, p belongs to $L^{\tilde{m}}(0,T;L^m(\Omega))$. Let $\frac{1}{r} = \frac{1}{m}+\frac{1}{k}$, then $r > 1$ and $ap \in L^1(0,T;L^r(\Omega))$. Since $\frac{N}{2j_1} < \frac{N}{2m}+\frac{N}{2k} = \frac{N}{2r}$, we have $j_1 > r$. Moreover, $\frac{N}{2r} = \frac{N}{2m}+\frac{N}{2k} < 1+\frac{N}{2j_1}-\frac{\xi_1}{2}$, and therefore π belongs to $L^1(0,T;W^{\xi_1,j_1}(\Omega))$.

Let $\tilde{\pi}$ be the solution to the equation

$$-\frac{\partial\tilde{\pi}}{\partial t}+A\tilde{\pi} = 0 \text{ in } Q, \qquad \frac{\partial\tilde{\pi}}{\partial n_A} = 0 \text{ on } \Sigma, \qquad \tilde{\pi}(T) = p_T \text{ in } \Omega, \tag{5.6}$$

Under condition (A9'), we can find (ξ, j) satisfying

$$0 \leq \xi < 2, \quad \frac{N(N-3)(\theta-1)}{2(N-1)} < 1+\frac{N}{2j}-\frac{\xi}{2} < \frac{1}{2}. \tag{5.7}$$

We can choose $\epsilon \leq j$ satisfying

$$\frac{N(N-3)(\theta-1)}{2(N-1)} < \frac{N}{2\epsilon} < 1+\frac{N}{2j}-\frac{\xi}{2}.$$

Then $\tilde{\pi} \in L^1(0,T;W^{\xi,j}(\Omega))$. Since $p = \pi+\tilde{\pi}$, we have shown that p belongs to $L^1(0,T;W^{\xi_1,j_1}(\Omega))+L^1(0,T;W^{\xi,j}(\Omega))$.

3 - Let (ξ, j) obey (5.7), and (ξ_1, j_1) obey (5.5). Then there exists α such that

$$1 > \alpha > 0, \quad \xi > \alpha+1+\frac{N}{j}, \quad \xi_1 > \alpha+1+\frac{N}{j_1}. \tag{5.8}$$

Thus $L^1(0,T;W^{\xi_1,j_1}(\Omega))+L^1(0,T;W^{\xi,j}(\Omega))$ is included in $L^1(0,T;C^{1,\alpha}(\overline{\Omega}))$. The estimate of p in $L^1(0,T;C^{1,\alpha}(\overline{\Omega}))$ in function of $\|p_T\|_{L^\epsilon(\Omega)}$ may be deduced from the analysis of Step 1 and Step 2. ∎

5.2. Derivative of the cost functionnal

First we give a convergence result.

Proposition 5.1 – *Let y_λ be the solution of (1.1) corresponding to u and γ_λ, and $\bar{y}$ be the solution of (1.1) corresponding to u and γ, then $(y_\lambda)_\lambda$ converges to $\bar{y}$ in $L^{\tilde{r}}(0,T;L^r(\Omega))$ for every $(\tilde{r},r)$ obeying*

$$\tilde{r} < \infty, \quad 1 \leq r < \frac{N-1}{N-3}. \tag{5.9}$$

Therefore, if we set $a_\lambda = \int_0^1 \Phi'(\bar{y} + \chi(y_\lambda - \bar{y}))d\chi$, the sequence $(a_\lambda)_\lambda$ tends to $a = \Phi'(\bar{y})$ in $L^{\tilde{k}}(0,T;L^k(\Omega))$ for every $(\tilde{k},k)$ satisfying (5.2).

Proof. The function $w_\lambda = y_\lambda - \bar{y}$ satisfies

$$\frac{\partial w}{\partial t} + Aw + a_\lambda w = 0 \text{ in } Q, \quad \frac{\partial w}{\partial n_A} = u(t)(\delta_{\gamma_\lambda} - \delta_\gamma) \text{ on } \Sigma, \quad w(0) = 0 \text{ in } \Omega.$$

Due to Proposition 2.6, the sequence $(w_\lambda)_\lambda$ is bounded in $L^{\tilde{r}}(0,T;L^r(\Omega))$ for every $(\tilde{r},r)$ obeying (5.9). Since the sequence $(u(\delta_{\gamma_\lambda} - \delta_\gamma))_\lambda$ converges to zero for the weak-star topology of $\mathcal{M}_\lfloor(\overline{Q} \setminus (\otimes \times \{T\}))$, due to Lemma 2.3.1, the sequence $(w_\lambda)_\lambda$ (or at least a subsequence) converges to zero in $L^1(Q)$. With Lemma 4.1 in [13], we can prove that $(w_\lambda)_\lambda$ converges to zero in $L^{\tilde{r}}(0,T;L^r(\Omega))$ for every $(\tilde{r},r)$ obeying (5.9). ∎

Proposition 5.2 – *Let p be the solution of (5.1) corresponding to a and p_T. Let $(a_\lambda)_\lambda$ be a sequence of nonnegative functions converging to a in $L^{\tilde{k}}(0,T;L^k(\Omega))$, for all $(\tilde{k},k)$ satisfying (5.2). Let $(p_T^\lambda)_\lambda$ be a sequence of functions converging to p_T for the weak topology of $L^\epsilon(\Omega)$, for all $\epsilon < \frac{N-1}{(N-3)(\theta-1)}$. Let p_λ be the solution to*

$$-\frac{\partial p}{\partial t} + Ap + a_\lambda p = 0 \text{ in } Q, \quad \frac{\partial p}{\partial n_A} = 0 \text{ on } \Sigma, \quad p(T) = p_T^\lambda \text{ in } \Omega. \tag{5.10}$$

Then the sequence $(p_\lambda)_\lambda$ converges to p in $L^1(0,T;C^1(\overline{\Omega}))$.

Proof. From Theorem 5.1, we know that the sequence $(p_\lambda)_\lambda$ is bounded in $L^1(0,T;C^{1,\alpha}(\overline{\Omega}))$ for some $0 < \alpha < 1$. The identity mapping from $C^{1,\alpha}(\overline{\Omega})$ into $C^1(\overline{\Omega})$ is compact. The sequence $(\frac{dp_\lambda}{dt})_\lambda$ is bounded in the space $L^1(0,T;(W^{1,\beta}(\Omega))')$ for some β big enough. From a compactness result ([28], Corollary 4), we deduce that the identity mapping is compact from $L^1(0,T;C^{1,\alpha}(\overline{\Omega})) \cap W^{1,1}(0,T;(W^{1,\beta}(\Omega))')$ into $L^1(0,T;C^1(\overline{\Omega}))$. Therefore the sequence $(p_\lambda)_\lambda$ converges to some function p in $L^1(0,T;C^1(\overline{\Omega}))$. Next, we can easily verify that p is the solution to equation (5.1). ∎

Lemma 5.1 – *Let p be the solution to equation (5.1) corresponding to a, p_T. Suppose that $(a_\lambda)_\lambda$ and $(p_T^\lambda)_\lambda$ satisfy the assumptions of Propositon 5.2. Let p_λ be the solution to equation (5.10), then we have*

$$
\begin{aligned}
lim_{\lambda\to 0}\ & \frac{1}{\lambda}\int_0^T \left(\int_{\gamma_\lambda} p_\lambda d\zeta - \int_\gamma p_\lambda d\zeta\right) u\, dt \\
&= \int_0^T \int_\gamma \nabla p \cdot G\, d\zeta\, u\, dt + \int_0^T \int_\gamma p(\zeta)\tau(\zeta)^T G'(\zeta)\tau(\zeta) u\, d\zeta\, dt,
\end{aligned}
$$

where $\tau(\zeta) = \tau(g(\vartheta)) = \frac{g'(\vartheta)}{|g'(\vartheta)|}$ is the unit tangent vector to γ at $\zeta = g(\vartheta)$.

Proof. We can write

$$
\frac{1}{\lambda}\left(\int_{\gamma_\lambda} p_\lambda d\zeta - \int_\gamma p_\lambda d\zeta\right) = J_1^\lambda + J_2^\lambda + J_3^\lambda,
$$

where

$$
J_1^\lambda = \frac{1}{\lambda}\int_I p_\lambda(g_\lambda(\vartheta))(|g_\lambda'(\vartheta)| - |g'(\vartheta)|)d\vartheta,
$$

$$
J_2^\lambda = \int_I \nabla p_\lambda(g(\vartheta)) \cdot (G(g(\vartheta)) + \varepsilon(\lambda,\vartheta))|g'(\vartheta)|d\vartheta,
$$

$$
\begin{aligned}
J_3^\lambda = \int_I\int_0^1 & \Big[\nabla p_\lambda(g(\vartheta) + \chi(g_\lambda(\vartheta) - g(\vartheta))) - \nabla p_\lambda(g(\vartheta))\Big] d\chi \\
& \times (G(g(\vartheta)) + \varepsilon(\lambda,\vartheta))|g'(\vartheta)|d\vartheta.
\end{aligned}
$$

First, observe that the sequence $(\frac{|g_\lambda'|-|g'|}{\lambda})_\lambda$ converges to $\left(\frac{g'}{|g'|}\right)^T \cdot G'(g) \cdot g'$ uniformly on I. Due to Proposition 5.2, the sequence $(p_\lambda)_\lambda$ converges to p in $L^1(0,T;C^1(\overline{\Omega}))$. Thus we have the following convergence in $L^1(0,T)$:

$$
\lim_{\lambda\to 0} J_1^\lambda = \int_\gamma p(\zeta)\tau(\zeta)^T G'(\zeta)\tau(\zeta)\, d\zeta.
$$

Secondly,

$$
\lim_{\lambda\to 0} J_2^\lambda = \int_I \nabla p(g(\vartheta)) \cdot G(g(\vartheta))|g'(\vartheta)|d\vartheta = \int_\gamma \nabla p \cdot G d\zeta\ .
$$

Due to Theorem 5.1, the sequence $(p_\lambda)_\lambda$ is bounded in $L^1(0,T;C^{1,\alpha}(\overline{\Omega}))$ for some $0 < \alpha < 1$. Therefore we have

$$
\begin{aligned}
&\left|\int_0^T u(t)J_3^\lambda dt\right| \le C\int_0^T |u(t)| \int_I\int_0^1 \mathcal{F}(\lambda)d\chi\, |g_\lambda(\vartheta) - g(\vartheta)|^\alpha d\vartheta dt \\
&\le C\|u\|_{L^\infty(0,T)}\|\nabla p_\lambda\|_{L^1(0,T;C^{0,\alpha}(\overline{\Omega}))} \sup_{\vartheta\in I}|g_\lambda(\vartheta) - g(\vartheta)|^\alpha \longrightarrow 0 \text{ as } \lambda \to 0,
\end{aligned}
$$

where

$$\mathcal{F}(\lambda)=\frac{|\nabla p_\lambda(g(\vartheta)+\chi(g_\lambda(\vartheta)-g(\vartheta)))-\nabla p_\lambda(g(\vartheta))|}{|g(\vartheta)+\chi(g_\lambda(\vartheta)-g(\vartheta))-g(\vartheta)|^\alpha}.$$

The proof is complete. ■

Theorem 5.2 – *The derivative of the functional F is given by*

$$lim_{\lambda\to 0}\frac{F(\gamma_\lambda)-F(\gamma)}{\lambda}=\int_0^T u(t)\int_\gamma \nabla p\cdot G\, d\zeta dt+\int_0^T u(t)\int_\gamma p\tau^T\cdot G'\cdot\tau\, d\zeta dt,$$

where p is the solution to

$$-\frac{\partial p}{\partial t}+Ap+ap=0 \quad in\ \ Q,\qquad \frac{\partial p}{\partial n_A}=0\quad on\ \ \Sigma,$$
$$p(T)=\theta|\bar y(T)-y_d|^{\theta-2}(\bar y(T)-y_d)\quad in\ \ \Omega, \tag{5.11}$$

$\bar y$ is the solution of (1.1) corresponding to u and γ, $a=\Phi'(\bar y)$.

Proof. We denote by y_λ the solution of (1.1) corresponding to u and γ_λ. Set $a_\lambda=\int_0^1\Phi'(\bar y+\vartheta(y_\lambda-\bar y))d\vartheta$. Thanks to Lemma 5.1, the sequence $(y_\lambda)_\lambda$ converges to $\bar y$ in $L^{\tilde r}(0,T;L^r(\Omega))$ for every $(\tilde r,r)$ obeying (5.9). Therefore, $(a_\lambda)_\lambda$ tends to $a=\Phi'(\bar y)$ in $L^{\tilde k}(0,T;L^k(\Omega))$ for every $(\tilde k,k)$ satisfying (5.2).

We set $z_\lambda=(y_\lambda-\bar y)/\lambda$ and we denote by z the solution to

$$\frac{\partial z}{\partial t}+Az+a_\lambda z=0 \text{ in } Q,\quad \frac{\partial z}{\partial n_A}=u(t)(\delta_{\gamma_\lambda}-\delta_\gamma)/\lambda \text{ on } \Sigma,\quad z(0)=0 \text{ in } \Omega.$$

From the convexity of the mapping $y\mapsto\int_\Omega|y-y_d|^\theta dx$, it follows that

$$\theta\int_\Omega|\bar y(T)-y_d|^{\theta-2}(\bar y(T)-y_d)z_\lambda(T)dx\le\frac{F(\gamma_\lambda)-F(\gamma)}{\lambda}$$
$$\le\theta\int_\Omega|y_\lambda(T)-y_d|^{\theta-2}(y_\lambda(T)-y_d)z_\lambda(T)dx\ .$$

Let p_λ be the solution to (5.10) when $p_\lambda(T)=\theta|y_\lambda(T)-y_d|^{\theta-2}(y_\lambda(T)-y_d)$, and $\tilde p_\lambda$ be the solution to the equation

$$-\frac{\partial\tilde p_\lambda}{\partial t}+A\tilde p_\lambda+a_\lambda\tilde p_\lambda=0 \text{ in } Q,\quad \frac{\partial\tilde p_\lambda}{\partial n_A}=0 \text{ on } \Sigma,$$
$$\tilde p_\lambda(T)=\theta|\bar y(T)-y_d|^{\theta-2}(\bar y(T)-y_d) \text{ in } \Omega.$$

Using the Green formula (4.87) when $h=0$ and $b=0$, we obtain

$$\int_0^T\frac{1}{\lambda}u\Big[\int_{\gamma_\lambda}\tilde p_\lambda d\zeta-\int_\gamma\tilde p_\lambda d\zeta\Big]dt\le\frac{F(\gamma_\lambda)-F(\gamma)}{\lambda}\le\int_0^T\frac{1}{\lambda}u\Big[\int_{\gamma_\lambda}p_\lambda d\zeta-\int_\gamma p_\lambda d\zeta\Big]dt.$$

We conclude with Lemma 5.1, by passing to the limit when $\lambda\to 0$. ■

References

[1] R. A. Adams, Sobolev spaces, Academic Press, New-York, 1975.

[2] H. Amann, Dual semigroups and second order linear elliptic boundary value problems, Is. J. Math., Vol. 45 (1983), 225-254.

[3] S. Anita, Optimal impulse-control of population dynamics with diffusion, in "Differential equations and control theory, V. Barbu Ed., Longman, Harlow, 1991, 1-6.

[4] S. Anita, Optimal control of parameter distributed systems with impulses, App. Math. Optim., Vol. 29 (1994), 93-107.

[5] P. Baras, Compacité de l'opérateur $f \mapsto u$ solution d'une équation non linéaire $f \in (du/dt) + Au$, C. R. Acad. Sc. Paris, Série A, Vol. 286 (1978), 1113-1116.

[6] P. Baras, M. Pierre, Problèmes paraboliques semilinéaires avec données mesures, Applicable Anal., Vol. 18 (1984), 111-149.

[7] H. Brezis, A. Friedman, Nonlinear parabolic equations involving measures as initial conditions, J. Math. Pures Appl., Vol. 62 (1983), 73-97.

[8] H. Brezis, W. Strauss, Semilinear Elliptic Equation in L^1, J. Math. Soc. Japan, Vol. 25, 1973, 15-26.

[9] M. Berggren, R. Glowinski, J. L. Lions, A Computational Approach to Controllability Issues for Flow-Related Models (Part 1), International Journal of Computational Fluid Dynamics, Vol.7 (1996), 237-252.

[10] M. Berggren, R. Glowinski, J. L. Lions, A Computational Approach to Controllability Issues for Flow-Related Models (Part 2), International Journal of Computational Fluid Dynamics, Vol. 6 (1996), 253-247.

[11] R. Dautray, J. -L. Lions, Analyse Mathématique et Calcul Numérique, Tome 8, Masson, Paris, 1987.

[12] E. J. Dean and P. Gubernatis, Pointwise Control of Burgers' Equation – A Numerical Approach, Computers and Mathematics with Applications, Vol. 22 (1991), No. 7, 93-100.

[13] J. Droniou, J. -P. Raymond, Optimal pointwise control of semilinear parabolic equations, Nonlinear An., T.M.A, Vol. 39 (2000), 135-156.

[14] R. Glowinski, J. L. Lions, Exact and approximate controllability for distributed parameter systems (Part 1), Acta Numerica (1994), 269-378.

[15] R. Glowinski, J. L. Lions, Exact and approximate controllability for distributed parameter systems (Part 2), Acta Numerica (1995), 269-378.

[16] H. Henrot, W. Horn, J. Sokolowski, Domain optimization problem for stationary heat equation, Appl. Math. and Comp. Sci., vol. 6, No. 2, 1-21.

[17] H. Henrot, J. Sokolowski, Shape Optimization Problem for Heat Equation, Rapport de recherche INRIA, Juin 1997.

[18] D. Henry, Geometric Theory of Semilinear Parabolic Equations, Springer-Verlag, Berlin/Heidelberg/New York, 1981.
[19] K. - H. Hoffmann, J. Sokolowski, Interface optimization problems for parabolic equations, Control and Cybernetics, vol. 23 (1994), No. 3.
[20] I. Lasiecka, R. Triggiani, Differential and algebraic Riccati equations with application to boundary/point control problems: continuous theory and approximation theory, Lecture Notes in Control and Information Sciences, 164, Springer-Verlag, Berlin, 1991.
[21] J. -L. Lions, Some Methods in the Mathematical Analysis of Systems and their Control, Science Press, Beijing, Gordon and Breach, 1981.
[22] J. -L. Lions, Pointwise control for distributed systems, in "Control and estimation in distributed parameters sytems", H. T. Banks Ed., SIAM, Philadelphia, 1992, 1-39.
[23] A. Pazy, Semigroups of Linear Operators and Applications to Partial Differential Equations, Springer-Verlag, Berlin Heidelberg, 1983.
[24] J. P. Raymond, Nonlinear boundary control of semilinear parabolic equation, Disc. and Cont. Dyn. Syst., Vol. 3 (1997), 341-370.
[25] J. P. Raymond, H. Zidani, Pontryagin's principles for state-constrained control problems governed by parabolic equations with unbounded controls, SIAM J. Control Optim., Vol. 36 (1998), No. 6, 1853-1879.
[26] J. P. Raymond, H. Zidani, Hamiltonian Pontryagin's principles for control problems governed by semilinear parabolic equations, Appl. Math. Optim., Vol. 39 (1999), 143-177.
[27] J. P. Raymond, H. Zidani, Time optimal problems with boundary controls, Diff. Int. Eq., Vol. 13 (2000), 1039-1072.
[28] J. Simon, Compact Sets in the Space $L^p(0,T;B)$, Ann. Mat. Pura Appl., 196 (1987), 65-96.
[29] J. Simon, Caractérisation d'un espace fonctionnel intervenant en contrôle optimal, Ann. Fac. Sciences Toulouse, Vol. 5 (1983), 149-169.
[30] H. Triebel, "Interpolation Theory, Functions Spaces, Differential Operators", North Holland, Amsterdam New York, Oxford, 1977.
[31] V. Vespri, Analytic Semigroups Generated in $H^{-m,p}$ by Elliptic Variational Operators and Applications to Linear Cauchy Problems, Semigroup theory and applications (Clemens et al., ed.), Marcel Dekker, New York, 1989, 419-431.

P. A. Nguyen and J. -P. Raymond
UMR CNRS MIP
UFR MIG
Université Paul Sabatier
31062 Toulouse Cedex 4
France

E-mail: nguyen@mip.ups-tlse.fr, E-mail:raymond@mip.ups-tlse.fr

Studies in Mathematics and its Applications, Vol. 31
D. Cioranescu and J.L. Lions (Editors)

Chapter 27

STABILITE DES ONDES DE CHOC DE VISCOSITE QUI PEUVENT ETRE CARACTERISTIQUES

D. SERRE

Résumé. Tout choc satisfaisant la condition d'Oleinik dans un sens strict admet bien sûr un profil de viscosité. On donne ici une démonstration simple de la stabilité asymptotique dans $L^1(\mathbb{R})$ d'un tel profil, qui n'utilise que le principe de contraction, c'est-à-dire une forme précisée du principe du maximum. Les hypothèses sont naturelles et bien plus faibles que ce qui est connu : la condition initiale prend ses valeurs entre ses deux limites u_g, u_d en $\pm\infty$, tandis que la perturbation est intégrable. En particulier, il n'y a pas d'hypothèse de petitesse. De plus, le choc (u_g, u_d) peut être caractéristique d'un coté ou même des deux cotés. A cause de cette généralité, on n'obtient pas d'estimation du taux de convergence quand t tend vers $+\infty$.

Abstract. Any choc satisfying the condition of Oleinik in a strict sense admits obviously a viscosity profile. We give here a simple proof of the $L^1(\mathbb{R})$−asymptotic stability of such a profile, by making use only of the contraction principle, i.e., a precise form of the maximum principle. The hypotheses we make are natural and are much weaker than those known in the litterature. They asks to the initial condition to take its values between two limits u_g, u_d at $\pm\infty$ while the perturbation has to be integrable. In particular, there is no "smalness" hypothesis. Moreover, the choc (u_g, u_d) can be characteristic on one side or even on both sides. Due to this generality, we do not obtain an estimate of the rate of convergence as $t \to \infty$.

1. Introduction

Considérons une loi de conservation scalaire

$$\frac{\partial u}{\partial t} + \frac{\partial}{\partial x} f(u) = 0, \tag{1}$$

où f est une fonction de classe $\mathcal{C}^2$. Les solutions admissibles de (1) sont classiquement les limites des solutions de l'équation perturbée

$$\frac{\partial u}{\partial t} + \frac{\partial}{\partial x} f(u) = \epsilon \frac{\partial^2 u}{\partial x^2}, \tag{2}$$

lorsque $\epsilon \to 0+$. Les solutions de (1) développent en général des discontinuités au bout d'un temps fini, même lorsque la donnée initiale est très régulière. L'archétype d'une discontinuité est une solution constante de part et d'autre d'une droite :

$$u(x,t) = \begin{cases} u_g, & x < st, \\ u_d, & x > st. \end{cases}$$

Le nombre s est la *vitesse* de la discontinuité. Celle-ci n'est admissible que si les conditions suivantes sont remplies :

$$f(u_d) - f(u_g) = s(u_d - u_g), \tag{3}$$

$$(u_d - u_g) f(u) \geq (u - u_g) f(u_d) + (u_d - u) f(u_g), \forall u$$
$$(u_d - u)(u - u_g) > 0. \tag{4}$$

La première condition est dite *de Rankine-Hugoniot* et exprime que u est une solution au sens des distributions pour (1). La seconde est la condition *d'Oleinik* et exprime une forme de stabilité. Lorsqu'elle est satisfaite avec une inégalité stricte, l'équation différentielle

$$v' = f(v) - f(u_g) - s(v - u_g) \tag{5}$$

possède, à translation près, une et une seule solution hétérocline allant de u_g vers u_d, appelée *profil de viscosité*. A cette solution correspond une onde progressive pour l'équation perturbée :

$$u^\epsilon(x,t) = v\left(\frac{x - st}{\epsilon}\right),$$

qui tend vers u lorsque $\epsilon \to 0+$.

Une question importante est l'uniformité de cette convergence par rapport à des perturbations. Une question connexe est, en raison des changements d'échelles possibles $((x,t,\epsilon) \mapsto (\epsilon x, \epsilon t, 1))$, la stabilité asymptotique d'un profil de viscosité. En clair, étant donnée une solution ϕ de (5) et une perturbation $\alpha : \mathbb{R} \mapsto \mathbb{R}$, disons intégrable, la solution (bien définie et régulière) du problème de Cauchy

$$\frac{\partial u}{\partial t} + \frac{\partial}{\partial x} f(u) = \frac{\partial^2 u}{\partial x^2}, \tag{6}$$

$$u(x,0) = \phi(x) + \alpha(x) \tag{7}$$

est-elle asymptotique à $U(x,t) := \phi(x-st-x_0)$ dans une norme convenable lorsque $t \to +\infty$? Si la réponse est positive pour la norme de $L^1(\mathbb{R})$, alors le décalage x_0 est entièrement déterminé par la condition initiale à cause de la conservation de la masse. En effet, u et U étant deux solutions ayant les mêmes valeurs en $x = \pm\infty$, on a

$$\begin{aligned}\int_{\mathbb{R}}(u(x,t) - U(x,t))dx &= \int_{\mathbb{R}}(u(x,0) - U(x,0))dx \\ &= \int_{\mathbb{R}}(\phi(x) - \phi(x-x_0) + \alpha(x))dx \\ &= (u_d - u_g)x_0 + \int_{\mathbb{R}} \alpha(x)dx.\end{aligned}$$

Ainsi,

$$x_0 = \frac{1}{u_d - u_g}\int_{\mathbb{R}} \alpha(x)dx.$$

Le premier résultat de stabilité a été obtenu par Il'in et Oleinik [1] lorsque $\inf_{\mathbb{R}} f'' > 0$, ce qui s'applique à l'équation de Burgers-Hopf. Comme dans ce cas toutes les discontinuités sont non caractéristiques (on a $f'(u_d) < s < f'(u_g)$), la convergence de u vers U est exponentiellement rapide par rapport au temps. Ceci est lié à une propriété du spectre de l'opérateur linéarisé

$$Lz := -\frac{d^2 z}{dx^2} + \frac{d}{dx}(f'(\phi)z).$$

Par un changement de variable $(x,t) \mapsto (x - st, t)$, on peut toujours se ramener au cas où s est nul, c'est-à-dire que ϕ est une solution stationnaire. On a alors $L\phi' = 0$ et $\phi' > 0$, ce qui montre que le spectre de L est positif, strictement à l'exception de la valeur propre 0. En ce plaçant dans un espace de type L^2 à poids, on peut obtenir que le spectre de L soit discret. Enfin, en se restreignant au sous-espace invariant défini par

$$\int_{\mathbb{R}}(z - \phi)dx = \int_{\mathbb{R}} \alpha dx,$$

le spectre de L devient minoré par un nombre strictement positif.

Le second résultat important est celui de Osher et Ralston [2], dans lequel le flux f est quelconque. La méthode employée est la bonne, qui utilise la propriété de contraction dans L^1 du semi-groupe associé à la résolution du problème de Cauchy pour (6). Mais, curieusement, trop de détails techniques en cachent la généralité et les auteurs n'obtiennent la convergence que pour un choc vérifiant la condition de Lax

$$f'(u_d) < s < f'(u_g), \tag{8}$$

au lieu d'inégalités larges. Ici, une restriction importante a lieu, qui est peut-être nécéssaire : la condition initiale $\phi + \alpha$ est à valeurs dans l'intervalle I d'extrémités u_g et u_d.

Enfin, un article récent de Matsumura et Nishihara montre la stabilité du profil d'un choc caractéristique (c'est-à-dire qui vérifie $f'(u_d) = s$ ou $f'(u_g) = s$). Cependant, la perturbation initiale est supposée petite. De plus, la méthode est basée sur des estimations d'énergie et la perturbation doit donc appartenir à un espace L^2 à poids, qui n'a pas de sens physique.

Nous présentons donc ici un résultat plus complet, sans hypothèse de petitesse sur α, dans l'espace naturel $L^1(\mathbb{R})$ et sans autre restriction que $\phi(x) + \alpha(x) \in I$ pour tout $x \in \mathbb{R}$.

Théorème 1.1 – *Soit ϕ un profil de viscosité entre u_g et u_d. Soit $\alpha \in L^1(\mathbb{R})$ une perturbation initiale telle que*

$$\phi(x) + \alpha(x) \in I,$$

pour presque tout $x \in \mathbb{R}$. Soit $x_0 = (u_d - u_g)^{-1} \int_{\mathbb{R}} \alpha(x)dx$ et soit enfin $U(x,t) = \phi(x - st - x_0)$. Alors

$$\lim_{t \to +\infty} \int_{\mathbb{R}} |u(x,t) - U(x,t)|dx = 0.$$

2. Rappels ; réduction à un cas particulier

La théorie des opérateurs monotones permet de construire un semi-groupe $(S(t))_{t \geq 0}$ qui résout le problème de Cauchy pour l'équation (6) lorsque la donnée initiale u_0 est dans $L^\infty(\mathbb{R})$. Comme l'équation (6) satisfait le principe du maximum, ce semi-groupe jouit des propriétés suivantes.

SG1 (régularité) Pour tout $a \in L^\infty(\mathbb{R})$, $S(t)a$ est indéfiniment différentiable et borné sur $\mathbb{R}$ pour tout $t > 0$.

SG2 (principe du maximum) Si $a \leq b$ presque-partout, alors $S(t)a \leq S(t)b$ pour tout $t > 0$.

SG3 (conservation de la masse) Si $a - b \in L^1(\mathbb{R})$, alors $S(t)a - S(t)b \in L^1(\mathbb{R})$ et on a pour tout $t > 0$

$$\int_{\mathbb{R}} (S(t)a - S(t)b)dx = \int_{\mathbb{R}} (a - b)dx.$$

SG4 (contraction) Sous les mêmes hypothèses qu'en (SG3), l'application

$$t \mapsto \int_{\mathbb{R}} |S(t)a - S(t)b|dx$$

est décroissante.

Il n'est pas nécessaire à la compréhension de ce rapport de démontrer les assertions ci-dessus. Elles sont classiques.

Le théorème se déduit aisément du lemme suivant :

Lemme 2.1 – *Soit ϕ un profil de viscosité qui joint u_g à u_d. Soit α une perturbation, fonction mesurable sur $\mathbb{R}$, telle que $\phi + \alpha$ soit compris entre deux translatés de ϕ (il existe $\gamma, \beta \in \mathbb{R}$ tels que $\phi(x+\gamma) \leq \phi(x) + \alpha(x) \leq \phi(x+\beta)$ presque partout). Alors la solution $u(t) = S(t)(\phi+\alpha)$ du problème de Cauchy pour (6) satisfait*

$$\lim_{t\to+\infty} \int_{\mathbb{R}} |u(t) - \phi(\cdot - x_0 - st)|dx = 0,$$

où $x_0 = (u_d - u_g)^{-1} \int_{\mathbb{R}} a(x)dx$.

En effet, si $\alpha \in L^1$ est tel que $\phi + \alpha$ prenne ses valeurs dans I, alors il existe, pour tout $\eta > 0$, une fonction $\alpha^\eta \in L^1$ et des nombres réels β, γ tels que

$$\begin{aligned} \phi(x+\gamma) \leq \quad & \phi(x) + \alpha(x) & \leq \phi(x+\beta) \\ & int_{\mathbb{R}}|\alpha^\eta(x) - \alpha(x)|dx & < \eta. \end{aligned}$$

Soit u^η la solution du problème de Cauchy dont la condition initiale est $\phi + \alpha^\eta$. Alors, pour tout $t > 0$, on a

$$\int_{\mathbb{R}} |u^\eta(x,t) - u(x,t)|dx < \eta,$$

d'après (SG4). Appliquant le lemme 2.1 à u^η :

$$\lim_{t\to+\infty} \int_{\mathbb{R}} |u^\eta(x,t) - \phi(x - st - x_0^\eta)|dx = 0.$$

De plus, $|x_0^\eta - x_0| \cdot |u_d - u_g| < \eta$. Finalement,

$$\limsup_{t\to\ +\infty} \int_{\mathbb{R}} |u(x,t) - \phi(x - st - x_0)|dx < 2\eta,$$

ce qui prouve le théorème 1.1.

3. Preuve du lemme 2.1

On notera $\|\cdot\|_1$ la norme de $L^1(\mathbb{R})$.

Tout d'abord, quitte à faire le changement de variables $(t,x)\mapsto(t,x-st)$, on peut supposer que le choc est stationnaire : $s=0$. Les fonctions $(t,x)\mapsto\phi(x+\gamma)$ et $(t,x)\mapsto\phi(x+\beta)$ sont donc des solutions stationnaires de (6). Utilisant le principe du maximum et l'hypothèse concernant la perturbation (qui assure que α est borné), on a $\phi(x+\gamma)\leq u(t,x)\leq\phi(x+\beta)$. Notons $v(t)=u(t)-\phi$. Alors $v(t)$ est compris entre deux fonctions intégrables qui ne dépendent pas du temps et reste donc dans un borné de $L^1(\mathbb{R})$. De plus, la propriété de contraction fournit l'inégalité $\|v(t,\cdot+r)-v(t)\|_1=\|u(t,\cdot+r)-u(t)\|_1\leq\|u(0,\cdot+r)-u(0)\|_1=\|\alpha(\cdot+r)-\alpha\|_1$ qui tend vers zéro avec r. D'après le théorème de compacité de Fréchet-Kolmogorov, la famille $(v(t))_{t\geq 0}$ est donc relativement compacte dans $L^1(\mathbb{R})$. L'ensemble ω-limite $A=:\phi+\cap_{s\geq 0}B_s$, où B_s est l'adhérence dans $L^1(\mathbb{R})$ de $\{v(t);t\geq s\}$, est donc non vide puisque $A-\phi$ est l'intersection décroissante de compacts non vides. Cet ensemble est celui de toutes les valeurs d'adhérence, pour la distance $d(z,w)=\|z-w\|_1$, des sous-suites $(u(t_n))_{n\in\mathbb{N}}$ où $t_n\to+\infty$.

L'ensemble ω-limite est invariant par le semi-groupe S puisque si $a\in A$, où $a=\lim_{n\to\infty}u(t_n)$, alors $S(t)a=\lim_{n\to\infty}u(t+t_n)$. Pour la même raison, $S(t)$ est surjectif sur A car on a aussi $a=S(t)b$ où b est une valeur d'adhérence de la suite $(u(t_n-t))_{n\in\mathbb{N}}$. La propriété de régularité [SG1] implique donc que A est inclus dans $\mathcal{C}^\infty$.

Soit maintenant $k\in\mathbb{R}$. La fonction $t\to\|u(t)-\phi(\cdot-k)\|_1$, décroissante, admet une limite notée $c(k)$ quand $t\to\infty$. Si $a\in A$, on en déduit que $\|a-\phi(\cdot-k)\|_1=c(k)$. Cependant, $S(t)a$ appartient encore à A. Il s'ensuit que $t\mapsto\|S(t)a-\phi(\cdot-k)\|_1$ est constant. Notons provisoirement $w(t)=S(t)a$ et $z(t)=S(t)a-\phi(\cdot-k)$. On a

$$0=\frac{d}{dt}\|z(t)\|_1=\int_{\mathbb{R}}z_t\text{sgn}\,z\,dx.$$

Or $z_t+(f(w)-f(\phi(\cdot-k)))_x=z_{xx}$, où on a utilisé l'équation de profil pour ϕ. Multipliant ceci par $\text{sgn}\,z$, on en déduit

$$|z|_t+((f(w)-f(\phi(\cdot-k)))\text{sgn}\,z)_x=z_{xx}\text{sgn}\,z,$$

ce qui donne après intégration sur $\mathbb{R}$:

$$\frac{d}{dt}\int_{\mathbb{R}}|z|dx=\int_{\mathbb{R}}z_{xx}\text{sgn}\,z\,dx.$$

Finalement,

$$0=\int_{\mathbb{R}}z_{xx}\text{sgn}\,z\,dx$$

et donc

$$0 = \int_{\mathbb{R}} a_{xx} \operatorname{sgn} a \, dx. \tag{9}$$

Cependant les estimations *a priori* faites lors de la construction du semi-groupe S montrent que w_{xx} est intégrable sur $\mathbb{R}$ et donc aussi a_{xx}. On a donc, d'après le théorème de convergence dominée,

$$0 = \lim_{\epsilon \to 0} \int_{\mathbb{R}} a_{xx} j'_\epsilon(a) \, dx$$

où $j_\epsilon(\tau) = \sqrt{\epsilon^2 + \tau^2}$. Intégrant par parties, il vient

$$0 = \lim_{\epsilon \to 0} \int_{\mathbb{R}} a_x^2 j''_\epsilon(a) \, dx.$$

Soit y_0 un point où a s'annule et soit $\delta > 0$ tel que $\delta |a_x(y_0)| < 1$. Pour $\epsilon > 0$ assez petit, on a $|a| < \epsilon$ sur $]y_0 - \delta\epsilon, y_0 + \delta\epsilon[$ puisque a est continument différentiable. Or $j''_\epsilon(\tau) = \frac{1}{\epsilon} J(\frac{\tau}{\epsilon})$ avec $J(\tau) = (1 + \tau^2)^{-3/2}$. Ainsi

$$\int_{\mathbb{R}} a_x^2 j''_\epsilon(a) \, dx \geq \frac{1}{\epsilon} \int_{y_0 - \delta\epsilon}^{y_0 + \delta\epsilon} J(1) a_x^2 dx,$$

dont le second membre tend vers $2\delta J(1) a_x(y_0)^2$ quand ϵ tend vers zéro. On en déduit que $a_x(y_0) = 0$. Finalement nous avons prouvé que

$$\forall a \in A, \forall k \in \mathbb{R}, \forall x \in \mathbb{R}, \quad (a(x) = \phi(x-k)) \Longrightarrow (a'(x) = \phi'(x-k)). \tag{10}$$

Pour conclure, on note d'abord que a est compris entre $\phi(\cdot + \gamma)$ et $\phi(\cdot + \beta)$ comme limite de telles fonctions, donc a prend ses valeurs strictement entre u_g et u_d. Aussi la fonction $x \mapsto k(x) =: x - \phi^{-1} \circ a(x)$ est-elle bien définie et régulière (rappelons que ϕ est strictement monotone). Par construction, $a(x) = \phi(x - k(x))$, ce qui donne en dérivant $a'(x) = \phi'(x - k(x))(1 - k'(x))$. Utilisant (10) avec $k = k(x)$, il reste

$$\phi'(x - k(x)) k'(x) = 0$$

et donc $k'(x) = 0$ puisque ϕ' ne s'annule pas. Finalement, k est une constante et $a = \phi(\cdot - k)$. Cependant, les éléments de A satisfont

$$\int_{\mathbb{R}} (a - \phi - \alpha) dx = 0,$$

ce qui fixe, on l'a vu, la valeur de k : on a $k = -y_0$.

On a donc prouvé que l'ensemble ω-limite est réduit à un seul élément (au moins un puisqu'il n'est pas vide et seulement celui-là) : $\phi(\cdot - y_0)$. Puisque la famille $(v(t))_{t\geq 0}$ est relativement compacte dans $L^1(\mathbb{R})$ et comme elle n'a qu'une seule valeur d'adhérence quand $t \to +\infty$, elle est convergente, c'est-à-dire que

$$\lim_{t\to+\infty} \|u(t) - \phi(\cdot - y_0)\|_1 = 0.$$

Ceci achève la preuve du lemme et donc celle du théorème.

Remerciements Je remercie Heinrich Freistühler pour l'intérêt qu'il a porté à ce travail, les discussions fructueuses et pour avoir attiré mon attention sur des travaux antérieurs. Je remercie aussi J.-L. Lions et ses collaborateurs pour leurs constants encouragements.

References

[1] A. M. Il'in, O. A. Oleinik, Behaviour of the solutions of the Cauchy problem for certain quasilinear equations for unbounded increase of time. AMS Translations, 42 (1964), 19–23.

[2] S. Osher, J. Ralston, L^1 stability of travelling waves with application to convective porous media flow. Comm. Pure and Applied Maths., 35 (1982), 737–751.

[3] A. Matsumura, K. Nishihara, Asymptotic stability of travelling waves for scalar viscous conservation laws with non-convex nonlinearity. Comm. Math. Physics, 165 (1994), 83–96.

Denis Serre
Unité de Mathématiques Pures et Appliquées
CNRS UMR 128
ENS Lyon
46, Allée d'Italie
69364 Lyon
France
E-mail: serre@umpa.ens-lyon.fr